Handbuch der Urologie
Encyclopedia of Urology · Encyclopédie d'Urologie
Gesamtdisposition · Outline · Disposition générale

HANDBUCH DER UROLOGIE

ENCYCLOPEDIA OF UROLOGY

ENCYCLOPÉDIE D'UROLOGIE

HERAUSGEGEBEN VON · EDITED BY
PUBLIÉE SOUS LA DIRECTION DE

C. E. ALKEN
HOMBURG (SAAR)

V. W. DIX
LONDON

W. E. GOODWIN
LOS ANGELES

H. M. WEYRAUCH†

E. WILDBOLZ
BERN

VII/1

SPRINGER-VERLAG BERLIN · HEIDELBERG · NEW YORK 1968

MALFORMATIONS

BY

A. D. AMAR · O. S. CULP · F. FARMAN
J. A. HUTCH · H. W. JONES, JR. · V. F. MARSHALL
J. W. McROBERTS · E. C. MUECKE · J. J. MURPHY · R. J. PRENTISS
TH. A. TRISTAN · K. WATERHOUSE

WITH 348 FIGURES

SPRINGER-VERLAG BERLIN · HEIDELBERG · NEW YORK 1968

ISBN 978-3-642-87401-7   ISBN 978-3-642-87399-7 (eBook)
DOI 10. 1007/978-3-642-87399-7

Titel-Nr. 5880

Table of Contents

Contributors to Volume VII/1

AMAR, ARJAN D., M.B.B.S., M.S., F.R.C.S.(C), F.A.C.S., Chief, Dept. of Urology, Kaiser Foundation Hospital, 1425 S. Main Street, Walnut Creek, Cal. 94596/ USA

CULP, ORMOND S., M.D., Consultant, Section of Urology, Mayo Clinic, Rochester, Professor of Urology, Mayo Graduate School of Medicine (University of Minnesota), Rochester, Minn., Section of Urology, Mayo Clinic, 200 First Street Southwest, Rochester, Minn. 55901/USA

FARMAN, FRANKLIN, A.B., M.D., F.A.C.S., Robert S. Fox Urological Foundation, P.O.B. 618, Whittier, Cal./USA

HUTCH, JOHN A., M. D., Asso. Clinical Professor of Urology, University of California, San Francisco, Cal. 94122/USA

JONES, HOWARD W., JR., M.D., Professor of Gynecology and Obstetrics, Johns Hopkins University, School of Medicine, Baltimore, Md./USA

MARSHALL, VICTOR F., M.D., Professor of Surgery (Urology), Cornell University Medical College, Attending Surgeon-In-Charge (Urology), The New York Hospital, 525 East 68th Street, New York, N.Y./USA

McROBERTS, J. WILLIAM, M.D., Resident in Urology, Mayo Graduate School of Medicine (University of Minnesota), Rochester, Minn., Mayo Clinic, 200 First Street Southwest, Rochester, Minn. 55901/USA

MUECKE, EDWARD C., M.D., Assistant Professor of Clinical Surgery (Urology) Cornell University, Assistant Attending Surgeon (Urology), The New York Hospital, Medical College, 525 East 68th Street, New York, N.Y./USA

MURPHY, JOHN J., M.D., F.A.C.S., Professor of Urology, School of Medicine, University of Pennsylvania, Suite W-310, University Hospital, 3400 Spruce Street, Philadelphia, Pa. 19104/USA

PRENTISS, ROBERT J., B.A., M. D., 3415, 6th Avenue, San Diego, Cal./USA

TRISTAN, THEODORE ATHERTON, B.S. Medicine, M.S. Anatomy, M.S. Radiology, M.D., Director, Division of Radiology, Polyclinic Hospital, Harrisburg, Pa. 17105/USA

WATERHOUSE, KEITH, M.A., M.B., B.Chir., F.A.C.S., Professor and Head, Urology, State University of New York, Downstate Medical Center, 450 Clarkson Avenue, Brooklyn, N.Y. 11203/USA

Vesicoureteral Reflux

John A. Hutch and Arjan D. Amar

With 36 Figures

A. Definition of Reflux

Reflux is defined in Dorland's Illustrated Medical Dictionary as "a backward or return flow". Vesicoureteral reflux, therefore, is the backward flow of urine from the bladder to the ureter. Such reflux in humans is thought to be pathologic, as has been reported by Gibson (1949), Iannaccone and Panzironi (1955), Jones and Headstream (1958), and Leadbetter, Duxbury and Dreyfuss (1960).

B. Anatomy of the Ureterovesical Junction

Vesicoureteral reflux and the ureterovesical junction are so intimately related that an understanding of reflux requires a knowledge of the anatomy, embryology, and physiology of this junction. It is formed by the union of totally dissimilar structures — the bladder, which arises from the endoderm and is controlled by the parasympathetic nervous system, and the ureter, which arises from the mesoderm and is controlled by the sympathetic nervous system. By this union is formed a valve which, when it functions correctly, prevents vesicoureteral reflux.

I. Contribution of the Ureter and Trigone to the Formation of the Ureterovesical Junction

To appreciate the part played by these structures in the formation of the ureterovesical junction it must be realized that the collecting tubules, calyces, renal pelvis, extravesical ureter, intravesical ureter, trigone, Bell's muscle, and crista urethralis form a functional and anatomical unit that is completely separate from the bladder. For convenience, this will be referred to as the mesodermal component.

When the human embryo has achieved the 5-mm stage and the caudal end of the wolffian duct has opened into the cloaca, a small outpouching (the ureteral bud) appears on the dorsal aspect of the wolffian duct a short distance above its junction with the cloaca (Fig. 1). During the 9—15 mm stage the ureteral bud pushes cranially into the primordial kidney and achieves an independent opening into the cloaca just dorsolateral to the opening of the wolffian duct. As development progresses, the opening of the wolffian duct (which will become the ejaculatory duct in the male) migrates downward and medially while the opening of the ureteral bud (which will become the ureteral orifice) migrates upward and laterally. As these 2 openings migrate further apart, the mesodermal tissue between them forms the vesical trigone. Thus the structures that eventually make up the mesodermal component all arise from the wolffian duct and are mesodermal in origin as opposed to the bladder which arises from endoderm.

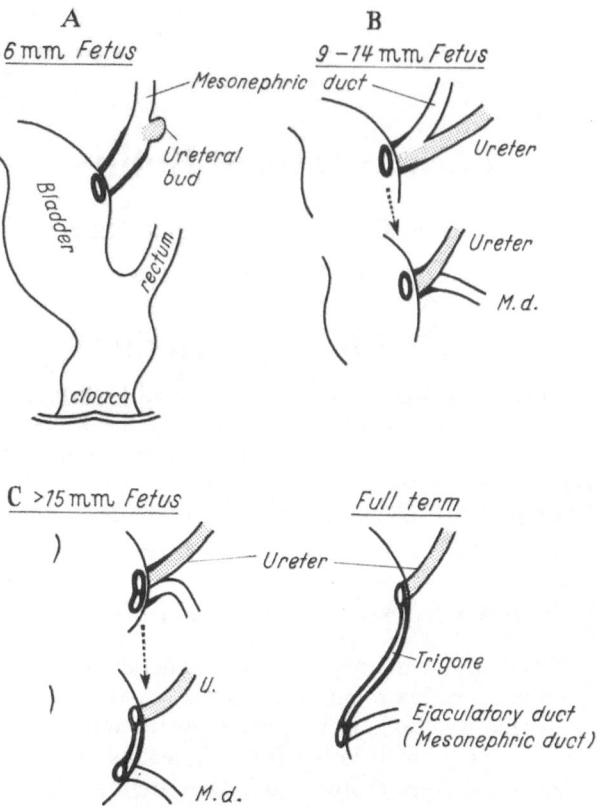

Fig. 1. Development of the ureter as a bud from the mesonephric duct which also gives rise to the ejaculatory duct. The segment of the mesonephric duct that is incorporated in the urogenital sinus expands to form trigonal tissue (TANAGHO and HUTCH, 1965)

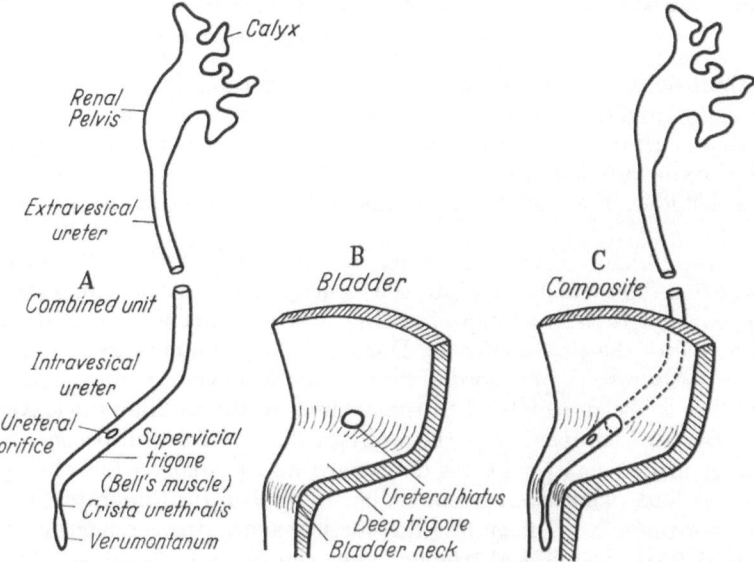

Fig. 2 A—C. Component parts of the urinary tract. A. Mesodermal component (called combined unit in drawing) arising from the wolffian duct. B. Endodermal component arising from the cloaca. C. Manner in which the 2 elements are combined to form the normal urinary tract (HUTCH, 1966 a)

The mesodermal component begins superiorly at the point where the collecting tubules join the nephron (Fig. 2A). At the verumontanum it becomes continuous with the ejaculatory duct. Histologically, the mesodermal component contains longitudinal smooth muscle fibers and connective tissue throughout its entire length. In the calyces, renal pelvis and extravesical ureter, circular smooth muscle bundles are applied around longitudinal muscle to aid in peristalsis. The mesodermal component has a lumen from its junction with the nephron to the ureteral orifice but is solid from the ureteral orifice to the verumontanum. The

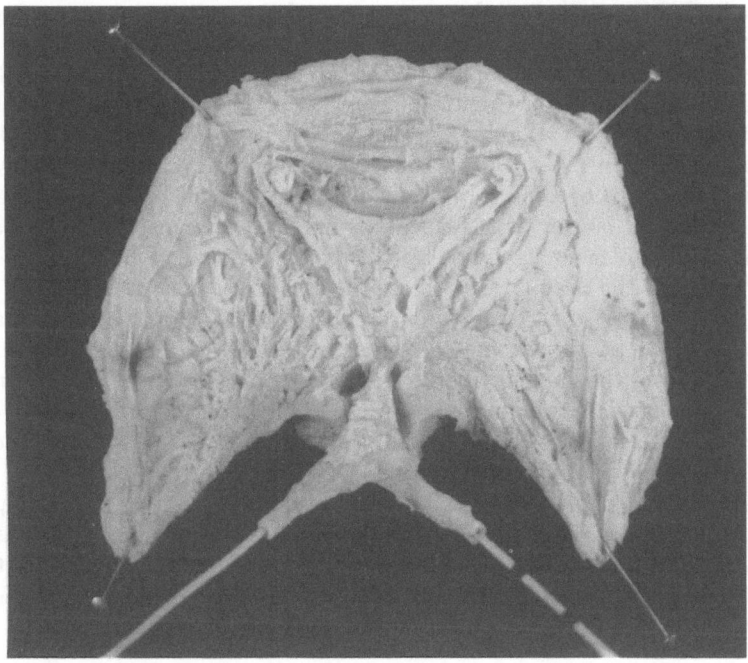

Fig. 3. Dissection of the trigonal area from human bladder. The mucosa and part of the inner longitudinal layer have been removed. The ureters have been cut off in their intravesical segment and reflected downward along with the superficial trigone which is continuous with the ureter. The deep trigone is exposed in its entirety. Note that at each ureteral hiatus the deep trigone terminates by forming a sheath around the ureter. This is Waldeyer's sheath

intravesical ureter is the point of union between its trigonal and extravesical parts. The intravesical ureter and trigone have no circular smooth muscle coat and therefore cannot convey peristalsis.

Throughout most of its course, the extravesical ureter comprises a lumen surrounded by a mucosa of transitional epithelium and 3 distinct smooth muscle layers: an inner longitudinal, a middle circular, and an outer longitudinal layer. About 15—20 mm above the ureteral orifice (in adults) the circular layer ends, so that, in the intravesical ureter, only longitudinally oriented smooth muscle is present. At the ureteral orifice, the lumen disappears, but this in no way disrupts the continuity of the mesodermal component which continues downward and medially to the bladder neck as Bell's muscle and the superficial trigone. (The deep trigone is a direct continuation of Waldeyer's sheath (Fig. 2B), and will be described later.)

In the superior half of its course between the ureteral orifice and the bladder neck, the superficial trigone lies on, but is not attached to, the deep trigone (Fig. 2C). Nearer the bladder neck, the 2 parts of the trigone are difficult to separate by dissection, but they remain distinct entities. The superficial trigone then dips over the bladder neck to pass downward to the verumontanum along the dorsal wall of the prostatic urethra. Along this passage it joins the contra-lateral superficial trigone to form the crista urethralis (see Fig. 4).

Tanagho and Pugh (1963) have shown that Waldeyer's sheath and the deep trigone are continuous (Fig. 3). About 2—3 cm above the point where the ureter enters the bladder (4—6 cm above the ureteral orifice) there begins a sheath of collagenous and smooth muscle fibers which completely surrounds the ureter. While this sheath is separate from the ureter and can be dissected free of the ureter, it is closely applied to the outer layer of the ureter and many fibers pass back and forth between the 2 structures. This relationship between the ureter and Waldeyer's sheath is maintained as these structures enter the ureteral hiatus. However, at the point in the ureteral hiatus where the intravesical ureter becomes submucosal, the roof of Waldeyer's sheath leaves the ureter. It divides into: 1. a lateral component that courses obliquely downward and laterally around the intravesical ureter to become continuous with the floor of Waldeyer's sheath and to form the lateral border of the deep trigone, and 2. a medial component that courses medially to meet its mate from the other side and to form part of the superior border of the deep trigone (Mercier's Bar). The floor of Waldeyer's sheath continues uninterrupted through the ureteral hiatus to emerge into the bladder as the deep trigone. It fans out medially to join its mate from the other side and, with support from the medial fibers of the roof, extends downwardand medially under the superficial trigone. The close approximation of tissues between Waldeyer's sheath and the extravesical ureter is lost after the roof of the sheath divides and leaves the intravesical ureter.

Despite disagreement in detail, most authorities agree in principle that Waldeyer's sheath is affixed to the ureter outside the bladder and that, inside the bladder, it leaves the ureter to become continuous with different parts of the bladder wall. Thus, the ureter and bladder wall are attached in such a way that the ureter is free to move in and out of the bladder through the hiatus.

II. The Vesical Component of the Ureterovesical Valve

1. The Mucosal Layer

The bladder is lined by a thin mucosal layer of transitional epithelium. This layer is continuous with the urethral mucosa distally and with the ureteral mucosa proximally.

2. The Inner Longitudinal Muscle Layer

The inner longitudinal muscle layer lies directly beneath the bladder mucosa. Over most of the bladder, the fibers of this layer are widely separated and run in many different directions. Nearer the bladder neck, they become organized in a longitudinal direction and pass over the bladder neck to become the inner longitudinal muscle layer of the urethra. Along the superior and the lateral borders of the trigone it fuses with the superficial trigonal layer.

3. The Middle Circular Muscle Layer

The middle circular muscle layer is prominent in all parts of the bladder wall but, as it approaches the bladder neck, it undergoes changes that may have sphincteric significance. First, it does not pass into the urethra as the other layers of the bladder do. Second, it thickens as it approaches the bladder neck and its fibers become more prominent. These fibers assume a circular position about the bladder neck, lying at right angles to the direction of the fibers of the inner and outer longitudinal layers. This thickening of the middle circular muscle layer around the bladder neck was first described by UHLENHUTH, HUNTER and LOECHEL (1953) who called it the "fundus ring". The interested reader is referred to their original work.

Except at the trigone, the fundus ring can be exposed during dissection by removal of the inner longitudinal muscle layer at the bladder neck. At the ventral

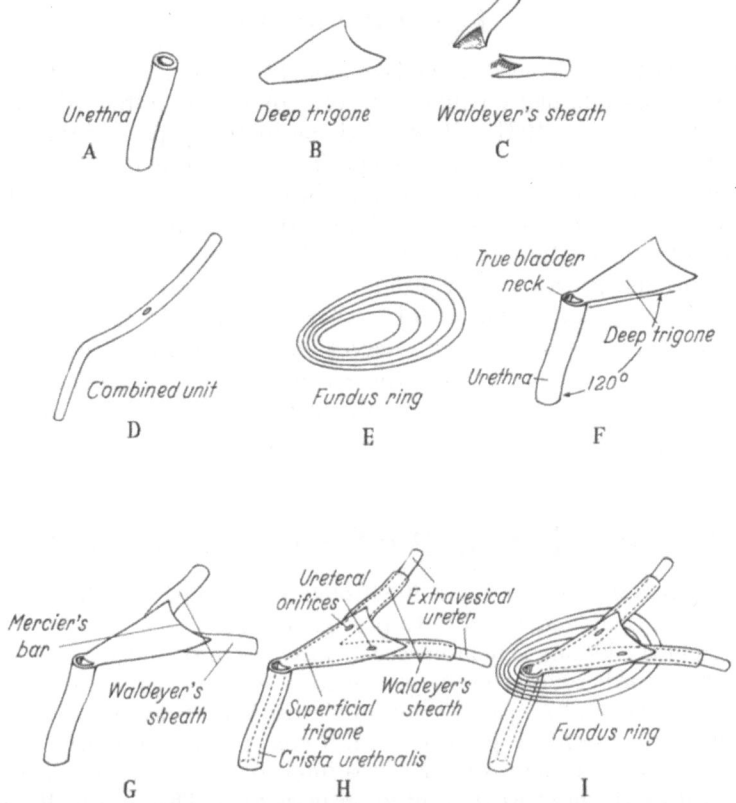

Fig. 4 A—I. Schematic representation of the anatomy of the bladder neck, trigone, and intravesical ureter. A (urethra), B (deep trigone), C (Waldeyer's sheath), D (mesodermal component) and E (fundus ring) are the component parts. F shows how the deep trigone is joined to the urethra. G shows Waldeyer's sheath as a continuation of the deep trigone. H shows how the mesodermal component passes inside Waldeyer's sheath as the intravesical ureter, lies on the deep trigone as the superficial trigone and passes down the urethra as the crista urethralis. I shows the relationship of fundus ring to these structures and how it forms the internal urinary sphincter (HUTCH, 1965)

and at both lateral portions of the bladder neck, the ring is easily identified and dissectable. The caudal-most ring (the true bladder neck) passes around the urethral orifice to insert into the apex of the trigone at the dorsal tip of the bladder neck. The cranial-most ring inserts into the ureteral hiatus, and the remaining rings insert into the trigone in corresponding positions between these 2 points. As these fibers approach the trigone, they fan out and, for the most part, break up into many minute fibrils so that their exact point of insertion into the trigone is difficult to determine. Some of these rings continue around beneath the trigone to meet their components from the other side, and thus form a complete loop (Fig. 4). At all times however, these muscular rings maintain their relationship to one another and to the urethral orifice.

4. The Outer Longitudinal Muscle Layer

The bladder is invested in a heavy outer smooth muscle layer, the fibers of which run, for the most part, longitudinally. This layer is best developed along its anterior and posterior aspects, and is quite thin laterally. As the outer longitudinal muscle layer approaches the bladder neck, part of it continues across the bladder neck to become the outermost smooth muscle coat of the urethra (circular) while other bundles insert into the dorsal side of the apex of the trigone.

C. Innervation

The innervation of the mesodermal component is from the sympathetic nervous system. No parasympathetic ganglia are present in the ureter, trigone or crista urethralis (LEARMONTH, 1931). Strips of tissue from any part of the mesodermal component contract in response to epinephrine or ergotamine, but do not respond to pilocarpine, physostigmine or urocholine. Stimulation of the hypogastric nerves causes contraction of the ureter, trigone and crista urethralis.

On the other hand, the motor nerves to the bladder are all parasympathetic. In the voiding reflex, the sensory stimuli initiated by stretch and tension in the bladder wall are carried to the sacral cord (S_2, S_3, and S_4) through the afferent fibers of the sacral parasympathetic nerves; the motor branch of this reflex arc is formed by the efferent fibers of these same sacral parasympathetic nerves. Although certain sensory impulses from the bladder (pain and thermal perception) are conveyed by sympathetic fibers, extensive sympathectomy does not alter vesical function.

D. Physiology of the Ureterovesical Valve

The ureterovesical valve has 2 functions: 1. to allow ureteral urine to pass unimpeded into the bladder and 2. to prevent reflux.

Urine from the kidney is held up at the junction of the intravesical and extravesical ureters until several cubic centimeters have accumulated in the lower ureter. Then a peristaltic wave is initiated which propels the urine downward with enough force to open the intravesical ureter, allowing the urine to pass into the bladder. The intravesical ureter closes immediately and remains closed until it is reopened by the next wave of ureteral peristalsis. Thus, normally, the intravesical ureter is constantly closed except during ureteral peristalsis when urine moves through it in one direction only — from the ureter to the bladder.

The position of the ureteral orifice in relation to the ureteral hiatus and the bladder neck is not fixed. As seen through the cystoscope, the ureteral orifice during the resting stage is a longitudinal slit on the ventral surface of the lateral border of the trigone and the intravesical ureter. Suddenly, and without apparent muscular activity in the trigone, bladder wall, or intravesical ureter, urine spurts from the ureteral orifice. (This movement is more easily seen if indigo carmine has been injected intravenously.) At the conclusion of this spurt of urine, muscular activity takes place in the intravesical ureter and the lateral border of the trigone, and the ureteral orifice moves upward along the intravesical ureter toward the ureteral hiatus. At the end of its maximal upward excursion, the orifice slowly settles back to its resting position to await the next spurt of urine. HUTCH (1954) feels that this upward movement of the ureteral orifice is primarily the result of active contraction of the longitudinal muscle of the ureter immediately behind the peristaltic wave.

The first study of the physiology of the ureterovesical junction was reported in 1965 by TANAGHO, HUTCH, MEYER and RAMBO. They attempted to answer 4 important questions. 1. What is the effect of interrupting the anatomical continuity between the ureter and the trigone? 2. What is the effect of cutting the nerves that supply the trigonal muscle? 3. Does active contraction of the trigonal muscle tighten or increase the resistance of the intravesical ureter? 4. During voiding, does the trigonal muscle actively contract to tighten and occlude the intravesical ureter?

Because of the importance of the ureterovesical junction in reflux, the answers to these questions will be presented in detail.

I. The Effect of Interrupting the Anatomical Continuity between the Ureter and the Trigone

Ten dogs that did not show reflux were chosen for the experiment. In 6 of these dogs, one side of the trigone was cut transversely 3 mm below the ureteral orifice. In the other 4 dogs, a 2—3 mm segment of the trigone was excised (Fig. 5).

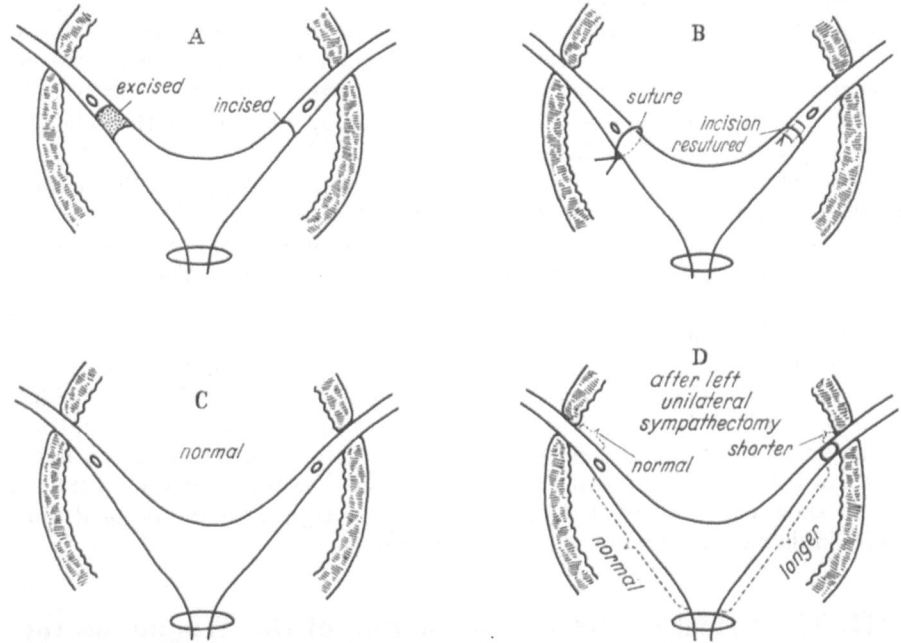

Fig. 5 A—D. Diagrammatic representation of the canine trigone and various surgical experiments done. A. Simple transverse incision 2—3 mm distal to the ureteral orifice on the left side and excision of segment of 2—3 mm length on the right side. B. Transverse incision and resuturing on the left side with control suture under the right trigonal muscle. C. Normal trigone. D. Postsympathectomy trigonal changes. Note lateral displacement of the ureteral orifice and shortening of the intravesical ureter on the left (side of sympathectomy) (TANAGHO, HUTCH, MEYERS and RAMBO, 1965)

Cystograms were performed on all dogs immediately postoperatively and at frequent intervals for 45 days thereafter.

In all 10 dogs, reflux occurred immediately postoperatively on the operated side; it persisted for a variable period and eventually stopped. The shortest time required for the reflux to stop was 10 days and the longest was 28 days, with a mean of 17 days. Once the reflux had stopped it did not recur. Cases with simple incision recovered earlier than cases with excision of a segment.

It was noted that when the lateral border of the trigone was incised transversely, the cut edges immediately spread apart for a distance of 3—4 mm. The
ureteral orifice and the intravesical ureter moved upward and the trigonal muscle
moved downward the bladder neck.

II. The Effect of Cutting the Nerves that Supply the Trigonal Muscle

It is known that the trigonal muscle is supplied by the sympathetic nerves
while the detrusor muscle is innervated by the parasympathetic fibers. Thus
TANAGHO, HUTCH, MEYER and RAMBO (1955) assumed that sympathectomy would
cause paralysis of the trigonal muscle on the operated side which would result
in vesicoureteral reflux. On 5 dogs examined beforehand and shown not to have
reflux, unilateral left lumbar sympathectomy (L_3, L_4, L_5) was done. In each dog,
the distance between the ureteral orifice and the bladder neck was measured at
the time of operation. Nothing was done locally to the ureterovesical junction. In
each of these dogs a cystogram was obtained immediately after the operation and
at frequent intervals thereafter.

None of the immediately postoperative cystograms showed reflux; however,
repeated cystograms showed reflux developing on the same side as the sympathectomy in all 5 dogs, and in 2 dogs, on the contralateral side as well. The
earliest appearance of reflux was 5 days after operation and the latest was
22 days, with a mean of 14 days.

Three of the dog were sacrificed 30 days postoperatively, and the others 60
and 90 days postoperatively. In all dogs, the distance between the ureteral orifice
and the bladder neck had increased on the side of sympathectomy when compared
with the measurements made at the time of the operation. The length of the
intravesical ureter on the side of the sympathectomy was consistently shorter
than that on the contralateral side.

In 1938, LANGWORTHY and KOLB had carried out a similar experiment in
which they did unilateral sympathectomies on 6 cats and sacrificed them at intervals of 6 to 46 days. Each cat developed a dilated ureter and a gaping ureteral orifice
on the sympathectomised side, while the contralateral ureter remained normal.
LANGWORTHY and KOLB's experiment and that of TANAGHO, HUTCH, MEYER and
RAMBO show conclusively that cutting the nerve supply to the trigonal muscle
results in damage to the ureterovesical junction.

III. The Effect of Active Contraction of the Trigone on the Resistance of the Intravesical Ureter

Several procedures were designed to record the effect of contraction of the
trigonal muscle on the rate of flow from the ureter. In 5 dogs, the lower ureter was
dissected, clamped, and cut approximately 10 cm above the bladder. A cannula
was fixed in the cut distal end of the ureter. The cannula was connected to an
infusion pump apparatus adjusted to give a flow of 0.76 cc of water per minute.
In the same circuit, a sensitive transducer was placed and connected to a Grass
Strain Gauge, Model 5B, polygraph machine (Fig. 6). Water was allowed to flow
down the cannula into the ureter, through the ureterovesical junction, and into
the open bladder. The resistance of the ureterovesical junction to this constant
flow was recorded. The trigonal muscle on the side with the cannulated ureter was
then stimulated electrically. Any change in the resistance to flow caused by this

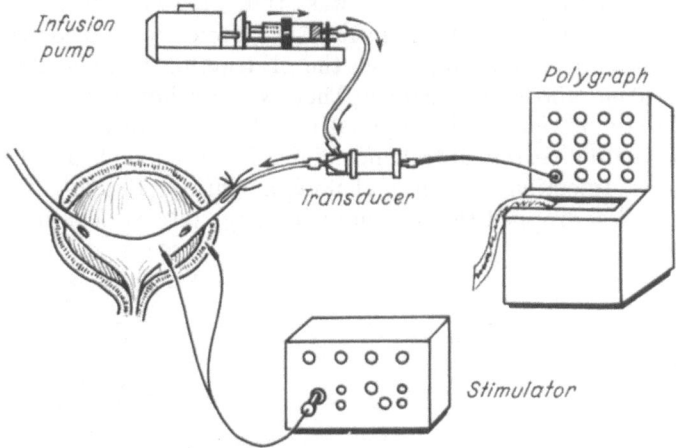

Fig. 6. Apparatus used to study resistance of the intravesical ureter to constant flow (generated by infusion pump) passing through the side arm of the transducer, then down the ureterovesical junction to the open bladder. Trigonal contraction was achieved by an electrical stimulator or by drugs (TANAGHO, HUTCH, MEYERS and RAMBO, 1965)

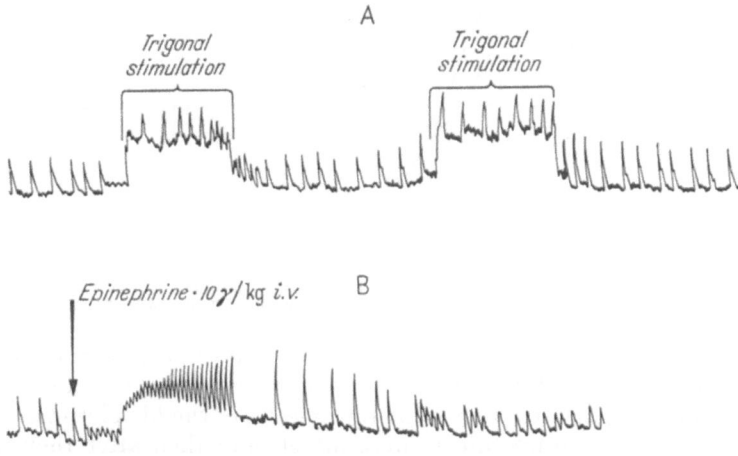

Fig. 7 A and B. Graphic recordings of changes in intraluminal pressure of the intravesical ureter. A. Note appreciable rise in pressure with electrical trigonal stimulation. B. Marked effect of intravenous injection of epinephrine (TANAGHO, HUTCH, MEYERS and RAMBO, 1965)

active trigonal contraction was recorded. Fig. 7 A shows that as soon as the trigonal muscle contracted, the resistance at the ureterovesical junction increased, manifest by a sharp and maintained rise in the resistance to flow. When the stimulation of the trigonal muscle ceased, the resistance to flow returned to its pre-stimulation level. This experiment was repeated several times and yielded identical results. A similar or even more pronounced effect was recorded after epinephrine (10 γ/kg) was injected intravenously (Fig. 7 B).

A similar experiment was designed to show even more directly that contraction of the trigonal muscle increases the resistance of the intravesical ureter and that this increased resistance is the result of active contraction of the walls of the intravesical ureter. Four dogs were studied by this method. A small rubber balloon (0.5 ml) was mounted on a 3F ureteral catheter. Through the opened bladder, this catheter was passed up the ureteral orifice until the balloon lay exactly within the intravesical ureter. By the use of a sensitive transducer connected to a strain gauge, changes in pressure inside the balloon could be recorded (Fig. 8). In this way, when the trigonal muscle was stimulated electrically, any changes in the intraluminal pressure of the intravesical ureter were recorded. To avoid the effect of urine accumulating above the obstructing balloon, the ureter was cut and tied 10 cm above the bladder. There was an immediate marked rise in pressure with stimulation of the trigonal muscle; pressure returned to baseline when the stimulation ceased. Intravenous injection of epinephrine (10 γ/kg) gave similar changes. Injection of isopropyl norepinephrine (10 γ/kg), which relaxes all smooth muscles, dropped the pressure within the intravesical ureter to a very ow level.

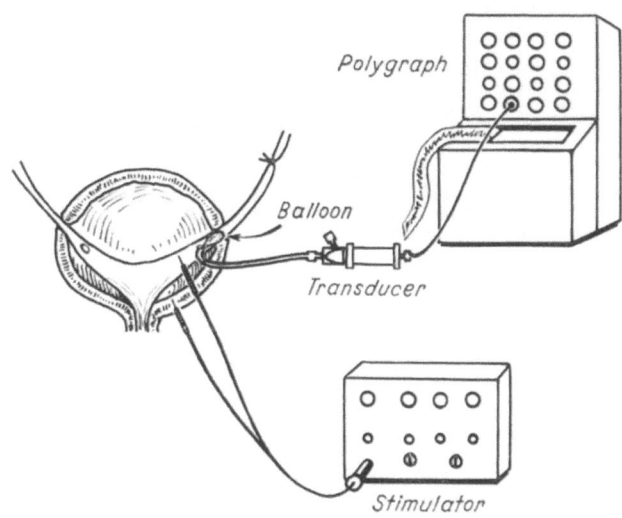

Fig. 8. Apparatus used to record changes in intraluminal pressure of the intravesical ureter using a small balloon lying within the intravesical ureter and connected to a sensitive transducer. Effect of trigonal contraction produced by electrical stimulation or drugs was recorded (TANAGHO, HUTCH, MEYERS and RAMBO, 1965)

To demonstrate the function of the posterior support of the intravesical ureter as a fulcrum against which the trigonal muscle compresses the ureteral walls, the 2 experiments were repeated, first while the muscles of the bladder that lie behind and support the intravesical ureter were intact, and then after they had been cleanly incised with a pair of surgical scissors. It was quite apparent that trigonal muscle contraction is much less effective after the posterior support of the intravesical ureter has been removed.

These tests showed conclusively that active contraction of the trigonal muscle increases the resistance of the intravesical ureter by bringing its walls into tighter apposition. They also showed that this action is seriously impaired if the continuity of the intravesical ureter and trigone is interrupted, or if the posterior support to the intravesical ureter is damaged.

IV. The Effect of Active Contraction of the Trigone in Tightening and Occluding the Intravesical Ureter during Voiding

This was difficult to determine because the action to be measured occurs during voiding when cystoscopic examination is impossible.

By studying multiple cystourethrograms taken from the true lateral projection during voiding (method of KJELLBERG, ERICSSON and RUDHE, 1957), it is possible to locate accurately the upper border of the trigone, as well as the bladder neck and the verumontanum. Such cystourethrograms show marked shortening of the distance between the upper border of the trigone and the verumontanum as micturition progresses, implying active contraction of the trigone during voiding.

Having learned that the trigone contracts during voiding and that trigonal contraction increases the resistance in the intravesical ureter, TANAGHO, HUTCH,

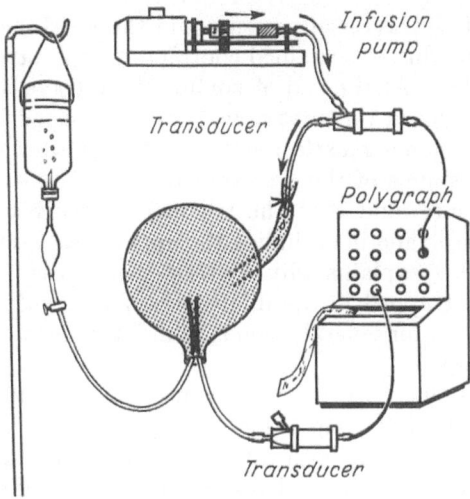

Fig. 9. Technique used to record changes in resistance of intravesical ureter to constant flow with various phases of bladder distension and voiding. The bladder was gradually filled by gravity through one catheter and intravesical pressure was recorded through the second catheter (TANAGHO, HUTCH, MEYERS and RAMBO, 1965)

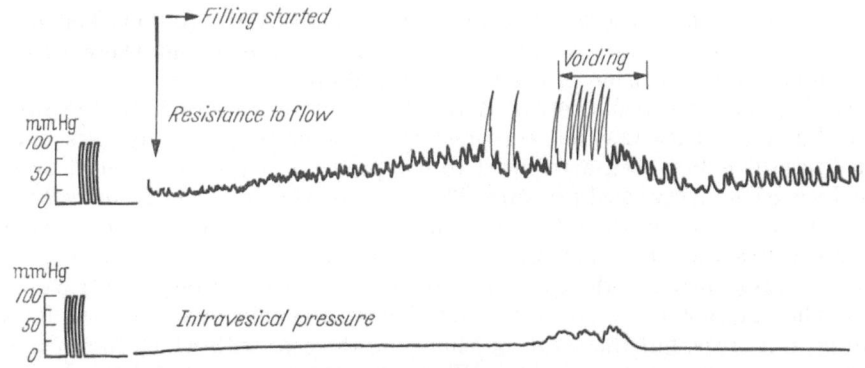

Fig. 10. Simultaneous graphic recording of intravesical pressure and resistance to flow through the intravesical ureter. Note marked rise in resistance of the intravesical ureter (due to trigonal stretch) to flow of fluid through it as the bladder fills, and minimal rise in the intravesical pressure. Also note the pre-voiding sharp rise which is also associated with a rise in intravesical pressure (due to maintained trigonal contraction during the act of voiding), and the gradual drop in resistance to flow after the end of voiding (due to gradual relaxation of trigonal muscle) (TANAGHO, HUTCH, MEYERS and RAMBO, 1965)

MEYER and RAMBO next attempted to determine if voiding increases the resistance in the intravesical ureter. To accomplish this, the ureter of a dog was cut, cannulated, and connected to a sensitive transducer and to the infusion pump described in the previous experiments. Two catheters were passed into the bladder through the urethra: one for filling the bladder, and one for recording intravesical pressure (Fig. 9). While recording the resistance of the intravesical ureter to a constant flow of water (0.76 cc/minute), the bladder was gradually filled until the dog voided around the catheters. Thus, were obtained simultaneous recordings of: 1. the intravesical pressure, and 2. the resistance of the intravesical ureter during bladder filling and voiding. The graph obtained (Fig. 10) shows a very slight rise in intravesical pressure during filling of the bladder, and a marked increase in the resistance of the intravesical ureter which was out of proportion to the degree of rise in the intravesical pressure. This was interpreted as the effect of distension which stretched the trigonal muscle and increased its tone.

With the contraction of the detrusor muscle and the start of voiding there was a rise in both the intravesical pressure and the resistance of the intravesical ureter, the latter being due to sustained contraction of the trigonal muscle during the whole act of voiding. At the end of voiding the intravesical pressure dropped sharply but the resistance at the intravesical ureter persisted for about 20 seconds, indicating that the trigonal muscle relaxes gradually after voiding ceases. This suggests that the resistance of the intravesical ureter is not solely dependent on the intravesical pressure, and that the trigonal muscle plays an active role in opening the bladder neck and in occluding the intravesical ureter during distension and voiding. This also explains why contrast medium seen in cinefluoroscopic voiding movies to be trapped in the dilated lower ureter during voiding cannot move on into the bladder for several seconds (sometimes as long as 15 to 20 seconds) after voiding has stopped.

In an effort to demonstrate that the detrusor muscle and the trigonal muscle function separately, the following experiment was carried out, again in a dog. With the bladder partly filled, both intravesical pressure and resistance to flow at the intravesical ureter were recorded under the effects of antagonistic stimuli. The intravenous injection of epinephrine ($10 \gamma/kg$) induced a marked rise in the resistance to flow without any change in the intravesical pressure. Mecholyl ($5 \gamma/kg$), a cholinergic drug which acts on the detrusor muscle alone and not on the trigone, was then given intravenously. This evoked a marked rise in the intravesical pressure without the slightest increase in the resistance to flow. The intravesical pressure was almost as high as that during voiding, but there was no change in degree of occlusion of the intravesical ureter.

It has been shown conclusively by Kiil (1957) and was confirmed by Tanagho, Hutch, Meyer and Rambo that the pressure in the ureter gradually rises as the bladder fills. This has been attributed to compression of the intravesical ureter by the increasing intravesical pressure. This last experiment suggests instead that the increase resistance in the intravesical ureter is the result of increased trigonal tone and contraction. Gradual distension of the bladder stretches the trigonal muscles, and this increases their tone and their occlusive effect on the intravesical ureter. When mecholyl was given, the detrusor muscle contracted and the intravesical pressure rose, but there was no change in the ease with which fluid passed down the ureter and into the bladder. When epinephrine was administered under identical conditions, there was no increase in intravesical pressure, but the trigonal contraction that resulted caused a marked increase in the resistance to the flow of fluid through the ureter.

E. Pathology of the Ureterovesical Junction

Proper closure of the ureterovesical valve is partly dependent on the length of the intravesical ureter. In general, reflux is less likely to occur through longer segments then through shorter ones although a minimum "competent" length has not been established. Paquin (1959) stated that the ratio of the length of the intravesical ureter to its width should be 4 or 5 to 1. Susceptibility to reflux

varies greatly among species. As many as 86 per cent of rabbits studied by
GRAVES and DAVIDOFF (1924) showed reflux, whereas the normal human bladder
almost never shows reflux. GRUBER (1929a, 1929b) showed experimentally that
the trigone is poorly developed and the intravesical ureter is short in animals
very susceptible to reflux; in humans and experimental animals in which reflux
is rare, the trigone is well-developed and the intravesical ureter is long. He also
showed that destruction of the intravesical part of the ureter in humans produces
reflux. From fresh cadavers he removed the bladder along with the lower ureters
and part of the urethra. After tightly suturing the urethra, he removed the roof
of one intravesical ureter through a small opening in the dome of the bladder.
In 8 of 9 specimens so treated, vesical distension produced reflux in that side.
Reflux never occurred in the undamaged side. Surgical procedures to correct
reflux are all based on this principle and their aim is to lengthen the intravesical
segment of the ureter.

The length of the intravesical ureter is affected by structural changes at the
ureterovesical junction, vesical edema, congenital anomalies, and maturation.

I. Structural Changes at the Ureterovesical Junction

Normally, in adults, the ureteral orifice lies 13—15 mm below the ureteral
hiatus. Reduction of this distance can be accomplished in 2 ways: 1. the ureteral
orifice can remain fixed and the ureteral hiatus can migrate downward; 2. the
ureteral hiatus can remain stationary and the ureteral orifice can migrate upward
toward it.

1. Conditions in which the Ureteral Hiatus Moves Downward in Relationship to the Ureteral Orifice

In any bladder that is undergoing severe trabeculation, the ureteral hiatus,
which is a weak point in the bladder wall, dilates. Invariably, the cause of the
trabeculation is either neurogenic disease or infravesical obstruction. The dilatation of the ureteral hiatus occurs at the expense of the supporting muscles that
iie behind and support the intravesical ureter. As the hiatus dilates, a saccule
forms (Figs. 11 and 12) and the length of the intravesical ureter shortens. In
severe cases, the intravesical ureter, the ureteral orifice, and the lateral border of
the trigone become engulfed in the saccule. These severe structural changes at
the ureterovesical junction almost always lead to reflux.

Saccules that form secondary to obstruction neurological disease are called
secondary saccules. Primary saccules occur in children with smooth-walled
bladders. They are an intermittent herniation of vesical mucosa through the
ureteral hiatus (Fig. 13). Apparently there is a weakness between the roof of the
ureteral hiatus and the roof of the intravesical ureter which allows this hernia to
form (HUTCH, 1961a). Although these saccules are quite different from the
saccules that occur in trabeculated bladders, both are equally harmful to the
ureterovesical junction.

2. Conditions in which the Ureteral Orifice Moves Upward in Relation to the Ureteral Hiatus

As discussed earlier, the position of the ureteral orifice on the mesodermal
component and particularly its position in relationship to the bladder wall is
dependent upon the tonus of the longitudinal muscle in the trigone pulling downward and the tonus in the ureter pulling upward. Upward movement of the

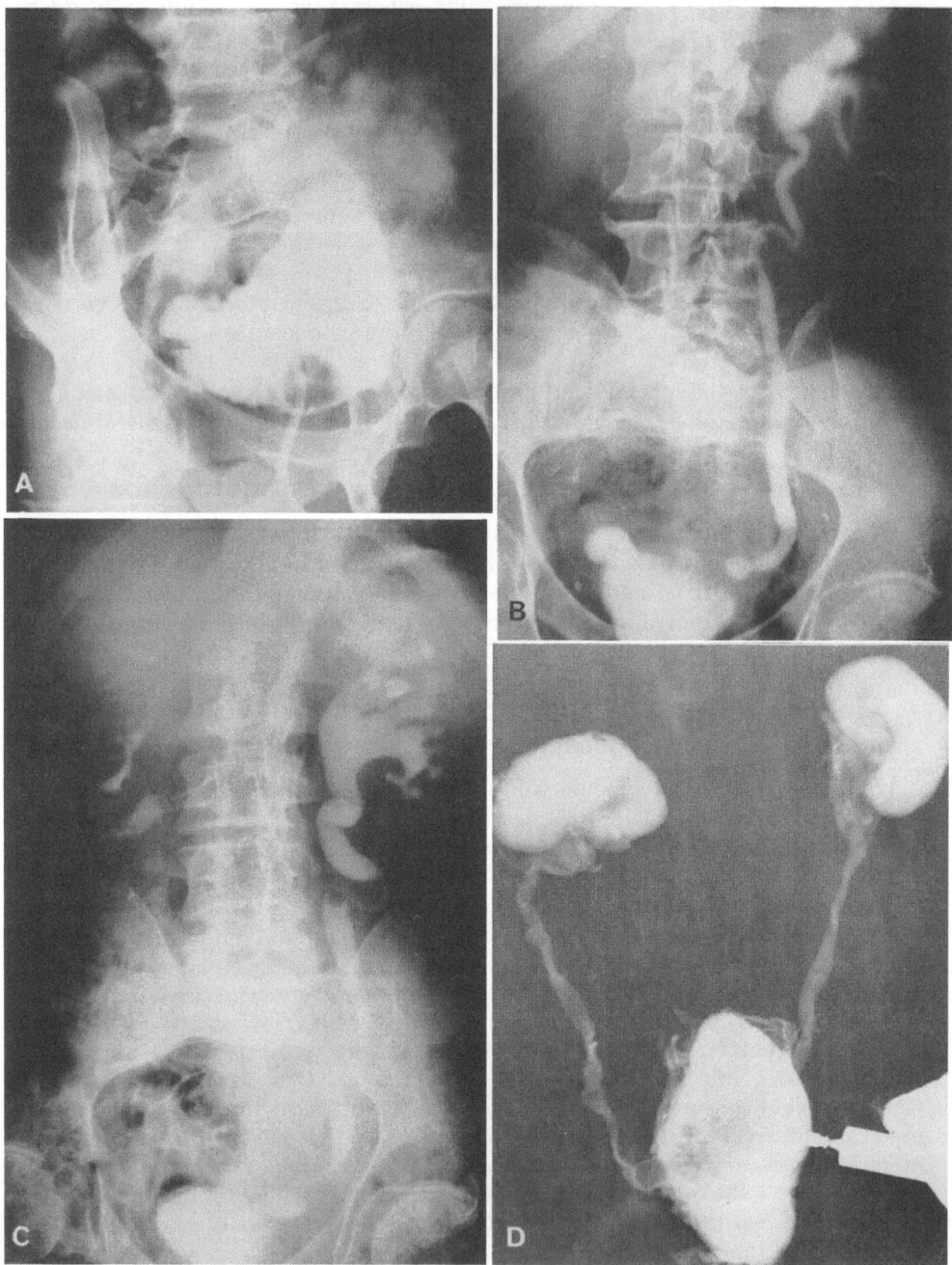

Fig. 11 A—D. Four cystograms demonstrating saccule formation at the ureterovesical junction in neurologic and obstructed vesical disease. A, B, and C are of patients with cord injury. D is a postmortem specimen of a patient with obstruction due to prostatic hypertrophy. The bladder and saccule at the right ureterovesical junction were filled with air (HUTCH, 1966 b)

ureteral orifice is usually due to weakness of the trigonal muscle and is believed by some to be the etiology of primary reflux (HUTCH and TANAGHO, 1965).

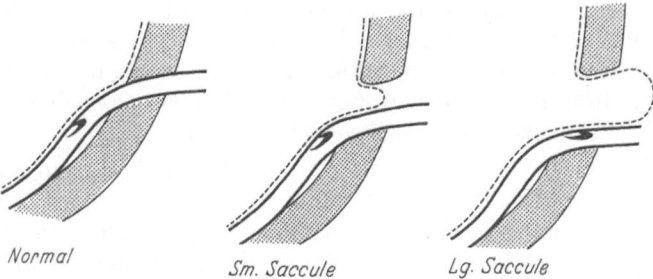

Normal   *Sm. Saccule*   *Lg. Saccule*

Fig. 12. Schematic drawing demonstrating the pathogenesis of saccule formation in neurogenic and obstructed bladders. The ureteral hiatus dilates allowing the saccule to form. The intravesical ureter and ureteral orifice drop into the saccule in advanced cases (HUTCH, 1966a)

Fig. 13. Primary saccules occur in smooth walled bladder due to a primary defect between Waldeyer's sheath and the vesical wall along the roof of the intravesical ureter (HUTCH, 1961a)

Primary reflux

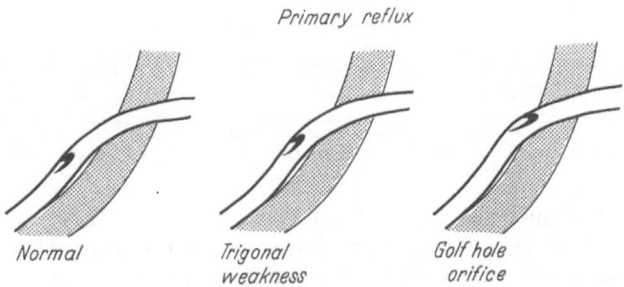

Normal   *Trigonal weakness*   *Golf hole orifice*

Fig. 14. Schematic drawing demonstrating the etiology of primary reflux. Note that there is a long intravesical ureter in the normal situation; as the trigone weakens, the orifice moves upward toward the ureteral hiatus shortening the intravesical segment. With a golf hole orifice, the ureteral orifice lies directly over the ureteral hiatus (HUTCH, 1966a)

In primary reflux, the "golf hole" orifice (one created when the ureteral orifice is superimposed on the ureteral hiatus) is a characteristic finding (Fig. 14) The concept of congenital lateral ectopia of the ureteral orifice advanced by McGovern, Marshall and Paquin (1960) and supported by Ambrose and Nicolson (1962) goes far toward proving that, in the development of a "golf hole" orifice, the ureteral hiatus remains fixed and the ureteral orifice does indeed migrate upward and outward to assume a position over the hiatus. These investigators noted that the trigone of a refluxing ureter appears large when compared with the trigones of normal patients of comparable age.

In patients with bilateral reflux, all the trigonal dimensions (right orifice to bladder neck, left orifice to bladder neck and right orifice to left orifice) are enlarged. In patients with unilateral reflux the trigone may appear asymmetrical with the distance from the refluxing orifice to the bladder neck being considerably greater than that from the normal orifice to the bladder neck.

The experiments by TANAGHO, HUTCH, MEYER and RAMBO, reported in section D, all indicated that downward pull of the trigone was part of the normal physiology of the ureterovesical junction. Thus, any weakness in the trigone would allow the orifice to assume a position above and lateral to its normal position. This would place the ureteral orifice abnormally close to the ureteral

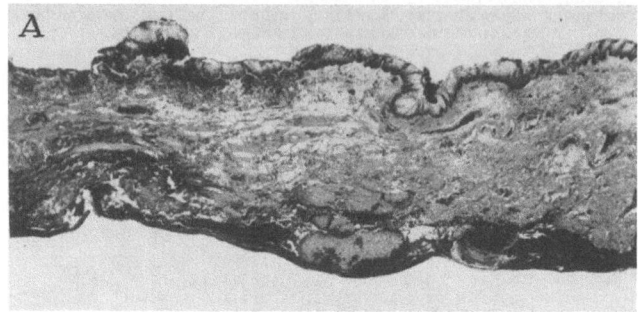

Fig. 15. Longitudinal histologic section of trigonal tissue. A. From case of primary reflux. B. From normal bladder of comparable age. Note the marked deficiency in trigonal musculature in primary reflux compared to the heavily muscular normal trigone. H. and E., ×10 (TANAGHO, HUTCH, MEYERS and RAMBO, 1965)

hiatus and shorten the intravesical segment of the ureter. Many variations in the degree of this pathologic process occur, from the ureteral orifice that still appears essentially normal with only a slight shortening of the intravesical segment to the gaping orifice where no intravesical ureter remains. In adults, an upward migration of the ureteral orifice of about 13 mm is required to create a "golf hole" orifice; in the newborn, it is created by an upward shift of only 5 mm.

That an abnormally high position of the ureteral orifice is caused by excessively powerful contractions of the longitudinal muscles of the ureter in the presence of a normal trigone must be considered. It is certainly logical. However, the conclusion of TANAGHO, HUTCH, MEYER and RAMBO that trigonal weakness is the etiology of such abnormality is based on their experience with 35 ureteric advancement operations during which the lateral border of the trigone was excised surgically and studied microscopically. Almost without exception, the trigones of these patients with primary reflux were attenuated thin structures with sparse smooth muscle fibers, as compared with normal trigones in patients of comparable age (Fig. 15).

II. The Effect of Vesical Edema on the Ureterovesical Junction

Many investigators of reflux have concluded that infection plays a part in its etiology. An abundance of clinical evidence supports this conclusion. Diseases that produce prolonged cystitis, such as tuberculosis or interstitial cystitis (Hunner's ulcer), are often accompanied by reflux. The presence of bladder stones, which produces marked edema of the vesical wall, is sometimes complicated by reflux and ureteral dilatation. After removal of these stones, the reflux stops and the dilated upper urinary tract returns to normal. CAMPBELL, in 1951, stated "With the clearing of the vesical infection and disappearance of edema and infiltration of the ureteral meatus and intramural segment, the normal function of the ureterovesical valve is usually restored and the reflux no longer occurs." Clinically, KJELLBERG, ERICSSON and RUDHE (1957), HANLEY (1962) and others have observed that reflux may disappear when infection has been overcome.

Experimental evidence that bladder infection produces reflux has been presented in the following studies. AUER and SEAGER (1937) produced reflux by injecting saline solution under the bladder mucosa around the ureteral orifice. GRAVES and DAVIDOFF (1924) found that insertion of stones into canine bladders increased the incidence of reflux. KJELLBERG, ERICSSON and RUDHE (1957) have been able to demonstrate roentgenologically a relationship between edema of the ureteral orifice and reflux. GRUBER (1929a, 1929b) as well as CAMPBELL (1951) suggested that infection converted the intravesical ureter into a rigid tube that could not be compressed.

A recent study of the anatomy of the ureterovesical junction by HUTCH, AYRES and LOQUVAM (1961) offers a plausible explanation of the manner in which vesical edema can cause reflux. At its caudal end (the ureteral orifice) the intravesical ureter is fixed to the vesical mucosa, and at its cranial end it is attached to the bladder wall through Waldeyer's sheath. Therefore, any edema occurring between the bladder mucosa and muscularis would cause the intravesical ureter to ride upward with the vesical mucosa. This would tend to straighten out or reduce the obliquity of the intravesical ureter. It is known that the normal intravesical ureter can move backward and forward within Waldeyer's sheath. Some authorities feel that this allows the ureter to adjust to changes in bladder filling and intravesical pressure. If the structures involved were swollen with edema, the intravesical ureter would lose its freedom of movement and this protective mechanism would be lost, making it easier for reflux to occur.

III. Congenital Anomalies of the Ureter

Congenital anomalies at the ureterovesical junction that may cause reflux include ureteral duplication and ectopically placed ureteral orifices. When 2 complete ureters are present on one side, both ureters usually pass through the same hiatus but the orifice of the ureter from the upper renal segment terminates on the trigone in a position caudal and medial to the orifice from the lower renal segment. The lower orifice is usually normally placed and has an intravesical ureter of normal length. Therefore, the upper orifice must have an abnormally shortened intravesical segment. In cases where complete ureteral duplication is complicated by ureterocele, the ureterocele almost always involves the lower orifice (Fig. 16). In such cases, the upper orifice may insert into the roof of the ureterocele and be incompetent. For an interesting discussion of this subject see AMBROSE and NICOLSON (1964).

The ureteral orifice may terminate ectopically at any point along the mesodermal component in the bladder, urethra, vagina, or seminal vesicles. Orifices opening abnormally low in the bladder or in the urethra often reflux, sometimes only during voiding. They frequently dilate and form ectopic ureteroceles. Some ureteral orifices terminate ectopically in a position cranial and lateral to their normal position (ectopy lateralis). This could result from an abnormally high

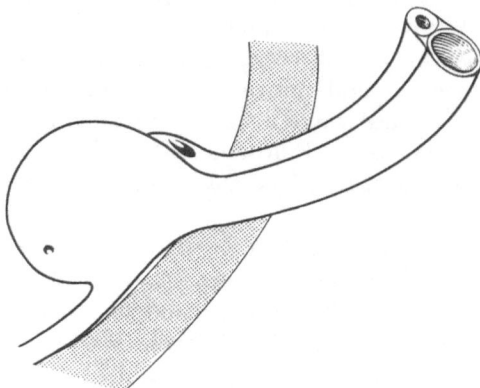

Fig. 16. Ureterocele involving the lower intravesical orifice. The upper orifice is normal but is supported by the wall of the ureterocele rather than by firm bladder muscle. Therefore the upper orifice often leaks in this situation (Hutch, 1966a)

entrance of the ureteral hiatus into the bladder, or from a weakness of the trigone that would allow the ureteral orifice to lie in an abnormally high and lateral position.

Thus, when complete ureteral duplication is present and both ureteral orifices reach the usual position, the upper orifice (from the lower renal segment) is often incompetent. When ureteral duplication is complicated by ectopy, the lower orifice (from the upper renal segment) is invariably the one that is ectopic, and it is often incompetent. When ureterocele complicates ureteral duplication, it invariably involves the lower orifice. The ureterocele refluxes rarely, but the other orifice, lying on the roof of the ureterocele, may reflux.

IV. Maturation of the Intravesical Ureter

The ureterovesical junction is much smaller in a newborn than it is in an adult (Fig. 17). By studying autopsy material, Hutch (1961b) found that the intravesical ureter in newborns varied from 4—6 mm in length, with an average of 5 mm. In adults it varied from 11—18 mm and averaged 13 mm. It takes about 10—12 years for the intravesical ureter to reach its adult length (Table 1).

V. Concept of the Marginally Competent Ureterovesical Junction

In the past, reflux has been regarded as being either demonstrable by cystogram and therefore present, or not demonstrable and therefore absent. Patients have been classified as having reflux or not having reflux on the basis of a single cystogram. The dynamic nature of reflux has not been appreciated. We have tended to consider a ureterovesical valve as either competent or incompetent, not appreciating the fact that a given valve may be competent most of the time and yet permit reflux at other times under different intravesical conditions. There

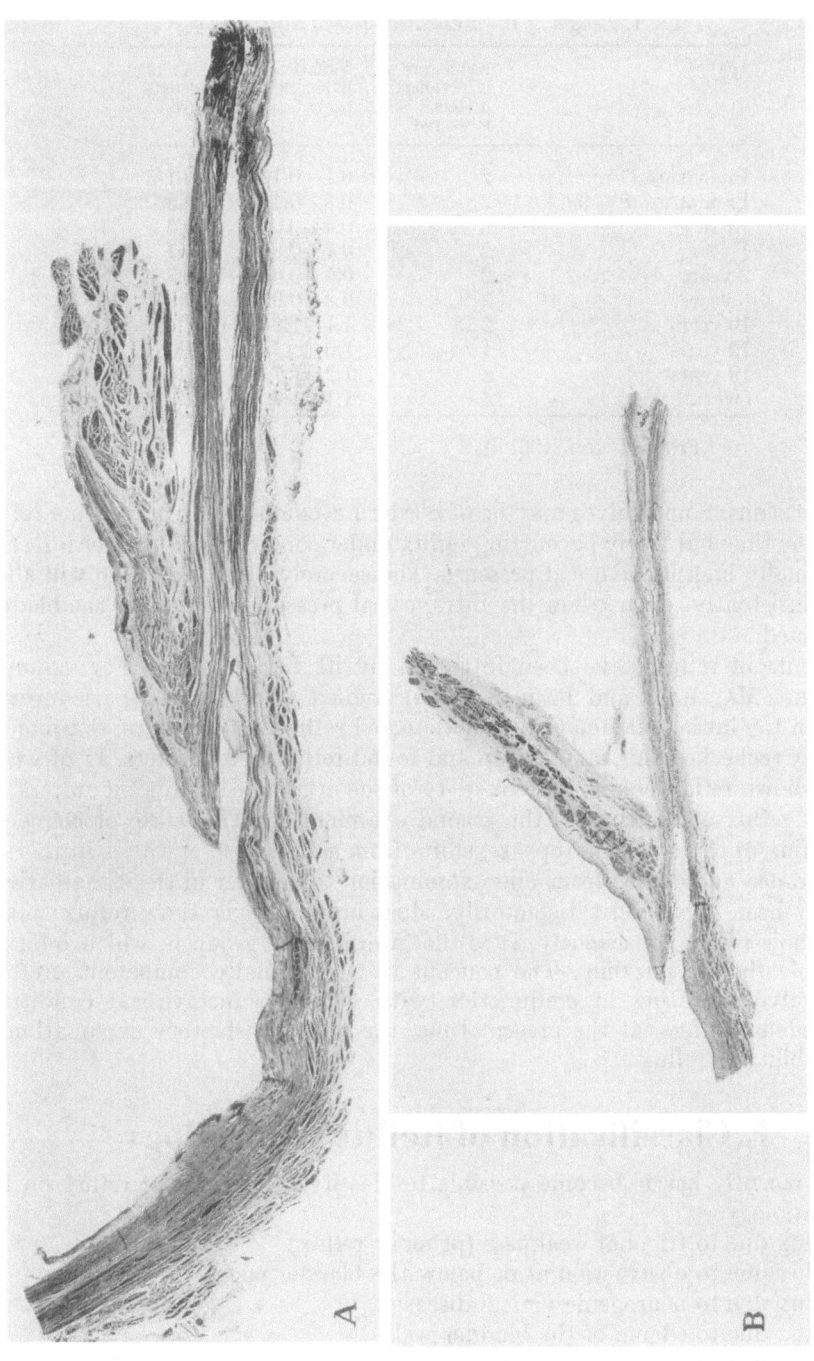

Fig. 17. A. Normal adult ureterovesical junction. B. Ureterovesical junction of normal newborn. Magnification the same in each photomicrograph (HUTCH, 1961 b)

may even be a marked variation in the degree or margin of competence of the valves in normal people. Undoubtedly many people have ureterovesical valves that, for a lifetime, meet the challenge of the changing situation within the bladder without allowing reflux. Others may have valves that allow reflux only on the

Table 1. *Length of intravesical ureters at different ages*[1]

Age	Number of intravesical ureters measured	Variation in length (cc.)	Average length (cc.)
Premature (7 mo.)	2	0.1—0.1	0.1
Premature (8$\frac{1}{2}$ mo.)	2	0.3—0.4	0.35
Birth	6	0.4—0.6	0.5
1 year	2	0.7—0.8	0.7
2 years	2	0.5—0.6	0.55
6 years	2	0.8—0.9	0.85
10 years	2	1.1—1.2	1.15
12 years	1	1.0	1.0
19 years	2	1.2—1.7	1.45
Over 21 years	14	1.1—1.7	1.3

[1] From Hutch (1961b).

rarest occasions. Some valves may be of borderline competence, preventing reflux most of the time but freely permitting reflux under conditions of bladder infection or abnormally high intravesical pressure. The severely damaged valve will allow reflux consistently, even when the intravesical pressure is low and the bladder is uninfected.

Students of reflux have been intrigued by its fickle nature. For example, McGovern, Marshall and Paquin (1960) studied a series of 43 patients with reflux. On the initial examination they detected reflux in 64 ureters. Six months later, they rechecked this same group and found reflux in 61 ureters, 11 of which had not shown reflux before. Moreover, 14 of the original 64 refluxing ureters did not show reflux at the time of the second examination. This state of change in which reflux appears and disappears, shifts from side to side or varies in its time of appearance and degree from one examination to another in the same patient, has never been explained satisfactorily. Most ureters never show reflux; many ureters show reflux consistently. It is the intermediate group in which reflux is so variable that is puzzling. The concept of a marginally competent uretero-vesical valve operating in conjunction with changing intravesical conditions (vesical edema) offers, at the present time, the most satisfactory explanation of the variability of reflux.

F. Classification of Reflux by Etiology

Only recently has it become possible to classify most cases of reflux on the basis of etiology:
 I. Reflux due to trigonal weakness (primary reflux)
 II. Reflux due to obstruction at or below the bladder neck
III. Reflux due to neurogenic vesical disease
IV. Reflux due to edema of the bladder wall
 V. Reflux due to congential ureteral anomalies
VI. Reflux due to iatrogenic causes

I. Reflux Due to Trigonal Weakness

By far, the commonest cause of reflux is trigonal weakness. Here, the defect in the valve is in the mesodermal component rather than in the bladder wall. As has been mentioned earlier, the position of the ureteral orifice in the mesodermal component depends on the longitudinal smooth muscle in the ureter pulling the orifice upward and the trigone pulling it downward. Trigonal weakness, therefore, results in an abnormally high position of the ureteral orifice and a short or absent intravesical ureter. Most cases of ectopy lateralis can be explained on the basis of trigonal weakness.

The term primary reflux was used to describe the type of reflux that resulted from primary disease of the ureterovesical valve as opposed to reflux secondary to obstructive or neurological disease. Now that the etiology of this type of reflux is known to be trigonal weakness, the term primary reflux is not longer a good one. Reflux caused by trigonal weakness is probably more exact.

II. Reflux Due to Obstruction at or below the Bladder Neck

Obstruction to the outflow of urine from the bladder may cause reflux. The ureteral hiatus is dilated because of changes occurring in the bladder wall (trabeculation and saccule formation) as it strains to void against the obstruction. As the ureteral hiatus dilates, the supporting tissue behind the ureter shortens and eventually breaks down, thereby shortening the intravesical segment.

Obstructive lesions that will produce these changes include: meatal stenosis; urethral stricture; urethral valves; fibroelastosis of Bodian; benign or malignant prostatic hypertrophy; and idiopathic stenosis of the bladder neck.

III. Reflux Due to Neurogenic Vesical Disease

In neurogenic vesical disease, the pathogenesis is identical to that of the obstructed bladder, in that the valve is damaged because of changes that occur in the bladder wall (trabeculation, saccule formation, alterations of tone). Under these influences, the ureteral hiatus dilates and the support behind the intravesical ureter is lost. This results in shortening of the intravesical ureter and reflux.

IV. Reflux Due to Edema of the Bladder Wall

Inflammation of the bladder wall sometimes causes reflux by causing a collection of edematous fluid between the vesical mucosa and muscularis interfering with the efficiency of the ureterovesical valve. Inflammatory reflux usually occurs in conjunction with some other kind of reflux. It serves as an added insult that will make a marginally competent valve reflux.

V. Reflux Due to Congenital Anomalies

Anomalies at the ureterovesical junction that may cause reflux include ureteral duplication and ectopically placed ureteral orifices.

VI. Reflux Due to Iatrogenic Causes

Iatrogenic reflux results from surgical procedures at the ureterovesical junction that shorten the intravesical ureter. Examples would be: ureteral meatotomy done for ureteroceles or to facilitate the passage of a stone from the ureter to the bladder; transurethral resection of the intravesical ureter, sometimes necessary to remove a bladder tumor; reimplantation of the ureter into the bladder in such a manner that reflux can occur.

G. Etiology of Nonocclusive Ureteral Dilatation (Megaloureter)

Past efforts to understand the cause of unobstructed dilated ureters have been directed toward an intensive study of the undilated segments since they were believed to represent obstructed or strictured area in the ureter. These "strictured areas" of ureter have been calibrated with probes and ureteral bougies and studied by all the tools of the pathologists, but still they remain an enigma. These methods of study have failed because they are concerned only with the structure and caliber of the ureter, and ignore the part played by the rate of flow of the urine through the ureter.

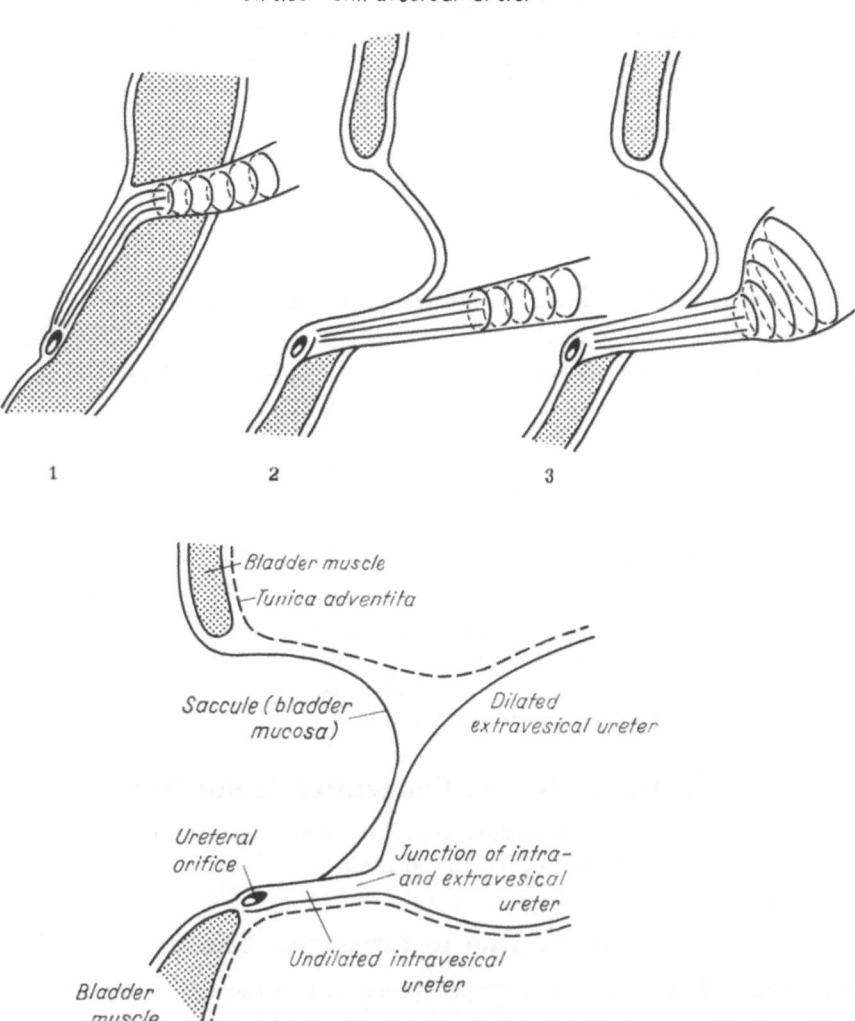

Fig. 18. Schematic drawing of the normal ureterovesical junction. Note that the intravesical ureter contains only longitudinal smooth muscle fibers whereas the rest of the ureter contains circular smooth muscle fibers as well. 2. The same junction as in 1 with saccule formation. Note that peristalsis can no longer reach the bladder because the intravesical ureter has become extravesical. 3. Dilation of the extravesical ureter beginning at the junction of the intravesical segment (SMITH, 1963)

In megaloureter, the dilatation begins at the junction between the intravesical and extravesical ureter, not at the ureteral orifice (HUTCH, 1954). In the early stages the intravesical ureter remains undilated and gives the appearance of being obstructed. This interesting segment of the ureter (intravesical segment) contains within its walls a number of features that distinguish it from the rest of the ureter (Fig. 18). First, it contains the built-in defense against reflux described in the section on physiology. Second, it lacks the expansibility present in the rest of the ureter, a point that will be discussed in detail later in this chapter. Third, its walls contain no circular smooth muscle fibers (SATANI, 1919) and therefore, the intravesical ureter cannot convey peristalsis like the rest of the ureter (a point easily confirmed by watching the intravesical ureter cystoscopically). Under normal conditions, the inability of the intravesical ureter to convey peristalsis is not caused by obstruction; when the intravesical ureter becomes extravesical as it does in most types of reflux, it constitutes a partial ureteral obstruction. The junction of the intravesical and extravesical ureter is now well outside the bladder (HUTCH, 1958). Because ureteral peristalsis ends at this point, the urine cannot be conveyed into the bladder in the normal way but must be forced through an inert intravesical ureter. Thus the intravesical ureter in an extravesical position is partially obstructed in spite of the fact that its lumen is not occluded.

In a tube that has a flow rate of 25 cc/hour and an emptying capacity of 50 cc/hour, no obstruction exists. If, however, the flow rate increases to more than 50 cc/hour, obstruction will occur. Thus, a tube will become obstructive if the flow rate through it exceeds its emptying capacity. In like manner, a ureter that is not obstructive under certain conditions may become obstructive without any change in its nature or its caliber if the flow rate through it is increased. In the human ureter, the points of lowest emptying capacity are the ureterovesical and the ureteropelvic junctions and for this reason, nonobstructive ureteral dilatation usually begins above one or both of these points.

I. Factors that Increase the Volume of Urine in the Ureter

1. Dilation of the Renal Pelvis and Ureter by Forced Intake of Fluids

In over 1,000 normal intravenous pyelograms done by the cinefluoroscopy technique, HANLEY 1959 found that 85 per cent of the renal pelves were of the open or funnel-shaped type and that 15 per cent were of the closed or box-shaped type. Both types of renal pelves could transport a normal flow of urine without dilation. After fluids were forced orally in patients with both types of pelves, the open or funnel-shaped pelves were found to transport the extra urine without dilating, but the closed or box-shaped pelves began to dilate.

2. Dilation of the Upper Urinary Tract Due to Diabetes Insipidus

Some patients with diabetes insipidus develop massive dilatation of the ureters and renal pelves. When one considers the tremendous urinary output of these patients (8—12 liters per day as an average and as high as 43 liters per day in some instances) (CECIL), it is surprising that the urinary tracts in any of these tatients fail to dilate. Whether dilatation will occur probably depends on the emptying capacity of the ureteropelvic and ureterovesical junctions. If HANLEY's observations are applied here, any patient with a combination of diabetes insipidus and closed-type renal pelvis should certainly develop hydronephrosis (obstruction of the ureteropelvic junction).

3. Dilation of the Pelvis of a Kidney Forced to Increase its Output after Unilateral Nephrectomy

In patients with unilateral hydronephrosis, the apparently normal kidney may become hydronephrotic after the diseased kidney has been removed. A pyelonephrotic left kidney was removed from a 44 year old patient. At that time, the right kidney was normal with a renal pelvis of the closed type. A pyelogram done 7 years later revealed a Grade IV hydronephrosis in the solitary kidney (Fig. 19). It is generally believed that hydronephrosis is inherently a bilateral disease:

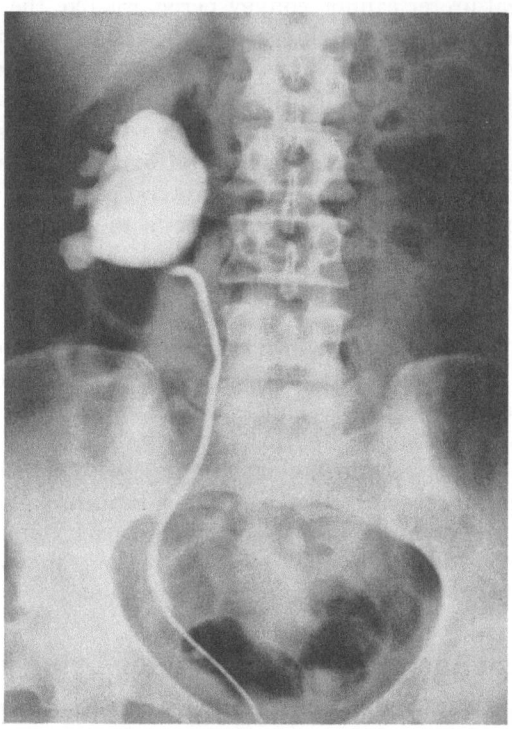

Fig. 19. Retrograde pyelogram of a 51-year-old woman whose left kidney was removed 7 years prior because of pyelonephrosis secondary to obstruction at the ureteropelvic junction. At time of nephrectomy, this was a normal kidney with a closed-type renal pelvis. Cystogram failed to reveal any reflux
(HUTCH and TANAGHO, 1965)

however, we believe that during the 44 years in which this kidney was excreting only part of the total urinary volume, it did its job without dilating, but when called upon to excrete the entire urinary volume it could not do so through that particular type of ureteropelvic function. Thus, in 7 years it became markedly dilated.

4. Reflux as a Cause of Ureteral Overfilling

The major source of ureteral overfilling is vesicoureteral reflux, and it is with this source that we are primarily concerned. The effect of this reflux depends on the type of ureter into which it is occurring. Take, for example, 3 ureters, each refluxing 50 cc of urine per hour. In one ureter, in which both the ureteropelvic and ureterovesical junctions have ample emptying capacities, the refluxed urine runs up and down the ureter without being trapped at any point. This will result in reflux into an undilated ureter (Fig. 20 A). In the second ureter with a ureteropelvic junction of the closed type, the volume of refluxed urine exceeds the emptying capacity of the ureteropelvic junction and hydronephrosis results (Fig. 20 B). In the third ureter with a ureteropelvic junction of the open type and a ureterovesical junction that cannot transport the additional urine, hydroureter will develop (Fig. 20 B). Thus, 3 ureters, each subjected to the same volume of refluxed urine, respond in different ways depending on the emptying capacities of their respective ureteropelvic and ureterovesical junctions.

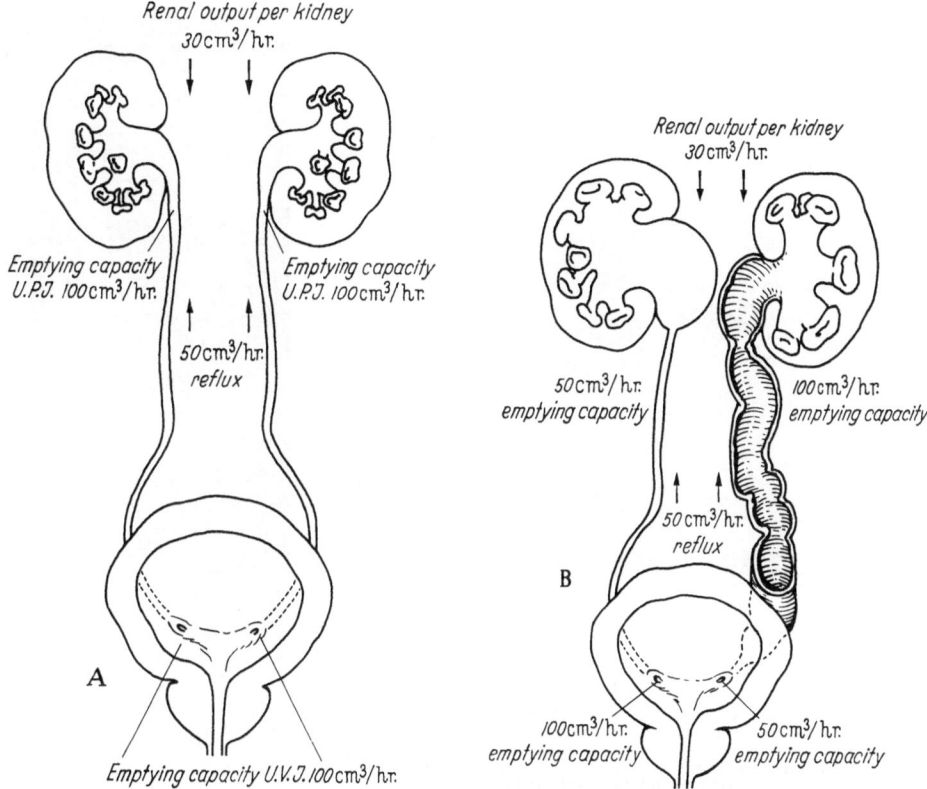

Fig. 20. A. Reflux into undilated ureters. Ureters do not dilate because the combined urinary output (renal output plus refluxed urine) does not exceed the emptying capacity of either ureter at any point. B. Hydronephrosis develops on the right side because combined urinary output exceeds the emptying capacity of the right ureteropelvic junction. Hydroureter develops on left side because combined urinary output exceeds the emptying capacity of the left ureterovesical junction (HUTCH and TANAGHO, 1965)

II. Emptying Capacity of Various Areas of the Ureter

Areas of physiologic narrowing in the ureter (ureterovesical and ureteropelvic junctions) contain more collagenous tissue than is present in the rest of the ureter. Since collagen is nonexpansible, it limits the expansibility of the ureter (Figs. 21 and 22). This, in turn, limits the maximum diameter and the emptying capacity of these narrow segments. Under normal conditions, they cause no impediment to the flow of urine down the ureter, but in situations that compel the ureter to transport a volume of urine that exceeds its emptying capacity, they become obstructive (HUTCH and TANAGHO).

III. Reflux and the Ureteropelvic Junction

Under physiologic conditions, the variations in size, shape, and emptying capacities of the normal renal pelves are insignificant because all of the different types can comfortably transport the volume of urine made by the kidney. Earlier, we discussed what happens in the various types of pelves when the flow of urine from the kidney is greater than the capacity of the upper ureter; we must now consider the pathologic changes that may be produced by an upward surge of large volumes of urine. Certainly, as refluxed urine sweeps up the ureter, the calyces and the renal pelvis are the end of the line. The urine has no place to go

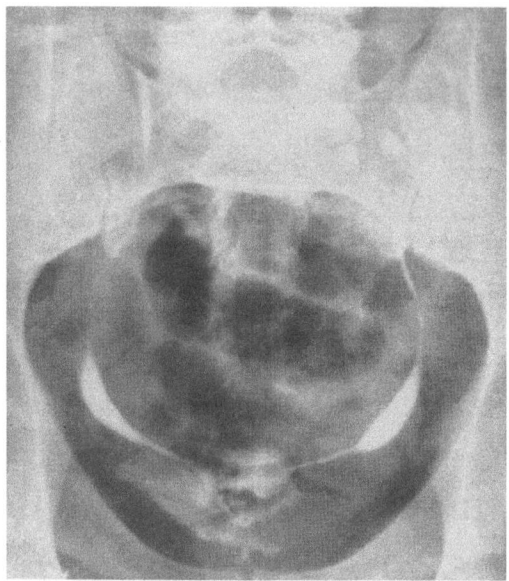

Fig. 21

Fig. 21. Withdrawal film of normal retrograde pyelogram shows ureteral orifices, undilated intravesical ureters, and slightly dilated lower spindles of extravesical ureters (HUTCH and TANAGHO, 1965)

Fig. 22. Ureterogram of ureter removed at autopsy and filled under manual pressure to produce distension. The ureter was tied off about 1 mm above the ureteral orifice. Note that the extravesical ureter distended whereas the intravesical ureter (arrow) resisted distension (HUTCH and TANAGHO, 1965)

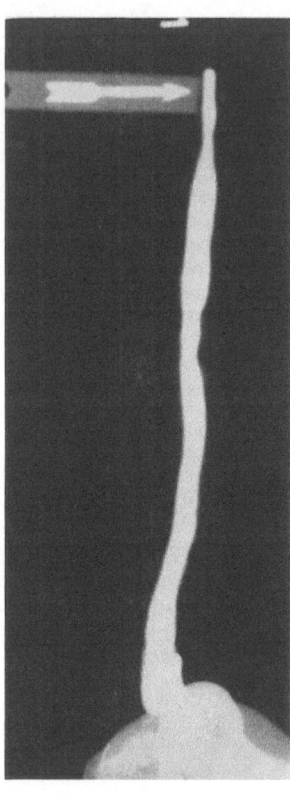

Fig. 22

except back downward, and if the road is blocked by a ureteropelvic junction that has been overwhelmed by this volume of urine, the calyces and the pelvis will have to dilate (Fig. 23). It is reasonable that this process, repeated over and over again, can result in permanent dilation of the renal pelvis (Fig. 24).

In 1962, HUTCH, HINMAN and MILLER studied voiding movies (cinefluoroscopy) and noted that some pictures of renal pelves overfilled by refluxed urine were identical to pictures characteristic of primary obstruction at the ureteropelvic junction. They postulated that the combination of reflux occurring concomitantly or at some time in the past and closed type of renal pelvis could result in the permanent dilation of the renal pelvis characteristic of obstruction at the uretero-pelvic junction.

Fig. 23. A. Retrograde pyelogram shows advanced bilateral ureteropelvic junction obstruction. B. Cystogram of same patient shows right reflux. Cystogram 6 months later failed to show any reflux (HUTCH and TANAGHO, 1965)

Fig. 24. Cystogram of 3-month old male with congenital absence of right kidney with refluxing ureter. The left side showed massive hydronephrosis and a relatively undilated ureter

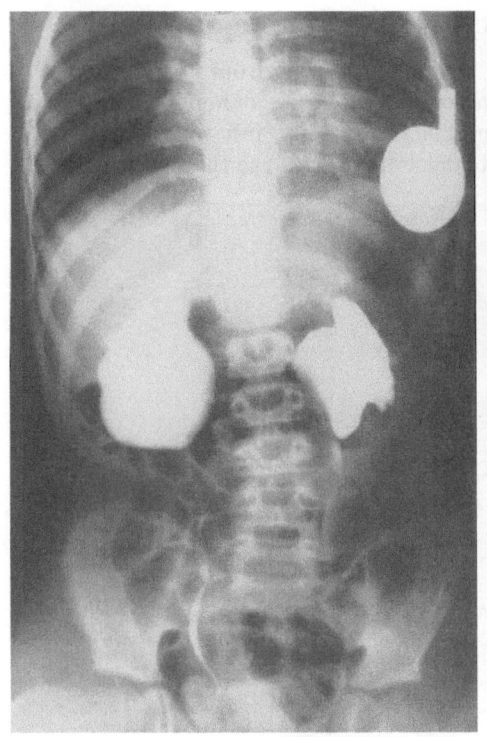 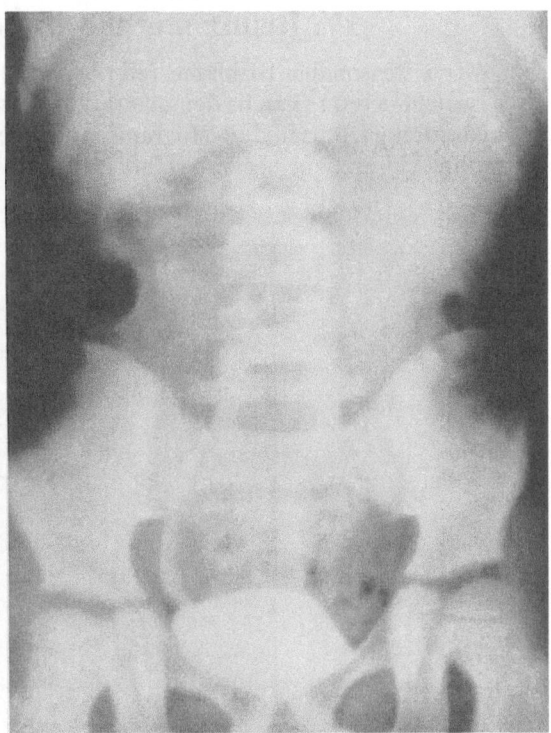

Fig. 23 A   Fig. 23 B

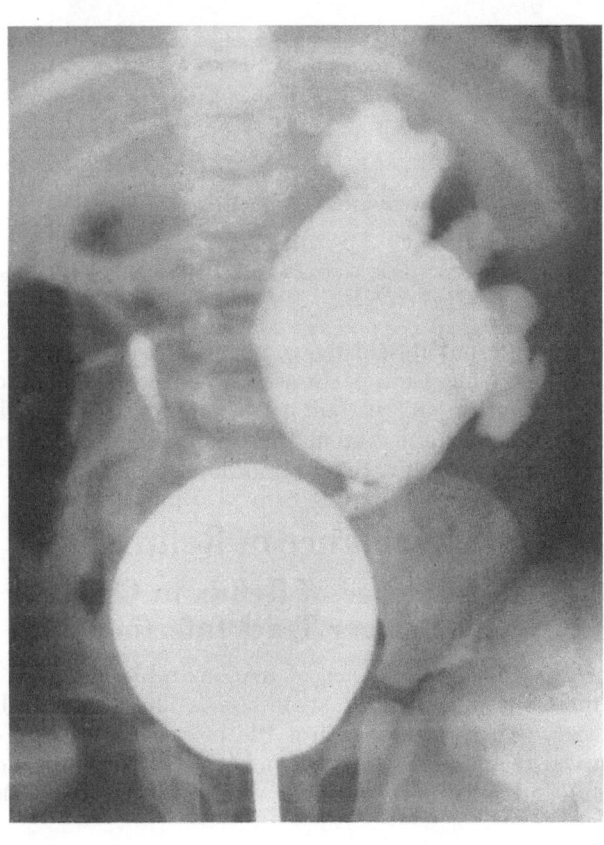

Fig. 24

IV. Reflux and the Ureterovesical Junction

It is reasonable to blame reflux for the accompanying hydroureter in those cases where reflux can be demonstrated. However, in many cases of nonobstructive hydroureter, repeated cystograms and even cinefluoroscopy fail to show any reflux. This may be because reflux is often intermittent (Fig. 25) so that a negative

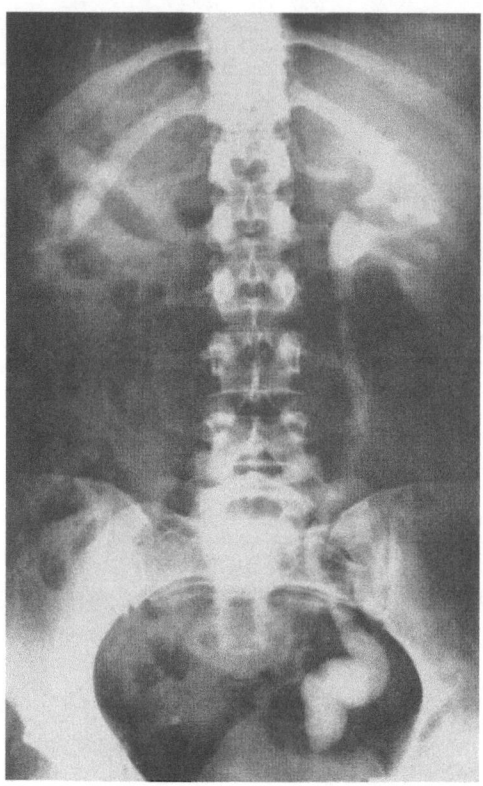

Fig. 25. Intravenous pyelogram of a 29- year-old woman with markedly dilated lower ureter who has failed to show reflux on several cystograms (HUTCH and TANAGHO, 1965)

cystogram does not rule out its existence, and because, in many people, reflux occurs during childhood but disappears as the ureterovesical junction matures. Therefore, when evaluating any patient with nonobstructive hydroureter, it is impossible to be certain that reflux did not play a role in its development, although cystograms at the time of examination are normal.

H. Incidence of Reflux

I. High Incidence of Reflux in Children with Urinary Tract Infection

The dramatic rise in the incidence of demonstrable reflux in children with urinary tract infections during the last decade is illustrated in Table 2 and is worth careful study. These figures reflect the increasing interest in the subject as well as improved techniques for the demonstration of reflux, i.e., the voiding cystourethrogram and cinefluoroscopy. At the University of California we re-

Table 2. *Incidence of reflux in children with pyelonephritis* *

Source and date	Number of cases	Percent reflux	Type of patients	Type of cystogram
CAMPBELL[1], 1951	722	12	Children with pyuria and enuresis	Immediate and after voiding
ST. MARTIN et al.[2], 1956	74	13.5	Children with bladder neck obstruction	Immediate, voiding and after voiding
KJELLBERG et al.[3], 1958	290	34	Children with urinary tract infection without obstruction or neurologic lesion	Voiding cysto-urethrography (multiple films during voiding)
FORSYTHE, WALLACE[4], 1958	51	39	Children with recurring urinary tract infection	Voiding cystogram (multiple films during voiding)
McGOVERN et al.[5], 1960	152	35	Symptomatic children	Multiple films, some during voiding
PALKEN, KENNELLY[6], 1960	53	37.5	Girls with recurring urinary tract infection	Voiding cysto-uretrography and cinefluoroscopy
GROSS, SANDERSON[7], 1961	83	47	Children with recurring urinary tract infection	Cinefluoroscopy
HUTCH et al.[8], 1962	190	48	Children and adults with urinary tract infection	Cinefluoroscopy

* From HUTCH, MILLER and HINMAN (1963a).
[1] CAMPELL, M.: Clinical pediatric urology. Philadelphia: W.B. Saunders Co. 1951. — [2] ST. MARTIN, E.C., J.H. CAMPBELL, and C. M. PASQUIER: J. Urol. (Baltimore) 75, 151 (1956). — [3] KJELLBERG, S. R., N. D. ERICSSON, and U. RUDHE: The lower urinary tract in childhood. Chicago: Year Book Publ. ,Inc. 1957. — [4] FORSYTHE, W. J., and I. R. WALLACE: Brit. J. Urol. 30, 297 (1958). — [5] McGOVERN, J. H., V. F. MARSHALL, and A. J. PAQUIN jr.: J. Urol. (Baltimore) 83, 122 (1960). — [6] PALKEN, M., and J. KENNELLY jr.: J. Urol. (Baltimore) 83, 745 (1960). — [7] GROSS, K. E., and S. S. SANDERSON: Radiology 77, 573 (1961). — [8] HUTCH, J. A., F. HINMAN jr., and E. R. MILLER: Unpublished data.

viewed the first 190 voiding movies made and found that reflux was present in 92 cases, for an incidence of 48 per cent (HUTCH, MILLER and HINMAN, 1963a). These patients were selected and many of them were known to have reflux before being sent to us. On the other hand, there were many patients in whom reflux had been demonstrated on previous cystograms who failed to show reflux during cinefluoroscopy. These patients were included in our statistics as not having reflux, but we know that on certain occasions, they did reflux. In GROSS and SANDERSON'S report, cinefluoroscopy was the initial diagnostic study. Thirty-nine cases of reflux were seen in 83 patients examined, for an incidence of 47 per cent.

If, by these most sensitive techniques, reflux can be demonstrated in nearly half of the children with recurring urinary tract infections, the true incidence of reflux in this group of patients must be even higher. For example, if reflux was demonstrated in 50 of 100 children with recurring urinary tract infections, later study of 50 who did not reflux on the first examination would undoubtedly reveal reflux in some of them on the second examination. Another indication that the true incidence of reflux is higher than our highest statistics show is the fact that reflux is more common when the bladder is infected then when it is uninfected. Thus, if these children were studied for reflux during the attacks of chills and fever, more reflux would be found than by our present technique in which the child is studied at a time when he is clinically well. On this assumption, it is conceivable that reflux is present intermittently in nearly all of these children.

II. Demonstration of Reflux in Adults with Nonobstructive Pyelonephritis

Let us now turn our attention to what is known about reflux in adults. In their study of a large group of adults demonstrating reflux, HUTCH, MILLER and HINMAN (1963a) found 22 patients with acute recurring or chronic pyelonephritis with normal-sized ureters and renal pelves (as judged by intravenous pyelography) in whom reflux was demonstrated by cystography, cinefluoroscopy or both. HODSON and EDWARDS reported 20 cases of reflux pyelonephritis revealed on cystograms. Eight of these cases were in adults. The roentgenographic appearance was that of nonobstructive pyelonephritis. They also investigated 10 patients under medical treatment for chronic nonobstructive pyelonephritis (unilateral or bilateral) and were able to demonstrate reflux into the involved kidney (or kidneys) in 9 of the 10 patients. WILLIAMS studied 16 adult patients with nonobstructive pyelonephritis and found reflux in three. MILLIEZ and coworkers investigated 95 patients with hypertension and found reflux in nine. The incidence of reflux was greatest in those patients with previous attacks of pyelonephritis, recent enuresis, double voiding or kidney pain associated with a full bladder and relief by voiding. Recently, ROSENHEIM reported a series of 31 adult patients with pyelonephritis and proven reflux.

In adults as well as in children, a normal cystoscopic examination and normal pyelogram do not rule out the possibility of reflux. Therefore, a urologic study that is complete by present standards may miss the reflux which is perpetuating the pyuria or bacteriuria. Even if we do careful cystogram and fail to demonstrate reflux, we cannot be certain that intermittent reflux is not the etiology of the urinary tract infection. The statement that reflux may be present when the urinary tract is infected and disappear after the infection is cured is just as valid in adults as in children. Thus, reflux may be missed by studying these patients when they are clinically well. Many children and some adults cannot void during the cystogram, yet in some patients reflux occurs only during voiding. Undoubtedly these factors cause us to miss many cases of reflux even when we look for it.

I. Effects of Reflux

Recent evidence suggests that reflux may play a role in the etiology of several urologic diseases not heretofore associated with reflux. These diseases include obstruction of the ureteropelvic junction, chronic pyelonephritis and atrophic pyelonephritis. We are also beginning to understand how infection can be perpetuated in a urinary tract in which the ureterovesical valves have become incompetent.

I. The Role of Reflux in Obstruction at the Ureteropelvic Junction

HANLEY's discovery in 1959 of the effect of flooding a "closed" or "box-shaped" renal pelvis by over-hydration has been discussed in a previous section. Briefly, such dilatation occurs when the emptying capacity of the ureteropelvic junction is exceeded by urine produced from above. If this occurs repeatedly, the

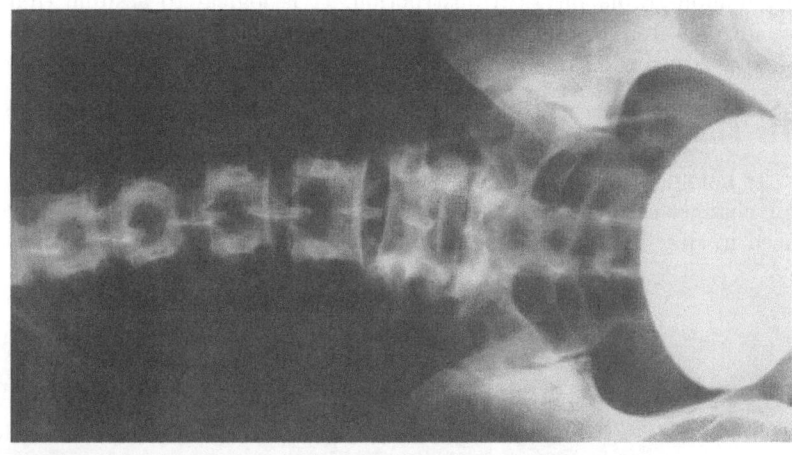
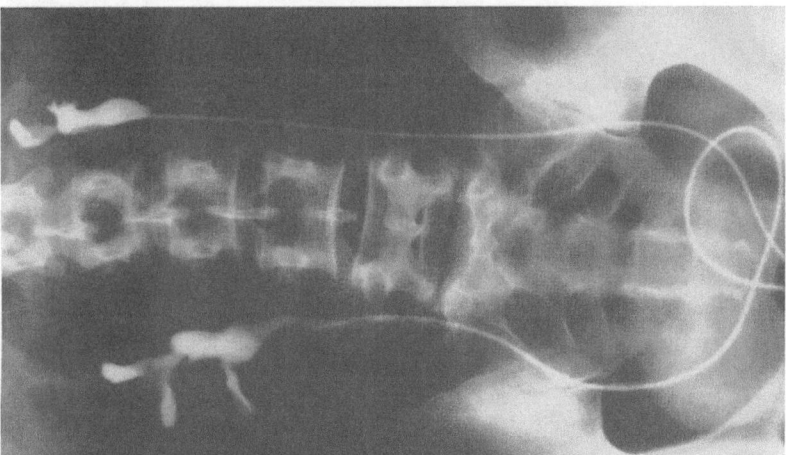
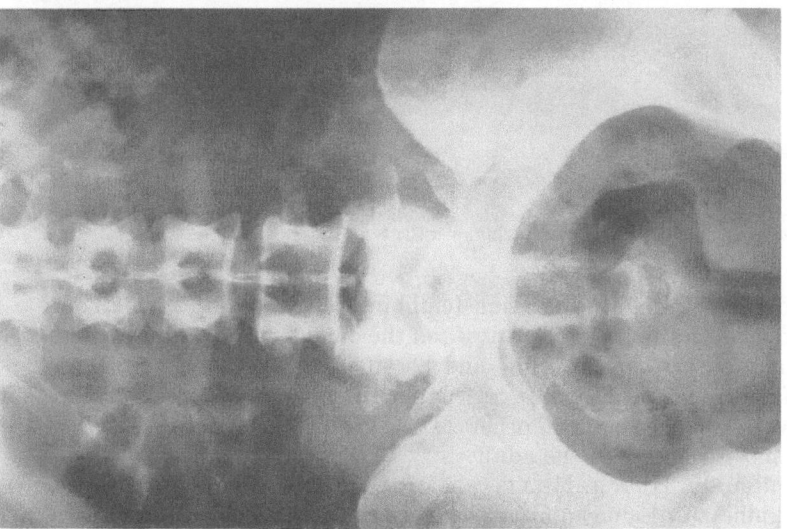

A   B   C

Fig. 26 A—C. Films of a 32-year-old woman who began to have intermittent attacks of chills, fever and pyuria at age 22. She had no associated history of attacks of cystitis and no definite kidney pain. The BUN was 16.7 mg-% and blood pressure was 120/88 mm Hg. Urine culture showed E. coli. Urinalysis revealed intermittent pyuria. Intravenous and retrograde pyelograms showed an atrophic chronic pyelonephritic kidney on the left and a chronic pyelonephritic kidney on the right. The cystogram revealed reflux into the left undilated ureter and kidney. A. Intravenous pyelogram showing poor visualization bilaterally. B. Retrograde pyelogram showing an atrophic chronic pyelonephritic (hypoplastic) kidney on the left and a chronic pyelonephritic kidney on the right. C. A cystogram reveals left reflux into an undilated ureter. Reflux could not be demonstrated on the right. The failure to demonstrate reflux does not rule out the existence of reflux at some time in the past (HUTCH, MILLER and HINMAN, 1963 a)

classical picture of obstruction at the ureteropelvic junction is produced in the absence of any demonstrable physical obstruction. It is logical to assume that reflux of urine into the renal pelvis could also overcome the emptying capacity of the ureter and produce the same pathologic condition.

II. The Kidney with Chronic Pyelonephritis

For years, radiologists and urologists have been interpreting certain types of pyelographic changes as being indicative of chronic pyelonephritis. The kidney becomes reduced in size. The caliceal infundibula become narrowed, which leads

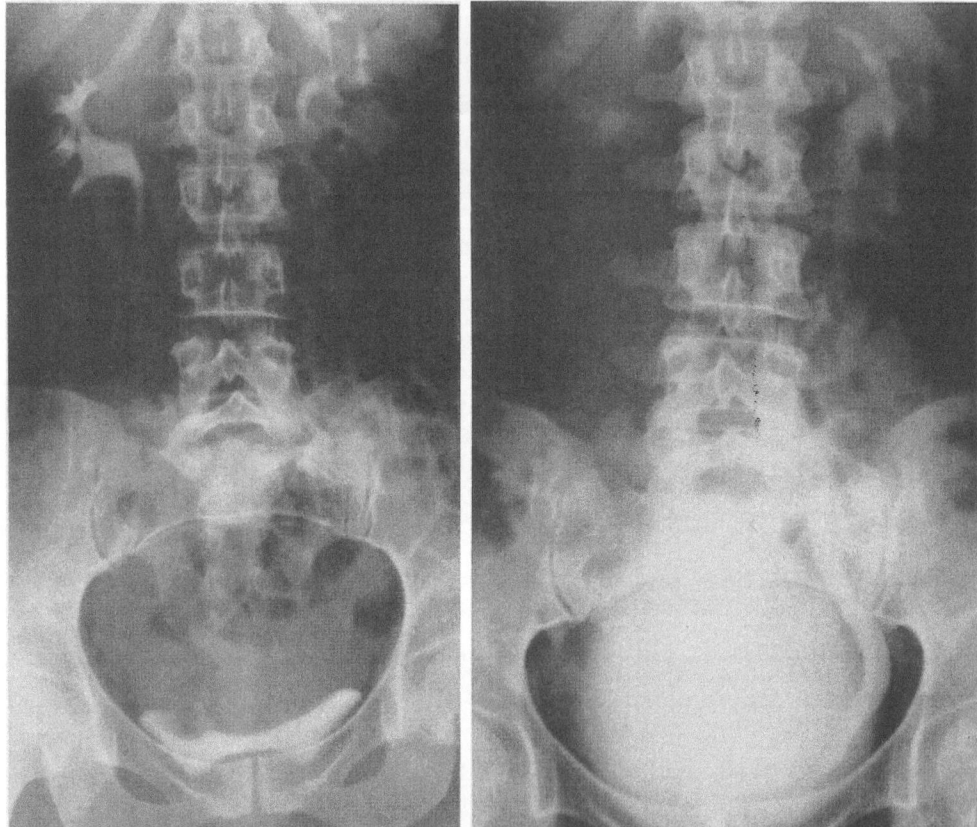

Fig. 27. A. Intravenous pyelogram of 40-year-old female with right atrophic pyelonephritis. B. Cystogram reveals reflux into slightly dilated right ureter (Hutch, Hinman and Miller, 1962)

to obstruction and therefore dilatation (clubbing) of the calyces. The thickness of the cortex between the tip of the calyx and the outside of the kidney is reduced. The calyx may lose its sharp outline and become fuzzy. There is no significant dilatation of the renal pelvis or ureter. Often these changes are pronounced in some calyces and almost absent in others. This type of kidney may be seen at any age but it is most common in the adult.

In a recent study, Hutch, Miller and Hinman (1963a) observed reflux into the infected kidney (or kidneys) of 22 adult patients with chronic nonobstructive pyelonephritis. Eleven of these patients had bilateral pyelonephritis and bilateral

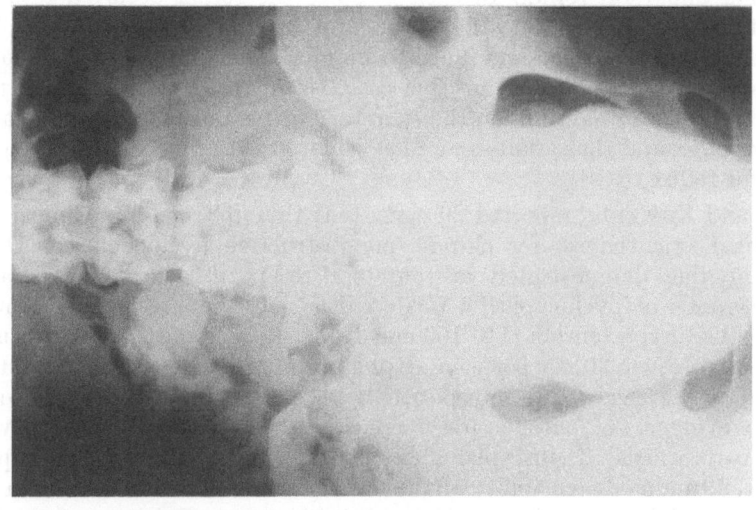

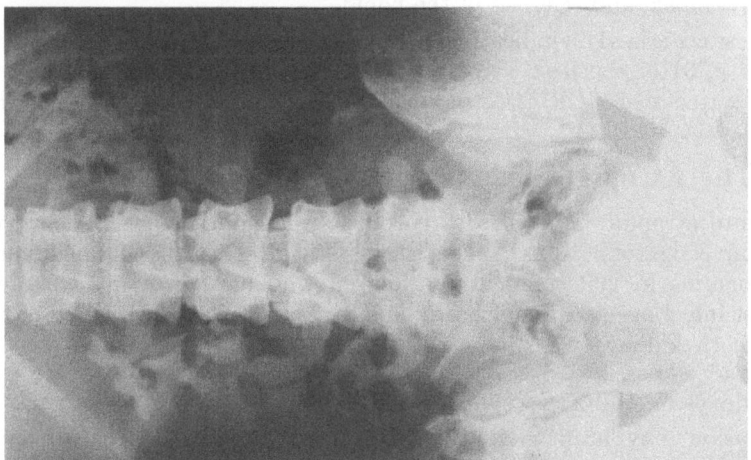

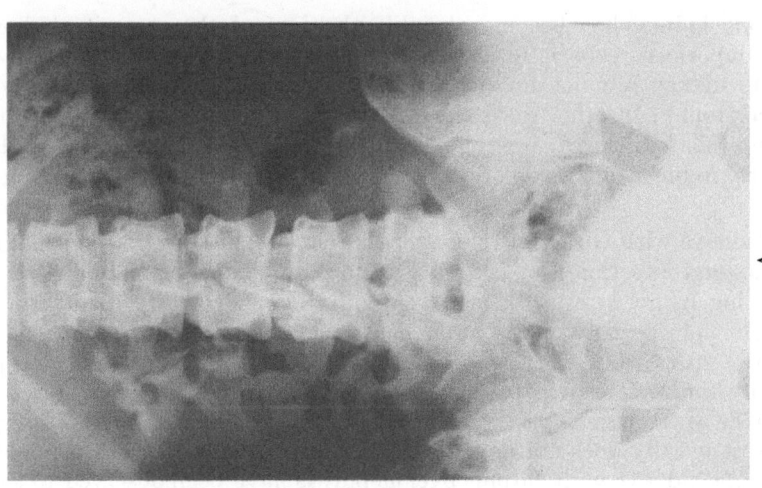

Fig. 28. A. Intravenous pyelogram of 30-year-old woman whose left kidney and ureter had been removed for pyelonephritis. The remaining kidney shows marked pyelonephritic changes. The BUN was 35 mg-%; blood pressure was normal. B and C. Cystograms of same patient demonstrating reflux into the right kidney (HUTCH, HINMAN and MILLER, 1962)

reflux; 11 had unilateral reflux. In each instance the reflux occurred into the pyelonephritic kidney and not into the normal appearing kidney (Figs. 26, 27, 28). Five patients had hypertension and four had uremia. Four kidneys were removed and pyelonephritis was confirmed in each case. It should be reemphasized that the absence of ureteral dilatation on the excretory urogram does not rule out the possibility of reflux and that adults as well as children with pyelonephritis should be checked for reflux.

Hodson and Edwards reported 20 cases that they felt met the clinical and radiologic diagnostic criteria for chronic nonobstructive pyelonephritis. Using cinefluoroscopy, they demonstrated vesicoureteral reflux in each of the 20 cases. In 4, the diagnosis of pyelonephritis was confirmed by nephrectomy. Eight of their patients had hypertension (140/100 mm Hg or higher) and 9 had blood urea nitrogen values of over 30 mg per cent. Hodson and Edwards concluded that reflux should be searched for in any patient with one or more of the following: 1. radiologic evidence of chronic pyelonephritis, 2. very small kidneys with generalized calycectasis, 3. unexplainable nonobstructive dilatation of upper urinary tract, 4. undue distensibility of the urinary tract, 5. residual urine in the bladder, 6. unexplained renal osteodystrophy (mixture of rickets, secondary hyperparathyroidism and sclerosis of the bone).

Using these criteria as an indication for cinefluoroscopy, Hodson and Edwards found reflux in 31 of the first 37 patients they examined (83 per cent). Reflux was demonstrated in 6 of 8 cases of renal rickets examined by these authors.

III. The Atrophic Kidney with Chronic Pyelonephritis

Reflux into atrophic kidneys with chronic pyelonephritis has been reported by several investigators. Kjellberg, Ericsson and Rudhe obtained voiding cystourethrograms in 290 consecutive children who had recurring episodes of urinary tract infection, and found reflux in 99 cases (35 per cent). In 18 of those 99, one or both kidneys were hypoplastic. Hinman and Hutch, Hodson and Edwards, and Gross and Sanderson have reported reflux in atrophic pyelonephritic kidneys in adults as well as in children.

Hypertension may be a complication of the atrophic kidney with chronic pyelonephritis and some of the most successful results in the surgical treatment of hypertension have come from the removal of this type of kidney.

When one kidney has become atrophic the fate of the opposite kidney is extremely important, especially since reflux is so often bilateral. Fortunately, the opposite kidney remains uninfected in many cases and undergoes compensatory hypertrophy. In other cases the opposite kidney undergoes pyelonephritic changes and bilateral reflux can be demonstrated. If both kidneys undergo atrophy early in life, renal rickets results and bilateral reflux can often be demonstrated.

These kidneys with chronic pyelonephritis that are smaller than normal may reflect reflux and infection that occurred at an early age when the kidneys were small resulting in an arrest of normal growth. The normal-sized kidney with chronic pyelonephritis may represent changes occurring in a kidney that achieved full or partial growth before the reflux and infection developed. This concept is supported by the fact that in patients whose spinal cord has been injured and whose urinary tract was apparently normal until the time of injury, atrophic pyelonephritis usually does not develop in spite of frequent reflux and infection. However, some kidneys with chronic pyelonephritis have been observed to shrink slowly to a state of marked atrophy. Thus, atrophic chronic pyelonephritis and chronic pyelonephritis may, in some instances, represent stages of the same process.

IV. The Normal Kidney

The concept that a normal excretory urogram rules out upper urinary tract disease has been disproved in recent years by the repeated demonstration of reflux with varying degrees of hydroureter and hydronephrosis on the cystograms of patients (particularly children) whose excretory urograms are normal.

V. Reflux as the Cause of Pyelonephritis of Pregnancy

It is recognized that the pregnancies of women with a history of pyelonephritis during childhood have a much greater than average chance of being complicated by pyelonephritis. The relationship between pyelonephritis of pregnancy and chronic pyelonephritis later in life is also well known. Since vesicoureteral reflux is being recognized as a major causative agent in pyelonephritis of childhood and adult pyelonephritis, it is reasonable tos uspect that such reflux also causes pyleonephritis of pregnancy. Dilatation of the urinary tract from the pelvic brim upward develops in almost all pregnant women. It is usually bilateral, although more pronounced on the right (Fig. 29). In about 2 per cent of these women, the static urine that results becomes infected and a severe type of pyelonephritis results. Apparently the kidney becomes infected by the direct reflux of this infected urine from the bladder.

McLane and Traut found that 6 per cent of pregnant women had B. coli in cultures of urine from the bladder, but less than 1 per cent of this same group had B. coli in the cultures of urine from the right kidney. Kass (1960) studied 4000 pregnant women and found bacteria in the urine of 6 per cent. Whether the patients were in the first, second or third trimester of pregnancy had no effect on this incidence. Certainly, if the bacteriuria were coming from the kidney and were caused by the hydroureter and hydronephrosis of pregnancy, the incidence of bacteriuria should increase progressively as pregnancy advanced. Kass also showed that the 2 per cent of women who will develop pyelonephritis will come from the 6 per cent who have significant bacteriuria early in pregnancy. Thus the bladder is the source of the bacteria; whether or not the kidneys will become involved depends upon the competency of the ureterovesical valves.

Hutch, Ayres and Noll did cystograms on 13 women during or shortly after acute episodes of pyelonephritis of pregnancy or the puerperium and were able to demonstrate reflux in 6 of the 13. These same authors reviewed the pregnancies of 24 women with recurring or chronic pyelonephritis and proved vesicoureteral reflux. Thex found that: 1. Fifteen of the 24 patients had had at least one attack of pyelonephritis either during pregnancy or the puerperium. 2. These 24 women had had 57 pregnancies. Twenty-seven pregnancies were complicated by pyelonephritis (24 during pregnancy and 3 during puerperium), an incidence of 47 per cent. Eight of the pregnancies had terminated in abortion; if they are excluded from the total pregnancies, 27 of 49 pregnancies were complicated by pyelonephritis. This is an incidence of 55 per cent as compared to the incidence of 2 per cent unselected pregnancies. 3. Five of the 57 pregnancies terminated prematurely; of these, one infant died following cesarean section, one infant was blind (retrolental fibroplasia), two (twins) were stillborn, and two babies were normal.

3*

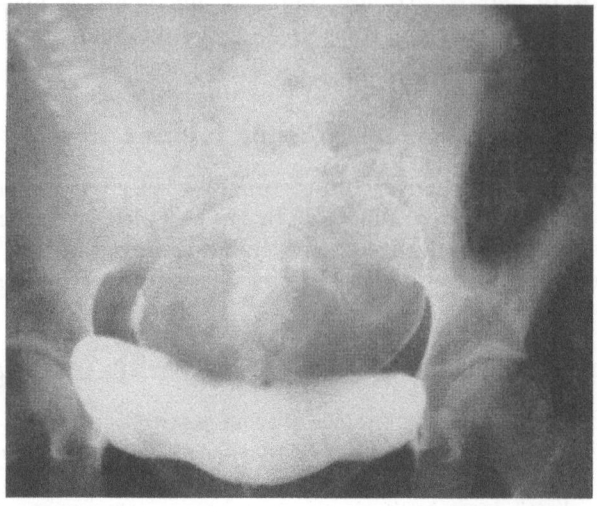

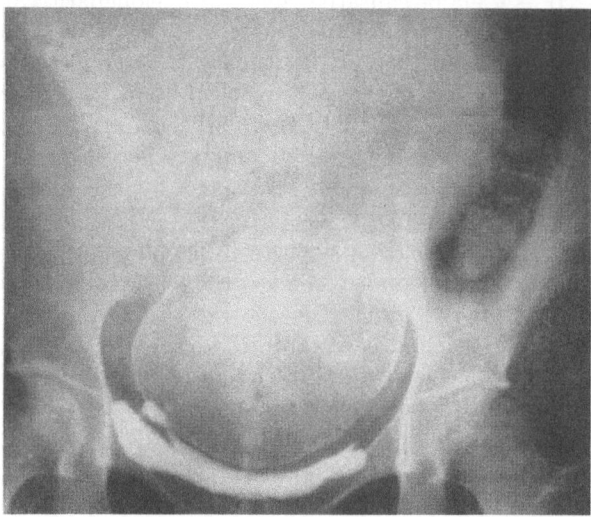

Fig. 29. Cystogram made during an attack of pyelonephritis of pregnancy showing a full-term pregnancy and right reflux before and ofter voiding (HUTCH, AYRES and NOLL, 1963)

4. Thirteen of the 24 patients gave a history of pyelonephritis during childhood. Dismal as this record is, it would have been even worse had not many of the pregnancies been protected by constant antibacterial therapy.

VI. Role of Reflux in the Perpetuation of Urinary Tract Infection

Once a bladder has become infected, its best defensive mechanism is complete voiding (Fig. 30). This action rids it of million of bacteria, retaining only those organisms contained in the small amount of urine that clings to the bladder mucosa after complete voiding. As the bladder begins to refill with sterile urine from the kidneys, the bacterial count in the bladder urine is greatly diluted. Cox and HINMAN have shown that there exists within the urinary tract some intrinsic defensive mechanism capable of overcoming the remaining bacteria. Therefore, the bladder defensive mechanism, with the aid of complete voiding, is able to rid a healthy bladder of invading bacteria. However, if ureterovesical reflux is present at the time of the infection, the situation is different. Now after voiding, even though the bladder may empty completely, there remains a residual of refluxed infected urine in one or both ureters. The bladder refills with infected urine rather than sterile urine and the intrinsic protective mechanism may be unable to cope with the additional infection returning to the bladder from the

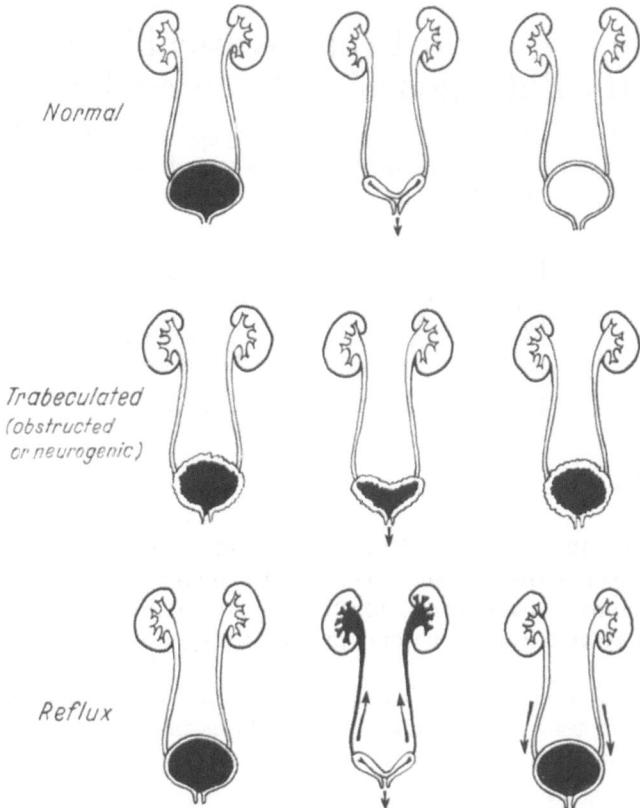

Fig. 30. Top. 3 drawings show how normal bladder rids itself of infection. Drawing on left shows normal bladder filled with infected urine (colored black). In center, almost all infected urine has been eliminated by complete voiding. On right, bladder has refilled with sterile urine. This is over-simplification; actually it takes several voidings and antibacterial activity by the vesical mucosa before the urine becomes completely sterile. Middle: 3 drawings illustrate how infection becomes chronic in a bladder that cannot empty itself completely. Drawing on left shows obstructed or neurogenic bladder filled with infected urine. After voiding (center drawing), significant infected residual urine remains (right drawing) and infection continues. Lower: 3 drawings illustrate how infection can become chronic in an obstructed bladder if vesicoureteral reflux is present. Drawing on left shows normal-appearing bladder filled with infected urine. With voiding (center), the bladder empties completely but reflux occurs leaving the ureters and renal pelves filled with infected urine. On right, the bladder has refilled with infected urine from the upper urinary tract (HUTCH, MILLER and HINMAN, 1963 b)

ureters. In this manner, the refluxed urine perpetuates the infection very much like true residual urine does. Of the 2 mechanisms, reflux is much more dangerous because it also supplies the route of access for the bacteria to reach the kidney (HUTCH, MILLER and HINMAN, 1963 b).

VII. Miscellaneous

Reflux plays a role in the spread into the upper urinary tract in certain miscellaneous conditions such as urinary tuberculosis, bladder tumors, and vesicle calculi. Tuberculosis arising in one kidney can produce bladder changes that would lead to reflux of the ureter of the uninvolved kidney allowing the tuberculous urine to infect the uninvolved kidney. A common cause of iatrogenic reflux is transurethral resection of bladder tumors in which it is necessary to resect part of the intravesical ureter. This may result in the transmission of the bladder tumor up the refluxing ureter. In paraplegic patients where small vesical calculi are common, these stones may be passed into the upper urinary tract through the refluxing ureteral orifice.

J. The Role of the Ureterovesical Junction in the Pathogenesis of Pyelonephritis

The natural history of urinary tract infections falls into 4 distinctive age groups or periods: 1. Childhood, ages 0 to 12 years. Urinary tract infections are common, mainly in females, and pyelonephritis predominates over cystitis.

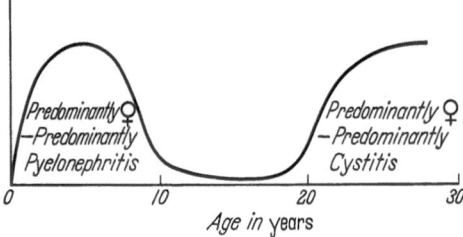

Fig. 31. Frequency of urinary tract infections. See text for explanation (HUTCH, 1961 b)

2. Teenage, ages 12 to 20 years. Urinary tract infections are rare. 3. Adulthood, 20 years and over. Urinary tract infections are common, mainly in females, and cystitis predominates over pyelonephritis (Fig. 31). 4. Prostatic age, 60 years or over. Urinary tract infections become predominantly a disease of males (HUTCH, 1962).

Each of these periods will be discussed separately, and an attempt will be made to explain their idiosyncrasies on the basis of changes occurring at the ureterovesical junction.

I. Childhood

The key to the natural history of urinary tract infection in childhood is the maturation of the intravesical ureter which has been discussed previously. Reflux does not occur in all children but the immature ureterovesical junction and the short intravesical ureter make them more vulnerable to reflux than adults. The margin of defense against reflux is smaller in childhood; the ureterovesical valve of a child, though competent under normal conditions, may be made incompetent more easily than its adult counterpart. This tendency to reflux does not necessarily manifest itself in clinical disease, nor do all children who develop urinary tract infections develop reflux. However, if this short intravesical ureter of childhood is combined with one or more pathological processes, clinical disease may result.

Let us consider the combination of a small intravesical ureter and cystitis. Undoubtedly cystitis is the commonest disease of the urinary tract. It is more prevalent in little girls because of the contamination of the urethral meatus by vaginal and fecal flora and the resulting upward migration of these bacteria along the mucosa of the short urethra (as proved by COX after a method developed by

HELMHOLZ). In childhood, particularly during the diaper stage, opportunities for such contamination are abundant. Cystitis in children is a twofold threat: 1. it produces the vesical edema required to convert a marginally competent intravesical ureter into an incompetent one. 2. it produces the infected urine necessary to make the reflux dangerous to the kidneys.

II. The Teenage Period

This is the period of relative freedom from urinary tract infections. Many individuals who had frequent episodes of pyelonephritis during childhood seem to outgrow their disease and to get well by themselves. New infections are rare. Only those who had a severe form of the disease in childhood and who sustained serious damage to the ureterovesical junction continue to have urinary tract infections in the teenage period.

Credit for this period of well-being must be attributed to the maximum development of the protective factors: 1. the ureterovesical junction has achieved its maximum development, 2. the urethra has lengthened, and 3. female hygiene is at its maximum. The days of vaginal infection, intercourse, childbirth, catheters and douche nozzles lie in the future.

III. The Adult Period

As the female patient begins to lead a life of sexual activity, urinary tract infections again become common. However, these are different from those in childhood. The state of maturity achieved by the intravesical ureter limits these infections to the bladder, with the result that cystitis is far more common than pyelonephritis in this age group. However, each patient carries into adult life her childhood legacy. Many women who had a mild form of reflux in childhood will probably never again experience reflux but others who had a more severe type of childhood reflux, particularly with some permanent shortening of the intravesical ureter, may experience reflux intermittently throughout their adult life. These patients may present a clinical pattern of recurring cystitis-pyelonephritis, manifest by flank pain, chills and fever associated with intermittent cystitis. Such patients may have chronic pyelonephritis with constant albuminuria, pyuria, bacilluria, hypertension and eventual renal failure. This infection may become resistant to therapy, not because the kidney cannot heal occasional bacterial invasion but because it is constantly subjected to reinfection until the damage is beyond repair. QUINN and KASS's observation that many asymptomatic patients have a significant number of bacteria in the bladder urine makes it possible to postulate the aforementioned sequence of events even in patients with no history of clinical cystitis.

During the adult period, pyelonephritis is predominantly a disease of women. Prostatitis resulting in prostato-cystitis and an ascending type of pyelonephritis may develop in men during this period but it is not common.

IV. Prostatic Period

Autopsy studies on the incidence of pyelonephritis show that as many men as women have pyelonephritis at the time of death (KIMMELSTIEL). In later life the incidence of pyelonephritis in men rises sharply because of their enlarging prostates and this balances the number of women who have continued having pyelonephritis from earlier adulthood. Prostatic enlargment causes vesical trabeculation and saccule formation throughout the vesical wall and produces the characteristic acquired (secondary) type of saccule at the ureterovesical junction. This damage to the efficiency of the valve, plus the high intravesical pressures required to initiate voiding, plus the bladder infection so often associated with residual urine constitute an ideal combination of circumstances for the production of pyelonephritis.

K. The Treatment of Reflux and its Complications
I. Medical Treatment

Antibacterial drugs are extremely important and will usually sterilize the urine and stop the chills and fever. However, the urinary infection and the symptoms of pyelonephritis usually recur when therapy is stopped because the basic defect in the urinary tract, the vesicoureteral reflux, has not been corrected. Such drugs may be administered intermittently to combat the clinical episodes of acute pyelonephritis or they may be given prophylactically over long periods of time. Many children with recurring attacks of pyelonephritis due to reflux have been kept clinically well for years with the daily administration of sulfonamides or nitrofurantoin. The treatment seems to work by maintaining a drug level in the bladder urine, thus aiding the bladder in repelling bacterial invasion from below.

Constant open catheter drainage (urethral or suprapubic) prevents filling of the bladder so that reflux does not occur even though the valves remain incompetent. The combination of constant bladder drainage and constant antibacterial treatment may work when drugs alone would fail.

Double or even triple voiding has been advocated as a treatment for reflux. The rationale for this therapy lies in the fact that patients with reflux into large dilated ureters retain a considerable quantity of infected urine in the ureters after voiding. Most of this "residual" urine returns to the bladder during the first 3—5 minutes after voiding and can be expelled if the patients void again. The effectiveness of double voiding is questionable, but it does afford an opportunity for parent participation in the long-range program.

Secondary reflux may be abolished by correcting the disease that is causing the reflux. For example, if reflux is due to spastic changes in the bladder wall secondary to spastic impulses from the spinal cord, the chain can be broken by interrupting the sacral parasympathetic nerves that carry the spastic impulses (Hutch, 1957; Misak; Ross). Or, if reflux is due to bladder trabeculation caused ba a true urethral obstruction (such as urethral strictures, congenital valves in the prostatic urethra and true bladder neck obstruction), correction of this obstruction may cure the reflux.

II. The Surgical Correction of Reflux
1. Types of Operation

The first operation designed to prevent reflux was introduced by Hutch in 1952 (Fig. 32). Prior to that time neoureterocystostomy was employed to anastomose the ureter to the bladder in cases where part of the bladder was resected for tumor, where the lower ureter was accidently injured or after excising the "strictured" area where the ureter entered the bladder in megaloureter. These operations often failed because of postoperative stricture and reflux. Hutch's antireflux operation (1952) had 3 main objectives: 1. that the ureter maintain its attachment to the trigone; 2. that a long intravesical ureter be created; and 3. that the bladder muscle lying behind and supporting the intravesical ureter be strong. These 3 objectives became the basis for subsequent ureterovesical surgery and, except for PAQUIN'S operation which forfeits the attachment of the ureter to the trigone, all modifications of the original operation incorporate them.

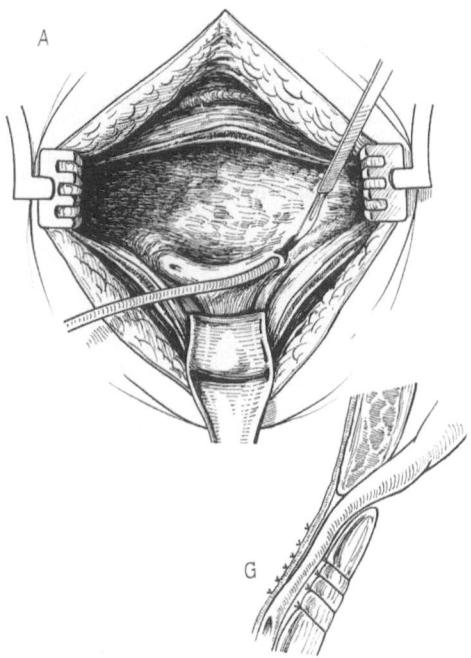

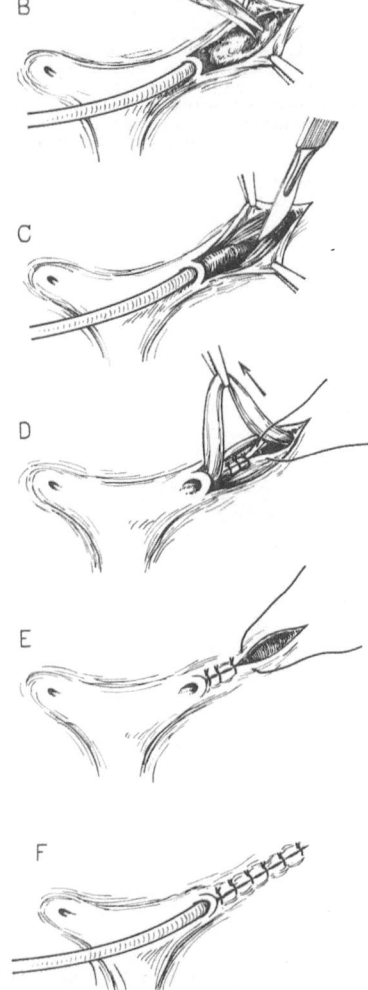

Fig. 32. The steps (A—G) in the Hutch I operation. The incision is made above the ureteral orifice completely through all the layers of the vesical wall along the anticipated course of the ureter for about 1—1½ inches. The ureter is identified in the floor of the incision and pulled into the bladder. The ureter is dissected free of any attachments until it lies freely and without tension in its new position. The attachment of the ureter to the trigone is preserved. The incision in the vesical wall is then closed under the newly-created intravesical ureter with 2 or 3 chromic catgut sutures which include all of the layers except the mucosa. Care is taken that the superior angle of the incision is not sutured too tightly as it is through this point that the ureter now leaves the bladder. The mucosa is either sutured over the transplanted ureter with plain catgut or left unsutured. The mucosa must not be sutured under the newly-created intravesical ureter as this may lead to obstruction. The ureteral catheter which is in place during the operation may be left in dwelling for a few days or it may be removed (Ethicon publication)

Two modifications are of great importance. 1. Because it is no longer considered necessary to free the lower ureter from behind the bladder, the operation can be done completely from within the bladder. This change has simplified the procedure. 2. Originally, the vesical mucosa as well as the bladder muscle was sutured under the newly formed intravesical ureter. This resulted in what is called the bridging defect in which the mucosa grows around the intravesical segment as well as under it, producing an intravesical ureter that stands up from the bladder wall like the handle of a suitcase (Fig. 33). When bridging occurs, the vesical mucosa tends to obstruct the ureter at the superior angle (the point where

42

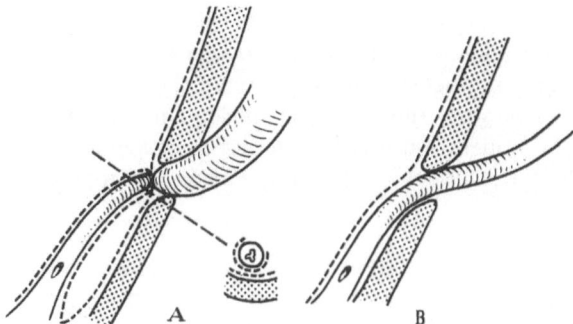

Fig. 33. A. Schematic drawing of undesirable result which sometimes occurs with Hutch I type of ureterovesico-plasty. Note that mucosa (dotted lines) has grown over the vesical muscle under the intravesical ureter and completely around the intravesical ureter. The intravesical ureter stands up from the vesical wall like the handle of a suitcase. When this defect, which is called "bridging", occurs, the ureter may be physically obstructed by vesical mucosa at the point where the intravesical ureter passes through the vesical wall at the superior angle of the transplant. B. Illustration of the desired result (HUTCH, 1963)

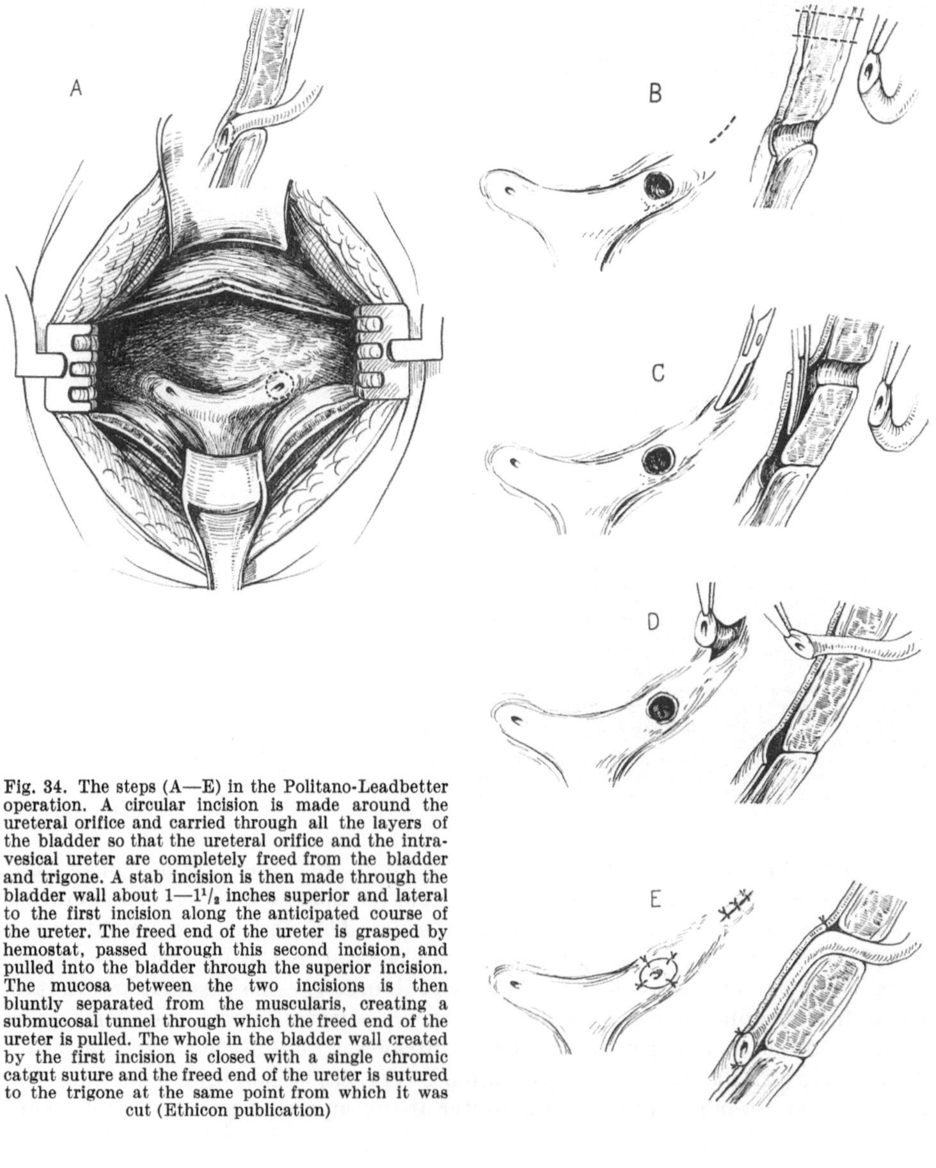

Fig. 34. The steps (A—E) in the Politano-Leadbetter operation. A circular incision is made around the ureteral orifice and carried through all the layers of the bladder so that the ureteral orifice and the intra-vesical ureter are completely freed from the bladder and trigone. A stab incision is then made through the bladder wall about 1—1$^1/_2$ inches superior and lateral to the first incision along the anticipated course of the ureter. The freed end of the ureter is grasped by hemostat, passed through this second incision, and pulled into the bladder through the superior incision. The mucosa between the two incisions is then bluntly separated from the muscularis, creating a submucosal tunnel through which the freed end of the ureter is pulled. The whole in the bladder wall created by the first incision is closed with a single chromic catgut suture and the freed end of the ureter is sutured to the trigone at the same point from which it was cut (Ethicon publication)

the intravesical ureter leaves the bladder). The mucosa forms around the circumference of the ureter almost like a suture, and partial obstruction of the ureter may result. "Nicking" this mucosal ring with a knife instantly frees the obstructed ureter. Most of the postoperative obstruction caused by the HUTCH I operation was due to this defect and can be prevented by simply suturing the mucosa over the intravesical ureter (JEWITT, 1955) or by not suturing the mucosa at all (AMBROSE and NICOLSON, 1962). On the basis of these observations, a fourth general principle of ureterovesical surgery should be added: that the vesical mucosa should never be sutured under the intravesical ureter.

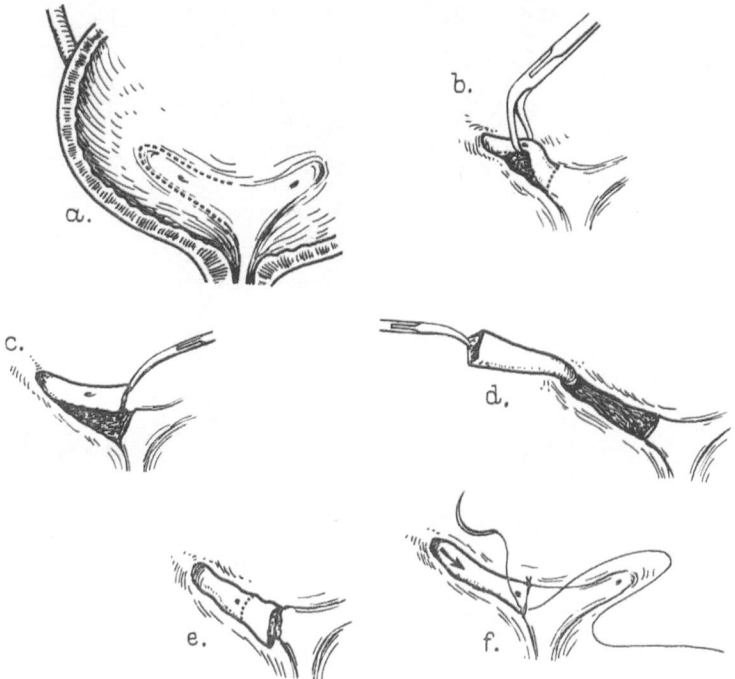

Fig. 35a—f. Ureteric advancement operation. a. Dotted lines represent incision made in the vesical mucosa. b. A towel clip is placed under the combined tube (intravesical ureter and lateral border of trigone) and dissection is continued until the combined tube is free from the vesical wall. c. The combined tube is cut from the trigone at its inferior end. d. The combined tube is reflected upward to facilitate dissection of the ureter from the vesical muscle at the ureteral hiatus. Muscle that supports the combined tube is exposed. e. The combined tube is cut off just below the ureteral orifice and the lateral border of the trigone is discarded. f. The intravesical ureter is sutured to the cut edge of the trigone. Note increased length of the intravesical ureter (HUTCH, 1963)

The Bischoff operation (BISCHOFF and BUSCH, 1961) is widely used in Europe It consists of lengthening the intravesical ureter by the creation of a tube of bladder mucosa inferior to the uretral orifice. It involves principles learned from repair of hypospadias.

The Politano-Leadbetter operation completely frees the ureter from the bladder wall by a circular incision made around the ureteral orifice (Fig. 34). A second incision is made through the bladder wall about 2—3 cm above and lateral to the ureteral orifice. The ureter, which has been freed from its original position, is brought into the bladder through the upper incision and carried downward through a submucosal tunnel to be reanastomosed with the trigone. This is an excellent operation and gives highly satisfactory results.

In Paquin's operation, the ureter is cut off outside the bladder and brought back into the bladder at a site higher up on the bladder wall. The ureter is then

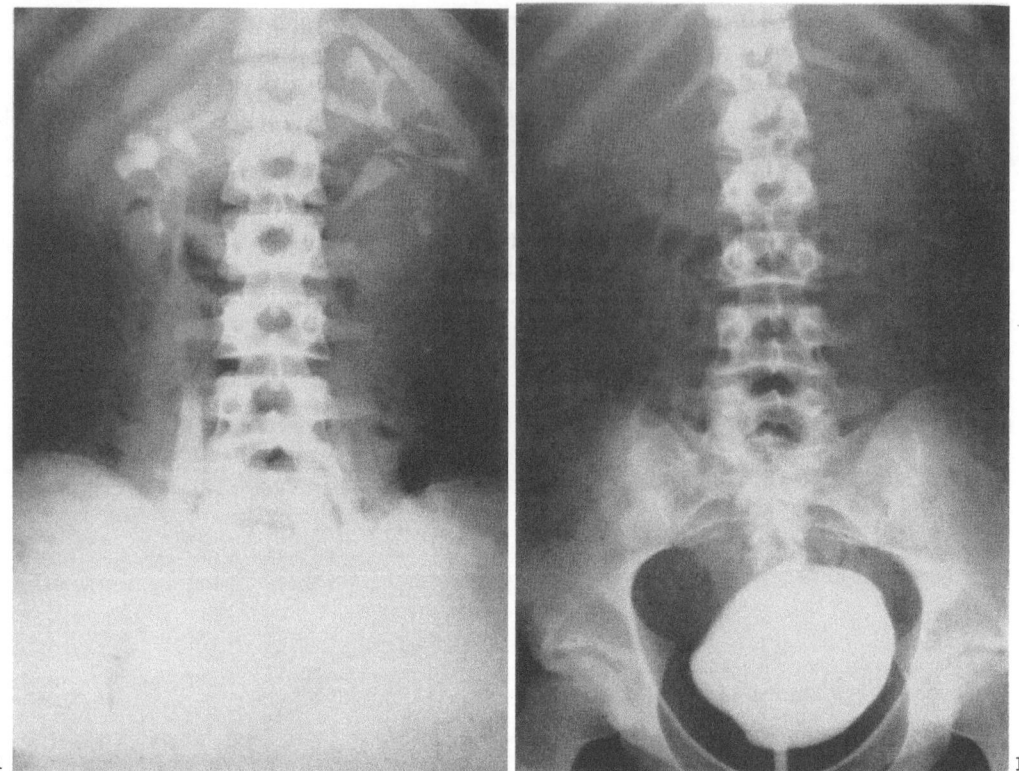

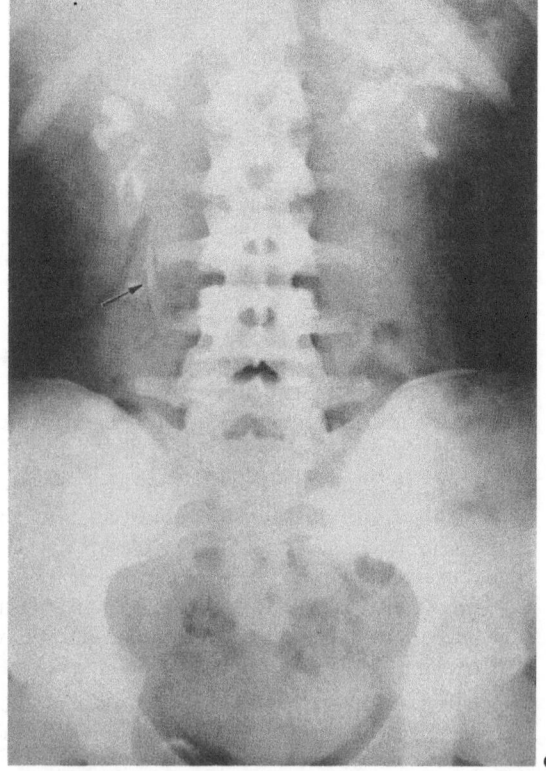

Fig. 36 A—C. Effect of Hutch II operation. A. Preoperative intravenous pyelogram of 6-year-old female with bilateral chronic pyelonephritis and
bilateral vesicoureteral reflux. B. Postoperative
cystogram showing that reflux was stopped. C. Postoperative intravenous pyelogram revealing reduction in hydroureter (arrow) and hydronephrosis

passed downward through a submucosal tunnel. To prevent stenosis of the cut end of the ureter, a nipple is formed as in the Vest operation.

The Lich, Howerton and Davis operation is done entirely outside of the bladder. The ureter is carefully dissected to its junction with the bladder wall. The bladder muscle is incised upward from the ureteral hiatus, with care being taken not to open the mucosa. The ureter is then placed into this incision against the unopened mucosa and the muscle is sutured under the ureter.

The Hutch II operation (HUTCH, 1963) is made possible by the fact that the intravesical ureter and lateral border of the trigone can be dissected free of the bladder wall beneath them. The lateral border of the trigone is excised and the intravesical ureter is advanced downward a distance equal to the length of lateral border of the excised trigone (Fig. 35). The Williams ureteric advancement operation (WILLIAMS, SCOTT and TURNER-WARWICK) is based on the same principles.

Thus, procedures to repair defects in the ureterovesical junction may be divided into 2 groups: suprahiatal including the Hutch I (1952), Politano-Leadbetter, Paquin, and the Lich, Howerton and Davis procedures, where the operation is done above the ureteral hiatus, and infrahiatal procedures such as the Bischoff (BISCHOFF and BUSCH), Williams ureteric advancement operation, and Hutch II (1963) where the operation is done below the ureteral hiatus.

2. Results of Operation

How successful is anti-reflux surgery? To be completely successful, ureterovesical surgery must meet 4 rigid criteria: 1. it must stop the reflux as determined by postoperative cystograms; 2. it must improve, or at last not damage, the upper urinary tract as determined by postoperative intravenous pyelograms (Fig. 36); 3. the attacks of recurring pyelonephritis must stop; and 4. the urine should become microscopically normal and sterile on culture even without antibacterial medication. In LEADBETTER and LEADBETTER's carefully controlled series of 28 patients and 45 ureters, these 4 criteria were met in this manner.

1. 41 of 45 operations stopped the reflux.

2. 43 of 45 operations caused no stricture of the operative site.

3. 27 of 28 patients were freed of their clinical symptoms (recurring pyelonephritis).

4. 19 of 28 patients achieved sterile urine (both microscopically and by culture).

Table 3 summarizes most of the large series published to date on the results of ureterovesicoplasties. Obstruction encountered in the early series was due to a combination of factors. First, the urologist had only the Hutch I operation, the disadvantages of which have been discussed. Second, most of the surgeons were unfamiliar with the operation. Finally, most of the patients operated upon had advanced hydroureter, often complicated by neurogenic disease. Even today, this is the type of case in which results are poorest. As technique and skill improved, and as the operations were used more to correct primary reflux and less to treat advanced neurogenic reflux, the results began to improve.

3. Indications for Operation

It is impossible to outline precise indications for surgery but the following features of each case carry weight in making the decision.

Table 3. *Analysis of published results of ureterovesicoplasies*

Source and date	Type of case	Type of operation	Number of ureters	Success	Miscellaneus
Hutch[1], 1952	Paraplegics	Hutch I	12	10	No obstruction
Hutch, Bunge and Flocks[2], 1955	Meningomyelocele and primary reflux	Hutch I	14	12	2 obstructions
Vinson[3], Hutch and Bunts, 1957	Paraplegics, meningomyelocele, obstruction and primary reflux	Hutch I	252	185	Failures due to reflux 26 obstr tions
Bunts[4], 1958	Paraplegics	Hutch I	51	38	Failures due to 8 recurrent ref 5 obstructions
Paquin[5], 1959	Primary reflux	Paquin	63	43	—
Leadbetter and Leadbetter[6], 1961	Primary reflux	Politano-Lead-better	45	41	Postoperative I.V.P. worse i 2 ureters
Palmer and Rooney[7], 1961	Primary reflux	Hutch I, Politano-Leadbetter and YV plasties	26	26	3 mild tempor dilations
Williams, Scott and Turner-Warwick[8], 1961	Primary reflux	Hutch I Politano-Leadbetter Williams ureteric advancement	35 8 16	23 8 12	No obstruction
Ambrose and Nicolson[9], 1962	Primary reflux	Hutch I	25	25	3 mild tempora dilations
Auvert and Zmerli[10], 1962	Primary reflux	Hutch I Bischoff Combined Hutch I and Bischoff	9 1 12	8 1 12	Obstruction in 2 cases (Hutch
Hutch[11], 1962	Primary reflux	Hutch II	17	13	No obstruction
Johnston[12], 1962	Primary reflux	Hutch I	21	16	5 failed to stop reflux
McGovern and Marshall[13], 1962	Primary reflux	Paquin	71	57	—
Politano[14], 1962	Primary reflux	Politano-Leadbetter	162	151	Almost no obstruction
Williams and Eckstein[15], 1965	Primary reflux	Bischoff Williams ureteric advancement Hutch I Paquin Politano-Leadbetter	21 14 48 136 57	5 12 44 113 55	
Johnston[16], 1966	Primary reflux	Politano-Leadbetter	69	64	Obstruction in 2 ureters
Scott[17], 1966	Primary reflux	Bischoff Williams ureteric advancement Paquin	11 8 66	4 7 66	
Garrett an.d Switzer[18], 1966	Primary reflux	Politano-Leadbetter	96	95	Obstruction in 4 ureters
Total			1366	1146	

References to Table 3, pag. 47.

a) The Severity of the Reflux

If the volume of the refluxed urine is large and if it occurs at low intravesical pressures, surgery is indicated. If the reflux occurs only in high intravesical pressures and the ureter is not much dilated, conservative treatment is advised.

b) The Presence of Saccules

If a saccule is present around the incompetent valve, surgery should be done. Structural changes of this magnitude will probably never heal spontaneously.

c) The Severity of the Clinical Course

The severity of the clinical course greatly influences the decision to operate. Patients who have recurring bouts of chills and fever in spite of constant anti-bacterial therapy are candidates for surgery.

d) The Conditions of the Kidneys

Progressive deterioration of the kidneys favors surgery. When reflux is present, the kidney may be damaged by the transmission of intravesical pressure into the upper urinary tract. This results in hydroureter and hydronephrosis. Only recently have we become aware of the harm that can also be done to the kidney by reflux into essentially undilated upper urinary tracts. Here the pressure factors are not significant but infection introduced into the kidney by the reflux of infected urine from the bladder produces multiple pyelonephritic scarring. The end result is chronic pyelonephritis, atrophic pyelonephritis, and even renal rickets (see section on effects of reflux). A kidney being damaged by pressure or infection or both should have the reflux stopped, if possible, to prevent further damage. It should be pointed out that when the ureter becomes greatly dilated, the chances of a successful ureterovesicoplasty diminish. A point can be reached where diversion is the only alternative.

e) The Age of the Patient

The age of the patient is of real importance in considering surgery. As pointed out in the section on maturation of the intravesical ureter, it is reasonable to hope for spontaneous improvement in children under the ages of 10—12 years. Therefore, if possible, conservative treatment should be carried out in this age group.

Actually, it is easier to outline the indications for conservative treatment than for surgery. For example, there are many children under 12 years who have

[1] HUTCH, J. A.: J. Urol. (Baltimore) 68, 475 (1952). — [2] HUTCH, J. A., G. BUNGE, and R. H. FLOCKS: J. Urol. (Baltimore) 74, 607 (1955). — [3] VINSON, C. E., J. A. HUTCH, and R. C. BUNTS: J. Urol. (Baltimore) 78, 611 (1957). — [4] BUNTS, R. C.: J. Urol. (Baltimore) 79, 733 (1958). — [5] PAQUIN jr., A. J.: J. Urol. (Baltimore) 82, 573 (1959). — [6] LEADBETTER jr., G. W., and W. R. LEADBETTER: J. Amer. med. Ass. 175, 349 (1961). — [7] PALMER jr., J. G., and D. R. ROONEY: J. med. Ass. Ga 50, 393 (1961). — [8] WILLIAMS, D. I., J. SCOTT, and R. T. TURNER-WARWICK: Brit. J. Urol. 33, 435 (1961). — [9] AMBROSE, S. S., and W. P. NICOLSON, III: J. Urol. (Baltimore) 87, 695 (1962). — [10] AUVERT, J., and S. ZMERLI: Read at Los Angeles Urological Research Congress, Los Angeles 1962. — [11] HUTCH, J. A.: J. Urol. (Baltimore) 89, 180 (1962). — [12] JOHNSTON, J. H.: Ann. roy. Coll. Surg. Engl. 30, 324 (1962). — [13] McGOVERN, J. H., and V. F. MARSHALL: Surg. Gynec. Obstet. 114, 143 (1962). — [14] POLITANO, V. A.: J. Urol. (Baltimore) 90, 696 (1963). — [15] WILLIAMS, D. I., and H. B. ECKSTEIN: Brit. J. Urol. 37, 13 (1965). — [16] JOHNSTON, J. H.: J. Pediat. Surg. 1, 145 (1966). — [17] SCOTT, J.: Arch. Dis. Childh. 41, 165 (1966). — [18] GARRETT, R. A., and R. W. SWITZER: J. Amer. med. Ass. 195, 636 (1966).

recurring urinary tract infections due to reflux. Many of them have essentially normal kidneys and can be kept free of acute pyelonephritis with intermittent or constant antibacterial therapy. Under these conditions conservative treatment seems indicated in anticipation of spontaneous remission in the teenage period. If these children begin to demonstrate renal damage or have recurring attacks of pyelonephritis in spite of long-term antibacterial therapy, or continue to reflux into the teenage period, surgery is indicated.

References

Advancing with surgery — surgical correction of ureterovesical reflux. Somerville, N. J.: Ethicon Inc. 1963.

AMBROSE, S. S., and W. P. NICOLSON: III: Vesicoureteral reflux secondary to anomalies of the ureterovesical junction: Management and results. J. Urol. (Baltimore) 87, 695—700 (1962).

— — Ureteral reflux in duplicated ureters. J. Urol. (Baltimore) 92, 439—443 (1964).

AUER, J., and L. D. SEAGER: Experimental local bladder edema causing urine reflux into ureter and kidney. J. exp. Med. 66, 741—754 (1937).

BETTEX, M.: Über den vesiko-ureteralen Reflux beim Säugling und Kind. Bern u. Stuttgart: Hans Huber 1965.

— Le reflux vésico-urétéral chez l'enfant. Urol. int. (Basel) 22, 97—123 (1967).

BISCHOFF, P.: Zur chirurgischen Behandlung des kindlichen Megaloureters. J. Urol. int. (Basel) 6, 12—49 (1958).

—, and H. G. BUSCH: Origin, clinical experiences and treatment of urinary obstructions of the lower ureter in childhood. J. Urol. (Baltimore) 85, 739—748 (1961).

BOEMINGHAUS, H.: Mega-Ureter (Betrachtungen zur Ätiologie und Therapie). Urol. int. (Basel) 4, 257—292 (1957).

BRUÉZIÈRE, J.: Le reflux vésico-urétéral chez l'enfant et le nourrisson (à propos de 70 observations). J. Urol. Néphrol. 71, 141—170 (1965).

CAMPBELL, M. F.: Clinical pediatric urology. Philadelphia: W. B. Saunders Co. 1951.

CECIL, R. L.: A textbook of medicine. Philadelphia: W. B. Saunders Co. 1939.

CHAUVIN, H.-F.: Le reflux vésico-rénal. Rapport de l'Assoc. française d'Urologie, 58e session, Paris 1964.

COX, C. E.: The urethra and its relationship to urinary tract infection. I. The flora of the normal female urethra. J. Urol. (Baltimore) (in press).

—, and F. HINMAN jr.: Experiments with induced bacteriuria, vesical emptying and bacterial growth on the mechanism of bladder defense to infection. J. Urol. (Baltimore) 86, 739—748 (1961).

GIBSON, H. M.: Ureteral reflux in the normal child. J. Urol. (Baltimore) 62, 40—43 (1949).

GRAVES, R. C., and L. M. DAVIDOFF: II. Studies on the ureter and bladder with especial reference to regurgitation of the vesical contents. J. Urol. (Baltimore) 12, 93—103 (1924).

GRÉGOIR, W.: Le reflux vésico-urétéral congénital. Acta urol. belg. 30, 286—300 (1962).

GROSS, K. E., and S. S. SANDERSON: Cineurethrography and voiding cinecystography with special attention to vesicoureteral reflux. Radiology 77, 573—585 (1961).

GRUBER, C. M.: I. A comparative study of the intravesical ureters (uretero-vesical valves) in man and in experimental animals. J. Urol. (Baltimore) 21, 567—581 (1929a).

— II. The ureterovesical valve. J. Urol. (Baltimore) 22, 275—292 (1929b).

HANLEY, H. G.: The pelvi-ureteric junction: A cine-pyelography study. Brit. J. Urol. 31, 377—384 (1959).

— Transient stasis and reflux in the lower ureter. Brit. J. Urol. 34, 283—285 (1962).

HELMHOLZ sr., H. F.: Determination of the bacterial count of the urethra: A new method with results of a study of 82 men. J. Urol. (Baltimore) 64, 158—166 (1950).

HINMAN jr., F., and J. A. HUTCH: Atrophic pyelonephritis from ureteral reflux without obstructive signs ("reflux pyelonephritis"). J. Urol. (Baltimore) 87, 230—242 (1962).

HODSON, C. J., and D. EDWARDS: Chronic pyelonephritis and vesico-ureteric reflux. Clin. Radiol. 11, 219—231 (1960).

HUTCH, C. A.: Vesico-ureteral reflux in the paraplegic: Cause and correction. J. Urol. (Baltimore) 68, 457—467 (1952).

— Nonobstructive dilatation of the upper urinary tract. J. Urol. (Baltimore) 71, 412—420 (1954).

— The treatment of hydronephrosis by sacral rhizotomy in paraplegics. J. Urol. (Baltimore) 77, 123—134 (1957).

— The ureterovesical junction. Berkeley: University of California Press 1958.

HUTCH, J. A.: Saccule formation at the ureterovesical junction in smooth walled bladders. J. Urol. (Baltimore) 86, 390—399 (1961 a).
— Theory of maturation of the intravesical ureter. J. Urol. (Baltimore) 86, 534—538 (1961 b).
— The role of the ureterovesical junction in the natural history of pyelonephritis. J. Urol. (Baltimore) 88, 354—362 (1962).
— Ureteric advancement operation: Anatomy, technique and early results. J. Urol. (Baltimore) 89, 180—184 (1963).
— A new theory of the anatomy of the internal urinary sphincter and the physiology of micturition. Invest. Urol. 3, 36—58 (1965).
— The etiology and treatment of vesicoureteral reflux. Bull. N.Y. Acad. Med. 42, 209—220 (1966 a).
— Vesicoureteral reflux. Chapter in the ureter, ed. by H. BERGMAN. New York: Hoeber Med. Div., Harper & Row, Publ. Inc. 1966 b.
— R. D. AYRES, and G. S. LOQUVAM: The bladder musculature with special reference to the ureterovesical junction. J. Urol. (Baltimore) 85, 531—539 (1961).
— —, and L. E. NOLL: Vesicoureteral reflux as cause of pyelonephritis of pregnancy. Amer. J. Obstet. Gynec. 87, 478—485 (1963).
— F. HINMAN jr., and E. R. MILLER: Reflux as a cause of hydronephrosis and chronic pyelonephritis. J. Urol. (Baltimore) 88, 169—175 (1962).
— E. R. MILLER, and F. HINMAN jr.: Vesicoureteral reflux — Role in pyelonephritis. Amer. J. Med. 34, 338—349 (1963 a).
— — — Perpetuation of infection in unobstructed urinary tracts by vesicoureteral reflux. J.Urol. (Baltimore) 90, 88—91 (1963 b).
—, and E. A. TANAGHO: Etiology of non-occlusive ureteral dilatation. J. Urol. (Baltimore) 93, 177—184 (1965).
IANNACCONE, G., and P. E. PANZIRONI: Ureteral reflux in normal infants. Acta radiol. (Stockh.) 44, 451—456 (1955).
JEWITT, H. J.: Upper urinary tract obstructions in infants and children. Diagnosis and treatment. Pediat. Clin. N. Amer. 2, 737—754 (1955).
JONES, B. W., and J. W. HEADSTREAM: Vesicoureteral reflux in children. J. Urol. (Baltimore) 80, 114—115 (1958).
KASS, E. H.: Bacteriuria and pyelonephritis of pregnancy. Arch. intern. Med. 105, 194—198 (1960).
KIIL, F.: The function of the ureter and renal pelvis. Philadelphia: W. B. Saunders Co. 1957.
KIMMELSTIEL, P.: Significance of chronic pyelonephritis. Chapt. in: QUINN and KASS, Biology of pyelonephritis, p. 215—224. Boston: Little, Brown & Co. 1960.
KJELLBERG, S. R., N. O. ERICSSON, and U. RUDHE: The lower urinary tract in childhood. Chicago: Yearbook Publ. 1957.
LANGWORTHY, O. R., and L. C. KOLB: Histological changes in the vesical muscle following injury of the peripheral innervation. Anat. Rec. 71, 249—263 (1938).
LAURET, G., et A. VIGNERON: Les malformations congénitales de la jonction urétéro-vésicale chez l'enfant. J. Urol. (Paris) 61, 15—32 (1955).
LEADBETTER jr., G. W., J. H. DUXBURY, and J. R. DREYFUSS: Absence of vesicoureteral reflux in normal adult males. J. Urol. (Baltimore) 84, 69—70 (1960).
—, and W. F. LEADBETTER: Ureteral re-implantation and bladder neck reconstruction. Four and one-half years' experience. J. Amer. med. Ass. 175, 349—353 (1961).
LEARMONTH, J. R.: A contribution to the neurophysiology of the urinary bladder in man. Brain 54, 147—176 (1931).
LEFÈVRE, J., J. SAUVEGRAIN, CL. MAITRE, M. SAVARY et M. ETHIER: Le reflux vésico-urétéral chez l'enfant. Ann. Radiol. 3, 173—196 (1960).
LICH jr., R., L. W. HOWERTON, and L. A. DAVIS: Recurrent urosepsis in children. J. Urol. (Baltimore) 86, 554—558 (1961).
MARCEL, J. E.: Le syndrome mégavessie — reflux cystopyélique. Presse méd. 60, 1793—1796 (1952.
MCGOVERN, J. H., V. F. MARSHALL, and A. J. PAQUIN jr.: Vesicoureteral regurgitation in children. J. Urol. (Baltimore) 83, 122—149 (1960).
MCLANE, C. M., and H. F. TRAUT: The relationship between infected urine and the etiology of pyelitis in pregnancy. Amer. J. Obstet. Gynec. 33, 828—834 (1937).
MILLIEZ, P., G. LAGRUE, P. SAMARCO, M. NOIX, and J. L. BINET: Functional signs which cause suspicion of ascending nephritis. Radiocinematographical study of reflux in ascending nephritis. Rev. Med. Moy. Or. 17, 71—74 (1960).
MISAK, S. J., R. C. BUNTS, J. L. ULMER, and W. M. EAGLES: Nerve interruption procedures in urologic management of paraplegic patients. J. Urol. (Baltimore) 88, 392—401 (1962).
NOIX, M.: La miction fractionnée dans les grands reflux cysto-urétéro-pyéliques. J. Radiol. Electrol. 6/7, 335—339 (1964).

Paquin jr., A. J.: Ureterovesical anastomosis: The description and evaluation of a technique. J. Urol. (Baltimore) 82, 573—583 (1959).

Politano, V. A., and W. F. Leadbetter: An operative technique for the correction of vesicoureteral reflux. J. Urol. (Baltimore) 79, 932—941 (1958).

Quinn, E. L., and E. H. Kass: Biology of pyelonephritis. Boston: Little, Brown & Co. 1960.

Rosenheim, M. L.: Problems of chronic pyelonephritis. Brit. med. J. 1963, No 5343, 1433—1440.

Ross, J. C.: Some complications of the neurogenic bladder. Brit. J. Urol. 33, 381—391 (1961).

Satani, Y.: Histologic study of the ureter. J. Urol. (Baltimore) 3, 247—267 (1919).

Smith, D. R.: General urology, 4th ed. Los Altos: Lange Med. Publ. 1963.

Tanagho, E. A., and J. A. Hutch: Primary reflux. J. Urol. (Baltimore) 93, 158—164 (1965).

— — F. H. Meyers, and O. N. Rambo jr.: Primary vesicoureteral reflux: Experimental studies of its etiology. J. Urol. (Baltimore) 93, 165—176 (1965).

—, and R. C. B. Pugh: The anatomy and function of the ureterovesical junction. Brit. J. Urol. 35, 151—165 (1963).

Uhlenhuth, E. de, W. T. Hunter, and W. E. Loechel: Problems in the anatomy of the pelvis. Philadelphia: J. B. Lippincott Co. 1953.

Vest, S.: Address delivered at Ninth Annual Urologic Research Congress, Los Angeles, California 1954.

Williams, D. I.: Megacystis and mega-ureter in children. Bull. N.Y. Acad. Med. 35, 317—327 (1959).

— J. Scott, and R. T. Turner-Warwick: Reflux and recurrent infection. Brit. J. Urol. 33, 435—441 (1961).

Anomalies of the Kidney

Franklin Farman

With 8 Figures

Introduction

Investigators have learned and casuistics confirm that 10% or more of the world population is born with some form of urogenital anomaly. Of this number of deformities one-half constitute developmental defects of the upper urinary tract. It is estimated that 40% of all diseases of the kidney are associated with a congenital anomaly. Many of these anomalies do not come to clinical recognition; they remain "silent" throughout life and are discovered first upon postmortem examination. In addition, anomalies of the urogenital system are accompanied frequently by maldevelopment of other organs in the body. This classification, based upon renal embryological theory, is adapted to anomalies by FARMAN:

Defect in Germ Plasm	failure of development of metanephros	complete incomplete defective	Agenesis Aplasia Hypoplasia Dysplasia
	alteration of tubulo-glomerulo structure	non-union separation obstruction	Solitary cyst Multiple cysts Multilocular cysts Polycystic disease
	disturbance of growth of ureteric buds and vessels	ascent rotation	Simple ectopia Crossed ectopia Malrotation Aberrant vessels
	failure of separation of metanephrogenic cell mass	partial complete	Horseshoe kidney Unilateral fused kidney Fused pelvic kidney

I. Agenesis

Agenesis signifies lack of development of the metanephros with complete absence of any renal structure. Both unilateral and bilateral renal agenesis occur. The former is the only type which is of clinical interest since the latter is incompatible with life.

Incidence: Aristotle was the first to observe the anomaly. The first scientific description of unilateral renal agenesis was made by Counsiliorium in 1609. By 1932, 581 cases of solitary kidney were reported in the literature.

At necropsy congenital absence of one kidney varies from 1:705 to 1:1,610 with a mean ratio of 1:1,000. In clinical practice the ratio is 1:1,500 cases. In children CAMPBELL (1957) found 88 cases in 47,409 postmortem examinations (1 in 528); WILLIAMS (London) considers this percentage too high for the general

---

The figures of the color-plates are redrawn from original drawings by WAYNE WILLIAMS, Medical Artist, Duke University.

Common Renal Anomalies

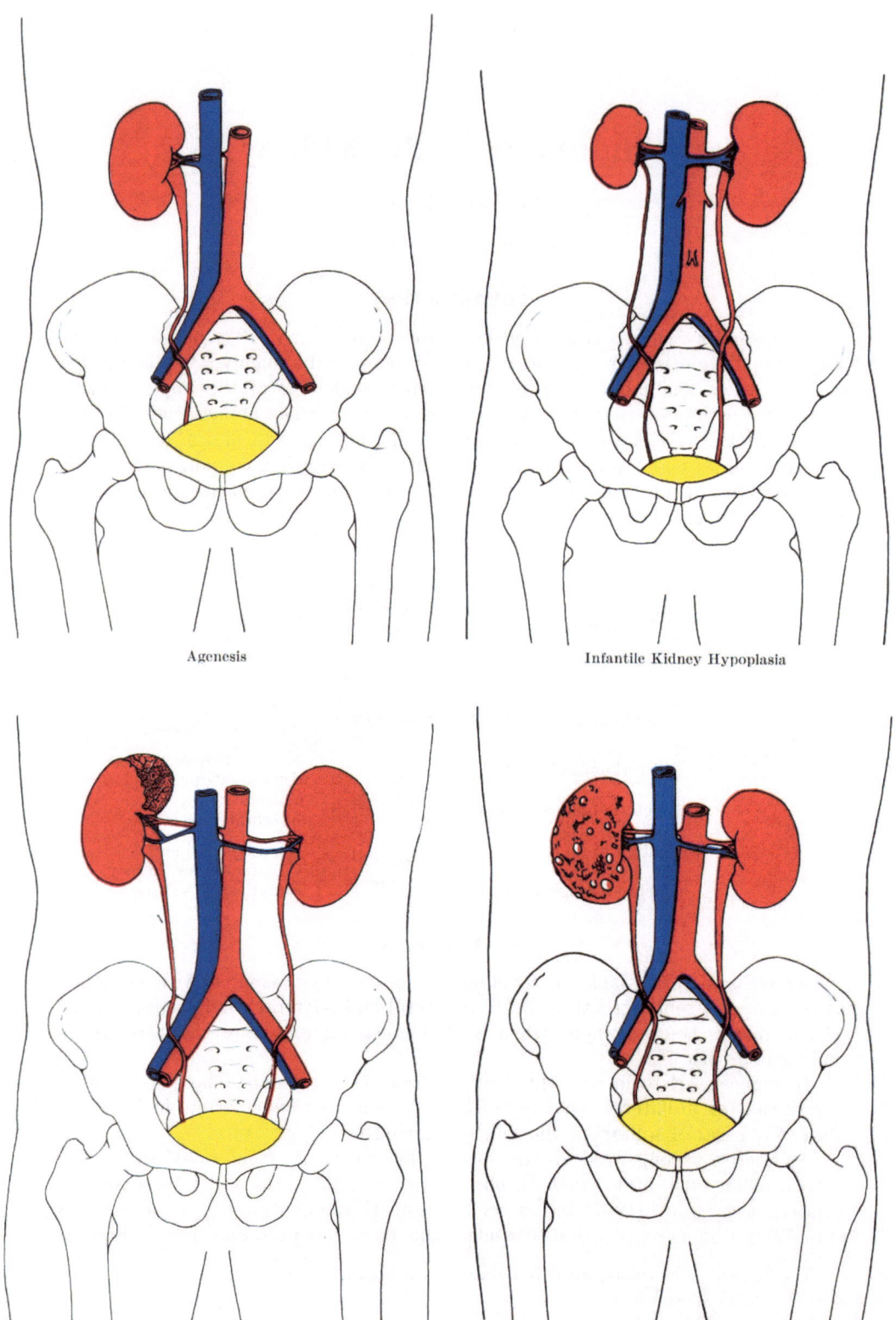

Agenesis

Infantile Kidney Hypoplasia

Simple Cyst (Solitary)

Polycystic Kidney

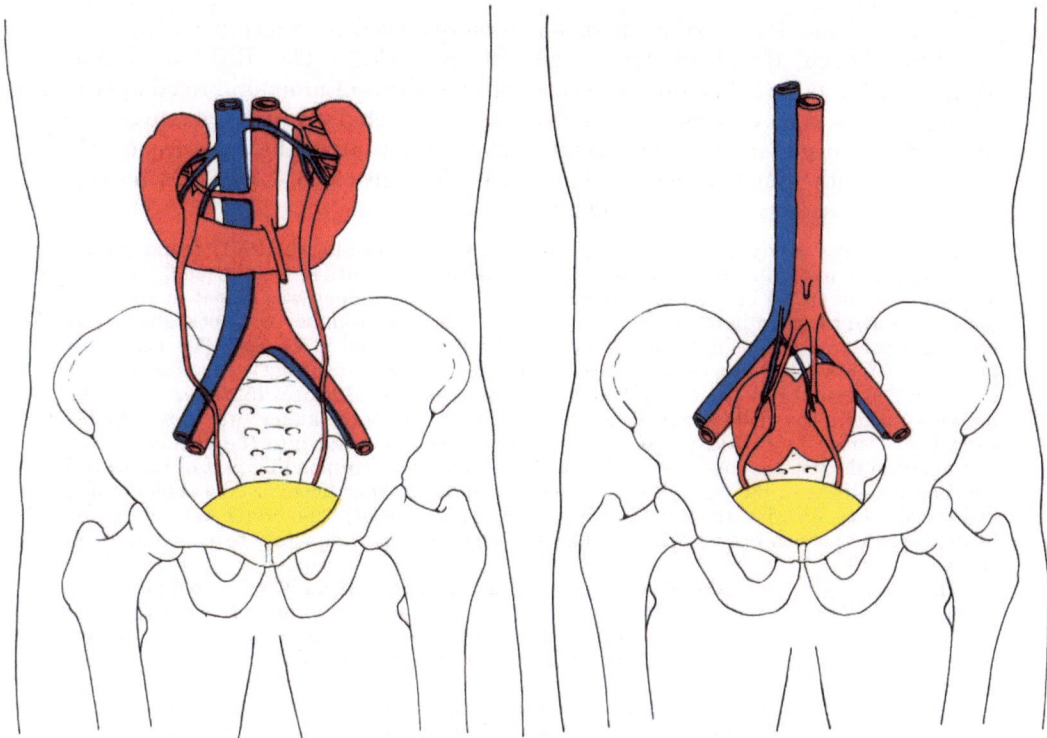

Horseshoe Kidney   Fused Pelvic Kidney

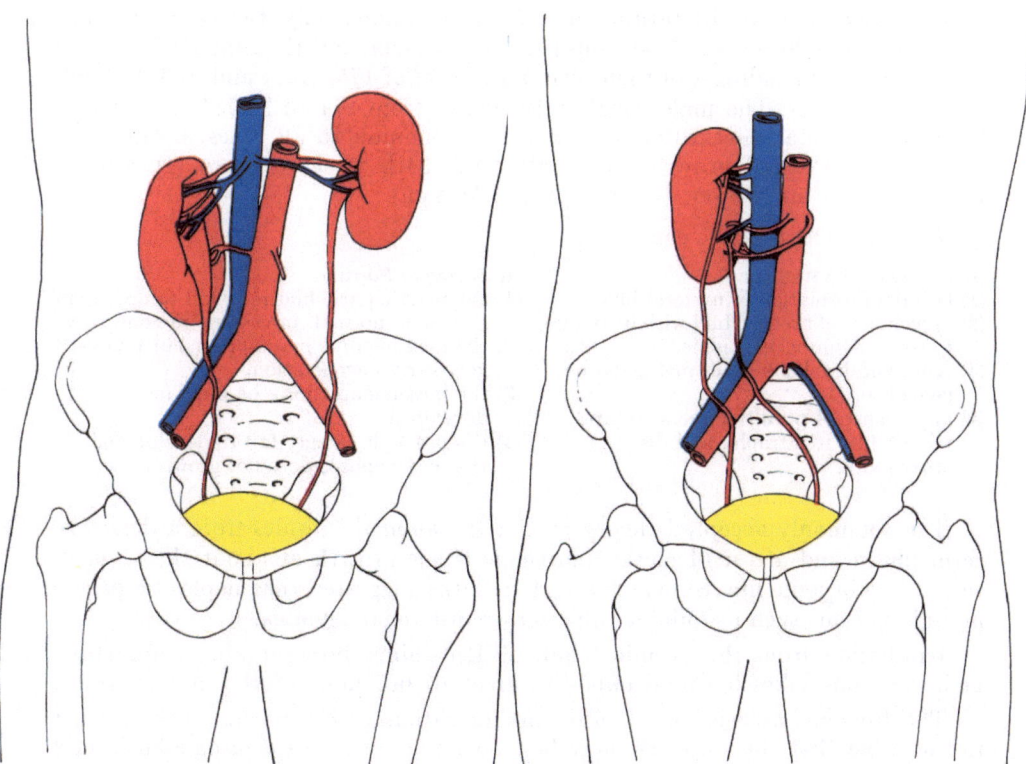

Simple Ectopia and Malrotation   Crossed Ectopia with Fusion

population. Unilateral renal agenesis has been observed in two siblings (Gorvoy et al.) although the hereditary factor is not clear. The Russian author, Den'mukhamedov (1957), observed only two instances of unilateral renal aplasia among 205 cases of various renal anomalies. Monoorchidism, unicornuate uterus and other urogenital and extra-genitourinary tract anomalies are frequently associated with renal agenesis (Howarth, Thompson-Walker, Campbell, Collins, Hennessey, Nation, Neerhut).

In a necropsy study of 1,498 cases, Doroshow and Abeshouse (1961) found the incidence to be 1 in 1,070. There were 801 males (64.1 per cent) and 448 females (35.9 per cent). The sex was not stated in 249 cases (16.6 per cent). The left kidney was absent in 769 cases (56.4 per cent); the right in 595 cases (43.6 per cent). The absent kidney was not stated in 134 cases (8.9 per cent). The single kidney was smaller than normal in 47 cases (7.4 per cent); normal in 37 cases (5.9 per cent); enlarged, in 548 cases (86.7 per cent); the size was not stated in 866 cases (57.4 per cent). The ureter was absent in 779 cases (52.0 per cent); the ureteral orifice in the bladder was absent on the side on which the kidney was absent in 284 cases (19.0 per cent). In 5 cases (0.33 per cent) the ureter crossed into the bladder on the side opposite the renal agenesis. One ureteral orifice and the corresponding part of the vesical trigone were absent in 136 cases (9.1 per cent). The ureter on the affected side was obliterated and remained as a cord of fibrous tissue in 19 cases (1.3 per cent). The ureter was obliterated in the cephalic portion only in 47 cases (3.1 per cent). It was completely patent in 22 cases (1.5 per cent).

The adrenal gland was absent on the side of the renal agenesis in 92 cases (6.5 per cent); present on the side of agenesis in 189 cases (14.3 per cent); not stated in 1,217 cases (81.2 per cent). There were 97 cases of congenital solitary pelvic kidney. The number of cases with associated genital anomalies was 27 (18.5 per cent). Extragenitourinary tract anomalies were present in 85 cases.

The condition of the solitary kidney was normal in 764 cases (51.0 per cent); diseased, in 450 cases (30.0 per cent); not stated, in 284 cases (19.0 per cent).

The youngest case was 2 hours; the oldest, 92 years of age; the average age was 37 years.

In contrast, the number of cases of bilateral renal agenesis is few. Hinman (1940) stated that the literature, since 1663, contained only 135 cases. Potter (1946) reported 20 cases in 5,000 autopsies. Sylvester and Hughes (1954) added 40 more cases, including 4 of their own, for a total of 175. Bain and Scott (1960) reported 23 cases. The male-female ratio varies from 2:1 to 3.07:1. Absence of intrauterine foetal micturition was a feature common to all cases, resulting in oligohydramnios and growth failure after the 34th week in utero (Duxbury, Hinman, Bain and Scott, Ruderman and Mayer).

Etiology. The possible causes are:

According to Hinman (Sr.)
(1) failure of formation of ureteral bud.
(2) appearance of ureteral bud with its failure to reach nephrogenic tissue.
(3) failure of development of metanephrogenic cap.
(4) appearance of primitive metanephros followed by its atrophy and disappearance.

According to Fortune
(1) the metanephric bud may fail to appear in site of a normal preceding mesonephros.
(2) the metanephros may appear but undergoes early degeneration.
(3) the mesonephros may be imperfectly developed.
(4) the pronephros may fail to develop and the mesonephros does not grow.

The commonly accepted theory is that the anomaly results from a defective germ plasm and arrest of metanephrogenic tissue growth at about the seventh week of embryonic life. Sylvester and Hughes suggested that nephrotic protoplasmic poisons such as quinine salts, may cause renal agenesis.

Irradiation from the atomic bomb of Hiroshima brought about numerous malformations following fetal exposure in utero but none of the urinary tract.

The frequent association of other malformation with renal agenesis has led to the belief that this anomaly may be associated with wide neuromesenchymal tissue defects. A contributory factor which has been suggested is increased intra-uterine pressure.

Based on 11 cases in which there was development of the kidney without the presence of a ureter ASHLEY and MOSTOFI believe the initiating factor is in the metanephric blastema and not the ureteric bud (BROWNE, BYRNES and BOELLAARD, DE SPA).

Clinical Considerations. Exact diagnosis may be difficult. The anomaly frequently comes to attention when the excretory pyelogram suggests absence of a functioning kidney on one side. WILLIAMS (London) states that asymmetry of the trigone and absence of the ureteric orifice may lead to suspicion of renal agenesis; however, he cautions that these features may also be observed in cases of renal "aplasia" and in ectopic ureter. Diagnosis may be confused with secondary atrophy or obstruction of one kidney due to disease. GILLASPY and RENTERGHEM (1962) cited an example of renal space replacement by lymphoid tissue.

Complications of the solitary kidney are as for a normal kidney — most commonly pyelonephritis, hydro- or pyonephrosis, calculi, and tumor.

The blood supply of an agenetic kidney may be underdeveloped or vestigial in size. The solitary kidney is usually in normal position. Malformation of the anatomical structure is common, especially of the calyceal system and vascular supply (BURFORD, HINMAN, OPOLSKI, DAREFER and BURON).

Usual delay of recognition of unilateral renal agenesis until adult life leads to the belief that this abnormality does not materially shorten "life span". DEES (1960) found "no history of any trouble" in 11 of 33 patients in support of compatibility of health with this anomaly.

EWERT (1964) observed an asymptomatic carcinoma in a solitary kidney. COPE (1964) in reporting a case of traumatic rupture of a solitary kidney states that such a kidney is more susceptible to injury because it projects below the rib cage. The kidney may be ectopic or excessively mobile. Should it fail to ascend, a solitary pelvic kidney is encountered.

The treatment of congenital solitary kidney is that of any complicating disease. Surgery is not contraindicated because the kidney is single but is restricted in extent because of the danger of irreversible renal failure. Careful pre- and postoperative study and availability of an artificial kidney are important to successful operation either on the acquired or congenital solitary diseased kidney. The most frequent operation is simple nephrostomy drainage which may be lifesaving in cases of "blocked" ureter or impacted stone in an inaccessible location. Heminephrectomy for tumor and excision of infected portions (tuberculous pyonephrosis) have been reported in which the importance of knowledge concerning variations of renal structure are emphasized (CAMPBELL, HINMAN, OPOLSKI, GOLDSTEIN).

a) Bilateral Renal Agenesis

Absence of both kidneys is usually accompanied by multiple malformations. In POTTER's series of 20 cases, there was hypoplasia of the lungs, spade-like hands and the characteristic facies of "premature senility". BAIN, BEATH and FLINT observed bilateral renal agenesis as a constant finding in sirenomelia (fusion of legs) and monomelia (one limb). Since the anomaly is compatible with intra-uterine life, about two-thirds are "live born". Death usually occurs a few hours after birth because of hypoplastic lungs, unable to sustain life. In one case a child with bilateral renal agenesis lived 39 days (DAVIDSON and ROSS, POTTER, WOOLF and ALLEN).

BENEVENTI reports three instances of nephrectomy in 44 cases of solitary pelvic kidney, all before the year 1911. Two instances of diagnostic error leading to nephrectomy for "solitary kidney" are recorded in the recent Russian literature (IVASHKO).

b) Solitary Pelvic Kidney

Solitary pelvic kidney is fixed in position within the bony pelvis and occupies a central rather than a lateral position. It is accompanied by an abbreviated ureter and anomalous blood supply. VUORINEN (1960) emphasizes the need for careful palpation and urological examination in all cases of pelvic masses. Surgical correction is not feasible.

II. Hypoplasia

a) Unilateral

Renal hypoplasia signifies failure of development of the kidney to attain normal size. This represents an arrest of growth. It should not be confused with atrophy or secondary contracture. The condition usually is unilateral. Defective blood supply, retarded maturation of the nephros or faulty union of the metanephric (Wolffian) duct with the metanephrogenic blastema have been presented as contributary influences. Hypoplasia applies to various degrees and forms of incomplete development. The ureter may or may not be patent. BOISSONNAT includes both quantitative and qualitative deficiency. PAETZEL prefers the term "dwarf kidney".

Hypoplasia differs from aplasia in that the latter represents a diminutive fibrous body containing only embryonic renal tissue. When secreting glomeruli are present, the condition should be classified as hypoplastic kidney. Dysplasia refers to a hypoplastic kidney which contains primitive renal elements or tissues foreign to the kidney, such as striated muscle or cartilage (renal osteodystrophy) (BAGGENSTOSS, EKSTROM, TEDESCHI and HOLTHAM, WILLIAMS).

Incidence. Since the literature is ambiguous as to what constitutes hypoplasia, frequency is difficult to ascertain. It is estimated to occur once in 800 births (THOMPSON). GRAHAM quotes the incidence as 4% of all renal anomalies (50 cases in 50,000 autopsies). Female patients predominate and the right side is involved more often than the left. Both GAMBLE and GROZS have given the frequency as one in 600 postmortem examinations and CAMPBELL has noted the incidence about equal to that of unilateral agenesis (1:527).

Gross appearance. Hypoplasia is almost always unilateral. As a general rule the kidney occupies a normal position; on rare occasions it is ectopic. A typical hypoplastic kidney weighs less than one-half the normal, varying from 30 to 100 grams. The common types of hypoplasia are 1) general or small kidney, 2) partial — in which one-half of a kidney with a double pelvis shows incomplete development, and 3) regional arrest — with localized calyces decreased in number, size and configuration in an otherwise normal kidney. The main renal artery usually is reduced in size. The contralateral kidney shows "compensatory hypertrophy", unless it is diseased (EKSTROM, HERBUT, LERUITTE).

Histological Appearance. In addition to its diminutive size, a unilateral hypoplastic kidney presents histologic changes. Glomeruli are fewer in number and decreased in size; tubules of the cortex tend to be dilated and cystic. There is increased fibrotic stroma; calcific deposits are observed. The medulla usually shows rudimentary pyramids. BELL and HERBUT consider a decreased number of pyramids to be one criterion of renal hypoplasia. The arteries may be normal or reduced in size; they become fibrotic or arteriosclerotic in the presence of infection.

It is difficult to distinguish congenital hypoplasia from severe chronic pyelo-nephritis with atrophy. In some cases hypoplastic kidneys have large glomeruli. Ciliated epithelium has been found in the hypoplastic kidney; it is thought to originate from the foetal epididymis (an embrogenic spillover) (BELL, ANDERSON et al., EKSTROM, GROZS, HERBUT, WILLIAMS).

Clinical Considerations. Increased knowledge concerning renal insufficiency and hypertension have focused attention upon the "hypoplastic kidney". In the unilateral type symptoms do not arise until complications appear — usually during adolescence or early adult life. In cases of bilateral hypoplasia the condition is recognized early. Characteristic clinical features are failure in growth, gain in

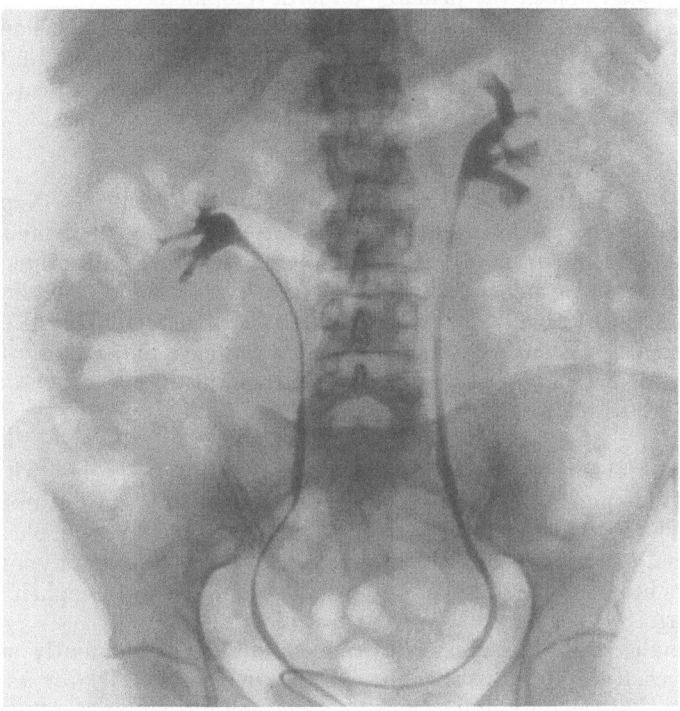

Fig. 1. Age 59. Hypoplasia right. Chronic pyelonephritis, hypertension relieved by nephrectomy of a 100-gram kidney; small kidney shadow and blunted or abbreviated calyces are characteristic

weight, hypertension and osteodystrophy (renal dwarfism). Early death results from renal failure; often there is complicating infection. Sometimes there is a long period of latency before renal functional impairment and hypertension develop, as observed in 40 cases reviewed by DEGOY and SCHULLER. In contrast to renal aplasia, usually discovered only by accident, a hypoplastic kidney which fails to produce symptoms is a rarity.

The symptoms of hypoplasia may be vague, nonspecific and often imitative of other illness. Sudden pain or mild recurring dull ache on the affected side are commonly observed. Infection and stone are frequent. The association of vascular changes in the brain has been noted, producing convulsions and hypotonicity (TROY, BYRNES and LINDAU, BACHER, BOISSONNAT, EKSTROM, SZIBERTH).

In differential diagnosis, consideration should be given other diseases which produce a reduction in size of the kidney. Most confusing is chronic pyelonephritis, chronic nephrosis, nephrosclerosis and arterionephrosclerosis. There is no uni-

form relationship between hypertension and disease of the unilateral hypoplastic
kidney. Reports of cure (permanent reduction in blood pressure) by nephrectomy
have been reported. All cases of persistent hypertension should be investigated
for the possible presence of a renal anomaly (CAMPBELL, EKSTROM, HUTCHINSON
and MONCRIEFF, PENSON, SOSZKA, VERGER, WERMUT et al.).

Retrograde investigation of the upper urinary tract in suspected cases is more
reliable than excretory urography. The characteristic urogram shows a small
kidney shadow on the affected side in which the calyces are small and abbreviated
(infantile) and blunted; a single calyx may be present. FORST suggests aortography as a diagnostic aid. Diagnosis may be difficult when other diseases and
secondary changes mask the primary underdevelopment.

Treatment. When the unilateral hypoplastic kidney has become infected,
obstructed or causes hypertension, the only form of treatment that will afford
permanent relief is nephrectomy. This procedure is clearly indicated providing
the opposite organ is normal.

b) Bilateral

The opposite is true for the rare case of bilateral hypoplasia. Nephrectomy is
contraindicated and prognosis is poor. WILLIAMS states, "even if the two kidneys,
though small, are structurally normal at birth it is inevitable that renal insufficiency will lead to gross tubular dilation and hypertrophy. Complicating pyelonephritis practically always occurs with scarring and distortion of the parenchyma.
There is often hypertension causing vascular changes, and at autopsy therefore
bilateral hypoplastic kidneys always present a complex appearance which may be
avoided in the unilateral" (CIBERT and COLLENET, EISENDRATH, MOMBAERTS,
PAETZEL).

Since arterial surgery is not feasible in the hypoplastic kidney with a vascular
abnormality, nephrectomy is the preferred treatment for cure of unilateral renal
ischemia. The secondary changes of pyelonephritis and hydronephrotic atrophy
are the usual cause of renal ischemia rather than the primary hypoplasia. Good
results have been noted in cases of unilateral hypoplastic pyelonephritis (BOEMING-
HAUS, NIETH, SMITH, VERGER, WERMUT et al.).

Since malignant hypertension during childhood is frequently secondary to
these changes, it is important that the cause be determined. DEGOY and SCHULLER
(1958) state that the only indications *sine qua non* for nephrectomy are unilateral
renal lesions and a child with full blown malignant hypertension which appears
redoubtable.

Failure to achieve satisfactory results following nephrectomy may be explained
on the basis of irreversible disease (chronic pyelonephritis or arteriosclerosis and
glomerulonephritis) in the remaining kidney. This emphasizes the need for early
recognition and evaluation of complications in the contralateral mate. Despite
apparent good function in one or the other, a small kidney or a diseased larger
kidney may be incapable of sustaining life unaided by its fellow (CYNOWSKI,
GOLDBLATT, HARTWICH, HINMAN, NEIMANN et al., SCHULTZE-JENA and HILLEN-
BRAND, WERMUT et al.).

III. Cystic Disease

Cystic disease includes anomalies of volume and structure of the kidney. As
in the most forms of congenital maldevelopment, the cause is considered to be an
embryonic defect in germplasm, in this instance leading to alteration of tubulo-
glomerulo structure. The alteration may take the form of non-union, separation
or obstruction of collecting tubules and the glomerulus secreting unit resulting

in cysts of the single, multiple, multilocular or polycystic type. Pyelogenic cyst, or calyceal diverticula, is a rare form which sometimes leads to entire replacement of the parenchyma. Renal dysplasia, an uncommon condition in itself, may produce small cysts within the underdeveloped metanephrogenic cap. Small retention cysts noted at autopsy or during surgical procedures are regarded generally as due to nephritis, infarction or inflammatory process. In this chapter other theories of etiology will be mentioned but only cyst formation of congenital origin will be considered in detail (RALL and ODEL, HILDEBRANDT, FISTER, WAKELEY et al., MOORE, WEYRAUCH, FLEMMING, HART et al.).

The following classification is based upon embryological theory and clinical observation:

Defect in Germ Plasm

alteration of	non-union
tubulo-glomerulo	separation
structure	obstruction

Congenital (embryological)	*Clinical (form and structure)*
single (solitary)	retention cysts
multiple	small
multilocular	large
polycystic	polycystic disease
	adult
	neo-natal
	pyelogenic (calyceal)
	peri-pelvic
	acquired
	para-pelvic
	nephritides
	inflammatory

a) Simple (Solitary) Cysts

To the pathologist the term "simple cyst" is preferred when describing isolated sacs filled with serous fluid. The world's literature, however, abounds with reports of "solitary cyst". In urologic practice the latter term is in common usage. Clinically a cyst presents as a globular mass from a few to several centimeters in diameter which usually projects from the surface of the kidney. Solitary cysts are usually unilateral, but may be bilateral.

Incidence. Renal cysts have been described since 1634. Reports place the incidence of solitary cyst at 1 per 3,500 persons. The average age incidence has been found to be about 50 years (FISH, NATVIG, GUTIERREZ, BRASCH and HENDRICK, LOWSLEY and CURTIS).

Etiology. HERBERT expresses the view that simple serous cysts are congenital. He believes their origin is similar to the same basis as of polycystic disease. When multiple, he states they may be indistinguishable from the latter disease. On the other hand, COLBY states that most evidence favors an acquired origin — a change within the kidney resulting from tubular obstruction and anemic degeneration of the parenchyma due to circulatory interference. According to FERGUSSON, present day opinion remains divided as to their origin from congenital or acquired causes. There is no indication of a hereditary tendency.

Clinical Manifestations. From a clinical point of view solitary cyst is quite different from polycystic disease. Simple cysts frequently are symptomless, many are discovered accidentally by intravenous pyelography. Large cysts may produce a dragging pain in the loin, with or without urinary symptoms. FISTER states the localizing symptoms may be lacking or cause lumbar pain, caused by pressure on renal tissue. An abdominal mass may be felt. Gastro-intestinal or gallbladder

disturbance may be simulated. Hypertension has been mentioned by KREUTZ-
MANN. Polycythemia has been reported in association with simple cysts. Evidence
of a casual relationship with erythropoietin activity has been remission after
removal of the cyst (KIER and YOUNG, ROSSEE, et al.).

Diagnosis is reached most commonly by radiographic means. Pyelograms
show distortion and displacement of the calyces, typical of a round, symmetrical
space consuming lesion of the kidney. ANDRESEN has placed location of the "cyst
mass" in the upper, middle or lower part of kidney in the proportionate ratio of
8—5—1. Calcification has been observed in the wall of cysts. Hydronephrosis and
displacement of the kidney may be present. Confusion arises because of the
similarity of simple cyst to neoplasm of the kidney. Renal angiography is helpful
to differentiate between the avascular area of the cyst and the vascular network
of opaque medium in neoplasms. Unfortunately both false positives and false
negatives have been observed with respect to this sign.

FERGUSSON advocates diagnosis by needle aspiration in selected cases, especially in elderly patients when operation may be contraindicated. This method
may lead to implanting neoplastic cells in the tract of the needle.

Since there is no absolute way to differentiate a cyst from malignancy, most
authorities advocate surgical exploration in suspected cases. The preferred
operation is uncapping the cyst or nephrectomy if little functional parenchyma
remains. Results generally are good.

b) Multiple Cysts of the Kidney

Although common to both sexes and all age groups the condition is most
frequently observed soon after birth, presenting as a unilateral upper abdominal
mass in infants.

Incidence. Only six cases of bilateral multicystic kidney have been recorded.
All appeared in neonates who died shortly after birth with other congenital
anomalies (CRAIG). Up to 1962 sixty-eight cases of the unilateral disease have been
described. The largest group of cases (19) was reported from the University
Central Hospital, Stockholm (PARKKULAINEN et al.). The age of the patients
varied from 1 hour to 1 year, 8 months. The incidence is given as 0.085% of
pediatric admissions. Autopsy was performed in fifteen and operation in four.

Etiology. There are various theories of pathogenesis, a common belief is that
there is partial or complete non-union of the collecting tubules with the glomerulus
which results in the formation of cysts. As described by VERMOOTEN there is lack
of metanephros differentiation. A concept advanced by HILDEBRANDT in 1894 —
incomplete fusion between the ureteral bud and metanephric blastema — is not
universally accepted. KAMPMEIER and McKENNA attribute the defect to persistence of primitive uriniferous tubules with subsequent cystic formation. NORRIS
and TYSON believe degenerative processes abnormally continue into tubular later
developmental generations of the kidney. ALLEN attributes the defect to a
distorted influence of "organizers" with failure of canalization at various levels
of the nephron.

Renal hypoplasia as the basic cause was suggested by BELL in 1935. More
recently AREY and ABESHOUSE have described advanced cystic hypoplasia of the
kidney, using the term "unilateral multicystic kidney disease". HERBST, APFEL-
BACH, and HAWES reported similar cases. The characteristic finding is abundant
interstitial connective tissue, occasionally associated with cartilage, a small or
absent renal pelvis, an atretic ureter and deficient renal vascularization (SPENCE,
IVKER et al., HINMAN, HEPLER and LYNCH).

Clinical Considerations. Since unilateral multicystic kidney is primarily a disease of early infancy, the condition usually is noted upon routine examination as a "silent" abdominal mass. If the mass is large, symptoms such as gastrointestinal disturbance may develop. A radiograph will outline a renal mass but not usually a renal pelvis. The contralateral kidney is usually normal. The finding of only one functioning kidney should arouse suspicion of the abnormality. Congenital hydronephrosis and Wilm's tumor are differentiated by retrograde pyelography.

Unlike bilateral multicystic kidney, unilateral multicystic disease generally occurs alone and rarely is associated with other congenital defects. MOE and CROFFORD (1960) reported the case of an ectopic unilateral multicystic kidney in an infant associated with tracheo-esophageal fistula and a duodenal obstruction due to annular pancreas (death followed the second operation). In all six reported cases of bilateral multicystic kidney other congenital anomalies, such as esophageal atresia, have been present (CRAIG, 1962).

Treatment consists of exploratory operation with biopsy, if necessary to establish the diagnosis. Nephrectomy is the procedure of choice for true unilateral multicystic kidney. Prognosis is good if the opposite kidney is normal. LYNCH and BRADHAM used a substitute ileal ureter in an infant with congenital hydronephrosis of the left kidney and congenital cystic hypoplasia of the right kidney. The latter is sometimes considered to be a form of multicystic kidney disease.

c) Multilocular Cysts

Multiloculated renal cysts are a form of "closed" cyst and should be differentiated from the single, solitary and the multiple type retention cyst. In contrast to multicystic disease usually seen in infants multilocular cysts are most

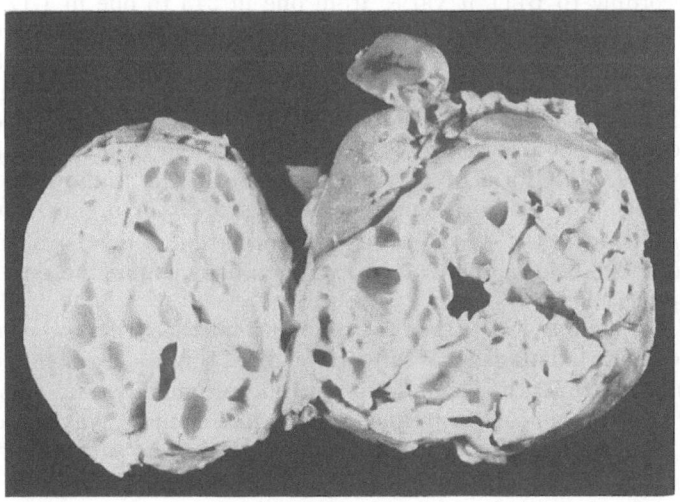

Fig. 2. Multilocular cyst of kidney; form of "closed" cyst

frequently found in adults. The pathogenesis is not clearly understood. The histological picture is one of cystic formation in which loculated septa contain embryological elements of neoplastic tissue. The latter are classified as cystadenoma. WEINGARTEN states about 7% of all cases of renal cystic disease are malignant.

Some cystadenoma are benign epithelial new growths confined to the cortex and periphery of the kidney. Heckel quotes Fahr and Lubarch (1925) who believed that such adenomas result from urinary tubules which have become separated from surrounding tissue. Some investigators have noted the close association of papillary cystic adenoma to hypernephroid growth, and have compared the development of large adenomas to carcinoma. Both Carver and Cristol et al. have statistically studied the rare occurrence or association (under 1%) of renal tumor with adenoma.

Clinical Considerations. The chief symptom of a large multilocular cyst is abdominal pain or discomfort caused by the renal mass. Hematuria usually signifies associated malignancy, although some cysts contain hemorrhagic fluid in association with benign epithelial growths. The "filling defect" of the pyelogram is indistinguishable between cyst-adenoma and renal tumor. Uson and Melicow (1963) reported four cases of intrapelvic herniation of the little recognized "daughter" cyst. They state that findings in the excretory and retrograde pyelogram, such as an expanding renal mass associated with hydronephrosis and a filling defect within the pelvis, suggest the possibility that the lesion is a multiloculated cyst with a herniated daughter cyst within the renal pelvis. Hypertension is sometimes produced by multilocular cysts.

Nephrectomy is generally necessary although resection of the kidney may be performed for benign adenoma if histologically proven at the time of operation. Weingarten and others advise surgical exploration in all cases to be certain a malignant tumor within the cyst or in another part of the kidney is recognized and removed.

d) Polycystic Kidney Disease

Incidence. The incidence of polycystic disease is usually based upon autopsy records. According to Bell it varies from one in 243 to one in 1,173 autopsies. In clinical cases Cameron (1961) lists the ratio at 1 to 230 and Melicow and Gile found 77 cases in 14,155 hospital admissions. The incidence of the bilateral type is 5 times that of the unilateral — 11 to 64 in 37,360 autopsies (Oppenheimer and Narins). Campbell states that one in 265 infants is born with congenital renal polycystic disease. The records of Cannon and Rall and Odel agree that the age incidence of the adult form of polycystic disease of the kidneys is from 35 to 60 years. Association with congenital anomalies of other organs is uncommon.

Etiology. The main theories regarding the formation of renal cysts are grouped by Rall and Odel into four categories: metabolic, inflammatory, neoplastic, and developmental.

Metabolic. Virchow was first of the opinion that the structural changes of the kidney were due to a deposition of salts in the renal tubules, leading to obstruction of the tubular lumen, with accumulation of fluid proximally, and cyst formation. Since arteriosclerosis is such a frequent accompaniment of polycystic disease of the kidney, Ritter and Baehr considered this to be the causative agent.

Inflammation. Later Virchow suggested that pyelitis in fetal life resulted in tubular fibrosis, obstruction, and cyst formation. Ribbert was of the belief that nonspecific inflammation during fetal life caused cystic disease — preventive union of uriniferous and collecting tubules.

Neoplasm. Because polycystic disease often presents as a vigorously growing tumor mass, the possibility of a neoplastic process was held by several observers. Brigidi and Severi were the first to suggest that the lesion represented a true neoplasia. Staemnier called the disorder cystadenoma fibroma.

Developmental Defect. Histologic studies suggested to MUTACH that cysts result from cessation of growth of renal analgen before union of uriniferous and collecting tubules. KAMPMEIER and McKENNA reconstructed serial sections of fetal kidneys and of one polycystic kidney. They concluded that the fundamental defect was failure of the second, third, and fourth generations of the uriniferous tubules to atrophy — that lack of atrophy or reunion of these generations of uriniferous tubules in the presence of active glomerular secretion results in dilation and cyst formation. NORRIS and HERMAN, also on the basis of reconstruction, confirmed KAMPMEIER's work. They suggested that in the polycystic kidney a greater amount of metanephros than normal is provisional, and failure to atrophy results in polycystic disease.

Pathology. The pathology of infantile and adult forms of polycystic disease is similar. The former is described by WILLIAMS (Vol. XV) and the latter by HERBUT

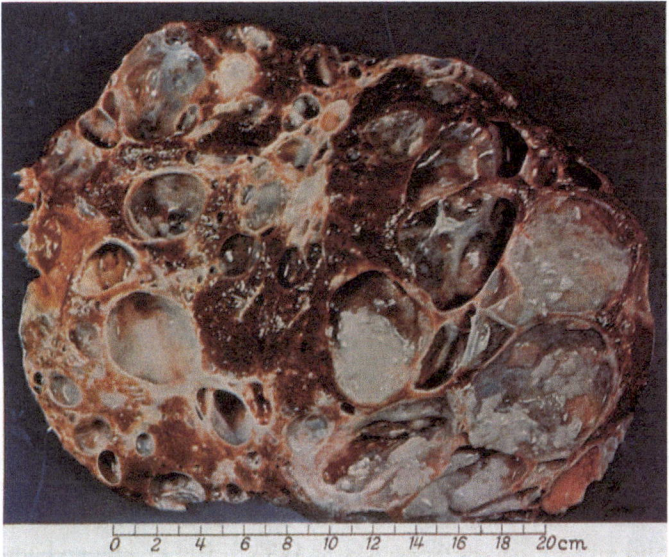

Fig. 3. Polycystic kidney. Gross specimen, polycystic kidney disease

(1957) in *Urological Pathology* (Vol. I). The bilateral type occurs twelve to fifteen times more frequently than the unilateral which is usually found only in adults.

Gross Appearance. The affected kidneys are much enlarged, the external surface being lobulated with closely-set cysts under tension. The kidneys generally occupy a normal position except for caudal displacement because of size. The cut surface presents usually a honeycomb formation filled with thousands of minute oval or elongated cysts from 2 to 3 cm. in diameter arranged radially throughout the parenchyma. Most cysts are under tension and filled with straw colored or blood tinged fluid. Occasionally the medullary and subcapsular zones are spared from involvement. In advanced cases there is no parenchyma remaining; if present it is distorted and compressed. In the adult form there usually is an abundant amount of connective tissue near the hilum.

Microscopic Appearance. The outstanding feature is the presence of cysts. LAMBERT states most cysts in infants are distended nephrons or portions of nephrons separated from their excretory tubules. Any portion of the nephros or even the collecting tubule may be affected. The number of cysts in each unit varies from one to three. Cysts connected with glomeruli are of two varieties (1) closed, without tubular connection, termed vesicles, (2) open, connected with a tubule. The glomerular cyst is usually lined with flat epithelial cells which may be cuboidal, rarely are they columnar. Tubules that leave the cyst upon reconstruction seem to end blindly or later dilate into a terminal sac. The vascular tuft is usually well formed but poorly differentiated. Occasionally cysts contain two or even three tufts. Tubular cysts may be in any part of a closed nephron. In adults secondary bossings are frequent. Excretory cysts are found in collecting tubules which open into the pelvis. According to LAMBERT the cystic nephrons of adults retain a considerable part of their functional ability.

Aside from the cysts, there is an over-all reduction of the number of nephrons estimated to be as low as 10% of normal. Connective tissue is abundant; in adults it may be loose or fibrous. It is seen around the cysts, occasionally mixed with muscle fibers and rarely with cartilage. In the medulla connective tissue may replace collecting tubules and resemble medullary fibroma. Scattered through the connective tissue are foci of round cells and isolated epithelial cells. In adults the arteries usually have a high degree of initial thickening.

Pathologic Physiology. MELAMED showed that the earliest manifestation of functional disturbance in polycystic kidney is a reduction in tubular secretion of cardiotrast (diotrast). PFAU and STAMEY, on the basis of differential renal function studies, conclude that filtration is derived mainly from nephrons with normal glomeruli; excessive reabsorption of sodium and water takes place in the distal tubules.

Complications. GAJOWNICZEK reported the unusual yet rare association of congenital renal, hepatic and possibly pancreatic polycystic disease. At the Massachusetts General Hospital (1963) the rare finding of small pancreatic cysts in a case of polycystic disease of the kidneys and liver was observed. VOROBTSOV describes the rare finding of polycystic disease in horseshoe kidney (heminephrectomy for hematuria). CAINUS (1928) and FAIRLEY et al. (1963) recorded the association of cystic renal disease with visual defects and suggested a developmental relationship between these two disorders. The occasional association of carcinoma of the kidney with polycystic disease was mentioned by WALTERS and BRAASCH (1936) and by MELICOW and GILE (1940), the incidence being under 1% (3 in 85, and 1 in 77).

Clinical Considerations. The infantile form of congenital polycystic disease differs from the adult form mainly in respect to life span — the pathology is similar. The majority of afflicted infants die within the first few months of birth from renal failure. In cases of large foetal polycystic kidneys, dystocia and stillbirth may ensue (FRANCIS, 1960).

In the adult form the patient usually reaches mid-life before diagnosis is made. A significant number of simultaneous congenital and hereditary anomalies elsewhere in the body have been reported. CAMERON cites a "pied en pince de homard" deformity (split and cloven feet) and also hand abnormality (the author has noted thumb deformity in a case of horseshoe kidney).

A familial history is useful but should not be taken as conclusive. GOLDSTON et al. (1963) have cited bilateral polycystic renal disease all associated with central nervous system defects in three of the six children of normal parents.

The clinical manifestations of nephritis (albuminuria) with or without elevation of blood pressure, and generalized weakness are primary symptoms in the adult form of polycystic disease. FRIEND et al. (1961) pointed out that erythrocytosis (polycythemia), occasionally found in association with severe cystic renal disease, may be due to the presence of a space-occupying lesion.

With advancing enlargement of one or both kidneys, palpable tumor, pain, pyuria and/or hematuria, gastrointestinal symptoms from compression and nephrotoxic changes are added. WILLOX (1962) reports a case of mechanical intestinal obstruction as a rare complication of polycystic kidney caused by sudden increase in size.

Diagnosis is based upon intravenous or retrograde pyelograms. Elongation, blunting, obliteration, and bizarre irregularity of the calyces and pelvis are characteristic. The recent addition of radioisotope renograms as developed by WINTER is confirmatory as a diagnostic test. Lumbar abdominal masses especially in infancy should be differentiated from: 1) Wilm's tumor, 2) neuroblastoma sympathicum, 3) hypernephroma, 4) unattached retroperitoneal embryoma or teratoma, and 5) congenital hydronephrosis.

All therapy should be directed toward the conservative management of nephritis and the "failing kidney". The urologic complications for which surgery may be required are: obstruction to urinary output, calculus, associated neoplasm, acute onset of hypertension and pain. Cyst puncture, drainage, or nephrectomy are temporizing measures which give relief in selected cases but do not necessarily prolong life. POLLASTRI following in a review of all forms of medical and surgical therapy concluded that treatment cannot be standarized and must be selected on the basis of morphologic appearance and evaluation of functional capacity of the involved kidneys. KATAMURA et al. (1962) advocate multiple sclerosing cyst puncture in nonadvanced cases.

MILAM et al. believe surgery is contraindicated for bilateral polycystic disease. GOLDSTEIN, on the other hand found in his series of cases (1960) the operative patients lived longer than the non-operative; longevity averaged 4.9 years in 38 cases against 2.5 years in 19 cases. A few adults have been known to live 10 years or longer after discovery of polycystic renal disease. The average prognosis depends upon the onset of "azotemia" which usually signifies termination within a 5-year period.

e) Peripelvic Cysts

Peripelvic cysts of developmental origin are a rare form of cyst formation situated in hilus of the kidney. These involve the renal collecting system; they are frequently referred to as "pyelogenic or calyceal cysts". The incidence is low; the etiology is obscure. The cysts are usually asymptomatic. Thirty-nine (39) cases have been reported up to 1962 (LLANOS).

Two theories of etiology have been advanced — one relating to congenital origin traced to a Wolffian duct defect, the other secondary to inflammation. The author favors the theory of acquired origin. The evidence, according to HENTHORNE, is just as good for congenital maldevelopment.

The cysts may be single or multiloculated, are found around the pelvis between elements of the pedicle and may partially replace the renal parenchyma. They are lined with endothelium supported by a thin wall of connective tissue. In about one-third of cases the lymphatics of the renal pelvis are blocked by hyaline thrombi, proximal to the kidney, resulting in lymphatic ectasia.

Clinical records of the few reported cases indicate occasional hypertension, hematuria, costo-vertebral tenderness or pain on the affected side.

Diagnosis is usually established by intravenous urography; in some suspected cases a retrograde pyelogram is required to outline clearly the cyst cavity.

Treatment, when necessary, consists of surgical removal by wedge resection of the kidney. Nephrectomy is to be avoided.

IV. Fusion Anomalies of the Kidney

The picturesque term "horseshoe kidney" describes one of the well known and commonest fusion anomalies of the urinary system. Before 1800 A. D. these strange anomalies were regarded as curiosities found only at post mortem examination. De Carpi, in 1522, was one of the first to report such a finding, and Morgagni (1820) was the first to give an anatomical description of the anomaly and associated pathology. It was not until the beginning of the twentieth century, following introduction of the roentgen ray, cystoscopy and urography that frequency of occurrence and clinical importance became known. As determined from autopsy statistics the incidence varies from one in 285 to one in 1000 post mortem examinations (Culp, Bell, Campbell, Glenn). In clinical practice the writer estimates the average frequency of some type of horseshoe kidney fusion at one per 500 population.

Embryological Theory and Classification. Fusion implies the union of two separate renal structures, each having its own collecting system. Failure of separation of the metanephrogenic cell mass results in partial or complete fusion leading to formation of (1) a horseshoe kidney, (2) a unilateral fused kidney or (3) a fused pelvic kidney. Ectopia and malrotation follow the fusion process in embryonic life and are part of the factors which determine the ultimate abnormality. Ninfe advanced the theory that vicarious migration or unknown cause might induce the ureteric bud to germinate in an anomalous direction by passing the median line, thus reaching the contralateral wolffian body. Fusion of the primitive renal cell masses takes place first at a very early stage of development (5 to 8 mm. stage) and according to Carleton deviation of point of origin of the umbilical artery may cause primary fusion by interfering with normal upward growth of the ureteric buds. Cases of early fetal horseshoe kidney confirm this observation (Jazuta, Boyden, Budde and Felix, cited by Zondek).

The fused kidney has an abnormal vascular supply, retaining primitive vessels, especially renal arteries from the middle sacral and common iliac vessels. Mechanical-like embryological influences apparently play a part in determining types and final position of the anomalous fused kidney. Fusion of the renal fundaments in the course of their development produces bizarre anomalies of form (Hinman).

Types of Fusion Anomalies

Bilateral fused	Unilateral fused (with crossed ectopia)	Fused pelvic
Horseshoe	end to end	lump
lower pole	end to side	clump
upper pole	sigmoid	cake
L shape		disc
		scutiform

a) Horseshoe Kidney

The horseshoe shape is attained by fusion of the upper or lower "polar" parts of the two renal anlages at a very early stage in foetal development (fourth to seventh week). Fusion prevents rotation of the kidneys, and ascent out of the bony pelvis is retarded. For this reason horseshoe kidneys are always situated lower than normal. The lower they are the more firmly are they fixed in the pelvis. The ureters are shorter than normal and the kidneys (separate halves) are nearly always smaller than normal. The average horseshoe kidney (total fused kidney) weighs 250 to 350 grams.

In 90% of the cases the fusion is in the lower poles. In the remaining 10% fusion is made by union of the upper poles. Fusion of the upper poles takes place before the fourth week of fetal life.

GERARD classified these usual types as "symmetric". An "asymmetric" type is seen in kidneys which assume different forms and shapes as governed by surrounding structures. Fusion of the lower poles causes the "long" kidney; central fusion, the "disc or caked" kidney; fusion of a larger with a small kidney, the "L" shaped or sigmoid kidney. Unusual forms of fusion have been reported by POTAMPA and CATLOW (the tandem horseshoe type) and by PETROVCIC and MILIC (asymmetrical horseshoe fusion with a solitary crossed ureter). VEROBTEEV reported polycystic diseases associated with horseshoe kidney. In 85% of fused kidneys the isthmus consists of parenchyma of the same character as that found throughout the organ; in 15% the isthmus is of fibrous tissue.

Symptoms. Individuals with fused kidneys tend, sooner or later, to experience pain and urinary retention, to develop secondary infection and calculi. FOLEY believes that this is due to pelvic dilation. The position of a fused kidney in itself may produce symptoms from pressure of the isthmus on nerve plexuses and interference with free renal circulation. GUTIERREZ has emphasized this point — he used the term "horseshoe kidney disease" to distinguish cases of the anomaly with vague abdominal, gastrointestinal and urinary complaints. A high percentage (80 per cent) of nephritis is found in SANGREE's et al. series of autopsy specimens and in KRETSCHMER's review of fused kidney.

The author advances the suggestion that more refined study of patients with so-called orthostatic albuminuria, medical nephritis, and essential hypertension may reveal degrees and types of renal anomaly not previously recognized. COLES (1961) reported the case of a girl, aged 2 years 10 months, with urinary excretion of T substance (a possible quinone or quinolone) in association with horseshoe kidney. GLENN (1959) states that his statistics of horseshoe kidney reveal an average age at diagnosis of 34 years.

Complications. A horseshoe kidney is more subject to complications than is the normally formed kidney. In a clinical group 90% of cases had associated renal disease (CULP). BELL's observation upon postmortem examination does not support the expressed opinion that horseshoe kidneys are more liable to disease than normal kidneys.

Common complications are hydronephrosis, infection, and calculus formation. The incidence of tumor is not greater due to the anomaly. A review of the literature by SHOUP, POLLACK, and DOU (1962) reveals only 47 cases of tumors in horseshoe kidneys including 21 instances of adenocarcinoma and 7 of nephroblastoma (WILM's). Only 2 cases of tumor involving the isthmus are reported.

Diagnosis. The symptoms of vague umbilical pain, frequency of urination, and pyuria and the palpation of a mass in the lower abdomen should lead to the suspicion of this anomaly. Diagnosis is established by intravenous or retrograde pyelography. The particular features of the roentgen picture are the relatively low level of the renal shadows, absence of the lower renal pole, downward convergence of the renal axes, medially located and nonrotated pelves, and the rather high departure of the ureters from the anterior or lateral aspects together with their short course near the midline (GAMMELGAARD). Aortography in final diagnosis and as an aid planning of surgery is advised by many (BOREHAM and GAMMELGAARD). FALOR and RUFFLO (1964) stress the value and limitations of aortograms and excretory urography in the complex association of abdominal aneurysm and horseshoe kidney.

5*

Differential Diagnosis. In the differential diagnosis, cases simulating acute intestinal obstruction have been reported by SALTER, CHESTERMAN, and GOHN. JEWELL (1964) emphasizes that retroperitoneal inflammatory lesions may mimic peritonitis caused by intra-abdominal lesions and that in the presence of paralytic ileus, radiographs may fail to disclose a renal anomaly. Ectopia and malrotation

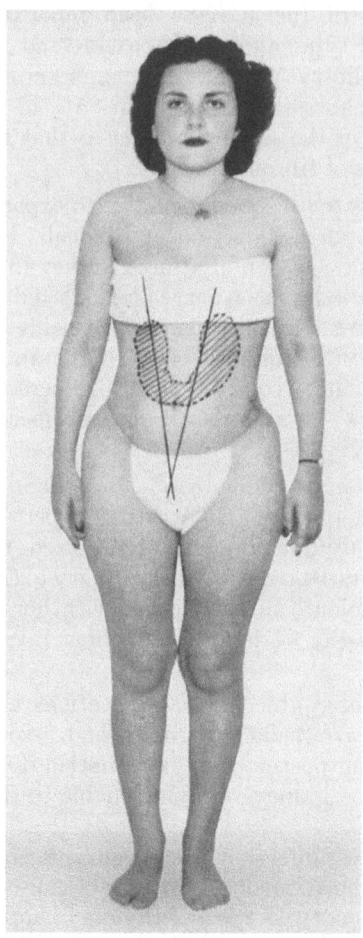 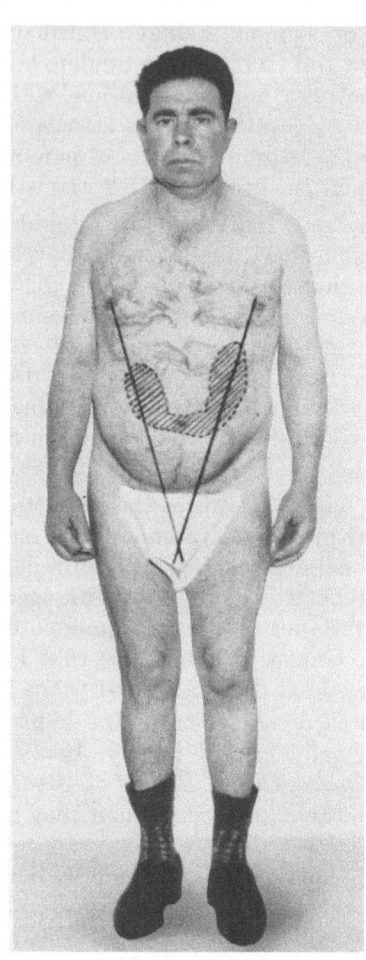

Fig. 4   Fig. 5

Fig. 4. Case I — Age 30. Illustration of body type. Shows characteristic downward convergence of renal axes

Fig. 5. Case II — Age 36. Portugese. Illustration of body type. Shows characteristic downward convergence of renal axes

in separate kidneys may similate the fused kidney. Sometimes distortion caused by a renal cyst or tumor causes confusion.

Treatment. The uncomplicated case of horseshoe kidney fusion will require no treatment. If discovered through routine examination reassurance to the patient and family are all that is necessary. Minor cases of urinary tract infection respond well to antimicrobial therapy. The group that warrants surgical intervention are those cases complicated by other diseases of the upper urinary tract such as hydronephrosis, repeated infection, calculous formation, and less frequently tumor. Surgery for relief of pain only has not proved satisfactory.

Surgery of the Horseshoe Kidney. Surgery of the horseshoe (fused) kidney embraces a wide field of renal operations and for that reason is of great interest and importance. Surgical advances in its cure have kept pace with improved diagnosis until today the "fused kidney patient" can be offered a life span almost equal to persons possessing two normal kidneys.

The extraperitoneal oblique lumbo-abdominal approach to operations upon the fused kidney is advocated by most urological surgeons; however, the transperitoneal incision has been used by some urologists and quite a number of general surgeons in the course of abdominal exploration for unexplained masses. Familiarity with both types of exposure is necessary to successfully correct these anomalies and their concomitant pathology.

Division of the Horseshoe Isthmus. Symphysiotomy, division of the fused isthmus, may be said to be the principal operative procedure for the cure of "horseshoe kidney disease". ROVSING (1911) was the first to advise division of the isthmus by crushing, through a transperitoneal approach. FOLEY (1940) advised nephropexy in addition to symphysiotomy. Division of the isthmus, partial resection of the kidney, or total heminephrectomy can be accomplished by various methods or techniques. Essentially they are the same — sharp dissection of the parenchymatous stump, and closure of the raw surface to insure complete hemostasis and prevent non-fistulous formation. The classical V-shaped incision may be used; a pad of fat placed within the wedge; and closure effected with mattress or overlapping sutures of plain, chromic, or ribbon catgut. FELTON and MILLER (1962) in reporting a successful surgical resection of abdominal aortic aneurysm associated with horseshoe kidney stated there appeared to be no particular hazard in division of the isthmus to gain access.

GREGOIR (1963) outlined the following steps in the surgery of the horseshoe kidney to (1) eliminate compression of the ureter, (2) eliminate all abnormal vessels, and (3) provide free pyelo-ureteral drainage:

(1) incision;
(2) dissection;
(3) section of the isthmus;
(4) polar nephrectomy;
(5) uretero-pyeloplasty or
(6) uretero-ileoplasty.

Nephropexy is useful to maintain the position of the separate units and prevent pressure on the upper portion of the ureters. One of the characteristics of fused (horseshoe) kidney is the low position it generally occupies and the complexity of the renal pedicle (ANSON). After division of the isthmus one must insure an adequate blood supply to both halves. This may preclude a high nephropexy. The principal point in fixation of the kidney is to insure proper drainage from the renal pelvis which almost always lies anteriorly. Any type of anchor suture may be used — transfixed to the intercostal, abdominal, or lumbospinal muscles. Drainage is necessary because of the danger of fistulae from the stump of the isthmus. Excellent results have been reported by FOLEY, GUTIERREZ, CULP, BOREHAM, FARMAN, and LACAL et al. by the combined operation of symphysiotomy and nephropexy. PARKER uses a transabdominal, retro-peritoneal incision to approach the isthmus for symphysiotomy; he does not recommend nephropexy.

Pyelolithotomy in Horseshoe Kidney. Lithiasis is common in horseshoe kidney. The incidence was one third ($^1/_3$) in the Rathbun series, and more than one-half ($^1/_2$) in the Walters-Priestley report. Pelviolithotomy is the operation of choice for removal of calculi. In most instances this can be readily performed because of the accessible anterior position of the renal pelvis. A drawback to pyelotomy is the frequent multiplicity of stones, a wedge-shaped arrangement (Sangree) and tendency to staghorn formation. Such formations may make it necessary to execute nephrotomy incisions for their removal.

Boreham reports a case of bilateral nephrolithiasis in which E. W. Riches successfully removed the calculi by a two stage division of the isthmus and nephropexy. The author has recorded a 12-year followup study of a similar case.

Heminephrectomy. Heminephrectomy may be said to be the operation of last resort in horseshoe (fused) kidney. Nevertheless it ranks high in the total number (almost 50%) of cases reported; this is evidence of late diagnosis with the development of a destructive lesion. Many heminephrectomies have been performed transperitoneally by general surgeons during the course of exploratory laparotomy for abdominal masses. There are some technical advantages to the transperitoneal route; however, most urological surgeons prefer the extraperitoneal lumbar approach. This provides good exposure and minimizes the danger of peritonitis.

The difference between nephrectomy of the non-fused and the fused kidney is in treatment of the isthmus, ligation of the aberrant renal vascular supply. In the latter there is increased danger of injury to the great vessels and the opposite ureter. Occasionally the opposite ureter curves across the vertebral column as has been observed by the author. In rare instances there is only one ureter (Sangree's series showed one ureter in four out of twenty-five autopsy cases). Persistent urinary fistulae are common as the resected "stump" tends to atrophy (Hess). Heminephrectomy should be undertaken only for obstructive calculous pyonephrosis, tumor, or other lesion incompatible with life and health.

Pyeloplasty in Horseshoe Kidney. Repair of the hydronephrotic pelvis often found in fused kidney is the operation of choice to restore normal drainage and to prevent infection and recurrent calculosis. The frequent high insertion of the ureter in an anterior position lends itself to re-implantation at a lower level (Atherton).

Renal transplantation now offers a new hope for the treatment of complicated fusion anomalies of the kidney. In development of surgery of renal fusion anomalies, credit should be given such early pioneers as Rovsing, Newman, Israel and Brawn.

b) Unilateral Fused Kidney

Unilateral fused kidney refers to displacement of one kidney to the opposite side of the retroperitoneum; it is commonly called crossed renal ectopia with fusion. This constitutes the second most common type of fusion anomaly. A total of 433 cases has been reported with an average autopsy incidence of 1 in 7,500 necropsies. In clinical reports the anomaly is most frequent in the young adult group. It is more common in the male sex (ratio of 3 to 2). There is a predilection for the right side (Abeshouse, Baggenstoss, Hinman, Wilmer).

Embryologic Development. Crossed renal ectopia with fusion involves vicarious growth of the ureteric buds and failure of separation of the metanephrogenic cell mass. The anomaly begins very early in embryonic life and is of unknown origin. Ninfe advanced the theory that error in migration might induce the ureteric bud to germinate in an anomalous direction, by-passing the median line and reaching the contralateral wolffian body. In this anomaly a wide variety of renal formations may be achieved, the most frequent being the elongated kidney in which the ectopic organ is fused by its upper pole to the lower pole of a normally placed kidney. Other variations are end-to-end, end-to-side, and L shaped kidneys. Unilateral fused kidney, as a rule, is placed lower than normal and is supplied by endless variety of aberrant blood vessels.

Complications occur in unilateral fused kidney from one-third to one-half of collected series. The displacement predisposes to hydronephrosis, infection and lithiasis. The clinical manifestations are those of the complication; usually localized pain, urinary and gastro-intestinal disturbance (HINMAN, LANGWORTHY and DREXLER, FOLEY and WILMER).

Diagnosis is more reached than in other anomalies since fused mass may be readily palpable and interpretation of pyelograms is more accurate. The nephrogram shows adjacent collecting systems on one side; one ureter crosses the vertebral column; both enter the bladder in a normal position.

Life span is not shortened by unilateral fused kidney. In operation for acquired complications a mortality rate higher than for similar operations on the nonanomalous kidney is shown (BELL, ABESHOUSE).

c) Fused Pelvic Kidney

Fused pelvic kidney is an extremely rare form of fusion ectopia. It differs from solitary pelvic kidney in that the fused pelvis has two distinct pelves and

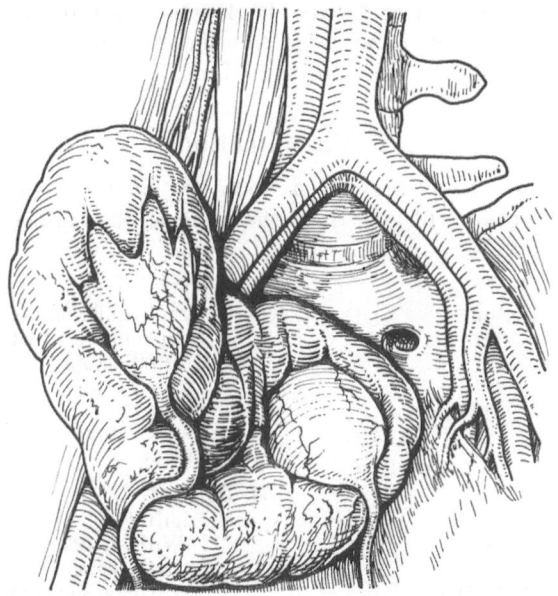

Fig. 6. Pelvic fused kidney. Sketch made at time of operation. (Courtesy Dr. J. F. GLENN)

ureters, and the solitary kidney only one pelvis with two ureters. It is caused by arrest of development after union of the ureteric buds with the renal blastema but before migration and rotation have taken place (10 to 14 mm. stage).

Grossly the fused pelvic kidney presents as a lump, cake, or disc kidney. It has a smooth posterior surface and a lobulated anterior surface from which two ureters emerge. The vascular supply retains its primitive derivation from the iliac and sacral vessels. SHILLER and WISMELL reported the tenth case on record. Some classifications include bilateral fused kidney as a form of crossed ectopia with fusion when the mass is located at the brim of the pelvis (ABESHOUSE, HERBERT).

Diagnosis. A pelvic mass on abdominal examination and absence of the renal shadows on urography suggests an anomaly. The diagnosis is made by cystoscopy, showing two ureteral orifices. Retrograde pyelography, and excretory urography if function is sufficient confirm the diagnosis.

Treatment. Treatment is predicated on any complicating disease — infection, stasis, stone. Surgical interference is technically difficult. Any attempt at division of the fusion or partial resection is not advised (CAMPBELL, GLENN).

V. Ectopia of the Kidney

1. Introduction

From the fourth to the eighth week of foetal life disturbances of embryologic growth may result in renal ectopia and malrotation. Normally the kidney reaches its permanent level opposite the second lumbar vertebrae at the end of the second month. Congenital misplacement theoretically attributes to (1) faulty development of the ureteric buds, (2) abnormal vascularization and (3) the rate of body growth (caudad) in relation to development of the entire uropoietic system. HINMAN advanced the theory that ectopic kidney is brought about by failure of the ureter to grow or the primitive vessels to degenerate, tending to draw the kidney downward. Another theory is that of LUCAS (1934) who suggested that overgrowth of the Wolffian body causes its caudal end to adhere to the metanephros and thus prevent the latter from detaching and ascending.

2. Classification

Classifications of ectopia are based upon the final form of renal anomaly, resulting from disturbance of growth of the ureteric buds and blood vessels.

HINMAN

1. Unilateral ectopy
2. Bilateral ectopy
3. Crossed ectopy
   a) with fusion
   b) without fusion

HARRIS

1. Simple unilateral ectopia
2. Simple bilateral ectopia
3. Crossed ectopia with fusion
4. Crossed ectopia without fusion
5. Pelvic ectopic solitary kidney

HOCHENEG

I. Unilateral (right or left)
   A. Sagittal
     1. Level of lumbar vertebrae
     2. Level of sacro-iliac synchodrosis
     3. Level of pelvis
   B. Transverse (crossed dystopia)
     1. With fusion
     2. Without fusion

II. Bilateral
   A. Sagittal
     1. Without fusion
     2. With fusion
       a) Horse-shoe kidney
       b) Clump kidney or cake (discoid)
       c) Long kidney or "langniere"
       d) S-shaped kidney (sigmoid)
   B. Transverse (bilateral crossed dystopia)
     1. Without fusion
     2. With fusion

a) Simple Ectopy

The term "simple ectopia" is used to classify malascent of the kidney without crossing the midline. The position may be lumbo-sacral, sacroiliac, or pelvic, according to the level of ascent (HINMAN). Bilateral ectopia without fusion is extremely rare.

The incidence is quoted at 1:1000 (HARRIS), 1:650 to 1:500 (NINFE) and 1:1190 (SVOTTOLOW). In clinical records the percentage increases to 1:500 (CULP, NATION, SMITH, and ORKIN). Upon intravenous pyelography in 17,229 examinations, ectopia was discovered only 30 times, 1 in 574 (Mallinckrodt Institute of Radiology).

The ectopic kidney may be located anywhere from a point above the normal location (thoracic) to the bony pelvis. The most common locations are lumbar or sacral. Approximately one-half of ectopic kidneys are found within the bony pelvis. Some degree of anterior malrotation is the usual accompaniment of ectopia. Pyelography will demonstrate an anteriorly placed pelvis and ureter. Ectopic ureteral openings, a bifid pelvis, and partial renal duplication are frequent associated anomalies.

The ectopic kidney usually functions normally and produces symptoms. In diseased states it is subject to usual pathological processes. Infection and stone are most frequent; hydronephrosis is common, neoplasm may be associated. In the diseased group pain, hematuria, polyuria, dysuria, or abdominal mass and gastrointestinal symptoms are characteristic. Pain in the lower abdomen may lead to such erroneous diagnosis as appendicitis, tumor of the large bowel, mesenteric cyst, and ovarian disease. Difficult labor also may result from the ectopic kidney. ANDERSON, RICE, and HARRIS concluded from a study of 34,206 deliveries that most women with pelvic kidney may be delivered vaginally, but if all renal tissue lies in the pelvis, as in bilateral ectopy or solitary fused pelvic kidney, the best method of delivery is Cesarean section. Pelvic ectopic kidney without complications does not constitute an indication for interruption of pregnancy or sterilization.

b) Bilateral Ectopia

Bilateral ectopia without fusion is exceptionally rare. It should not be confused with fused pelvic kidney. Its chief interest concerns differential diagnosis from nephroptosis. The level of fixity may be at any point below the normal renal position to the sacro-iliac region. Differential diagnosis is made by pyelography. The movable kidney is easily demonstrated by shift of position. Operation is not advised except for complications.

c) Crossed Renal Ectopia

Crossed renal ectopia is a congenital displacement in which the two kidneys lie adjacent to each other on one side of the lumbar spine or more rarely in a prevertebral position.

Transposition of an ectopic kidney to the opposite side, commonly results in fusion of the two renal structures. Rarely this occurs without fusion. Still more rarely, a single fused ectopic kidney is located in the pelvis. The cause may be traced to arrest of one of the ureteric buds during its growth into metanephrogenic cell mass, coupled with interference with migration and ascent. Abnormality of the genital tract, as absence of an ovary, uterine body, cervix, and vagina in females and abnormality of the females and abnormality of the vas deferens in the male are present in many cases.

The anatomical varients of crossed renal extopia are many:

    I. Crossed renal ectopia with fusion
       A. Ectopic kidney inferior (most frequent)
       B. Ectopic kidney superior (rare)
       C. Bizarre shapes caused by growth with fusion
          1. sigmoid
          2. disc
          3. shield-shape
          4. L-shape
          5. lump or cake
          6. horseshoe

II. Crossed renal ectopia without fusion
    A. Ectopic kidney below the uncrossed kidney
    B. Ectopic kidney above the uncrossed kidney
    C. Bilateral crossed renal ectopia.

Crossed ectopia with fusion occurs more often in males than females (ratio 3 to 2) and is commonly observed on the right side. In crossed ectopia without fusion the anomaly (ectopic component) is more common on the left side (ABES-

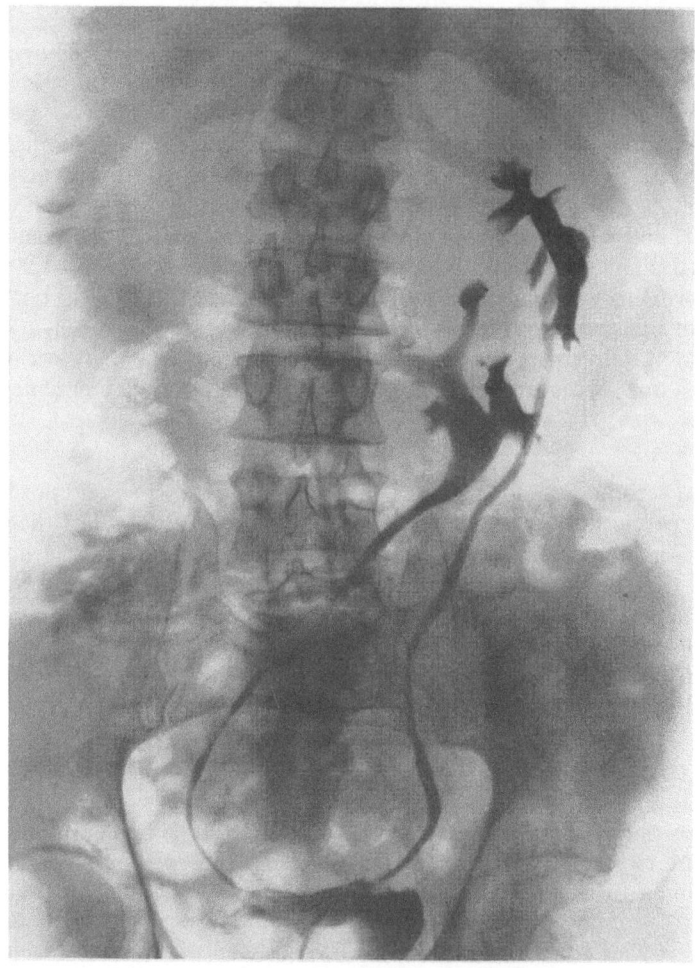

Fig. 7. Crossed renal ectopia. Pyelogram showing bilateral crossed renal ectopia, probably with fusion. (Not operated.) (Courtesy of Dr. B. K. BROCK)

HOUSE, SHIH HSI-EN). Crossed renal ectopia (fused and unfused) may be discovered at any age; it is most frequently diagnosed in the third decade of life.

The symptoms depend, not on the anomaly, but on any complication — abdominal or back pain, palpable mass, gastro-intestinal disturbance, and urinary symptoms, as frequency hematuria and pyuria. The usual acquired complications are infection, hydronephrosis, and stone which occur in fully one-third of the cases.

Diagnosis. Diagnosis of crossed ectopia is made by cystoscopy and pyelography. In differential diagnosis from other abdominal and retroperitoneal lesions the question of fusion always presents a problem. Poor excretory function and poorly defined visualization of pelvis and calyx are findings suggesting union of the transposed ectopic component to the more normally placed renal body.

Lack of clear cut symptoms and infrequency of occurrence, contribute to the difficulty of diagnosis. In review of cases of crossed ectopia the correct diagnosis was not made until after autopsy in about 35% and at operation in about 25% of cases. A history of pain on the affected side with a palpable mass may suggest an anomaly. In differential diagnosis the following conditions are more commonly encountered: congenital solitary kidney, hydronephrosis, renal tumor, appendiceal abscess, cholecystitis, intestinal obstruction, intra-abdominal tumor, retroperitoneal neoplasm, uterine myoma, ovarian cyst, salpingitis and aortic aneurism. Retroperitoneal air injection and aortography are helpful diagnostic aids.

Treatment. When ectopia is discovered accidentally, without symptoms and in the absence of complications, treatment generally is not required.

Medical and surgical measures should be carried out for any complication as indicated. In one-third of the reported cases treatment has been necessary. In the presence of infection antibiotic therapy has been successful in some cases of crossed ectopia without fusion (SHIH HSI-EN, et al.). Pyelotomy for removal of calculi has been reported by MAYERS; transposition and nephropexy was carried out in one case by Diaz, transperitoneal nephrectomy by HARRIS. Extraperitoneal nephrectomy for diseased crossed ectopic kidney is the surgical procedure of choice according to BEER. WINRAM and WARD-McQUAID in reporting the tenth case of crossed renal ectopia without fusion carried out nephrectomy to cure a functionless ectopic iliac kidney containing a cortical abscess.

The method of surgical approach is determined by the location of the ectopic kidney. Wide exposure and careful dissection is necessary since the relationship of the two abnormally placed kidneys and ureters to adjacent structures and anatomical variations in blood supply cannot always be determined beforehand. Arteriography is a help. A low abdominal or high lumbar extraperitoneal approach generally is preferable to the transperitoneal, so as to avoid the complication of peritonitis.

d) Ectopic Pelvic Kidney

Single ectopic pelvic kidney is a rare anomaly. It should not be confused with the bilateral fused pelvic kidney or the solitary crossed ectopic kidney. In solitary ectopic pelvic kidney there is only one renal mass and one ureter; in bilateral pelvic kidney there are two renal structures with separate ureters. In solitary crossed ectopia one renal mass may lie above or below the crest of the ileum.

The first author to describe solitary pelvic kidney was HEMUT (1830). According to STEVENS (1937) not more than 25 cases were published in the 100-year period following; he added two under his personal observation. COMUZZI (1956) added one case to the series of BORELL and FERNSTROM (1952) making a total of 69 cases In 1959 BURWELL and KENT reviewed 86 cases of solitary ectopic pelvic kidney. Dr. BURTON K. BROCK, a former associate, has observed 2 cases which emphasizes that the published incidence of single ectopic pelvic kidney may be low. The author agrees with COMUZZI (1956) who cited the need to rectify the casuistics table of the world literature.

Solitary ectopic kidney implies interference with growth factors governing migration and ascent during embryological development. In addition there is agenesis (absence) of the opposite kidney caused by arrest of development of one ureteric bud and its related metanephrogenic cell mass. Abnormalities of the genital tract are in frequent association, as absence of the uterine body or vagina in the female and of the vas deferens in the male. Burwell and Kent's analysis of congenital abnormalities in the female genital tract includes 39 cases.

Absent organ	Number of cases		Per cent
Ovary, one	7		
Ovary, two	2	9	23.1
Uterine tubes, one	6		
Uterine tubes, two	4	10	25.6
Uterine body	15		
Uterine cervix	1	16	40.1
Vagina	20		
Clitoris	1	21	51.3

The majority of pelvic kidneys of both the solitary and "bilateral fused" type have been discovered during exploratory surgical operations or at necropsy. The diagnosis is not difficult if the existence of such an abnormality is kept in mind. All retroperitoneal and low abdominal masses should be differentiated. Diamantis reports the case of a pelvic kidney mistaken for "appendicitis" and Sokolski (1960) states that in all cases of difficult labor (dystocia) this condition, though rare, should be considered. He added the eighth case of pregnancy complicated by solitary pelvic kidney.

Treatment is indicated only for complications- as hydronephrosis, hydro-ureter, and calculi. Successful nephrostomy and pyeloplasty are reported (Burwell and Kent).

e) Thoracic Kidney (Congenital Superior Ectopia)

High dystopia of the kidney is a rarity of clinical importance in the differential diagnosis of masses at the posterior lung base.

The first report was by Mikulicz (1922). Radecki described a congenital diaphragmatic hernia involving the major portion of the right kidney in a seven months stillborn female. Berlin, Stein, and Poppel reported two cases of superior ectopy of the left kidney and collected 10 similar cases from the literature. Paul, Uragoda, and Jayewardene reported the eighth case of congenital herniation of the kidney into the thoracic cage.

In high ectopia the kidney ascends through a gap, bulge, or foramen in the diaphragm. The high placed kidney has an elongated ureter and blood vessel supply and is associated with some degree of malrotation. Diagnosis is by pyelography and chest x-ray. Wolfromm was the first to identify superior dystopia of the kidney by such means. Treatment rarely is indicated as the high ectopic kidney usually is normal in size and shape and the intra-thoracic site does not diminish its functional value (Franciskovic and Martincic).

VI. Anomalies of Rotation

1. Introduction

Weyrauch defines anomalous rotation as a "congenital abnormality which is manifest by an atypical location of the hilum renale". Jonsson and Olsson say that "malrotation is incomplete rotation". The condition assumes clinical importance only when complications arise such as renal infection, formation of calculi and hydronephrosis. Malrotation is accompanied frequently by other abnormalities of the kidney, particularly fusion, caudal ectopia, and accessory vessels. Supplementary arteries always start fairly far away from the main renal artery, either from the distal part of the lumbar aorta or from the iliac or sacral artery. The position of the hilum and placement of the ureter determine the type of rotation.

2. Embryology

Normal movement of the kidney in the early embryo is as follows: (1) ascent along the posterior body wall (2) lateral movement away from the midline (3) axial deflection so that, in contrast to the early embryo in which the caudal poles are closer together in the adult the cephalic poles are closer than the caudal (4) internal rotation through a right angle so that the renal pelves, at first anterior, now lie medial to the renal mass (PARTON).

The types of malrotation are:
1. ventral (failure of)
2. ventromedial (incomplete)
3. lateral (reverse)
4. dorsal (excessive).

3. Incidence

Few cases have been recorded in the literature, a total of 44 having been collected up to 1959 (HERBUT; CAMPBELL). GERD BOLL (1957) describes a case of lateral rotation of the right kidney; T'UNG SHANG-T'AI, TS'AO CHIEN and HU CHI' JUNG (1959) reported a case of transversely situated right kidney and a laterally malrotated left kidney. JONSSON and OLSSON (1962) recorded a case of renal ectopia with malrotation and vascular anomalies causing renal pain which was relieved by resection of the proximal pole. NATHAN (1963) reports the case of a laterally rotated right kidney.

4. Etiology

Normally the kidney ascends from the pelvis to its usual position in the lumbar region. This ascent is due to differences in growth of the body rather than to a true migration. Change in the position of the pelvis is likewise a manifestation of differential regional growth rather than actual rotation. It represents gradual intrarenal displacement of the pelvis. From embryological studies it is known that normally tubules from the second and third order on usually send out two branches ventrally to each branch dorsally. Multiplication of metanephrogenic tissue to cover these is correspondingly irregular resulting in a relative displacement of the pelvis medially. Abnormal irregular proliferation of tubules results in displacement of metanephrogenic tissue and abnormal rotation, that is, no rotation, incomplete rotation, excessive rotation to as much as 360° and reverse rotation (WEYRAUCH, quoted by HERBUT).

5. Symptoms and Diagnosis

The condition assumes clinical importance only when complications arise — as renal infection, calculi and hydronephrosis. The outstanding symptom is pain. Stereoscopic and lateral pyelograms and angiographs establish the diagnosis.

6. Treatment

Malrotation does not require treatment unless complicated by associated renal lesions. For advanced hydronephrosis nephrectomy is indicated providing function of the other kidney permits — at other times pyeloplasty is the operation of choice. In performing nephropexy care should be taken not to produce angulation or torsion of the ureter and blood vessels. Resection of part of the kidney may be necessary.

VII. Anomalies of the Vessels

1. Introduction

In ascent from the hollow of the sacrum to the lumbar region the metanephrogenic cell mass derives its vascular supply from a ladder type exchange of blood vessels apparently arranged for mechanical convenience. This provides opportunity for variation and may account for the observation that in nearly 25%

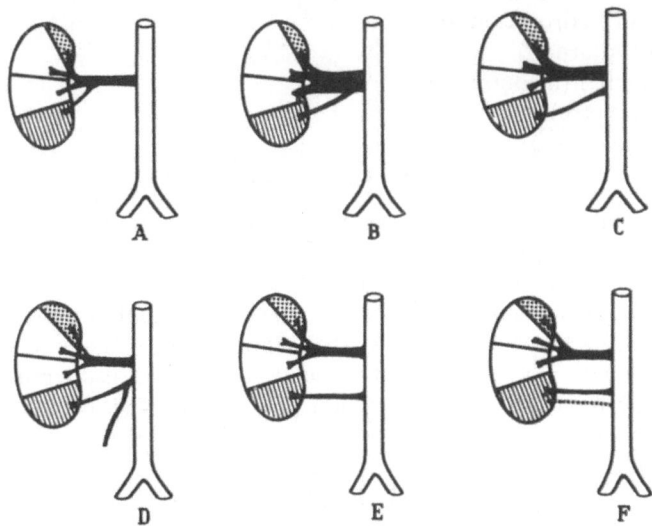

Fig. 8. A diagram illustrating the various origins of the artery to the lower segment. (After GRAVES)

of all individuals the renal blood supply shows some abnormal development. The original renal blood supply commonly is derived from the middle sacral, iliacs or lower aortic vessels. Retention of any of these vascular connections may explain one cause of renal ectopy as well as persistence of accessory vessels to the kidney and ureter. The final renal artery generally is a single vessel bifurcating into two main branches in region of the kidney pelvis; however, multiple renal arteries are found in one out of three kidneys and may supply important portions of the renal parenchyma. Anomalous deviation from the normal arterial pattern to the fully ascended kidney may be the cause of obstruction at the ureteropelvic junction (ANSON et al., BREMER, JEWETT, McDONALD et al., GRAVES).

Accessory vessels generally arise from the main vascular pedicle, the aorta, vena cava, or adjacent branches of the great vessels (supra renal and iliac). GRAVES cites the various origins of the artery to the lower segment as:

1. artery arising at the hilum
2. from the main stem at its junction with the aorta
3. from the aorta close to the main stem
4. from the aorta, the testicular or ovarian artery arising from the artery to the lower segment
5. the lower segment artery arising from the aorta at some distance from that of the main stem and

6. the lower segment artery and its posterior branch arising separately from the aorta.

Aberrant vessels which course anterior or posterior to the renal pelvis are referred to as polar vessels because they supply the upper or lower halves of the kidney. Accessory vessels which may interfere with normal urinary outflow from the kidney more commonly are found in front of the ureter at or near the uretero-pelvic junction. ABOWITZ stated that abnormal vessels supplying upper pole of the kidney are incapable of producing obstruction.

2. Production of Hydronephrosis by Aberrant Vessels

Slowly developing hydronephrosis may result from compression or distortion of the ureter by an anomalous vessel. These abnormal vessels usually compress the upper part of the ureter at, or near, its junction with the pelvis. As a rule these obstructing aberrant vessels go to the lower pole of the kidney or to the inferior margin of the hilum, and may pass either anterior or posterior to the ureter. ALBARRAN (cited by YOUNG) states that "whether the abnormal vessels pass in front or behind the ureter, one is unable to understand how it is able to kink the ureter in order to cause hydronephrosis. On the contrary, a hydronephrotic kidney may descend and the ureter become encroached upon by vessels which cross either in front or behind it". The factors of mechanical compression of the ureter and pelvis associated with malrotation and ptosis of the entire kidney, and true intrinsic disease of the ureter such as stricture, dilation or diminished peristalsis all play a part in causing interference with normal urinary excretion (HERBUT, JEWETT, WILLIAMS).

3. Incidence

Although the incidence of anomalous renal vessels is high, vascular obstruction producing disease of the renal pelvis and kidney is, by comparison, small. The Mayo Clinical in 1909 reported twenty-seven cases of surgically treated hydronephrosis; in twenty there were accessory vessels. To prove the causative relationship, in thirteen of the twenty cases, the vessels were severed and ligated, followed by a cure of the condition. KLUKOW, in 1912, collected fifty-seven cases of hydronephrosis due to abnormal vessels. In 1922 KUMMER collected 56 cases in which ligation of aberrant obstructing arteries cured the condition. FARMAN in 1927 reviewed the literature on surgery of anomalous vessels and concluded that the ligation of accessory renal vessels is indicated to relieve ureteral obstruction, to aid in mobilization of the kidney, and may be done without serious, if any, impairment of renal function.

About this time WILDBOLZ of Berne, Switzerland made a plea to avoid division of aberrant renal arteries. Many reports appeared regarding the role of the anomalous vessel in the production of hydronephrosis. Among them may be mentioned the reports of ARNEMANN (1959) on 26 cases, in 24 of which he carried out types of ureteroplasty; of CHAPMAN (1959) who obtained good results by displacing and fixing the aberrant artery to a position close to the main artery by fixing it with sutures to the renal pelvis; and of ADAMS (1959) using a method of nephroplication in 6 cases with clinical success. There are still those who resort to use of simple division of aberrant vessels in selected cases, including WINSBURY-WHITE, STEFFENS-KREBS, and STIRLING who in addition to ligation of the lower pole vessels advocate resection of the area supplied by these vessels as shown in the angiogram.

4. Symptoms

The symptoms of vascular ureteral obstruction are similar to those of other types of hydronephrosis. Intermittent dull flank pain, attacks of colic in the loin, periodic fever with pyuria and associated bladder disturbance may arise in cases of infection.

5. Diagnosis

An anomalous renal vessel should be suspected in cases of hydronephrosis produced by obstruction at the ureteropelvic junction which cannot be ascribed to ureteral stone, stricture, tumor, or nephroptosis. Diagnosis is made by pyelography, employing both the delayed pyelogram and ureterogram. A thin line of compression or a transverse filling-like defect may be demonstrated where the vessel crosses the ureter. Angiography may be helpful in differentiation and follow-up, confirming STIRLING's statement that a kidney is only as good as its blood supply. Operation clarifies the true cause of the obstruction. Many patients reach young adulthood before recognition of the basic difficulty (CAMPBELL).

6. Treatment

Nephrectomy has been performed in about one-half of the cases due to advanced renal damage. The need for early recognition and corrective surgery is apparent.

WINSBURY-WHITE (1959) states that the obstructing blood-vessel is the condition which calls for remedy by division, thereby giving immediate relief of symptoms and reduction in size of the hydronephrosis. The danger of producing renal infarction by division of any branch of the renal artery is apparent. In clinical practice such infarction and parenchyma scarring generally is endured without inconvenience and no decrease in total renal function results due to the factor of early compensatory lower polar circulation. In the author's experience ligation and severence of the aberrant vessel causing obstruction has been satisfactory in many cases. It is easily performed and a safe surgical procedure. The spontaneous resumption of function in a severely hydronephrotic kidney is reported by STEFFENS-KREBS (1959) following removal of an aberrant vessel obstructing the pyeloureteral junction.

Many types of plastic and preserving techniques for the relief and cure of hydronephrosis caused by abnormal vessels have been presented during the past two decades. The most frequent types of modern conservative surgery are modifications by pyelo-uretero anastomoses to provide unobstructed urine flow without division of vessels. Complete or partial resection of the ureter and re-implantation into the hydronephrotic sac, plication of the dilated pelvic wall or transposition of vessels to relieve pressure of kinking are the principal techniques employed by LeRoy, PERLET, MICHALOWSKI and MODELSKI, GIBSON. The novel method of HAMILTON STEWART in which the kidney is moulded bringing the two poles and consequently the two arteries together is applicable in some cases. Modifications of the techniques of ureteroplasty and nephroplication have tended to replace simple division of aberrant renal vessels.

VIII. Supernumerary Kidney
1. Introduction and Incidence

The supernumerary kidney is a free accessory renal organ with its own ureter. The first case was recorded in 1677. Since then 66 cases have been described, the latest by PHOKITIS in 1964. Some authors do not accept all of these cases as fulfilling the criteria for true extra kidney. Collected series by PARIN (1924) 24 cases; KRETSCHMER (1915, 1929) 30 cases; GEISINGER (1936) 40 cases; and CARLSON (1946 and 1950) 51 cases have enhanced knowledge of the supernumerary kidney. FRATKIN (1963) recorded an instance of adenocarcinoma in an extra kidney and referred to one other similar case reported by EXLEY and HOTCHKISS in 1944. TEMELKOFF (1957) reported upon the unusual operative finding of four independent kidneys in a male 25 years of age, two of which were extirpated.

2. Pathogenesis

The pathogenesis of supernumerary kidney is similar to double kidney, the only difference being a matter of embryological stage. The first anomaly develops from complete separation of the blastema having its only excretory apparatus and metanephrogenic cap (renal parenchyma) and the latter or duplicated kidney (pelvis) results from a growth variance or cleavage of the ureteral bud (mesonephros) without renal parenchymal involvement.

The anomaly may occur on either side and is found equally in the male and female. It generally but not always lies below the ipsilateral normal kidney. Usually the supernumerary kidney is underdeveloped both as to size and function. One case of bilateral formation has been reported. A supernumerary kidney may vary in size and shape and usually occupies a lumbar position; occasionally it is found as low as the iliopelvic region. The ureter may be independent with a separate intravesical orifice not necessarily following the Weigert-R. MEYER law as in duplication of pelvis — lower orifice always drains upper pelvis. Occasionally it becomes a branch of a bifurcated ureter or it may have an ectopic orifice.

3. Symptoms

The clinical signs that have led to discovery of the anomaly vary widely and arise from pathological change within the normal and supernumerary kidney or simultaneous involvement. Associated urinary tract and abdominal organ disease is frequent. Only five of the sixty-four cases were correctly diagnosed before operation.

4. Diagnosis

Inadequate methods of examination and indistinguishable symptoms render diagnosis of the supernumerary kidney difficult. The anomaly usually is discovered accidentally during operation or at necropsy.

The finding of two ureteric openings in close proximity upon cystoscopy may indicate also upper urinary tract malformation. By retrograde pyelography the two kidney pelves may be outlined. In STEFAN's case retroperitoneal air insufflation aided visualization. Excretory urography may have the disadvantage of poor dye output. Aortography may be useful to confirm the presence of the two independent organs.

5. Treatment

Surgical treatment of the supernumerary kidney should take into consideration both the extra kidney, the ipsilateral normal kidney and the contralateral kidney. The following tables are from the report of PHOKITIS (1964):

Table 1. *Operations upon the supernumerary kidney*

Kind of operation	Disease	Name of author
Nephrectomy	Hydronephrosis	Bacon
Nephrectomy	Lithiasis, pyonephrosis	Boretti
Nephrectomy	Papillary cystadenoma	Cobb and Gibbings
Nephrectomy	Nephroptosis	Depage
Nephrectomy	Cancer	Exley and Hotchkiss
Nephrectomy	Undiagnosed vague pain	Fenker
Nephrectomy	Pyonephrosis	Fischer and Rosenloecher
Nephrectomy	Chronic pyelonephritis with secondary contraction of the kidney	Holmes
Nephrectomy	Solitary simple cyst	Kuksinkaya
Nephrectomy	Undiagnosed vague pain	Lebedeff
Nephrectomy	Undiagnosed vague pain	Linberg
Nephrectomy	Hydronephrosis	Munro and Goddard
Nephrectomy	Undiagnosed vague pain	Parin
Hydronephrosis		Stewart
Two operations: a) nephrostomy b) nephrectomy	Pyonephrosis, ectopic ureter	Israel
Total ureteronephrectomy	Hydroureter, ectopic ureter, renal atrophy	Gayet
Total ureteronephrectomy	Ectopic ureter, hypoplastic kidney	Hoffmann and MacMillan
Total ureteronephrectomy	Ectopic ureter	Samuels, Kern and Sachs
Total ureteronephrectomy	Ectopic ureter	Shane
Subtotal ureteronephrectomy	Pyonephrosis; ectopic ureter	Author's case

Table 2. *Operations upon both the supernumerary and the ipsilateral normal kidney*

Kind of operation	Disease	Name of author
Nephrectomy of both kidneys	Pyonephrosis of the supernumerary kidney	Brunner
Nephrectomy of both kidneys	Infected hydroureter	Martin du Pan and Christen
Nephrectomy of both kidneys	Injury	Vatschensky
Nephrectomy of both kidneys	Lithiasis, pyonephrosis	Kretschmer
Nephrectomy of both kidneys	Tumor	Rolnick
Nephrectomy of both kidneys	Injury	Rutschinsky
a) Nephrostomy b) Nephrectomy c) Nephrectomy of the non-diseased supernumerary kidney	Pyonephrosis	Bernasconi and Bernard
a) Nephrostomy b) Nephrectomy of the supernumerary kidney c) Nephrectomy of the ipsilateral kidney because of post-operative hydronephrosis and suppurative nephritis	Infected hydronephrosis	Kretschmer
Nephropexy (of both kidneys)	Nephroptosis	Calabrese

IX. Cystic Disease of the Renal Pyramids

(Sponge Kidney)

1. Definition and Historical

The term "cystic disease of the renal pyramids" is included in this chapter because most authorities now regard the condition as of congenital origin. It has been shown to be a distinct disease entity and variously referred to as medullary sponge kidney, multiple cysts of the renal medulla, cystic dilation of the renal collecting tubules, renal tubular ectasia, or simply sponge kidney.

The original account of this uncommon lesion was made by CACCHI and RICCI (1949). Credit however, is given to LENARDUZZI (1939) who has first to recognize the radiological features of "sponge kidney".

2. Incidence

Medullary sponge kidney is rare. The largest number of cases are recorded by ABESHOUSE (1960) who collected 131 and added 5 personal cases. In reporting 2,465 intravenous urograms, 14 examples of the disease were found (PALUBINSKAS). The condition is seen most frequently in the fourth, fifth and sixth decades of life and is observed more commonly in males than females (2.5:1). Bilateral involvement predominates.

3. Etiology

Medullary sponge kidney is a developmental defect caused by alteration of the tubulo-glomerulo structure which form changes within the kidney. It has been called polycystic disease of the renal pyramids, in contrast to polycystic involvement of the entire kidney.

LOWEN and SMYTHE summarize the theoretical considerations:

(a) damage to collecting tubules during foetal life, produced by deposits of uric acid crystals (VIRCHOW, 1869; 1892)

(b) disturbance of fusion of the ducts formed from nephrogenic tissue with those originating from the Wolffian ducts (HANAU, 1890; DETTMER, 1903; RUCKERT, 1903)

(c) persistence of early generations of tubules (McKENNA and KAMPMEIR, 1934)

(d) an abnormal chemical factor (organizer) acting on the embryo.

KAMPMEIR points out that every human individual during fetal life normally passes through a period characterized by the presence of numerous cystic renal tubules.

4. Pathology

The basic pathology of this lesion is the presence of multiple small cystic cavities throughout the pyramids. They surround the terminal calyces in a cluster-like arrangement, limited to the medullary zone and impart a porous or spongy appearance, hence the term sponge kidney. Minute or microscopic calculi may be present in some or all of the cystic cavities from retention and precipitation of urinary salts. Round cell infiltration is a common feature (ANTON, ABESHOUSE). CACCHI and RICCI describe two types of cysts in sponge kidney: (a) an irregular type lined by cylindrical epithelium and consisting of large cells of an embryonic nature and (b) spheroid cysts lined by flattened or thin epithelium. Extent of the disease may vary from involvement of a single pyramid in one kidney to involvement of all the pyramids of both kidneys.

5. Clinical Considerations

Cystic disease of the renal pyramids is subject to long periods of latency and does not impair functional capacity of the kidney until late. The condition may remain asymptomatic until secondary renal disease develops, such as infection and stone formation. A typical case may present in adult life with mild intermittent renal pain, recurring urinary infection and hematuria.

6*

Alteration of clearance tests occurs in advanced cases. CACCHI and RICCI found tubular deficiency in 5 of 12 patients; and LEVIN et al. produced evidence that the cystic change interferes with ammonium excretion.

Cystic disease of the renal pyramids does not induce hypertension.

6. Diagnosis

Symptoms of urinary infection are the most common presenting complaints. The frequent association of sponge kidney with lithiasis may render differentiation from nephrolithiasis and nephrocalcinosis difficult. Intravenous pyelography is the best means of showing the cystic spaces within the pyramids; there is little or no filling with contrast medium using retrograde pyelography. The shape of the cysts is variable: they may be elongated, round, oval or irregular; when grouped have the appearance of a cluster of grapes (ANTON). The renal outline may be enlarged. Hypercalciuria was found in 14 of 36 patients reported by EKSTROM et al. (1959).

Differential diagnosis should exclude (1) pyelogenic cyst (2) calyceal diverticula (3) pyelosinus reflux and (4(cystic cavities associated with renal tuberculosis, necrotizing papillitis and nephrocalcinosis.

7. Treatment

Medical measures are advised in the therapy of the uncomplicated case of medullary sponge kidney. Although the disease is subject to long periods of quiescence, successful treatment of cases complicated by infection and stone formation is very difficult. CACCHI and RICCI have stated that one can never speak of curing sponge kidney since regression of the cystic process has not been known to occur.

Pelvic lavage has been tried without success; nor has surgical drainage of the infected, stagnated urine from the multiple sacculations proven feasible. In a case of ISHIHARA and KAWAMOTO relief of symptoms followed injection of the renal pelvis with 10% renacidin solution in the presence of multiple small calculi.

8. Prognosis

Prognosis is good in the majority of patients. Progressive renal damage and uremia are to be expected in the severely complicated case.

X. Anomalies of the Renal Pelvis
Reduplication or double kidney

1. Incidence

Reduplication of the renal pelvis, so-called double kidney, is a ureteral rather than a renal anomaly. However, it occurs with such frequency, and in such common association with other renal anomalies that it is included here. Duplication of the renal pelvis and ureters is the most common anomaly of the upper urinary tract. The incidence is from 3 to 4% of the total population (HERBUT, NORDMARK, COLBY). DEES (1941) found duplications in 2.7% of 1,610 consecutive pyelograms; PAYNE found 83 cases of reduplication in 5,000 urograms studied between the years 1953 and 1956. NATION (1944) in an analysis of 230 cases concluded that ureteral duplication is less common than most of the statistics in the literature indicate (0.7 per cent of 16,000 autopsies).

2. Embryological Development

Anomalies of the pelvis result from a growth variance or cleavage of the ureteral bud before it is encompassed by the nephrogenetic cap. Complete reduplication occurs when two ureteral buds form on the same Wolffian duct. Incomplete reduplication results from bifurcation of the renal pelvis and ureter when the former divides into major calyces. There are innumerable variations in number, shape and size of the pelvis and calyces — deviating from the normal which is a single pelvis with three major calyceal branchings (superior, medial and inferior).

Degrees of reduplication vary from the simple bifid pelvis to formations with two separate pelves and ureters. In most cases the duplicated kidney resembles a normal kidney externally; in some there is a cleft separating the two fused segments, the upper, as a rule, being smaller than the lower. Reduplicated pelves are placed one above the other, never in front or behind. When two ureteral orifices are seen in the bladder, according to the Weigert-R. MEYER law the lower always drains the upper of the two renal pelves. A reversal of this pathological truism has been reported four times (LUND, ST. MARTIN et al.). In partial reduplication the two ureters may join to form one ureter at any level between the kidney pelvis and bladder, but is observed more commonly in the lower segment.

Differentiation must be made from the rare supernumerary kidney which is a completely free organ developed from a separate blastema with its own ureter opening into the bladder. Extra kidneys almost always lie below the normal one. They are usually small, situated deeply, and have their own fibrotic capsule, pelvis and ureter, and arteries which stem from the abdominal aorta (TEMELKOFF).

3. Clinical Considerations

TOLLE, in a study of double kidney, concluded that anomalies of this sort frequently occur without symptoms. PAYNE, in a review of 141 cases, listed the presenting symptoms as pain, infection phenomena, haematuria, pyrexia, enuresis, calculi and associated congenital anomalies including multiple malformations of the urinary tract. Hypertension occurs no more commonly with ureteral duplication than without (NATION).

The diagnosis of reduplication is usually made without difficulty by means of an excretory urogram. Sometimes it is difficult to distinguish an abnormal calyceal system and interpretation must be coupled with other evidence of disease or anomaly. A double kidney is suggested by an elongated renal shadow and variation in the characteristic shape of the pelvis. The retrograde pyelogram though useful can not be depended upon in diagnosis because of incomplete filling of the duplicated pelvis and ureters. Duplication of the renal pelvis and ureter tend to make diagnosis of associated pathologic and anomalous conditions more difficult, hydronephrosis and hydro-ureter being the most common associated findings.

4. Differential Diagnosis

MARQUARDT (1956) reported two cases of double kidney both explored because of suspicion of renal tumor. MUKHIN reported a case of retroperitoneal tumor (liposarcoma) associated with double kidney. Ross (1964) presents a case of hemipyonephrosis in a duplicated kidney unsuspected before operation; nephrectomy was required because of infection.

ALEXANDER LIVADITIS et al. (1964) report upon the rare condition of ureteropelvic triplication demonstrated clearly by simultaneous retrograde pyelography and excretory urography. Excision of an ectopic ureter draining the middle pelvis and anastomosis to the lower ureter cured the child of incontinence and infection.

5. Treatment

In treatment only associated disease of the kidney requires attention, infection and obstruction being the most common. Persistent infections are indications for surgical intervention. Heminephrectomy is the preferred operation for isolated hydro- or pyo-nephrosis with or without stone formation in double kidney. SETH-SMITH and PARK state the partial nephrectomy appears justifiable when the disease is localized to one almost separate segment. It also is indicated when the ureter has an ectopic opening. In cases of tuberculosis or malignancy total nephrectomy is indicated providing the contralateral kidney has adequate function.

Acknowledgments

The author gratefully acknowledges the assistance of: J. E. MARKEE, Ph. D., Professor of Anatomy, Duke University, for consultative advice on the embryological basis for abnormalities of the kidney; WAYNE WILLIAMS and MARY E. DUNN, Department of Anatomy, Duke University School of Medicine, for illustrations; The Armed Forces Institute of Pathology, Washington, D. C., for pathological photographs; BELLE FARMAN, M. A., for copy editing; MARGARET BAILES, A. A., secretary, for bibliographic research and to The Robert S. Fox Urological Foundation, Whittier, California, for study grants.

References

Renal Agenesis

ALLEN, A. C.: The kidney. Medical and surgical disease. New York: Greene & Stratton 1951.

ASHLEY, D. J. B., and F. K. MOSTOFI: Renal agenesis and dysgenesis. J. Urol. (Baltimore) **83**, 211 (1960).

BAIN, A. D., M. D. BEATH, and W. F. FLINT: Sirenomelia and monomelia with renal agenesis and amonium nodosum. Arch. Dis. Childh. **53**, 250 (1960).

—, and J. S. SCOTT: Renal agenesis and severe urinary tract dysplasia. Brit. med. J. **1960** I, 841, No 5176.

BROWNE, E. L.: Congenital absence of one kidney. Canad. med. Ass. J. **68**, 489 (1953).

BURFORD, C. E., and E. H. BURFORD: Congenital solitary kidney. Sth. med. J. (Bham, Ala.) **48**, 934 (1955).

BYRNES, R. L., and J. W. BOELLAARD: Renal agenesis and meningocerebral angeiomatosis. Arch. Path. **66**, 23 (1958).

CAMPBELL, M. F.: Principles of urology. Philadelphia: W. B. Saunders Co. 1957.

CASTLEMEN, B., and B. U. KIBBEE: Case records of the Mass. General Hosp. Presentation of a case. New Engl. J. Med. **260**, 1031 (1959).

COLLINS, D. C.: Congenital unilateral renal agenesis. Ann. Surg. **95**, 715 (1932).

COMUZZI, U.: Rectification of statistics on the congenital solitary pelvic kidney. Acta urol. belg. **24**, 58 (1956).

COPE, J. C.: Traumatic rupture of a congenital solitary kidney. J. Urol. (Baltimore) **92**, 377 (1964).

DASELER, E. H., and B. J. ANSON: Unilateral renal agenesis: anatomical description of a specimen. J. Urol. (Baltimore) **50**, 155 (1943).

DAVIDSON, W. M., and G. I. M. ROSS: Bilateral absence of the kidneys and related congenital anomalies. J. Path. Bact. **68**, 459 (1954).

DEES, J. E.: Prognosis of the solitary kidney. J. Urol. (Baltimore) **83**, 550 (1960).

DEN'MUKHAMEDOV, S. R.: Levostoronnii gidronefroz pri aplazii pravci pochki. Urologija **22**, 59 (1957).

DOROSHOW, L., and B. S. ABESHOUSE: Congenital unilateral solitary kidney: report of 37 cases and a review of the literature. Urol. Surv. **11**, 219 (1961).

DOURMASHKIN, R. L., and I. LIGHT: Fallacies in diagnosis of renal agenesis. Amer. J. Surg. **50**, 348 (1940).

DUXBURY, J. H.: Bilateral agenesis of the kidneys. Canad. med. Ass. J. **78**, 123 (1958).
ELIAHOU, H. E., H. BOICHIS, and E. EDEN: Traumatic renal infarction in a solitary kidney. J. Urol. (Baltimore) **90**, 16 (1963).
EWERT, E. E.: Carcinoma of the kidney: incidental discovery in a solitary kidney. Lahey Clin. Bull. **13**, 198 (1964).
FORTUNE, C. H.: Pathological and clinical significance of congenital one-sided kidney defect with presentation of three new cases of agenesia and one of aplasia. Ann. intern. Med. **1**, 377 (1927).
FROST, I. F.: Case report of a patient with a true unicornuate uterus with unilateral renal agenesis. Amer. J. Obstet. Gynec. **75**, 210 (1958).
FRUMKIN, J., and R. MARZ: Bilateral renal agenesis. J. Urol. (Baltimore) **71**, 268 (1954).
GILLASPY, C. C., and O. J. RENTERGHEM: Unilateral renal agenesis associated with a recto-urethral fistula: the anatomical description of a specimen. Brit. J. Urol. **34**, 270 (1962).
GOLDSTEIN, A. E., and R. B. GOLDSTEIN Management of the diseased solitary kidney. J. Urol. (Baltimore) **90**, 361 (1963).
GORVOY, J. D., J. SMULEWIEZ, and H. ROTHFELD: Unilateral renal agenesis in two siblings. Pediatrics **59**, 270 (1962).
HENNESSEY, R. A.: Congenital solitary kidney. J. Urol. (Baltimore) **21**, 193 (1929).
HERBUT, P. A.: Urological pathology. Philadelphia: Lea & Febiger 1952.
HINMAN jr., F.: Congenital bilateral absence of the kidneys. Surg. Gynec. Obstet. **71**, 101 (1940).
HOWARTH, V. S.: Solitary kidney, a report of three cases. Brit. J. Surg. **37**, 237 (1950).
IVASHKO, L. M.: Sluchai podkozhogo razryva edinstvennoi gidroefroticheskoi pochki. Urologija **22**, 43 (1957).
LEBEDEV, A. P.: Kamen' rasshcheplennogo mochetochnika edinstvennoi udvoennoi pochki. Urologija **22**, 58 (1957).
MUIR, C. S.: Bilateral renal aplasia: a case in a chinese infant. Brit. J. Urol. **32**, 39 (1960).
NATION, E. F.: Renal agenesis. Surg. Gynec. Obstet. **79**, 175 (1944).
— Renal aplasia: a study of sixteen cases. J. Urol. (Baltimore) **51**, 579 (1944).
NEERHUT, K. J.: Aust. N. Z. J. Surg. **24**, 137 (1955).
OPOLSKI, W.: Przypadek czesciowego wyciecia jedynej nerki gruzliczej. Urol. pol. **9**, 185 (1956).
OSTRY, J.: Right renal agenesis, aneurysm of renal artery and left rudimentary unicornuate uterus. J. Urol. (Baltimore) **63**, 424 (1950).
POTTER, E. L.: Bilateral renal agenesis. J. Pediat. **29**, 68 (1946).
RUDERMAN, R. L., and J. M. MAYER: Unilateral renal agenesis with unicornuate uterus. Canad. med. Ass. J. **87**, 235 (1962).
SPA, V. DE: Documents cliniques sur l'etiologie de certaines malformations reno-ureterales. Sem. Hôp. Paris **34**, 2273 (1958).
SYLVESTER, P. E., and D. R. HUGHES: Congenital absence of both kidneys. Brit. med. J. 1944 I, 77.
THOMPSON-WALKER: Surgical diseases and injuries of the genito-urinary organs. London: Cassel & Co. 1936.
VUORINEN, P.: Single pelvic kidney. Brit. J. Radiol. **33**, 129 (1960).
— Anomalies and malformation of the upper urinary tract detected at intravenous urography in a series of 2,000 patients. Ann. Chir. Gynaec. Fenn. **50**, 407 (1961).
WEBER, L. L., and S. L. ISRAEL: Renal agenesis and oligohydramnios. Obstet. and Gynaec. **12**, 375 (1958).
WILLIAMS, D. I.: Urology in childhood. Encyclopedia of Urology. Berlin-Göttingen-Heidelberg: Springer 1958.
WOOLF, R. B., and W. M. ALLEN: Congenital malformations. Amer. J. Obstet Gynec **2**, 236 (1953).

Hypoplasia

ANDERSON, J. C., R. A. PAYNE, and J. RIGHE: Unilateral renal hypoplasia with vestigial epididymal tubules. Brit. J. Urol. **32**, 273 (1960).
BACHER, E.: Vollkommene Aplasie einer Niere. Z. Urol. **49**, 563 (1956).
BAGGENSTOSS, A. H.: Congenital anomalies of the kidney. Med. Clin. N. Amer. **35**, 987 (1951).
BELL, E. T.: Renal disease. Philadelphia: Lea & Febiger 1950.
BOEMINGHAUS, H.: Nierenhypoplasie und Harndruck. Z. Urol. **51**, 323 (1958).
BOISSONNAT, P.: What to call hypoplastic kidneys. Arch. Dis. Childh. **37**, 142 (1962).
BURKLAND, C. E.: Clinical considerations in aplasia, hypoplasia and atrophy of the kidney. J. Urol. (Baltimore) **71**, 1 (1954).
CAMPBELL, M. F.: Principles of urology. Philadelphia: W. B. Saunders Co. 1957.
CECIL, A. B.: A case of renal hypoplasia with ureter opening into the vagina. J. Urol. (Baltimore) **70**, 835 (1953).

Cibert, J., et J. Collenet: Nephrectomies pour aplasie ou atrophie renale. J. Urol. Néphrol. 58, 807 (1952).

Degoy, A., et C. Schuller: Hypertension maligne chez une enfant de dix ans avec hypoplasie renale unilaterale, guerie par nephrectomie. J. Méd. Bordeaux 135, 971 (1958).

Eisendrath, D. N.: Clinical importance of congenital renal hypoplasia. J. Urol. (Baltimore) 33, 331 (1935).

Ekstrom, T.: Renal hypoplasia. Acta chir. scand., Suppl. 203, 1 (1955).

Emmett, J. L., J. J. Alvarez-Ierena, and J. R. McDonald: Atrophic pyelonephritis versus congenital renal hypoplasia. J. Amer. med. Ass. 148, 1470 (1956).

Fanconi, G., E. Caldelare, H. Menano, and R. Cramer: Congenital hypoplasia of both kidneys with symptoms of Cushing syndrome and Lightwood, Albright disease, case. Helv. paediat. Acta 7, 350 (1952).

Forst, H.: Aortographic diagnosis of unilateral renal aplasia. Z. Urol. 50, 366 (1957).

Gamble, P. G.: Hypoplastic kidney. Sth. med. J. (Bgham, Ala.) 28, 887 (1935).

Giordano, G.: High lumbar ectopy of hypoplastic micropolycystic kidney with double pelvis and bifid ureter, pyelographic diagnosis. Radiol. med. (Torino) 37, 353 (1951).

Goldblatt, H.: The renal origin of hypertension. Springfield (Ill.): Ch. C. Thomas 1948.

Graham, A. P.: Hypoplastic kidney, bilateral: case report. J. Urol. (Baltimore) 60, 581 (1948).

Grozs, F.: Angeborene einseitige Nierenaplasie mit verkalkter zystischer Degeneration. Zbl. Chir. 82, 1801 (1957).

Hartwich, A.: Die Beziehungen zwischen Niere und Blutdruck im Tierexperiment. Verh. dtsch. Ges. inn. Med., Kongr.-Bd. 41, 187 (1929).

— Demonstrationen zum Klinischen. Referat über Kreislaufwirkungen körpereigener Stoffe. Verh. dtsch. Ges. inn. Med., Kongr.-Bd. 44, 76 (1932).

Herbut, P. A.: Urological pathology. Philadelphia: Lea & Febiger 1952.

Hinman, F.: Principles and practice of urology. Philadelphia: W. B. Saunders Co. 1935.

Huffman, L. F.: Aplastic kidney with ectopic ureteral orifice in the prostatic urethra. J. Urol. (Baltimore) 68, 673 (1952).

Hutchinson, R., and A. A. Moncrieff: Care of primary hypertension in a child. Brit. J. Child. Dis. 27, 201 (1930).

Leruitte: Le diagnostic des petits reins. Acta belg. Arte med. pharm. milit. 109, 183 (1956).

Mombaerts, J.: Les petits reins. Acta urol. belg. 23, 24 (1955).

Neimann, N., M. Pierson, G. Rauber, G. Lascombes, and F. Lehodey: Hypertension arterielle par malformation de l'artere renale. Arch. franç. Pédiat. 15, 688 (1958).

Nieth, H.: Einseitige Nierenerkrankung und Hochdruck. Ärztl. Wschr. 12, 983 (1957).

Oehlecker, F.: Unusual abnormality: contribution to study of Wolffian body and anlage of antanephros. Chirurg 22, 157 (1951).

Paetzel, W.: Zwergnieren. Chirurg 27, 444 (1956).

Penson, J., i. L. Cynowski: Nadcisnienie tetnicze zlosliwe objawowe. Pol. Arch. Med. wewnet. 25, 107 (1955).

Perkel, H.: Acute hypertension probably to infarction within congenitally hypoplastic kidney. J. Pediat. 36, 254 (1950).

Ritter, J. S., and S. E. Kramer: Hypoplastic kidney and atrophic polynephritic kidney. J. Urol. (Baltimore) 63, 48 (1950).

Schultze-Jena, B. S., and H. J. Hillenbrand: Einseitige Nierenhypoplasie und Hochdruck im Kindesalter. Kinderärztl. Prax. 24, 433 (1956).

Slungaard, R. K., and J. L. Jaeck: Bilateral renal hypoplasia. J. Lancet 79, 236 (1959).

Smith, H. W.: Hypertension and urologic diseases. Amer. J. Med. 4, 724 (1948).

Soszka, A.: Nadcisnienie tetnicze z jedostronnymi zmianami w nerkach. Pol. Arch. Med. wewnet. 20, 279 (1950).

Sziberth, K.: Hammaturie bei Nierenhypoplasie. Z. Urol. 49, 225 (1956).

Tedeschi, C. G., and W. H. Holtham: Renal ossification: report of occurrence in an aplastic kidney. J. Urol. (Baltimore) 67, 50 (1952).

Thompson, G. C. V., and A. S. Feddersen: Congenital anomalies of clinical interest. Med. J. Aust. 1, 360 (1950).

Trolle, D.: Ectopic hypoplastic kidney-ureter system opening into cystically dilated Gartner's duct in a 17 year old woman. Acta obstet. gynec. scand. 32, 335 (1953).

Troy, J. M.: Renal hypoplasia. J. Indiana med. Ass. 55, 1769 (1962).

Verger, P.: Hypertension arterielle maligne et hypoplasie renale unilaterale chez l'enfant. Pédiatrie 42, 693 (1953).

Wermut, W., S. Swica i Z. Wajda: Przypadek nadcisnienia objawowego wyleczony za pomoca usuniecia hipoplastycznej nerki. Pol. Tyg. lek. 13, 1635 (1958).

Williams, D. I.: Urology in childhood. Encyclopedia of urology. Berlin-Göttingen-Heidelberg: Springer 1958.

Cystic Disease

ABESHOUSE, B. S.: Congenital renal aplasia with calcified cystic degeneration. J. int. Coll. Surg. **26**, 283 (1956).

ALLEN, A. C.: The kidney, medical and surgical diseases. New York: Grune & Stratton 1956.

ANDRESEN, K.: Beitrag zur Röntgenologie der solitaren Nierencyste. Röntgenpraxis 8, 505 (1936).

ANNAMUNTHODO, H.: Multicystic disease of the kidney in the newborn: report of two cases. Brit. J. Urol. **32**, 34 (1960).

AREY, J. B.: Cystic lesions of kidney in infants and children. J. Pediat. **54**, 429 (1959).

BAXTER, TH. J.: Observations on a polycystic kidney by microdissection. Aust. N.Z. J. Surg. **30**, Nov. (1961).

BEDFORD, P. W.: Bilateral polycystic disease of kidney complicated by intestinal obstruction. Brit. med. J. **1958** II, No 5105, 1146.

BELL, E. T.: Cystic disease of the kidneys. Amer. J. Path. **11**, 373 (1935).

BELTRAN, J. C.: Congenital unilateral multicystic kidney in infancy. J. Urol. (Baltimore) **81**, 602 (1950).

BLOCK, CL., and H. M. PECK: Radiological notes: congenital cyst of the right kidney with gastric obstruction. J. Mt Sinai Hosp. **31**, 234 (1964), Case No 237.

BRASCH, W. F., and J. A. HENDRICK: Renal cysts, simple and otherwise. J. Urol. (Baltimore) **51**, 1 (1944).

BRIGIDI and SEVERI: Quoted from RALL and ODEL.

BROSCH, W.: Enormous, nonfunctioning hydronephrosis on the left with stoneformation in the lower kidney pole, where a calcified solitary cyst was suspected. Z. Urol. **49**, 442 (1956).

BURFORD, E. H., and C. E. BURFORD: Congenital polycystic kidney disease. Missouri Med. **59**, 1126 (1960).

CAMERON, J. R.: Bilateral "hereditary" polycystic disease of the kidneys associated with bilateral teratodactyly of the feet. Brit. J. Urol. **33**, 473 (1961).

CAMPBELL, W. F.: Principles of urology. Philadelphia and London: W. B. Saunders Co. 1957.

CANNON, J. F.: Hereditary polycystic kidney. Amer. int. Med. **26**, 610 (1947).

CARVER, J.: Renal adenoma. Brit. J. Med. **7**, 229 (1935).

Case records of the Massachusetts General Hospital. Case 18-1963. New Engl. J. Med. **268**, 601 (1963).

COLBY, FL. H.: Essential urology, 3rd ed. Baltimore: Williams & Wilkins Co. 1956.

CHOVNICK, ST. D., and D. S. SILVERT: Infected solitary cyst of the kidney. J. Urol. (Baltimore) **83**, 7. (1960).

CRAIG, R. D. P.: Unilateral multicystic disease of the kidney. Brit. J. Urol. **34**, 19 (1962).

CRISTOL, D. S., J. R. MCDONALD, and J. L. EMMETT: Renal adenomas in hypernephromatous kidney. J. Urol. (Baltimore) **55**, 18 (1946).

EKSTROM, T.: Renal hypoplasia. Acta chir. scand., Suppl. **203**, 1 (1955).

FAHR: Quoted in HECKEL.

FAIRLEY, K. F., P. W. LEIGHTON, and PRISCILLA KINCAID-SMITH: Familial visual defects associated with polycystic kidney and medullary sponge kidney. Brit. med. J. **1963** I, 1060.

FERGUSSON, J. E.: Cysts of the kidney. Ann. roy. Coll. Surg. Engl. **26**, 115 (1960).

FINE, M. C., and E. BURNS: Unilateral multicystic kidney: report of 6 cases and discussion of the literature. J. Urol. (Baltimore) **81**, 42 (1959).

FISH, G. W.: Large solitary serous cysts of the kidney. Report of 32 cases. J. Amer. med. Ass. **112**, 514 (1939).

FISTER, G. M.: Simple serous cysts of the kidney. J. Urol. (Baltimore) **49**, 408 (1943).

FRANCIS, W. G.: Foetal polycystic disease and ruptured uterus. Report of a case. J. Obstet. Gynaec. Brit. Emp. **67**, 126 (1960).

FRAZIER, T. H.: Multilocular cysts of the kidney. J. Urol. (Baltimore) **65**, 351 (1951).

FRIEND, D. G., R. G. HOSKINS, and M. W. KIRKIN: Relative erythrocythemia (polycythemia) and polycystic kidney disease, with uremia. New Engl. J. Med. **264**, 17 (1961).

FRIENDRICHS, P.: A cystic, calcified kidney abnormality. Z. Urol. **52**, 35 (1959).

GAJOWNICZEK, K.: Przypadek torbielowatesci watreby, nerek i trzustki (Case of hepatic, renal and pancreatic polycystic disease). Pol. Tyg. lek. **13**, 560 (1958).

GALL, H., and I. J. KURTZ: Simple renal cyst in child. Amer. J. Dis. Child. **96**, 314 (1958).

GIBSON, TH. E.: Multilocular cyst of the kidney: case report. Trans. Amer. Ass. gen.-urin. Surg. **53**, 53 (1961).

GOLDSTEIN, A. E., and R. B. GOLDSTEIN: Polycystic renal disease. J. Urol. (Baltimore) **84**, 268 (1960).

GOLDSTON, ANNE S., E. C. BURKE, A. D'AGOSTINO, and W. T. E. MCCAUGHEY: Neonatal polycystic kidney with brain defect. Amer. J. Dis. Child. **106**, 484 (1963).

Gutierrez, R.: Large solitary cyst of the kidney. Arch. Surg. 44, 279 (1942).

Hart, L. B., Beckley, E. I. Dobos, and R. P. Forbes: Congenital cystic dilation of single kidney calix. J. Pediat. 16, 206 (1940).

Heckel, N. J., and H. V. Gould: Papillary cystadenoma of the kidney. J. Urol. (Baltimore) 44, 200 (1940).

Hepler, A. B.: Etiology of multilocular cysts of the kidney. J. Urol. (Baltimore) 64, 206 (1940).

Herbst, R. H., and H. J. Polkey: Solitary renal cysts. J. Urol. (Baltimore) 37, 490 (1937).

Herbut, P. A.: Urological pathology, vol. I. Philadelphia: Lea & Febiger 1952.

Higgins, C. C.: Adenoma of the kidney. Amer. J. Surg. 65, 3 (1944).

Hildebrandt, A.: Weiterer Beitrag zur pathologischen Anatomie der Nierengeschwülste. Langenbecks Arch. klin. Chir. 48, 343 (1894).

Hinman, F.: Principles and practice of urology. Philadelphia and London: W. B. Saunders Co. 1935.

Hyams, J. A., and H. R. Kenyon: Localized obliterating pyelonephritis. J. Urol. (Baltimore) 46, 380 (1941).

Ivker, M., and S. Keesal: Loin masses due to infant and juvenile unilateral multicystic kidney disease. Arch. Surg. 81, 798 (1960).

Jordan, W. P.: Peripelvic cysts of the kidney. J. Urol. (Baltimore) 87, 97 (1962).

Katamura, E., T. Kitayama, and M. Kuze: Treatment of the polycystic kidney with multiple sclerosing punctures. Acta urol. (Kyoto) Japanica, 8—1 (3—11) (1962).

Kelley, T., and G. H. Guin: Clinical pathological conference. Presentation of a case (polycystic kidney disease). Clin. Proc. Child. Hosp. (Wash.) 19, 19 (1963).

Kier, J. H., and J. M. Young: Polycythemia and renal cyst. J. Tenn. med. Ass. 56, 181 (1963).

Kreutzmann, H. A. R.: Hypertension associated with solitary renal cyst. J. Urol. (Baltimore) 57, 467 (1947).

Kurtz, C. W., and W. L. Vlk: Congenital unilateral multicystic kidney disease in infancy. Ariz. Med. 16, 475 (1959).

Lambert, G. P.: Polycystic disease of the kidney. Arch. Path. 44, 34 (1947).

Llanos, M. A.: Peripelvic renal cysts. Calif. Med. 96, 201 (1962).

Lowsley, O. S., and M. S. Curtis: Surgical aspects of cystic disease of the kidney. J. Amer. med. Ass. 127, 1112 (1945).

Lubarch: Quoted from Heckel.

Lynch, K. M., and R. R. Bradham: Substitute ileal ureter in an infant with congenital hydronephrosis of the left kidney and congenital cystic hypoplasia of the right kidney. Surgery 49, 278 (1961).

Mayers, M. M.: Treatment of deep renal cortical cysts. J. Urol. (Baltimore) 82, 10 (1959).

McKenna, Ch. M., and O. F. Kampmeier: A consideration of the development of polycystic kidney. J. Urol. (Baltimore) 32, 37 (1934).

Melamed, S. B.: The functional state of polycystic kidneys. Urologiya 23, No 6 (9—13) (1958).

Melicow, M. M., and H. H. Gile: An hypernephroma in a polycystic kidney. J. Urol. (Baltimore) 43, 767 (1940).

Milam, J. H., J. H. Magee, and R. C. Bunts: Evaluation of surgical decompression of polycystic kidneys by differential renal clearances. J. Urol. (Baltimore) 90, 144 (1963).

Moe, P. J., and W. L. Crofford: Ectopic unilateral multicystic kidney in infant. Amer. J. Dis. Child. 99, 35 (1960).

Moore, T.: Unilateral cystic kidneys. Brit. J. Urol. 29, 3 (1957).

Natvig, P.: Two cases of solitary renal cyst with communication to the renal pelvis. Acta radiol. (Stockh.) 22, 732 (1941).

Norris, R. F., and L. Herman: Pathogenesis of polycystic kidneys. J. Urol. (Baltimore) 46, 147 (1941).

Oppenheimer, G. J., and L. Narins: Unilateral polycystic kidney disease. J. Urol. (Baltimore) 61, 866 (1949).

Parkkulainen, K. V., L. Hjelt, and K. Sirola: Congenital multicystic dysplasia of the kidney. Acta chir. scand., Suppl. 244, 1 (1959).

—, and J. K. Visakorpi: Congenital multicystic dysplasia of the kidney: analysis of cyst fluid. Urol. Int. 8, 204 (1959).

Pfau, A., and Th. A. Stamey: Some functional characteristics of polycystic renal disease. A differential renal function study of a patient with "unilateral" polycystic disease. Invest. Urol. (Baltimore) 1, 593 (1964).

Pollastri, S.: Terapia del rene policistica. Urologia (Venice) 24, 399 (1957).

Prat, V.: Hypertense u nemocnych polycystickmi ledviami. Vnitřni Lek. 6, 413 (1960).

RALL, J. E., and H. M. ODEL: Congenital polycystic disease of the kidney. Amer. J. med. Sci. **218**, 399 (1949).
RAVITCH, M. M., and M. O. SANDFORD: Unilateral multicystic kidney in infants. Pediatrics **4**, 769 (1949).
RIBBERT: Quoted from RALL and ODEL.
RITTER and BAEHR: Quoted from RALL and ODEL.
ROSSI, C.: Su di un caso di cisti ematica del rene. Acta ital. uro. **15**, 265 (1938).
SANDER, E.: Case report on persistent pronephros. Z. Urol. **49**, 654 (1956).
SCHENCKER, B., and P. ZANCA: A rational approach to the diagnosis of renal masses. Med. Tms (New York) **92**, 685 (1964).
SCHNEIDERMAN, F. J.: Familial infantile polycystic disease of the kidney. J. Amer. osteopath. Ass. **62**, 1004 (1963).
SPENCE, H. M.: Congenital unilateral multicystic kidney. An entity to be distinguished from polycystic kidney disease and other cystic disorders. J. Urol. (Baltimore) **74**, 693 (1955).
STAEMNIER: Quoted from RALL and ODEL.
STAUBITZ, W. J., TH. C. JEWETT, and R. J. PLETMAN: Renal cystic disease in childhood. J. Urol. (Baltimore) **90**, 8 (1963).
TRINKLE, J. K.: Polycystic kidney disease. G P, Kansas City, Mo., U.S.A. **27**, 92 (1963).
USON, A. C., MELICOW, and M. MEYER: Multilocular cysts of kidney with intrapelvic herniation of a "daughter" cyst: report of 4 cases. J. Urol. (Baltimore) **89**, 341 (1963).
VELLIOS, F., and R. A. GARRETT: Congenital multicystic disease of the kidney. Amer. J. clin. Path. **35**, 244 (1961).
VERMOOTEN, V.: The kidneys: neoplasms and cystic disease. Postgrad. Med. **3**, 369 (1963).
VIRCHOW: Quoted from RALL and ODEL.
VOROBTSOV, V. I.: Secondary heminephrectomy in hemorrhage from a polycystic horseshoe kidney. Urologiya **21**, 49 (1956).
WAKELEY, C. P. G.: Cystic kidney in an infant. Proc. Soc. S. Med. **23**, 547 (1929/30).
WALTERS, W., and W. F. BRAASCH: Surgical aspects of polycystic kidney. Trans. Amer. Ass. gen.-urin. Surg. **26**, 285 (1933).
WATKINS, K. H.: Cysts of kidney due to hydrocalycosis. Brit. J. Urol. **11**, 245 (1939).
WEINGARTEN, CH. J.: Renal cysts — a clinical enigma. Henry Ford Hosp. Bull. **9**, 559 (1961).
WEYRAUCH, H. M., and H. E. FLEMING: Congenital hydrocalycosis. J. Urol. (Baltimore) **63**, 582 (1950).
WILLIAMS, D. I.: Urology in childhood. In: Encyclopedia of urology. Berlin-Göttingen-Heidelberg: Springer 1958.
WILLOX, S. W.: Intestinal obstruction caused by polycystic disease of the kidneys. Brit. J. Urol. **34**, 267 (1962).
WINTER, C. C., M. H. MAXWELL, R. E. ROCKNEY, and C. R. KLEEMAN: Results of the radio-isotope renogram and comparison with other kidney tests among hypertensive persons. J. Urol. (Baltimore) **82**, 674 (1959).

Fusion Anomalies

ABESHOUSE, B. S.: Crossed ectopia with fusion. Amer. J. Surg. **73**, 657 (1947).
ADAN, R.: A descriptive study of a new technique relating to the surgical treatment of tuberculosis of the small bladder and ureter in the case of a single kidney. J. Urol. (Baltimore) **62**, 491 (1956).
ANSON, B. J., J. W. PICK, and E. W. CAULDWELL: The anatomy of commoner renal anomalies: ectopic and horseshoe kidneys. J. Urol. (Baltimore) **47**, 112 (1942).
ATHERTON, L.: Horseshoe kidney and its clinical management. Sth. Surg. (Brit.) **11**, 173 (1942).
BAGGENTOSS, A. H.: Congenital anomalies of the kidney. Med. Clin. N. Amer. **35**, 987 (1941).
BECK, W. C., and A. E. HLIVKO: Wilm's tumor in the isthmus of a horseshoe kidney. Arch. Surg. **81**, 803 (1960).
BELL, E. T.: Cystic disease of the kidneys. Amer. J. Path. **11**, 373 (1935).
BOREHAM, P.: Horseshoe kidney. Arch. Middx Hosp. **4**, 46 (1954).
BRIDGE, A. C.: Horseshoe kidneys in identical twins. Brit. J. Urol. **32**, 32 (1960).
BURCKHART, T., and A. SCHMITT: Complete symmetrically fused pelvic kidney. Zbl. Chir. **80**, 149 (1955).
CAMPBELL, M. F.: Principles of urology. Philadelphia and London: W. B. Saunders Co. 1957.
CARLETON, A.: Crossed ectopia of the kidney and its possible cause. J. Anat. (Lond.) **71**, 292 (1937).
CHESTERMAN, J. T.: Hydronephrosis of horseshoe kidney presenting an acute intestinal obstruction. Brit. J. Urol. **13**, 163 (1941).

Coles, H. M. T.: T substance anomaly with horseshoe kidney. Proc. roy. Soc. Med. **54**, 330 (1961).

Comiti: A propos d'une anomalie renale. J. Urol. med. chir. **57**, 213 (1951).

Culp, O. S.: Renal ectopia. J. Urol. (Baltimore) **52**, 420 (1940).

Doyle, A. D., and G. C. Schofield: Unilateral renal agenesis, genital abnormalities and hypertension. Med. J. Aust. **2**, 145 (1964).

Edward, N., G. P. G. Young, and M. Macleod: Fibrinolytic activity in plasma and urine in chronic renal disease. J. clin. Path. **17**, 365 (1964).

Falor, W. H., and R. A. Rufflo: Horseshoe kidney complicated by abdominal aortic aneurysm. J. Urol. (Baltimore) **91**, 131 (1964).

Farman, F.: Bilateral nephrolithiasis in horseshoe kidney. J. Urol. (Baltimore) **51**, No 5 (1944).

—, and B. K. Brock: Review of the horseshoe kidney. Urol. cutan. Rev. **52**, 217 (1938).

—, L. Kamens, and B. K. Brock: A follow-up study of horseshoe kidney surgery. Scientific Exhibit, Amer. Urol. Ass. 50th Annual Meeting 1955.

Felton, W., and J. Miller: Surgical resection of abdominal aortic aneurysm associated with horseshoe kidney. J. Okla. med. Ass. **55**, 480 (1962).

Foley, F. E. D., and H. A. Wilmer: Surgery of the unilateral fused kidney. Int. Abstr. Surg. **70**, 155 (1940).

Gammelgaard, A.: Horseshoe kidney in a child of 16 months. Acta chir. scand. **107**, 168 (1954).

Gerard, G. J.: Les anomalies congènitales du rein chez l'homme; essai de classification d'après 527 cas. J. Anat. (Paris) **41**, 241, 411 (1905).

Gibson, G. R.: Ruptured horseshoe (fused) kidney: a review and report of a case with traumatic renal hypertension. J. Urol. (Baltimore) **92**, 374 (1964).

Glenn, J. F.: Fused pelvic kidney. J. Urol. (Baltimore) **80**, 7 (1958).

— Analysis of 51 patients with horseshoe kidney. New Engl. J. Med. **261**, 684 (1959).

Gohn, G.: Ileus caused by large hydronephrosis of portion of horseshoe kidney. Zbl. Chir. **55**, 1426 (1928).

Gregoir, W.: Conservative surgery in horseshoe kidney. Urol. int. (Basel) **16**, 129 (1963).

Gutierrez, R.: The clinical management of horseshoe kidney. Amer. J. Surg., part I, **14**, 657 (1931); part II, **15**, 132 (1932); part III, **15**, 345 (1932).

Herbut, P. A.: Urological pathology. Philadelphia: Lea & Febiger 1952.

Hess, E.: Surgical horseshoe kidney, report of an unusual case. J. Urol. (Baltimore) **12**, 627 (1924).

Hinman, F.: Principles and practice of urology. Philadelphia and London: W. B. Saunders Co. 1935.

—, H. T. Langworthy, and L. A. Drexler: Carcinoma in crossed renal ectopia. J. Urol. (Baltimore) **47**, 776 (1942).

Jewell, J. H. A.: Spontaneous rupture of calix of solitary kidney due to stone. Brit. J. Surg. **51**, 392 (1964).

Kretschmer, H. L.: Unilateral fused kidney. Surg. Gynec. Obstet. **40**, 360 (1935).

Kron, S. D., and D. R. Meranze: Completely fused pelvic kidney. J. Urol. (Baltimore) **62**, 278 (1949).

Lacal, F., and B. Singer: Symphysiotomy in therapy of painful horseshoe kidney. Sem. méd. (B. Aires) **104**, 478 (1954).

Lathem, J. E., and K. H. Smith: Wilm's tumor in a horseshoe kidney: a surviving case. J. Urol. (Baltimore) **88**, 25 (1962).

Newman: Cited from Kretschmer 1935.

Ninfe, S.: Sacral ectopy of the left kidney. Acta chir. ital. **12**, 466 (1956).

Parker, G.: The surgical approach to a horseshoe kidney. Brit. J. Urol. **28**, 447 (1956).

Petrovcic, F., and N. Milic: Horseshoe kidney with crossed ureter condition after right nephrectomy. Brit. J. Radiol. **29**, 114 (1956).

Potampa, P. B., M. D. Hyman, and C. E. Catlow: An unusual renal anomaly: combined tandem and horseshoe kidney. J. Urol. (Baltimore) **6**, 340 (1949).

Rathburn, N. P.: Notes on the clinical aspects of horseshoe kidney. J. Urol. (Baltimore) **12**, 611 (1924).

Riches, E. W.: Urology, modern trend series. New York: Paul B. Hoeber, Inc. 1953.

Rovsing, T.: Symptomatology, diagnosis and treatment of horseshoe kidney. Z. Urol. **5**, 586 (1911).

Salter, C. L.: A case of horseshoe kidney presenting as an intestinal obstruction. J. med. Ass. Ala **26**, 116 (1956).

Sangree, H., D. Morgan, T. Klein, and R. Trasi: Horseshoe kidney and the relation of nephritis and calculous formation to anomalous circulation. J. Urol. (Baltimore) **32**, 648 (1934).

SHILLER, W. R., and O. B. WISMELL: A fused pelvic (cake) kidney. J. Urol. (Baltimore) 78, 9 (1957).

SHOUP, G. D., H. M. POLLACK, and JUNG HAE DOU: Adenocarcinoma occurring in a horseshoe kidney. Arch. Surg. 84, 413 (1962).

SIDAWAY, M. E.: Wilm's tumour in a horseshoe kidney. Brit. J. Radiol. 35, 341 (1962).

STEWART, M. J., and W. O. LODGE: Unilateral fused kidney and allied renal malformations. Brit. J. Surg. 11, 27 (1923).

THOMPSON, G. C. V., and A. S. FEDDERSEN: Congenital anomalies of clinical interest. Med. J. Aust. 1, 360 (1950).

VOROBTSOV, V. I.: Secondary heminephrectomy in hemorrhage from a polycystic horseshoe kidney. Urologiya 21, 49 (1956).

WALTERS, W., and J. B. PRIESTLY: Horseshoe kidney. J. Urol. (Baltimore) 28, 271 (1932).

WILMER, H. A.: Unilateral fused kidney. A report of five cases and a review of the literature. J. Urol. (Baltimore) 40, 551 (1938).

ZONDEK, T.: Notes on the topography of the fetal horseshoe kidney. Brit. J. Urol. 24, 201 (1952).

Ectopia of the Kidney

ABESHOUSE, B. S.: Crossed ectopia with fusion. Amer. J. Surg. 73, 658 (1947).

—, and I. BHISTKUL: Crossed renal ectopia with and without fusion. Urol. int. (Basel) 9, 63 (1959).

ANDERSON, G. W., G. G. RICE, and B. A. HARRIS: Pregnancy and labor complicated by pelvic ectopic kidney. J. Urol. (Baltimore) 65, 760 (1951).

BEER, E., and W. L. F. FERBER: Crossed renal ectopia. J. Urol. (Baltimore) 38, 541 (1937).

— — Crossed renal ectopia. J. Urol. (Baltimore) 39, 479 (1938).

BERLIN, H. S., J. STEIN, and M. H. POPPEL: Congenital superior ectopia of the kidney. Amer. J. Roentgenol. 78, 508 (1957).

BOISSONNAT, P.: Intervention plastique creant un rein droit et un rein gaudre independants. J. Urol. med. chir. 61, 408 (1955).

BOWLES, W. T., and K. R. SMITH jr.: Renal ectopia. Report of a case of fused pelvic renal ectopia. Surgery 46, 539 (1959).

BURWELL, R. G., and S. G. KENT: The solitary ectopic pelvic kidney: case report with a review. Brit. J. Urol. 31, 254 (1959).

CAIRRE, M.: Crossed renal ectopia without fusion. Brit. J. Urol. 28, 257 (1956).

CAMPBELL, W. F.: Renal ectopy. J. Urol. (Baltimore) 24, 187 (1930).

— Principles of urology. Philadelphia: W. B. Saunders Co. 1957.

CULP, O. S.: Renal ectopia. J. Urol. (Baltimore) 52, 420 (1944).

DIAMANTIS, S.: A single ectopic pelvic kidney. J. Urol. méd. chir. 56, 95 (1950).

FALKINBURG, L. W., M. M. KAY, and W. S. KLUTZ: Crossed renal ectopia without fusion. Amer. J. Dis. Child. 99, 86 (1960).

FRANCISKOVIC, V., and N. MARTINCIC: Intrathoracic kidney. Brit. J. Urol. 31, 156 (1959).

HANLEY, G. H., and N. A. STEEL: The solitary ectopic pelvic kidney. Brit. J. Surg. 34, 402 (1947).

HARRIS, W.: Renal ectopia — special reference to crossed ectopia without fusion. U. Urol. (Baltimore) 42, 1051 (1939).

HEARD, G.: Renal ectopia in sisters. Brit. J. Surg. 44, 154 (1956).

HILL, J. E., and R. C. BUNTS: Thoracic kidney: case reports. J. Urol. (Baltimore) 84, 460 (1960).

HINMAN, F.: Principles and practice of urology. Philadelphia: W. B. Saunders Co. 1935.

JOHN, H. T.: Large abdominal cyst associated with an ectopic pelvic kidney. Brit. J. Urol. 32, 148 (1960).

KEUSENHUFF, W.: Gekreuzte Nierendystopie. Z. Urol. 49, 114 (1956).

LAUGHLIN, V.: Renal dystopia (ectopia): a reprint of an interesting case and a brief review of the literature. J. Urol. (Baltimore) 47, 632 (1942).

LUCAS, C.: Coalesced kidneys in rabbit and associated anomalies in circulatory and nervous systems. J. Anat. (Lond.) 68, 270 (1934).

MADISSON, H.: Über das Fehlen beider Nieren (Aplasia renum bylateralis). Zbl. allg. Path. path. Anat. 60, 1 (1934).

MALAMENT, MAXWELL, and G. MAJNARICH: Giant hydronephrosis in crossed renal ectopia without fusion. J. Urol. (Baltimore) 83, 542 (1960).

MAYERS, M. M.: Crossed renal ectopia. J. Urol. (Baltimore) 36, 111 (1936).

MAYS, H. B.: Pelvic single kidney. J. Urol. (Baltimore) 56, 619 (1946).

MIKULICZ-RADECKI, F.: Contribution to congenital intra-thoracic dystrophy of kidney. Zbl. Gynäk. 44, 1718 (1922).

94 F. FARMAN:

Moe, P. J., and W. L. Crofford: Ectopic unilateral multicystic kidney in infant. Amer. J. Dis. Child. **99**, 35 (1960).
Nation, E. F.: Renal ectopia. Amer. J. Surg. **68**, 67 (1945).
Ninfe, G.: Sacral ectopy of the left kidney. Acta chir. ital. **12**, 466 (1956).
Paul, A. T. S., C. G. Uragoda, and F. L. W. Jayewardene: Thoracic kidney with report of a case. Brit. J. Surg. **47**, 395 (1960).
Purpon, I.: Crossed renal ectopy with solitary kidney: a review of the literature. J. Urol. (Baltimore) **90**, 13 (1963).
Shih Hsi-En, F., S. Wu-Hou, and C. Hung-Knang: Crossed unfused renal ectopia. Chin. med. J. **75**, 841 (1957).
Smith, E. C., and L. A. Orkin: A clinical and statistical study of 471 congenital anomalies of the kidney and ureter. J. Urol. (Baltimore) **53**, 11 (1945)
Sokolski, E. J.: Pregnancy complicated by solitary pelvic kidney. Presentation of a case and review of the literature. Obstet. and Gynec. **16**, 365 (1960).
Svottolow: Quoted from Ninfe.
Thompson, G. T., and T. M. Pace: Ectopic kidney. Surg. Gynec. Obstet. **64**, 935 (1937).
Wilmer, H. A.: Unilateral fused kidney. J. Urol. (Baltimore) **40**, 551 (1938).
Winram, R. G., and J. N. Ward-McQuaid: Crossed renal ectopia without fusion. Canad. med. Ass. J. **81**, 481 (1959).
Wolfromm, G.: Situation du rein daus/inventration diaphramatique droite. Main. Acad. Chir. **66**, 41 (1940).
Young, H. H., and I. M. Davis: Practice of urology, vol. 5, p. 12. Philadelphia: W. B. Saunders Co. 1926.

Anomalies of Rotation

Boijsen, E.: Angiographic studies of the anatomy of single and multiple renal arteries. Acta radiol. (Stockh.) Suppl. **183**, 1 (1959).
Boll, G.: A faulty rotation of the kidney. Z. Urol. **50**, 91 (1957).
Campbell, M.: Principles of urology. Philadelphia: W. B. Saunders Co. 1957.
Chauvin, E., and H. F. Chauvin: Congenital abnormalities of renal orientation. J. Urol. méd. chir. **56**, 481 (1950).
Herbut, P. A.: Urological pathology, vol. I, p. 512. Philadelphia: Lea & Febiger 1952.
Hinman, F.: Principles and practice of urology. Philadelphia and London: W. B. Saunders Co. 1935.
Jonsson, G., and O. Olsson: A case of renal ectopia with malrotation and vascular anomalies causing renal pain. Acta chir. scand. **123**, 447 (1962).
Nathan, H.: Aberrant renal artery producing developmental anomaly of kidney associated with unusual course of gonadal (ovarian) vessels. J. Urol. (Baltimore) **89**, 560 (1963).
Parton, L. I.: Renal fusion. N. Z. med. J. **61**, 506 (1962).
T'ung Shang-T'ai, Ts'ao Chien, and Hu Chi'Jung: Bilateral renal anomalies. A case report. Chin. med. J. **78**, 233 (1959).
Weyrauch, H. M.: Anomalies of renal rotation. Surg. Gynec. Obstet. **69**, 183 (1939).

Anomalies of the Vessels

Abowitz, J.: Obstructive hydronephrosis produced by aberrant blood vessels and diagnosed by intravenous urography. Radiology **48**, 33 (1947).
Adams, A. W.: Hydronephrosis due to the inferior polar artery: late results after nephroplication. Appendix of recent cases. Brit. J. Urol. **31**, 461 (1959).
Anson, B. J., J. W. Pick, and E. W. Cauldwell: The anatomy of commoner renal anomalies. Ectopic and horseshoe kidneys. J. Urol. (Baltimore) **47**, 112 (1942).
Arcadi, J. A., and F. Farman: Experimental studies and clinical aspects of the renal circulation. J. Urol. (Baltimore) **62**, 756 (1949).
Arnemann, W.: Surgery of aberrant renal vessels. Z. Urol. **52**, 429 (1959).
Beringer, A., E. Deutsch, and H. Thaler: Zusammenhänge zwischen Anatomie und Funktion der Niere. Wien. klin. Wschr. **72**, 456 (1960).
Bremer, J. L.: The origins of the renal artery in mammals and its anomalies. Amer. J. Anat. 18, 179 (1915).
Campbell, M. F.: Urology. Philadelphia: W. B. Saunders Co. 1954.
Chapman, T. L.: Hydronephrosis due to aberrant renal artery, general discussion on hydronephrosis. Brit. J. Urol. **31**, 394 (1959).
Derrick, J. R., and Ch. A. Hooks: Surgical significance of vascular variations in systemic hypertension, with especial reference to aberrant renal arteries. J. Urol. (Baltimore) **87**, 273 (1962).
Farman, F.: Aberrant renal artery. Calif. west. Med. 28, No 5 (1928).

GAGARINOV, V. W.: Rupture of hydronephrotic kidney. Urologiya **24**, 60 (1959).

GIBSON, H. M.: Ureteral reflux in the normal child. J. Urol. (Baltimore) **62**, 40 (1949).

GRAVES, F. T.: Anatomy of the intra-renal arteries. Brit. J. Surg. **42**, 132 (1955).

HERBUT, P.: Urological pathology. Philadelphia: Lea & Febiger 1952.

HINMAN, F.: Principles and practice of urology. Philadelphia: W. B. Saunders Co. 1935.

IPPOLITO, J. J., and H. H. LeVEEN: Treatment of renal artery aneurysms. J. Urol. (Baltimore) **83**, 10 (1960).

JEWETT, H. J.: Accessory renal vessels. Surg. Gynec. Obstet. **68**, 666 (1939).

KHOURY, E. N.: Chief variations in the intrarenal distribution of the multiple renal arteries. J. Urol. (Baltimore) **76**, 149 (1956).

KREBS-STEFFENS, D.: Functionless, severely hydronephrotic kidney. Z. Urol. **52**, 153 (1959).

LeRoY, A.: Report on hydronephrosis from an abnormal tube: a new preserving technique. Mém. Acad. Chir. **83**, 37 (1957).

McDONALD, D. R., and J. M. KENNELLY: Intrarenal distribution of multiple renal arteries. J. Urol. (Baltimore) **81**, 25 (1959).

MICHALOWSKI, E., u. W. MODELSKI: Verlagerung der pogfaesse-transpositio vasorum. Z. Urol. **51**, 569 (1958).

MICHON, M. L.: Hydronephrosis from an abnormal tube: a new preserving technique by M. ANDRE LeRoY. Mém. Acad. Chir. **83**, 16 (1957).

PERLET, F.: Cura chirurgica conservatuia in alcuni con di idronefrose. Urologica (Venice) **24**, 146 (1957).

RICHARDSON, E. H.: Diverticulum of the ureter. J. Urol. (Baltimore) **47**, 335 (1942).

SMITH, E. C., and L. A. ORKIN: A clinical and statistical study of 471 congenital anomalies of the kidney and ureter. J. Urol. (Baltimore) **53**, 11 (1945).

STEFFENS-KREBS, D.: The differential diagnosis of contrast medium filling defects in the kidney pelvis region. Z. Urol. **52**, 446 (1959).

STEWART, H. H.: New operation for treatment of hydronephrosis in association with lower polar (or aberrant) artery. Brit. J. Surg. **35**, 51 (1947).

STIRLING, W. B.: "Angiography in hydronephrosis", general discussion on hydronephrosis. Brit. J. Urol. **31**, 294 (1959).

TRUETTA, J., A. E. BARCLAY, P. M. DANIEL, K. J. FRANKLIN, and P. M. PRICHARD: Studies in the renal circulation. Springfield (Ill.): Ch. C. Thomas 1947.

WILLIAMS, J. I.: Urology in childhood. Encyclopaedy of urology. Berlin-Göttingen-Heidelberg: Springer 1958.

WINSBURY-WHITE: "Aberrant vessels", general discussion on hydronephrosis. Brit. J. Urol. **31**, 391 (1959).

WULFF, HELGE G., and ARNE MALM: Coarctation of the aorta and renal abnormalities. Surg. Univ. Acta chlr. scand., Suppl. **283**, 200 (1961).

YOUNG, H. M.: That aberrant artery. Urol. cutan. Rev. **46**, 66 (1942).

Supernumerary Kidney

CARLSON, H. E.: Supernumerary kidney. A summary of fifty-one cases. J. Urol. (Baltimore) **64**, 224 (1950).

FRATKIN, L. B., H. W. JOHNSON, and B. B. MOSCOVICH: Adenocarcinoma in supernumerary kidney. Canad. J. Surg. **6**, 195 (1963).

GEISINGER, J.: Supernumerary kidney. J. Urol. (Baltimore) **38**, 331 (1937).

KRETSCHMER, H. L.: Supernumerary kidney. J. Amer. med. Ass. **65**, 1447 (1915).

— Supernumerary kidney. Surg. Gynec. Obstet. **49**, 818 (1929).

PARIN, B. V.: Case of a third, supplementary kidney. Vestn. Khir. **16**, 118 (1929).

PHOKITIS, P.: The supernumerary kidney. Urol. int. (Basel) **17**, 265 (1964).

RABINOWITSCH: Cit. in PHOKITIS.

STEFAN, H.: A clinically diagnosed supernumerary kidney. Z. Urol. **49**, 343 (1956).

TEMELKOFF, S.: Unusual kidney abnormality, four kidneys independent from each other. Z. Urol. **50**, 18 (1957).

Cystic Disease of the Renal Pyramids

ABESHOUSE, B. S., and G. A. ABESHOUSE: Sponge kidney: a review of the literature and a report of five cases. J. Urol. (Baltimore) **84**, 252 (1960).

ANTON, H. C.: Medullary sponge kidney. Brit. J. Radiol. **35**, 144 (1962).

CACCHI, R., and V. RICCI: Sponge kidney. J. Urol. méd. chir. **55**, 498 (1949). Abstracted from Italian. J. Urol. (Baltimore) **84**, 246 (1960).

EKSTROM, T., B. ENGFELDT, C. LAGERGREN, and N. LINDVALL: Medullary sponge kidney. Nord. Med. **63**, 678 (1960).

FENNA, L. R.: Cystic disease of the renal pyramids. Brit. J. Urol. **33**, 34 (1961).

ISHIHARA, T., and M. KAWAMOTO: A case of sponge kidney successfully treated with renacidin for the associated renal lithiasis. Acta urol. (Kyoto) **7**—**12**, 1050 (1961).

LAGERGREN, C., and N. LINDVALL: Medullary sponge kidney and polycystic diseases of the kidney: distinct entities. Amer. J. Roentgenol. **88**, 153 (1962).

LENARDUZZI, G.: L'indagine radiologica nelle occlusioni intestinali da ascaridi. Arch. ital. chir. **52**, 645 (1938).

LEVIN, N. W., B. ROSENBERG, S. ZWI, and F. P. REID: Medullary cystic disease of the kidney, with some observations on ammonium excretion. Amer. J. Med. **30**, 807 (1961).

LINDVALL, NILA: Roentgenologic diagnosis of medullary sponge kidney. Acta radiol. (Stockh.) **51**, 193 (1959).

LOWEN, W., and A. D. SMYTHE: Cystic disease of the renal pyramids: "medullary sponge kidney". Med. J. Aust. **2**, 60 (1962).

PALUBINSKAS, A. J.: Medullary sponge kidney. Radiology **76**, 911 (1961).

PENNISI, S. A., and R. C. BUNTS: Sponge kidney. J. Urol. (Baltimore) **84**, 246 (1960).

ROSSE, W. F., TH. A. WALDMANN, and PH. COHEN: Renal cysts, erythropoietin and polycythemia. Amer. J. Med. **34**, 76 (1963).

ROWLING, J. TH.: Cystic disease of the renal pyramids. Brit. J. Urol. **33**, 38 (1961).

TURNER, T. A., J. I. WALLER, C. A. HELLWIG, and E. N. McCUSKER: Medullary sponge kidney. An uncommon urological disorder, probably congenital in origin, and curable. J. Kans. med. Soc. **64**, 303 (1963).

Anomalies of the Pelvis

AREY, L. B.: Developmental anatomy, ed. 4. Philadelphia: W. B. Saunders Co. 1942.

BRAASCH, W. F., and A. J. SCHOLL: Cited by NATION.

BURT, J. C., CL. M. LANE, and J. L. HAMILTON: Unusual anomaly of upper urinary tract. J. Urol. (Baltimore) **46**, 235 (1941).

CAMPBELL, M. F.: Principles of urology. Philadelphia: W. B. Saunders Co. 1957.

CAMPBELL, MEREDITH: Surgical treatment of anomalies of upper urinary tract in children. J. Amer. med. Ass. **106**, 193 (1936).

COLBY, F. H.: Essential urology. Baltimore: Williams & Wilkins Co. 1956.

DEES, J. E.: Clinical importance of congenital anomalies of upper urinary tract. J. Urol. (Baltimore) **46**, 569 (1941).

EISENDRATH, D. N.: Double kidney. Ann. Surg. **77**, 450 (1923).

EVERETT, H. S.: Cit. by NATION.

FOWLER, B. J.: Bifid renal pelvis as a factor in upper pole hydronephrosis following operation for congenital hydronephrosis. Brit. J. Urol. **31**, 154 (1959).

GOYANNA, R., and L. F. GREENE: Pathologic and anomalous conditions associated with duplication of renal pelvis and ureter. J. Urol. (Baltimore) **54**, 1 (1945).

GRAUHAN, M.: Über Wachstum und Form der Hydronephrosen. Langenbecks Arch. klin. Chir. **180**, 517 (1934).

GUTIERREZ, R.: J. Amer. med. Ass. **106**, 183 (1936).

HARPSTER, CH. M., T. H. BROWN, and H. A. DELCHER: Cit. by NATION.

HAWTHORN, A. B.: Cit. by NATION.

HERBUT, P. A.: Urological pathology. Philadelphia: Lea & Febiger 1956.

HINMAN, F.: Principles and practice of urology. Philadelphia: W. B. Saunders Co. 1935.

JEWETT, H. J.: Accessory renal vessels. Surg. Gynec. Obstet. **68**, 666 (1939).

KEIBEL, FR., and FR. P. MALL: Manual of human embryology, vol. 2, The development of the urogenital organs (W. FELIZ). Philadelphia: J. B. Lippincott Co. 1912.

KOMAROVICH, F. P.: Redkii variant anomalii pochechonoi lokhanki. Urologiya **22**, 44 (1957).

LAU, FR. T., and R. B. HENLINE: Ureteral anomalies: report of a case manifesting three ureters on one side with one ending blindly in aplastic kidney and bifid pelvis with single ureter in other side. J. Amer. med. Ass. **96**, 587 (1931).

LIVADITIS, A., K. MAURSETH, and P. A. SKOG: Unilateral triplication of the ureter and renal pelvis. Acta chir. scand. **127**, 181 (1964).

LUND, A. J.: Uncrossed double ureter with rare intravesical orifice relationship: Case report with review of literature. J. Urol. (Baltimore) **62**, 22 (1949).

MARQUARDT, H. D.: Diagnosis and therapy of double kidneys. Z. Urol. **49**, 736 (1956).

MARTIN, E. C. ST., H. A. O'BRIEN, and J. D. MITCHELL: Double kidney: reversal of the R. Weigert-Meyer Law, a case report. J. Urol. (Baltimore) **66**, 486 (1951).

MILLS, J. C.: Cit. by NATION.

MUKHIN, I. V.: Sluchai liposarkomy paranefral 'noi Kletchatki udvoennoi i distopirovannoi pochki. Urologiya **22**, 46 (1957).

NATION, E. F.: Duplication of kidney and ureter: statistical study of 230 new cases. J. Urol. (Baltimore) **51**, 456 (1944).

NORDMARK, B.: Double formations of the pelvis of the kidneys and the ureters. Acta radiol. (Stockh.) **30**, 267 (1948).

PAYNE, R. A.: Clinical significance of reduplicated kidneys. Brit. J. Urol. **31**, 141 (1959).

ROSS, K. F.: A case of hemipyonephrosis in a duplicated kidney presenting in middle age. Brit. J. Urol. **36**, 191 (1964).

ROTHAUGE, C. FR.: Heminephrektomie einer Doppelniere wegen Steinbildung im oberen hydronephrotischen Nierenbecken. Z. Urol. **50**, 35 (1957).

SETH-SMITH, A. B., and W. D. PARK: Duplication of the renal pelvis and ureter: an unusual case. Brit. J. Urol. **31**, 150 (1959).

SHANE, J. H.: Supernumerary kidney with vaginal ureteral orifice. J. Urol. (Baltimore) **45**, 344 (1942).

TEMELKOFF, S.: Seltene Nierenanomalie vier voneinander vollkommen unabhängiger Nieren. Z. Urol. **50**, 18 (1957).

TOLLE, R.: Zur Kasuistik der Doppelniere. Z. Urol. **49**, 501 (1956).

TRUSZ, F.: Zur Klinik der Doppelniere. Med. Klin. **28**, 1215 (1957).

WEIGERT: Cit. by EISENDRATH.

WILLIAMS, D. I.: Encyclopedia of urology. Berlin-Göttingen-Heidelberg: Springer 1958.

WOODRUFF, ST.: Complete unilateral triplication of ureter and renal pelvis. J. Urol. (Baltimore) **46**, 376 (1941).

Anomalies of the Ureter

ARJAN D. AMAR and JOHN A. HUTCH

With 41 Figures

The following clinical classification of ureteral anomalies places the emphasis of primacy upon the most common and important of these entities. Anomalies that are frequently associated have been grouped together.

A. Multiplication, Ectopia and Ureterocele
    I. Duplication, Triplication
    II. Ectopia
    III. Ureterocele
B. Ureteropelvic Junction Obstruction
C. Unusual Positions of the Ureter
    I. Retrocaval (Post-Caval; Circumcaval) Ureter
    II. Retroiliac Ureter
    III. Herniation of the Ureter
D. Agenesis, Aplasia, Blind Ending
    I. Agenesis
    II. Aplasia
    III. Blind Ending Ureters
E. Congenital Dilatation of the Ureter
    I. With Obstruction
      1. Congenital Ureteral Stricture
      2. Distal Ureteral Atresia
      3. Congenital Ureteral Valves (or Folds)
    II. Without Obstruction (Discussed in Chapter on Reflux)
F. Twists, Kinks, Congenital Diverticula, Blockage by Vessels
    I. Spiral Twist or Torsion of Ureter
    II. Ureteral Kinks
    III. Diverticulum of the Ureter
    IV. Blockage of Ureter by Blood Vessels.

Because of the nature of ureteral embryonic development, certain anomalies cannot be clearly discussed without occasional mention of others that are frequently associated with them. For example, ureteral ectopia, ureteral duplication, and ureterocele are so closely related that consideration of one will inevitably entail mention of the others.

A. Multiplication, Ectopia and Ureterocele
I. Ureteral Duplication (and Triplication)
1. Definitions

Duplication of the ureter means longitudinal segmentation of the ureter into two tubes. *Triplication* is segmentation into three tubes. (For further discussion of triplication see pp. 122ff.). *Multiplication*, a general term, connotes segmentation into two or more tubes. Unless otherwise specified, information and comments in the following text apply to both duplication and triplication of the ureter, although the term "ureteral duplication" is used throughout the discussion because it is by far the commoner form of ureteral multiplication.

Although the term, "reduplication", has been used by some authors to designate this anomaly, it would appear not to be justified, since in strict definition, the *reduplication* of any object is its "duplication *again*", or, in actuality, its multiplication into *four* (quadruplication).

Ureteral duplication is the term applied specifically to segmentations of the conduction system that originate distal to the ureteropelvic junction (GREENE, 1944; GOYANNA and GREENE, 1945). Segmentations of the conducting system that originate proximal to the ureteropelvic junction are termed *bifid* (or *trifid*) *pelvis*.

Ureteral duplication may be *unilateral* or *bilateral*.

With few exceptions, the two renal pelves from which duplicated ureters descend lie one above the other. Rarely, one of the two pelves in such a kidney lies in front of the other or in some other peculiar relationship (EMMETT).

Ureteral duplication may be *partial* or *complete*. In partial duplication, the lower or distal portion of the ureter is a single tube. The duplicated portion of the ureter is in its upper or proximal part: it consists of two (or more) tubes leading from two (or more) renal pelves. Occasionally, one of the ureteral segments has a blind superior end, and is not attached to any discernible renal tissue.

Ureteral duplication should not be confused with ureteral ectopia (see pp. 124 ff.), although the two are frequently associated. Ureteral ectopia is the occurrence of the ureteral orifice elsewhere than in its normal site, the vesical trigone.

2. Incidence (Table 1)

Duplication of the renal pelvis or of the renal pelvis and its attached ureter is the most common anomaly of the upper urinary tract. Segmentation of the ureter into three or more parts, however, is rare. The clinical incidence of ureteral duplication (which is, for all practical purposes, tantamount to its urographic incidence) is much higher than that noted in autopsy series (Table 1). Thus CAMPBELL (1963) noted 342 cases of ureteral duplication among 51,880 autopsy subjects, an incidence of 1:160, while in addition to the incidences noted in the table, NORDMARK found 138 cases among 4,774 patients studied radiographically, an incidence of 1:35. Among patients with urologic disease who were admitted to the University of Michigan Hospital (THOMPSON and AMAR) we noted a 6% urographic incidence of ureteral duplication. SWENSON and RATNER found radiographic evidence of ureteral duplication in 62 of 4,000 urographic examinations of children, an incidence of 1:64.

CAMPBELL (1963) observed the anomaly twice as often in males as in females in the clinical situation, but found an approximately equal incidence in the two sexes at autopsy. NATION, on the contrary, noted that even in autopsy subjects ureteral duplication was 27% more common in the female than in the male; while in his clinical experience the relative incidence was 63% in females, 37% in males. Seventy-seven per cent were unilateral, 23% bilateral; 53% were on the right, 47% on the left. The incidence of complete duplication was greater on the right than on the left. In females, complete and incomplete duplication occurred with nearly equal frequency; in males, the majority of duplications were incomplete. Among 100 consecutive cases seen in our clinic during the past four years, 42% were in males; 58% in females. A similar, though slighter, preponderance of females over males in cases seen clinically was reported by

Table 1. *Relative incidence of ureteral duplication, and of clinically significant subcategories, in four series*

Type of study	CAMPBELL, 1963 Autopsy	NATION, 1944 Autopsy	NATION, 1944 Clinical	PAYNE, 1959 Clinical	AMAR, 1961—1965[1] Clinical	
Total subjects examined	51,880	16,00	121	5,000	100	
Ratio Duplications:subjects	1:160	1:160		1:160		
	%	%		%	%	
Adults	82			(10 mos.-	72	
Infants and children	18			80 yrs.; max. incl. 20—40 yrs.)	28	
Male	50	49	37	38	42	
Female	50	51	63	62	58	
Incomplete	70	59		55	71	
Complete	30	48		32[2]	40	
Bilaterally complete	1	10			8	
Unilateral	84	77		85	83	
Right		(50)		(53)	(49)	(43)
Left		(50)		(47)	(51)	(57)
Bilateral	16	23		15	17	
Site of junction						
Upper[3]	29	27			24	
Middle[3]	23	45			24	
Lower[3]	24	28			23	

[1] Unpublished data.

[2] Bifid pelvis only in 18 cases.

[3] Upper = superior 4 cm. of ureter; lower = within 8 cm. of bladder; middle = between these two sites.

Nordmark. In contrast to Campbell's experience, therefore, the observations of the last three authors thus support the concept that, when ureteral duplication is present, the tendency to development of urinary tract disease is greater in the female than in the male.

There appeared to be a familial tendency to duplication of the urinary tract in a family reported by Girsh and Karpinski, and in five other families cited by those authors from the literature. The families of 39 patients with renal and ureteral duplication were studied recently by Whitaker and Danks for data on the inheritance of such anomalies. Of the 123 excretory urograms of relatives of the propositi, 16 showed unsuspected duplication.

3. Embryology and Physiology

a) Embryology

Comprehension of the usual and unusual anatomic relationships in ureteral multiplication depends upon an understanding of the course of embryonic development. The embryology of the urinary tract has been dealt with extensively in a separate chapter of this book, and by a number of other authors: Chwalla (1927 a, b); Felix (1912); Hawthorne (1936); Mertz (1918, 1920); Meyer (1946); Pohlman (1905). A few of the pertinent points are briefly recapitulated here.

The earliest discernible rudiments of the ureteropelvic conduit system appear during the fourth or fifth week of intrauterine life (Fig. 1), when a frontal fold (the urorectal septum) emerges to divide the cloaca into two parts: a dorsal section, which will develop into the rectum; and a ventral section, the urogenital segment. The lower lateral portion of the urogenital segment contains the opening of the two wolffian ducts. When the embryo is about 5 mm. in length, a ureteral bud arises in the medial dorsal side of each wolffian duct, near its opening into the urogenital segment. Each ureteral bud subsequently migrates from the medial dorsal to the lateral side of the duct, while also descending slightly in a caudal direction so that the *inferior portion* of the developing ureter approaches the side wall of the bladder.

At about the 10 mm. stage, the common end portion of the wolffian duct dilates and becomes divided by a common septum into two openings (wolffian

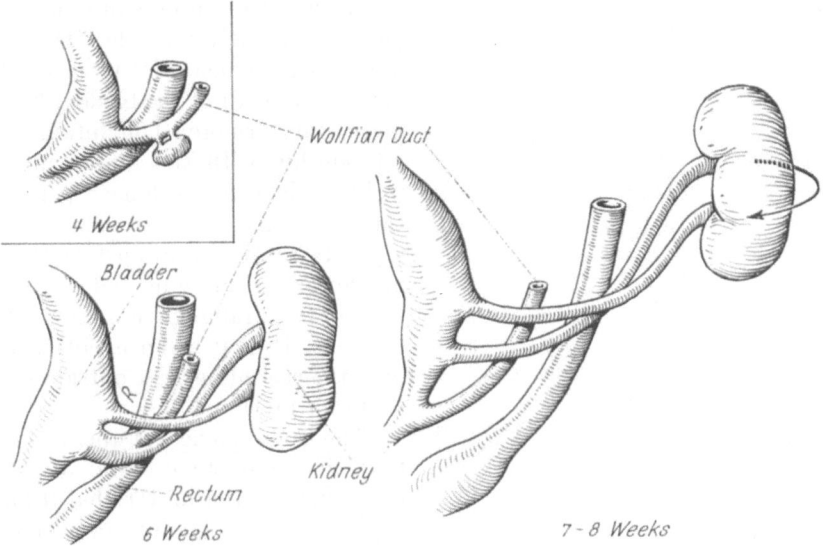

Fig. 1. Development of a kidney with duplicated ureter. *4-Week stage.* Double ureter starting from the wolffian duct. *6-Week stage.* Through expansion of the lateral portion of the allantois, the lower wolffian duct becomes dilated and the lower ureter is the first to reach the bladder. The wolffian duct, carrying the upper ureter with it, shifts in a downward direction. When the upper ureter becomes implanted, it is implanted further down and more medially than the ureter belonging to the lower renal segment. *7—8-Week stage.* The wolffian duct continues downward and, in the male, becomes permanently lodged at the bladder neck; in the female, it continues still further down. This diagram represents the final arrangement seen in the adult. (Redrawn from KELLY, H. A., and C. F. BURNAM: Diseases of the Kidneys, Ureters, and Bladder, Vol. 2, New York: D. Appleton and Co., 1914)

duct and ureteral lower end) into the bladder instead of the single wolffian duct. When the embryo is approximately 12 mm. in length, the opening of the wolffian duct lies medial and cranial to the ureteral opening; but it soon migrates caudally to a position inferior and medial to the ureteral orifice. This downward movement progresses until the wolffian ducts terminate side by side in the posterior urethra, forming the ejaculatory ducts in the male, and Gärtner's ducts in the female.

Meanwhile, the bladder is growing rapidly. Most authors believe that the bladder trigone is formed from wolffian duct mesoderm. The trigone normally contains the ureteral orifices, which are drawn cranially and laterally with the bladder growth, and move away from the wolffian duct openings. According to MEYER the two contralateral ureteral orifices are equidistant from the urethra.

While the lower end of the ureter is being separated from the wolffian duct, its upper end continues to proliferate and finally enters the renal blastema, where it forms the urinary collecting system within the kidney. Although it is not certain whether, in cases of complete ureteral duplication, one or two ureteral buds are originally present on the affected side, there is probably a short period during which the wolffian duct does contain two such buds. It is believed that one of these, located cranially to the other, will enter the superior part of the renal blastema; the other, more caudally located, will enter the lower part of the renal blastema.

As the two ends of the wolffian ducts merge into the bladder floor, the lower of the two duplicated ureteral buds reaches the bladder first, and the rapid growth of the bladder pulls this lower ureteral bud with it, so that the ureteral orifice rises cranially and laterally, away from the opening of the wolffian duct. The lower end of the ureter draining the superior renal pelvis reaches the bladder later, after much of the upward and outward growth of the bladder has taken place; furthermore, it maintains a close relationship with the lower end of the wolffian duct for a longer period. For these reasons, when it does ultimately separate from the wolffian duct, it remains in a lower and more medial position than the ureter draining the inferior renal segment. It may even eventually open near the vesical neck or outside it, in the posterior urethra.

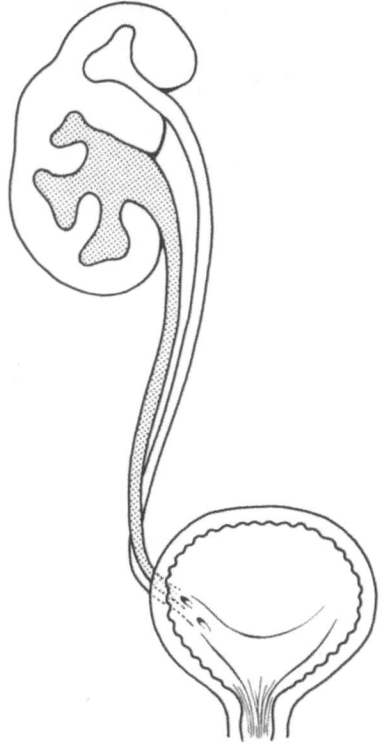

Fig. 2. Schema of ureteral duplication, depicting the Weigert-Meyer law: the final arrangement of completely duplicated ureters, as seen in the majority of adults with this anomaly. The orifice of the ureter belonging to the lower renal segment is in the bladder, above and lateral to the opening of that belonging to the upper renal segment

The consequence of these events — that, in complete ureteral duplication, the orifice of the ureter that drains the superior renal pelvis emerges at a point lower in the bladder than does that of the ureter that drains the inferior renal pelvis — has been called the *Weigert-Meyer law* (Fig. 2). Rare exceptions to this law have been noted (LUND).

The early intimate relationship of the ureteral buds and the wolffian derivatives provides an explanation for such unusual types of ectopia of the ureteral orifice as, in the male, openings into the seminal vesicle or, in the female, into the vagina, cervix, or uterus. In either sex, the intimate relation of the ureter to the cloaca during the early developmental phase makes possible ultimate ectopic opening of the ureter into the rectum.

Crossing of the duplicated ureters, according to HAWTHORNE, is caused by two factors: (1) Because the orifice of the ureter from the superior renal segment is caudal and medial to that of the ureter from the inferior renal segment, the ureters frequently cross immediately above the bladder. (2) Because the ureter from the lower renal segment is more redundant, it tends to form a loop or loops which cross the ureter from the upper renal segment. It is also frequently dilated. Both the redundancy and the dilatation may be congenital, and may also be caused or increased by obstruction, infection, or associated reflux.

b) Physiology

The caliber of the two limbs (above the bifurcation) of an incompletely duplicated ureter may be normal, but it often shows dilatation when compared to the normal caliber of the common stem. If fluid that has been colored by indigo carmine is introduced into one of two renal pelves attached to an incompletely duplicated ureter, the fluid can be seen to regurgitate from that limb into the

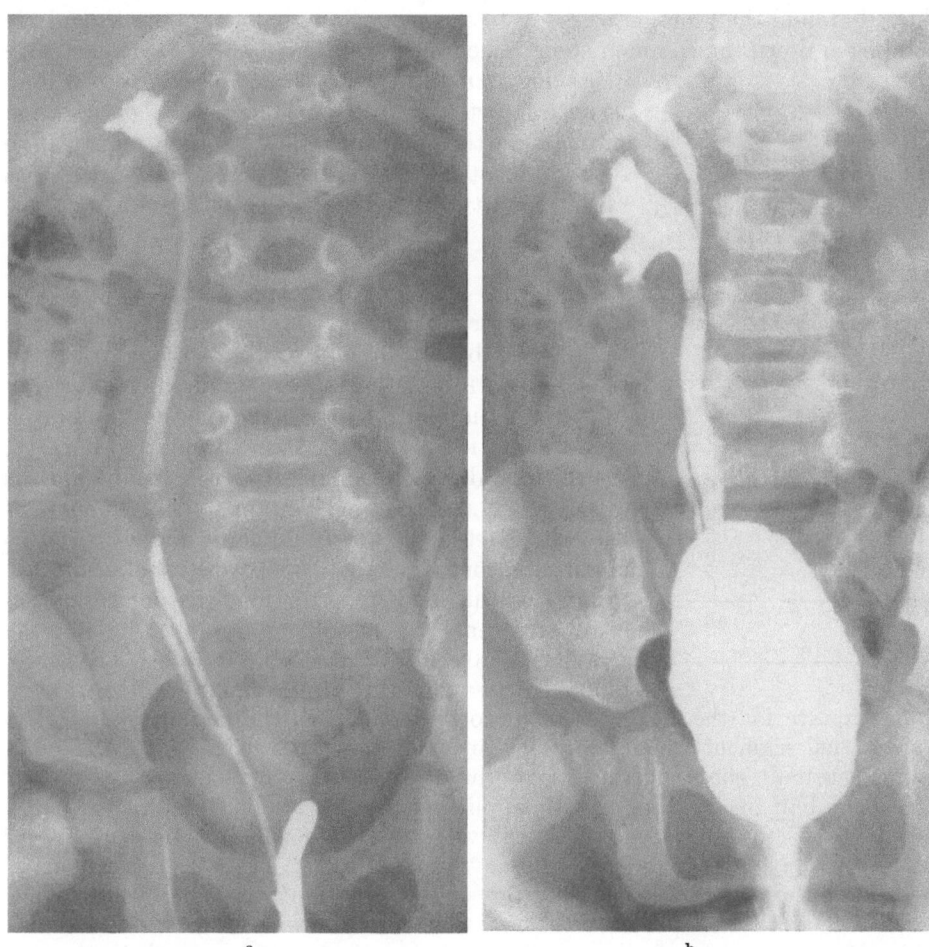

a   b

Fig. 3. (a) Ureter-to-ureter reflux demonstrated by retrograde urography in a 4-year-old girl with extravesical ureteral junction. (b) In the same patient, vesicoureteral reflux outlines the duplicate ureters with corresponding renal pelves

other limb around the site of junction. This regurgitation has also been observed by the use of cineradiography. LENAGHAN noted that this regurgitation caused a delay in emptying the ureter, from a normal emptying time of a few seconds, to periods ranging from 10 minutes to $1^{1}/_{2}$ hours. The length of the delay period was directly related to the caliber of the dilated limbs.

Ureter-to-ureter reflux was noted by STEPHENS in six cases of duplicated ureters with extravesical junction. CAMPBELL (1951a) observed that ureter-to-ureter reflux could be demonstrated by retrograde urography in most bifid ureters with extravesical junction (Fig. 3): the two limbs of the bifid ureter are usually outlined by dye which flows around the junction of the Y. Such reflux was not noted if the junction lay within the bladder wall.

Lenaghan observed ureteral peristalsis at the time of surgical exploration of bifid ureters. He noted that the peristaltic waves in the two limbs of incompletely duplicated ureters were asynchronous. Peristaltic asynchronicity in incompletely duplicated ureters was similarly demonstrated by electromanometric pressure recordings. In a 7-year-old girl with bifid ureter, observed during operation, contractions in one limb of the incompletely duplicated ureter were often conducted around the junction of the Y to the other limb, but were only irregularly conducted down the common stem. When, however, one limb was occluded with a bulldog clamp, all waves from the unclamped limb were conducted down the stem. This observation provides further evidence that excision of one limb of a duplicated ureter with extravesical junction is rational treatment for uretero-ureteral reflux.

Table 2. *Differences in length between nondiseased bifid kidney and contralateral normal nonbifid kidney* (From Amar and Scheer: New Engl. J. Med. **273**, 211 (1965)]

Length difference between bifid and nonbifid kidney millimeters	Cases No.
< —4	2
—4— 0	1
1— 5	8
6—10	6
11—15	16
16—20	5
21—25	2
> 25	5
Total	45
Average 13	

4. Surgical Anatomy

The kidney belonging to duplicated ureters has been found by Amar and Scheer (1965) to be, on the average, 13 mm. longer (as measured on the x-ray film) than the contralateral nonduplicated kidney in the same patient (Table 2). In nonduplicated kidneys, the average difference in radiographic renal length is 3 mm. (Moëll, Amar and Scheer, 1965). In view of the many factors that may influence kidney size, including hydronephrosis, pyelonephrosis, pyelonephritis, urinary calculi, and hypertension, the determination of relative renal size may be important in judging the presence or absence of disease. Within the duplicated kidney, we and others (MacAlpine; Ross) have found it rare for the upper segment to be larger than the lower segment. The upper renal segment usually contains one major calyx; whereas the lower renal segment usually contains two or more major calyces.

About half of all kidneys with duplicated ureters studied by Boijsen were supplied with multiple renal arteries; whereas the incidence of multiple renal arteries supplying nonduplicated kidneys was only 20%. In the series reported by Derrick and Hooks, the incidence of hypertension was found to be higher in patients with multiple renal arteries than in those with single renal arteries. This observation suggests the possibility that patients with ureteral duplication may have a higher incidence of hypertension.

As the two duplicated ureters approach the bladder, they are usually contained within a single sheath, and their blood supply is often closely associated, a point of great surgical significance since the possibility of injury to the blood supply of the nondiseased ureter must be borne in mind during surgical excision of a diseased bifid ureter.

The junction of the two segments may be at any level distal to the renal pelves, including the intramural portion (Fig. 4). The relative frequency of possible sites of junction, as noted by various observers, is indicated in Table 1.

The majority of completely duplicated ureters cross one another once or twice between the renal pelves and the bladder, although Lund, after a careful analysis

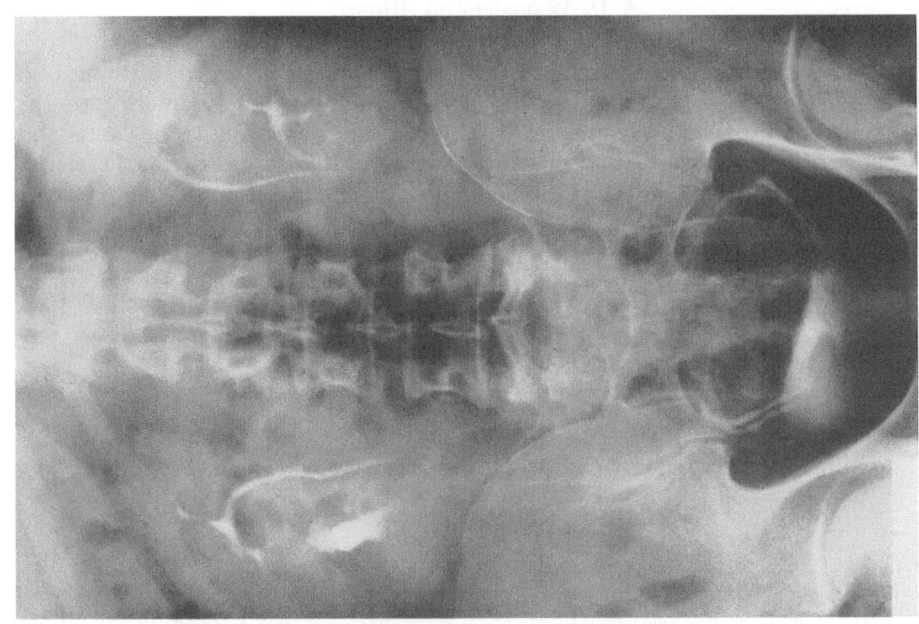

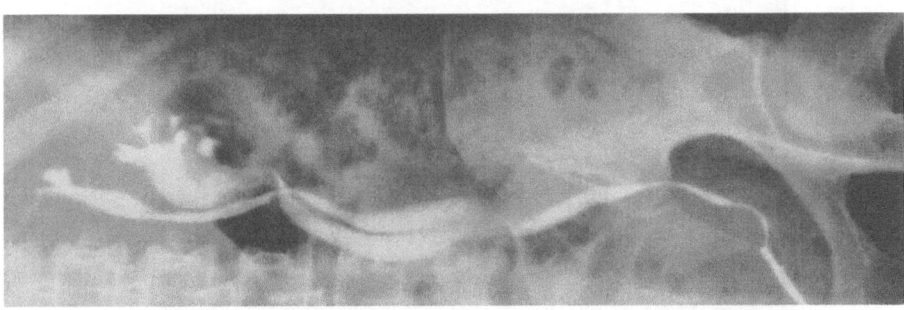

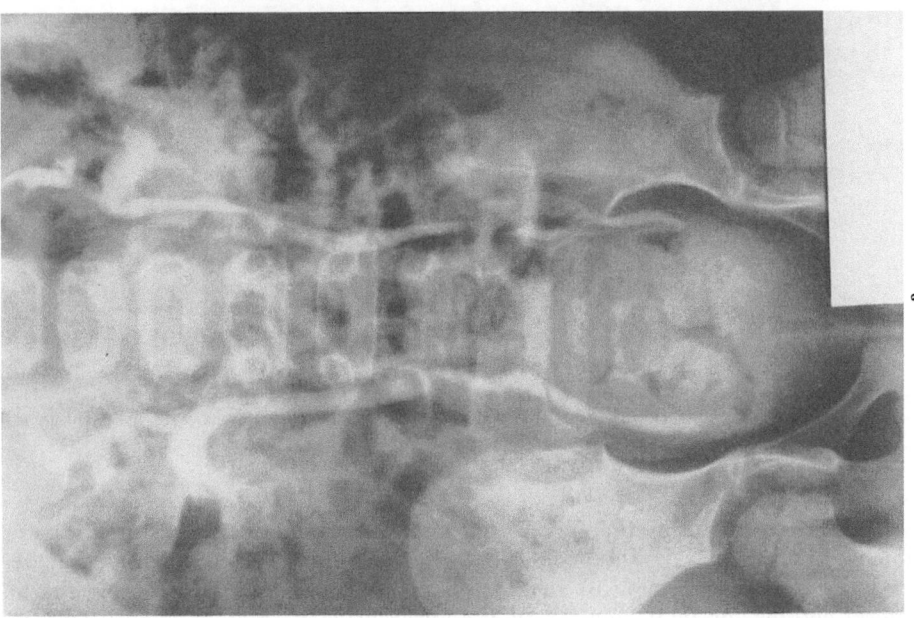

Fig. 4a—c. Levels of ureteral junction: (a) Bifid renal pelvis (right). (b) Ureteral junction at level of sacrum; below this point the ureter is single. (c) Bilateral complete ureteral duplication

of published reports, estimated that from 8 to 13% of completely duplicated ureters are uncrossed. Crossing occurs most often immediately inferior to the renal pelves and immediately superior to the bladder. According to BRAASCH and HAGER, crossing is only apparent in some cases, and will disappear after a catheter is inserted into one of the ureteral segments. WEIGERT suggested that embryologic development probably influences the site of crossing.

With rare exceptions, the orifice of the ureter that leads from the lower renal pelvis is superior and lateral to the orifice of the ureter leading from the upper renal pelvis. The commonly accepted explanation for these relative locations is given under Embryology.

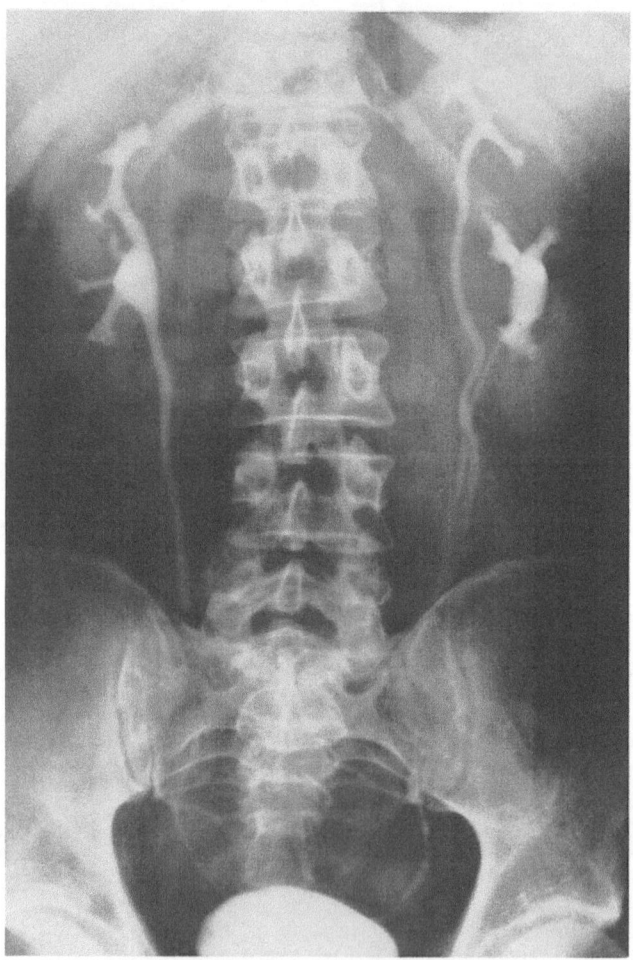

Fig. 5. Complete (left) ureteral duplication, with malrotated lower segment, in a 35-year-old man with prostatic inflammation. The single right kidney and ureter show no abnormality. Symptoms were unrelated to the duplication and malrotation

5. Associated Anomalies

Duplication of the ureter may be accompanied by almost any other type of anomaly of the urinary tract, such as malrotation (Fig. 5), solitary kidney, horseshoe kidney, hypoplasia (Fig. 6), fusion, renal ectopia (Fig. 7), ureterocele (Fig. 8), ureteral ectopia, or malformation of the lower urinary tract or genital system.

The urographic possibilities of such combinations are almost endless. Other anomalies of the urinary tract were present in 27 (or 12%) of the 230 patients with ureteral duplication studied by NATION. Fifteen of the 27 associated anomalies were on the same side as the duplication; 9 were on the opposite side; 3 were bilateral. The 27 associated anomalies were renal ectopia (in 7 patients), stenosis of one duplicated ureter (5 patients), renal hypoplasia (5), agenesis (3), polycystic disease (2), aplasia (1), malrotation (1), partial triplication (1), and partial quadruplication and quintuplication (1 each). In 4 cases (3%), the orifice of one or more of the completely duplicated ureters was ectopic. Two female

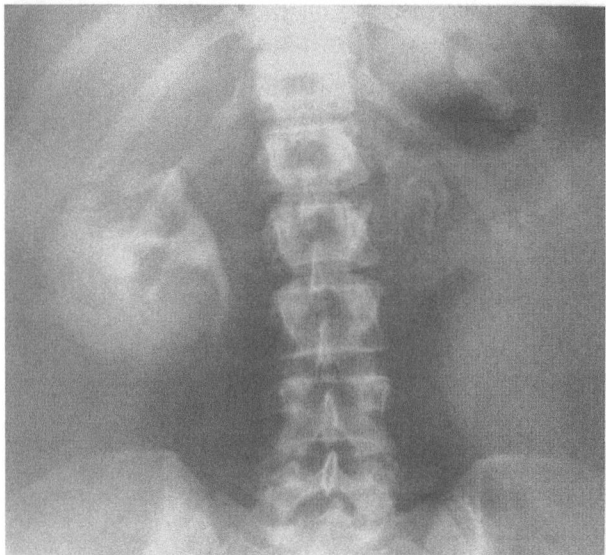

a

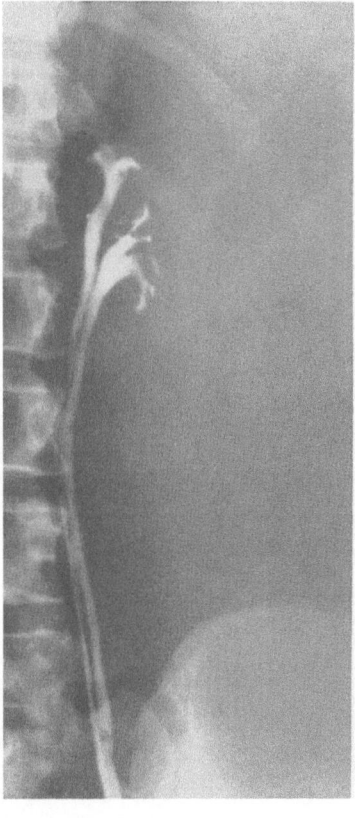

Fig. 6a and b. 39-Year-old woman with recurrent urinary infection of long duration. (a) Excretory urogram: Small, hypoplastic left kidney; compensatory hypertrophy of right kidney. (b) Retrograde urogram: Small, hypoplastic left kidney with ureteral duplication. There has been no urinary infection since left nephrectomy for chronic pyelonephritis, 5 years ago

b

patients had associated anomalies of the internal genitalia. Among 29 children studied by LENAGHAN, 12 had other anomalies of the urinary tract. The majority of the associated anomalies involved the ureters. Included among them were two instances of bifid ureter with blind superior ending. CAMPBELL (1963) stated that associated anomalies may be found in at least a fourth of patients with ureteral duplication. GROSS (1953) noted other malformation of the urinary tract in association with double ureters in nearly 20% of children reported by him. He observed that ureterocele is an especially likely accompaniment of bifid ureter, and that such an associated ureterocele is almost always of the ureter that is attached to the superior renal segment, which opens lower than the other ureter in the trigone, or into the bladder neck or urethra.

Because other anomalies are so frequently associated with ureteral duplication, the physician is obliged to examine the urinary tract in its entirety, by all relevant means including thorough urographic study, in all patients whose symptoms and signs suggest ureteral duplication. Such duplication may be only a contributory factor in the total disease.

An unusual, and perhaps unique, configuration was reported by Berman and Sidorenko in 1959: together with two perfectly formed right and left kidneys having complete ureters, a left accessory kidney with complete ureter was attached by an isthmus to the lower pole of the right kidney, forming a horseshoe kidney.

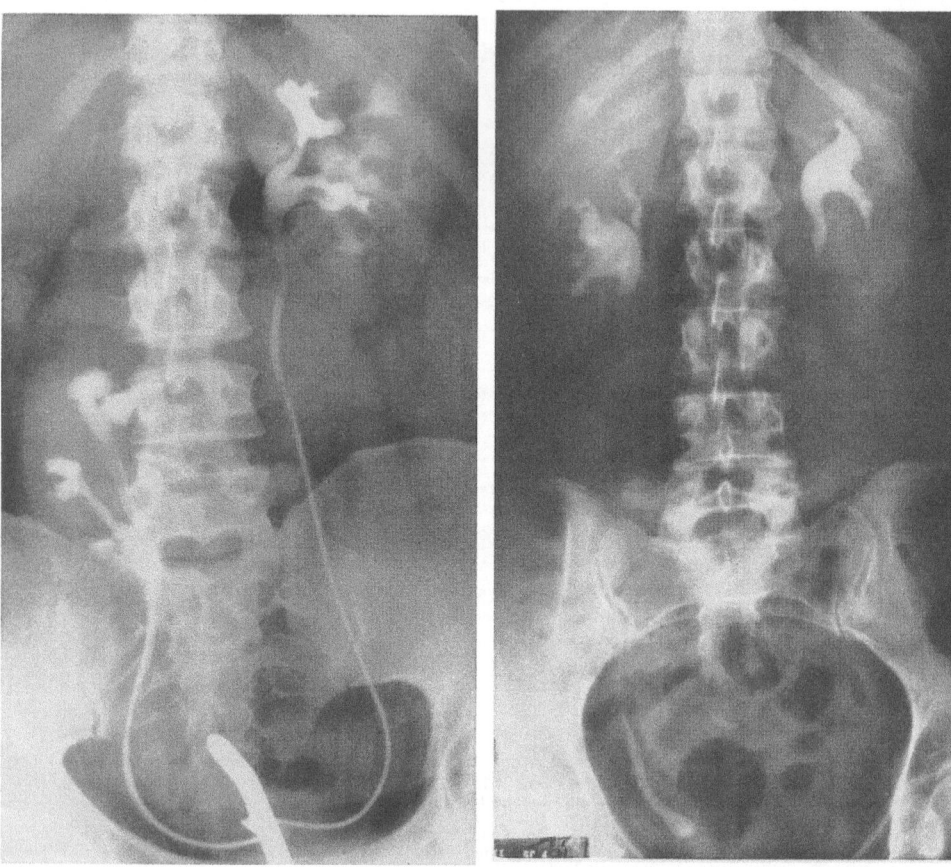

Fig. 7   Fig. 8

Fig. 7. Bilateral retrograde urogram: Ectopic right kidney with bifid pelvis

Fig. 8. Excretory urogram: Right complete ureteral duplication with ureterocele belonging to ureter of upper renal segment

The orifices of the two ureters from the horseshoe kidney were side by side, 4 or 5 cm apart. The ureter of the normal left kidney crossed in front of the accessory kidney, and was believed to be the cause of the presenting intermittent colic with a palpable abdominal mass. Surgical correction appears to have been followed by relief of symptoms.

6. Associated Reflux
(See also, chapter on Reflux, pp. 1—50)

Either or both of two types of reflux may be associated with incomplete duplication: ureteroureteral, and ureterovesical. Only ureterovesical reflux can exist with complete duplication. There is a high incidence of vesicoureteral reflux accompanying complete duplication, and of ureteroureteral reflux with incomplete duplication. The embryologic basis for the high incidence of associated vesicoureteral reflux has been discussed by AMBROSE and NICOLSON.

According to LENAGHAN, the chance that vesicoureteral reflux will be present is greater in patients with incomplete ureteral duplication than in persons with

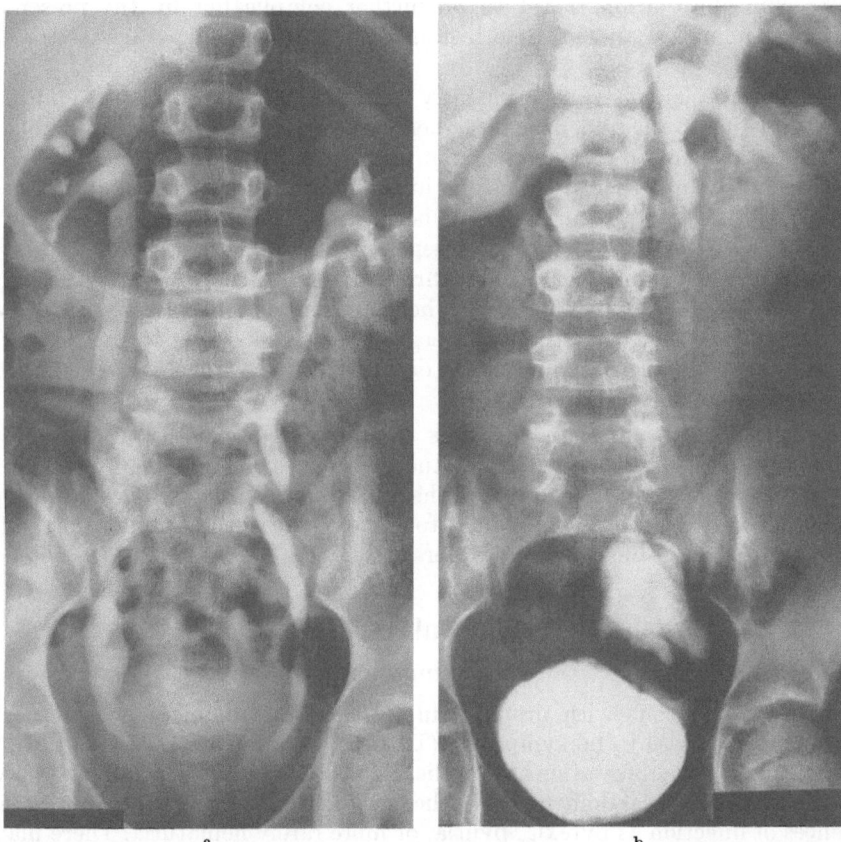

a   b

Fig. 9a and b. 8-Year-old boy with recurrent abdominal pain. (a) Excretory urogram: Bilateral ureteral dilatation; the left kidney is smaller than the right; ureterocele of the upper renal segment ureter (left), causing bladder neck obstruction and bilateral ureteral dilatation, was proved by voiding cystourethrograms (b). (From AMAR, 1966b)

nonduplicated ureters. He noted vesicoureteral reflux into the stem of the incompletely duplicated ureter, or into the ipsilateral completely duplicated ureter, or into the opposite single ureter, in more than one third of patients investigated by him for reflux. The urinary stasis produced by reflux fosters infection. In bifid ureters of normal caliber, with junction in the upper one third, ureteroureteral reflux usually does not occur. In dilated duplicated ureters which join in their lower portion, ureteroureteral reflux and stasis are likely and infection may supervene.

7. Associated Disease

Some of the diseases that are commonly associated with ureteral duplication are caused by the nonphysiologic position of the orifice of the upper renal segment ureter, even though that orifice may be within the bladder wall. Thus, when the orifice is in the vesical neck or in the urethra, obstruction and infection may be present without urinary incontinence. Obstruction and back pressure are frequently associated with opening of the ureter into a seminal vesicle, uterus, or fallopian tube, and infection may result. If the ureter opens into the vestibule, vagina, or rectum bacterial infection may involve the kidney. The infection is often difficult to treat effectively because obstruction tends to induce ureteral atonia, irrespective of the presence of infection.

The clinical situation is apt to be further complicated by the presence of associated anomalies, and in some instances, is complicated by disease traceable to them.

The ureterocele that is so frequently found, usually at the lower end of the ureter belonging to the upper renal segment, may cause obstruction of the bladder neck and of the ureter belonging to the nonduplicated kidney (Fig. 9).

It has not been sufficiently stressed in the literature that the duplicate ureter belonging to the lower renal segment, because its orifice is more lateral in the bladder wall, has a short intravesical segment and is therefore prone to reflux, with its attendant danger of pyelonephritis involving the lower renal segment.

According to CAMPBELL (1963), when ureteral ectopia is associated with duplication, in 9 of 10 cases the upper segment is the one that is diseased. When there is no ectopia, and only half of the organ is diseased, then the lower segment is the more commonly affected.

Obstruction, congenital or other, is common in duplicated ureters; it was found in 14 of 61 duplicated ureters studied at autopsy of 19,046 children and infants (CAMPBELL, 1963). In adults, of the 281 autopsies with ureteral duplication, obstruction was found in 14, of which stones were present in 8 and stricture in 4, and an undesignated blockage in 2 others.

8. Clinical Aspects

a) Symptoms

In 60% of persons with ureteral duplication, the anomaly does not cause disease and gives rise to no symptoms (THOMPSON and AMAR). In the remaining 40%, the clinical manifestations of duplication are highly variable. There may be pain throughout the abdomen, backache, or referred pain. There may be such evidences of infection as pyrexia, pyuria, or more rarely hematuria. There may be such bladder symptoms as frequency, dysuria, nocturia. In females with ectopic ureteral opening into the urethra or genital tract there may be urinary incontinence. In some patients the presenting manifestations may be those of urinary calculi. The incidence of these presenting manifestations in the series of 141 cases reported by PAYNE was as follows: pain, 48 cases; infection, 44; enuresis, 7; hematuria, 5; pyrexia, 4; symptoms due to epididymitis, 2; symptoms due to hypernephroma, 1. In 41 cases the symptoms were not relevant to the duplication and the finding was coincidental. Ten of the 141 patients had urinary calculi, six of which were in the lower renal segment or its ureter. Forty-nine of the 141 patients were pregnant women, five of whom dated their symptoms from the time of pregnancy; in 41 of the pregnant women gestation had no detectable effect on the status of the urinary system.

CAMPBELL (1963) has summarized the probabilities by stating that the manifestations of ureteral duplication are those that are typical of a normally formed kidney that is particularly prone to infection or obstruction. In addition, the duplicated collecting system is as prone to other types of disease as is the normal urinary tract. Finally, it is our clinical impression that a duplicated kidney and ureter are more vulnerable to injury than one of normal configuration, apparently because the greater size of the duplicated organ, and possibly its more frequent involvement in disease, expose it to trauma.

b) Diagnosis

The majority of diagnoses of ureteral duplication are made during investigation of the urinary tract for disease unrelated to this anomaly, or as an incidental finding either during life or at routine autopsy. In other instances, however, the duplication is of clinical importance because of obstruction, urinary infection, or both that may be caused by the anomaly.

The clinical manifestations of ureteral duplication with or without ectopia are often puzzling. Symptoms and signs of urinary tract disease that cannot be explained on any other basis are more frequently attributable to ureteral duplication and ectopia than is generally recognized. Since the manifestations may be nonspecific, it is often difficult to predict whether the anomaly will be found even by thorough investigation. The following order of procedure is recommended:

α) Personal History

Ureteral duplication and ectopia must always be considered when urinary symptoms or signs persist and cannot be definitively traced to other cause. Intermittent leakage of urine with intervals of normal voiding, may be the presenting evidence of ureteral duplication with ectopia. Ureteral ectopia usually causes incontinence in the female, but this diagnostic sign is not always present. In a 20-year-old women treated by THOMPSON and AMAR, leakage of urine had not occurred until after the delivery of her first child; two extravesical ureteral orifices were then found. LEVACK reported a similar course in a women who had never been incontinent until after her only pregnancy at age 23 years, when manifestations typical of ureteral duplication and ectopia developed. A tortuous, dilated duplicated ureter was found, which opened into the vulva. Leakage of urine may well be prevented in this and other instances by (a) compression of an elongated and ectopic ureter, either at a single site or throughout its course through the vesical neck and vaginal septa, (b) plugging of the lumen by debris from ingrown vulvovaginal epithelium, or (c) the amount of functioning renal tissue attached to this ureter may be small, and may therefore produce little or no urine.

β) Family History

In view of the increased incidence of malformations of the urinary tract in certain families (GIRSH and KARPINSKI), if a patient has symptoms referable to the urinary system, and a member of the patient's family is known to have ureteral duplication, all appropriate attempts should be made to determine whether a similar anomaly is the basis for the present patient's disease.

γ) Physical Examination

Routine inspection of the urethra and bladder should include a close search for an ectopic ureteral orifice. If such an opening is found, it should be filled with radiopaque material through a bulb catheter. In the female, physical examination must include close inspection of the external genitalia for one or more ectopic orifices. At endoscopy of the urethra, any oddly situated urethral diverticulum should be unroofed, and a voiding cystogram should be made. The presence of an abdominal mass should suggest the possibility of a tremendously dilated, duplicated ureter. Urethroscopic and vaginal examination are then important, and should include the application of gentle pressure to the abdominal mass. The ectopic opening may become apparent if urine or pus can thus be caused to exude from the dilated ureter.

δ) Laboratory Study of the Urine

Laboratory study of the urine, for evidence of infection or microscopic hematuria, is mandatory.

ε) Excretory Urography

The most common and usually the first, definitive test for the presence of ureteral duplication is excretory urography. As a preliminary clue, the lengths of the left and right kidney are compared. As Amar and Scheer reported in 1965,

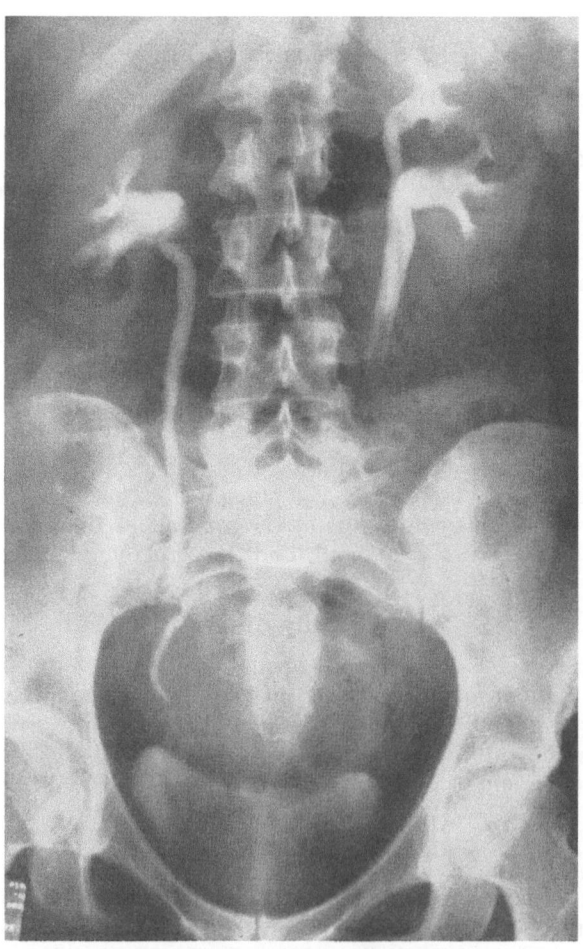

Fig. 10. 42-Year-old woman with left ureteral duplication. Excretory urogram: The left kidney is longer than the contralateral nonduplicated kidney

in 45 normotensive patients the unilateral bifid kidney averaged 13 mm longer than its nonbifid counterpart (Fig. 10); whereas the average difference between left and right single kidney in their controls, and in the normal kidneys measured by MOËLL in 1956, was 3 mm. If both segments of the duplicated kidney are functioning, excretion of the dye and outlining of the duplicated system permit prompt diagnosis. But one or the other segment of a duplicated kidney may function poorly; it may excrete little or no dye, and the ureter attached to the nonfunctioning segment may therefore not be visualized. The upper renal segment is usually the one that fails to function, because it contains insufficient tissue, or

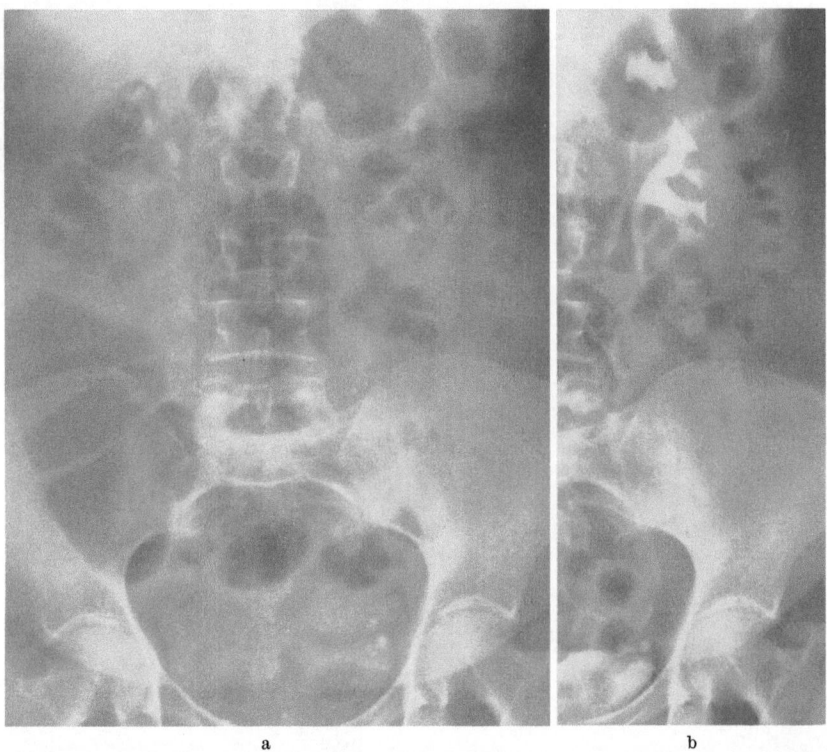

a   b

Fig. 11a and b. 62-Year-old woman with left flank pain. (a) Single-dose urogram: The left collecting system could not be visualized. (b) Double-dose urogram, 15-minute film: Duplicated left kidney and ureter were promptly visualized. Lower left ureteral calculus proved to be the cause of symptoms

because of longstanding obstruction at the lower end of its tortuous, ectopic ureter. When the upper segment fails to function, and the duplication is unilateral, the lower renal segment alone appears as a small kidney opposite the normal kidney. This appearance of such a renal segment, with or without associated pyelonephritis, has occasionally been misdiagnosed as a hypoplastic kidney.

Whenever we suspect ureteral duplication, we anticipate the probability of segmental renal hypofunction, and double the standard dose of contrast medium (AMAR, 1964b), then make delayed films at one and two hours or if necessary up to 24 hours. This technique has established the diagnosis of ureteral duplication in a number of instances where single-dose excretory urography had been unrevealing. Our incidence of visualization of duplicated collecting systems with this method has been much higher than in our previous experience with the single dose (Fig. 11).

In patients with incomplete ureteral duplication, and one poorly functioning renal segment, if ureteroureteral reflux is present it may provide an effective aid to diagnosis. Only on a delayed film, one may see reflux of the contrast medium up the ureteral limb that is attached to the nonfunctioning renal segment.

When all components of a duplicated collecting system do not clearly visualize, the presence of a nonvisualized renal segment is suggested by a discrepancy

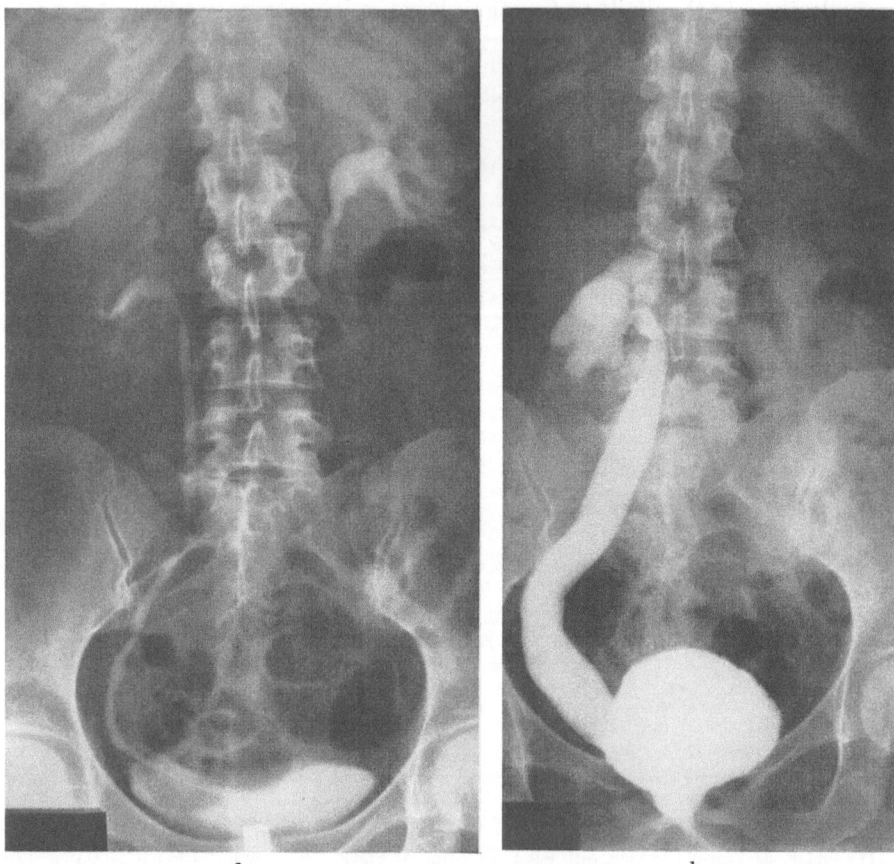

a   b

Fig. 12a and b. 42-Year-old woman with persistent urinary infection. (a) Excretory urogram: Left kidney appears normal. The right kidney appears to droop. Right duplication unsuspected. (b) Voiding cystogram: Upper renal segment (right) and its ureter filled with contrast medium by reflux. This ureter had an ectopic opening into the proximal urethra, missed on previous cystoscopic examinations

(usually in the superior renal segment) between the indicated volume of the collecting system and the indicated mass of the kidney parenchyma; or by displacement inferiorly and laterally of the visualized collecting system, in a kidney that does not appear deformed (Fig. 9). Duplication should also be suspected if the kidney appears normal but lacks an upper major calyx. In addition, it should be borne in mind in the differential diagnosis of kidneys whose radiographic contours do not closely reflect one another.

That the usual urographic features of nonvisualized ureteral segments are not diagnostically infallible is illustrated by the following two cases: In a 42-year-old woman, urographic and cystoscopic examinations had repeatedly failed to reveal an abnormality; yet urinary infection persisted. Eventually, the orifice of a duplicated ureter was found in the bladder neck (Fig. 12). In another instance, a urethral diverticulum was believed to account for recurrent urinary infection. Because the diverticulum was at the vesical neck, it was unroofed rather than excised. Postoperatively, the patient complained of pain in the flank during micturition. Cystographic study showed reflux up a duplicated ureter. Improved drainage of the upper renal segment ureter was established. Subsequent intravenous urography demonstrated return of renal function (Fig. 13).

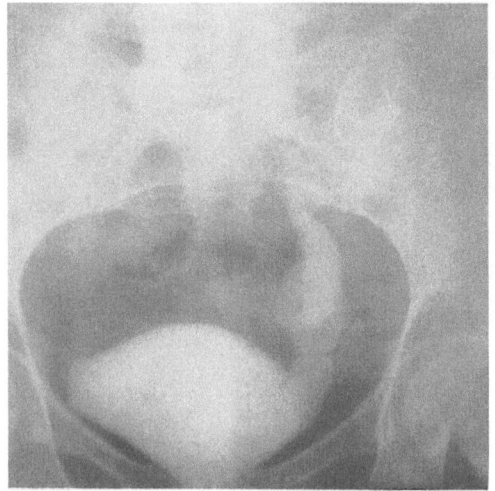

a

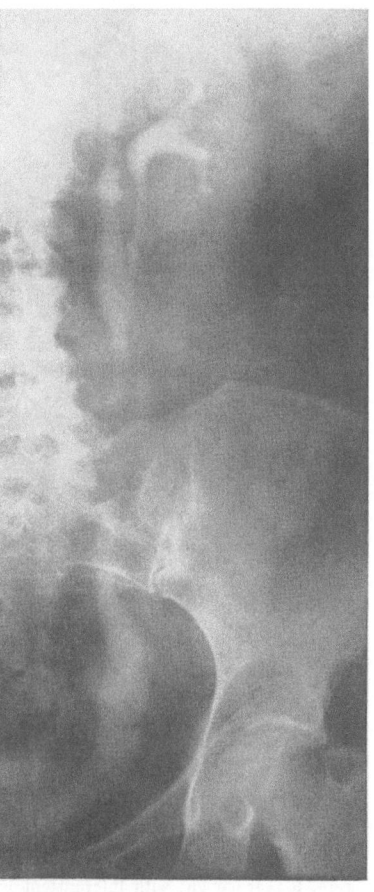

b

Fig. 13a and b. 43-Year-old woman with recurrent urinary tract infections, believed to be due to a urethral diverticulum; this structure was unroofed. (a) Cystogram: Left ureteral reflux. (b) Excretory urogram: Bilateral ureteral duplication. Left upper renal segment with ureter was visualized for the first time after unroofing of the structure that had appeared to be a diverticulum

ζ) Cystoscopy

If excretory urography fails to reveal ureteral duplication, but otherwise unexplained signs persist, special cystoscopic techniques are helpful: (1) *Dual dye technique* (AMAR, 1966a). Administer orally, for 24 hours in divided doses, a strongly colored dye (e.g., pyridium). Then fill the bladder with sterile water to which a dye of a contrasting color (e.g., indigo carmine) has been added. Allow the patient to expel the fluid from the bladder. The fluid coming from the bladder will be blue. If a ureter from a functioning renal segment has an ectopic orifice outside the bladder, orange urine will be seen to come separately from that opening. In an 8-year-old girl an ectopic ureteral opening of the upper renal segment into the vagina was found by this method, and proved by simultaneous retrograde and excretory urography (Fig. 14). (2) *Indigo carmine test* (AMAR,

1964a; 1966b) (Fig. 15). We have found the indigo carmine test helpful not only in the diagnosis of vesicoureteral reflux in general, but also specifically in finding ureteral duplication, which is so often accompanied by reflux. Indigo carmine in sterile water (10:200) is introduced into the bladder through a catheter. When the bladder is filled to capacity the patient is allowed to void. The cystoscope is then introduced immediately. If the anatomic situation permits reflux, voiding pressure

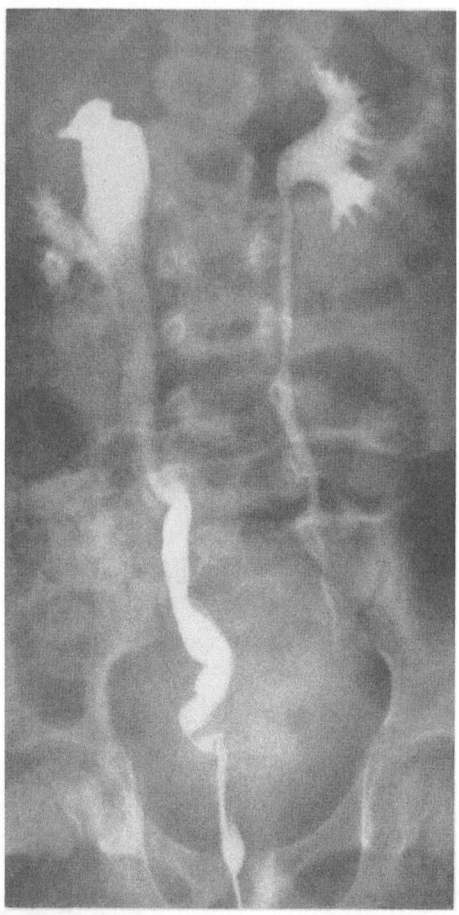

Fig. 14. 9-Year-old girl with intermittent urinary incontinence. (Right ureteral duplication was not seen on excretory urogram or cystourethroscopic examination. The ectopic ureteral opening, in the vagina, was revealed by dual-dye technique.) Simultaneous retrograde and excretory urography prove right ureteral duplication. (From Amar, 1966b)

will have produced it. Blue fluid can then be seen to issue from the refluxing ureter. Retrograde urography, voiding cystoureterography, or both will then prove the duplication (Fig. 16). In 4 patients recently studied in our clinic, all other tests had failed to demonstrate ureteral duplication. The indigo carmine cystoscopic test alone not only made possible the diagnosis of duplication, but also showed the precise site of the orifice of the upper renal segment ureter.

η) Vaginoscopy and Urethroscopy

Vaginoscopy and urethroscopy are frequently necessary in the search for a duplicated ureter.

9) Other Radiographic Techniques

Occasionally, retrograde urography is required. According to PAYNE, a 10-minute delayed film is an important part of the retrograde urogram. A voiding cystourethrogram, with or without cineradiographic control, occasionally delineates a duplicated ureter not demonstrated by other means, because a ureter with ectopic orifice may fill more readily by this technique than by other methods. Similarly, retrograde urethrography may be of value and is occasionally necessary.

a) *Normal urinary tract, without reflux*

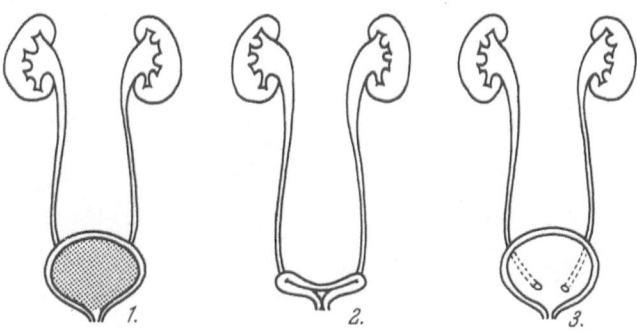

b) *Abnormal urinary tract, with reflux*

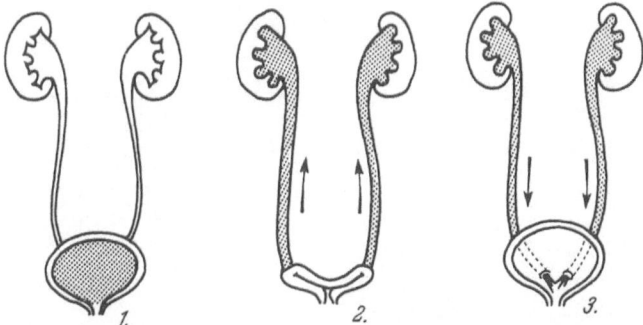

Fig. 15. Indigo carmine technique for demonstration of vesicoureteral reflux. A. (1) Normal bladder filled with blue fluid (stippled area). (2) Complete voiding has eliminated the blue fluid. (3) Postvoiding cystoscopic examination; clear urine seen coming from the ureteral orifices. B. (1) Normal-appearing bladder filled with blue fluid. (2) on voiding, bladder empties but vesicoureteral reflux causes ureters and renal pelves to fill with blue fluid. (3) On postvoiding cystoscopic examination, blue fluid effluxes from ureteral openings. (From AMAR, 1964a)

A duplicated ureter that appears radiographically and urographically to be incomplete may be complete. When the clinical manifestations cannot be satisfactorily explained on the basis of incomplete duplication, further investigation should be made, including, if necessary, cystoscopy with retrograde catheterization, and attempts to determine whether reflux is present.

Finally, in every patient with unilateral ureteral duplication, the probability of bilateral duplication is greater than in the general population, and a search for duplication of the contralateral ureter is mandatory.

In summary, one should attempt to demonstrate duplication if any of the following are encountered:

(1) Repeated, unexplained urinary tract infection.

(2) An oddly situated urethral diverticulum.

(3) Failure of the two contralateral kidneys to show a "mirror image" of one another on the urogram.

(4) Unilateral duplication.

The order of procedure recommended for diagnosis is:

(1) Personal history.

(2) Family history.

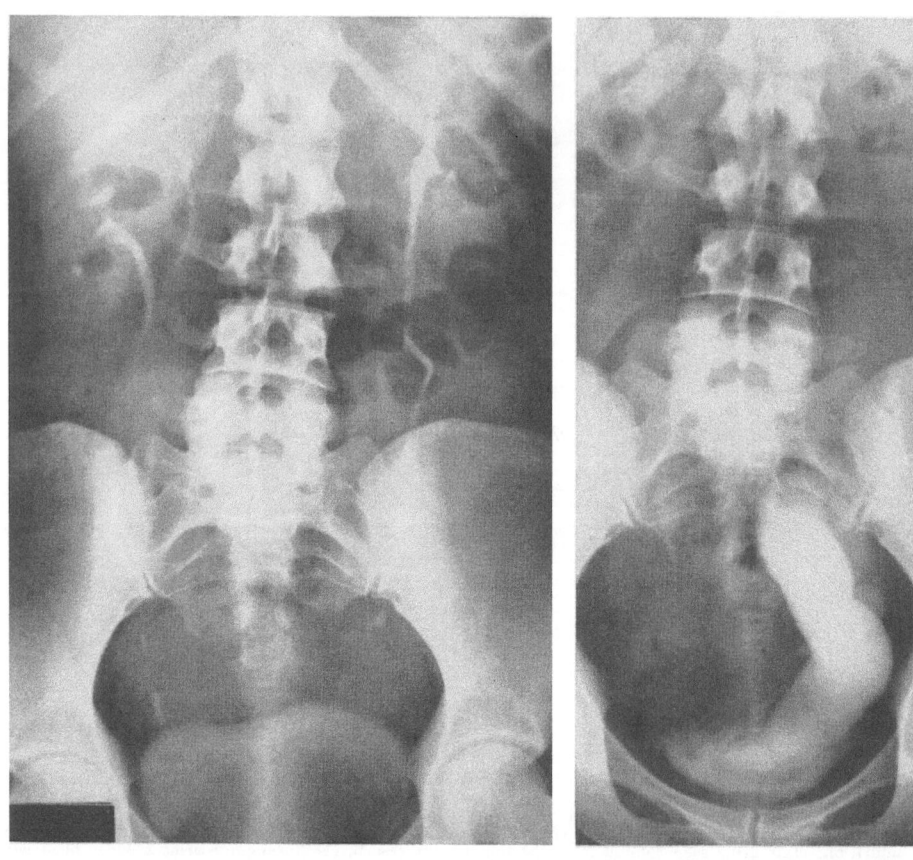

a   b

Fig. 16a and b. 28-Year-old woman who had had toxemia and pyelonephritis during recent pregnancy. (a) Excretory urogram: Right ureteral duplication demonstrated; left ureteral duplication unsuspected. Cystourethroscopic examination was unrewarding. On indigo carmine test, blue fluid exuded from an opening in the proximal urethra, which had been hidden by mucosal folds at time of urethroscopy. (b) Retrograde catheterization of this opening proved left duplication with markedly dilated upper segment ureter. (From AMAR, 1966b)

(3) Physical examination, including inspection of female external genitalia, rectum in both sexes, and palpation of the abdomen for masses.

(4) Urinalysis.

(5) Excretory urography, double-dose technique.

(6) Cystoscopy including dual dye technique and indigo carmine test.

(7) Vaginoscopy, urethroscopy.

(8) Retrograde urography; voiding cystourethrography with or without cineradiographic control; retrograde urethrography.

c) Principles of Management

There is no set plan for the treatment of all cases of ureteral duplication. One must consider the problem as it exists in the patient, and plan the management accordingly. The necessary first step is to heighten one's awareness of the possibility and to establish the diagnosis. The plan of treatment is usually clearly suggested once the extent of the problem is known. Factors to be considered are obstruction, infection, calculi, presence or absence of associated reflux, and the status of the total urinary tract, including renal function. The treatment of associated ureterocele and ectopia is discussed under the appropriate headings.

The treatment of ureteral duplication is aimed primarily at the elimination of urinary infection. Stasis of urine fosters infection; therefore, when infection cannot be controlled by chemotherapy directed against the separate attack, the anomaly that is producing the stasis must, if possible, be corrected.

α) Incomplete Ureteral Duplication

Incomplete ureteral duplication may produce much, little, or no urinary stasis depending on the amount of ureteroureteral reflux. The duplicated ureter of normal caliber with the junction of the Y in its superior third, produces only slight reflux and is not usually a source of major infection. If the junction of the Y is in the lower part of the ureter, a greater quantity of urine tends to collect at this junction. If ureteroureteral reflux is present, it will foster infection. If urinary infection is a frequent, persistent problem that cannot be controlled by other means, one limb of the partially duplicated ureter, together with the renal segment (usually the upper) to which it is attached, must be removed surgically. As LENAGHAN has pointed out, if the stem of a duplicate ureter with ureterectasis is dilated, surgical removal is less likely to eliminate infection. In such a case, although ureterectomy does away with ureteroureteral reflux, a difference in caliber persists between the upper and lower portions of the resultant tube, and causes relative obstruction.

GIBSON's technique (described in the following section on treatment of complete duplication) may be useful in the therapy of incomplete duplication.

When vesicoureteral reflux is the cause of stasis and infection, the patient is instructed to void two or three times at short intervals, on each occasion of need.

Antimicrobial therapy is given concurrently. LENAGHAN has suggested that if this regimen is continued for years, vesicoureteral reflux may disappear, but ureteroureteral reflux is likely to persist.

β) Complete Ureteral Duplication

Complete ureteral duplication is not amenable to treatment by removal of the involved renal segment. Either GIBSON's technique or ureteroureterostomy may be helpful in individual situations. GIBSON in 1957 (and SANDEGÅRD in 1958), in order to preserve a normal-appearing renal parenchyma in a 17-year-old girl incontinent since birth, with anomalous renal pedicle which entered the kidney near the upper pole and so rendered partial nephrectomy infeasible, removed as much of the ectopic ureter as could be reached, then anastomosed the upper pelvis to the lower pelvis (Fig. 17). At the time of GIBSON's report, the patient had remained continent and free of infection for longer than one year. The most commonly performed surgical treatment is removal of the upper renal segment and its attached ureter (upper segment nephroureterostomy). During this procedure, the ureter belonging to the lower segment is particularly vulnerable to injury, because of the common sheath and blood supply of the two ureters. To

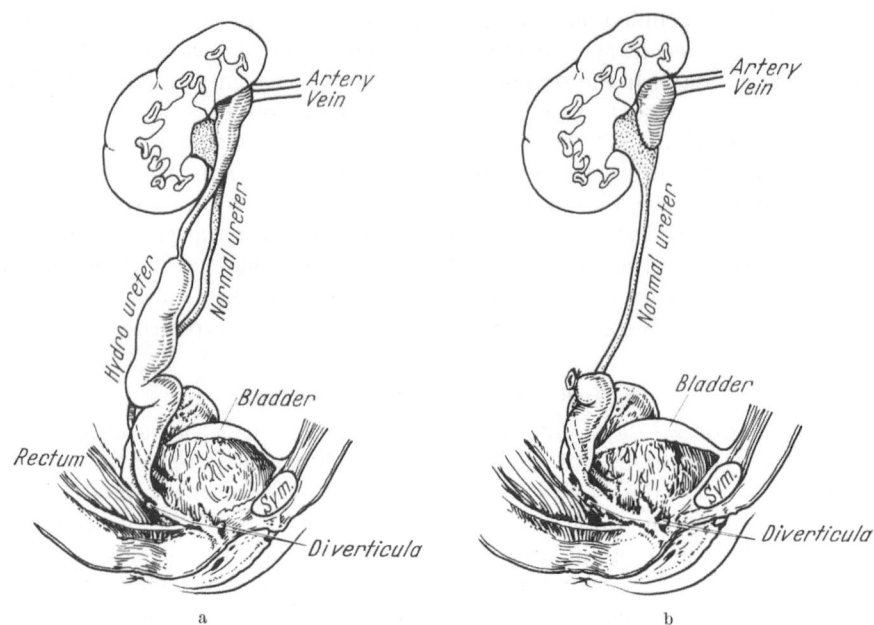

Fig. 17a and b. GIBSON'S technique for treatment of ureteral duplication. (a) Ectopic ureteral opening in vestibule. Composite sketch of findings. (b) The upper renal pelvis is anastomosed to the lower renal pelvis. Most of the ureter belonging to the upper renal segment is removed. (From GIBSON, 1957)

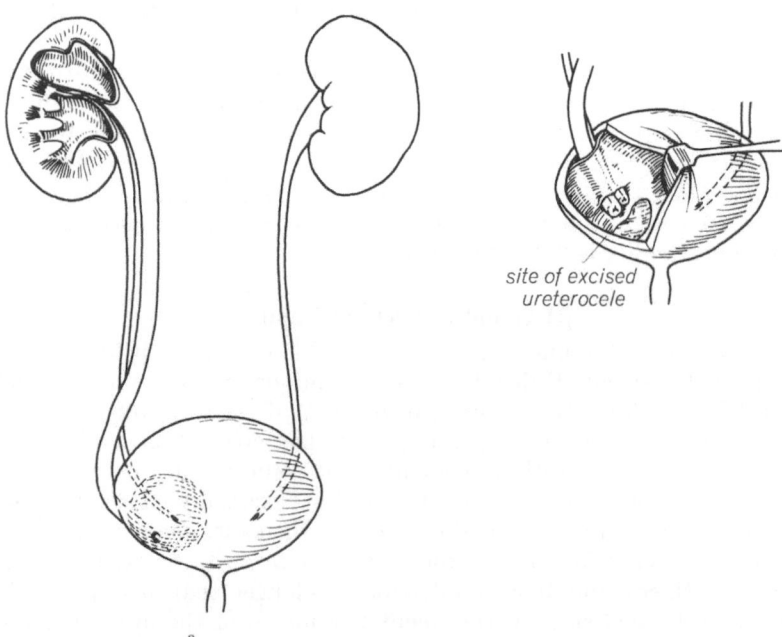

Fig. 18. PAQUIN'S technique for treatment of ureteral duplication. (a) Ureterocele associated with complete ureteral duplication. (b) The ureterocele is excised, and both ureters are reimplanted in the bladder through a common submucosal tunnel. (From PAQUIN, 1964)

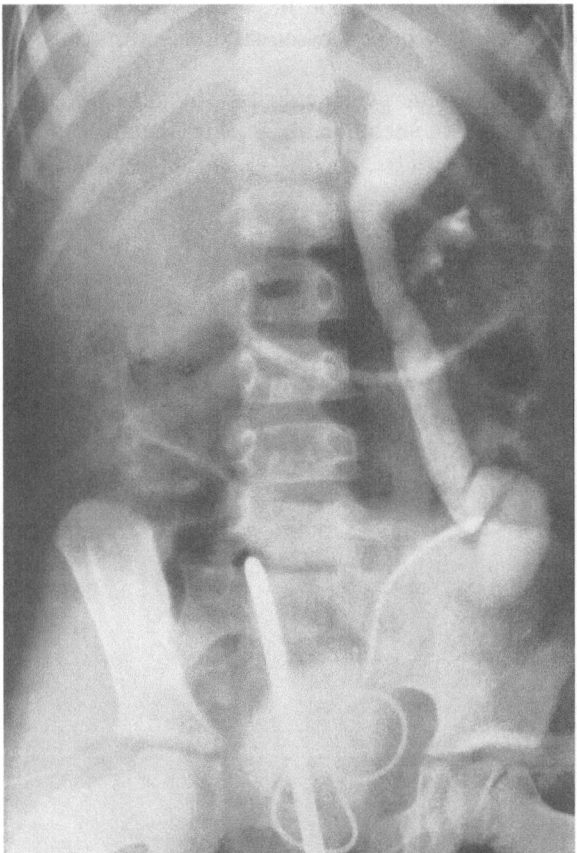

a

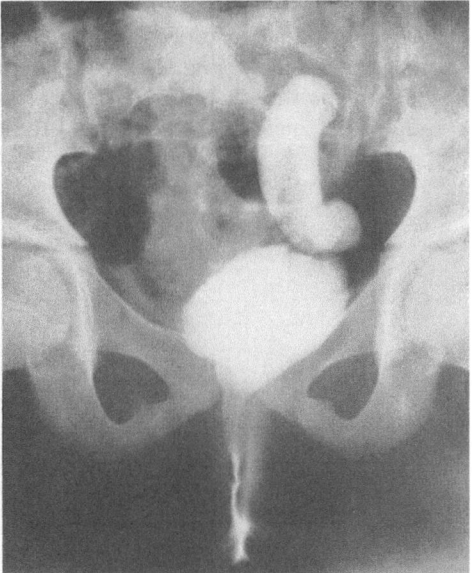

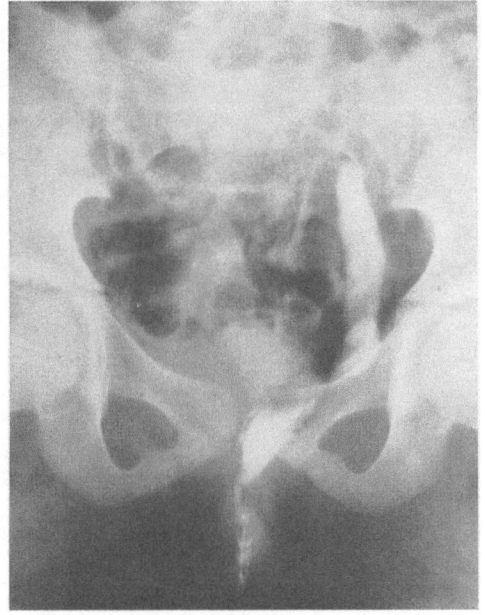

b

Fig. 19a and b. 6-Year-old girl with chronic urinary tract
infection. (a) Retrograde urogram: Marked hydronephrosis
of upper renal segment (left); wide dilatation of corresponding ureter. Lower renal segment and its ureter appear
normal. Upper renal segment and superior 17 cm. of its
ureter were removed because of hydronephrosis and pyelonephritis. (b) Six years later, voiding cystogram shows
reflux into dilated ureteral stump remaining from previous
operation. The stump had acted as a reservoir of urinary
infection, which was brought under control only after its
removal. (From AMAR, 1964c)

avoid this type of injury, PAQUIN has suggested that reflux can be prevented by
reimplanting both ureters so that each orifice is in a new site, and the distal end
of each ureter traverses a tube constructed of bladder wall mucosa (Fig. 18).
If the ureter is markedly dilated, the reimplantation procedure may have to be
preceded by preliminary nephrostomy of the dilated renal segment.

The treatment of vesicoureteral reflux associated with duplication is discussed
in the chapter on Reflux.

Following nephrectomy with or without partial ureterectomy, reflux into the remaining ureteral stump may persist, and the site may then become a reservoir for continuous or repeated urinary infection. If this occurs, removal of the stump is recommended (Fig. 19) (AMAR, 1964c).

When a lower renal segment is the seat of chronic pyelonephritis, this may be due to associated vesicoureteral reflux (Fig. 20). In such a case, if other attempts

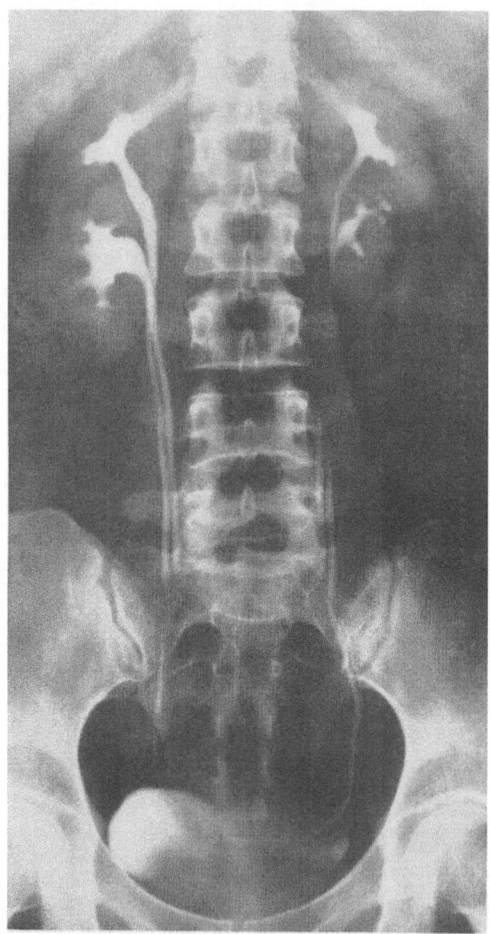

Fig. 20. 45-Year-old woman with recurrent and chronic urinary infection. Excretory urogram: Bilateral complete duplication. On the left, lower renal segment is hypoplastic, and seat of chronic pyelonephritis; its ureter was subject to vesicoureteral reflux

to control the persistent or recurrent pyelonephritis are unsuccessful, the lower renal segment and its attached ureter may have to be surgically removed; or ureteral reimplantation may be tried.

The treatment of urinary calculi (Fig. 21), cysts (Fig. 22), tumors, and other diseases of the duplicated ureter does not differ essentially from appropriate management of these diseases in the single ureter.

γ) Triplication or Quadruplication

Segmentation of the kidney and ureter into three (or more) complete or partial collecting systems (*ureteral triplication, quadruplication,* etc.) is unusual. Although

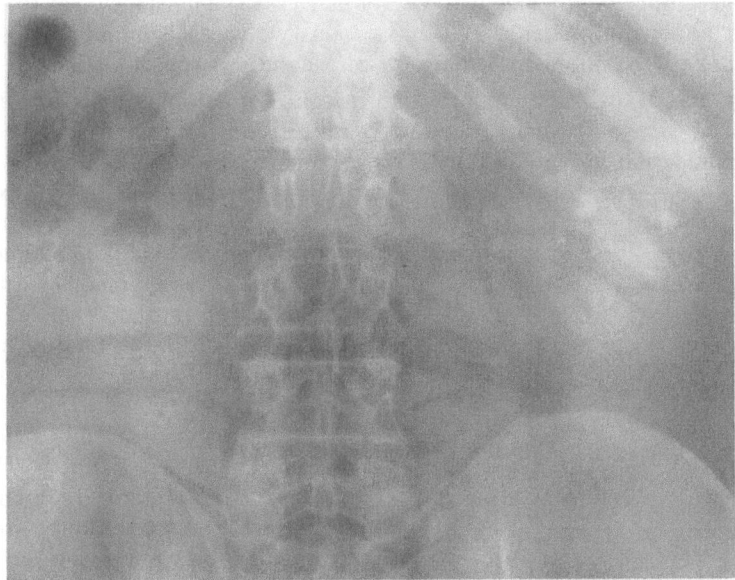

a

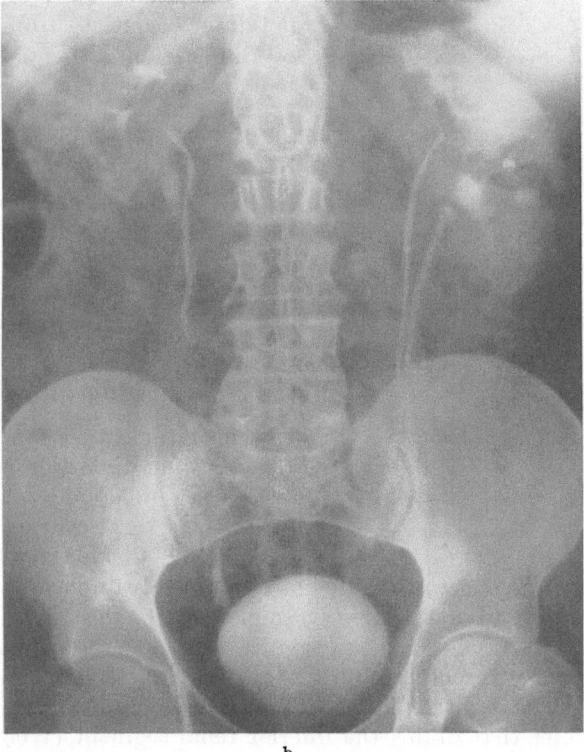

b

Fig. 21a and b. 55-Year-old man with history of chronic pyelonephritis and urinary calculi. (a) Preliminary
film: Calcifications in renal area bilaterally. (b) Excretory urogram: Changes of chronic pyelonephritis in lower
renal segments bilaterally. Vesicoureteral reflux was found in left and right lower renal segments, which contained
the calculi

fewer than twenty cases have been reported, the condition is probably somewhat less rare than this might suggest. CAMPBELL (1963) mentioned thirteen cases, three from his own experience. We have seen two cases. Undoubtedly sporadic instances have been noted by urologists who have not reported them.

RINGER and MACFARLAN reported an instance of complete triplication of the ureter in a 27-year-old woman; LIVADITIS and his coworkers observed complete ureteral triplication in a child with incontinence. In both cases, the middle ureter opened ectopically into the urethra. In the woman, who had a history of urinary infections in childhood and whose complaint was "honeymoon cystitis" and right flank pain, surgical correction was performed according to GIBSON's technique: the middle segment ureter was brought into communication with the lower renal pelvis by end-to-side anastomosis. The ureter distal to the anastomosis was not removed. In the child, pyeloureterostomy to the lower ureter was performed, and the middle ureter distal to the surgical juncture was resected. In both cases, excellent results were obtained.

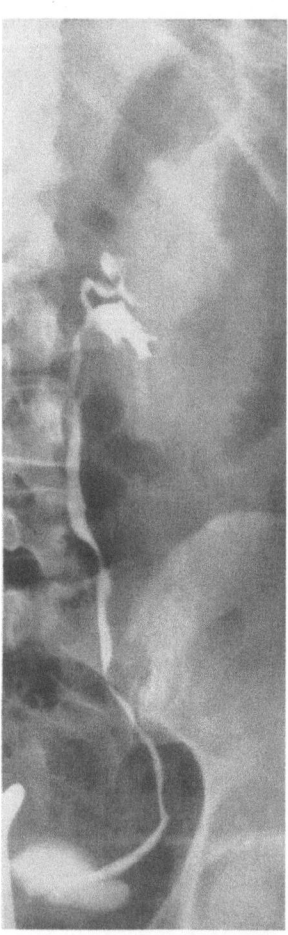

The symptoms, signs, diagnosis, and management of ureteral triplication do not differ essentially from those of ureteral duplication, although treatment may be more complex.

II. Ureteral Ectopia

Note: Because of the nature of their embryonic development, ureteral ectopia, ureteral duplication, and ureterocele are frequently associated. It will therefore be necessary to mention these anomalies occasionally during the discussion of ectopic ureter.

1. Definition and Description

The term *ectopic ureter* is applied in the following discussion to ureters whose orifice is outside the bladder — in the internal urethral orifice, the urethra, or completely outside the urinary tract.

Fig. 22. 48-Year-old woman. Retrograde urogram: On the left, a cyst is seen in the upper segment of the duplicated kidney. Proved on exploration

From 70% to 80% of ectopic ureters are completely duplicated; the remaining 20% to 30% are partially duplicated or single.

It is of medical and surgical importance that, in ureteral duplication, with rare exceptions, the orifice of the ureter that drains the superior renal segment is, as a result of its embryonic development, lower and closer to the midline than the orifice of the ureter that drains the inferior renal segment (Weigert-Meyer law; see Ureteral Duplication: Embryology). If this lower point is within the bladder wall, the associated ureter is by definition not ectopic; if it is still lower, outside the bladder, it is by definition ectopic. Because the ureter arises from the wolffian duct, its orifice in the male will, even if ectopic, be above the verumontanum and therefore proximal to the external urethral sphincter. (For the sites of ectopic ureteral orifices in men, seen below.) In the female, in whom the wolffian ducts normally regress, an ectopic ureteral orifice may be either within or outside of urinary sphincter control. A further source of clinically significant ectopia in the female is the close proximity of the persistent müllerian duct system. The ureteral orifice is occasionally incorporated during embryonic growth into the structures derived from the müllerian ducts, and the ureter may then empty into a fallopian tube, the uterus, the vagina, or its external vestibule. In either sex, the early relationship between the developing ureter and the cloaca may result in a ureteral opening into the large intestine or rectum.

2. Sites of Ectopic Orifices

The sites of the orifices of ectopic ureters that had been reported in the literature up to 1957 were tabulated in two reviews as follows:

Males	BURFORD et al., 1949 THOM, 1928	*Females*	BURFORD et al., 1949
Prostatic urethra	54%	Vestibule	38%
Seminal vesicle	28%	Urethra	32%
Vas deferens	10%	Vagina	27%
Ejaculatory duct	8%	Uterus	3%

ALLANSMITH found ectopic ureter terminating in a seminal vesicle three times as often on the left side as on the right, in autopsy specimens; and twice as often on the left as on the right in clinical cases. No explanation for this predominance was suggested. When an ectopic ureter opens into a seminal vesicle, the kidney may be absent on this side, as was true in a case observed by FARR, who in 1960, reported what he believed to be the sixteenth clinically treated case of ectopic ureteral opening into a seminal vesicle, diagnosed as a cyst of the seminal vesicle. The associated ureter had no attached kidney. The chief manifestations were sacrococcygeal pain and epidydimitis. Other case reports of ectopic ureter entering the seminal vesicle include those of ENGEL (1948), HAMILTON and PEYTON (1950), MEISEL (1952), GOLDSTEIN and HELLER (1956), DICKINSON (1963).

3. Incidence

Reviewing the literature up to 1949, BURFORD and his colleagues found reports of 425 cases of ureteral ectopia. DESGREZ and his coworkers in 1964 reviewed 185 cases of extravesical ureteral openings in men, 87 of whom showed pathologic manifestations. The incidence is undoubtedly higher than these figures suggest, as increasing awareness of the entity and advancing diagnostic techniques will surely prove. PARKKULAINEN and REJMAN in 1961 estimated the incidence of ectopic ureter in Finland to be 1 in 80,000 live births, and 1 in 4,700 hospital admissions.

Ectopic ureter is diagnosed more often in the female than in the male, and it is possible that the absolute incidence is higher. The fact that the diagnosis is made more frequently in girls and women is, however, attributable to the typical incontinence in the female, and to the ease with which the female urinary tract becomes infected, since either of these situations leads to investigation.

Bilateral single ectopic ureter, a relatively rare entity, was reported by PARKKULAINEN and REJMAN in 3 of their 17 patients with ectopic ureters; MOORE noted that 6 cases had been published prior to his report of an additional case, in 1952.

4. Associated Anomalies

Twenty-seven of the 45 cases of ureteral ectopia observed by KJELLBERG and his colleagues were associated with ureterocele. Ten of the remaining 18 ectopic ureters without ureterocele (14 in girls, 4 in boys) were completely duplicated; all of these 10 were in girls, and in each of the 10 duplications the ectopic ureter was that which belonged to the upper renal segment. Other congenital anomalies, such as posterior urethral valve, may also be associated with ureteral ectopia.

In the female, if two nonduplicated ureters open into the urethra, so that the bladder contains no trigone, the mechanism for sphincter control does not exist. Such a case has been reported by CHUN and BRAGA; two additional cases were seen by COX and HUTCH.

5. Clinical Features

Ureteral ectopia, like a variety of other major anomalies of the urinary tract, does not necessarily cause symptoms. Two of five patients with ectopic ureter reported by H.T. THOMPSON had no clinical manifestations such as recurrent pyuria, fever, or interference with general good health; in the remaining three cases the anomaly was discovered during search for the cause of recurrent pyelonephritis. The urine may show no evidence of pyuria or infection at the time the patient is examined.

The clinical manifestations that do occur have their basis in either of two types of anatomic situation: incontinence, or stasis and back pressure due to obstruction of urinary flow. Incontinence and obstruction may, however, be concomitant. A third clinical manifestation may be urinary infection added to incontinence or obstruction, or both.

a) Incontinence

In the male, ureteral ectopia does not cause incontinence of urine. In the female, the ectopic ureteral orifice may be situated outside the control of the external sphincter; if so, incontinence is likely to result. The type of incontinence that is characteristic of ureteral ectopia in the female is continual dribbling of urine. Incontinence of this type was noted in 8 of 14 girls with ureteral ectopia studied by KJELLBERG et al.; 5 of the 8 had been treated elsewhere for enuresis.

Where incontinence is due to unilateral ectopia, the bladder may fill normally with urine from the contralateral ureter. In the female, if there is no other abnormality, normal voiding and continuous leakage of urine may therefore coexist. When ureteral ectopia is associated with ureteral duplication, a similar situation can exist because the bladder fills with urine from the lower renal segment ureter, which opens into the bladder.

In patients with dilatation of the lower end of the ectopic ureter, there may be orthostatic incontinence, since the dilated portion of the ureter may act as a reservoir while the patient is recumbent. When the patient sits or stands, sudden leakage may occur as the reservoir empties. In an excellent study of the clinical manifestations of ectopia, LANE pointed out that in two of four cases observed by him, leakage occurred only during the day. He also noted that an ectopic orifice within the vagina may cause purulent vaginitis, especially in an infant; and that leakage of urine may be recognized by excoriation of the inner aspects of the thighs. Rarely, incontinence does not arise even though the orifice is in the vulva or the urethra (outside urethral sphincter control), until some event causes a shift in anatomic relationships. Thus, a 20-year-old woman reported by THOMPSON and AMAR was not incontinent until after the delivery of her first child; two extravesical ureteral orifices were then found. Similarly, a 23-year-old woman reported by LEVACK had never been incontinent until her first pregnancy. A tortuous, dilated, duplicated ureter was then observed to open into the vulva. In such cases, it is possible that incontinence was prevented earlier by obstruction and possibly compression of the ureter.

b) Obstruction and Dilatation

The anatomic situation of the ectopic orifice is quite likely to cause urinary obstruction and stasis, with their attendant high incidence of infection. An ectopic

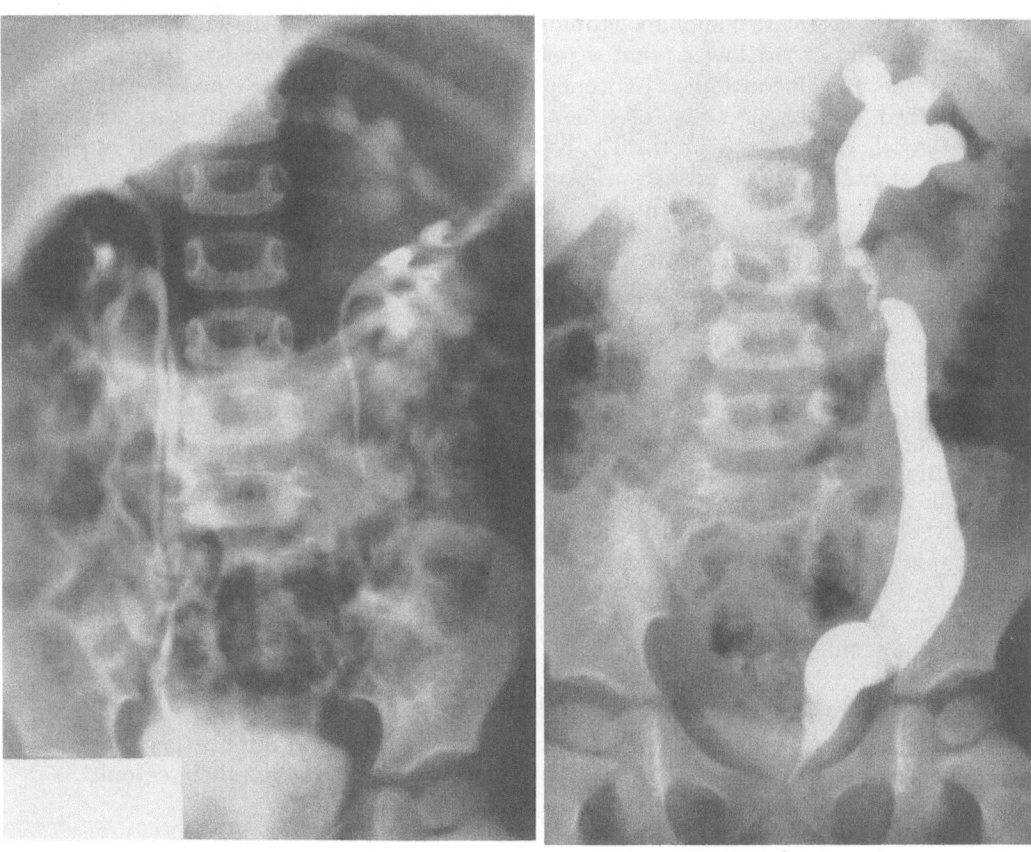

a   b

Fig. 23a and b. 4-Year-old girl with recurrent urinary infection. (a) Excretory urogram: Bilateral ureteral duplication. Function of left upper renal segment is impaired and its ureter is not outlined. (b) Voiding cystogram: Reflux into dilated upper renal segment and its ureter; ectopic ureter opens into proximal urethra

ureter is usually dilated, particularly at its inferior end, and the function of the corresponding kidney is impaired (Fig. 23). Where the ectopic orifice is in the urethra, ureteral dilatation may be due partly to obstruction of the ureter as it passes through the sphincter apparatus, and partly to reflux from the urethra into the ectopic ureter, which may occur only during voiding. Such reflux is often not easily demonstrable, but investigation for it must be adequately pursued. Definite stenosis of the ectopic orifice has been observed in a few cases.

In patients with duplication and associated ureteral ectopia, the upper renal segment, to which the ectopic ureter is attached, is often small, hydronephrotic, and infected (LANE). DESGREZ and his coworkers, in their review of 185 extravesical ureteral openings in men, noted that the kidney with a single ureter was usually hypoplastic and in some instances was ectopic, absent, or fused. Like LANE, they observed that the superior segment of a kidney with ureteral duplication was subject to infection, and that it generally presented a hydronephrotic or pyonephritic mass with flaccid walls. The ectopic ureter was dilated, and in some cases reached the caliber of an intestinal loop. In half of the patients studied by them, excretory urography failed to reveal the anomaly directly because the associated kidney was hypoplastic and nonfunctioning; or, where there was duplication of the collecting system, the superior renal portion did not excrete sufficient contrast medium to afford visualization of collecting structures distended by urinary stasis. Dilatation of an ectopic duplicated ureter may also be accompanied by dilatation of the lower renal segment ureter with intravesical orifice, on the same side. In some instances hydronephrosis is bilateral, and may occasionally be explained by the presence of other associated congenital anomalies such as ureterocele or posterior urethral valves. In some instances the lower end of the ureter showed a cystic dilatation, constituting an ectopic ureterocele. In 10% of male children with ectopic ureter, compression of the vesical neck resulted in dysuria, or in acute or chronic retention with distention of the bladder. Stones, occasionally multiple and bilateral, were observed in 12% of the patients, and were associated with stasis and infection.

c) Infection

The importance of urinary tract infection as a sign of ureteral ectopia has been underestimated. The symptoms of recurrent or chronic urinary infection are usually the most prominent among the complaints that cause the patient to seek medical aid. Thirteen of 17 patients with ureteral ectopia reported by PARKKU-LAINEN and REJMAN named recurrent or chronic urinary infection as their chief complaint.

In the male, urinary infection is the principal manifestation of ureteral ectopia. Epidydimitis may be the presenting symptom when the ectopic orifice is in the seminal vesicle.

6. Diagnosis

Ureteral ectopia is so commonly associated with ureteral duplication that all investigations pertinent to duplication must be carried out when ectopia is considered (see Ureteral Duplication: Diagnosis).

a) History

In the female, the diagnosis of ureteral ectopia is suggested by a history of incontinence, recurrent urinary infection with or without pyelonephritis, or a combination of incontinence and urinary infection.

In the male, the presenting symptoms may be urinary infection, sterility, pollakiuria, hematuria, epidydimitis, low backache, or sacrococcygeal pain.

b) Family History

A family history of ureteral ectopia or ureteral duplication suggests the possibility that ureteral ectopia may exist in the patient under study for symptoms mentioned above.

c) Physical Examination

In the female, physical examination must include a thorough inspection under a bright light of the external genitalia, including vaginoscopy. In the male, it must include rectal examination for a seminal bulge or cyst, above the prostate.

d) Urinalysis

Pyuria in a clean-voided specimen, from a patient in whom the catheter specimen shows no pyuria, strongly suggests the possibility of an ectopic ureteral orifice.

e) Urographic Procedures

All investigations and special procedures described under Ureteral Duplication are appropriate to the diagnosis of ureteral ectopia with or without ureteral duplication. When ectopia is suspected, it is essential to follow the course of the ureters, especially that of the upper renal segment, to their termination.

α) Excretory Urography

In the experience of other authors, the findings at excretory urography have been of limited diagnostic value because the renal segment to which the ectopic ureter is attached is usually nonfunctioning. The double-dose technique (AMAR, 1964 b) has improved the value of this examination. Duplication of the upper portion of the urinary tract can be clearly demonstrated by the double-dose technique in some patients, but in others it may not be possible to establish by this means that the supernumerary ureter is ectopic.

β) Micturition Urethrocystography

Micturition urethrocystography has been of great value in our experience and in that of others because, in the majority of cases in which the ectopic opening is into the urethra or the internal urethral orifice, urethro-ureteral reflux occurs during micturition, allowing visualization of the ectopic ureter and of the associated renal pelvis. Such reflux was observed in all of five patients with an ectopic opening in the urethra, observed by us, and in 8 of 10 patients with similar ectopia noted by KJELLBERG et al. Where the ureter had no connection with the urethra, the micturition views showed no abnormality, and search for ectopic opening outside the urethra led to discovery of the ectopic orifice.

γ) Vasography

Vasography in the male will demonstrate an enlarged seminal vesicle with a cyst and may help to show an ectopic ureteral orifice opening into the seminal vesicle.

f) Endoscopy

On cystoscopic examination, a ridge-like elevation of the mucosa may be noted in the trigonal area on the involved side. Such an elevation is caused by the presence of the submucosal portion of the ureter as it transcends the bladder wall toward its ectopic orifice, whether that orifice is within or outside of the urethra. Where a nonduplicated ureter is ectopic, there is no distinct trigone in the corresponding half of the bladder. Where both of two bilaterally single ureters are ectopic, the trigone is completely missing. Where ectopia of one ureter is combined with complete ureteral duplication, a normal trigone is seen.

Visualization of the ureteral orifice may be facilitated by intravenous injection of indigo carmine; however, since the function of the corresponding renal segment is likely to be poor, this test may fail to produce the desired effect. We have found helpful a test recently described by AMAR (1964a) in which the bladder is filled with blue fluid (2, 5-cc. ampules of indigo carmine in 200 cc. of sterile water); then cystourethroscopic examination is performed immediately after voiding. If urethroureteral reflux is present, blue fluid will have entered the ureter during voiding, and can be seen effluxing from the ectopic ureteral opening.

An orifice in the urethra can be observed in most instances through the urethroscope. It usually lies in the posterior urethral wall in the midline; but mucosal folds may cover and hide the opening.

In the male: It is difficult to find an ectopic ureteral orifice in the ejaculatory duct or seminal vesicle. A projection may be seen on the posterior bladder wall, caused by a seminal vesicle cyst which may need to be unroofed to establish the diagnosis. If the ectopic orifice is in the prostatic urethra, it is well to catheterize this opening in order to determine whether the ectopia is associated with ureterocele. If ureterocele is present, it may be the cause of bladder neck obstruction. Retrograde urethrography and micturition cystourethrography will help to establish the diagnosis provided urethroureteral reflux outlines the ectopic ureter.

In the female: If the ectopic ureteral opening is in the vestibule, it is usually seen on inspection of this area; but location of an ectopic ureteral orifice in the uterus or vagina presents special problems. The ectopic ureter may present as a cyst in the vaginal wall. Needle puncture of the cyst, and filling it with contrast medium, may reveal its nature. If the ectopic orifice can be identified on inspection, an attempt should be made to introduce a catheter into it. This cannot be done in every case. When it can be achieved, the ureter and renal pelvis may easily be filled with contrast medium and depicted in their entirety. Reflux of contrast medium from the vagina has permitted detection of an abnormal ureteral opening into the vagina (CENDRON; KATZEN and TRACHTMAN).

The dual-dye technique (AMAR, 1966b) (Fig. 14) may be employed to determine whether there is an ectopic orifice in the urethra distal to the urethral sphincter or in the vagina or vestibule (see Ureteral Duplication). If a urethral diverticulum appears to be present, it should be unroofed and catheterized; contrast medium should then be injected. The "diverticulum" may prove to be an ectopic ureter.

In some cases, surgical exploration may be necessary to establish the diagnosis of ectopic ureter, at which time excision of the ectopic segment is performed as treatment.

7. Treatment

An ectopic ureter must be corrected surgically. The objectives of treatment are: (1) to abolish incontinence, (2) to eradicate infection, and (3) to eliminate a dilated ureteral stump, since such a stump with or without reflux may become a reservoir of infection (AMAR, 1964c).

For single ectopic ureter, ureterovesical implantation — so performed as to prevent future reflux — may be selected, provided the function of the kidney to which the ectopic ureter is attached is sufficiently good to make it worth preserving. PAQUIN's (1964) operation is the standard procedure for this situation.

However, the incontinence that results from absence of urethral sphincter control in bilateral ectopic nonduplicated ureters (CHUN and BRAGA; COX and HUTCH) cannot be cured by reimplantation of the ureters into the bladder, since the reason for the absence of sphincter control (nonexistence of the vesical trigone) would not be altered by repositioning of the ureters. If the function of a kidney drained by a single ectopic ureter is poor while that of the contralateral kidney is good, nephrectomy on the involved side is indicated. If vesicoureteral or urethroureteral reflux is demonstrated preoperatively on the side of ectopia, nephroureterectomy is necessary in order to prevent the refluxing ureter from acting as a reservoir of infection at a later date (AMAR, 1964c). If there is obstruction at the distal end of the ureter, which prevents proper drainage, then whether reflux is present or not ureteronephrectomy must be performed. Where the ureteral orifice lies in the urethra, and there is no incontinence, a large meatotomy may be performed to improve drainage of this ureter. Reflux may occur postoperatively. If infection supervenes, further treatment is required.

For completely duplicated ureter, if the function of the associated renal segment is poor, ureteroheminephrectomy is the procedure of choice. If the function of the associated renal segment is good and the contralateral kidney is defective, the recommended procedure is excision of the distal part of the ectopic ureter followed by anastomosis between the renal pelves and/or ureters, according to the technique of GIBSON.

III. Ureterocele
1. Definition, Description, Types

A ureterocele is a congenital or acquired cystic dilatation of the distal end of the ureter. The shape, characteristic of the majority of ureteroceles in adults, is often described as "cobra head" or "spring onion". Within recent years, comprehensive articles on ureterocele have been published by ERICSSON; by CAMPBELL (1951c); GROSS and CLATWORTHY; KJELLBERG and his associates; USON, LATTIMER and MELICOW; WERSHUB and KIRWIN, and by THOMPSON and KELALIS.

A ureterocele may be either *simple* or *ectopic*. Either a simple or an ectopic ureterocele may arise in a single or a duplicated ureter.

A *simple ureterocele* is one that (a) arises in a ureter whose orifice opens into the urinary bladder in a normal or nearly normal position, and (b) is entirely contained within the bladder. Most but not all simple ureteroceles are small (Fig. 24). A small simple ureterocele is one that is small in relation to the bladder lumen. It causes only a mild degree of ureteral obstruction and never obstructs the bladder neck. A large simple ureterocele is large in relation to the bladder lumen; it may fill a major portion of the bladder cavity (Fig. 25). Such a ureterocele may obstruct the ureter(s) on one or both sides and may also obstruct the bladder neck. Approximately 75% of large simple ureteroceles are associated with ureteral duplication.

An ectopic ureterocele, first described as a distinct entity by ERICSSON in 1954, arises in a ureter that has an ectopic orifice situated in or extending into the urethra. The frequent association of ectopic ureterocele with ureteral duplication in infants and children (Fig. 26) has led most authors to agree that this type is almost certainly congenital.

An ectopic ureterocele is invariably large in relation to the bladder lumen (Fig. 27). Its orifice may be round like that of a normal ureter (except that it opens into the urethra instead of into the bladder); or it may be a slit, as much as

9*

a centimeter in length, and may involve the bladder neck and the proximal urethra.

The orifice of an ectopic ureterocele is always proximal to the external urethral sphincter. This location accounts for the fact that incontinence does not occur even though the ureter is, by definition, ectopic. Occasionally a small, round orifice well within the bladder lumen cannot be found on endoscopic examination, so that the ureterocele appears to be a blind sac. In an infant or

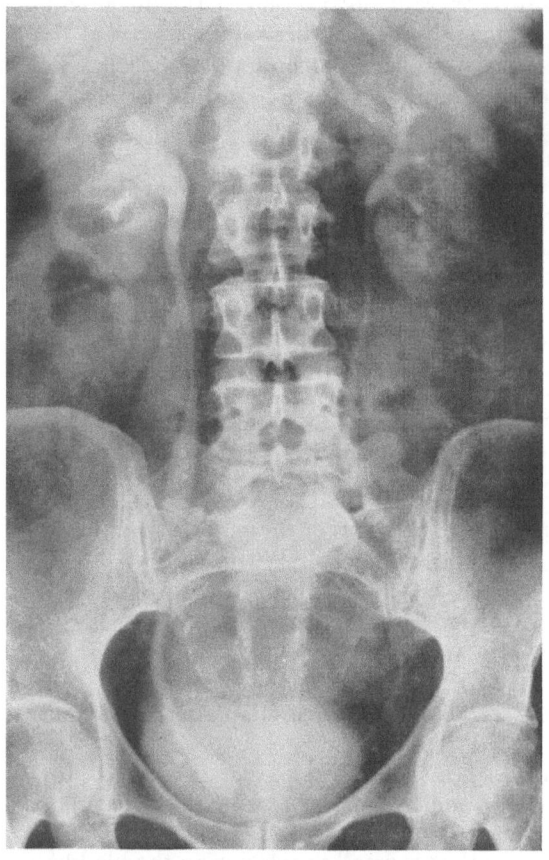

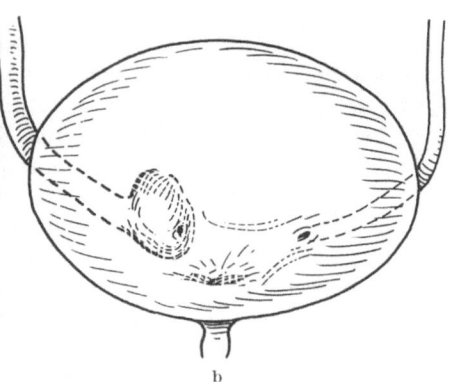

Fig. 24a and b. 37-Year-old woman. (a) Excretory urogram: Small ureterocele on right. (b) Schema of a small simple ureterocele. [Redrawn from THORNBURY, J. R.: Amer. J. Roentgenol. *90*, 15 (1963)]

small child, such an anatomic configuration makes it difficult to demonstrate the orifice even at operation.

Identification of the type of ureterocele may be aided in such cases by noting the configuration of its base. That of a simple ureterocele is usually a pedicle; that of an ectopic ureterocele is broad and extends into the urethra.

Both the size and the shape of an ectopic ureterocele depend largely on the course that the ureter traverses through the wall of the urinary bladder. If the intramural portion of the ureter is submucosal, its intravesical part may protrude as a large, typical ectopic ureterocele; if its intramural portion is more peripheral, away from the bladder lumen, the inner layers of the bladder wall offer greater resistance to intravesical protrusion. The ureterocele then presents only a shallow bulge within the bladder lumen.

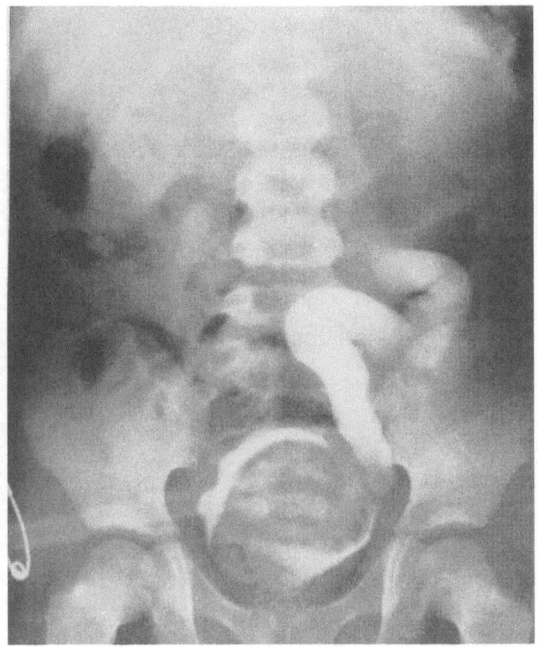

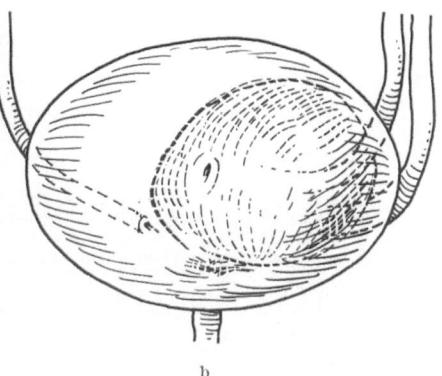

b

Fig. 25a and b. 8-Year-old girl. (a) Retrograde cystogram: Large simple ureterocele almost fills bladder lumen, causing bladder neck obstruction which necessitated suprapubic cystostomy. Vesicoureteral reflux outlines the ureter of the lower renal segment (left); ureterocele belonged to upper renal segment. (b) Schema of a large simple ureterocele. (Redrawn from THORNBURY, 1963)

a

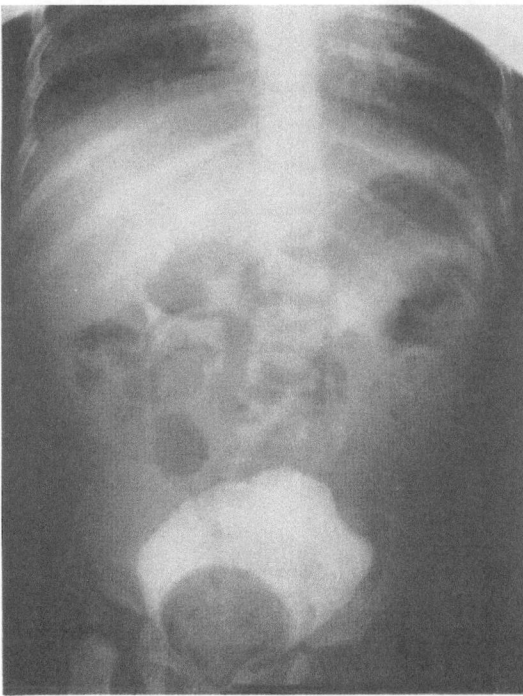

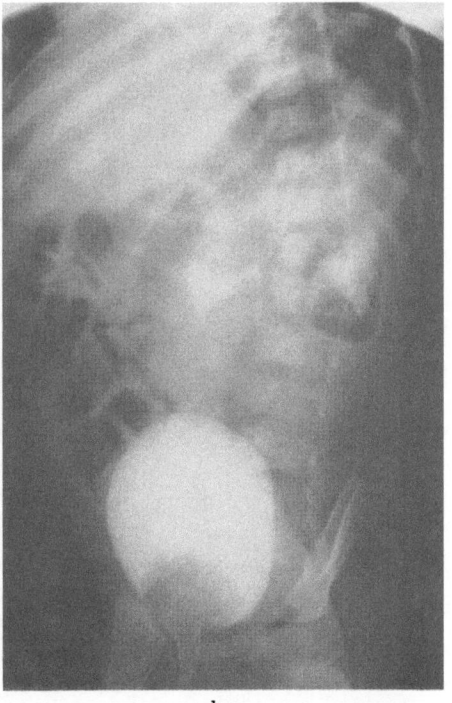

a   b

Fig. 26a and b. 2-Month-old infant. (a) Excretory urogram, A/P view: Large ureterocele in bladder. Right duplication is suggested by the incomplete collecting structures on the right; upper renal segment is not visualized. Left side is normal. (b) Oblique view again shows the lower renal segment with typical incompleteness of collecting system

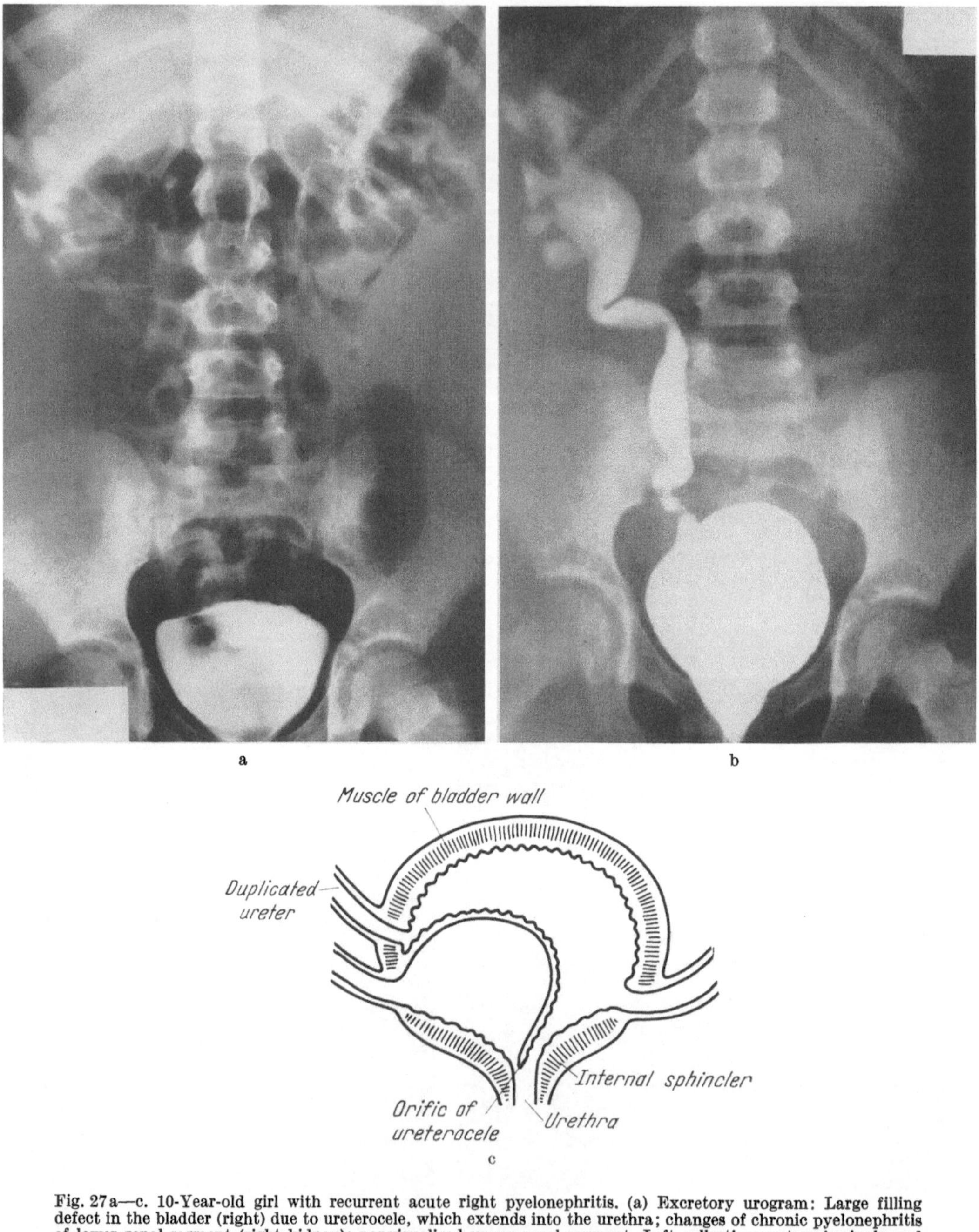

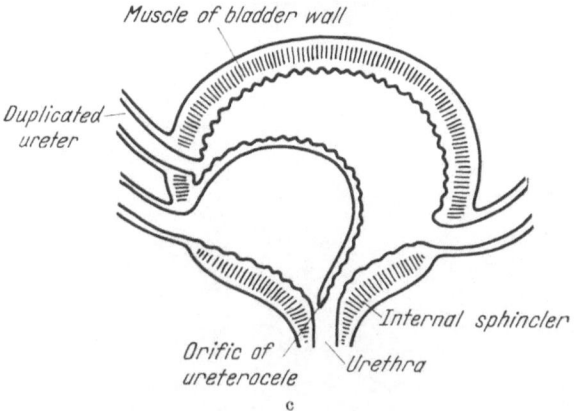

Fig. 27a—c. 10-Year-old girl with recurrent acute right pyelonephritis. (a) Excretory urogram: Large filling defect in the bladder (right) due to ureterocele, which extends into the urethra; changes of chronic pyelonephritis of lower renal segment (right kidney); nonvisualized upper renal segment. Left collecting system is single and normal. (b) Voiding cystogram: Reflux into lower renal segment and its ureter (right); wide dilatation of renal pelvis and ureter. (c) Schema of anatomic relationships in an ectopic ureterocele. (Redrawn from THORNBURY, 1963)

A ureterocele does not occur in a ureter that has an ectopic opening outside the urinary tract, as in the vestibule or the vagina, since such a ureter lacks the intimate relationship to the bladder wall that is requisite to the formation of a ureterocele.

2. Incidence

Ureteroceles of all types are more common in the female than in the male. Although the phenomenon is of greater clinical significance in infants and children than in adults, the patient is usually an adult at the time the diagnosis is made.

Simple ureteroceles are seen more often in adults than in children; they are, in fact, rather common in adults, but only 10% of the ureteroceles seen in children are of the small, simple type.

Ectopic ureterocele is more frequently seen in children than in adults. The incidence of this type in females far exceeds that in males, ranging as high as 13:1 in the series reported by ERICSSON.

Among 33 ureteroceles noted in infants and children by KJELLBERG and his coworkers, 28 were ectopic; 5 simple. In one of the patients with ectopic ureterocele, this anomaly was present bilaterally; in the remaining patients the distribution with respect to right or left side was even.

3. Pathogenesis

The cause of ureterocele is not known. Many authors have believed that its chief cause is probably stenosis of the ureteral orifice; but others do not share this view (HELLSTRÖM; HIGGINS and coworkers; JUHL; PETILLO).

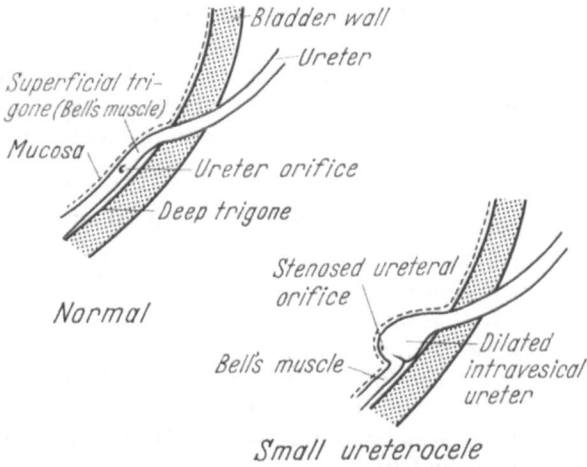

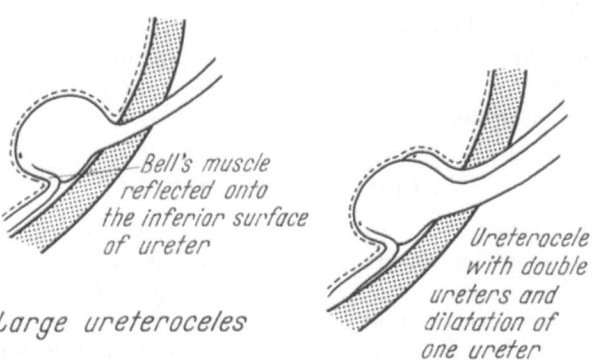

Fig. 28. Schema elucidating the boundaries of small and large ureterocele. (From HUTCH and CHISHOLM)

The position of the orifice in the bladder may be of some consequence to the formation of a simple ureterocele. The fact that simple ureteroceles are seen more often in adults than in children has led some observers to consider that acquired factors may contribute to their pathogenesis.

The following outline of the probable sequence of events in the formation of a simple ureterocele is based on the assumption that a congenital ureterocele results from obstruction at the ureteral orifice. Obstruction at that point would cause a rise in hydrostatic pressure within the immediately adjacent portion of the ureter, which is within the bladder wall. The increased pressure would result in a tendency of the intravesical ureter to dilate.

The intravesical portion of the ureter has four boundaries: ventrally, the vesical mucosa; dorsally, that portion of the detrusor muscle which lies posterior to the intravesical ureter; superiorly, the ureteral hiatus, and inferiorly, the trigone (Fig. 28). Of these tissues, the bladder mucosa is the least resistant to pressure; therefore, dilatation of the intravesical ureter in ureterocele is chiefly ventral, into the bladder lumen. In addition, a lesser degree of dilatation occurs inferiorly because the superior half of the superficial trigone (Bell's muscle), which is immediately inferior to the ureteral orifice, is only loosely attached to the deep trigone. As the ureterocele expands, it lifts the upper half of the superficial trigone away from the deep trigone and reflects it ventrally onto the inferior surface of the ureterocele. The ureteral orifice also migrates ventrally. Its downward excursion is limited by the fact that the deep and superficial trigones are fused below a point midway between the ureteral orifice and the bladder neck. As a result of this limitation to the available free length of the superficial trigone, the ureteral orifice in a simple ureterocele (usually in the inferior wall of the ureterocele) is separated by a distance of 1—2 cm from the inner surface of the bladder.

4. Associated Anomalies and Disease

Duplication of the kidney and ureter is very frequently associated with ectopic ureterocele. The upper urinary tract was completely duplicated in 27 of the 28 children with ectopic ureterocele noted by Kjellberg and his coauthors; in 20 of the 27 cases the duplication was on the side of the ureterocele. The duplication was bilateral in 12 cases. The kidney and ureter were duplicated in two of five patients with simple ureterocele. In patients with ectopic ureterocele, the upper portion of the ipsilateral urinary system was invariably dilated, and dilatation of the lower renal pelvis and its ureter on the side of a simple ureterocele was common. In many cases, abnormality was present in other parts of the urinary tract. Hydronephrosis and hydroureter on the contralateral side were not unusual. In some cases there was considerable reduction in renal parenchyma. Bilateral hydronephrosis caused by unilateral ectopic ureterocele was seen in four of the 28 patients. Stones may be found in the kidney or at any point in the ureter including the ureterocele itself.

5. Clinical Manifestations

The clinical manifestations of ureterocele result from obstruction of the ureter, bladder neck, or even the urethra, depending on the type and size of the lesion. Small ureteroceles may cause no symptoms or signs and are at times discovered incidentally during investigation of the urinary tract for unrelated disease. Ureterocele may be discovered during study of duplicated ureter and kidney; it is frequently found in the upper renal segment ureter.

Recurrent urinary tract infection with or without urinary calculi is the most frequent presenting manifestation. Other signs and symptoms include inability to void, interruption of urinary stream, urinary calculi, hematuria. With advanced obstruction of the bladder neck there may be uremia. In the female, prolapse of the ureterocele through the external urethral orifice occurs in rare instances. In two of our patients, a boy and a girl, ureterocele was found during investigation for abdominal pain of undetermined origin. Pain during micturition may be a presenting symptom. Localization of pain in the flank is due to increased intra-ureteral pressure caused by compression of the ureterocele transmitted throughout the ureter, during micturition; or by associated reflux.

6. Reflux

In simple, small ureteroceles, reflux is rare. AMAR and SCHEER (1964) have recently reported a case of simple ureterocele with associated vesicoureteral reflux (Fig. 29). When an ectopic ureterocele involves a duplicated ureter that is attached

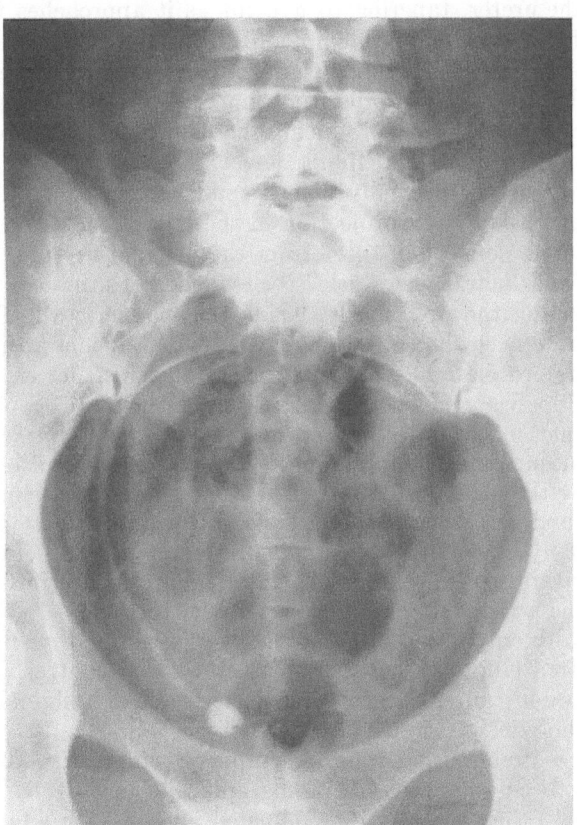

Fig. 29. Small simple ureterocele with reflux: Post-voiding cystogram shows contrast medium in right ureter and simple ureterocele. (From AMAR and SCHEER, 1964)

to the upper renal segment, reflux into the ipsilateral ureter from the lower renal segment is common (see Fig. 27). Reflux into the ectopic ureterocele and its ureter may also occur. If so, it can usually be demonstrated by delayed and voiding retrograde cystography, provided the orifice in the ureterocele is large enough to permit entrance of the contrast medium. Frequently, reflux can be demon-

strated for the first time after resection or incision of the ureterocele. There may
have been sufficient obstruction to impede function of the involved renal segment;
after correction of the obstruction, renal function in the involved kidney segment
may improve. This kidney may now be observed on excretory urography, whereas
it was not visualized before the obstruction was relieved.

7. Diagnosis

The diagnosis of ureterocele depends chiefly on urographic study and endoscopy.

a) Urography

α) Excretory Urography

Although excretory urography is the procedure of choice for the demonstration
of a ureterocele, it must be borne in mind that either a large or a small ureterocele
may not be depicted by this means. Typically, a ureterocele appears on the
ureteral portion of the excretory urogram as an irregular dilatation of the
lower third of the ureter, tapering to a point as it approaches the ureteral orifice. Such a deformity is not pathognomonic of ureterocele, since it may also be
observed in ureterovesical obstruction arising from any of several causes. In the
bladder, the dilated, cystic, terminal portion of the ureter projects inward and
impinges on the bladder lumen, producing the picture classically described as the
"cobra-head" or "spring-onion" filling defect.

A simple ureterocele, especially if it is small, is usually so demonstrable. It
appears as a smooth-bordered projection in the lateral corner of the trigone and
the contiguous area, somewhat apart from the internal urethral orifice. In both
anteroposterior and lateral views, one sees evidence of urine containing contrast
medium surrounding the ureterocele, inside the bladder wall. The depiction of
other forms of ureterocele, however, may be hampered by factors intrinsic to them.
Thus, a large ureterocele tends to impair or totally obstruct excretion from the
associated renal pelvis and so materially alters the radiographic appearance.
Such a malformation, whether simple or ectopic, then appears as a smooth-
bordered radiolucent defect seen against the intravesical pool of urine that has
been excreted by the other kidney. If the large ureterocele is simple, it generally
appears as a round or elliptical defect surrounded by contrast material. If it is
ectopic, it is usually larger than the simple ureterocele; its site of contact with the
bladder floor covers a larger area; the internal urethral orifice is likely to be
involved, and the wall of the ureterocele may lie very close to the anterior wall
of the bladder, with only a thin intervening layer of contrast medium. The ectopic
ureterocele is often elliptical or hemispherical, and no contrast medium is seen
at its base. In rare instances, large simple ureteroceles are bilateral; one then sees
heart-shaped filling defects on the bladder floor.

Although large ureteroceles, whether simple or ectopic, have been reported by
various authors to be demonstrable radiographically in 30% to 90% of children
(KJELLBERG et al.) they can rarely be depicted in adults. In adults and children,
delayed excretory roentgenograms up to two hours are often required to allow
enough contrast medium to reach the bladder to delineate the lesion. Gas within
the rectum may produce a misleading radiolucency; the distinction may be made
by oblique views of the bladder. In a number of our patients (AMAR, 1964b), the
use of a double dose of contrast medium in excretory urography has been helpful
in demonstrating ureterocele where the single dose has failed. Combining the
double-dose injection technique with delayed bladder films improves markedly
the cystographic quality (AMAR, 1965).

Demonstration of associated duplication of the ureter and kidney depends upon excretion of the contrast material by the duplicated segments; or upon non-visualization of the upper segment, the lower segment showing the typical incompleteness (Fig. 26).

β) Retrograde Urethrocystography

When excretory urography has been indeterminate, retrograde urethrocystography may lead to visualization of a ureterocele; however, this technique has a number of disadvantages. A bladder defect that can be clearly seen on excretory urography may be obscured by dense opacification in the retrograde examination. Weakness of the bladder wall at the site of the ureterocele may cause it to be pushed outward, away from the bladder cavity, as a diverticulum is pushed out during cystography. No bulge into the bladder cavity will then appear; instead, the ureterocele shadow may be seen as a pouch outside the bladder cavity. Increased intravesical pressure during cystography may, in other cases, cause the ureterocele to collapse and empty its contents into the dilated, atonic ureter above it so that the ureterocele is not demonstrated. A ureterocele is often diminished in volume, and may be completely compressed, during micturition.

If there is reflux at the ureterovesical junction, which is also the site of the ureterocele, contrast medium from the bladder may enter the ureterocele and ureter above it. This possibility is enhanced by voiding cystography. A post-voiding film, seen in Figure 29, shows a simple ureterocele and the lower portion of the ureter filled with contrast medium. On cystoscopic examination, a small simple ureterocele without associated bladder diverticulum was seen.

On retrograde urethrocystography, the filling defect in the bladder which represents a large simple or ectopic ureterocele is a smooth-bordered, radiolucent defect in the bladder that is otherwise filled with contrast medium. A large simple ureterocele generally appears as a round or elliptic defect surrounded by contrast material. An ectopic ureterocele is usually elliptic or hemispheric, without contrast material at its base.

b) Endoscopy

Cystoscopic examination is the only sure way of demonstrating a small simple ureterocele. Large simple ureteroceles are usually readily demonstrated radiographically. The chief endoscopic problem thus pertains to the ectopic ureterocele, which is most often large and so prominent that it is difficult to maneuver the cystoscope to a point from which the situation can be surveyed. A large ectopic ureterocele is less sharply demarcated from the bladder wall than is a simple or smaller cyst; in addition, it is compressible; or may collapse in response to distention of the bladder, empty its contents into the dilated ureter above, and so be easily overlooked. Rarely, a ureterocele may flatten or even evert. The best technique for overcoming these difficulties is to view the ectopic ureterocele from the posterior urethra and bladder neck, where the prominence is most distinct. In addition, it is usually possible to identify an opening in the urethra, where the orifice is often large. A visible opening into the urethra may be lacking, however. If an opening is found, it should be catheterized and retrograde urograms made to confirm the diagnosis.

A few large ureteroceles have been misdiagnosed as submucosal or extravesical tumors. This confusion is less likely to occur when the abnormality is in a child than when it is in an adult because tumors of this type are very rare in the young while ectopic ureterocele is common.

8. Treatment

a) Simple Ureterocele

Many simple ureteroceles require little therapy. Aas and Nilson studied 15 ureteroceles in 11 women and 10 ureteroceles in 7 men. The largest of these ureteroceles was about 3 cm. in diameter. The authors used conservative treatment only in 13 of the ureteroceles that did not have associated urinary calculi; the ureteroceles remained stationary. Twelve ureteroceles were treated surgically; of these 5 had no associated urinary calculi, and 3 of the 5 ureteroceles were treated by electroresection. The 7 cases of urinary calculi were treated surgically; one by transvesical electroresection; and among the 6 cases in which the transurethral approach was used, 4 were treated by electroresection, 2 by electro-

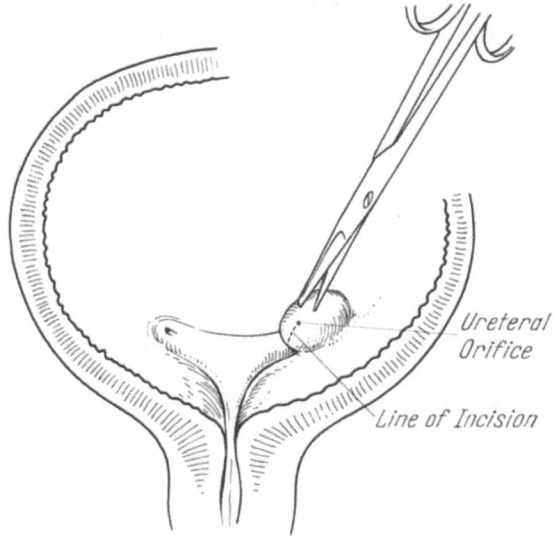

Fig. 30. Schema of inferior meatotomy. (From Hutch and Chisholm)

coagulation. In no instance had a small ureterocele led to a urographically demonstrable diminishment of kidney function. Among the 12 ureteroceles treated surgically, there was 1 recurrence, and postoperative ureterovesical reflux occurred in 1 patient. Periods of observation were from 9 months to 18 years. Thus, conservative therapy may be adequate for the small ureterocele in a number of cases. In the woman, frequent urethral dilatation may be needed; sporadic urinary infections are treated as they occur. When stones are present, they may need to be removed. It is sometimes possible to enlarge the ureteral orifice in a ureterocele with a cold scissor or a diathermy knife; but in a few patients that we have seen, such procedures have led to reflux that had not existed previously. The reflux gave rise to typical clinical manifestations, and had to be corrected by ureteral reimplantation. Transurethral electrocoagulation or transvesical incision has been used for treatment of a simple ureterocele. To obviate postoperative reflux and damage to the roof of the intravesical ureter, we have devised the following form of treatment for simple ureterocele: An inferior ureteromeatotomy is performed at open surgery, and a 6F or 7F ureteral catheter is left indwelling for a week postoperatively to keep the new orifice open. In both of the two cases so treated, at cystoscopic examination six weeks postoperatively, the enlarged meatus admitted a 7F catheter with ease. Some have advocated transurethral transverse meatotomy for the same reasons, but we feel that only inferior meatotomy should be used even though it necessitates open surgery. The technique of inferior meatotomy, as described by HUTCH and CHISHOLM, is shown diagrammatically in Fig. 30. Operation is indicated only where there is dilatation of the upper urinary tract; or where there is infection, pain, or lithiasis.

b) Ectopic Ureterocele

All patients with ectopic ureterocele are operated upon. Mere incision of an ectopic ureterocele is a therapeutic error. The form of surgical treatment must be selected for the individual case; it depends upon the degree of function of the associated kidney, the degree of dilatation of the involved ureter, the function of the contralateral kidney, the presence or absence of ureteral duplication, of infection, and of calculi. Primary excision may be partial or total.

Partial excision. The ureterocele alone may be excised. Close, long-term observation must follow. The transvesical approach is usually used for partial excision. The entire ureterocele, including the portion that extends into the urethra, must be removed in order to avert postoperative urethral obstruction. After such excision, there is usually improvement in symptoms but infection is seldom eradicated. Dilatation of the renal pelvis and ureter usually diminishes but rarely regresses entirely.

Total excision. The ureterocele, ureter, and associated renal segment or entire kidney may be removed either primarily or, if severe infection persists after ureterocele excision, as a secondary operation to eradicate infection. The best results have been obtained by primary total excision: 16 of 17 patients so treated by KJELLBERG et al. remained in good health throughout long terms of observation.

Ureteroneocystostomy. If reflux has been initiated by excision of a ureterocele, reimplantation of the ureter(s) may be considered. For this purpose, HUTCH and CHISHOLM have advocated the following operative technique: The ureterocele is exposed through any convenient incision, and the ureterocele is grasped firmly with sutures or an instrument for purposes of retraction. The mucosa is incised with Potts arterial scissors, beginning at the superior edge of the base of the ureterocele, and continuing around its base on both sides of the trigone (Fig. 31). While traction is maintained on the ureterocele, the intramural portion of the ureter is dissected from the bladder muscle as it passes through the ureteral hiatus. The base of the ureterocele is easily detached from the vesical wall; the extravesical portion of the ureter can be freed to a distance of 4 or 5 cm. Dissection enlarges the hiatus sufficiently to permit several centimeters of ureter to be drawn into the bladder with ease. Inferiorly, the ureterocele is freed by cutting the superficial trigone. Superiorly, the ureterocele is cut away from the ureter by an elliptical incision, which leaves a mucosa-lined ureteral orifice. Care is taken to insure that the bladder floor that will lie beneath the new intravesical ureter is strong, healthy muscle some 3 cm. long. At this time, it is frequently necessary to narrow the enlarged ureteral hiatus to a diameter conformable with that of the ureter, by means of sutures placed in the bladder muscle, meticulously avoiding placement of sutures in the mucosa, as this might result in obstruction. If a longer intravesical ureter is necessary, the ureteral hiatus may be enlarged by cutting with scissors superiolaterally, along the projected line of the lateral edge of the trigone. It is then repaired under the ureter. The severed elliptical end of the ureter is then sutured to the superficial trigone with three sutures of 000

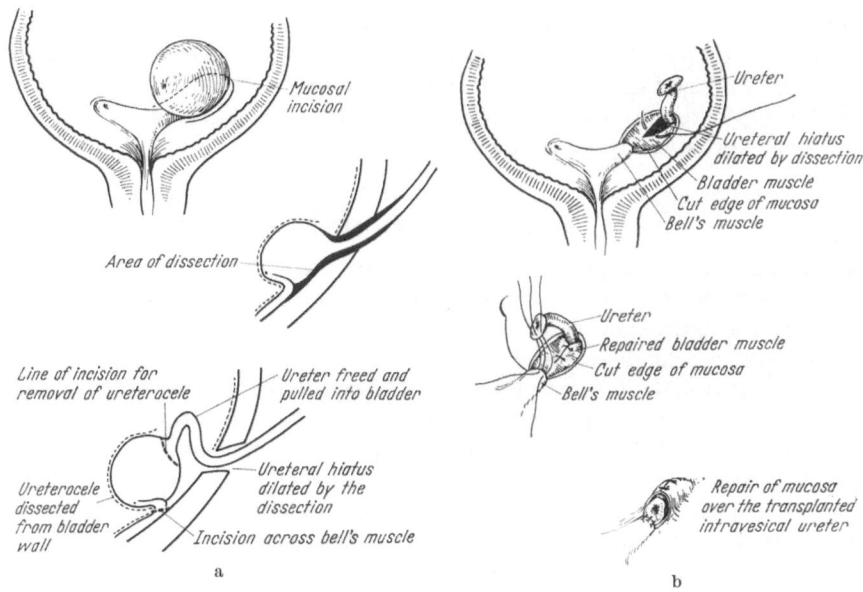

Fig. 31a, b. Technique of ureterocele excision and ureteroneocystostomy. (From Hutch and Chisholm)

chromic catgut, and the mucosa is sutured over the transplant. An indwelling splinting catheter, left in the ureter, is optional.

Nephrectomy or heminephrectomy without excision of the ureterocele, used by a few workers, has not led to good results. It is important to note that a residual ureteral stump after nephrectomy may become a reservoir of infection (Amar, 1964c).

B. Ureteropelvic Junction Obstruction

Congenital ureteropelvic junction obstruction derives clinical significance from the hydronephrosis to which it gives rise. The broad subject of hydronephrosis has been reviewed by Campbell (1936, 1951a, b) and by Kretschmer, but hydronephrosis due to congenital ureteropelvic junction obstruction forms a distinct clinicopathologic entity and has been discussed by Williams (Vol. XV of this Encyclopedia). It will therefore be only briefly summarized here.

1. Definition and Description

Ureteropelvic obstruction is impediment to the flow of urine out of the renal pelvis into the ureter. Such obstruction results first in hypertrophy of the muscle of the renal pelvis, then in its dilatation. The extrarenal portion is affected first, then the total renal pelvis. Subsequently, the calyces also dilate and show similar changes. This is followed by progressive destruction of functioning renal parenchyma as a result of direct hydrostatic pressure, and by vascular obstruction and obliteration. Secondary calculi, infection, and pyelonephritic changes are common complications. Ureteropelvic obstruction tends to be bilateral, but when first noted it is often more prominent on one side, becoming apparent on the contralateral side usually after nephrectomy on the side first affected.

2. Incidence

Ureteropelvic junction obstruction is relatively common. It is seen with equal frequency in patients of either sex, and may be discovered at any age, particularly in children older than five years and in adults at any decade. The incidence is high in malrotated, ectopic, and horseshoe kidneys. For an unknown reason, the left kidney is more frequently involved than is the right. Rarely, a familial incidence has been reported (RAFFLE). Hydronephrosis due to congential uretero-pelvic obstruction is not seen in the upper pelvis of double kidneys that lack a clearly defined ureteropelvic junction. It is most likely to occur in kidneys that have a well-developed pelvis outside the hilum.

3. Causes

a) Aberrant Renal Blood Vessels

There is some doubt whether an aberrant blood vessel alone can obstruct the flow of urine at the ureteropelvic junction, or whether such a vessel merely exacerbates an obstruction that is due to an intrinsic lesion. An aberrant vessel contributing to ureteropelvic junction obstruction may be a branch of the main renal artery to the lower pole, or may be one of the multiple renal arteries arising from the aorta. Veins also may participate in the obstruction, with or without an associated artery.

These vessels cross most often behind, but occasionally in front of the uretero-pelvic junction. In the majority of cases they cross the proximal ureter at a site 1 to 2 cm. distal to the junction, and may contribute to the binding of this part of the ureter to the dilated pelvis. Among 78 children with hydronephrosis due to ureteropelvic junction obstruction studied by NIXON, blood vessels were thought to be a contributory cause in 25.

b) Adhesions and Kinks

When patients with ureteropelvic obstruction are explored surgically, the proximal ureter is frequently found to be bound down to the surface of the distended renal pelvis by dense adhesions, so that the ureter is acutely kinked at the ureteropelvic junction. The adventitial sheath may fix, but does not cause these contortions. Most authors agree that the adhesions are not inflammatory in nature. They are commonly associated with vessels crossing the ureteropelvic junction area, but may be present without such vessels. Surgical division of the adhesions alone does not usually allow complete emptying of the renal pelvis and so does not relieve hydronephrosis. The kinks and adhesions therefore are not regarded as a primary cause of obstruction in the majority of cases; but they accentuate obstruction from other causes.

c) High Insertion of Ureter

Insertion of the ureter high in the renal pelvis has been believed by some authors to be a cause of ureteropelvic obstruction. A high position of the uretero-pelvic junction is a common finding in ureteropelvic junction obstruction; but since it is otherwise rare (except in malrotated kidneys), in this situation it is probably an effect rather than a cause. It appears likely that the high position of the junction results from unequal distention of the various portions of the renal pelvis: the lower portion dilates more than does the upper, and protrudes beneath the junction, so that the site of insertion of the ureter appears to be high.

d) Stenosis

It is commonly believed that the primary cause of the hydronephrosis associated with ureteropelvic obstruction is narrowing of the ureteral lumen immediately distal to the ureteropelvic junction. In this view, such abnormalities as ureteral kinks and vascular aberration are secondary to the ureteral narrowing. Such a hypothesis does not explain the cause of the narrowing. The stricture is seldom fibrotic in nature; indeed, at the affected site the muscular wall tends to be thinner than normal, and muscle fiber may be totally lacking. This striking finding has led MURNAGHAN to suggest that the primary lesion is interruption of the ureteral circular musculature. Others have believed that the narrowing is caused by spasm.

e) Functional Disorders

In some cases, no obvious cause for ureteropelvic junction obstruction can be found at operation. Since contraction waves in the distended pelvis are not propagated down the ureter, it has been suggested that the cause of obstruction is failure of conduction at this point; but ANDERSON showed that distention easily provokes ureteral contraction, and considered that the conduction failure is probably a result of the fact that no urine passes from the renal pelvis into the ureter. MURNAGHAN attributed the failure of propulsion of urine from the renal pelvis into the ureters to deficiency of circular muscles at the ureteropelvic junction.

f) Vesicoureteral Reflux

The association of ureteropelvic junction obstruction with reflux is discussed in the chapter on Vesicoureteral Reflux. It should be briefly mentioned here, however, that any tendency to obstruction at the junction will be accentuated by vesicoureteral reflux because urine that is returned into the renal pelvis from the ureter must be ejected into it once more via the ureteropelvic junction and finally into the bladder. In additon, vesicoureteral reflux may, by causing the renal pelvis and calyces to dilate disproportionately larger than the ureter, give rise to the false impression that there is obstruction at the ureteropelvic junction (Fig. 32). HUTCH, HINMAN, and MILLER in 1962 first suggested that many cases of hydronephrosis that appear to be secondary to ureteropelvic junction obstruction are actually due to vesicoureteral reflux. They reached this conclusion on the basis of the following facts: Any tube will become obstructed if the flow of fluid through it exceeds the emptying capacity of its narrowest point. In the ureter, the narrowest points are the ureteropelvic junction and the ureterovesical junction. There is a wide variation in the shapes of normal ureteropelvic junctions. In about 85% of patients studied by HANLEY, this junction was funnel-shaped, and had a large emptying capacity. Some 15% of normal ureteropelvic junctions were box-shaped, and their emptying capacity was limited. If a box-shaped, though normal, renal pelvis must expel a quantity of urine that exceeds its emptying capacity, obstruction ensues and the renal pelvis dilates. In patients with vesicoureteral reflux, the reflux is the cause for excess filling of the urinary tract. Many patients who appear to have ureteropelvic junction obstruction will be found on investigation to have reflux. Even in patients in whom reflux is not demonstrable at the time the diagnosis of ureteropelvic junction obstruction is made, reflux may have been present in early childhood, and may have disappeared as the intravesical ureter matured, leaving the hydronephrosis as an aftermath.

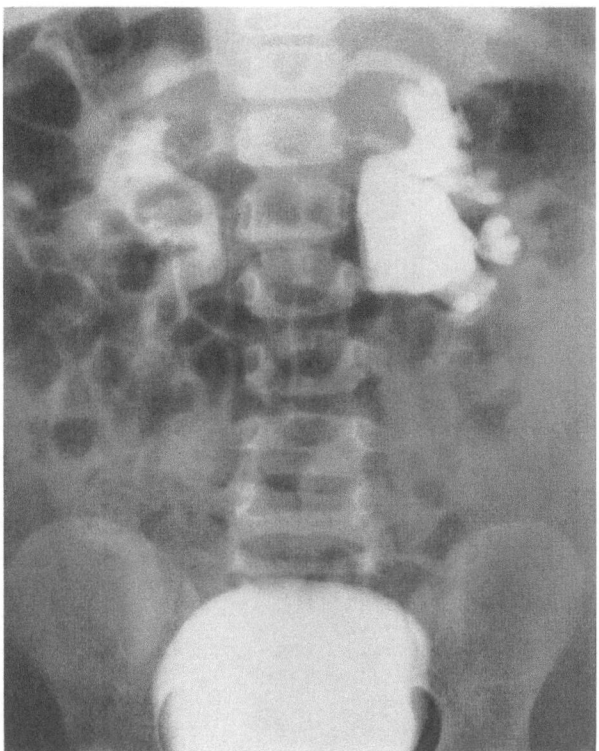

Fig. 32. 3-Year-old girl with recurrent urinary infection. Excretory urogram: Left renal pelvis dilated; calyces clubbed. Left vesicoureteral reflux was found to explain the apparent ureteropelvic obstruction

4. Clinical Aspects

The renal pelvis and calyces may undergo slow, progressive distention for a long time without causing symptoms. Such asymptomatic dilatation may be discovered incidentally, during routine examination or investigation for unrelated disease, as a large mass within the abdomen. On further investigation, advanced hydronephrosis may be found (Fig. 33). In infants, the enlarged kidney is easily palpated and must be distinguished urographically from renal neoplasm or cystic kidney.

The clinical manifestations that may occur are as follows:

a) Superimposed Intermittent Acute Obstruction

In children, the acute attack is characterized by loin pain, vomiting, and pallor, of several hours' duration; or vomiting alone may be the predominant sign. In adults there may have been discomfort or aching of long duration, in the flank or in the upper portion of the abdomen, which becomes more severe during the acute attack. It is usually necessary to make a differential diagnosis between ureteropelvic obstruction and ureteral colic. Because the pain is often not clearly localized, and may be referred to the central portion of the abdomen, an erroneous diagnosis of appendicitis has occasionally been made and a normal appendix has been removed. Pain in the lower portion of the abdomen may arise from hydronephrosis of a pelvic ectopic kidney. On examination during the acute attack, the

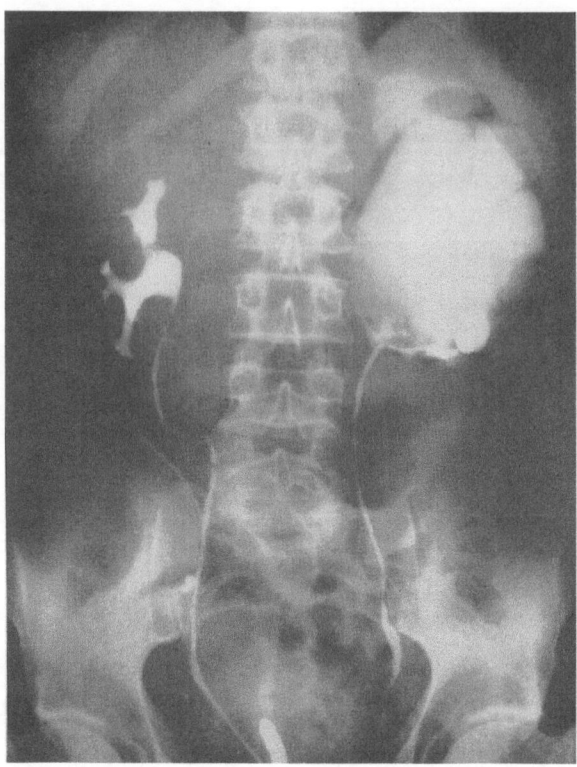

Fig. 33. 53-Year-old man with asymptomatic abdominal mass (left), found during routine physical examination. Retrograde urogram: Advanced left hydronephrosis. At operation, dilated renal pelvis contained 2500 cc. of urine. Congenital ureteropelvic junction obstruction had caused the hydronephrosis

kidney may be tender and palpable. In intervals between such episodes, the patient usually appears well and no evidence of abnormality is found on physical examination.

b) Recurrent Pyelonephritis with Loin Pain

Recurrent pyelonephritis with loin pain may be the presenting sign of hydronephrosis, and may instigate investigation leading to the discovery of ureteropelvic junction obstruction. Pus cells in the urine may be due to the presence of stones; finding of such cells may stimulate investigation in the asymptomatic patient.

c) Renal Calculi

Ureteropelvic junction obstruction frequently causes the formation of renal calculi. Clinically, the calculi may relate to the obstruction in any of three ways: (1) The stones may, by causing symptoms, produce the first evidence of the obstruction. (2) The calculi may be associated with urinary infection superimposed upon ureteropelvic junction obstruction; they may result from such infection, or they may contribute to it. (3) Renal calculi caused by ureteropelvic obstruction may give rise to no, or to negligible symptoms. They may then be discovered during investigation for the hydronephrosis associated with the obstruction.

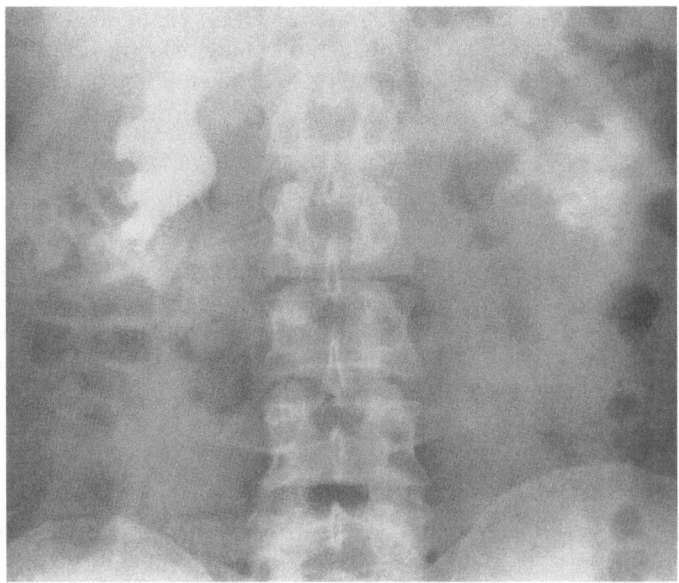

a

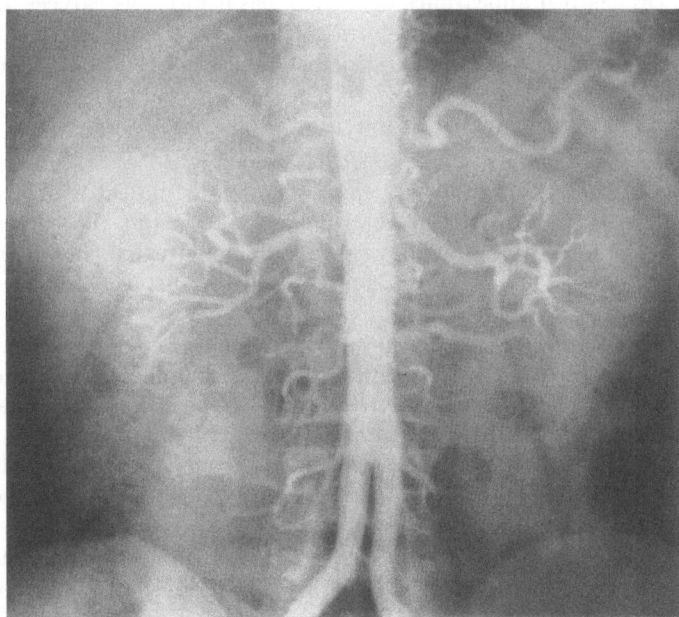

b

Fig. 34a and b. 43-Year-old man with hypertension. (a) Excretory urogram: Right ureteropelvic obstruction with hydronephrosis. (b) Transfemoral aortogram: Left renal artery obstruction, at its junction with the aorta

d) Hematuria

Hematuria with or without infection may be the presenting sign of ureteropelvic junction obstruction. The cause of the hematuria is presumed to be rupture of pelvic veins due to pressure; it may be precipitated by trauma.

10*

e) Trauma

Rarely, trauma to the hydronephrotic kidney leads to the discovery of ureteropelvic junction obstruction. The size of such a kidney, which causes it to protrude beyond the protective surrounding structures, makes it particularly vulnerable to injury.

f) Associated Hypertension

An occasional patient with uncomplicated ureteropelvic junction obstruction is hypertensive, but the hydronephrosis is often not the cause of the hypertension. In a 43-year-old man with elevated blood pressure, investigation revealed unilateral ureteropelvic junction obstruction with hydronephrosis which had caused no symptoms (Fig. 34). There was renal artery stenosis of the contralateral kidney, which was thought to be the cause of the hypertension.

5. Diagnosis

The most important aspect of investigation for ureteropelvic obstruction is intravenous urography. Treatment should not be undertaken without this study. The characteristic appearance on excretory urography is bulging of the renal pelvis and failure of the ureter to fill during the early stage of the examination. The delicate cup-shaped appearance of the calyces is lost; the calyces are clubbed. When the superior portion of the ureter is outlined by the radiopaque medium in later films of the series, a small filling defect is usually seen immediately below the renal pelvis; inferior to this defect, the ureter usually presents a normal appearance, but some degree of atonic dilatation may be noted in its lumbar portion.

The shape of the renal pelvis is a more significant indicator of ureteropelvic obstruction than is its size. If the renal pelvis is large but the calyces are well cupped and continuous filling of the ureters is demonstrated, there is no ureteropelvic obstruction, although such a kidney may be predisposed to the subsequent development of hydronephrosis.

The pelvis of a malrotated kidney ordinarily appears large in the anteroposterior view, and must be differentiated from hydronephrosis. For the diagnosis of ureteropelvic junction obstruction, dilatation should also be evident in the lateral view. If, in addition, the calyces are dilated, true hydronephrosis is present. Where intravenous urography shows no excretion of dye, but reveals only the nephrogram, as in renal colic, there may be acute exacerbation of obstruction in an otherwise functioning kidney.

When hydronephrosis is known or suspected to be present, we employ double-dose excretory urography (Amar, 1964b) making delayed films and using half-size films after identification of the affected side. Excellent visualization is usually afforded by this means and we have had to resort to retrograde kidney study in progressively fewer cases.

Retrograde urography may yield helpful information but it carries the risk of trauma and infection. This examination is preferably made in the operating room; if evidence of ureteropelvic obstruction is found, one can proceed with definitive surgical treatment. Although it may not be possible to fill the renal pelvis with radiopaque medium during retrograde study, an important advantage is gained with visualization of the entire ureter below the site of obstruction;

indeed, if obstruction is present at any additional site distal to the ureteropelvic junction, this must be known before operation.

Investigation for vesicoureteral reflux must also be carried out before treatment is instituted. Voiding cystography, supplemented by the indigo carmine test at the time of cystoscopic examination (AMAR, 1964a), may prove that reflux is present. The surgical treatment indicated will in that case differ considerably from procedures recommended if no reflux is found.

Aortography is not necessary either for diagnosis or for treatment of ureteropelvic obstruction except in rare circumstances; e.g., ectopic or horseshoe kidney, or in patients with hypertension. In one of our patients with asymptomatic unilateral hydronephrosis and elevated blood pressure, aortography demonstrated renal artery obstruction of the contralateral kidney which was considered to be the cause of the hypertension.

6. Treatment

Any operation that restores a straight conical outlet from the renal pelvis will lead to improvement. The recommended operative procedures are described in detail elsewhere; only certain principles will be mentioned here.

The aim of treatment is to preserve and improve the functioning of renal tissue. Operation is recommended in all cases where definite obstruction is demonstrated. The failure rate for surgical treatment of ureteropelvic junction obstruction is about 10—20%. Where such failure occurs, the operation must be repeated.

Conservative operation is obligatory in patients with bilateral involvement, and in those with unilateral obstruction whose contralateral kidney has a large, square pelvis suggesting the possibility that it also may later undergo the development of hydronephrosis. Nephrectomy is required for unilateral obstruction which has resulted in extensive destruction of the renal parenchyma, or which is complicated by severe pyelonephritis. For painful, unilateral, anomalous hydronephrosis, nephrectomy has been recommended as the treatment of choice.

Malrotated or ectopic kidneys do not respond satisfactorily to conservative plastic operation. For horseshoe kidney, we perform pyeloplasty, symphysiectomy, and nephropexy as recommended by CULP. For very low kidneys, HESS has performed direct anastomosis between the renal pelvis and the bladder, with good results.

Since all renal arteries are end arteries, we do not recommend ligation of a vessel crossing the ureteropelvic area if it supplies too large a renal segment. Where infection is present, an attempt is made to sterilize the urine through the preoperative administration of antibiotics.

Nephrostomy drainage may be helpful postoperatively, but is not essential in all cases. Where acute obstruction is associated with infection, it may be necessary to institute nephrostomy drainage for a preliminary period before plastic operation of the ureteropelvic area. Mere lysis of adhesions, and straightening of the area of obstrcution at the junction it not adequate treatment for ureteropelvic junction obstruction. The site of obstruction must be opened, and the lumen of the renal pelvis and of the junction must be enlarged from within, by any of the various pyeloplastic operations that have been devised. Selection of the specific procedure for the individual case depends upon the degree of obstruction present, and upon the length of the narrowed segment or hypoplastic ureter found distal to the ureteropelvic junction.

C. Unusual Positions of the Ureter

I. Retrocaval (Post-Caval; Circumcaval) Ureter

1. Definition

A retrocaval ureter, instead of pursuing a normal paravertebral course in its descent from the renal pelvis, is directed towards the midline. In the region of the third and fourth lumbar vertebrae, the ureter lies against the vertebral column, passing posterior to the vena cava, describing a spiraling semicircle that courses first medially, then anteriorly, and finally laterally to the vena cava. The lower portion of the ureter is directed toward the bladder at a normal angle. The total curve, from renal pelvis to bladder, has been called "sickle-shaped".

All published instances of retrocaval ureter have been on the right side. The single exception, that on the left noted by Brooks, was in a patient with situs inversus.

2. Incidence

Harrill, in 1940, made the first accurate preoperative diagnosis of retrocaval ureter. In all of the 27 cases published before that time, the diagnosis had been made at autopsy or at operation. The second preoperative diagnosis was reported by Greene and Kearns in 1946. As cognizance of the anomaly has sharpened during the last two decades, retrocaval ureter has been noted with increasing frequency (Duff; Laughlin). Publishing in January, 1965, Qureshi and Mulvaney had found 129 reported cases and added 2; several have been reported since that time.

3. Embryology

A retrocaval position of the ureter is secondary to anomalous vascular development (Fig. 35). The metanephros, as it ascends into the thoracic region from its early position near the cloaca, passes through an elongated "ring" or loop of blood vessels. The dorsal limb of this vascular loop normally persists into fetal life, to form the vena cava. Its ventral limb consists principally of the posterior cardinal and subcardinal veins, which subsequently disappear or become incorporated in other vessels, leaving the vena cava posterior to the ureter. Abnormally, the ventral limb of the loop may persist, as the vena cava, while the dorsal limb does not; or the dorsal limb may be absent from the start. (A third possible abnormality — persistence of both the ventral and the dorsal limbs — would leave the ureter threaded through the loop.) Either of these events would leave the ureter behind the anomalous vena cava: a retrocaval ureter.

4. Clinical Aspects

a) Symptoms and Signs

The clinical manifestations of retrocaval ureter are caused by obstruction due to pressure upon the ureter as it passes behind and around the vena cava. Such manifestations have occasionally been noted in children (Creevy; Freire; Marcel; Parks and Chase). Retrocaval ureter may be suspected in any patient with right pyelectasis and dilatation of the superior portion of the right ureter. Infection, urinary calculi, or both may supervene.

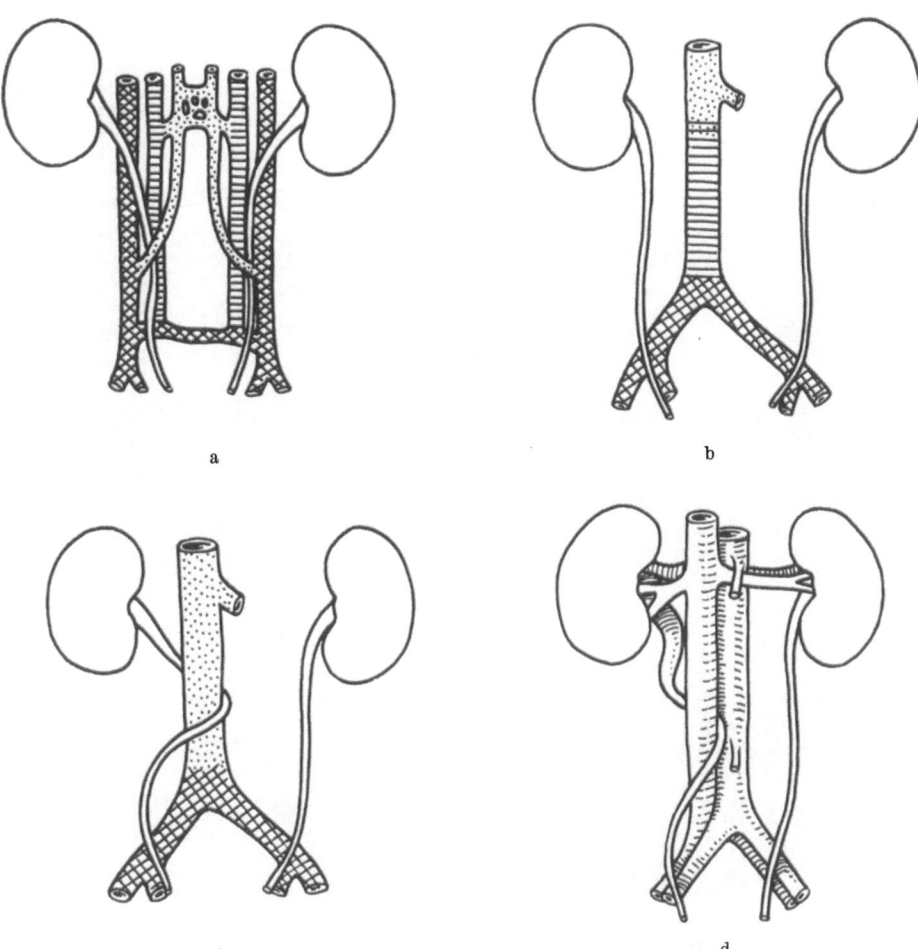

Fig. 35a—d. Schema showing relation between embryonic development of the infrarenal portion of the inferior vena cava and retrocaval ureter: (a) A primitive stage; the ureter winds among the three cardinal veins; (b) the usual development of the vena cava, from the right supracardinal vein (dorsal to the ureter), has left the ureter free; but in (c) the subcardinal vein, ventral to the ureter, has formed the main portion of the vena cava; (d) mature retrocaval ureter, as clinically seen. (From HOLLINSHEAD, W. H.: Anatomy for Surgeons, Vol. II. New York: Hoeber Medical Division, Harper and Row 1956)

b) Diagnosis

The urographic diagnosis of retrocaval ureter is frequently obscured by failure of such a ureter to fill with contrast medium throughout its length. In two of our patients, on excretory urography with a double dose of contrast medium, both the upper and the lower portions of the ureter were visualized, but that part of the ureter which was posterior to the vena cava could not be seen (Fig. 36a). A ureteral catheter was then inserted, and a venacavagram was obtained. In the lateral projection, the catheter in the ureter was seen to lie behind the vena cava (Fig. 36b). This technique, described by ROWLAND, BUNTS and IWANO in 1960, is the only sure way of proving the retrocaval position of a ureter before operation.

c) Treatment

When the kidney that is drained by a retrocaval ureter is well-preserved, a conservative surgical procedure is worth while. The dilated renal pelvis is transected. The ureter is dissected free, and is drawn through into a normal location.

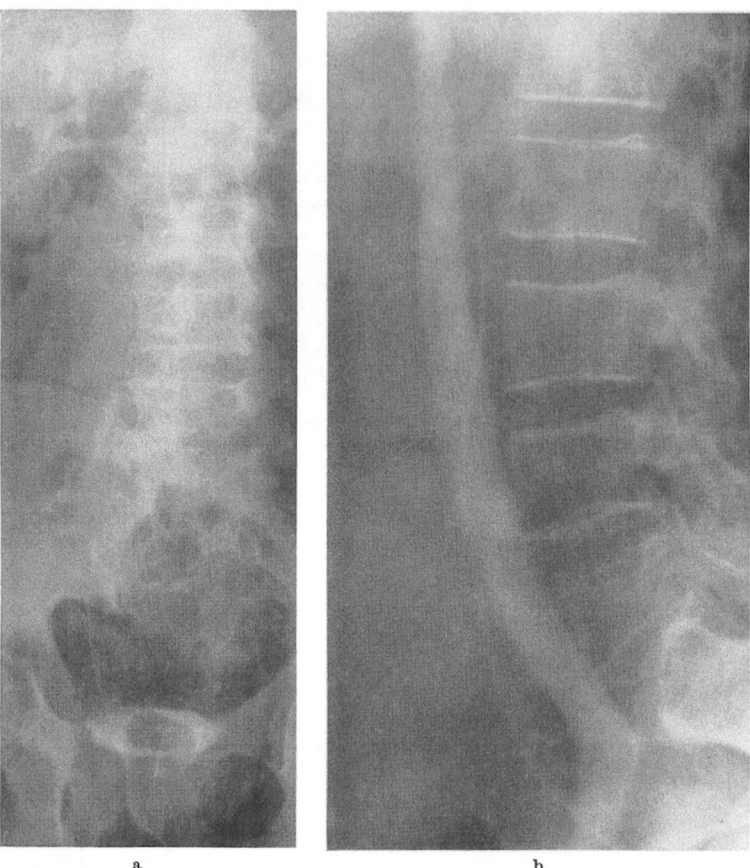

a   b

Fig. 36a and b. Retrocaval ureter (47-year-old man). (a) Excretory urogram: Upper and lower portions of ureter
are visualized; the ureteral segment that is behind the inferior vena cava could not be seen. (b) Ureteral catheter,
seen posterior to the inferior vena cava on lateral view, proves retrocaval ureter

Continuity of the ureter is re-established. This is the most frequently used procedure. It is based on logical grounds, and gives good results; however, a long
segment of ureter is dissected, and care should be taken that its blood supply
is not impaired during this maneuver. Goodwin, Burke and Muller transected
the vena cava; freed the ureter from its anomalous position; brought it forward,
then repaired the vena cava by end-to-end anastomosis. This procedure, although
it appears not to be generally acceptable, may be of value in rare cicrumstances.
Vittori, Moine, and Quinard preferred to section the ureter at its point of
crossing the aorta, to bring it into good position and perform anastomosis. They
felt that other techniques were less logical, and sometimes dangerous. After any
form of correction of retrocaval ureter, especially if hydronephrosis has been
advanced, some time may elapse before ureteral tonus is regained.

II. Retroiliac Ureter

Retroiliac ureter was reported for the first time by Corbus, Estrem, and
Hunt in 1960. The patient was a 25-year-old multiparous woman who had
recurrent episodes of urinary tract infection throughout life, made worse by each

pregnancy, and recent painful hematuria. There was early hydronephrosis and moderate hydroureter. The diagnosis was not made until surgical exploration revealed that the ureter was compressed behind the iliac artery at the level of the fifth lumbar vertebra, lateral to the ovarian vein. The ureter was widely dilated above this point. The ureter was divided obliquely and reanastomosed in front of the iliac artery. Ureteral dilatation was relieved; the patient had remained free of urinary tract infection for four years at the time of the report, and had undergone two uneventful pregnancies.

III. Herniation of the Ureter

Herniation of the ureter is rare; but it is still more rarely detected in the living patient. The problem was reviewed in 1937 by DOURMASHKIN, discussing 27 inguinal, 33 femoral, and 11 scrotal herniations of the ureter — a total of 71 cases.

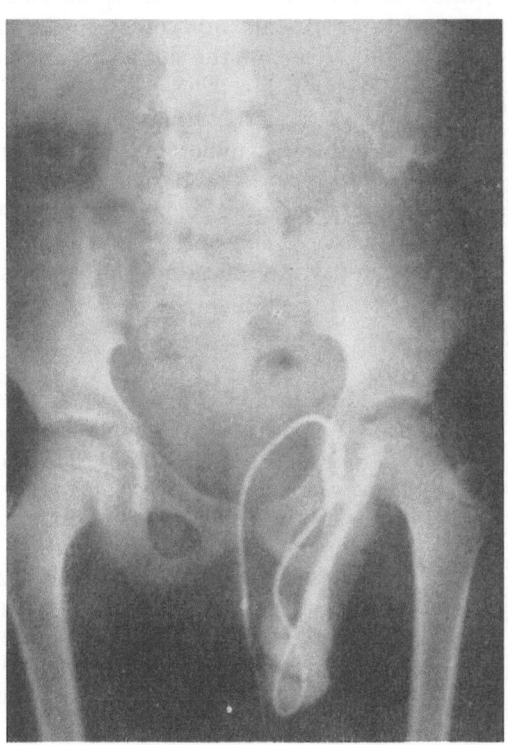

Fig. 37. Dilated ureteral loops containing ureteral catheter, visualized in left scrotum; 20 cm. of ureter was resected. (From JEWETT and HARRIS)

JEWETT and HARRIS, in 1953, reported a 9-year-old boy in whom the left ureter descended into the scrotum through the internal inguinal ring (Fig. 37). There was no hernial sac. The scrotal ureter was attached to a solitary hydronephrotic kidney. A total of 26 cm. of ureter was excised, and oblique end-to-end anastomosis was performed.

The symptoms of ureteral herniation are due to the obstruction which it causes. Infection is often present. Pain in the kidney, with or without pyuria, has been reported as the most frequent manifestation.

The diagnosis is urographic. If the herniated ureter fills with dye throughout its length, it will be seen to describe a loop within the scrotum or other site. Since obstruction may prevent the free flow of the contrast medium, particularly into the herniated loop, this unusual diagnosis is easily missed.

Treatment is selected on the basis of the renal status. If the corresponding kidney retains sufficient function to be salvageable, resection of the redundant portion of the ureter, with ureteroplasty, is carried out. Nephroureterectomy is necessary if this kidney has been irretrievably damaged by obstruction, with or without superimposed infection.

D. Agenesis, Aplasia, Blind Ending
I. Agenesis

Ureteral agenesis is part of a total anomaly which derives its primary importance from the failure of development of the ipsilateral kidney. The total anomaly is due to incomplete or absent embryonic growth of the ureteral bud.

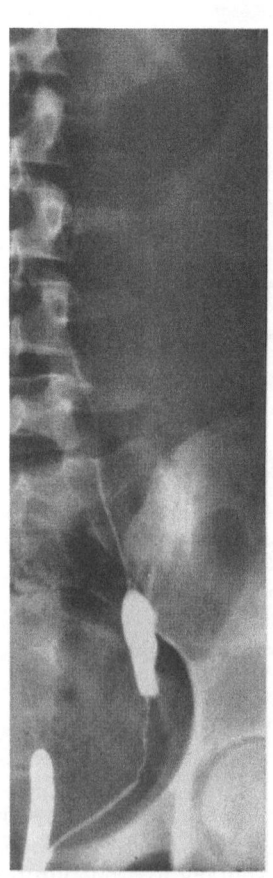

The trigone on the side of the anomaly may be missing or hypoplastic. There is often no ureteral orifice on that side; or there may be a blind dimple.

Clinically, unilateral ureteral agenesis has the same significance as ureterorenal agenesis (q.v.). Bilateral ureteral agenesis, being assoiated with bilateral renal agenesis, is incompatible with life, and is found at necropsy in certain nonviable monsters.

II. Aplasia

Like ureteral agenesis, ureteral aplasia is associated with anomalies of the kidney (described in the chapter on renal anomalies). The most frequent among these is renal agenesis (total absence of the kidney), but the kidney may be present yet lack normal development (renal hypoplasia). In some instances, an aplastic ureter is a thin, cord-like structure, with or without a small lumen, attached to one segment of a double kidney. An aplastic ureter frequently fails to open into the bladder; or its orifice may be smaller than normal. Aplasia obviously occludes the flow of urine, and a cyst may consequently develop at one or more sites along its course. Congenital multicystic kidney may be present, with complete obliteration of the ureteric lumen, and of the associated renal pelvis and calyces.

Fig. 38. Blind-ending ureter (17-year-old girl), seen on retrograde ureterogram. Ureteral lumen is hypoplastic and narrow, with beading. The apparent dilatation seen in the middle segment represents extravasation of contrast medium

III. Blind Ending Ureters

A ureter with blind ending usually presents an extremely small orifice, although occasionally its opening is of normal size. In most instances, the entire ureter is hypoplastic. The structure may entirely lack a lumen; or there may be a narrow lumen which is either continuous or interrupted by alternating areas of constriction and dilatation. At retrograde urography, only a small caliber ureteral catheter can be inserted.

A blind ending ureter may be one of a duplicated pair; its ipsilateral counterpart may then be normal, and may drain a functioning kidney. Instances of bilateral blind ending ureters have been reported; 11 such cases were noted by CHWALLA in 1927a, b. Large cysts may form in the blind-ending ureter, producing a mass within the abdomen which must be differentiated from a neoplasm. The *diagnosis* is made by retrograde ureterography, provided the lower end of the blind-ending ureter is patent so that contrast medium can be injected into it (Fig. 38). Occasionally the diagnosis is made at the time of surgical exploration. *Treatment* is not required in most cases, since a blind-ending ureter usually does not give rise to symptoms. If a large cyst forms, it may become necessary to remove the ureter surgically.

E. Congenital Dilatation of the Ureter

I. With Obstruction

1. Congenital Ureteral Stricture

Although the terms "ureteral stricture" and "ureteral atresia" are occasionally used interchangeably, the more precise restriction of "atresia" to absence of the lumen, and of "stricture" to narrowing of the lumen would reduce the number of occasions on which restatement of these definitions must be made. The term "ureteral stricture" properly denotes narrowing of the ureteral lumen. The ureteral wall is thicker than normal. Congenital ureteral stricture is present chiefly at the junction of the ureter and the renal plevis. The resultant obstruction gives rise to hydronephrosis (described under that heading). Congenital stricture at the ureterovesical junction causes a ballooning of the ureter resembling megaureter. Stricture rarely occurs between these two junctions.

The few cases of "stricture" of the ureter between the ureteropelvic and the ureterovesical junction have appeared to be different in character from the thick-walled narrowing of the lumen that involves those junctions. Narrowing along the length of the ureter is usually a thread-like area with thin, defective muscular wall. Such a defect is also called *ureteral hypoplasia*. A relatively long hypoplastic segment of ureter often totally lacks a lumen, and is frequently accompanied by multicystic or dysplastic kidney. A short segment usually has a narrow lumen; hydroureter or hydronephrosis is then only partial. In some affected ureters, there is more than one site of hypoplasia. Involvement may be bilateral.

Opinions differ about the incidence of true ureteral stricure. Most authors consider it rare. CAMPBELL (1963), however, cited an incidence of 0.6% in autopsy material and noted 123 cases among 19,046 children studies post mortem, 51% in the ureterovesical area and 34% in the ureteropelvic region (15% of the "strictures" were in the body of the ureter). The reason for this difference of opinion may be a simple one. Ureteral dilatation due to any cause affects only its extravesical portion (see chapter on Reflux). As a result, the intravesical ureter usually appears narrow by contrast. This appearance has frequently led to the erroneous diagnosis of ureteral stricture. In most of these situations, from the anatomic point of view there is no stricture, since the junction will admit passage of a ureteral catheter of adequate caliber.

The cause of ureteral stricture is unknown. OSTLING, in 1942, noted that most of the "ureteral strictures" encountered by him in cadavers were merely areas of redundant mucosa, in which there were no adventitial changes, but which produced intraureteral folds and narrowings that might function like a stricture. When the ureteral adventitial sheath was divided, however, and the mucosa was drawn out longitudinally, the fold was found to represent a true narrowing. ÖSTLING observed that some of the strictures were sharply localized, that their length was usually 1 to 3 cm., but some exceeded this measurement; and that their calibers ranged from broad to impassable.

The clinical manifestations of ureteral stricture express the resultant obstruction which predisposes to infection. Symptoms may represent the obstruction directly, or the infection, or both.

The diagnosis of ureteral stricture is made by excretory urography and cystoscopy, and retrograde urography.

Treatment is excision of the area of stricture. Where the lesion involves the vesical end of the ureter, the remaining ureter is implanted into the bladder in such a manner as to prevent vesicoureteral reflux. If much of the functioning renal tissue has been destroyed, nephrectomy may be necessary provided the function of the other kidney is adequate to sustain life. In individual circumstances, other appropriate methods of treatment are to be considered.

2. Distal Ureteral Atresia

Atresia (from the Greek *atrētos* [not perforated]) of the ureter is absence of a ureteral lumen. Atresia of the distal ureter is anomalous absence of a lumen within the distal third of the ureter, preventing the flow of urine from the upper two thirds of the ureter into the bladder. The embryologic sequence of events is such that distal ureteral atresia must be present before the fifth week of intrauterine life (GORDON and REED).

If the patent portion of the ureter drains a functioning kidney, hydrostatic pressure from the accumulated urine will gradually result in dilatation of the ureter proximal to the site of atresia. The sedentary column of urine is susceptible to bacterial infection, but since it does not empty into the bladder, the urine in and from the bladder is not contaminated from this source. When dilatation of the proximal ureter is sufficiently gradual, and infection of the sedentary column is minimal or absent, distal ureteral atresia may give rise to no symptoms or signs. This state may persist throughout childhood, and in some instances throughout life. Even when clinical manifestations are present, the diagnosis is frequently not made prior to surgical exploration.

A preoperative diagnosis may be made in a larger number of cases if certain clinical and roentgenologic features are kept in mind. In infancy, the ballooning of the ureter proximal to the site of atresia gives rise to the chief symptoms and constitutes the only clinical and radiographic sign. The cystic dilatation may become so large as to suggest a tumor within the abdomen. No other ureteral anomaly, and no disease of the genitourinary system including neoplasms, produces so large a mass as that which may be attained by the dilated ureter proximal to such atresia. There may be episodes of mild fever without localizing indications pointing to a site of infection. In most patients, these episodes respond to antibiotic therapy, but if recurrent infection of the sedentary urine leads, as it often does, to pyoureteronephrosis, the fever persists despite treatment. There may be concurrent, cramping abdominal pain. In the adult in whom distal ureteral atresia has not given rise to manifestations necessitating definitive study, the anomaly

is sometimes discovered accidentally during investigation for other disease; alternatively, an abdominal mass, unexplained persistent fever, or both, may lead to investigation for the first time during adult life.

The roentgenologic aspects of distal ureteral atresia have been well demonstrated by GORDON and REED.

The most significant urographic findings are those on the lateral film on excretory urography. The proximal cystic dilatation is there seen as a mass between the bladder and the rectum. Since ureteral atresia may be accompanied by other anomalies of the urinary system, one should be alert to this possibility when there is duplication of the ureter. If the ipsilateral duplicated ureter is patent, both it and the bladder will be displaced forward by the mass.

Treatment is excision of the involved ureter and kidney.

3. Congenital Ureteral Valves (or Folds)

A ureteral valve is a transverse fold of redundant mucosa which projects into the lumen of the ureter, usually within 3 cm. of the ureterovesical junction. The true ureteral valve, as suggested by WALL and WACHTER in 1952, contains a band of smooth muscle, although in a few instances the so-called "valve" has been an iris-diaphragm, confirmed surgically (PASSARO and SMITH). ÖSTLING, in 1942, reported a detailed study of the embryogenesis of this anomaly. The anomaly was held to be extremely rare by WALL and WACHTER, who found in the literature only 7 cases reported previous to their own, that satisfied their criteria for true congenital valves. Other authors have estimated that they are present in from 5 to 20% of fetuses and newborn infants dying from other causes.

Congenital ureteral valves tend to disappear as the patient becomes older. At no age are they likely to cause obstruction. In the fetus and newborn, the lumen of the ureter is relatively large, and folds suggesting valves are seen rather often near the ureterovesical junction, in both sexes and on either side. Congenital ureteral valves rarely cause obstruction, although ROBERTS encountered, in an infant that died 40 minutes after birth, a valve that totally obstructed the right ureter 5 cm. above the ureterovesical junction. The diagnosis is rarely made before operation. A ureterogram may suggest the condition. The treatment is chiefly directed toward the renal complications. Nephroureterectomy has been performed in a number of instances. In two cases the valve itself was resected, with good results.

II. Without Obstruction

Discussed in chapter on Reflux, pp. 1—50.

F. Twists, Kinks, Congenital Diverticula, Blockage by Vessels

I. Spiral Twist or Torsion of Ureter

That torsion of the ureter is extremely rare is indicated by CAMPBELL's (1963) finding of only 2 cases among 20,080 autopsies of children. The embryologic explanation usually given is that the ureter fails to rotate with the ascending kidney, but OSTLING considered that a ureteral twist is a phase of redundancy of the fetal mucosa. If torsion causes pronounced ureteral obstruction, hydronephrosis will follow. No clinical instance of this anomaly has been seen by us.

II. Ureteral Kinks

While kinking of the ureter is a rather frequent byproduct of angulation of the ureter secondary to the dilatation that occurs proximal to a site of obstruction, as by a calculus, a true congenital ureteral kink is rare.

Many ureteral kinks never give rise to symptoms. Those that do cause symptoms are usually in the superior portion of the ureter, between the kidney and the point at which the ureter is fixed to the peritoneum. The kidney, to cause kinking, must be abnormally mobile: as such an organ sags, its ureter becomes kinked and obstruction results. DIETL'S crisis is believed to be caused by such blockage.

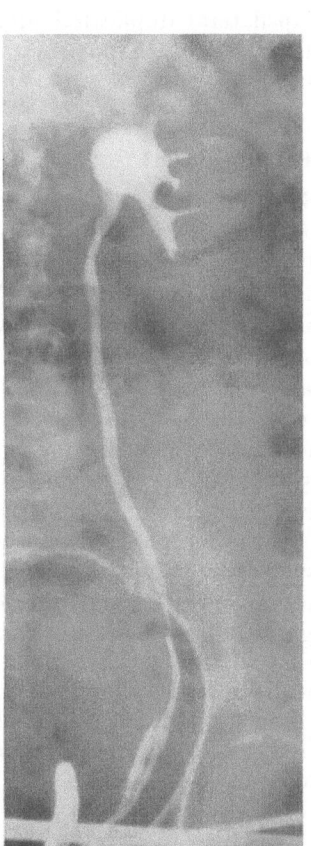

The diagnosis is most often made after pain or urinary infection has led to investigation of the urinary tract. Stereoscopic ureterography is performed with the patient in both the upright and the Trendelenburg positions. A ureteral kink remains fixed throughout these maneuvers.

Where a kink is secondary to ureteral obstruction, eradication of the cause (e.g., removal of a ureteral stone) often leads to a cure. A belt or other support for the hypermobile kidney has occasionally been advised. In some instances, surgical removal of a band or adhesion which held the kidney and ureter in malposition has corrected the kink. During surgical exploration, an area of ureteral hypoplasia may be found at the site of kinking; this may be excised and the ureteral ends anastomosed; or intubated ureterotomy may be performed.

Fig. 39. Blind-ending branch of a bifid ureter (left), seen on retrograde urogram. (Courtesy of Dr. OWEN DAVIES)

III. Diverticulum of the Ureter

A true congenital diverticulum of the ureter appears to be rare. CAMPBELL (1963) accepted only 13 cases reported prior to 1963 as genuine instances of this anomaly. The etiologic aspects of ureteral diverticula were discussed in 1960 by RANK, MELLINGER and SPIRO. In a number of instances the term has been applied erroneously to an acquired diverticulum, or to the blind-ending branch of a bifid ureter (Fig. 39). According to MILLER and TREMBLAY, 30 cases were reported between 1947 and 1964.

Like a blind-ending bifid ureter, a true congenital ureteral diverticulum results from premature cleavage of the ureteral bud, or from multiple budding, and its wall includes all coats (muscular and mucosal) that make up the ureteral wall. Unlike a bifid ureter, which is characteristically at least twice as long as its greatest width, a true congenital diverticulum is round or oval. The lumen of a blind-ending bifid ureter forms an acute angle with that of the main ureter at their

point of junction. The lumen of a true congenital diverticulum communicates with that of the ureter through a well-defined stoma in the ureteral wall.

A true diverticulum may be small or large; in some instances it has become a hugely dilated pouch. CULP (1947) reported a ureteral diverticulum at the ureteropelvic junction, which contained 1600 ml. of urine (Fig. 40). The anomaly has been noted in both male and female patients. It may occur on any aspect of

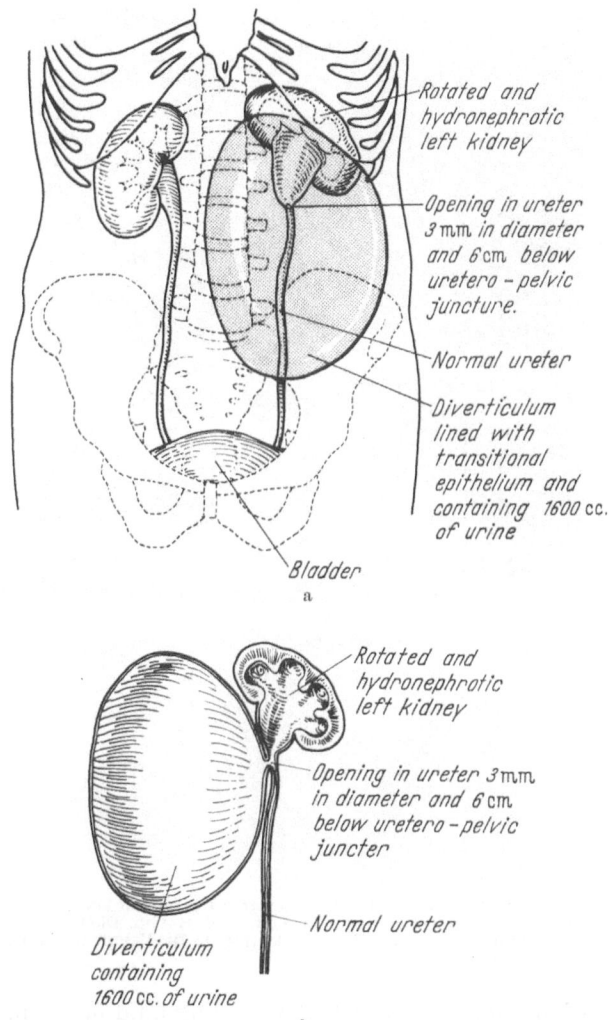

Fig. 40. (a) Congenital diverticulum of ureter (left). (b) General relationships of the diverticulum seen in (a). (Redrawn from CULP, 1947)

the ureter, and at any ureteral level. In the majority of reported cases it has been near the ureterovesical junction, but McGRAW and CULP (1952) observed a true diverticulum from the superior third of a ureter.

This entity differs distinctly from a so-called "acquired" ureteral diverticulum. CAMPBELL found 40 published instances of this lesion prior to 1963. The acquired lesion is an area in which some — but not all — coats of the ureteral wall have yielded to pressure resulting from obstruction distal to the site of ballooning. Such obstruction may be caused by a calculus, or may result from urinary extravasation following ureterostomy, ureterolithotomy, or injury. Outpouchings of this type are always a seat of urinary infection, and treatment consists in relieving the obstruction that has given rise to them.

A true congenital ureteral diverticulum presents no distinguishing clinical features. It may contain calculi. Most diverticula produce hydronephrosis, and

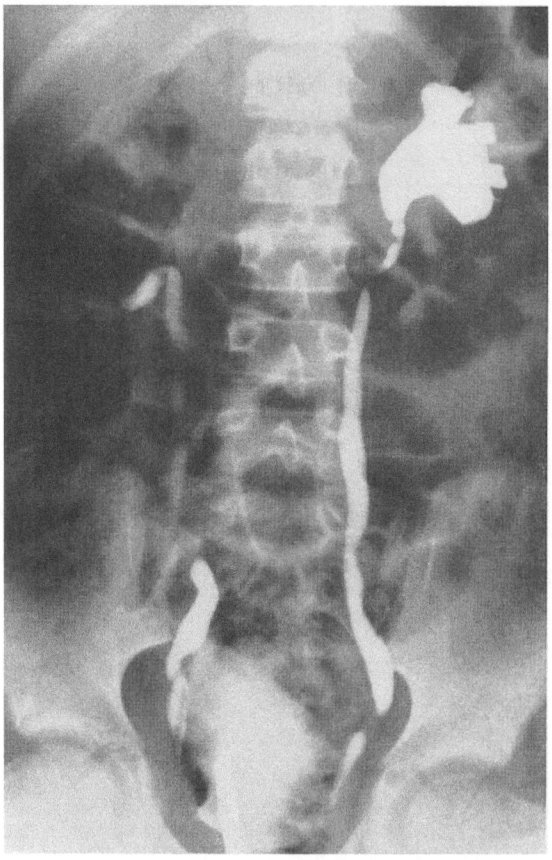

Fig. 41. 8-Year-old boy. Retrograde urogram: Bilateral vascular obstruction of lower ureters (arrows). Posterior urethral valves were thought to be the primary cause of ureteral dilatation. Dilatation did not improve until after the obstructing vessels were divided. (Courtesy of Dr. FREDERICK BURRELL)

may be the seat of urinary infection. A blind-ending bifid ureter may cause pain, as reported by MILLER and TREMBLAY.

In the majority of reported cases of congenital ureteral diverticulum, surgical removal has been necessary. The choice of procedure is influenced by the size and location of the diverticulum, by the size of its opening into the ureter, and by the presence or absence of disease. Depending upon these factors, it may be necessary to reimplant the ureter into the bladder, to perform end-to-end ureteral anastomosis, or to remove the involved ureter with its kidney. McGRAW and CULP (1952) reported an additional authentic case; the diverticulum, one of the largest on record, had destroyed the kidney, and was unique in that it contained numerous calcified plaques. The patient recovered after nephrectomy and diverticulectomy.

IV. Blockage of Ureter by Blood Vessels

Blockage of the ureter by anomalous renal vessels that obstruct the uretero-pelvic junction and superior portion of the ureter is discussed in Section B, Ureteropelvic Junction Obstruction. That the lower end of the ureter may be obstructed by blood vessels was emphasized for the first time by HYAMS, in 1929, who reviewed four cases and added one of his own. GREENE and his coworkers (1954) reported a similar condition in two patients, and five cases were reported in 1965 by YOUNG and KISER. CAMPBELL (1963) felt that primary vascular obstruction of the lower ureter is difficult to prove, but that secondary blockage by vessels (compression of the dilated ureter against the vessel) is not rare in this region.

HYAMS' observations and review of the pertinent anatomy led him to the conclusion that a variation in the course of vessels normally present in the area, or the presence of abnormal vascular branches might result in compression of the lower ureter. The umbilical branch of the hypogastric artery, coursing medially from a point lateral to the ureter, usually crosses in front of its lower portion. Compression by this artery is most likely to be within 2 or 3 cm. of the ureterovesical junction. Abnormal branches of the obturator artery have been thought to contribute to obstruction in some instances. In an 8-year-old boy, bilateral vascular obstruction accentuated ureteral dilatation that was primarily due to posterior urethral valves; ureteral dilatation and hydronephrosis did not subside until after the vessels were excised (Fig. 41).

DREYFUSS (1949) reported a 61-year-old man in whom the ureter was obstructed by a collateral vein of the vena cava. Severance of the obstructing anomalous vein resulted in relief of the ureteral obstruction.

In most instances of vascular compression of the ureter, a mistaken diagnosis of ureterovesical junction obstruction has been made, on the basis of the roentgenographic appearance. Particularly when unilateral hydronephrosis is not traced to another cause, vascular compression should be suspected.

In one of the five cases reported by YOUNG and KISER, obstruction was relieved by dividing the vessel. In three cases, ureteroneocystostomy was performed in addition to division of the involved vessel.

References

AAS, T. N., and A. E. NILSON: Ureterocele in adults: clinical and roentgenographic follow up. Acta chir. scand. 116, 263 (1959).

ALLANSMITH, R.: Ectopic ureter terminating in seminal vesicle; unilateral polycystic kidney: report of a case and review of literature. J. Urol. (Baltimore) 80, 425 (1958).

AMAR, A. D.: Demonstration of vesicoureteral reflux without radiation exposure. J. Urol. (Baltimore) 92, 286 (1964a).
— Double dose contrast medium in excretory urography. Surg. Gynec. Obstet. 118, 1083 (1964b).
— Refluxing ureteral stump: reservoir of urinary infection. J. Urol. (Baltimore) 91, 493 (1964c).
— Improved excretory urograms with double dose of contrast medium. Urol. Dig. 4, 21 (1965).
— Cystoscopic demonstration of vesicoureteral reflux: evaluation in 250 patients. J. Urol. (Baltimore) 95, 776 (1966a).
— Improved methods in the diagnosis of ureteral duplication and ectopia. Canad. med. Ass. J. 95, 813 (1966b).
—, and C. W. SCHEER: Ureterocele with associated vesicoureteral reflux. J. Urol. (Baltimore) 92, 197 (1964).
— — Comparative length of unilateral bifid kidney and its single counterpart. New Engl. J. Med. 273, 211 (1965).

AMBROSE, S. S., and W. P. NICOLSON: Ureteral reflux in duplicated ureters. J. Urol. (Baltimore) 92, 439 (1964).

ANDERSON, J. C.: Abnormal function of the upper urinary tract. Proc. roy. Soc. Med. **44**, 925 (1951).
— In: Modern trends in urology (E. W. RICHES, ed.). London: Butterworth & Co. 1954.
BERMAN, N. A., and L. N. SIDORENKO: Redkii variant anomalii mochevyvodiashchikh putei. Urologiya **23** (6), 51 (1958).
BOIJSEN, E.: Angiographic studies of the anatomy of single and multipe renal arteries. Acta radiol. (Stockh.), Suppl. **183**, 1 (1959).
BRAASCH, W. F., and B. H. HAGER: Urography, ed. 2. Philadelphia: W. B. Saunders Co. 1927.
BROOKS jr., R. E.: Left retrocaval ureter associated with situs inversus. J. Urol. (Baltimore) **88**, 484 (1962).
BURFORD, C. E., J. E. GLENN, and E. H. BURFORD: Ureteral ectopia: a review of the literature and two case reports. J. Urol. (Baltimore) **62**, 211 (1949).
CAMPBELL, M. F.: Vascular obstruction of ureter in children. J. Urol. (Baltimore) **36**, 366 (1936).
— Clinical pediatric urology. Philadelphia: W. B. Saunders Co. 1951a.
— Hydronephrosis in infants and children. J. Urol. (Baltimore) **65**, 734 (1951b).
— Ureterocele: a study of 94 instances in 80 infants and children. Surg. Gynec. Obstet. **93**, 705 (1951c).
— Anomalies of the ureter. In: Urology, ed. 2, vol. 2. Philadelphia and London: W. B. Saunders Co. 1963 (three vol.).
CENDRON, J.: Sur six cas d'urétérocèle. Arch. franç. Pédiat. **13**, 20 (1956).
CHUN, D., and C. BRAGA: Ectopic ureters with congenital absence of urethral sphincters (report of a case). Brit. J. Urol. **37**, 320 (1965).
CHWALLA, R.: Über die Entwicklung der Harnblase und der primären Harnröhre des Menschen mit besonderer Berücksichtigung der Art und Weise, in der sich die Ureter von den Urnierengängen trennen, nebst Bemerkungen über die Entwicklung der Müllerschen Gänge und des Mastdarms. Z. Anat. Entwickl.-Gesch. **83**, 615 (1927a).
— Process of formation of cystic dilatations of vesical end of ureter and of diverticula at the ureteral ostium. Urol. cutan. Rev. **31**, 499 (1927b).
CORBUS, B. C., R. D. ESTREM, and W. HUNT: Retro-iliac ureter. J. Urol. (Baltimore) **84**, 67 (1960).
COX, C. E., and J. A. HUTCH: Bilateral single ectopic ureter. J. Urol. (Baltimore) **95**, 493 (1966).
CREEVY, C. D.: Recognition and surgical correction of retrocaval ureter: a case report. J. Urol. (Baltimore) **60**, 26 (1948).
CULP, O. S.: Ureteral diverticulum: classification of literature and report of an authentic case. J. Urol. (Baltimore) **58**, 309 (1947).
DERRICK, J. R., and C. A. HOOKS: Surgical significance of vascular variations in systemic hypertension, with especial reference to aberrant renal arteries. J. Urol. (Baltimore) **87**, 273 (1962).
DESGREZ, H., F. HEITZ, B. VASSELLE, J. CORNU, and J. BARROIS: Radiologic study of ectopic ureters opening into posterior urethra. J. Radiol. Électrol. **45**, 308 (1964).
DICKINSON, K. M.: Ectopic ureter entering a seminal vesicle. Brit. J. Surg. **50**, 858 (1963).
DOURMASHKIN, R. L.: Scrotal hernia of ureter, associated with a unilateral fused kidney: a case report. J. Urol. (Baltimore) **38**, 455 (1937).
DREYFUSS, W.: Anomaly simulating a retrocaval ureter. J. Urol. (Baltimore) **82**, 630 (1959).
DUFF, P. A.: Retrocaval ureter: case report. J. Urol. (Baltimore) **63**, 496 (1950).
EMMETT, J. L.: Clinical urography: An atlas and textbook of roentgenologic diagnosis, ed. 2. Philadelphia: W. B. Saunders Co. 1964 (two vol.).
ENGEL, W. J.: Ureteral ectopia opening into seminal vesicle. J. Urol. (Baltimore) **60**, 46 (1948).
ERICSSON, N. O.: Ectopic ureterocele in infants and children: clinical study. Acta chir. scand., Suppl. **197**, 1 (1954).
FARR, J. L.: Ectopic ureteral opening into seminal vesicle. J. Urol. (Baltimore) **83**, 108 (1960).
FELIX, W.: The development of the urogenital organs. In: Manual of human embryology, vol. 2 (F. KEIBEL and F. P. MALL, eds.). Philadelphia: J. B. Lippincott Co. 1912 (two vol.).
FREIRE, J. G. DE C.: Uretère rétro-cave et rein hypoplasique. J. Urol. méd. Chir. **59**, 868 (1953).
GIBSON, T. E.: A new operation for ureteral ectopia: a case report. J. Urol. (Baltimore) **77**, 414 (1957).
GIRSH, L. S., and F. E. KARPINSKI jr.: Urinary-tract malformations: their familial occurrence, with special reference to double ureter, double pelvis and double kidney. New Engl. J. Med. **254**, 854 (1956).
GOLDSTEIN, A. E., and E. HELLER: Ectopic ureter opening into seminal vesicle. J. Urol. (Baltimore) **75**, 57 (1956).

GOODWIN, W. E., D. E. BURKE, and W. H. MULLER: Retrocaval ureter. Surg. Gynec. Obstet. 104, 337 (1957).

GORDON, M., and J. O. REED: Distal ureteral atresia. Amer. J. Roentgenol. 88, 579 (1962).

GOYANNA, R., and L. F. GREENE: The pathological and anomalous conditions associated with duplication of renal pelvis and ureter. J. Urol. (Baltimore) 54, 1 (1945).

GREENE, L. F.: Duplication of renal pelvis and ureter. Surg. Clin. N. Amer. 24, 910 (1944).

—, and W. M. KEARNS: Circumcaval ureter: report of a case with a consideration of preoperative diagnosis and successful plastic repair. J. Urol. (Baltimore) 55, 52 (1946).

—, J. T. PRIESTLEY, H. B. SIMON, and R. H. HEMPSTEAD: Obstruction of the lower third of the ureter by anomalous blood vessels. J. Urol. (Baltimore) 71, 544 (1954).

GROSS, R. E.: Surgery of infancy and childhood. Philadelphia: W. B. Saunders Co. 1953.

—, and H. W. CLATWORTHY: Ureterocele in infancy and childhood. Pediatrics 5, 68 (1950).

HAMILTON, G. R., and A. B. PEYTON: Ureter opening into seminal vesicle complicated by traumatic rupture of only functioning kidney. J. Urol. (Baltimore) 64, 731 (1950).

HANLEY, H. G.: The pelvi-ureteric junction: a cine-pyelographic study. Brit. J. Urol. 31, 377 (1959).

HARRILL, H. C.: Retrocaval ureter: report of a case with operative correction of defect. J. Urol. (Baltimore) 44, 450 (1940).

HAWTHORNE, A. B.: Embryologic and clinical aspect of double ureter. J. Amer. med. Ass. 106, 189 (1936).

HELLSTRÖM, J.: Zur Kenntnis der Ureterocele. Acta chir. scand. 71, 339 (1932).

HESS, E.: Pyelocystostomy in crossed renal dystopia. J. Urol. (Baltimore) 22, 667 (1929).

HIGGINS, T. T., D. I. WILLIAMS, and D. F. E. NASH: The urology of childhood. London: Butterworth & Co. 1951.

HUTCH, J. A., and E. CHISHOLM: Surgical repair of ureterocele. J. Urol. (Baltimore) 96, 445 (1966).

—, F. HINMAN jr., and E. R. MILLER: Reflux as a cause of hydronephrosis and chronic pyelonephritis. J. Urol. (Baltimore) 88, 169 (1962).

HYAMS, J. A.: Aberrant blood vessels as factor in lower ureteral obstruction: preliminary report. Surg. Gynec. Obstet. 48, 474 (1929).

JEWETT, H. J., and A. P. HARRIS: Scrotal ureter: report of a case. J. Urol. (Baltimore) 69. 184 (1953).

JUHL, S.: Om ureterocele. Nord. Med. 33, 235 (1947).

KATZEN, B., and B. TRACHTMAN: Diagnosis of vaginal ectopic ureter by vaginogram. J Urol. (Baltimore) 72, 808 (1954).

KJELLBERG, S. R., N. O. ERICSSON, and U. RUDHE: The lower urinary tract in childhood; some correlated clinical and roentgenologic observations. Chicago: Year Book Publ., Inc. 1957.

KRETSCHMER, H. L.: Hydronephrosis in infancy and childhood. Surg. Gynec. Obstet. 64, 634 (1937).

LANE, V.: Ectopic ureter: elusive cause of urinary incontinence in the female. Lancet 1962 I, 937.

LAUGHLIN, V. C.: Retrocaval (circumcaval) ureter associated with solitary kidney. J. Urol. (Baltimore) 71, 195 (1954).

LENAGHAN, D.: Bifid ureters in children: an anatomical, physiological and clinical study. J. Urol. (Baltimore) 87, 808 (1962).

LEVACK, J. E.: Ectopic ureter syndrome commencing during pregnancy. Brit. J. Urol. 32. 152 (1960).

LIVADITIS, A., K. MAURSETH, and P. A. SKOG: Unilateral triplication of ureter and renal pelvis: report of a case. Acta chir. scand. 127, 181 (1964).

LUND, A. J.: Uncrossed double ureter with rare intravesical orifice relationship: case report with review of literature. J. Urol. (Baltimore) 62, 22 (1949).

MACALPINE, J. B.: Cystoscopy and urography, ed. 2. Bristol: John Wright & Sons 1936.

MARCEL, J. E.: Grosse hydronéphrose par uretère rétrocave. Arch. franç. Pédiat. 10, 274 (1953).

McGRAW, A. B., and O. S. CULP: Diverticulum of the ureter: report of another authentic case. J. Urol. (Baltimore) 67, 262 (1952).

MEISEL, H. J.: Ectopic ureter opening into a seminal vesicle. J. Urol. (Baltimore) 68, 579 (1952).

MERTZ, H. O.: A review of the subject of multiple ureters with a study of 16 unpublished cases. Urol. cutan. Rev. 22, 553 (1918).

— Bilateral duplication of the ureters with a compilation of recorded cases. Urol. cutan. Rev. 24, 636 (1920).

MEYER, R.: Normal and abnormal development of ureter in human embryo — mechanistic consideration. Anat. Rec. 96, 355 (1946).

MILLER, E. V., and R. E. TREMBLAY: Symptomatic blindly ending bifid ureter. J. Urol. (Baltimore) 92, 109 (1964).

11*

Moëll, H.: Size of normal kidneys. Acta radiol. (Stockh.) **46**, 640 (1956).

Moore, T.: Ectopic openings of the ureter. Brit. J. Urol. **24**, 3 (1952).

Murnaghan, M. F.: Quoted from D. I. Williams: Op. cit.

Nation, E. F.: Duplication of the kidney and ureter: a statistical study of 230 new cases. J. Urol. (Baltimore) **51**, 456 (1944).

Nixon, H. H.: Hydronephrosis in children — a clinical study of 78 cases with special reference to the role of aberrant blood vessels and results of conservative operations. Brit. J. Surg. **40**, 601 (1953).

Nordmark, B.: Double formations of the pelves of the kidneys and the ureters: embryology, occurrence, and clinical significance. Acta radiol. (Stockh.) **30**, 4 (1948).

Östling, K.: The genesis of hydronephrosis. Acta chir. scand. **86**, Suppl. **72** (1942).

Paquin jr., A. J.: Considerations for the management of some complex problems for uretero-vesical anastomosis. Surg. Gynec. Obstet. **118**, 75 (1964).

Parkkulainen, K. V., and F. Rejman: Extravesical ectopic ureter in infants and children: analysis of seventeen cases, including three cases with bilateral ectopy of the single ureters. Ann. Paediat. Fenn. **7**, 290 (1961).

Parks, R. E., and W. E. Chase: Retrocaval ureter: report of two cases diagnosed pre-operatively in childhood. Amer. J. Dis. Child. **82**, 442 (1951).

Passaro jr., E., and J. P. Smith: Congenital ureteral valve in children: a case report. J. Urol. (Baltimore) **84**, 290 (1960).

Payne, R. A.: Clinical significance of reduplicated kidneys. Brit. J. Urol. **31**, 141 (1959).

Petillo, D.: Ureterocele: clinical significance and process of formation: report of four cases. Surg. Gynec. Obstet. **40**, 811 (1925).

Pohlman, A. G.: Abnormalities in the form of the kidney and ureter dependent on the development of the renal bud. Bull. Johns Hopk. Hosp. **16**, 51 (1905).

Qureshi, M. A., and W. P. Mulvaney: Retrocaval ureter: report of two cases. Amer. Surg. **31**, 50 (1965).

Raffle, R. B.: Familial hydronephrosis. Brit. med. J. **1955 I**, 580.

Rank, W. B., G. T. Mellinger, and E. Spiro: Ureteral diverticula: etiologic considerations. J. Urol. (Baltimore) **83**, 566 (1960).

Ringer, M. G., and S. M. MacFarlan: Complete triplication of ureter: a case report. J. Urol. (Baltimore) **92**, 429 (1964).

Roberts, R. R.: Complete valve of the ureter: congenital urethral valves. J. Urol. (Baltimore) **76**, 62 (1956).

Ross, J. A.: Unusual variant of duplication of ureter. Brit. J. Urol. **20**, 125 (1948).

Rowland jr., H. S., R. C. Bunts, and J. H. Iwano: Operative correction of retrocaval ureter: a report of four cases and review of literature. J. Urol. (Baltimore) **83**, 820 (1960).

Sandegård, E.: The treatment of ureteral ectopia. Acta chir. scand. **115**, 149 (1958).

Stephens, F. D.: Double ureter in the child. Aust. N.Z. J. Surg. **26**, 81 (1956).

Swenson, O., and I. A. Ratner: Pyeloureterostomy for treatment of symptomatic ureteral duplications in children. J. Urol. (Baltimore) **88**, 184 (1962).

Thom, B.: Harnleiter und Nierenverdoppelung mit besonderer Berücksichtigung der extra-vesikalen Harnleitermündungen. Z. Urol. **22**, 417 (1928).

Thompson, G. J., and P. P. Kelalis: Ureterocele: clinical appraisal of 176 cases. J. Urol. (Baltimore) **91**, 488 (1964).

Thompson, H. T.: Problems in diagnosis and treatment of ectopic ureteral orifices. Surgery **45**, 593 (1959).

Thompson, I. M., and A. D. Amar: Clinical importance of ureteral duplication and ectopia. J. Amer. med. Ass. **168**, 881 (1958).

Uson, A. C., J. K. Lattimer, and M. M. Melicow: Ureteroceles in infants and children: a report based on 44 cases. Pediatrics **27**, 971 (1961).

Vittori, J., D. Moine, and J. Quinard: A case of reterocaval ureter. J. Urol. Néphrol. **68**, 440 (1962).

Wall, B., and H. E. Wachter: Congenital ureteral valve: its role as a primary obstructive lesion: classification of the literature and report of an authentic case. J. Urol. (Baltimore) **68**, 684 (1952).

Weigert, C.: Quoted by H. O. Mertz, Op. cit. 1918.

Wershub, L. P., and T. J. Kirwin: Ureterocele, its etiology, pathogenesis and diagnosis. Amer. J. Surg. **88**, 317 (1954).

Whitaker, J., and D. M. Danks: A study of the inheritance of duplication of the kidneys and ureters. J. Urol. (Baltimore) **95**, 176 (1966).

Williams, D. I.: Urology in childhood, vol. XV. Encyclopedia of urology. Berlin-Göttingen-Heidelberg: Springer 1958.

Young, J. D., and W. S. Kiser: Obstruction of the lower ureter by aberrant blood vessels. J. Urol. (Baltimore) **94**, 101 (1965).

Congenital Abnormalities of the Bladder

Victor F. Marshall and Edward C. Muecke

With 39 Figures

A. Development of the Bladder

In man the bladder is derived from the ventral portion of the cloaca and is not completely separated from the intestinal tract until the 16 mm stage is reached (6 weeks' gestation). Careful reconstruction of developmental events in early embryos by the use of wax models has shown that the allantoic structures are not incorporated in the evolving bladder anlage (CHWALLA, FELIX, KEIBEL). The following summary of developmental events in the formation of the lower urinary tract is based on the detailed description of CHWALLA.

I. The Cloaca

In the early developmental stages of the embryo, the cloaca is a blind pouch, which on cross section appears oval. Its long axis lies in a dorso-ventral position. Superiorly, the cloaca receives the hindgut and the allantoic duct. Its anterior wall is formed by the cloacal membrane, which consists of two layers of epithelium, a stronger ectodermal and a weaker entodermal one (Fig. 1). Seen from without, the membrane forms the floor of a shallow, rhomboid-shaped groove (Fig. 2) between two lateral folds that originally contained the umbilical vessels (KEIBEL). In an embryo of 4.84 mm in length, the cloacal membrane reaches superiorly almost to the allantoic duct; in later stages, the cloacal membrane shortens progressively as the mesoderm pushes itself wedge-like between the ecto- and endodermal layers of the membrane. Because of this superior shortening of the cloacal membrane, in the 6 mm embryo it forms the anterior wall of only the lower two thirds of the cloaca. The persistence of an abnormally large cloacal membrane at this developmental stage would prevent the mesoderm from reinforcing the essentially membranous infra-umbilical abdominal wall and thus play a major role in the genesis of the exstrophy-epispadias complex (MARSHALL and MUECKE).

The cloaca becomes divided into the hindgut and ventral cloaca by the crescent-shaped urorectal septum or fold. It can be recognized from the outside by lateral indentations, and it grows downward faster at the lateral walls of the cloaca than it does in the middle. These lateral folds have been called "the plicae urorectales" and it has been suggested that the urorectal septum could be formed by the mid-line fusion of these lateral folds; however, the absence of a mid-line raphe, such as would be formed by the union of two such folds, makes the description of the septum urorectale as unpaired more correct (FELIX). As the septum grows, it gains in antero-posterior dimension and pushes the rectum away from the ventral cloaca. In a 6.4 mm embryo, the septum has divided the cloaca into the hindgut and ventral cloaca down to the level of the openings of the Wolffian ducts. The ventral cloaca is now the anlage of the bladder and the primary urethra; that is, the female urethra or the cranial portion of pars prostatica urethra in the male to the openings of the ejaculatory ducts. The part that lies distal to the openings of the Wolffian ducts is the urogenital sinus. In an 8 mm embryo, the lateral limbs of the crescent-shaped urorectal fold shorten and change into a transverse fold in the 10 mm embryo.

II. The Genital Tubercle

By this developmental stage, the character of the anterior abdominal wall has undergone marked changes. The proliferating mesoderm in the immediate infra-umbilical region has encroached upon the membranous portion of the ventral cloacal wall (cloacal membrane) and the cloacal tubercle arises in that region. In the 7.8 mm embryo, the tubercle encompasses the area from the umbilicus down to the caudal end of the cloaca. Between the lateral limits of the tubercle and the medial limits of the lower extremities, the raised genital folds appear in the 8 to 9 mm embryo. With the continued growth of the cloacal tubercle, the cloaca is drawn into the tubercle and the cloacal membrane rotates further downward, forming the base of the cloacal tubercle and the caudal end of the undivided cloacal portion (Fig. 3).

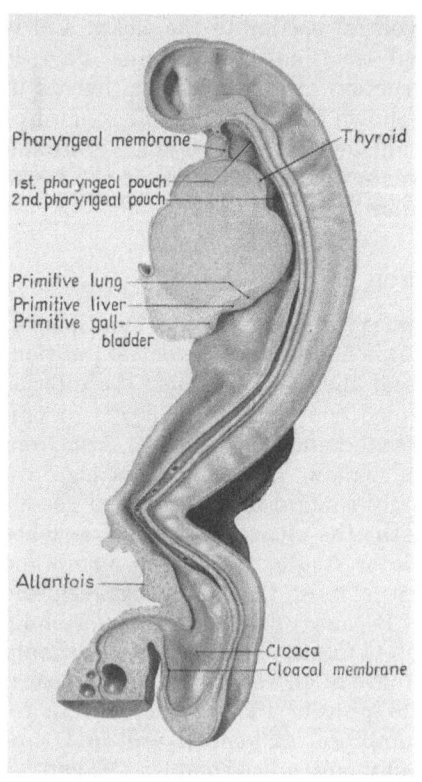

Fig. 1. Wax model reconstruction of a 30 day-old embryo (Carnegie embryo no. 2053). Mid-sagittal view on left. (From Davis)

III. Separation of the Bladder from the Intestinal Tract

As the division of the cloaca continues, the urogenital sinus lengthens distally. The communication between rectum and urogenital sinus becomes restricted to a small connecting passage, the cloacal duct of Reichel. Should development be temporarily inhibited at this stage, then this cloacal duct may persist, resulting in recto-urethral or recto-vaginal fistulae with imperforate anus.

In the 12 mm embryo, the urogenital sinus can be differentiated into the distal "pars phallica" and the proximal "pars pelvina." In cross sections, the pars pelvina is a flattened tube with many folds in the epithelial lining and the pars phallica is almost triangular in shape. Finally, in the 15.85 mm embryo, the urorectal septum meets and fuses with the cloacal plate and thus divides it into the urogenital membrane, or plate, anteriorly and the anal membrane, or plate, posteriorly. At the same time, the urogenital membrane dehisces and forms the urogenital orifice. Thus, the primitive perineum is formed. The cloacal tubercle is now designated as the genital tubercle or phallus (Fig. 4).

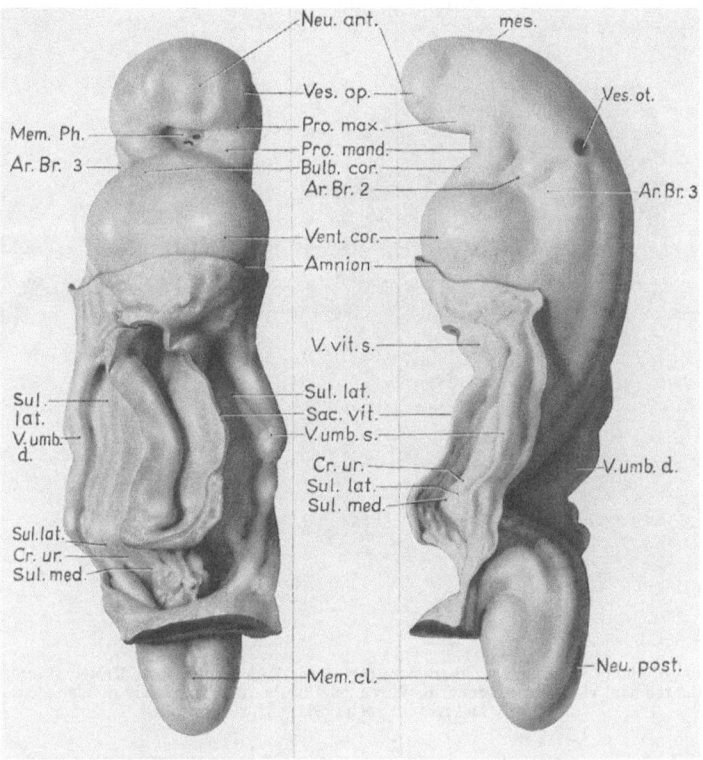

Fig. 2. External view of the 30-day embryo shown in Fig. 1. The prominent cloacal membrane (Mem. Cl.) occupies the central portion of the infra-umbilical abdominal wall flanked by ridges of mesoderm. (From Davis)

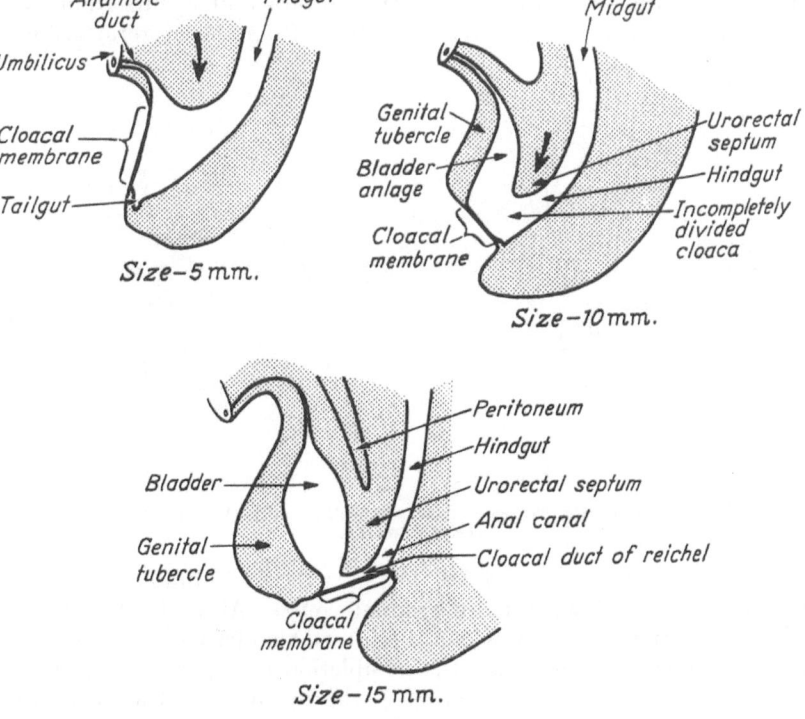

Fig. 3. Developmental changes of the cloaca and cloacal membrane. (Drawings are not to scale)

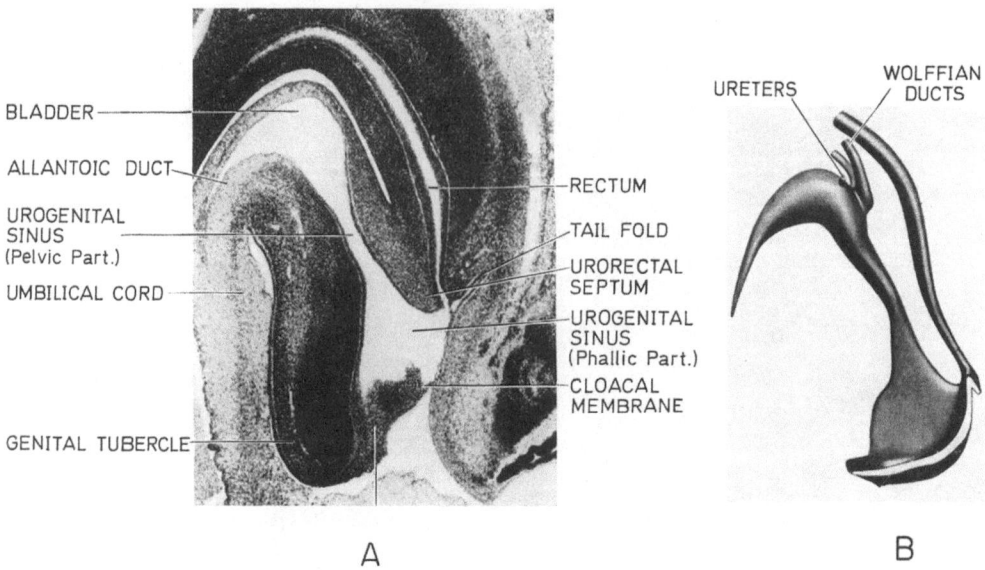

BLADDER

ALLANTOIC DUCT

UROGENITAL
SINUS
(Pelvic Part.)

UMBILICAL CORD

GENITAL TUBERCLE

RECTUM

TAIL FOLD

URORECTAL
SEPTUM

UROGENITAL
SINUS
(Phallic Part.)

CLOACAL
MEMBRANE

URETERS

WOLFFIAN
DUCTS

A

B

Fig. 4. Separation of bladder from intestinal tract in the 16 to 17 mm embryo. A, Urorectal septum in process of fusion with the remnant cloacal membrane. B, Wax model of the now completely divided cloaca into bladder anteriorly and rectum posteriorly. (After CHWALLA)

IV. Formation of the Trigone

After the division of the cloaca is completed, further development of the bladder anlage is characterized by the rapid growth and development of the bladder neck and the pars prostatica of the urethra that results from the rapid and increasing separation of the Wolffian ducts from the ureteral orifices. The

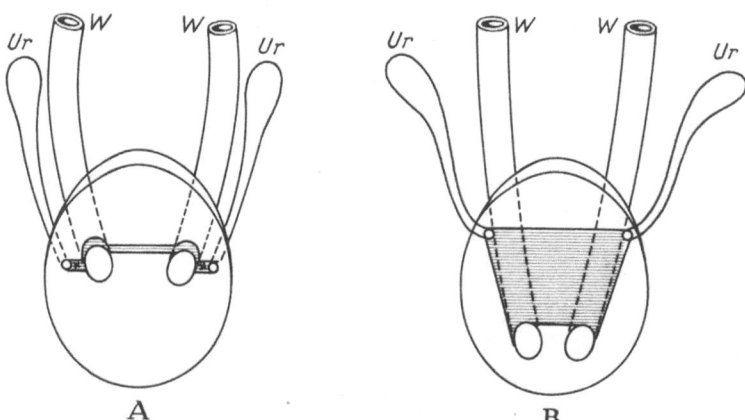

A

B

Fig. 5. Schemata of trigone formation. Trigone forms as the Wolffian ducts (W) separate from the ureters (Ur). (From CHWALLA)

trigone also is formed by this process. The growth of the fundus of the bladder, on the other hand, follows a more gradual course. At first, the bladder and urethra form a flattened tube in which the future bladder neck can be recognized as a narrowed section. Soon after the completion of separation of the ureters from the Wolffian ducts (12 mm embryo), the openings of the Wolffian ducts commence to migrate caudally and, at the same time, to approach each other.

Thus, the bladder trigone is formed. According to CHWALLA, the separation of the orifices of the ureters and Wolffian ducts is facilitated by the appearance and rapid growth of a new segment of posterior bladder wall that forms the trigone and the posterior wall of the primitive urethra (pars prostatica). The anlagen of this new segment were the epithelial margins of the original Wolffian duct openings that fused with the ureteric septa (mesoderm) and the narrow strip of posterior bladder wall between the two openings (entoderm) (Fig. 5). It is therefore held that the trigone of the bladder is in part a mesodermal derivative.

V. Regression of the Allantoic Duct

The allantoic duct, which originally communicated with the bladder vertex, shows obliterative changes of the distal lumen in the 10.6 mm embryo. The duct is almost completely fused in the 13 mm embryo and here the vertex of the bladder reaches to the umbilicus. In the 17 mm embryo, the obliteration and regression of the allantoic duct is complete. As development progresses, the bladder retains its umbilical attachment but its upper part narrows more and more and forms the urachus (BEGG). Therefore, the urachus is derived from the vertex of the bladder and not from the allantois (FELIX).

VI. Development of the Bladder Musculature

The musculature of the bladder varies in its development. FELIX observed in a 22.5 mm embryo the appearance of the first muscle layer as a longitudinal musculature extending from the apex to about the ureteral orifices. According to CHWALLA, the musculature of the bladder is first seen in the 20 mm embryo as longitudinal smooth-muscle bundles. A few transverse fibers are also seen in the ventral wall. The muscle layer appears separated from the bladder epithelium by a zone of loosely arranged embryonal connective tissue, the anlage of the tunica propria. The longitudinal muscle coat always appears stronger on the dorsal than on the ventral surfaces and surrounds the epithelium in wide curves, so that broad sheets of mesenchyme persist. In this loose embryonal connective tissue, a layer of circular muscle fibers develops in the 26 mm embryo, again appearing first in the upper portion of the bladder. In the older embryo, a third and last layer, an inner longitudinal musculature, is formed, and both longitudinal layers give off oblique bundles to the intervening circular layer. In the 80 to 90 mm embryo, the muscle layers lose their distinctiveness and the entire muscle coat appears to be tightly interwoven. The trigone of the bladder appears to be still devoid of musculature in the 50 mm stage according to CHWALLA although the anterior anlage of the bladder neck appears to be forming. The bladder neck was well formed in a 68 mm embryo according to CHWALLA but found to be somewhat later (in a 90 mm embryo) by FELIX. The musculature of the trigone is first seen as slender, loosely arranged muscle bundles in the 100 mm embryo. This single circular muscle syncitium is limited superiorly by the denser circular muscle layer forming the interureteric ridge and inferiorly by the densely grouped muscle fibers of the bladder neck.

VII. Fetal Descent of the Urinary Tract

A descent of the bladder and urethra already has occurred at the end of the second trimester. The cause of descent of the urinary tract probably is the growth of the bony pelvis and the deepening of the pelvic bowl. In the process of descent the pouch of DOUGLAS shortens and the posterior wall of the urogenital sinus and Müllerian tubercle lose their peritoneal cover. The vertex of the bladder continues to narrow into a stalk-like structure, thus forming the urachus. *Theoretically*, failure of descent could result in an excessively long urethra, thus causing increased resistance to urine flow and increased intravesical pressure, as well as in a continued high attachment of the bladder to the umbilicus, with a resulting vesico-cutaneous fistula at the time of separation of the cord.

VIII. Summary of important Early Developmental Events

A brief summary of important anatomic events in the embryogenesis of the bladder is presented in Tabulation 1. The time period in question is the first 6 weeks of the developing embryo. The anatomic details and embryo size are from CHWALLA; the timetable is based on STREETER.

Tabulation 1

At 3 wk. (2—3 mm in size)	Tail portion of embryo develops and cloaca forms as a cul-de-sac. Wolffian ducts from the primitive kidneys reach to the cloaca but do not communicate.
	Bilaminal cloacal membrane forms ventral wall of cloaca.
At 3$^1/_2$ wk. (4—5 mm in size)	Cloaca has formed two funnel-shaped evaginations in direction of distal ends of Wolffian ducts.
	Ureteric buds can be seen along the dorso-medial wall of each Wolffian duct.
	Cloacal membrane shows greatest dimension and superiorly reaches almost to the allantoic duct.
At 4 wk. (5—8 mm in size)	Urorectal septum grows caudally and begins to separate ventral bladder from hindgut.
	Wolffian ducts communicate with cloacal lumen.
	Ureteric buds are well developed along dorsal walls of Wolffian ducts and are capped by dense metanephric blastoderm.
	Cloacal membrane begins to regress superiorly as mesoderm sandwiches itself between the ecto- and endodermal layers.
	Cloacal tubercle forms.
	Allantoic duct begins to become obliterated.
At 5 wk. (8—12 mm in size)	Cloacal tubercle enlarges, thereby progressively obliterating the cloacal membrane superiorly.
	Only the caudal third of the ventral wall of the cloaca is still membranous.
	The urorectal septum has almost completely divided the bladder from the hindgut; the narrow communication between rectum and urogenital sinus is called "cloacal duct of Reichel".
	The Wolffian ducts appear to have shifted caudally; this, however, is due to the rapid growth of the bladder in a superior direction.
	The ureters are well demarcated from the Wolffian ducts and, having grown into long, tubular structures, course in a gentle curve along the Wolffian ducts.
	The allantoic duct is sharply differentiated from the apex of the bladder and forms an almost solid cord.
At 6 wk. (12—16 mm in size)	Division of the cloaca is complete as the urorectal septum fuses with the cloacal membrane.
	The cloacal membrane dehisces and establishes the urogenital opening.
	The genital tubercle is very prominent.
	Separation of ureters from the Wolffian ducts is completed but the ureters are temporarily closed by epithelial membranes.
	The Wolffian ducts appear to migrate in a caudal direction and the trigone begins to form.
	The allantoic duct is a solid cord.
	The urogenital sinus can be differentiated into a narrow, waist-like "pelvic" part and the distal, triangular-shaped "phallic" part.

B. Congenital Abnormalities of the Bladder

In this chapter, the primary anomalies of the urinary bladder will be described. Many of these congenital abnormalities are quite rare and only a few cases have been recorded in the literature. Abnormalities of the distal ureters and the uterovesical junction, as well as those of the bladder neck are not included but are described elsewhere (Chapters 1, 3 and 5).

I. Agenesis of the Bladder

Complete absence of the urinary bladder is very rare and appears to be more of anatomic than of clinical interest. When the ectopic ureters are unobstructed this congenital anomaly can be compatible with life and surgical correction or diversion is necessary. Twenty-five cases have been described in the world literature (CAMPBELL, ENEREN, GLENN, GRUBER, IGNATESCU et al., LEPOUTRE, MILLER).

1. Incidence

It is difficult to obtain any meaningful figures in such an unusual condition. CAMPBELL stated an incidence of seven cases in 19,046 autopsies, and GLENN described one child seen with this anomaly among 600,000 new hospital admissions.

2. Pathology

It appears that the abnormal embryogenesis is one of arrest or delay in the strict timetable of early development. LEPOUTRE described a persistent cloaca in a 4-month-old boy. Here, apparently, the ureters entered the rectal wall. They were unobstructed and an excretory urogram was normal. This case would be

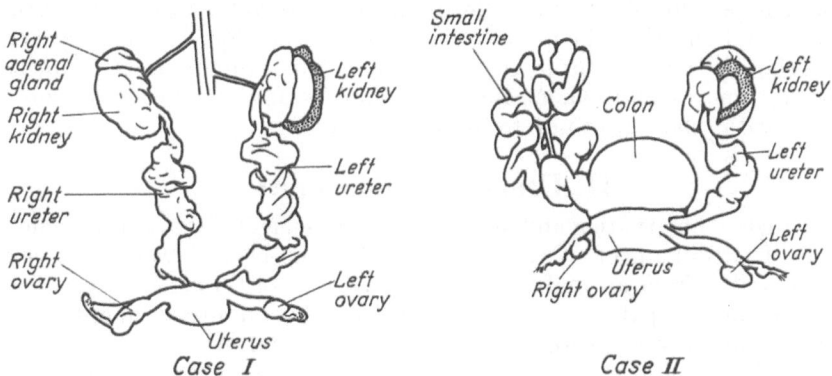

Fig. 6. Drawings of the urinary tracts in a pair of stillborn twins with bladder agenesis as reported by IGNATESCU et al. In each case, the ureters terminated in the uterus and the kidneys were hydronephrotic

an example of arrest or failure of formation of the urorectal septum. The other, and apparently more common mechanism, is the delay or failure of separation of the ureters from the Wolffian ducts. Thus, the ectopic ureteral orifices come to lie in the structures derived in part from the Wolffian duct system. In the case described by ENEREN the ureters opened into the posterior urethra. MILLER reported two ectopic ureters in the anterior vaginal wall of a 27-year-old woman with absence of urethra and bladder. The patient had been totally incontinent since birth and both upper urinary tracts were mildly obstructed. IGNATESCU et al.

reported two cases of bladder agenesis in stillborn twins in which the ureters terminated in the uteri. The upper tracts were grossly hydronephrotic (Fig. 6). In the case reported by GLENN the ectopic ureters opened into the urethro-vaginal septum, from which urine dribbled continuously (Fig. 7). An intravenous pyelogram showed bilateral hydronephrosis and duplication of the left upper urinary tract.

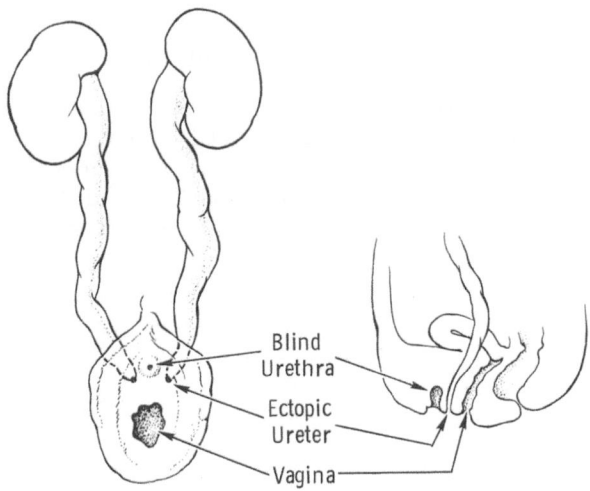

Fig. 7. Agenesis of the bladder as reported by GLENN. Composite drawing shows blind urethra and ectopic ureteral orifices

3. Treatment

Treatment should be directed toward improved urinary drainage to relieve obstruction; secondarily, to relieve incontinence. Should the degree of obstruction be severe and the child in marked renal decompensation, preliminary nephrostomies should be done prior to any diversionary procedure in order to ameliorate the grave clinical condition. As final diversion, we favor the construction of an ileal loop with bilateral ileo-ureterostomies.

II. Urachal Abnormalities

The original observation and description of a congenitally patent urachus were made by BARTHOLOMAEUS CABROLIUS in 1550. The patient was an 18-year-old girl who voided from an umbilical opening as well as through the urethra. His attempt to cure the patient by ligation of the fistula and dilatation of the urethra was successful (HERBST, CHERRY, NIX et al.). Urachal abnormalities have been reported in the veterinary literature. COOK, SALVISBERG, and CULLEN stated that the process of tearing off the cord frequently leads to umbilical fistulae in foals. Apparently, the formation of the urachus in horses is a late (neonatal) development and at birth the cavity of the bladder extends to the umbilicus. Because of this, urachal anomalies have been considered by some authors to represent an atavistic occurrence (BEGG).

1. Incidence

BEGG, in 1927, collected 72 cases of patent urachus from the world literature. Of these, he was able to study 58 cases in detail, the majority of which (68%) had been diagnosed before the age of 6 years. TRIMINGHAM and McDONALD reported 14 cases of urachal anomalies from the Mayo Clinic, and HINMAN described 18 cases of urachal disorders seen at the University of California School of Medicine in San Francisco. NIX et al., in 1958, reviewed 100 cases of patent urachus from the literature and reported an incidence of three cases among 200,000 admissions to the Children's and Infants Hospital in Boston and three cases among 1,168,760 admissions to the Charity Hospital of New Orleans.

2. Pathology

In embryos 10 to 24 mm in size, the bladder reaches to the umbilicus (BEGG, 1927). As development progresses, the apical portion of the bladder begins to narrow and the urachus forms. As the bony pelvis develops, the bladder begins its descent. At birth, the bladder apex is 4 cm above the symphysis (BEGG, 1929) and it retains its attachment with the umbilical structures by means of the well-defined tubular urachus. Anatomic studies based on 35 specimens from premature and term fetuses showed the urachus to be a muscular tube in all specimens, with the well-developed muscle coat directly continuous with that of the bladder dome (HAMMOND et al.). Fifty per cent of the specimens had a communication between the bladder cavity and the urachal lumen, albeit the opening was always minute (less than 1 mm in diameter). Earlier studies by LUSCHKA had shown that a central epithelial canal was readily dissectable even in adult specimens. Often cyst-like dilatations of the canal were seen (LUSCHKA's lacunae) that usually occurred singly and were filled with debris and amorphous concretions. Dissection of adult cadavers showed that the neonatal relationship of the urachus and umbilical arteries was maintained essentially unchanged (HAMMOND et al.). The varying degree of atrophy of the urachus found is probably secondary to the growth of the abdominal wall and the descent of the bladder in the neonatal period. In about one quarter of the cadavers, the urachus appeared as a tubular structure with a muscular wall; in another one quarter, it appeared as a smaller canal of fibrous tissue; and in the remaining one half, the lower half of the urachus was a muscular tube that gradually tapered into a fibrous cord in its upper half. The length of the urachus varied from 3.4 cm to 12 cm. In 10% of the adult specimens, a gross continuity between bladder cavity and urachal lumen was demonstrable, and in all cases in which the urachus was a muscular tube, the central epithelial canal was readily dissectable.

Thus, in summary, the following developmental events shall be reiterated: (1) Although the descent of the urinary tract commences toward the end of the second trimester as the pelvis deepens, the bladder is not entirely a pelvic organ at birth but lies high enough that its apex may reach the umbilicus. (2) The apical portion of the bladder remains attached to the umbilicus but narrows progressively into a narrow muscular tube, the urachus, which persists in the adult as a fibromuscular appendage at the bladder apex. (3) The epithelial canal of the urachus does not become entirely obliterated even in the adult. Consequently, patent urachus is principally a disorder of the neonatal period, at which time the bladder has yet to descend and the urachus has not been fully formed. On the other hand, urachal cysts occur principally in the adult patient after descent and maturation of the urachus have taken place (HINMAN).

3. Classification

HINMAN, in 1961, proposed the following classification based upon that of VAUGHAN, which had been followed in the older literature (Fig. 8).

(a) Congenital patent urachus — the urachus remains patent or the apex of the bladder fails to form a urachus. A urinary fistula is present at birth ("completely patent").

(b) Vesico-urachal diverticulum — the urachus is patent only at its vesical termination ("blind internal").

(c) Umbilical cyst or sinus — the subumbilical portion of the urachus remains patent ("blind external").

(d) Alternating urachal sinus — both ends of the urachus are potentially patent, thereby allowing infection to spread centrally to the bladder, peripherally to the umbilicus, or to both areas simultaneously ("blind").

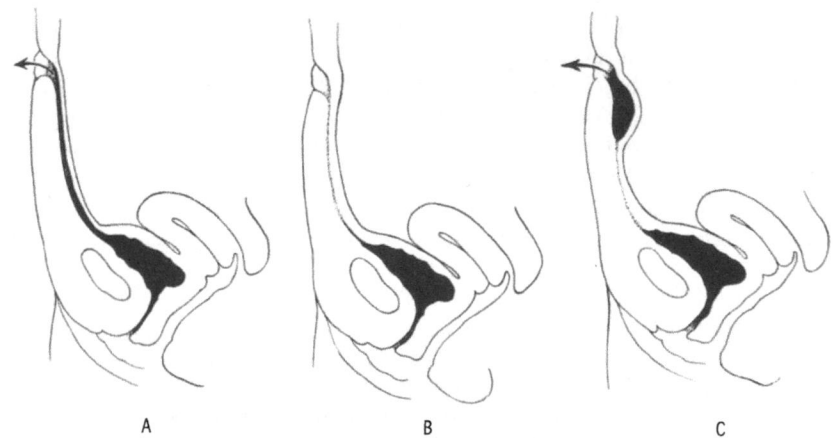

Fig. 8. Illustrations of the three major types of congenital urachal abnormalities. A, Congenital patent urachus. B, Vesico-urachal diverticulum. C, Umbilical cyst and sinus. (From HINMAN)

4. Clinical Features

The symptomatology of urachal anomalies varies according to the degree of patency of the urachal lumen. Most of the clinical manifestations of umbilical disorders of urachal origin should be considered congenital although symptoms may not have appeared until adulthood. Re-opening of the urachal canal in normal individuals has been described as impossible (BEGG, 1927) for more than a short distance (5 cm) above the bladder, although it is possible that a persistent, albeit a very minute, urachal canal could dilate in the face of prolonged urinary-tract obstruction and infection (category "d" in HINMAN's classification). Urachal lesions must be distinguished from other disorders of the umbilicus, such as oomphalitis and patent oomphalomesenteric duct, and at times a radiographic study of the upper intestinal tract is essential in differential diagnosis.

a) Congenital Patent Urachus

The presence of an unusually large umbilical cord at birth is commonly associated with a patent urachus. Reflux of urine through the urachus into the cord can cause cystic dilatation of the cord prior to separation (McCAULEY and LICHTENHELD). The diagnosis is readily made by observing leakage of urine from the umbilicus; however, usually a few hours or even days are necessary to allow the umbilical cord to slough. The umbilicus in these infants may be relatively normal in appearance or may show a tumor-like process consisting of the granulating stump of the cord, which is being nourished by the urachal arteries. Commonly, excoriation of the surrounding skin appears. Severe types of patent urachus have been described as "persistent fetal bladders" and have been the cause of fetal dystocia (SIMON and BRANDEBERRY, ROSENBERG). Simple diagnostic procedures that might be employed to prove the patency of the urachus have been described: (1) passage of catheters via the fistula into the bladder; (2) analysis of the draining fluid that shows it to be urine; (3) injection of methylene blue into the fistula and observance of blue dye in the voided specimen; (4) intravenous injection of indigocarmine and observance of discoloration of the umbilical fluid. In all cases of persistent urachal fistula in the newborn the possibility of lower urinary-tract obstruction must be ruled out. An excretory urogram and a voiding cystogram should be obtained in all children with this anomaly prior to any attempt at treatment, for there appears to be a high incidence of associated obstructive anomalies. HINMAN reported that three of seven cases of congenital patent urachus were associated with lower urinary-tract obstruction, and in HERBST'S series of 67 children with urachal anomalies, 12 (18%) had other anomalies, usually those producing urethral obstruction. In 50% of children with anterior abdominal muscle deficiency (the "prune-belly syndrome"), the urachus is patent and an outlet obstruction of the bladder appears to be present (COOPER and KINTZEN).

b) Vesico-urachal Diverticulum

This is commonly considered an "acquired" disorder, for most cases have been described in adults and usually have been associated with lower urinary-tract obstruction. However, HAMMOND et al. found a patent distal urachal lumen communicating with the bladder cavity in 10% of adult cadavers that were carefully dissected. Hence, this should be considered a congenital disorder that is asymptomatic until an obstructive feature is superimposed. In the cases that have been described, the symptomatology is that of lower urinary-tract infection. An interesting case reported by DREYFUSS and FLIESS was that of an 11-year-old girl. She had had an umbilical hernia repaired 2 hours after birth; thereafter, urine-like fluid drained from the umbilicus for 1 year. She failed to thrive. At 5 years of age she had pyuria and passed a stone per urethra. At 11 years of age, shortly after passing another stone, she was hospitalized with fever and convulsions and died of sepsis 36 hours after admission. At autopsy, the bladder was pear-shaped and extended to the umbilicus, where the fundus was adherent. The upper portion of the bladder showed severe cystitis and a 4×2 cm uric acid calculus was lodged in the apex of the bladder. Other cases of concretions in a vesico-urachal diverticulum have been reported by WARD, SIDDALL, LADD and GROSS, and WINTZELL. The diagnosis of a vesico-urachal diverticulum, when symptomatic, is made on cystoscopy and a voiding cystogram. More commonly, however, this condition in the asymptomatic patient is found incidentally at autopsy (HINMAN).

c) Umbilical Cyst and Sinus

The first clinical report of cyst formation of the urachus to be found in the English literature was contributed by LAWSON TAIT in 1886. Infected umbilical cysts are usually more common in adults, yet the underlying cause is a remnant of the urachal duct communicating externally in the region of the umbilicus. When infection occurs in this remnant a purulent discharge may be present, or, if the external opening of the sinus is occluded, para-umbilical induration and pain may be present. Tenderness to palpation is elicited, the overlying skin may be inflamed, and generalized symptoms of infection may be the presenting complaints. When the cyst ruptures it usually drains through the umbilicus. The largest cyst on record is that in Rippmann's case, which contained 52 liters of fluid.

d) Alternating Urachal Sinus

This term has been used by Hinman to describe those cases of urachal anomalies that present with clinical findings of both umbilical and urinary-tract infection. It is believed that the path of infection is the patent urachal canal, that the infection probably arises in the canal filled with epithelial desquamate and from there spreads both externally to the umbilicus and internally to involve the lower urinary tract. The diagnosis is confirmed when cystoscopic examination shows an inflammatory area in the dome of the bladder and a purulent discharge occurs from the inflamed umbilicus.

5. Treatment

Many modalities of treatment have been advised. Table 1 summarizes the therapy of congenital patent urachus in 124 cases reported by Nix and co-workers. Surgical intervention in the newborn with a patent urachus should not be delayed

Table 1. *Summary of treatment of congenital patent urachus in 124 cases (from Nix et al.)*

Type treatment	Tot. no. pt.	Result, no. pt.					
		Clos.	Non-clos.	Re-open.	Devel. ca.	Devel. cyst	Died
Ligation umb. tumor	6	4	1	1	0	0	0
Ligation and excision umb. tumor	4	3	1	0	0	0	0
Ligation and caut. umb. tumor	3	2	1	0	0	0	0
Caut. umb. tumor	13	3	6	0	3	1	0
Incision and caut. umb. tumor	1	1	0	0	0	0	0
Simple suture umb. tumor	5	2	1	2	0	0	0
Umb. tumor sutured and strapped	2	2	0	0	0	0	0
Umb. tumor taped or strapped	5	2	3	0	0	0	0
Circumcision only	5	2	3	0	0	0	0
Urachus excised (intraperitoneally)	12	11	0	0	0	0	1
Urachus excised (extraperitoneally)	31	31	0	0	0	0	0
Abd. surgery; urachus untreated	3	0	0	0	0	0	3
Spontaneous closure	3	2	0	1	0	0	0
No treatment given	21	0	0	0	0	0	5
Total	124	65	16	4	3	1	9

once the diagnosis is made unless delaying factors, such as infection, obstructive uropathy, poor renal function, multiple congenital anomalies, and prematurity, are present. Best results are obtained with complete surgical removal of the urachus, preferably through an extra-peritoneal approach. Small draining sinuses at the umbilicus may be treated with cautery although surgical extirpation of the sinus tract and urachal remnant is a better form of treatment. When the urachus is excised, the dome of the bladder should be closed in layers and the bladder drained by catheter for at least 5 days postoperatively.

III. Duplications of the Bladder

Duplication of the urinary bladder and related anomalies are rare and only a few cases have been described in the literature. Most of the early cases reported were based upon autopsy findings and many descriptions are vague as to the exact nature of the bladder anomaly encountered. There are a number of other malformations that might suggest a duplication, such as bladder diverticula, ectopic ureteroceles, and dilated posterior urethras in the cases of urethral valves. WEHRBEIN described a case of bilocular bladder in a 4-year-old child that is more correctly classified as a ureterocele; similarly, SENGER and SANTARE reported their findings in a case of congenital multilocular bladder in a 16-year-old girl in which the second loculus was really a cystic dilatation of the lower end of the ureter occurring in the bladder cavity. As diagnostic methods have improved,

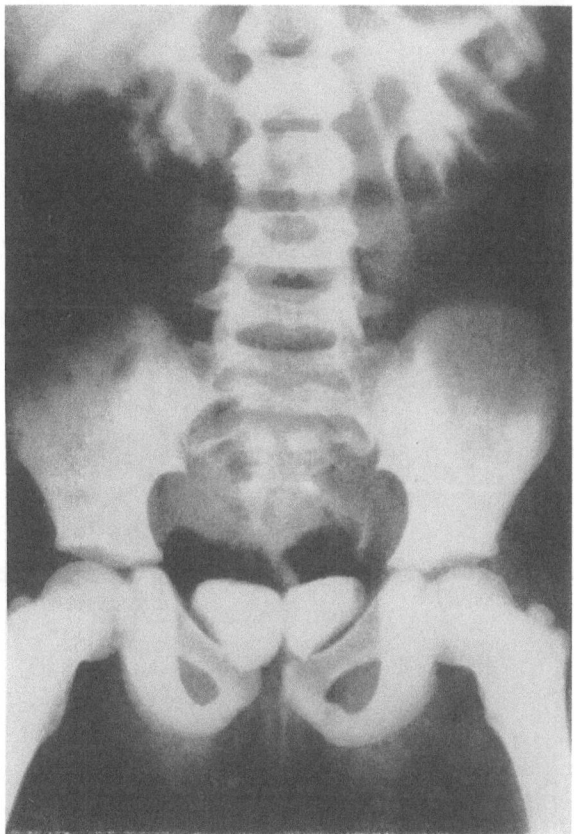

Fig. 9. Roentgenographic demonstration of a double bladder following intravenous urography. (From NESBIT and BROMME)

more of the cases of bladder duplication have been diagnosed in living persons and surgical correction, when necessary, has been attempted. The first radiographic diagnosis of true duplication of the bladder and urethra was reported by NESBIT and BROMME in 1933 (Fig. 9).

1. Classification

BURNS et al., in 1947, reviewed the literature and arranged the cases that were indeed either complete or incomplete bladder duplications into groups according to the pattern into which they seemed to fit best. A more recent review by ABRAHAMSON follows the classification of BURNS et al. (Fig. 10).

1. *Complete duplication*
(anterior view).

2. *Incomplete duplication*
(anterior view).

3. *Complete sagittal septum*
(anterior view).

4. *Incomplete sagittal septum*
(anterior view).

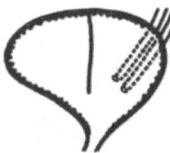

5. *Incomplete frontal septum*
(lateral view).

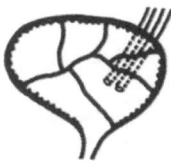

6. *Multiseptate bladder*
(lateral view).

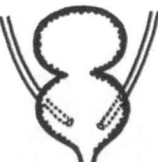

7. *Hour-glass bladder*
(anterior view).

Fig. 10. Various forms of bladder duplication. (From Abrahamson)

a) Complete Duplication

Two separate bladders are present, each quite distinct, lying side by side, and separated by a peritoneal fold of varying depth. The bladder walls are quite distinct and made up of the intact epithelial and muscular coats. Each bladder receives a ureter from the ipsilateral kidney and each bladder empties through a separate urethra. The external genitalia may or may not be diphallic.

Eighteen cases of this interesting anomaly have been reported (Abrahamson, Swenson and Oeconomopoulos, Wojewski and Kossowski). Other abnormalities of cloacal development are frequently present and congenital anomalies of the upper urinary tracts may occur in association with this as with other types of bladder anomalies. Of the 18 cases reported, nine (50%) had duplication of the lower gastrointestinal tract. Six had complete duplication from the ileocecal region, and three were duplicated more distally. RAVITCH, in 1953, reviewed 20 cases with hindgut duplication and found in 12 (60%) a definite description of major anomalies of the genito-urinary tract; in eight (40%) there was a double bladder present. Duplication of the external genitalia occurred in 16 (90%) of the reported cases of bladder duplication. Double penes were present in eight (44%). Abnormal cloacal development occurred in six cases (33%), the most common type being ano-rectal atresia with recto-vesical, recto-urethral, or recto-vaginal fistula. Duplication of the lower vertebral column was present in two cases.

b) Incomplete Duplication

The bladder is bilobed, separated by a peritoneal fold of varying depth. The two bladders have a single bladder neck and urethra.

Five such cases have been reported in the literature (ABRAHAMSON), the most recent being that of a 4-year-old boy (BOISONNAT). Cystogram and retrograde urethrogram showed a double bladder with right vesico-ureteral reflux. On cystoscopy, the instrument passed readily into both bladders and the septum was visualized. Both ureteral orifices were seen. Associated anomalies were incompletely rotated, low-lying kidneys. In the other cases reported (BURNS et al.) few if any other associated deformities were described, a marked contrast to complete duplication of the bladder.

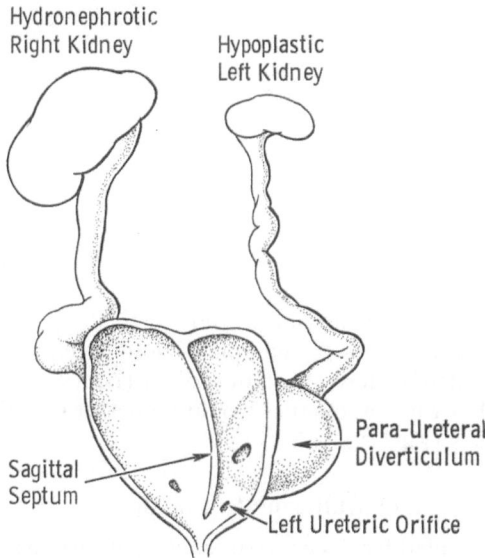

Fig. 11. Complete sagittal septum of the bladder in a 9-year-old girl. (From ABRAHAMSON)

c) Sagittal Septum

Both complete and incomplete sagittal septa have been described. Should the septum be complete, the bladder is divided into two equal or unequal chambers only one of which drains per urethra. Each chamber receives a ureter; if the blind chamber drains a functioning kidney, it distends and the increased intravesical pressure results in massive hydronephrosis and hydroureter on the ipsilateral side. Such changes may occur before or after birth. Abrahamson described such a case in a 9-year-old girl (Fig. 11) and listed five other cases from the literature, one of which, however, appears to have been an ectopic ureterocele.

In the cases of incomplete sagittal septum of the bladder, the attachment of the septum stops short at the lower end so that two bladder chambers communicate with one another and the urethra. However, the communication may be extremely small. Eight such cases have been recorded and in only one (Menton and Denny) were there associated anomalies in the form of duplicated appendix and colon.

d) Frontal Septum

A septum in the frontal plane divides the bladder, completely or incompletely, into two chambers lying antero-superior or infero-superior to one another. The

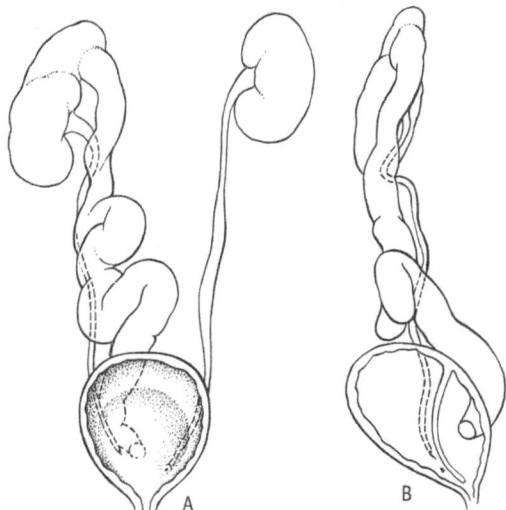

Fig. 12. "Bilocular" bladder with duplicated right upper urinary tract as described by Wehrbein. A, Front view. B, Side view. This case is more correctly classified as a ureterocele

thickness of the septum varies but most frequently it appears to be composed of bladder epithelium. Three cases of "complete" and seven cases of "incomplete" frontal septum of the bladder have been described (Burns et al., Abrahamson); however, it appears that these anomalies would be better classified as ureteroceles and ectopic ureteroceles (Fig. 12).

e) Multiseptate Bladder

In this anomaly, the bladder cavity is divided by multiple septa into numerous blind and communicating chambers. Only one such case has been described, that in a 10-day-old male infant (Kohler). The diagnosis was established at autopsy. Histologically, the septa were composed of fibro-muscular layers covered by bladder epithelium. Multiple other anomalies were present. The infant had died of renal failure.

f) Hourglass Bladder

The bladder is divided by a constriction in the horizontal plane into an upper and lower compartment. The bladder wall of each chamber is essentially normal

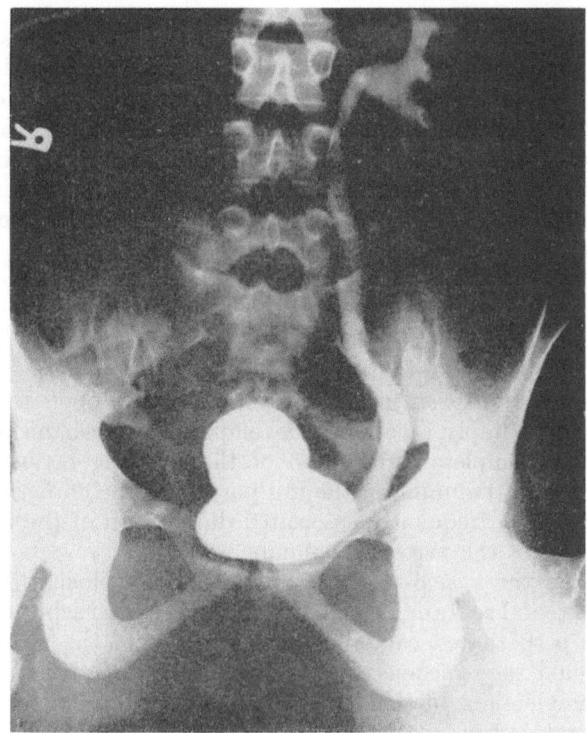

a

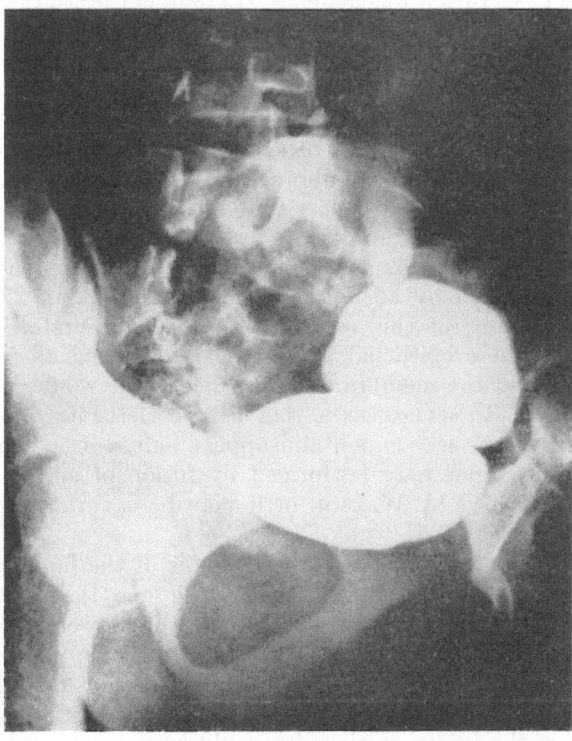

b

Fig. 13. Hourglass deformity of the bladder. (From KEARNS and TURKELTAUB)

except at the ring-like constriction, or waist, where an increased amount of fibro-muscular tissue is present. This condition must be carefully distinguished from that in which there are large apical diverticula (vesico-urachal diverticula) or from that in which the posterior urethra has dilated enormously behind urethral valves. Twenty-two cases of hourglass bladders have been described (Zeller-mayer and Carlson). In most of the cases described, the ureters entered the lower chamber. Although this condition has been described as congenital, it has not been recorded in the young child (Williams).

2. Pathology

Despite several attempts to explain the embryogenesis of this group of anom-alies, a comprehensive description is still lacking. Although these variations of bladder duplication have been grouped together, based upon their clinical de-scriptions, it does not imply a common developmental mechanism.

In the cases of complete duplication of the bladder, Ravitch and Scott proposed that a partial twinning of the tail portion of the embryo had occurred. That would explain the frequently associated duplication of the lower intestinal tract and anomalies of the vertebral column.

If a splitting of the vesico-urethral anlage (ventral cloaca) occurred at the time that the uro-rectal septum formed, then a mid-sagittal septum of the ventral cloaca would be perpetuated caudally as an element of the uro-rectal septum. Each bladder would then receive the ipsilateral ureter (Satter and Mossman). The degree of completeness of this sagittal division would depend upon the time when the uro-rectal septum acquired the sagittally oriented fold. The occurrence of supernumerary uro-rectal septa has been postulated by Burns et al.

An interesting observation was made by Watson in the study of fetal bladders. He demonstrated that on cross sections of these bladders many epithelial folds were fusing into definite septa, all of which appeared to occur in the sagittal plane. Hence, the formation of sagittal septa might be secondary to coalescence of mucosal folds of the fetal bladder.

It is conceivable that the origin of some bipartite bladders is related to a faulty resorption of Chwalla's membrane and thus has a similar causation to that of ureteroceles. Indeed, it appears that most of the cases earlier classified as "biloculated" bladders or frontal septa are more properly grouped as uretero-celes. As Chwalla has shown, as the ureters terminate in the early bladder anlage, they are occluded by an epithelial membrane (12 to 29 mm embryos). Just prior to dehiscence of this membrane, the distal ureters appear dilated immediately above these epithelial membranes, indicating an increased hydro-static pressure. Should the membrane fail to dehisce, it would balloon outward, forming a ureterocele. In severe cases, this ureterocele could conceivably fill the greater portion of the vesical lumen and appear indeed as a separate chamber.

A multiseptate bladder may be formed by fusion of epithelial folds of the fetal bladder as described by Watson or it may be a variation of a very large ureterocele.

A reasonable explanation of the embryogenesis of the hourglass bladder has not been proposed. Krasa and Paschkis held the view that it was an atavistic phenomenon, for it appears to occur normally in some lower animals. It appears more likely that the fault lies with the development of the bladder musculature, especially in view of the fact that the posterior aspect of the bladder (trigone) receives its muscular wall rather late in development. At that time the muscle wall at the level of the interureteric ridge could have condensed into a ring-like constriction.

3. Clinical Features

There is no diagnostic clinical feature by which these anomalies may be recognized. There should always be a high index of suspicion present of the existence of anomalies of the urinary tract when external abnormalities of the genitalia, anus, or lower spine are observed at birth. At times the presence of two urinary streams, as the child voids, may lead to the recognition of complete duplication of the bladder and urethra. Difficulties in micturition, with poor urinary stream, may be present when a bladder septum or ureterocele-like partition partially occludes the bladder outlet. A history of double or triple voiding may be obtainable when the communication between two bladder chambers is small, as, for example, in hourglass bladders. The features of upper urinary-tract infection may draw attention to some of these anomalies. The majority of patients who present with symptoms show only the various combinations of urinary-tract obstruction and the accompanying symptoms of urinary-tract infection.

4. Diagnosis

Since the clinical features are nonspecific, one is dependent upon special investigations to establish the presence and exact nature of the anomaly. As in all children with signs of urinary-tract infection, a systematic study should be undertaken to determine the kind and nature of the congenital anomaly that may be present.

When an apparently distended bladder persists after catheterization, the presence of an hourglass bladder or a septate bladder should be suspected. The possibility of a duplicated urethra that has only a pinpoint opening should be looked for (RAVITCH and SCOTT). A voiding cystogram and retrograde urethrogram will show most of the primary bladder anomalies, such as bilobulation or septum formation, by the heart-shaped outline of the bladder and the filling defect in the dome of the bladder where the septum is attached. Double shadows may lead to the diagnosis of ureteroceles or loculated bladders. Should complete duplication be present, both urethras have to be catheterized and contrast medium injected. At times the outward ballooning of a bladder septum can best be seen on a ciné cystographic study.

The upper urinary tracts should be studied by intravenous and often retrograde pyelography. Signs of obstruction, absence or duplication of the kidneys and ureters may lead one to suspect anomalies of the bladder and ureterovesical junction.

If at all possible, cystoscopy should be carried out. In complete duplication, the bladder chamber should be unremarkable except for the presence of only one ureteral orifice. With incomplete duplication, the sagittal septum should be visualized and the instrument should pass into each chamber. Cystoscopic examination of the bladder with various septate partitions may be difficult to interpret as is the case with ureteroceles. The presence of loose folds, bulging walls, apparently absent ureteral orifices, trabeculated walls in the presence of an hourglass bladder neck should raise the index of suspicion. The presence of an hourglass bladder may be disguised by the presence of a diverticulum-like opening in the dome of the bladder.

5. Treatment

In general, the aim of treatment is to remove obstruction and to restore the urinary tract to as great a degree of anatomic and physiological normalcy as possible. Preliminary nephrostomies may have to be done prior to any reconstructive efforts especially when obstruction has been severe and renal function compromised. Following prolonged drainage of dilated and infected upper urinary tracts, a more realistic assessment of renal function may be obtained and various forms of treatment instituted.

In the child with prolonged and marked urinary-tract obstruction and reduced renal function that shows no improvement after nephrostomy drainage, bilateral cutaneous ureterostomies or diversion by means of an ileal loop is the treatment of choice. When obstruction has been unilateral and the kidney shows poor or no function, unilateral nephrectomy may remove the source of chronic infection. This is especially true when one kidney drains into a blind bladder chamber or a poorly draining one. In the cases of double collecting systems and ureteroceles ("biloculated bladder"), a heminephrectomy and ureterectomy will remove a nonfunctioning and irreparably dilated component.

In dealing with the actual bladder anomaly, treatment should be designed not only to remove the obstructive feature and allow proper emptying of the bladder but also to prevent vesico-ureteral reflux. Hence, it is not enough merely to remove a septum, partition, or ureterocele; the ipsilateral ureter must be reimplanted if vesico-ureteral reflux is present (McGovern et al.).

Several congenital duplications are compatible with normal life and require no treatment. Thus, the patient with complete duplication reported by Satter and Mossman required no reconstructive surgery except for the removal of a short vaginal septum. In males, the presence of a double penis with a functioning urethra in each ipsilateral penis may be of concern to the parents but may present no threat to the life of the patient. Surgery is best postponed until the function of each organ can be fully assessed. Should one of the penes be small and atrophic, and obviously of no use sexually, then reconstructive surgery may be instituted earlier. Wojewski and Kossowski recommended removal of one bladder, reimplantation of the ipsilateral ureter, and a wedge resection of the bladder neck. Amputation of the ipsilateral phallus was undertaken 3 months later. In cases of incomplete duplication and in those of incomplete sagittal septa, open resection of the septum may be all that is necessary. Boisonnat reported an excellent postoperative result in a child with partial duplication and recurrent urinary-tract infection by resection of the septum, that is, the medial half of each bladder, suprapubically. A cystogram several months later showed no reflux and no bladder residuum. In cases of partial sagittal or frontal septa with double ureters (a form of ectopic ureterocele), the unobstructed or functioning ipsilateral ureter frequently will have to be reimplanted following removal of the bladder septum and heminephrectomy and ureterectomy on that side. The treatment of hourglass bladder will have to be determined by the type of constrictive band and the location of the ureteral orifices. Of the 22 cases reported, 11 of the patients had operative intervention (Zellermayer and Carlson). In six, the constrictive band was cut with scissors; in two, the entire band was excised. One operation was attempted transurethrally. In two cases, the base of the bladder was resected. It seems that the most logical treatment is a suprapubic excision of the band together with an anti-reflux procedure when indicated.

IV. Bladder Mucosal Redundancy (Trigonal Curtains)

Several cases have been reported in the literature in which the bladder outlet appeared to be obstructed by loose mucosal folds or valves. In two cases reported

(ESPINOZA et al., MORTON et al.), the redundant mucosa prolapsed through the urethral meatus. In all cases, the mucosal redundancy involved the trigone of the bladder (Fig. 14). The condition has been termed "trigonal curtain," "bladder neck valves," or "mucosal redundancy." Fifteen cases have been reported (BEER, WALLACE, CAMPBELL, 1932, 1963; HARRIS, LEARMONTH and WATKINS, POOLE-WILSON, KOOK et al., ESPINOZA et al., MORTON et al.). Several theories as to etiology have been proposed but none appears to be entirely satisfactory. KOOK et al. stated that these trigonal curtains might be incompletely resorbed epithelial folds that resulted from the fusion of the Wolffian ducts and urogenital sinus. This might explain the formation of valve-like flaps but does not include the more common occurrence of loose, redundant trigonal mucosa.

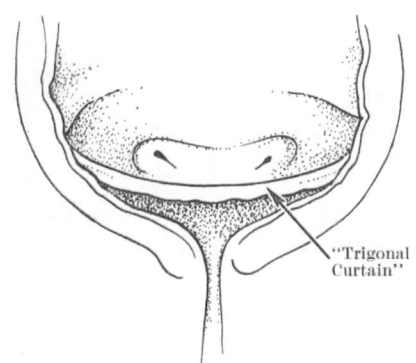

Fig. 14. Trigonal curtain. (From CAMPBELL, 1932)

1. Clinical Features

The physical appearance of the pro-truding redundant mucosa may be similar to a prolapsed ectopic ureterocele. Indeed, the presence of a protruding vaginal mass was the main complaint in two cases and led to the correct diagnosis (ESPINOZA et al., MORTON et al.). Usually the clinical features are much more insidious and in some of the reported cases the diagnosis was established only at autopsy. The trigonal folds act as valves, causing intermittent obstruction to the flow of urine. Thus, the most common presenting complaint is difficulty in micturition. The flow of urine is initiated normally but after a short while the stream is cut off completely by a valve, fold, or trigonal curtain floating over the internal urethral meatus. The harder the patient strains, the less likely is he to resume his urinary stream. As he relaxes, frequently he will be able to void again. Thus, the bladder is emptied with difficulty and only after repeated and frequent attempts. Catheterization will often yield a large residual after voiding. Therefore, the common combination of obstruction and infection is present. Upper urinary tracts may show dilatation, obstruction, and infection secondary to long-standing increased intravesical pressure and vesico-ureteral reflux. A standard cystogram may not be of diagnostic help except to document trabeculation of the walls and reflux, and cystoscopy may be ambiguous as the mucosal folds partially obstruct the field of vision. A ciné study of the act of micturition often may give the clue to an intermittent bladder-neck obstruction. Most of the time, however, diagnosis will be established at exploratory operation, as happened in one of our cases recently. It is important to remember that these mucosal folds may be ectopic ureteroceles, and a search for possible double urinary tracts should be made.

2. Treatment

The treatment of choice is excision of the mucosal fold or valve after the bladder has been opened suprapubically. The cut margins of the excised fold should be sutured to the underlying muscularis. Should reflux be present, reimplantation of the ureter should be undertaken. The bladder should then be drained suprapubically for about 7 days.

V. Congenital Cyst of the Bladder

1. Clinical Features

Another rare anomaly of the bladder is the congenital cyst arising in the trigonal area and causing obstruction to the bladder outflow. Michon et al. described a case in a 3-year-old girl in which a cyst-like protruberance arose in the posterior trigone below the interureteric ridge, causing obstruction at the bladder outlet. The child had a history of persistent pyuria. She was found to be uremic, had bilateral hydronephrosis and vesico-ureteral reflux on cystogram, with a persistent filling defect in the bladder. At the time of operation, a soft

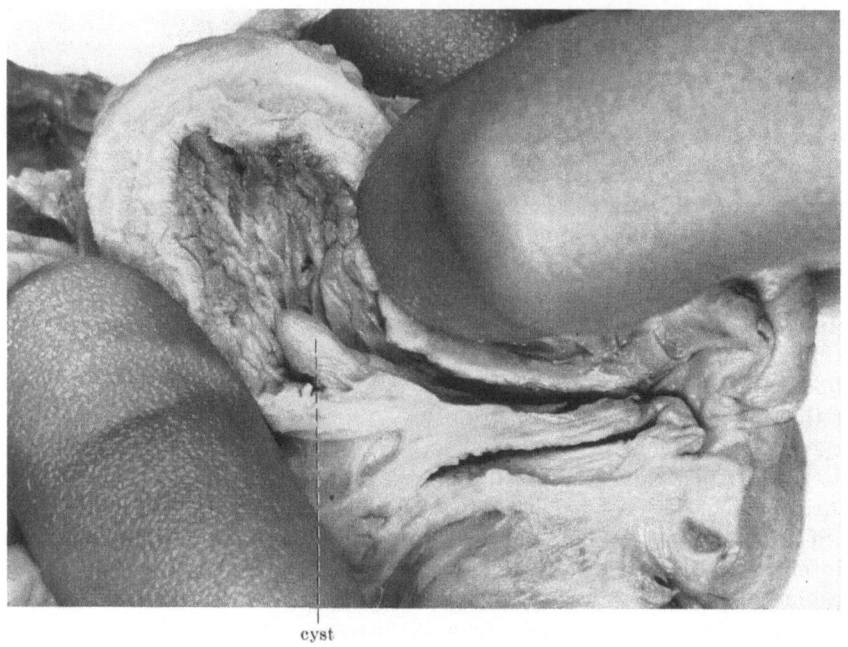

cyst

Fig. 15. Cyst of the trigone. Postmortem specimen from a 5-week-old female infant shows a thick-walled bladder. Bladder-neck obstruction was caused by a serous cyst arising in the bladder trigone. (From Williams, 1958)

mass was seen arising in the posterior trigone inferior to the interureteric ridge. When torn with forceps, purulent fluid ran out of the cyst cavity, which was found to be multiloculated. The cyst was unroofed, and the patient had an uncomplicated postoperative course. Follow-up cystogram, however, continued to show reflux on the left side. The authors cited 17 other cases of "cysts" reported in adults; in some of these, it would appear that the cyst was an ectopic ureterocele. Two additional cases of congenital cysts have been reported. Boisonnat and Bouteau described an "orange-sized" bulge in the bladder of a 3-year-old boy that compressed the right ureteral orifice. When opened, the cyst contained colorless liquid. Extra-vesical exploration did not reveal a double ureter on the right side. Examination of the cyst wall revealed epithelial lining on both sides of the cyst. Follow-up urograms revealed no reflux and some function of the contracted right kidney. The third case, described by Williams, was a postmortem study. The child died of urinary-tract obstruction and infection; the trigonal cyst was found to be lined with simple flattened epithelium and was filled with clear fluid (Fig. 15).

2. Pathology

Several theories have been proposed regarding the etiology of trigonal cysts (MICHON et al.). The hypothesis of "dysembroplasia" has been proposed; a more likely cause may be cystic transformation of glands (BRUNNER's glands) in the trigone.

3. Diagnosis and Treatment

Essentially, the diagnostic steps are the same as in cases of ureteroceles; indeed, the label of "trigonal cyst" should not be attached until the possibility of ureteroceles has been truly investigated and excluded. A good history is very important. The child may have difficulty in initiating and maintaining an adequate urinary stream. Persistent urinary-tract infection and bouts of fever may be the presenting complaints. A voiding cystogram may show a filling defect in the base of the bladder and massive vesico-ureteral reflux is usually present. An intravenous pyelogram may show hydronephrosis, hydroureters, or an absence of renal function on the more involved side. Should marked hydronephrosis be present, as well as chemical documentation of impaired renal function, preliminary nephrostomies are indicated. At the time of exploratory surgery, the cyst may be simply unroofed. Should reflux be present and renal function adequate, an anti-reflux procedure should be done at this time. Should renal function be markedly impaired, even after prolonged nephrostomy drainage, then a form of urinary diversion should be undertaken as the definitive procedure.

VI. Congenital Bladder Diverticulum

1. Incidence

The congenital diverticulum of the bladder once was described as a rare condition. In 1934, KRETSCHMER reviewed the world literature and reported 15 cases in children less than 12 years of age. In 1959, FORSYTHE and SMYTH added another 25 cases from the literature and reported 13 more from the Royal Belfast Hospital for Sick Children. MACKELLAR and STEPHENS described 23 children with congenital diverticula from Australia. CAMPBELL, in 1963, stated that true congenital diverticula constitute about 5% of all cases. It appears, therefore, that this condition is found more frequently than once thought, probably reflecting the increased awareness among physicians of possible anatomic abnormalities in children with urinary-tract difficulties, infections, or enuresis, and better diagnostic facilities.

2. Etiology

It is now usually assumed that vesical diverticula, with the exception of the urachal remnants, are formed only where there is obstruction and that if no abnormality is found at the urethra, then a bladder-neck obstruction must be present (WILLIAMS). This concept is supported by BADENOCH's observations that simple excision of the diverticulum was followed by progressive ureteral dilatation unless a bladder-neck revision was also carried out. Thus, some obstructive factor in the region of the bladder neck appears to be present. KRETSCHMER stated that children with bladder diverticula always have some type of bladder-neck or urethral obstruction and that those patients without obstruction were insufficiently studied or the obstruction was overlooked.

Of the many etiological concepts reviewed, two general categories may be mentioned. One is that some obstructive feature was present in intra-uterine life, resulting in increased vesical pressure and a ballooned-out weak spot in the bladder wall. The preponderance of occurrence of diverticula in the region of the trigone may reflect the late formation of trigonal (detrusor) musculae in relation to the remainder of the bladder. Englisch, in 1903, postulated a temporary occlusion of the urethra during fetal life as an etiological factor, and Chwalla believed that temporary obstruction of the ureteral orifice was responsible for congenital diverticula, especially the ones in which the ureter terminates in the diverticulum. The possibility that aberrant ureteral buds might cause weakness in the muscle wall was suggested by Joly and by Caulk. Williams described one case that seemed to fit into Caulk's explanation. A narrow, elongated sac, opening behind the ureteral orifice, ran upward in a common adventitial sheath with the ureter.

The other etiological concept is that the diverticula are not present at birth but are represented by muscular weakness in the wall (Lurz). Should urinary-tract obstruction develop, or even as the result of the force in normal urination, the diverticulum may appear. Thus, some authors believe that congenital diverticula may be present although no anatomic obstruction in the region of the bladder neck and urethra may be demonstrable (Johnston, MacKellar and Stephens).

3. Pathology

Although multiple diverticula of the bladder may occur, they are usually single in children. Their size varies from a few centimeters to that of the whole bladder and they are found most commonly at the base near the ureteral orifices

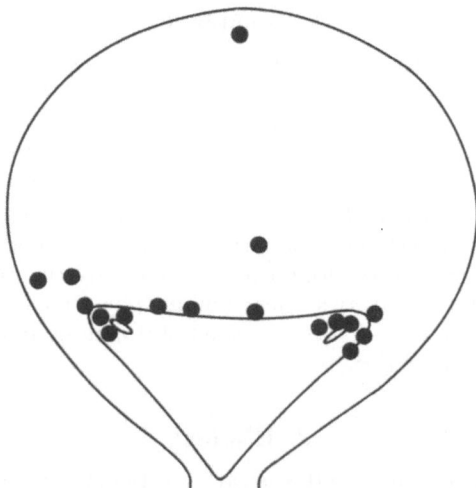

Fig. 16. Distribution of bladder diverticula in children as reported by Stephens. Note the preponderance around the trigone

(Fig. 16). Essentially, a diverticulum is a bladder mucosal herniation through a defect in the muscle coat and is usually covered by a thin coat of muscle fibers. Whether or not an obstructive feature is demonstrable, some local weakness in the detrusor coat must be presumed responsible for the herniation. Once the diverticulum has been formed, bladder contracture will distend it with urine and with each micturition effort the diverticulum will be gradually enlarged. HUTCH described such a progression of diverticulum formation in the region of the ureteral orifices where a weakness in WALDEYER's sheath has been postulated. Every time the diverticulum distends with urine, WALDEYER's sheath weakens further and the saccule dilates more, pushing supporting muscle elements downward (Fig. 17). Eventually, the ureter is carried into the saccule, the obliquity of its intramural course is altered, and reflux occurs. Fig. 18 shows the relationship

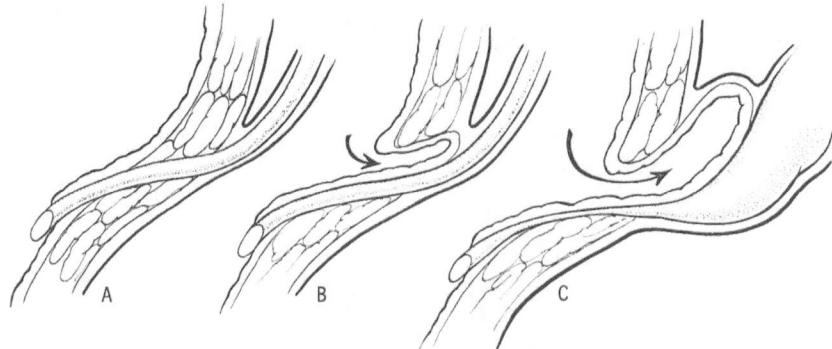

Fig. 17. Illustration of saccule formation in a smooth-walled bladder. A, Normal uretero-vesical junction. B, WALDEYER's sheath is defective and ureteral hiatus is dilated, leaving the intra-mural ureter without support. C, More advanced stage of saccule formation. (From HUTCH)

of diverticula and ureteral orifices in eight children reported by MACKELLAR and STEPHENS and the relationship of this location to vesico-ureteral reflux.

One additional causative factor that is commonly overlooked must be mentioned. That is the neurogenic bladder. We believe that the abnormality in neuro-muscular co-ordination of the detrusor coat during micturition is responsible for coarse trabeculation, eventual weaknesses in the muscle coat, and subsequent diverticula formation.

4. Diagnosis

There is no sign or symptom pathognomic of bladder diverticulum. The abnormality commonly is found radiologically during investigation of urinary-tract infection, difficulties in micturition, or enuresis.

During clinical examination any palpable enlargement of the bladder should be noted. A neurological examination may reveal a saddle anesthesia or poor anal-sphincter tone on rectal examination. The act of voiding should be observed for hesitancy or double voiding. The character of the stream may indicate distal obstruction.

The voiding cystogram is the best single diagnostic tool. The bladder is filled with contrast material via a urethral catheter at 15 cm and 30 cm gravity pressure. The filling phase may be monitored, using the image-intensification technique and television. Exposures are then made with the patient supine and the pelvis in a slightly oblique position. The act of voiding is then monitored and recorded on film with multiple spot exposures. It is important to record how the diverticulum empties, its location, and whether or not reflux or vesico-ureteral obstruction is present (Fig. 19). Should a large residuum remain in the bladder at the end of the voiding period, the bladder should be drained via catheter and another exposure obtained to be certain that the diverticulum empties.

5. Treatment

The method of treatment depends upon the radiographic findings as well as the neurological examination. Should the child have a neurogenic bladder secondary to a spinal-cord lesion, then urinary diversion eventually is necessary, and no local treatment should be attempted. A single diverticulum without ureteral reflux or ureteral obstruction should be excised and a bladder-neck plasty carried

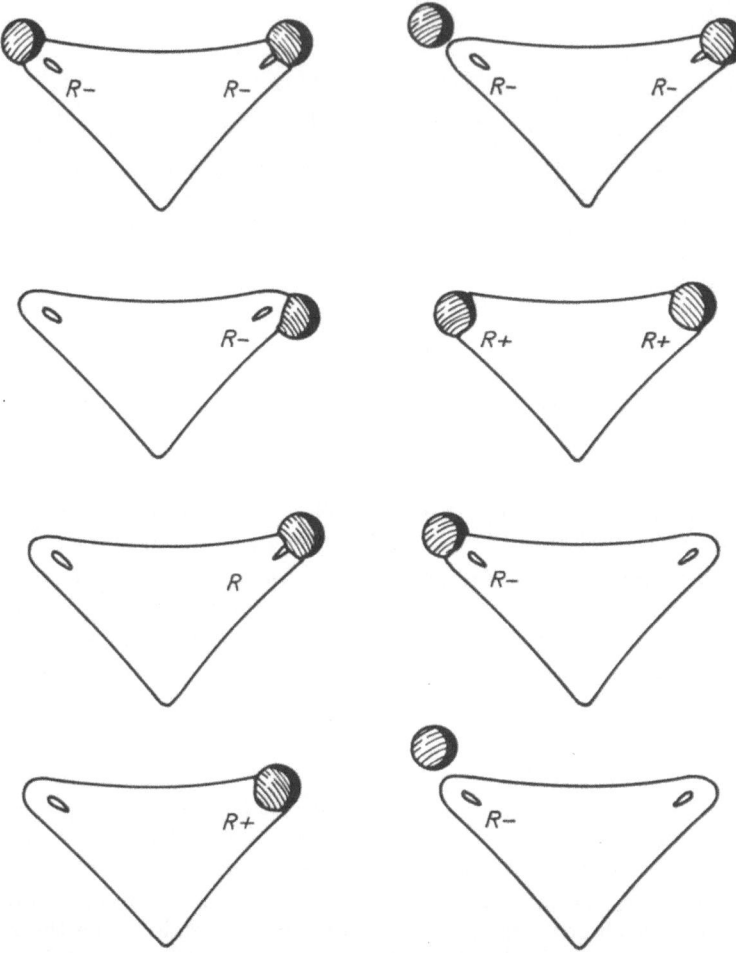

Fig. 18. Trigones of eight children described by Mackellar and Stephens to show the relationship of diverticulum, ureteral orifice, and vesico-ureteral reflux in each case. $R-$ No reflux; R Slight reflux; $R+$ Free reflux

out at the same time. Should reflux be present, the ureter will have to be re-implanted in addition to diverticulectomy and bladder-neck plasty. Should the intramural ureter be obstructed by the diverticulum, preliminary nephrostomy should be done to relieve the hydronephrosis. In some cases in which there is marked ipsilateral hydronephrosis and reduction of renal parenchyma and function, nephroureterectomy, diverticulectomy, and bladder-neck plasty give excellent results.

We should like to re-emphasize that a possible obstructive factor must be excluded in all cases of congenital diverticula. Distal obstructions, such as valves, meatal stenosis, or urethral strictures, must be removed before a diverticulectomy

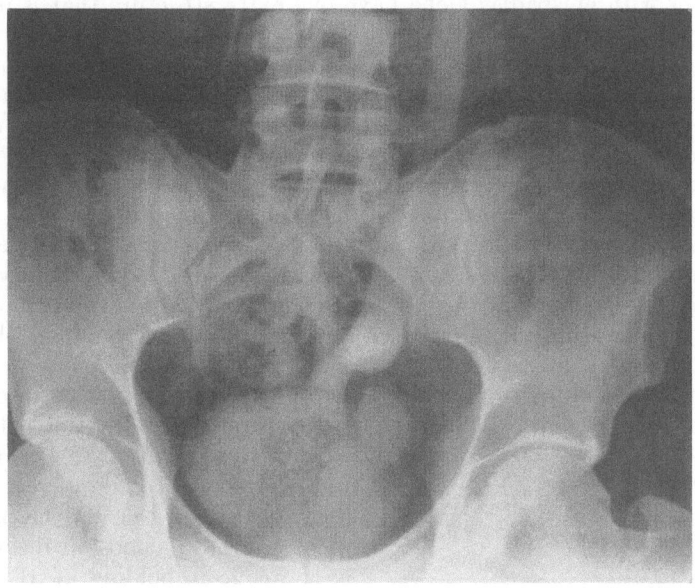

Fig. 19. Congenital diverticulum in a young adult. The left ureter lies in the wall of the saccule

is carried out. Notwithstanding reports in the literature that bladder-neck and urethral obstructions are not necessarily associated with bladder diverticula, we feel that an obstructive feature is responsible in the etiology of this congenital anomaly and that a Y-V plasty of the bladder neck is indicated in all cases.

VII. The Exstrophy-epispadias Complex
1. Pathology

Even among deformities generally, exstrophy of the bladder and its variants are exotic conditions. They not only involve the genito-urinary tract but routinely affect the musculoskeletal structures of the lower abdomen. In some cases, the lower gastro-intestinal tract is also involved, thus presenting a complex anatomic derangement that includes the urinary, genital, musculoskeletal, and intestinal systems in one continuum.

Historically, the first truly complete description was that of Mowatt in 1747. Rickham gave reference to an Assyrian clay tablet of about 2,000 B.C. that is believed to describe ectopia vesicae, and other authors (Hall et al., Connell) referred to descriptions of this anomalous complex dating from the sixteenth and seventeenth centuries. Perhaps the most famous patient with this congenital defect was Matthew Ussem (Duncan), who more than a century ago exhibited himself in many European medical centers and thereby earned his living. Most patients with untreated exstrophy, however, lead wretched, piteous lives (Cooper), and many authors have reported that about one half of the patients with exstrophy will die by their tenth year and two-thirds by their twenty-first year (Campbell, 1963; Higgins).

Many theories as to the faulty embryogenesis have been proposed (von Geldern). We have been impressed by the musculoskeletal abnormalities so commonly associated with the exstrophy-epispadias complex. Most patients with exstrophy or with epispadias alone have one extra structure that is not present in the normal person; that is, the triangular portion of the abdominal wall lying between the rectus sheaths with its apex toward the umbilicus but not occupied by the everted bladder or vesical outlet. This is a widely spread linea alba; a fascial layer devoid of muscle fibers. It appears as if this thin, fibrous, wedge-shaped structure held apart the developing anterior abdominal wall and probably is a derivative of the cloacal membrane that occupied the infra-umbilical abdominal wall during the first 6 to 8 weeks of embryonic life.

The cloacal membrane, a thin, bilaminar structure of ecto-endoderm, forms the infra-umbilical anterior abdominal wall in the very young embryo (Fig. 2). During normal development, mesodermal elements progressively encroach upon and obliterate this thin membrane. Thus, a prominent mesodermal plate, sandwiched between the original ecto-, endodermal membrane, comes to lie in the infra-umbilical abdominal wall, which subsequently organizes itself into the anterior abdominal musculature and osseous structure of the anterior pelvic girdle.

Table 2. *Variations of exstrophy-epispadias as reported by* MARSHALL *and* MUECKE *(1965) from The New York Hospital, New York*

Variety	No. cases
Spade penis only	1
Epispadias with continence	
Balanic	2
Penile	6
Epispadias with incontinence	
Subsymphyseal	1
Penopubic	17
Classical exstrophy	44
Cloacal exstrophy	4
Superior vesical fissure	1
Duplicate exstrophy	1
"Pseudo-exstrophy"	2
Total	79

Should the primitive cloacal membrane be overdeveloped, however, it could delay or hinder this mesodermal movement, prevent its mid-line fusion, and act as a wedge in holding apart developing musculoskeletal elements of the anterior abdominal wall. Thus, the recti muscles would be kept divergent yet intrinsically normal and the bony structure of the anterior pelvis would be separated. The remnant membranous cloacal membrane normally ruptures in the 15 to 16 mm embryo and establishes the urogenital sinus; should a large cloacal membrane dehisce, then the anterior wall of the developing bladder would be lacking and the bladder would evert. Should the cloacal membrane dehisce before the uro-rectal septum has fused with the primitive perineum, then the lower gastrointestinal tract would also evert. Thus, depending upon the size and extent of the non-obliterated cloacal membrane at the time of dehiscence, variations in the final product are to be expected.

Therefore, the underlying defect common to all cases of the exstrophy-epispadias complex is a large cloacal membrane that acts as a mechanical barrier to mesodermal movement during the first 6 weeks of embryonic life. POHLMAN, in 1911, described an abnormal embryo in which the cloacal membrane was unusually large, extending well into the umbilical stalk superiorly, and he commented that this embryonic defect could have "... a decided bearing on bladder exstrophy and epispadias." In 1964, MUECKE reported the successful experimental production of cloacal exstrophy in the chick. He showed that in the chick embryo a minute plastic disc introduced into the tail-bud region during early development interfered with normal cloacal membrane regression and initiated a defect in the infra-umbilical wall that persisted until the hatching stage and resulted in exstrophy strikingly analogous to human cloacal exstrophy.

In 1965, MARSHALL and MUECKE reported their series of patients with exstrophy-epispadias anomalies (Table 2) and observed that the most common type is classical or typical exstrophy (56%). Other varieties occur with decreasing frequency as they diverge from the classic form of this congenital defect (Fig. 39). Except for the rare cases in which secondary fusion appears to have taken place, all final anatomic varieties can be derived by simple embryonic progression from the starting point of an abnormally persistent or overdeveloped cloacal membrane.

2. Pseudo-exstrophy of the Bladder

The presence of characteristic musculoskeletal defects of the exstrophy-epispadias complex without a major defect in the urinary tract has been reported and termed "pseudo-exstrophy." About eight cases of this deformity have been described (MacKenzie, Higgins, Kittredge and Bradburn, Hejtmancik et al.,

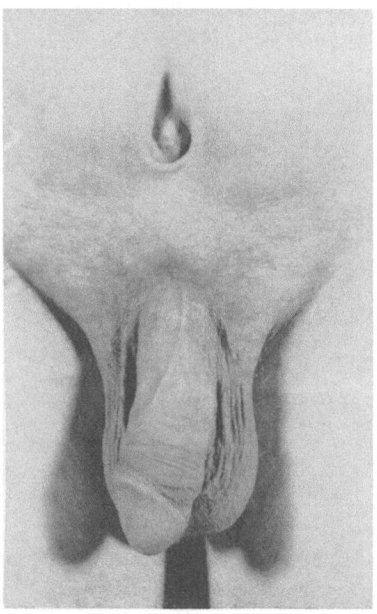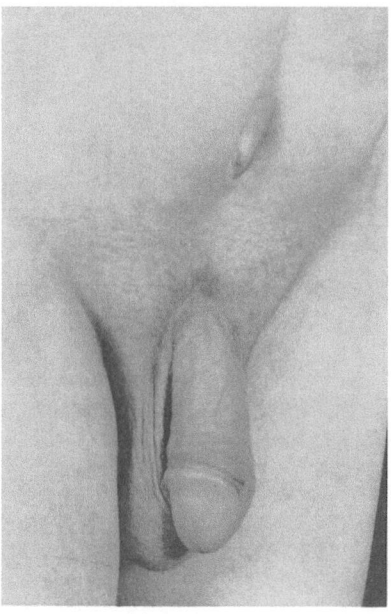

Fig. 20. Pseudo-exstrophy. Patient has the musculo-skeletal deformity of exstrophy with an intact urinary tract

Uson and Roberts, Marshall and Muecke, Van Buskirk et al.). In this deformity, the recti diverge from just above the umbilicus to attach on the separated pubic bones (Fig. 20). The umbilicus is low-set, deep, and elongated. Frequently the bladder will bulge outward as a ventral hernia, and the presence of inguinal herniae, with or without undescended testes, has been reported. The urinary tract is intact. This defect is related to the exstrophy anomaly in that it has the same embryonic origin. The wedge effect of an abnormally large cloacal membrane exerted itself in holding apart developing mesodermal elements of the anterior abdominal wall. Complete eversion of the bladder did not occur; perhaps, because the dehiscence of the cloacal membrane was incomplete. Thus, it persisted as a triangular-shaped linea alba. This defect frequently needs no repair; should a large ventral hernia be associated with it, we favor primary closure by mobilization of fascial flaps of anterior rectus sheath and external oblique aponeuroses. Osteotomies have been performed for closure of the anterior pelvic defect (Van Buskirk et al.); this, in our opinion, is not necessary to obtain an adequate hernial repair.

3. Classical Exstrophy of the Bladder
a) Pathology

Exstrophy of the bladder in its classical form is the most frequently encountered malformation of this anomalous complex. USON and co-workers reported 66 cases of classical exstrophy in a series of 72 cases. Of 158 cases of exstrophy reported by HIGGINS, all but five were of the classical form. In our series of 52 cases of various forms of exstrophy, 44 were of the classical variety. Exstrophy

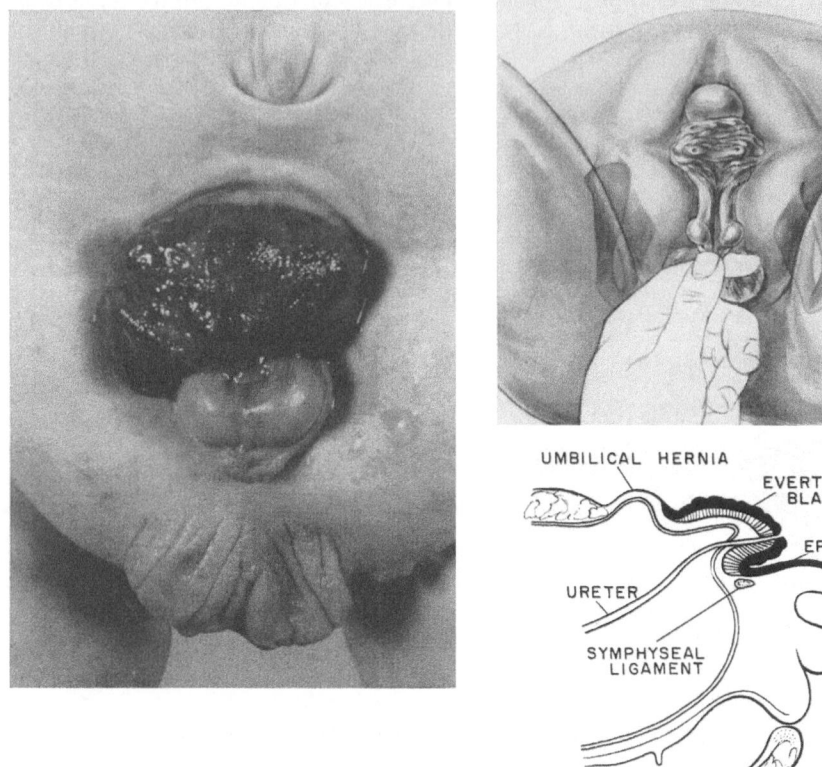

UMBILICAL HERNIA

EVERTED BLADDER

EPISPADIAS

URETER

SYMPHYSEAL LIGAMENT

RECTUM

Fig. 21.
Classical exstrophy of the bladder in a male child

of the bladder is observed more commonly in males than in females. HARVARD and THOMPSON reported a sex ratio of 2.1 males to 1 female in a large series of 198 cases; BAXTER and MORSE had a ratio of 1.8 males to 1 female in their series of 22 cases; STAGNER and HODGES reported a ratio of 2.9 to 1; and STURDY found a ratio of 2 to 1 in favor of males.

The malformation does not appear to be familial. HIGGINS observed only twice the occurrence of exstrophy in members of the same family, and GLASER and LEWIS reported one family in which two successive male children were born with exstrophy. CHISHOLM, in 1961, reported a patient with two siblings (one of the opposite sex) who both had exstrophy. Exstrophy of the bladder has appeared on at least one occasion in both members of a set of identical twins (HIGGINS). The twins reported by COATES and those reported by RUKSTINAT and

BALFOUR were of the same sex but only one twin of each pair had exstrophy; a similar group of "identical" twins was reported by ANSELL. In our twins of opposite sex, only the boy had exstrophy but both had a teratomatous development in the retroperitoneal tissues. The majority of children with this malformation are born into families without histories of previous urinary-tract anomalies. At least 33 children have been born to 26 mothers who had exstrophy without the deformity reappearing. There does not seem to be any racial factor involved nor has there been any relationship established to specific illnesses during pregnancy (USON et al.).

It has been estimated that about 100 children with exstrophy are born each year in the United States (HIGGINS), or an incidence of about 1 in 30,000 to

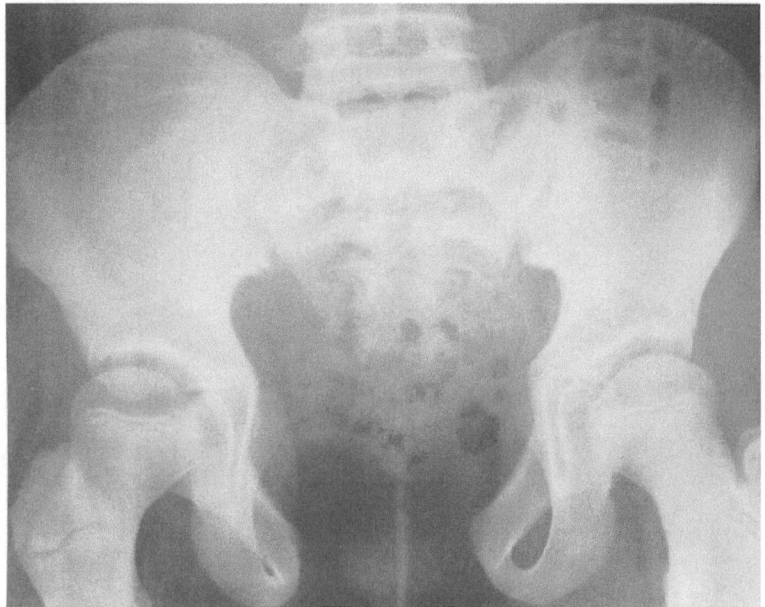

Fig. 22. Antero-posterior pelvic roentgenogram of a patient with exstrophy of the bladder

40,000 livebirths. At the Mayo Clinic, 198 cases of exstrophy were seen in 1.5 million admissions (HARVARD and THOMPSON). KIMBALL and DRUMMOND were able to find records of 18 cases of exstrophy during a 5 year period in Oklahoma. RICKHAM, in 1961, reported that between 1941 and 1953, 16 patients with exstrophy were treated in the Children's Hospitals of the Liverpool Region, or an incidence of 1 per 40,000 births. Since 1953, however, 27 additional children with exstrophy were seen and treated at the two Liverpool Children's Hospitals, thus raising the incidence to 1 per 10,000 livebirths. Hence, exstrophy of the urinary bladder is not so rare an anomaly as it was once described to be.

b) Anatomy

In the male child, the bladder lies everted on the lower abdominal wall (Fig. 21) with the moist mucosal surface exposed. The characteristic epispadiac urethra forms a mucosal surface covering the dorsal penis. The ureteral orifices are usually easily recognizable by the two little streams of effluxing urine. The pubic bones are separated on palpation and a fibrous structure can usually be felt just below the everted bladder neck, representing a symphyseal band of fibrous tissue. The umbilicus is low-set and elongated and a thin fibrous sheet lies between the two divergent recti muscles, occasionally bulging outward as a form of umbilical hernia. Bilateral inguinal herniae are frequently present and the testes may be undescended.

The penis is short and stubby and tends to be upturned. Although actual measurements have shown that the corpora cavernosa and crura are not deficient nor shorter than the normal penis, because of the marked separation of the anterior

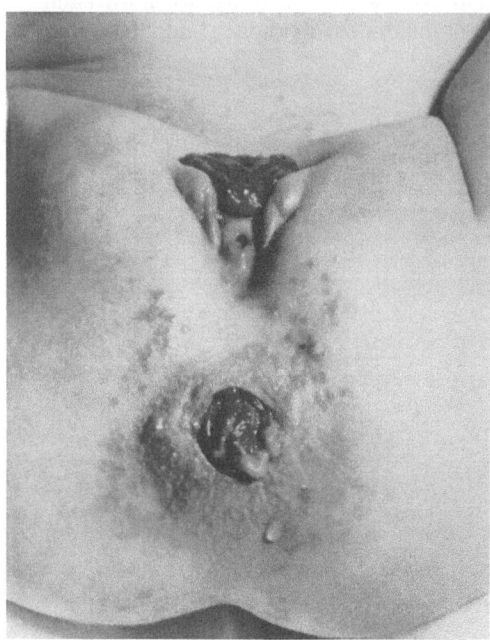
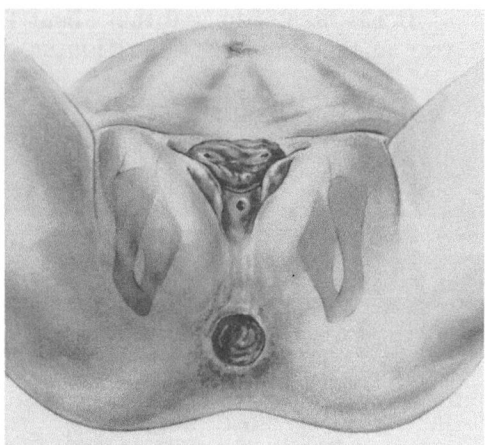
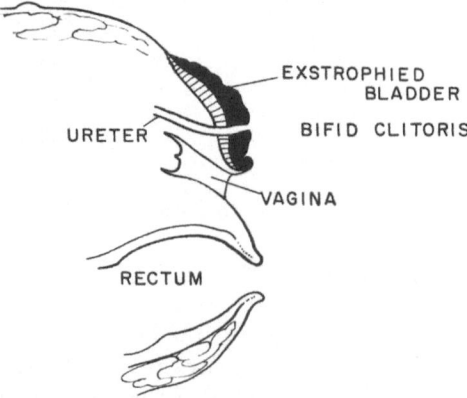

Fig. 23. Female child with exstrophy of the bladder
and patulous anus

pelvic girdle as well as outward rotation of the pubo-ischial rings, the fused pendular portion of the two corpora is short. The short tuft of prepuce lies ventrally. The frenum and penile raphe are normal. The scrotum is often flattened and may be hypoplastic. Incomplete testicular descent is common.

An antero-posterior roentgenogram of the pelvis shows the characteristic bony abnormality of this disorder (Fig. 22). The symphysis is markedly widened — in our series, the width of the symphysis ranged from 3 to 17 cm. There is an outward rotation of the innominate bones, relative to the sagittal plane of the body, along the sacro-iliac joints. In addition, there is an outward rotation or eversion of the pubic bone at its junction with the ischium and ilium. A third component, present in the more severe cases, consists of varying degrees of lateral separation of the innominate bones inferiorly with the fulcrum at the iliosacral joint. Because of the outward rotation and lateral displacement of the innominate bones, the distance between the acetabular fossae is variously increased, a finding that accounts for the increased distance between the hips and the waddling gait of the more severely affected patient. The roentgen anatomy of the bony abnormalities so characteristic of the exstrophy-epispadias complex has important bearing on the method of functional closure; for although the anterior pelvic girdle can be closed by bilateral posterior iliac osteotomies, the degree of outward rotation

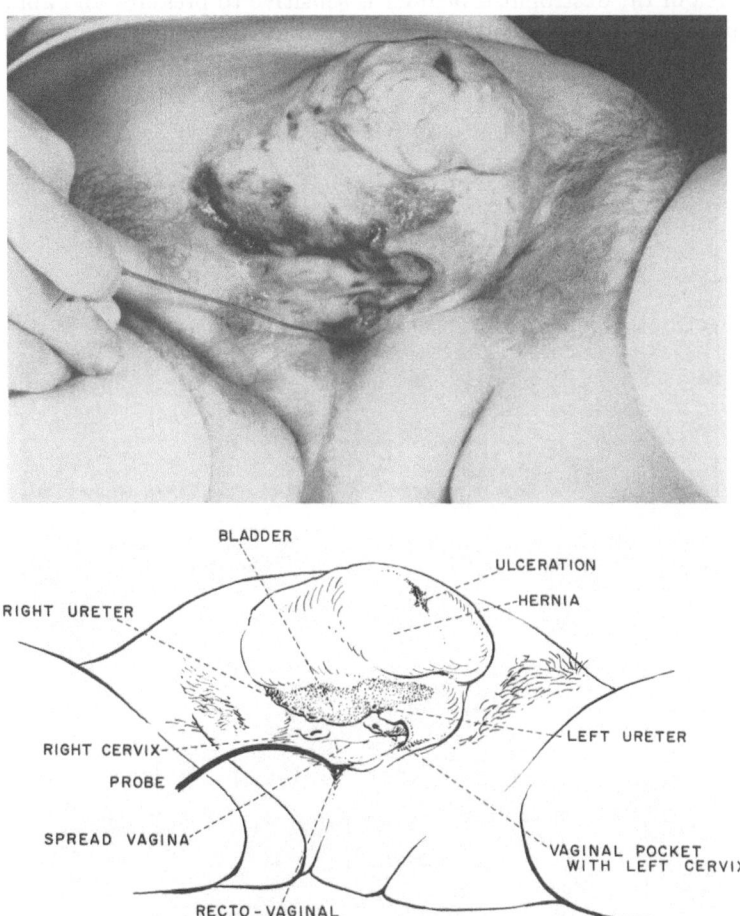

Fig. 24. Extreme exstrophy of the bladder in a young woman with recto-vaginal fistula (marked by probe)

of the pubo-ischial ring is thereby increased and the already stubby penis is likely to be further shortened as the crura separate even further.

In the female child (Fig. 23), the bladder lies open and everted on the surface of the anterior abdomen as in the male. The urethral strip is rather short and frequently almost indistinguishable from the edematous and polypoid bladder mucosa. The labia are spread apart, and the clitoris is bifid. The vaginal orifice is narrowed and lies immediately below the bladder. Usually, the vagina, uterus, and Fallopian tubes are normal; however, in cases of severe exstrophy and markedly separated pubic bones, the vagina may lie splayed open, and the cervix and uterus may be double (Fig. 24).

The anus is frequently abnormal in children with the exstrophy anomaly. Lax anal sphincter with patulous anus was encountered in 16 patients in a series of 40 cases of typical exstrophy (Marshall and Muecke); in 13 of these, the condition was such as to permit rectal prolapse. Imperforate anus with recto-vaginal, rectovesical, or rectoperineal fistula has been described (Jensen et al.).

c) Clinical Features

The area of the exstrophied bladder is sensitive to pressure and abrasion. The hyperemic mucosa bleeds easily. The continual discharge of urine tends to irritate

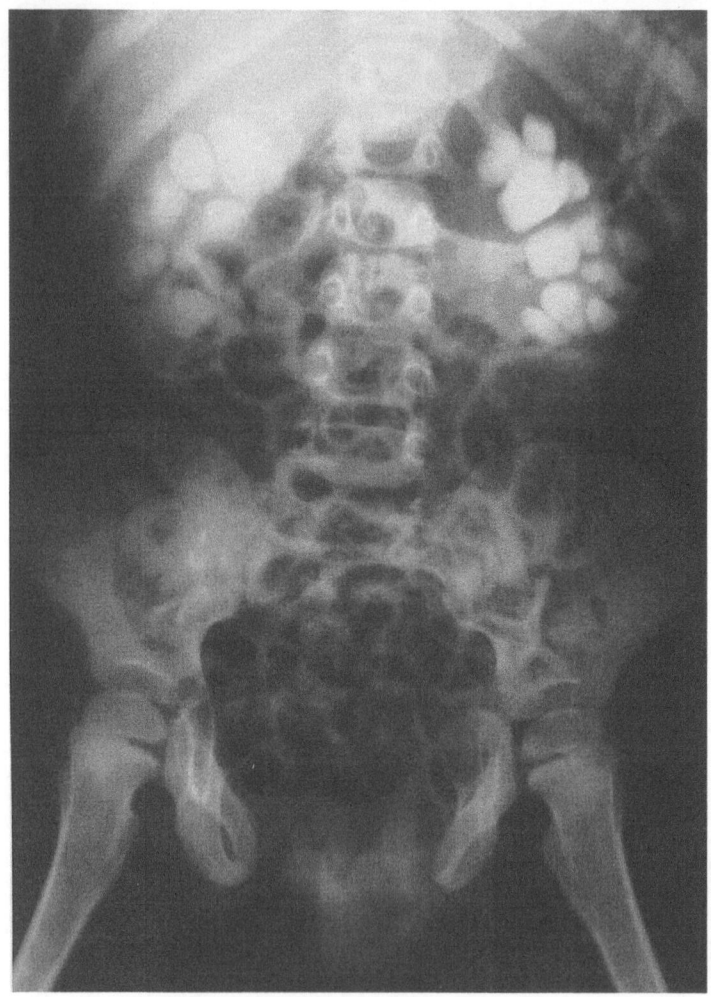

Fig. 25. Excretory urogram of an untreated 3-year-old boy with exstrophy of the bladder. Severe bilateral hydronephrosis is evident

the surrounding skin of the abdomen with resultant ammoniacal cutaneous rash. Therefore, the young child with exstrophy requires good hygienic care with frequent diaper change and skin care. A vaseline gauze pad covering the exposed bladder mucosa tends to reduce the painful irritation by diapers and the resultant bleeding. As the child grows older, the mucosa becomes less sensitive as squamous metaplasia and fibrosis of the mucosa occur. Anal anomalies, so commonly associated with this disorder, may require attention during early infancy, and the frequently occurring anal prolapse during straining and crying may have to be reduced manually by the mother.

Although the ureteral orifices lie exposed, frequently the upper urinary tracts show signs of obstruction, and symptoms of acute pyelonephritis may occur in the untreated patient (Fig. 25). In 17 of 50 patients reported by MALONEY and co-workers, unilateral or bilateral hydrouretero-nephrosis was present when the patients were first seen. Studies showed that active peristalsis was present throughout the length of the ureters. The point of obstruction seems to be the ureterovesical junction, partly because of the partial prolapse of the ureters in the exstrophied bladder base and partly because of the fibrosis and metaplasia occurring in the untreated everted bladder mucosa.

Associated malformations often demand attention before surgical closure of the exstrophied bladder can be undertaken. Imperforate anus with or without fistulous communication with the bladder, vagina, or perineum will have to be corrected first. Frequently, the commonly associated inguinal herniae will incarcerate and have to be repaired.

Because of the abnormal bony pelvis with increased separation of the acetabular fossae and abnormal outward rotation of the innominate bones and puboischial rings, the child will walk with a waddling, broad-based gait. In later life, the opened anterior pelvis appears to be of little handicap to locomotion, and special orthopedic treatment is not required.

d) Treatment

There are two alternative methods of definitive treatment of exstrophy of the urinary bladder: urinary diversion, usually with cystectomy, or functional reconstruction of the bladder, bladder neck, and urethra. Urinary diversion is the orthodox method and in the experience of most surgeons is the treatment of choice. In recent years, however, attempts at surgical restoration to the normal anatomic state have been reported. We favor this approach, partly as a progressive investigation; but should the result be unsatisfactory, urinary diversion can still be undertaken, the abdominal-wall defect having been repaired at the time of functional closure.

α) Urinary Diversion

The standard method of treatment of exstrophy of the bladder since the time of MAYDL (1894) has been ureterosigmoidostomies. It is, therefore, the only method for which any large series of adequately followed cases is available. A number of children with exstrophy surviving 30 years or longer after ureterosigmoidostomies have been reported (GOVERNEUER, STEPHENS, TURNER), but other long-term studies (GARRETT and MERTZ, HARVARD and THOMPSON, WILLIAMS and JOLLY, SANDERUD) presented less optimism. BOYCE and VEST, in 1952, were unable to find a single report of a patient having ureterosigmoidostomy in infancy who achieved a normal life expectancy. One of the longest survivors was our epispadiac patient who died of renal failure 44 years and 8 months after urinary diversion to the rectum (MARSHALL and GARDNER). CROSS and BARBER concluded a study of survivorship by saying that "... while a few patients may be treated and obtain satisfactory results, in the main they are the exception rather than the rule." In 1962, HIGGINS presented a follow-up study of 158 cases of exstrophy of the bladder treated by uretero-intestinal anastomosis. Of these, 99 patients survived for periods from 1 to 30 years. In 39 patients, immediate postoperative complications developed, and in 107 patients, complications were found after discharge from the hospital. The major complication of ureterocolic anastomoses is pyelonephritis; and broad-spectrum antibiotics have not greatly influenced the development of chronic pyelonephritis and the associated deterioration of renal function although they can abort acute attacks. One additional hazard to ureteral transplantation to the sigmoid colon in cases of exstrophy is the frequently associated weakness of the anal sphincter, with or without rectal prolapse. Hence, ureterocolic diversion may lead to incontinence and nocturnal enuresis of the fecal stream diluted with urine.

The separation of urinary and fecal streams offers some improvement provided that the anal sphincter is intact. Boyce and Vest, in 1952, presented a method of closure of exstrophy and epispadias, establishing a permanent vesicorectal fistula and diversion of the fecal stream by means of a permanent colostomy. Although some difficulty may be encountered in maintaining the vesicorectal fistula, the idea of separation of urinary and fecal streams has considerable merit. We have one rather long-term survivor treated by the Boyce-Vest technique who is well and without clinical evidence of pyelonephritis or impaired renal function. Radiographically, his upper urinary tracts show some pyelonephritic changes that have remained stable over the past 5 years.

Modification of the Gersuny procedure have been described in the hope of providing voluntary control, yet avoiding fecal contamination of the urinary tract (Kinman et al., Lowsley and Johnston, Snyder, Thompson, Powell). In this method of urinary diversion, the terminal colon and rectum are isolated and the bowel divided. The ureters are implanted in the isolated rectal pouch and the divided distal colon is brought down under the external anal sphincter in a pull-through procedure. Voluntary control of both the urinary and fecal streams is thus obtained. In another modification of this surgical technique, the urinary tract is diverted through an isolated segment of the sigmoid colon, which was anastomosed to the previously closed bladder. The distal end of this sigmoid colon conduit then emerges under the external sphincter anterior to the rectum. It seems unlikely that the sphincter would provide adequate continence to both the urinary and fecal streams but early reports are encouraging. Again, a weak anal sphincter, observed in more than half of our children with exstrophy, militates against success by these methods.

Simple cutaneous ureterostomies (Cordonnier) or transplantation of the vesical trigone to the skin (Sturdy) are less serious operations than most of the reconstructive or diversionary procedures and may be applicable in children with serious renal failure.

In recent years the construction of an ileal conduit as a means of urinary diversion has gained great popularity. The method was introduced by Bricker in 1950 for urinary diversion in patients with carcinoma of the bladder and it probably represents the best urinary diversion yet available, except that voluntary control is not obtained and an external receptacle has to be worn by the patient. The value of the ileal loop has been confirmed by a number of published series (Bricker et al., Burnham and Farrer, Parkhurst and Leadbetter, Stamey and Scott). Ureteric transplants into an isolated ileal segment has a number of advantages over ureterocolic anastomosis. In most cases, the contents of the ileal loop become sterile within a short period after operation. A mucosa-to-mucosa suturing largely eliminates the risk of obstruction at the site of uretero-intestinal anastomosis and hydronephrosis. There is little danger of electrolyte imbalance provided the loop is not too long and there is no stagnation of urine within it (RICKHAM). One of the major complications is stenosis of the ileal stoma, which may require revision. In our experience, the incidence of pyelonephritis and hyperchloremic acidosis has been minimized by this procedure. The surgical technique and recent modifications are well described by CORDONNIER and we refer to this author for a comprehensive description and discussion.

β) Functional Closure of Exstrophy

In 1906 TRENDELENBURG reported "... for more than twenty years I have endeavored to aid the direct union of the freshened edges in cases of ectopia by producing a separation of the pelvic bones at the sacro-iliac synchondrosis in order to provide for a closer approximation of the two halves of the pelvis anteriorly at the symphysis and consequently of the edges of the defect .... As regards my cases of bladder ectopy which were operated upon years ago, I desire to say that

Tabulation 2

Good (at ≧1 yr.) — Good kidneys (no or much less hydronephrosis; no or barely detectable reflux; no clinical pyelonephritis).
Good control (routinely 2 hr.; occasional dampening only; enuresis to 12 yr.).
No residuum, pyuria, stricture, or fistula.

Fair (at ≧1 yr.) — Good kidneys (Grade I hydronephrosis and unilateral mild reflux allowed; no clinical pyelonephritis).
Good control (routinely 1 hr., often 2+ hr.; nocturnal enuresis and rare day wetting allowed; good stream).
No residuum, pyuria, stricture, or fistula.

Poor — Diversion necessary; or renal deterioration; or only gravity control; or dehiscence.

there are 2 patients living whom I have had under observation almost continuously and a third who has been seen occasionally .... Retention of urine is not complete in any of my three cases .... These young men wear a contrivance supplied with a small spring which compresses the urethra at the root of the penis." TRENDELENBURG believed that iliac osteotomies were necessary to bring the widely separated pubic bones together in order to obtain satisfactory closure of the abdominal-wall defect and improve sphincter control; however, YOUNG, in 1942, achieved a very satisfactory plastic restoration by rotating fascial flaps only in covering the surgical inversion of the bladder. Since 1950 there has been a resurgence of interest in functional closure of the exstrophied bladder and varying degrees of success in achieving continence and physiologically normal upper urinary tracts. In 1964 we surveyed reported cases of surgical restoration of the lower urinary tract. Using the criteria given in Tabulation 2, we reviewed 277 cases (Table 3). In only 15 cases (5%) could the results be considered good and in only 40 (14%) could they be considered fair. Notwithstanding the poor results as regards continence (80% failures), the child treated by operation usually derives other great benefits (elimination of the painful bladder, less urinary irritation, and improved appearance), so that we believe that a trial of functional closure is justified in most cases — it might be successful, and only by effort can the ideal be reached.

No standard operative procedure can yet be recognized; indeed, it appears that each new case closed functionally represents a slight variation in technique from the previous one, and all surgeons experimenting with this method have made individual variations in each case. Our own method was reported in 1962; since then we have changed the anti-reflux technique to the tunnel-and-cuff

Table 3. *Review of functional closure of exstrophy*

Author	Year published	Result, no. cases		
		Good	Fair	Poor
von Trendelenburg	1906	—	—	3
Young, H. H.	1942	—	—	1
Michon	1948	1	1	4
Schultz	1958	1	—	2
Jonas	1959	1	—	9
Boyce	1961	—	—	17
Lapides	1961	—	2	2
Pytl	1961	—	1	3
Rickham	1961	—	1	17
Sturdy	1961	—	6	36
Thompson	1961	1	3	5
Young, B. W.	1961	—	—	3
Chisholm	1962	11	6	10
Cook et al.	1962	—	—	3
Higgins	1962	—	2	4
Baxter and Morse	1963	—	—	11
Landau and Lattimer	1963	—	2	48
Stagner and Hodges	1963	—	2	11
Swenson et al.	1963	—	14	33
Total		15 (5%)	40 (14%)	222 (80%)

ureteral reimplantation method of Paquin and have adopted Young's bladder-neck plasty in the reconstruction of the vesical neck. Our approach consists of a two-stage operation: first, bilateral nephrostomies are done as a preliminary procedure; then, as a second operation, the exstrophy and epispadias are closed. The time interval between the two operations varies between 2 and 4 weeks. The optimum age for this procedure is between 9 and 18 months of age. The child is hospitalized for bilateral nephrostomies and usually remains in the hospital until the exstrophy is closed. Urinary diversion by nephrostomies aids in the preoperative skin care and has proved to be of value in attaining an essentially uncomplicated postoperative course. Our method of exstrophy closure consists of incising circumferentially the mucocutaneous junction of the exstrophied bladder, then subcutaneously carrying the incision laterally to the upper medial portion of the separated pubic bones. By incising the superficial cartilage of the pubic bones, the attachments of structures of bladder base and vesical neck are freed from the pubes. These incisions are carried completely down the pubic rami into the ischiorectal region and accordingly the uppermost attachments of the levator ani muscles are free. This dissection permits the vesical neck and prostate to be readily depressed deep inside the pelvis. The lateral bladder walls are then dissected free, and the two ureters are divided at the point where they enter the bladder wall. Each ureter is then reimplanted, using the tunnel-and-cuff technique of Paquin. The epispadiac urethra is now incised and closed, following the general plan of Young. As the mucosa of the urethra is rolled into a tube over a no. 10 French catheter, a posterior bladder-neck plasty is also done. The tissues that were freed from the pubes are now folded over the reconstructed urethra; actually, the cartilaginous attachments themselves can be easily overlapped. The bladder is closed, and a small Malecot catheter is left indwelling as a cystostomy. The whole urinary structure is now readily displaced down into the pelvis. Large flaps of rectus sheath and external oblique aponeurosis are then raised on either side and turned downward across the defect to be sutured to the opposite pubic bones. Skin and subcutaneous tissue can be easily mobilized and drawn over the defect. Appropriate drains are left in place. After about 2 to 3 weeks, pressures are measured in each renal pelvis (WALZAK and PAQUIN) by connecting each nephrostomy tube to a water manometer overnight. Should no obstruction be present in the reimplanted ureters, as shown by low intrapelvic pressures, the nephrostomy tubes are removed, and the child is sent home after the cystotomy tube has been removed.

In the past 8 years we have attempted functional closure of the exstrophied bladder in 16 children. In all but two, closure of the abdominal defect by means of fascial flaps has been uniformly successful. Four children subsequently had urinary diversion by means of an ileal conduit: two because of failure of complete closure of the abdominal defect (in one child, the flaps became necrotic; in the other, exstrophy recurred when the child was re-operated upon in an attempt to tighten the bladder neck). The other two children who had diversion were totally incontinent. Two children have not been followed closely in the past 2 to 3 years but when last seen were essentially incontinent. Two others are known to be incontinent. Four children have some degree of control; that is, they void frequently and in small amounts (20 to 50 cc.) and are dry between voidings. Four children have good control and good capacity. Thus, we can summarize our experience with functional closure of the exstrophied bladder: one-fourth are failures and have had diversion; one-fourth are incontinent; one-fourth show some semblance of continence; and one-fourth have fair to good control and good capacity.

4. Cloacal Exstrophy

a) Pathology

Cloacal exstrophy is a rare and curious anomaly that involves not only the genito-urinary system but the intestinal tract as well. Another term that is employed in describing this disorder is "vesico-intestinal fissure," introduced by SCHWALBE in 1909. In a recent review article, SOPER and KILGER described 52 cases reported in the German and English literature since 1868 and added five cases from the State University of Iowa Hospitals. The works of BRAKELEY, DAVIES, McFARLAND, RUSSELL, and HALL et al. also contain extensive bibliographies relating to this complex congenital anomaly.

The incidence of this disorder is estimated to be approximately 1 in 200,000 to 1 in 250,000 livebirths. There seems to be no sex preponderance; SOPER and KILGER reported 29 male and 24 female children in their review. It is interesting to note that most infants born with this defect are premature. Chromosomal karyotype of one patient reported by these authors was normal, suggesting that there is no cytogenetic abnormality responsible for this disorder.

The embryological pathogenesis involves the cloacal membrane — its large size and the time of its dehiscence. Basically, the defect occurs as the abnormally large cloacal membrane divides before the rectal septum has fully developed and has fused with the primitive perineum; thus, the embryonic cloaca itself eventrates (Fig. 26). Before the urorectal septum has divided the cloaca horizontally (coronally), the vesical anlage must consist of the paired lateral portions of the two upper sides of the cloaca as well as the ventral or median portion that is the cloacal membrane itself. As the cloacal membrane dehisces (or splits) before the urorectal septum has separated the ventral portion of the cloaca from the future hindgut (posterior cloaca), the posterior cloaca would eventrate, carrying with it

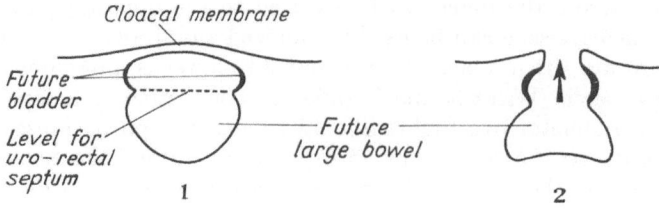

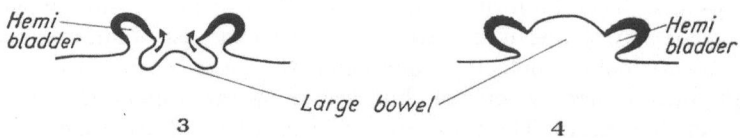

Fig. 26. Diagrams of eventration of the cloaca in the formation of cloacal exstrophy

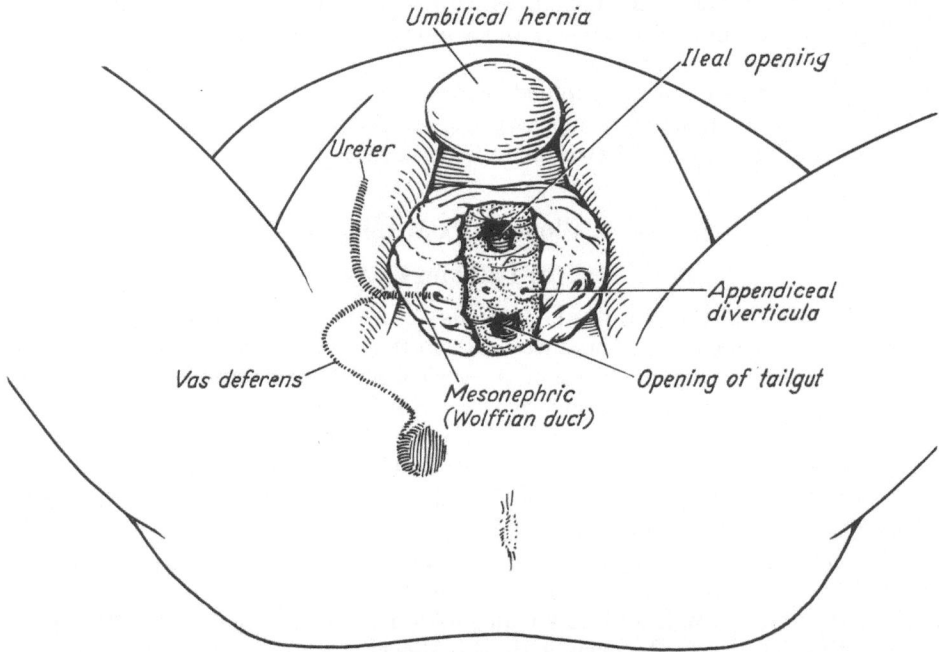

Fig. 27. Diagram of cloacal exstrophy. (After von Geldern)

the two lateral halves of the remaining ventral cloaca. Hence, bowel mucosa would lie in the middle of the abdominal defect, flanked on either side by a hemibladder. As the urorectal septum would already have some slight but significant beginning in almost every embryo before the cloacal membrane dehisced, the resulting bladder structure would be in a somewhat horseshoe shape, being joined in the upper portion. As early as 1913 Johnston wrote that if some observers could demonstrate vesical mucosa between the bowel and the umbilicus in any case of cloacal exstrophy, "... the case for an abnormally early and excessive rupture of the cloacal membrane will be established beyond doubt." Von Geldern made exactly this demonstration in 1924 (Fig. 27).

Experimental support was given to this embryological theory by MUECKE. An early embryonic stage in the chick was chosen, and the cloacal membrane anlage was surgically altered into a nonregressing entity by the insertion of a small bit of plastic into this region. Theoretically, this small foreign body would be expected to act as a mechanical barrier to mesodermal movements in the infra-umbilical abdominal wall, thus simulating a large, nonregressing cloacal membrane, and would initiate an abdominal defect and the subsequent formation of cloacal exstrophy. Fig. 28 shows an experimental chick with cloacal exstrophy. The small bit of plastic is still in place just superior to the open cloacal pouch. The separated pubic bones are evident and medial to each is a membranous hemi-

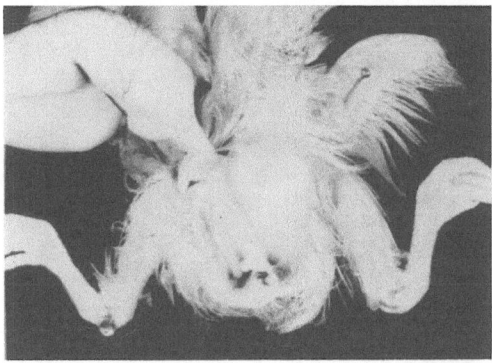

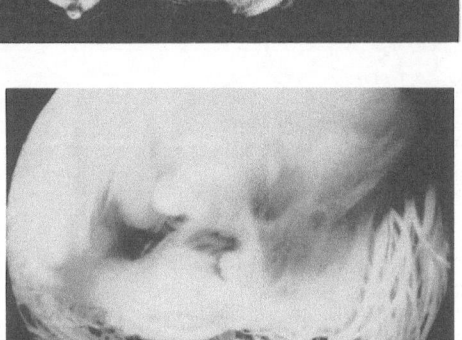

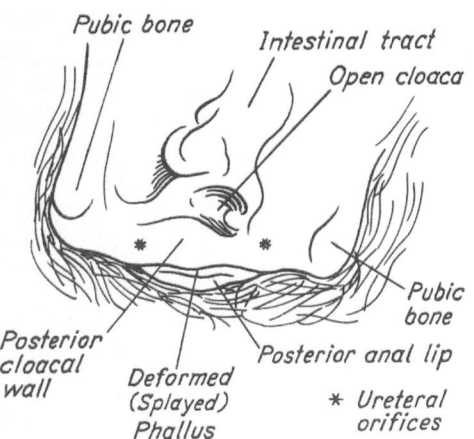

Fig. 28. Experimental cloacal exstrophy in a chick. The pubic bones are widely separated and the cloaca lies open. Asterisks mark the ureteral orifices

allantois into which the developing embryo excreted its mesonephric urine. The anal opening and phallus are splayed. The coiled intestines lie immediately superior to, and are continuous with, the open cloaca, the mucosal folds of which are clearly seen. Fig. 29 compares the anatomic features of an experimental chick with those described by VON GELDERN. Striking similarities are evident.

One interesting comment was annotated to the experimental study just described. In mammals, other than man, the exstrophy complex is rare. A possible explanation for this observed variation in incidence according to species may lie in the varying importance to embryonic survival of structures adjacent to the cloacal membrane, particularly the allantois. In domestic animals, such as the pig, horse, dog, and sheep, the allantoic vesicle is of vital importance in implantation (MOSSMAN). Thus, an early defect in the tail-bud allantoic region could well have fatal consequences for the embryo. In man, however, the allantois is a vestigial structure. Abnormalities in the neighboring cloacal area might thus be better tolerated, and a significant incidence of the exstrophy complex in live births could be expected.

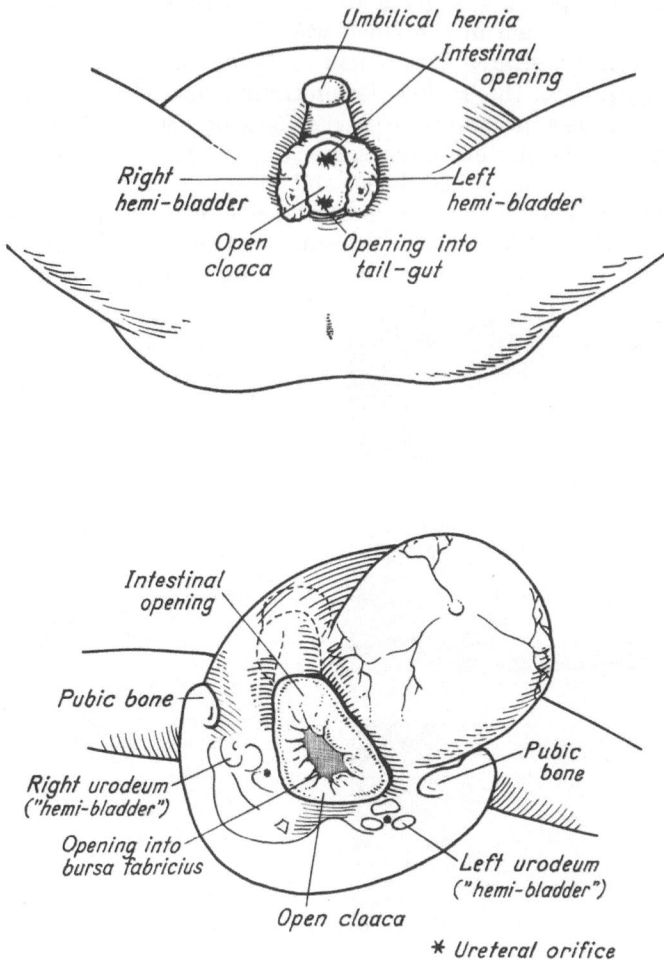

Fig. 29. Human cloacal exstrophy compared with experimental exstrophy in the chick. Striking similarities
are present

b) Anatomy and Clinical Features

The characteristic anatomic features of this congenital anomaly are the presence of two hemibladders, each with its own ureter, separated by an area of intestinal mucosa (Fig. 30). The exposed intestinal field is surfaced by mucosa histologically compatible with colon; this mucosal surface is readily distinguishable from bladder epithelium by its deeper red hue. This exstrophied intestine is thought to be the cecum. An attached remnant of tailgut is present in many cases of cloacal exstrophy; this gut opens as a blind intestinal tube from an ostium near the mid-line toward the posterior or caudal part of the exstrophied large bowel. A pair of bowel sacs near the upper or cephalic portion of the exstrophied large bowel are frequently found and thought to be appendiceal diverticula. These appendix-like diverticula have varied greatly in character in the reported cases and have been missing in a few but their tendency to be paired and their large-bowel character are real. The ileum is present and opens into the upper portion of the exstrophied bowel as an orifice closely resembling the usual ileocecal valve. Frequently the ileum is found to be prolapsed through this junction. The

ureter and vas normally become completely separated before 6 weeks [between the 9 and 14 mm stages (WILLIAMS)]. If eventration occurred well before this, according to the embryological theory presented, then ureter and vas should not infrequently be found to be joined in a common, actually mesonephric, duct. Several such cases have been reported (JENSEN et al., VON GELDERN, KINDRED, VEAL and McFETRIDGE) (Fig. 27). Many minor variations of the exstrophied structures have been described (MARSHALL and MUECKE). Epispadias is invariably present. According to SOPER and KILGER, in 29 male infants determinate for sex,

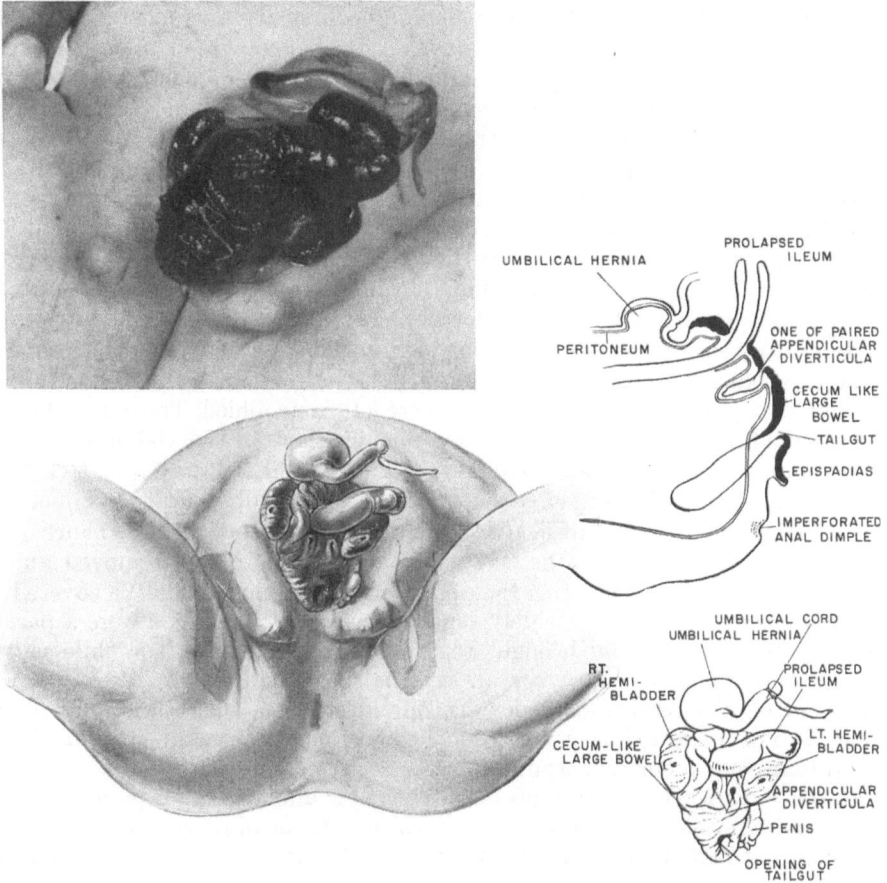

Fig. 30. Cloacal exstrophy in the newborn

the penis was absent in eight and the scrotum, in 19. Fifteen of the patients had diphallus, and eight had a bifid scrotum, with a bridge of noncorrugated skin separating each hemiscrotum. In the 24 female infants reviewed, the clitoris was bifid in seven and absent in 14. The vagina was duplicated in 14 and described as absent in six. Fifteen of the 53 infants with cloacal exstrophy described by these authors had a prominence in each groin overlying the separated pubic tubercles. These were referred to as labial or groin "swellings" and were composed histologically of vascular, erectile tissue. The vertebral column is frequently malformed, 38 of the 53 children in the series had bony abnormalities. Lumbar spina bifida was present in 32, and, of those, 24 had associated myelocele or meningomyelocele.

Finally, there are often recorded other accompanying anomalies not directly associated with the rump of the infant. These seem to be unrelated from an embryonic standpoint.

c) Treatment

In our experience, the greatest problem in the clinical management of these children is their prematurity. If neglected, these infants usually die in a short period of time, usually from pulmonary complications, electrolyte loss and dehydration secondary to the congenital ileostomy. Treatment is justified in those infants in whom congenital anomalies such as cardiac and neurological abnormalities do not preclude an inevitable early demise.

In 1900 von Steinbüchel documented the first attempt at staged surgical salvage of an infant with cloacal exstrophy. He had operated upon a newborn male infant in 1894, correcting the omphalocele at that time. Two days later the child underwent exploratory surgery because of intestinal obstruction, and a "pull-through" of a segment of atretic bowel to the perineum was done. The baby died in the postoperative period from a traumatic enema that resulted in intestinal perforation. In 1960 Rickham reported the cases of three infants with cloacal exstrophy in whom various corrective procedures were attempted. Two of the children died postoperatively but the third one was alive and thriving 5 years later. This child, a boy weighing $5^1/_2$ lb, had a medium-sized exomphalos, a split penis, and a small sacral myelomeningocele. At 1 day of age he was found at exploratory surgery to have only one kidney. This solitary kidney was situated in the pelvis and had a short ureter joining the left hemibladder. The two hemibladders were joined at the mid-line and were left exstrophied. The exposed cecum was closed. It was found that there was enough slack in the abdominal wall to allow complete repair of the exomphalos. The child did well postoperatively; the only complication was an attack of mechanical intestinal obstruction at $2^1/_2$ months of age. At age 17 months, an ileal conduit was constructed and the short ureter reimplanted. At age 19 months, the exstrophied bladder was removed and by mobilization of the skin margins the defect in the abdomen could be covered. At the same operation, the two widely separated crura of the penis were separated from the pubic arches and brought together in the mid-line. The child is well, is incontinent of feces, and wears an external ileal-conduit bag. He walks well, talks, and is of normal intelligence. To our knowledge, this is the only patient reported to be living so long after surgical correction of this complex anomaly and indeed represents a surgical triumph.

In our experience, four infants born with this anomaly were premature, and death resulted primarily from their prematurity. In one of these infants, a girl, we demonstrated that the tailgut was able to absorb water and electrolyte solutions. The ileum was therefore detached from the exstrophied cecum and anastomosed end-to-end to the rather long and prominent tailgut, which was then detached from the exstrophied structures and brought out through the skin as a colostomy. This improved the child's nutritional state to a marked degree by apparently reducing the water and electrolyte loss from the ileal stoma. The child died as the result of respiratory difficulties and pneumonia before further surgical correction could be undertaken.

5. Variants of Exstrophy

A great variety of anomalies associated with the exstrophy-epispadias complex are theoretically possible, all having in common a basically similar pathogenesis. The focal point is the abnormal cloacal membrane; depending upon its size, the extent of dehiscence, and the critical timetable of associated structural development, variations in the final product are to be expected.

a) Superior Vesical Fissure

The characteristic musculoskeletal deformity of the exstrophy complex is present with wide separation of the pubic bones. A vesical opening is present just below the umbilical hernia (Fig. 31). The lower, closed area of the abdominal

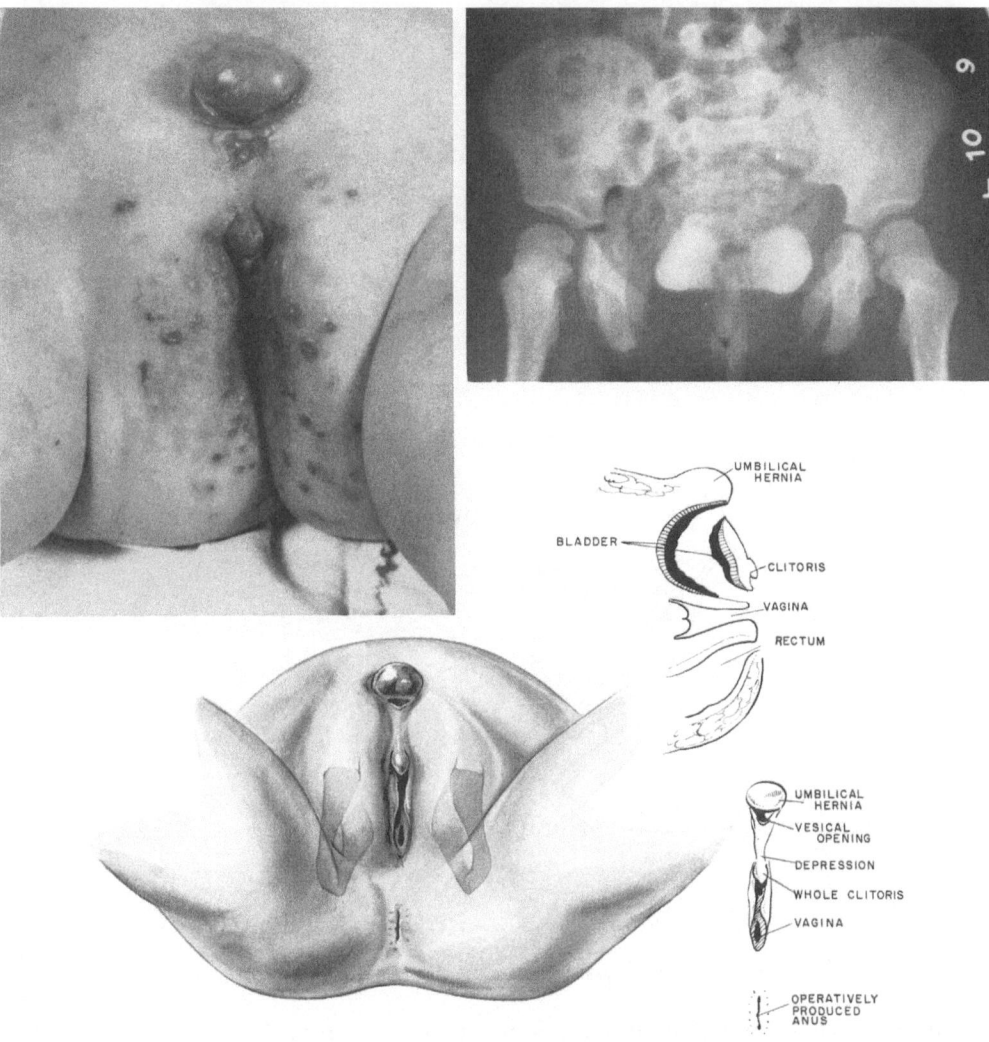

Fig. 31. Superior vesical fissure in a girl. Musculo-skeletal deformities are those typical of exstrophy of the bladder

wall has a composition compatible with what would be expected of a persistent cloacal membrane; that is, a fibrous sheath devoid of muscle fibers interpositioned between skin and anterior bladder wall. In this case, the clitoris was normal and the patient was continent. Closure of the vesical fissure was all that was necessary. The child had no reflux, good upper tracts, and no persistent urinary-tract

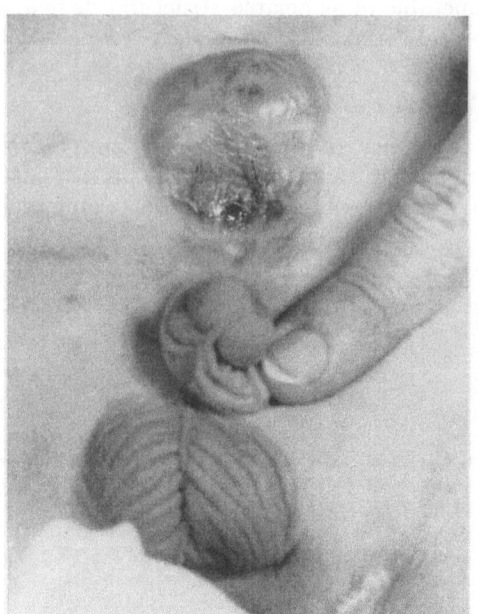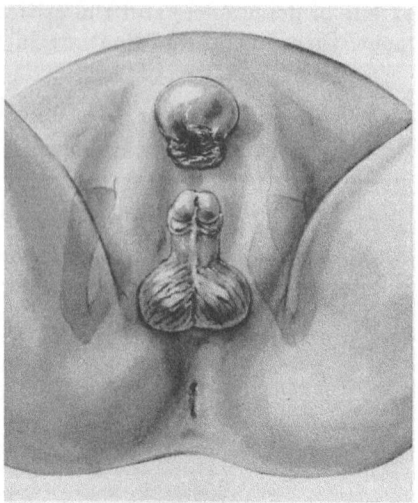

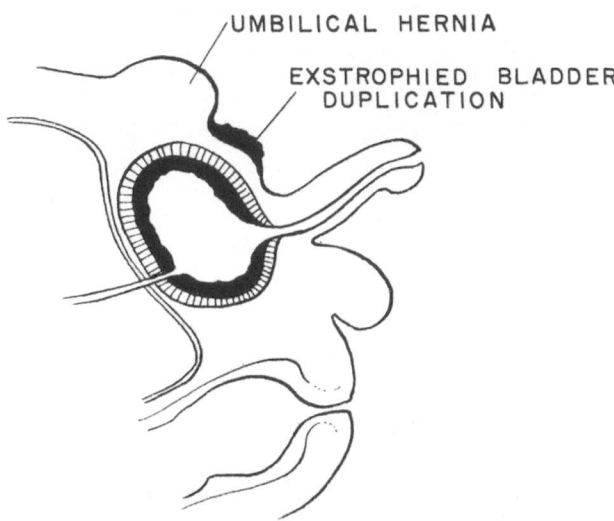

Fig. 32. Duplicate exstrophy in a boy with an intact urinary tract

infection. USON and ROBERTS presented two cases in their series of "pseudo-exstrophy," in which the patients had slight urinary leakage from a small opening just below the umbilicus. Superior vesical fissure is not merely an example of patent urachus, for the musculoskeletal abnormalities are present in the former and never in the latter.

b) Duplicate Exstrophy

If a superior vesical fissure should occur and then fuse, bladder elements might remain on the outside. Such an occurrence would be expected to be even more rare than is superior vesical fissure. About six cases have been reported in the literature (KITTREDGE and BRADBURN, GROSS and CRESSON, CAMPBELL, POWELL, MARSHALL and MUECKE). Our case seems to be typical of duplicate exstrophy (Fig. 32). This child was referred to us by his pediatrician who reported a patient with exstrophy but with apparently normal urinary control! Anatomically, the boy was exactly the same as those with superior vesical fissure except that a closed and normally functioning urinary tract was present internally and a rudimentary exstrophy presented externally in the middle of the lower abdomen with the characteristic musculoskeletal deformity. A possible transition between superior vesical fissure and duplicate exstrophy is the case of KITTREDGE and BRADBURN, in which a tiny fistula did connect the internal and external bladder mucosae. Treatment of this interesting anomaly is simple excision of the ectopic bladder elements.

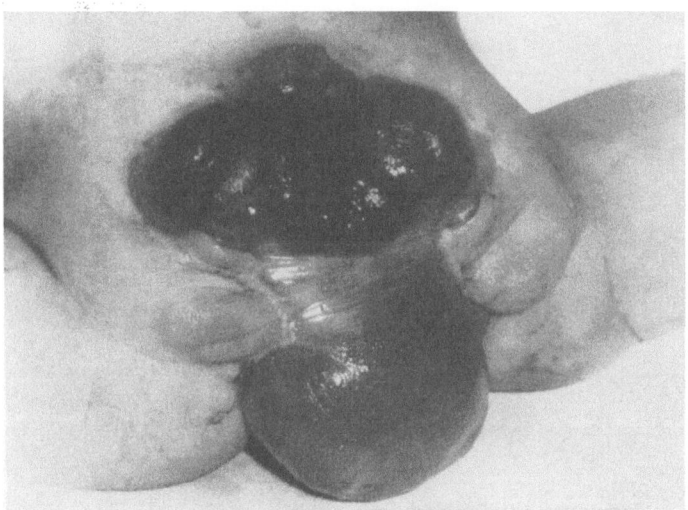

Fig. 33. Incomplete cloacal exstrophy. The eventrated cloacal duct of REICHEL covers the perineum and connectes the exstrophied bladder to the prolapsed rectum. (Photo courtesy of Mr. J. H. JOHNSTON, Liverpool)

c) Inferior Vesical Fissure

The deformity of inferior vesical fissure is mentioned several times in the literature as a variant of the exstrophy-epispadias complex. SORRENTINO and LEONETTI titled an illustration of penopubic epispadias as "fissura vesical inferior." Campbell equated inferior vesical fissure with classical exstrophy, and one unique case reported by HIGGINS, in which exstrophy was present except for an intact, fused penis, was given as an example of inferior vesical fissure. Epispadias with an open vesical neck and possible mucosal prolapse might conceivably be labeled "inferior fissure," but most authors considered this to be more correctly classified as complete epispadias: penopubic in the male, subsymphyseal in the female child.

d) Incomplete Cloacal Exstrophy

One interesting photograph of a child with incomplete cloacal exstrophy was sent to us by Mr. J. H. JOHNSTON from the Alder Hey Children's Hospital in Liverpool (Fig. 33). Here, the exstrophied bladder was joined by a bridge of intestinal mucosa that stretched over what should have been the perineal body to a prolapsed but otherwise intact colon. The anterior portion of the anal sphincter was deficient. The defect occurred at the time when the urorectal septum had not yet fused with the primitive perineum and the cloacal duct of Reichel was still open. Thus, the exstrophic bridge of intestinal-type mucosa may be considered the matured epithelial lining of the duct of Reichel, and thus we have the transition between classical exstrophy and complete cloacal eversion.

6. Epispadias

a) Pathology

The various forms of epispadias represent, as a group, the lesser degree of the basic embryological defect of the entire exstrophy complex. The embryogenesis of epispadias is that of vesical exstrophy, the difference between the

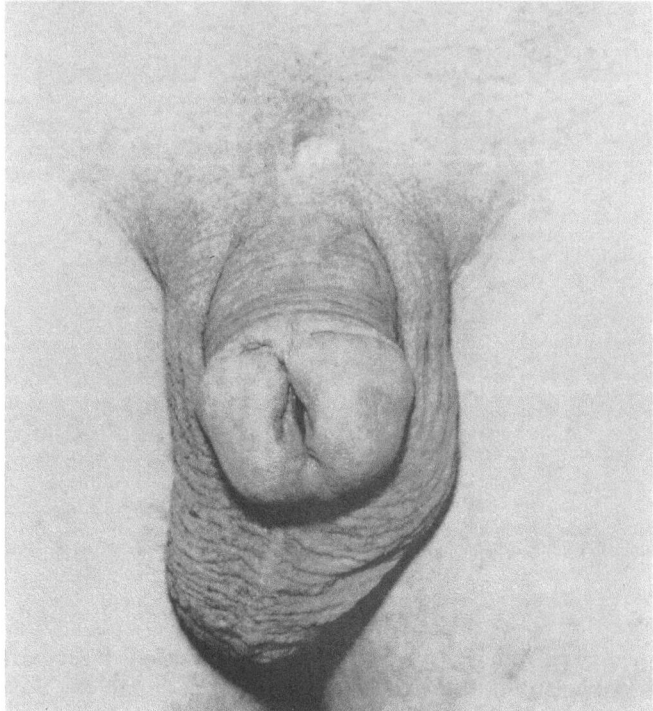

Fig. 34. Balanic epispadias

lesions being one of degree. An abnormal cloacal membrane remains on the dorsum of the genital tubercle and as the primitive phallus develops it roofs the ventrally fusing phallic anlagen. Thus, the primordium of the urethra lies in the dorsal aspect of the phallus. Mesoderm, however, successfully sandwiches itself between the membranous layers superiorly, so that when the split occurs, only the dorsum of the developing phallus is affected. The urethral ending may be located anywhere along the dorsal aspect of the penis; in the most severe, and commoner forms, the bladder neck may lie exposed and the patient is totally incontinent.

The incidence of this anomaly has been said to be 1 in 30,000 individuals (Campbell), but Dees, in a summary of 5,292,212 hospital admissions in several medical centers, reported only 45 cases of epispadias in male patients (1 in 117,604) and 11 in female patients (1 in 481,110). Sex incidence favors males 5 to 1 as compared to females (Gross and Cresson).

The musculoskeletal deformity, consisting of separated pubic bones and divergent recti muscles, is a common but not constant feature of epispadias alone. Of all patients with mere epispadias seen at The New York Hospital (27 patients), roentgenograms of the pelves showed the width of the symphysis increased in 12 and normal in seven; in eight, the films were unavailable. The pubic separation

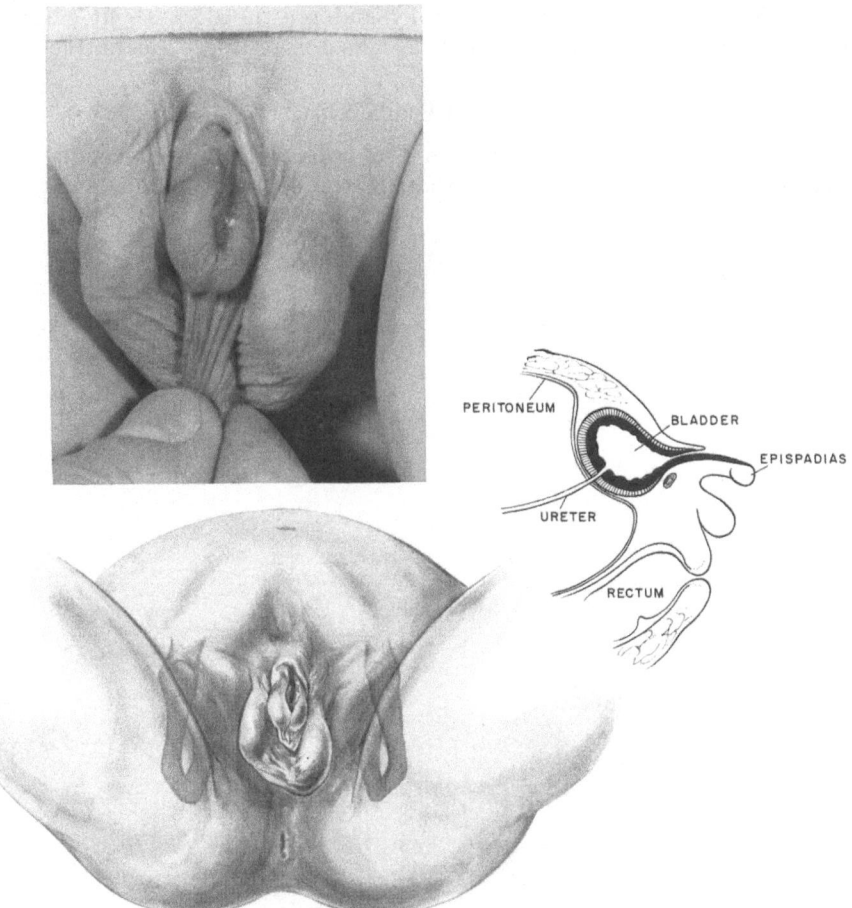

Fig. 35. Penile epispadias with continence

occurs in almost all cases of complete epispadias but is reduced or absent is some cases of the intermediate, lesser type; that is, the amount of pubic separation when present, seems to correlate with the severity of the epispadias. Thus, in epispadias alone, the musculoskelatal deformity often is less severe than in exstrophy of the bladder.

b) Description

In the male, there are three forms of epispadias: balanic, penile, and penopubic.

α) Balanic Type

The balanic type, or glandular epispadias, is the rarest form as well as the one with the least defect. The defect extends dorsally from the meatus to the corona

of the glans penis (Fig. 34). The glans appears flattened and the penis is stubby.
The patient is fully continent and the symphysis pubis is normal. In our series
of 27 patients with epispadias only, there were but two examples of the balanic
type. GROSS and CRESSON saw it only three times in 18 epispadiac patients and
DEES likewise found it to be the least common form. We have observed an even
rarer form of this mild degree of epispadias in a boy with a "spade penis," in
whom the urethral opening was only elongated but essentially in the tip of the
glans. The penis was short and stubby and protruded at about a right angle
from the pubis even in the flaccid state. The dorsal aspect was flattened.

β) Penile Type

The penile type is the next degree of gross anatomic defect in which the
urethral meatus lies in the dorsal shaft of the penis. If the glandular corona is

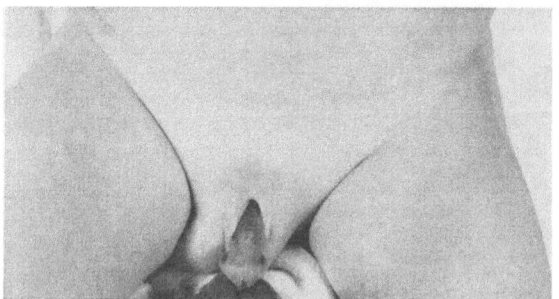

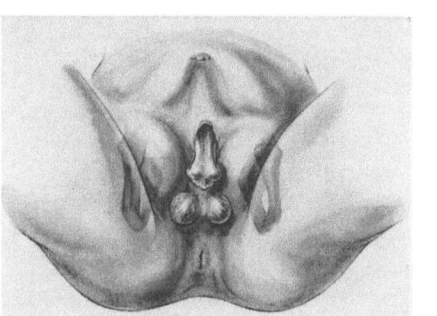

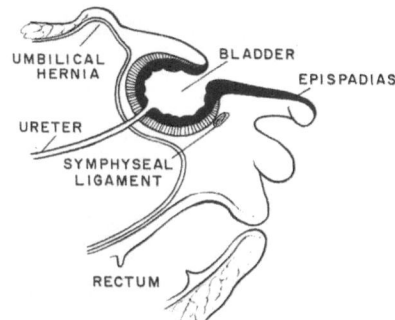

Fig. 36. Penopubic epispadias

divided, we consider the epispadias to be more than merely balanic. The penis is
usually short, flattened, and curved upward. A large portion of the pendulous
urethra lies open on the dorsum of the phallus, forming an epithelium-lined groove
between the corpora (Fig. 35). We have six such patients in our series of 27; all
were fully continent. The musculoskeletal defect, characterized by a widened
symphysis pubis, was not a constant feature in these patients.

γ) Penopubic Type

In the penopubic type, or complete epispadias, incontinence is the rule rather
than the exception (Fig. 36). It is the most common type of epispadias in the
absence of vesical exstrophy: 18 of our 27 cases without exstrophy; 11 of 18 in
GROSS and CRESSONS series; and the most common in DEES' collection. Some

patients with penopubic epispadias are not totally incontinent, again showing a borderline or transitional variation of the complex. In others, the urethral cleft was so extensive as to represent a minimal degree of exstrophy, with bladder mucosa prolapsing at times through the open vesical neck. Sometimes these cases have been referred to in the literature as "inferior vesical fissure;" in our opinion, these are transitions between a pronounced epispadias and a mild degree of exstrophy.

c) Clinical Features

The clinical findings and symptoms vary with the type of defect present. The male child with balanic or penile epispadias usually has good urinary control.

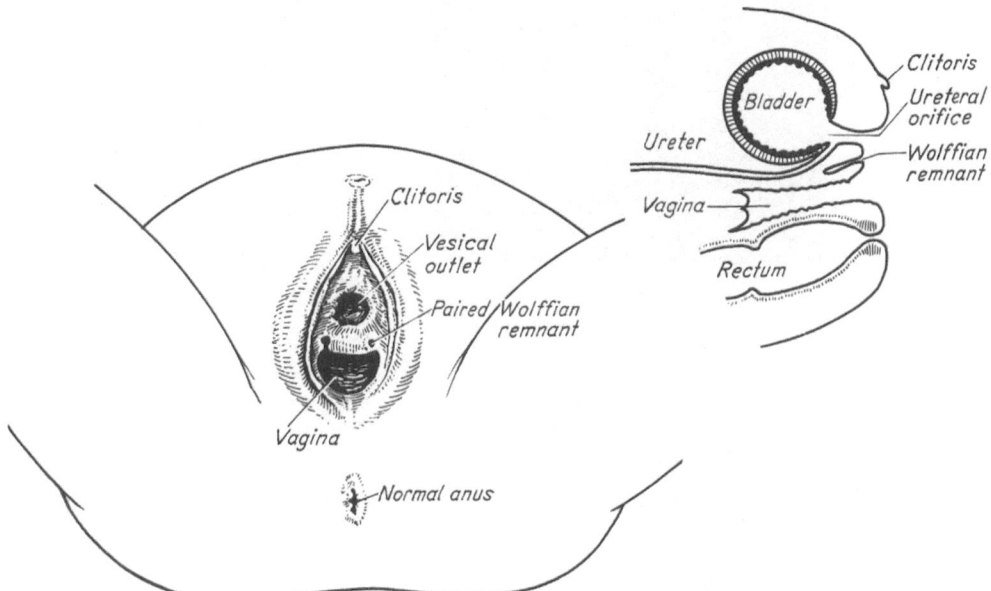

Fig. 37. Incontinent girl with a somewhat mild variety of subsymphyseal epispadias

It is the appearance of the deformed penis that brings the child to the physician's attention. In the uncircumcised penis, the mild or balanic form of epispadias may be completely hidden from view. It is important to observe the child during micturition, having him start and stop the urinary stream on command, to ascertain if the sphincteric mechanism is intact. In the penopubic type, in which the dorsal, groove-like opened urethra extends the entire length of the penis, incontinence may be the presenting complaint. On examination, a continuous dribble of urine is observed to come from the open bladder neck. Often the pendulous prepuce that hangs from the ventral surface has to be grasped in order to pull down the flattened, upward-turned penis, thereby exposing the open prostatic urethra and patulous bladder neck. The urinary opening in the lower abdominal wall often is large enough to admit the tip of the examiner's finger; some portion of bladder mucosa may actually prolapse. The typical musculoskeletal defect of exstrophy is almost always present; the symphysis pubis is widened — but not to the degree seen in classical exstrophy. The bladder is apt to be small and contracted. As a rule, the upper urinary tracts are not hydronephrotic, and reflux is not present on cystograms.

In female children, the same three general types do occur, but the first two lesser degrees of epispadias are difficult to detect and rarely come to the physician's attention. The urethra is shortened and patulous, and the clitoris may be bifid. Urinary control is usually good although the child may have some degree of stress incontinence. These children do not require surgery.

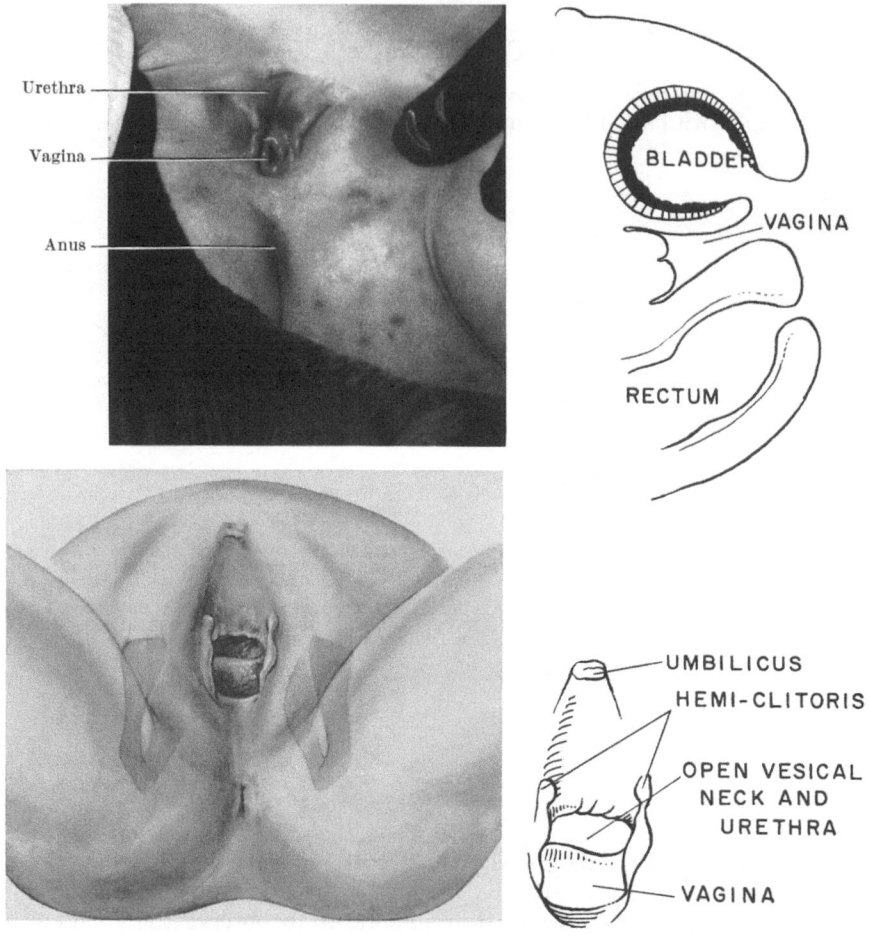

Fig. 38. Girl with subsymphyseal epispadias as reported by BROWNE

The subsymphyseal epispadias without vesical exstrophy does present a challenge in management. It is a rare disorder; we saw only one case in our series of 27 epispadiac patients (Fig. 37). The urethra was patulous and incontinent. The subsymphyseal area was widened, with divergent recti muscles, and the fatty mons was deficient. Two small openings, thought to represent GARTNER's ducts, were found in the lateral aspects of the urethrovaginal septum. In 1892 ALEXANDER described and illustrated such a case, and BALLANTYNE, in 1896, claimed to have collected as many as 33 examples. BROWNE presented photographs of a similar case in 1951 (Fig. 38). He preferred to call it "pubo-vesical cleft." In all these cases (ALEXANDER, BALLANTYNE, STILES, BROWNE, MAKINS), the clitoris has been bifid but it would seem possible that in milder varieties the clitoris might be whole and even the symphysis undivided.

d) Treatment

The objectives of surgical repair are both cosmetic and functional. In the male patient, an attempt must be made to restore continence and to improve the delivery of urine and semen. We should like to mention, however, that even the most severely epispadiac patient may be fertile and functional in the coital act, for one of our epispadiac patients sired five normal children! In the female patient, the urethra should be reconstructed and vulvar abnormalities, if obvious, corrected by plastic revision.

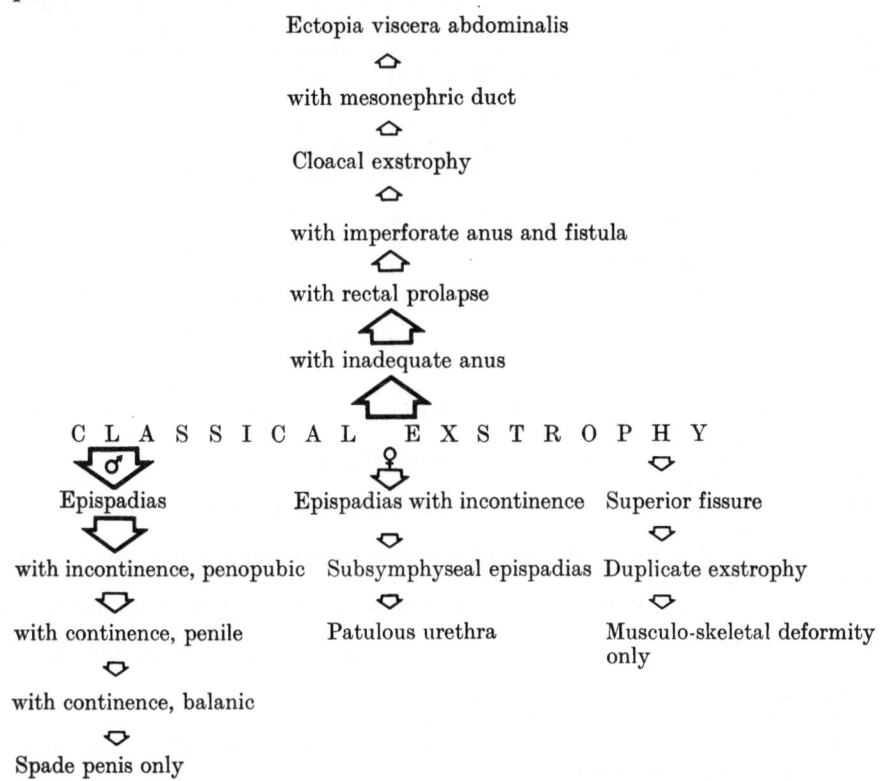

Fig. 39. Scheme of developmental relationships of variations within the exstrophy-epispadias complex. The most common deformity is classical exstrophy. Other variations within the complex diverge from classical form with decreasing frequencies as suggested by the size of the arrows. Progression shown above classical exstrophy is increasingly to exotic, while variations below approach normality

At the time of surgery, the patient should be younger than school age, preferably between 3 and 5 years, for attempted continence procedures; plastic revisions of the lesser degrees of urethral defects can be done more easily in the older child as structures are more developed.

For a comprehensive review of different surgical methods, we refer to a review by GROSS and CRESSON. In essence, we employ H. H. YOUNG's technique as modified by GROSS and CRESSON in the plastic reconstruction of the urethra, with excellent results in the male child without incontinence. We have found that in the penopubic-type defect a second operation is necessary; namely, an internal bladder-neck revision, as described by DEES, before an adequate degree of continence is achieved. In two patients we had to resort to reimplantation of ureters, as described by LEADBETTER, in order to gain additional length in the reconstruction of the bladder neck.

In the female patient, a similar operation may be performed (Davies, Dees). In our patient with subsymphyseal epispadias, plastic procedures failed to produce adequate urinary control, and the ureters had to be transplanted into an ileal conduit.

References

Abrahamson, J.: Double bladder and related anomalies: clinical and embryological aspects and a case report. Brit. J. Urol. **33**, 195 (1961).

Alexander, S.: Congenital deformity of external female genitals, entire absence of urethra, spontaneous dislocation of the hip as the result of efforts to retain urine; correction of deformity and restoration of urethra by plastic operation with some observations upon the technique of ureteral catheterism. J. Cutan. and Genito-Urin. Dis. **10**, 253 (1892).

Ansell, J.: Personal communication 1964.

Babcock, W. W.: Exstrophy of the bladder: plastic reconstruction versus ureterointestinal transplantation. Trans. Amer. Ans. Obstet. Gynec. **60**, 240 (1949).

Badenoch, A. W.: Congenital obstruction at the bladder neck. Ann. Roy. Coll. Surg. Engl. **4**, 295 (1949).

Ballantyne, J. W.: So-called epispadias in woman. Edinb. Hosp. Rep. **4**, 249 (1896).

Baxter, C. F., and T. S. Morse: Exstrophy of the bladder — evaluation of operative results. Ohio St. med. J. **59**, 149 (1963).

Beer, E.: Chronic retention of urine in young boys due to obstruction at the neck of the bladder. Ann. Surg. **79**, 264 (1924).

Begg, R. C.: The urachus and umbilical fistulae. Surg. Gynec. Obstet. **45**, 165 (1927).

— The urachus: its anatomy, histology and development. J. Anat. (Lond.) **64**, 170 (1929).

Boisonnat, P.: Vessie double (Vesica duplex) avec urètre unique chez un garcon de 4 ans: opération plastique; excellent resultat anatomique et fonctionnel. J. Urol. méd. chir. **59**, 883 (1953).

—, and P. Bouteau: Volumineux kyste du trigone chez un garcon de 3 ans. J. Urol. méd. chir. **60**, 688 (1954).

Boyce, W. H.: Risultati della chiusura Funzionale dell'estrofia Vesciale congenita. [Disc.] Urologia **28**, 101 (1961).

—, and S. A. Vest: A new concept concerning treatment of exstrophy of the bladder. J. Urol. (Baltimore) **67**, 503 (1952).

Brakeley, E.: Exstrophy of the bladder; complicated by other congenital anomalies. Amer. J. Dis. Child. **43**, 931 (1932).

Bricker, E. M.: Bladder substitution after pelvic evisceration. Surg. Clin. N. Amer. **30**, 1511 (1950).

— H. Butcher, and C. A. McAfee: Late results of bladder substitution with isolated ileal segments. Surg. Gynec. Obstet. **99**, 469 (1954).

Browne, D. I.: Some congenital deformities of rectum, anus, vagina and urethra. Ann. roy. Coll. Surg. Engl. **8**, 173 (1951).

Burnham, J. P., and J. Farrer: A group experience with uretero-ileal cutaneous anastomosis for urinary diversion: results and complications of the isolated ileal conduit (Bricker procedure) in 96 patients. J. Urol. (Baltimore) **83**, 622 (1960).

Burns, E., H. Cummins, and J. Hyman: Incomplete reduplication of the bladder ith congenital solitary kidney: report of a case. J. Urol. (Baltimore) **57**, 257 (1947).

Buskirk, K. E. van, G. C. Sanders, and D. T. Schamber: Exstrophic musculoskeletal defects with ectopic bladder in ventral hernia; report of a case and review of the literature. J. int. Coll. Surg. **40**, 251 (1963).

Campbell, M. F.: Trigonal curtain obstruction of the bladder outlet. J. Urol. (Baltimore) **27**, 157 (1932).

— Exstrophy of the bladder. In: Urology, 1st ed., vol. 1, sect. 4. Philadelphia and London: W. B. Saunders Co. 1954.

— Personal communication 1961.

— Anomalies of the genital tract. In: Urology, 2d ed., vol. 2, chapt. 37. Philadelphia and London: W. B. Saunders Co. 1963.

— Exstrophy of the bladder. In: Urology, 2d ed., vol. 2, chapt. 37. Philadelphia and London: W. B. Saunders Co. 1963.

— Anomalies of the bladder. In Urology, 2d ed., vol. 2, chapt. 36. Philadelphia and London: W. B. Saunders Co. 1963.

Caulk, J. R.: The ureter as a possible origin of certain diverticula of the bladder. J. Urol. (Baltimore) **21**, 23 (1929).

Cherry, J. W.: Patent urachus: review and report of a case. J. Urol. (Baltimore) **63**, 693 (1950).

CHISHOLM, T. C.: Exstrophy of the urinary bladder. In: Pediatric surgery. New York: Year-book Publ. 1962.
— Exstrophy of the urinary bladder. Amer. J. Surg. 101, 649 (1961).
CHWALLA, R.: Über die Entwicklung der Harnblase und der primären Harnröhre des Menschen mit besonderer Berücksichtigung der Art und Weise, in der sich die Ureteren von den Urnierengängen trennen, nebst Bemerkungen über die Entwicklung der Müllerschen Gänge und des Mastdarms. Z. Anat. Entwickl.-Gesch. 83, 615 (1927).
— The process of formation of cystic dilatations of the vesical end of the ureter and of diverti-cula at the ureteral ostium. Urol. cutan. Rev. 31, 499 (1927 b).
COATES, W. H.: Case of a remarkable conformation of the urinary and genital organs in a female child. Edinb. med. J. 1, 39 (1805).
CONNELL, F. G.: Exstrophy of the bladder. J. Amer. med. Ass. 36, 637 (1901).
COOK, C. E.: Pervious urachus: joint or navel illness; case report. Amer. Vet. Rev. 47, 618 (1915).
—, J. T. LESLIE, and E. W. BRANNON: Preliminary report: a new concept of abdominal closure in infants with exstrophy of the bladder. J. Urol. (Baltimore) 87, 823 (1962).
COOPER, A.: Case of malformation of the urinary and genital organs in a female. Edinb. med. J. 1, 132 (1805).
COOPER, E. A., and W. KINTZEN: Patent urachus associated with abdominal muscle deficiency. Canad. med. Ass. J. 87, 27 (1962).
CORDONNIER, J. J.: Surgery of the ureter and urinary conduits. In: Urology, ed. by M. F. CAMPBELL, 2d ed., vol. 3, chapt. 54. Philadelphia and London: W. B. Saunders Co. 1963.
CROSS jr., R. R., and K. E. BARBER: Exstrophy of the bladder: follow-up of survivors. J. Urol. (Baltimore) 82, 333 (1959).
CULLEN, T. S.: Congenital patent urachus. In: Embryology, anatomy and diseases of the umbilicus together with diseases of the urachus. Philadelphia and London: W. B. Saunders Co. 1916.
DAVIES, D. V.: Ectopia vesicae. Brit. J. Urol. 14, 1 (1942).
DAVIS, C. L.: Description of a human embryo having twenty paired somites. Contrib. Embryol. Carnegie Inst. of Washington. 15, ills. 21—22, p. 53—54 (1923).
DEES, J. E.: Congenital epispadias with incontinence. J. Urol. (Baltimore) 62, 513 (1949).
DREYFUSS, M. L., and M. M. FLIESS: Patent urachus with stone formation. J. Urol. (Baltimore) 46, 77 (1941).
DUNCAN, A.: An attempt toward a systematic account of the appearances connected with that malformation of the urinary organs. Edinb. med. J. 1, 43 (1805).
ENEREN, B.: Absence de la vessie et abouchement direct des deux urétères dans l'urètre postérieur. Bull. Soc. franc. Urol. (Paris) 201 (1939).
ENGLISCH, J.: Über eingesackte Harnsteine. Wien. med. Wschr. 53, 1194 (1903).
ESPINOZA, R. R., R. L. BLANCO, and B. V. PICORNELL: Prolapsing trigonal mucosa in infant. J. Pediat. 53, 446 (1958).
FELIX, W.: The development of the urogenital organs. In: Manual of human embryology, ed. by F. KEIBEL and F. MALL. Philadelphia: J. B. Lippincott Co. 1912.
FORSYTHE, I. W., and B. T. SMYTH: Diverticulum of the bladder in children; a study of 13 cases. Pediatrics 24, 322 (1959).
GARRETT, R. A., and J. H. O. MERTZ: Follow-up studies of bladder exstrophy with ureterosig-moidostomy. J. Urol. (Baltimore) 71, 299 (1954).
GELDERN, C. E. v.: The etiology of exstrophy of the bladder. Arch. Surg. 8, 61 (1924).
GLASER, K. H., and A. P. R. LEWIS: A case of familial incidence of ectopia vesicae. Brit. med. J. 2, 1333 (1961).
GLENN, J. F.: Agenesis of the bladder. J. Amer. med. Ass. 169, 2016 (1959).
GOVERNEUER, R.: Technique et resultats eloinges des transplantations de l'uretere. VII. Congr. Internat. Soc. Urol. Rep. (pt. 1) vol. 7, p. 373. (1939).
GROSS, R. E., and S. L. CRESSON: Exstrophy of bladder; observations from eighty cases. J. Amer. med. Ass. 149, 1640 (1952).
— — Treatment of epispadias: a report of 18 cases. J. Urol. (Baltimore) 68, 477 (1952).
GRUBER, G.: In: LUBARSCH-HENKE, Handbuch der Speziellen Pathologischen Anatomie und Histologie. Berlin: Springer 1928.
HALL, E. G., A. E. McCANDLESS, and P. P. RICKHAM: Vesico-intestinal fissure with diphallus. Brit. J. Urol. 25, 219 (1953).
HAMMOND, G., L. YGLESIAS, and J. E. DAVIS: The urachus, its anatomy and associated fasciae. Anat. Rec. 80, 271 (1941).
HARRIS, A.: Congenital vesical neck obstruction in a female child due to cup-valve formation: open operation: complete recovery. Amer. J. Surg. 20, 64 (1933).

Harvard, B. M., and G. T. Thompson: Congenital exstrophy of the urinary bladder: late results of treatment by the Coffey-Mayo method of uretero-intestinal anastomosis. J. Urol. (Baltimore) 65, 223 (1951).

Hejtmancik, J. H., W. B. King, and M. A. Magid: Pseudo-exstrophy of bladder. J. Urol. (Baltimore) 72, 829 (1954).

Herbst, W. P.: Patent urachus. South. med. J. (Bghm, Ala.) 30, 711 (1937).

Higgins, C. C.: Transplantation of the ureters into the rectosigmoid for exstrophy of the bladder; review of 41 cases. J. Urol. (Baltimore) 57, 693 (1947).

— An evaluation of cystectomy: for exstrophy, for papillomatosis, and for carcinoma of the bladder. J. Urol. (Baltimore) 80, 279 (1958).

— Exstrophy of the bladder: report of 158 cases. Amer. Surg. 28, 99 (1962).

Hinman jr., F.: Surgical disorders of the bladder and umbilicus of urachal origin. Surg. Gynec. Obstet. 113, 605 (1961).

— Urologic aspects of the alternating urachal sinus. Amer. J. Surg. 102, 339 (1961).

Hutch, J. A.: Saccule formation at the ureterovesical junction in smooth walled bladders. J. Urol. (Baltimore) 86, 390 (1961).

Ignatescu, M., H. Slobozianu, and E. Athanasiu-Vergu: Absence de la vessie et abouchement des uretères dans l'utérus chez deux jumelles. J. Urol. méd. chir. 45, 51 (1938).

Jensen, O. J., H. E. Eggers, A. H. Bill, and D. R. Dillard: Urinary and fecal incontinence due to congenital abnormalities in children: management by transplantation of ureters to an isolated ileostomy. J. Urol. (Baltimore) 73, 321 (1955).

Johnston, J. H.: Vesical diverticula without urinary obstruction in childhood. J. Urol. (Baltimore) 84, 535 (1960).

Johnston, T. B.: Extroversion of the bladder complicated by the presence of intestinal openings on the surface of the extroverted area. J. Anat. Physiol. 48, 89 (1939).

Jonas, K. C.: Results of surgical reconstruction in exstrophy of the bladder. Arch. Surg. 78, 146 (1959).

Kearns, W. M., and S. M. Turkeltaub: Hourglass deformity of the urinary bladder. J. Urol. (Baltimore) 29, 729 (1933).

Keibel, F.: Zur Entwicklungsgeschichte des menschlichen Urogenitalapparates. Arch. Anat. Physiol. 55, (1896).

Kimball, G. H., and N. R. Drummond: Exstrophy of the bladder; an experimental study and a clinical study of cases. J. Okla. med. Ass. 33, 2 (1940).

Kindred, J. E.: Eventration of the abdominal viscera associated with umbilical hernia, hemipenes and hydromyelomeningocele in a newborn infant. Anat. Rec. 128, 379 (1957).

Kinman, L. M., D. Sauer, V. T. Houston, and W. F. Melick: Substitution of the excluded rectosigmoid colon for the urinary bladder: preliminary report. Arch. Surg. 66, 531 (1953).

Kittredge, W. E., and C. Bradburn: Incomplete exstrophy of the bladder: case report. J. Urol. (Baltimore) 72, 38 (1954).

Kohler, H. H.: Septal bladder with multiple genito-urinary anomalies and uremia. J. Urol. (Baltimore) 44, 63 (1940).

Kook, H., B. Kamhi, and H. B. Hermann: Trigonal curtain obstruction in a female child. J. Urol. (Baltimore) 73, 1026 (1955).

Krasa, F. C., and R. Paschkis: Trigon of bladder in mammals. Z. urol. Chir. 6, 1 (1921).

Kretschmer, H. L.: Diverticulum of the bladder in infancy and in childhood. Amer. J. Dis. Child. 48, 842 (1934).

Ladd, W. E., and R. E. Cross: Rare conditions of abdominal wall. In: Abdominal surgery of infancy and childhood. Philadelphia and London: W. B. Saunders Co. 1941.

Landau, S. J., and J. K. Lattimer: Functional closure of bladder exstrophy; a review of fifty cases. Pediatrics 31, 433 (1963).

Lapides, J.: Risultatic della chiusura Funzionale dell'estrofia vescicale congenita. Urologia 28, 99 (1961).

Leadbetter jr., G. W.: Surgical correction of total urinary incontinence. J. Urol. (Baltimore) 91, 261 (1964).

Learmonth, J. R., and K. H. Watkins: A rare type of valvular obstruction of the neck of the bladder; record of two cases. Brit. J. Surg. 22, 879 (1935).

Lepoutre, C.: Sur un cas d'absence congénitale de la vessie (persistance du cloaque). J. Urol. méd. chir. 48, 334 (1939/40).

Lowsley, O. S., and T. H. Johnston: A new operation for the creation of an artifical bladder with voluntary control of urine and feces. J. Urol. (Baltimore) 73, 83 (1955).

Lurz, L.: So-called congenital diverticula of bladder. Z. urol. Chir. 18, 287 (1925).

Luschka, H.: Über den Bau des menschlichen Harnstranges. Virchows Arch. path. Anat. 23, 1 (1862).

MACKELLAR, A., and F. D. STEPHENS: Vesical diverticula in children. In: Congenital malformations of the rectum, anus and genito-urinary tracts, ed. by F. D. STEPHENS. Edinburgh and London: E. & S. Livingstone Ltd. 1963.

MACKENZIE, L. L.: Split pelvis in pregnancy. Amer. J. Obstet. Gynec. 29, 255 (1935).

MAKINS, G. H.: The treatment of epispadias in the female and hypospadias in the male. Lancet 1894/II, 1140.

MALONEY jr., P. K., D. M. GLEASON, and J. K. LATTIMER: Ureteral physiology and exstrophy of the bladder. J. Urol. (Baltimore) 93, 588 (1965).

MARSHALL, V. F., and J. S. GARDNER: An evaluation of the Coffey I method for uretero-intestinal anastomosis. Surg. Gynec. Obstet. 81, 559 (1954).

—, and E. C. MUECKE: Variations in exstrophy of the bladder. J. Urol. (Baltimore) 88, 279 (1962).

— — Variations in exstrophy of the bladder. J. Urol. (Baltimore) 88, 766 (1962).

McCAULEY, R. T., and F. R. LICHTENHELD: Congenital patent urachus. Sth. med. J. 53, 1138 (1960).

McFARLAND, J.: Exstrophy of the bladder with imperforate anus, absence of the greater part of the small and large intestines, continuity of the duodenum with the colon, absence of the left testis, epididymis, and cord, and enormous hydroureter. Amer. J. Path. 14, 509 (1958).

McGOVERN, J. H., V. F. MARSHALL, and A. J. PAQUIN jr.: Vesicoureteral regurgitation in children. J. Urol. (Baltimore) 83, 122 (1960).

MENTON, M. L., and H. E. DENNY: Duplication of the vermiform appendix, the large intestine and the urinary bladder: report of a case. Arch. Path. 40, 345 (1945).

MICHON, L.: Conservative operations for exstrophy of the bladder, with particular reference to urinary continence. Brit. J. Urol. 20, 167 (1948).

—, BOYET, and CAMMENOS: Kyste du trigone chez une fillette de 3 ans. J. Urol. méd. chir. 59, 529 (1953).

MILLER, H. L.: Agenesia of the urinary bladder and urethra. J. Urol. (Baltimore) 59, 1156 (1948).

MORTON, H. G., T. THOMPSON, and M. M. SIMMONS: Prolapsing trigonal mucosa in infant. J. Fl. med. Ass. 47, 162 (1960).

MOSSMAN, H. W.: Comparative morphogenesis of the fetal membranes and accessory uterine structures. Contr. Embryol. Carneg. Inst. 26, 129 (1937).

MOWATT, J.: An account of a child born with the urinary and genital organs preternaturally formed. Medical Essays and Observations 3, 220 (1747). Publ. by the Edinburgh Society.

MUECKE, E. C.: The role of the cloacal membrane in exstrophy; the first successful experimental study. J. Urol. (Baltimore) 92, 659 (1964).

NESBIT, R. M., and W. BROMME: Double penis and double bladder; with report of a case. Amer. J. Roentgenol. 30, 497 (1933).

NIX, J. T., J. G. MENVILLE, M. ALBERT, and D. L. WENDT: Congenital patent urachus. J. Urol. (Baltimore) 79, 264 (1958).

PAQUIN jr., A. J.: Ureterovesical anastomosis; the description and evaluation of a technique. J. Urol. (Baltimore) 82, 573 (1959).

PARKHURST, E. C., and W. F. LEADBETTER: A report on 93 ileal loop diversions. J. Urol. (Baltimore) 83, 397 (1960).

POHLMAN, A. G.: The development of the cloaca in human embryos. Amer. J. Anat. 12, 1 (1911).

POOLE-WILSON, D. S.: Congenital valvular obstruction of the neck of the bladder. Brit. J. Urol. 15, 11 (1943).

POWELL, T. O.: Surgery of the exstrophied bladder: preliminary report. J. Urol. (Baltimore) 74, 67 (1955).

PYTL, A.: Risultati della chiusura Funzionale dell'estrofia Vescicale Congenita. [Disc.] Urologia 28, 100 (1961).

RAVITCH, M. M.: Hind gut duplication — doubling of colon and genital urinary tracts. Ann. Surg. 137, 588 (1953).

—, and W. W. SCOTT: Duplication of the entire colon, bladder, and urethra. Surgery 34, 843 (1953).

RICKHAM, P. P.: The treatment of ectopia vesicae. Brit. J. plast. Surg. 10, 300 (1958).

— Vesico-intestinal fissure. Arch. Dis. Childh. 35, 97 (1960).

— The incidence and tratment of ectopia vesicae. (Abridged) Proc. roy. Soc. Med. 54, 389 (1961).

RIPPMANN, G.: Eine seröse Cyste in der Bauchhöhle, mit einem Inhalt von 50 l Flüssigkeit. Dsch. Klin. 22, 267 (1870).

ROSENBERG, M. Y.: Fetal dystocia due to urachal cyst and ascites: report of a case. Obstet. and Gynec. 16, 227 (1960).

RUKSTINAT, G. J., and J. D. BALFOUR: Hypogastroetroschisis, spina bifida, diaphragmatic hernia, and left renal agenesis in a viable fullterm child. Arch. Path. 6, 48 (1928).

Russell, K. F.: A case of complicated exstrophy of the bladder presenting many unusual features. Brit. J. Urol. 11, 31 (1939).

Salvisberg: Die Behandlung der Urachusfistel beim Fohlen. Schweiz. Arch. Tierh. 44, 228 (1902).

Sanderud, A.: Exstrophy of the bladder; a follow-up examination of 17 cases. Acta chir. scand. 106, 117 (1953).

Satter, E. J., and H. W. Mossman: A case report of a double bladder and double urethra in the female child. J. Urol. (Baltimore) 79, 274 (1958).

Schwalbe, E.: Die Morphologie der Mißbildungen des Menschen und der Tiere. Jena: Gustav Fischer 1909.

Senger, F. L., and V. J. Santare: Congenital multilocular bladder: a case report. J. Urol. (Baltimore) 68, 283 (1952).

Shultz, W. G.: Plastic repair of exstrophy of bladder combined with bilateral osteotomy of ilia. Urologists News Letter Club, June 1965.

Siddall, A. C.: Cyst of urachus with calculus formation. Chin. med. J. 46, 894 (1932).

Simon, H. E., and N. A. Brandeberry: Anomalies of the urachus: persistent fetal bladder. J. Urol. (Baltimore) 55, 401 (1946).

Snyder, C. C.: A new therapeutic concept of the exstrophied bladder. Plast. reconstr. Surg. 22, 1 (1958).

Soper, R. T., and K. Kilger: Vesicointestinal fissure. J. Urol. (Baltimore) 92, 490 (1964).

Sorrentino, F., and P. Leonetti: Terapia Estrofia Vescical. Naples ESI, Naples (1958).

Stagner, R. V., and C. V. Hodges: Experiences with exstrophy of the bladder. J. Urol. (Baltimore) 89, 53 (1963).

Stamey, T. A., and W. W. Scott: Ureteral ileal anastomosis. Surg. Gynec. Obstet. 104, 11 (1957).

Steinbüchel, R. v.: Über Nabelschnurbruch und Bauchblasenspalte mit Cloakenbildung von Seiten des Dünndarmes. Arch. Gynäk. 60, 465 (1900).

Stephens, A. R.: Longevity following uretero-intestinal anastomosis. J. Urol. (Baltimore) 46, 57 (1941).

Stiles, H. J.: Epispadias in the female, and its surgical treatment; with a report of two cases. Surg. Gynec. Obstet. 13, 127 (1911).

Streeter, G. L.: Developmental horizons in human embryos: description of age groups XV, XVI, XVII and XVIII, being the third issue of a survey of the Carnegie Collection. Contr. Embryol. Carneg. Inst. 32, 133 (1948).

Sturdy, D. E.: New concepts in the initial management of exstroversion of the bladder. Brit. J. Urol. (Baltimore) 33, 296 (1961).

Swenson, O., G. H. Moussatos, and J. H. Fisher: Results of repair of exstrophy of the bladder. Surg. Clin. N. Amer. 43, 151 (1963).

—, and C. T. Oeconomopoulos: Double lower genitourinary systems in a child. J. Urol. (Baltimore) 85, 540 (1961).

Tait, L.: Twelve cases of extraperitoneal cysts. Brit. gynaec. J. 2, 348 (1886/87).

Thompson, I. M.: Management of exstrophy of the urinary bladder by primary closure. South. med. J. Bghm, Ala.) 54, 1069 (1961).

Trendelenburg, F. v.: The treatment of ectopia vesicae. Ann. Surg. 44, 281 (1906).

Trimingham, H. L., and J. R. McDonald: Congenital anomalies in the region of the umbilicus. Surg. Gynec. Obstet. 80, 152 (1945).

Turner, G. G.: Transplantation of the ureters into the bowel. In: Textbook of genitourinary surgery, ed. by H. P. Winsbury-White. Baltimore: Williams & Wilkins Co. 1948.

Uson, A. C., J. K. Lattimer, and M. M. Melicow: Types of exstrophy of urinary bladder and concommitant malformations .Pediatrics 23, 927 (1958).

—, and M. S. Roberts: Incomplete exstrophy of urinary bladder: a report of two cases. J. Urol. (Baltimore) 79, 57 (1958).

Vaughan, G. T.: Patent urachus; review of the cases reported: operation on a case complicated with stones in the kidneys; a note on tumors and cysts of the urachus. Trans. Amer. surg. Ass. 23, 273 (1905).

Veal, J. R., and E. M. McFetridge: Exstrophy of the bladder (persistent cloaca) associated with intestinal fistulas; with a brief analysis of 36 cases of anal and rectal anomalies from the records of Charity Hospital in New Orleans. J. Pediat. 4, 95 (1934).

Wallace, W. J.: Unusual bladder obstruction. J. Urol. (Baltimore) 15, 325 (1926).

Walzak jr., M. P., and A. J. Paquin jr.: Renal pelvic pressure levels in management of nephrostomy. J. Urol. (Baltimore) 85, 697 (1961).

Ward, W. G.: Suppurative cyst of the urachus, with concretion. Ann. Surg. 69, 329 (1919).

Watson, E. M.: The developmental basis for certain vesical diverticula. J. Amer. med. Ass. 75, 1473 (1920).

Wehrbein, H. L.: Double kidney, double ureter, and bilocular bladder in a child. J. Urol. (Baltimore) 43, 804 (1940).

WILLIAMS, D. I.: The development of the trigone of the bladder. Brit. J. Urol. **23**, 123 (1951).
— Congenital abnormalities of the lower urinary tract. In: Urology in childhood. Handbuch der Urologie, Bd. 15. Berlin-Göttingen-Heidelberg: Springer 1958.
—, and H. R. JOLLY: Long term results of transplantation of the ureters for ectopia vesicae. G. Ormond Str. Hosp. J. No **3**, 9 (1953).
WINTZELL, K.: Ein Fall von Urachus-Cyste mit Konkrementen. Radiologe **3**, 425 (1963).
WOJEWSKI, A., and W. KOSSOWSKI: Total diphallia: a case of plastic repair. J. Urol. (Baltimore) **91**, 84 (1964).
YOUNG, B. W., R. L. MILLS, and J. NESBIT: Artificial urinary sphincter. Surg. Gynec. Obstet. **113**, 62 (1961).
YOUNG, H. H.: A new operation for epispadias. J. Urol. (Baltimore) **2**, 237 (1918).
— An operation for the cure of incontinence associated with epispadias. J. Urol. (Baltimore) **7**, 1 (1922).
— Exstrophy of the bladder: the first case in which a normal bladder and urinary control have been obtained by plastic operations. Surg. Gynec. Obstet. **74**, 729 (1942). Cit. by W. W. BABCOCK.
ZELLERMAYER, J., and H. E. CARLSON: Congenital hourglass bladder. J. Urol. (Baltimore) **51**, 24 (1944).

Anomalies of the Bladder Neck

John J. Murphy and Theodore A. Tristan

With 34 Figures

Anomalies of the Bladder Neck

The bladder neck may be defined as the portion of that viscus which marks the junction of the bladder cavity with the urethral lumen. In a static phase this consists of the distal portion of the trigone and adjacent distal detrusor, actually

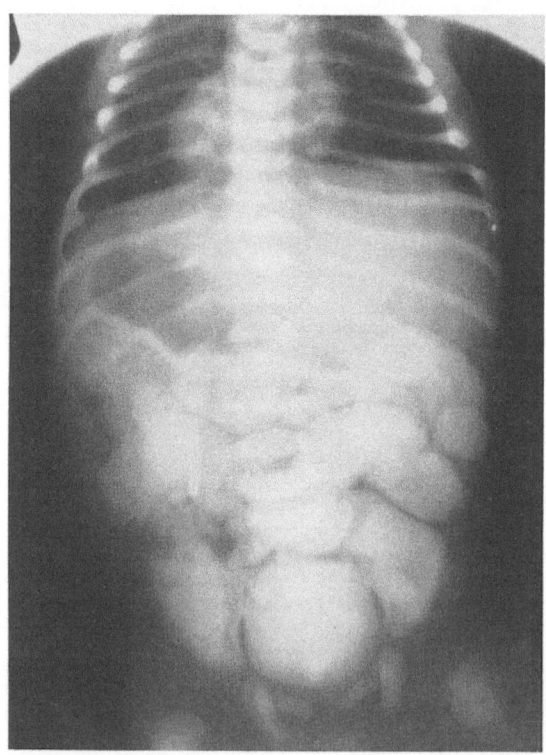

Fig. 1. Urogram of newborn male showing advanced hydroureter and hydronephrotic atrophy of kidneys due to bladder outlet obstruction by congenital anomaly at neck

apparent accumulations of smooth muscle bundles, some of which extend into the proximal urethra. The inherent tone of the smooth muscle augmented by abundant intermingled elastic tissue provides for urinary continence. Contraction of the smooth muscle of the detrusor results in shortening and widening of the proximal urethra to permit normal voiding [1, 2]. Anomalous development of this area involves abnormalities in smooth muscle, connective and elastic tissue. Some of these abnormalities produce gross and severe organic obstruction to the bladder outlet, so that there is advanced damage to the urinary tract apparent at birth (Fig. 1). Varying degrees of obstruction or dysfunction results from others. These may become apparent only later in life when infection or signs and symptoms of renal failure supervene (Fig. 2).

According to NANSON [3] the first description of bladder neck obstruction due to smooth muscle hypertrophy was by GUTHRIE in 1834 and further observations on the subject were made by MERCIER (1844) and CIVALE (1850). MARION presented his classic description of the problem with recommendations as to surgical

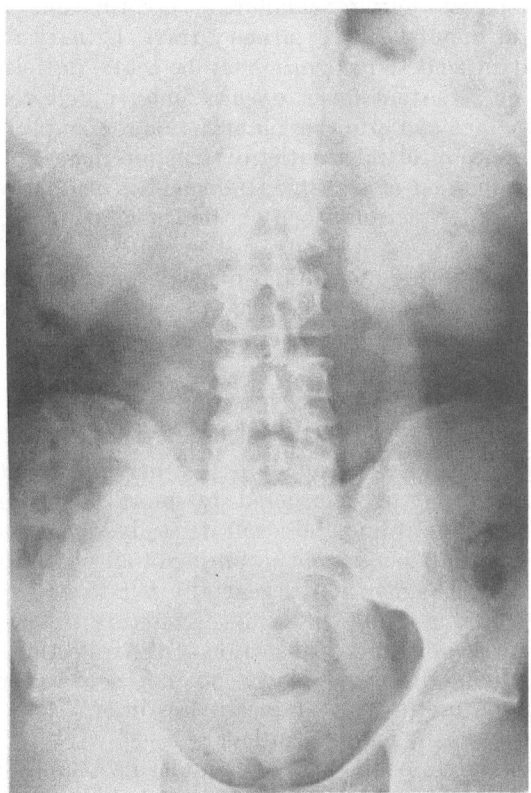

Fig. 2. Urogram of a 52 year old man showing massive hydronephrosis on left, non-visualization of hydronephrotic right kidney due to bladder outlet obstruction by bladder neck anomaly

management in 1933 [4], and this abnormality, when found in males under 50 years of age, has been called Marion's disease since that time. Actually MARION described three types of abnormality: (1) Hypertrophy of smooth muscle at the bladder neck, (2) with elements of inflammation and fibrosis, and (3) adenomatous formation.

In 1957 BODIAN [5] described an abnormality characterized by excessive amounts of elastic tissue in the posterior urethra which resulted in bladder outlet obstruction leading to severe damage to the upper urinary tracts. This has been accepted as an abnormality of elastic tissue development and has been called fibroelastosis. LEADBETTER [6] suggested that the anomaly was a fault of dissolution of mesenchyme or inclusion of abnormal amounts of non-muscular connective tissue at the bladder neck and that the obstruction resulted from smooth muscle hypertrophy, fibrotic contracture or changes due to chronic inflammation.

The latter is likely to be secondary although there are those who believe that infection alone can produce changes in the tissues in this area which result in obstruction. Hurst [7] described a case of severe obstructive uropathy and suggested that the bladder outlet obstruction was similar to achalasia as seen in the esophagus. Murphy [8] ascribes to this view and points out the fact that a good therapeutic response may be obtained by dilation or myotomy as in esophageal achalasia.

Symptoms

Obstruction or dysfunction of the bladder outlet may produce no symptoms recognizable by the patient until renal failure occurs. Enuresis or true incontinence may focus attention upon the lower urinary tract. Urinary retention, although usually gradual and difficult to recognize, may be acute, especially when infection occurs. Recurrent or persistent fever, dysuria, interrupted voiding, straining to void, flushing of the face and grunting on urination, or simply "failure to thrive" may signal the presence of bladder outlet obstruction due to this anomaly.

Signs include abdominal masses due to distended bladder or hydronephrosis, pallor, weak or interrupted stream, cloudy or foul smelling urine, and occasionally, hematuria.

Laboratory findings include bacilluria, pyuria, low hemoglobin level, elevated blood urea nitrogen and creatinine, decreased carbon dioxide combining power, increased serum chlorides and potassium, decrease in serum calcium and an increase in serum phosphorus and uric acid.

Diagnosis

The diagnosis of vesical neck anomaly is best made by radiologic techniques. Voiding cystourethrography is considered by most people interested in this problem to be the most rewarding and accurate study. Cinefluorographic voiding cystourethrography permits permanent recording of all phases of micturition and since this is a dynamic phenomenon is generally felt to be superior to multiple static films, though this is refuted by some. Excretory urography may suggest outlet obstruction by demonstrating dilatation of the collecting system or ureters, bladder trabeculation and diverticula, but is not sufficient to accurately delineate this lesion. Retrograde urethrocystography is less informative than the voiding technique and it is obviously more hazardous.

Endoscopic examination of the urethra, bladder neck and bladder is useful to determine the presence or absence of secondary signs of bladder outlet obstruction (trabeculation, cellules and diverticula) and to rule out other causes of obstruction such as calculi, bladder tumors, ureteroceles, benign and malignant enlargements of the prostate gland, posterior urethral valves and urethral strictures. Endoscopic recognition of fibrous contracture of the neck is not difficult, but simple hypertrophy of the smooth muscle in this area and dysfunction may be difficult to diagnose by this means.

Measurement of intravesical pressures during voiding is helpful in confirming the diagnosis of bladder outlet obstruction [8, 9] and accurate measurement of rates of urine flow have been described [10, 11, 12], but usually require such complicated apparatus as to exclude their general use.

Urethral calibration utilizing sound and bougie a boule is useful to detect meatal stenosis and obstructions in the distal portion of the urethra. Care must be exercised lest these instruments lead to the false impression that obstruction exists when in reality the normal caliber of the urethra is merely grossly exceeded by the size of the instrument.

The Normal Bladder Neck

In order to appreciate variations from the normal, one must first be familiar with what is normal and be aware of the alterations in appearance of normal induced by infection, neurological defects and psychological influences. A great deal depends upon how the studies are conducted. Uniformity of approach is essential to minimize the number of potential variables. Attempts to study toilet trained females by voiding cystourethrography, for example, prove not only frustrating and time consuming but also confusing because of artifacts introduced by straining, sphincter spasm and other effects of the unnatural situation faced by the patient. Males, including children, are usually able to void on command and in unusual situations. Nevertheless, for purposes of uniformity, it is recommended that cinefluorographic voiding cystourethrography be accomplished with the use of general anesthesia.

Technique

The bladder is emptied of urine with a small Foley catheter. The radiopaque material (Hypaque 50%)[1] is instilled by gravity by pouring the radiopaque into a 50 cc. syringe barrel which is attached to the catheter. This permits accurate determination of the volume of radiopaque material instilled and the pressure at which reflux occurs. During filling of the bladder the renal areas, abdomen and pelvis are surveyed fluoroscopically and recorded on motion picture film to detect reflux. When the bladder is full the skin of the suprapubic area is prepared by scrubbing it with Phisohex and the bladder is punctured percutaneously using a 17 gauge thin walled needle inserted in the mid-line just above the symphysis. A polyethylene catheter is introduced through the needle into the lumen of the bladder and the needle is removed. The polyethylene catheter is connected to a PR23-2 D-300 Statham Transducer. The pressure in the bladder is converted by the strain gauge to an electrical signal that is displayed on an oscilloscope and simultaneously recorded on paper by means of a seven channel recorder. The record obtained is marked simultaneously with a signal derived from the synchronous motor that drives the motion picture camera mounted on the image intensifier. The child is now permitted to recover from anesthesia, and awakening with a full bladder in an obtunded mental state, usually voids spontaneously. When the intravesical pressure begins to rise removal of the urethral catheter provides sufficient stimulus to initiate micturition. The roentgen appearance of the entire act of voiding is recorded by the motion picture camera through the image intensifier. Finally, the renal areas, abdomen and pelvis are survyed to detect late reflux, residual urine, diverticula, etc.

Cinefluorographic Appearance of Normal Bladder Outlets During Micturition

The appearance of the bladder neck changes during voiding but the configuration depicted in Fig. 3 is characteristic of a normal female bladder outlet during full voiding. The bladder neck itself is a tapered funnel leading into a urethra which is of practically uniform caliber to the urethral meatus. A long, thin, but normal, bladder neck is shown in Fig. 4, while a short, wide urethra and bladder neck are depicted in Fig. 5. Another minor variation of the relatively wide bladder neck and urethra is shown in Fig. 6.

---

[1] Winthrop Lab. N.Y. N.Y.

15*

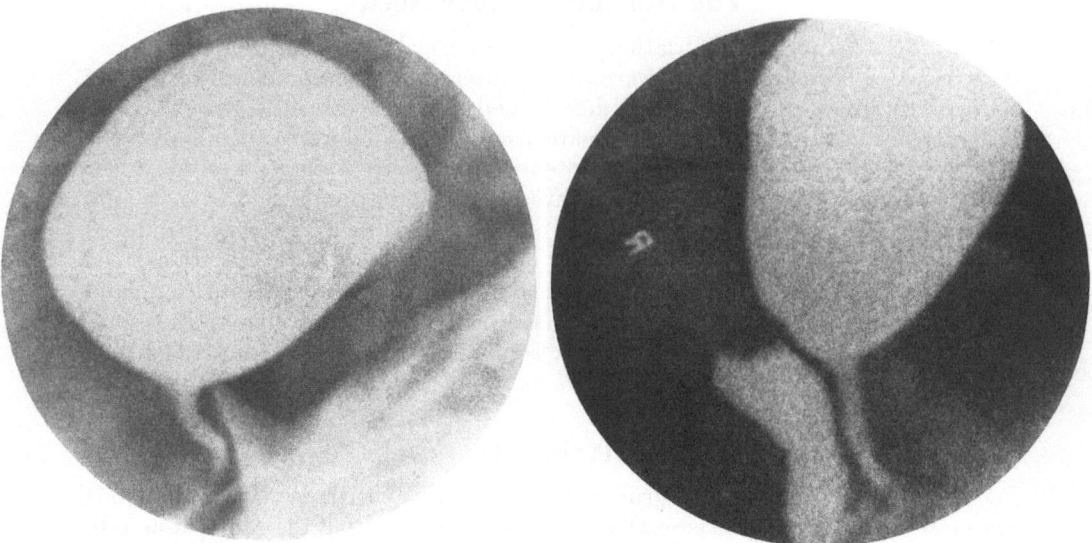

Fig. 3. Appearance of bladder neck and urethra in a typical normal female child during full voiding

Fig. 4. Long, thin normal bladder neck and urethra

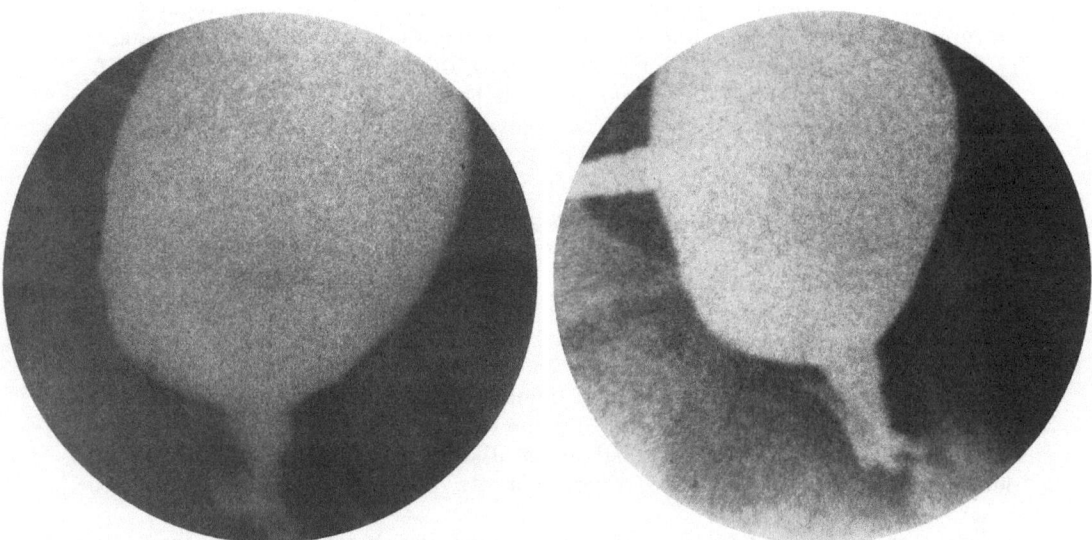

Fig. 5   Fig. 6

Fig. 5. Short, wide normal bladder neck and urethra

Fig. 6. Variation in short, wide normal bladder neck and urethra

The variations in the appearance of the bladder neck at different stages in voiding is demonstrated in Fig. 7a and b. A demonstrates what appears to be an anterior bulging of the urethral wall just below a normal neck at the beginning of micturition while B demonstrates a tapered funnel with normal caliber urethra at a later stage in voiding.

A very wide bladder neck with wide caliber urethra in a three year old white female is shown in Fig. 8. Vesicoureteral reflux is apparent on the right side. The

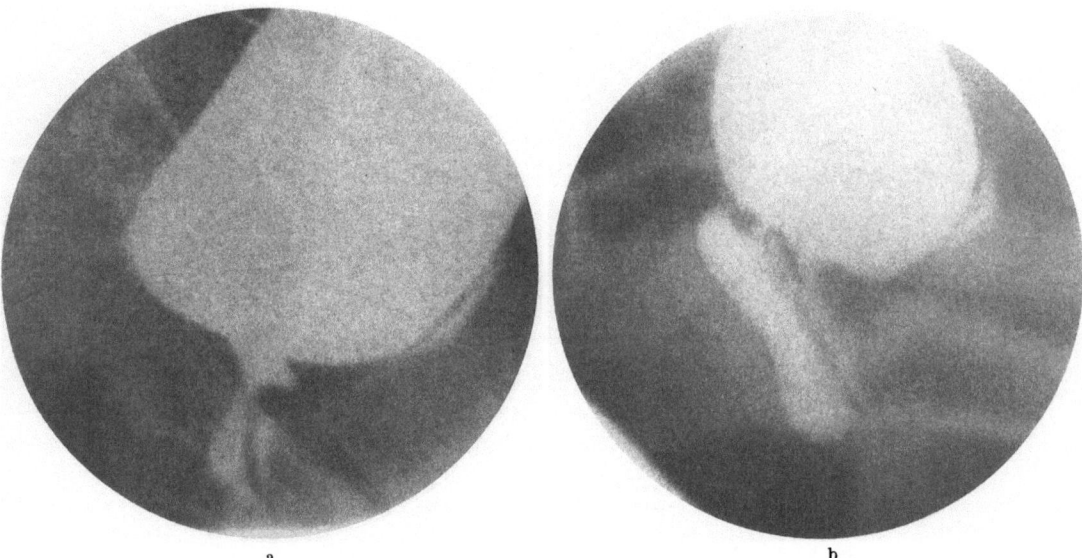

a   b

Fig. 7.a) Anterior bulge of urethral wall at beginning of micturition. b) Normal appearance at later stage

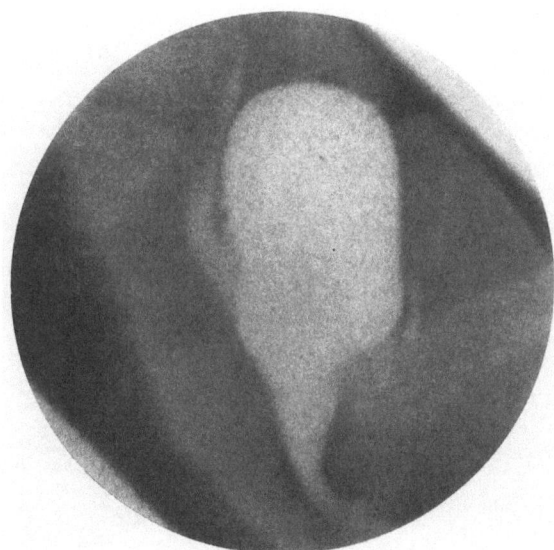

Fig. 8. Wide bladder neck and urethra in 3 year old female with right vesicoureteral reflux

appearance of this bladder neck is similar to that seen after surgical revision of the outlet.

A rather narrow, but normal, bladder neck and urethra is demonstrated in Fig. 9a. There is vesicoureteral reflux on the right side. The same patient studied two years later shows a more widely tapered bladder neck — right vesicoureteral reflux persists, Fig. 9b.

A normal appearing bladder neck in a two year old white female is demonstrated in Fig. 10. There is vesicoureteral reflux in the right side.

Fig. 11 demonstrates a wide bladder neck and urethra, another variation of normal, in a four year old white female. The radiopaque material seen on the right

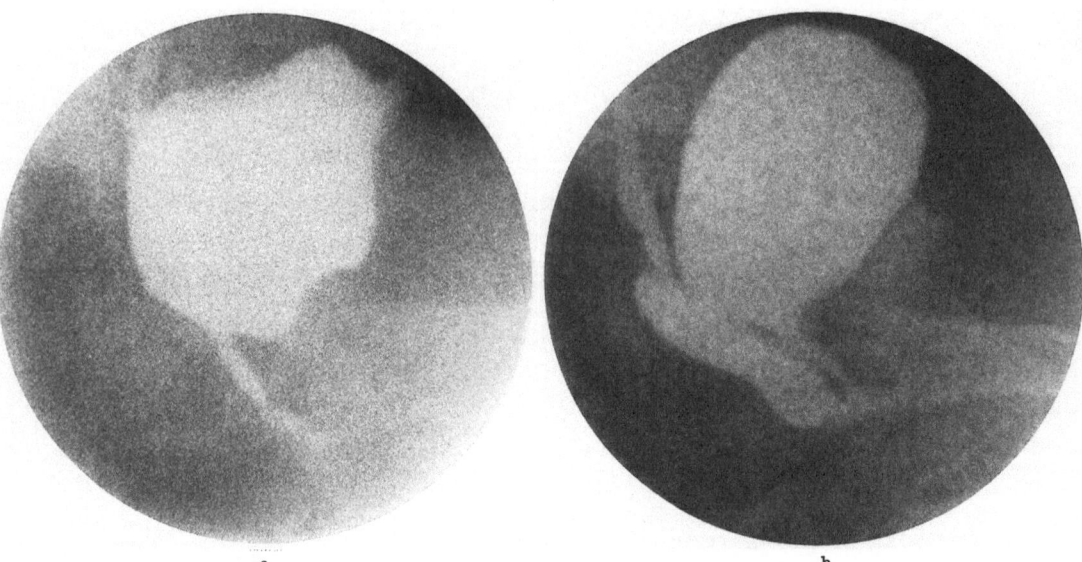

a   b

Fig. 9. a) Normal, narrow neck — right vesicoureteral reflux. b) Same child two years later. Normal neck — reflux
persists

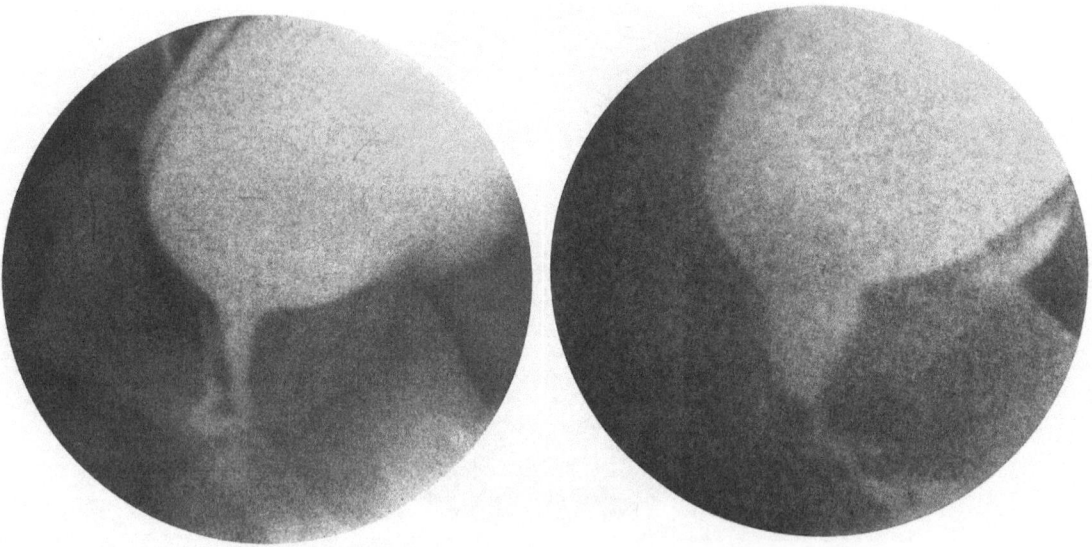

Fig. 10   Fig. 11

Fig. 10. Normal neck with right reflux

Fig. 11. Wide neck and urethra (normal) showing extravasation of radiopaque around catheter used to measure
pressures

side of the print is due to extravasation of the radiopaque liquid around the
catheter used to measure intravesical pressures.

Dilatation of the Urethra Distal to Wide Bladder Necks

Failure of the bladder neck to open widely and synchronously during contraction of the detrusor may be regarded as varying degrees of dysfunction of the
muscle at the bladder neck. Apparently minor degrees of this phenomenon have

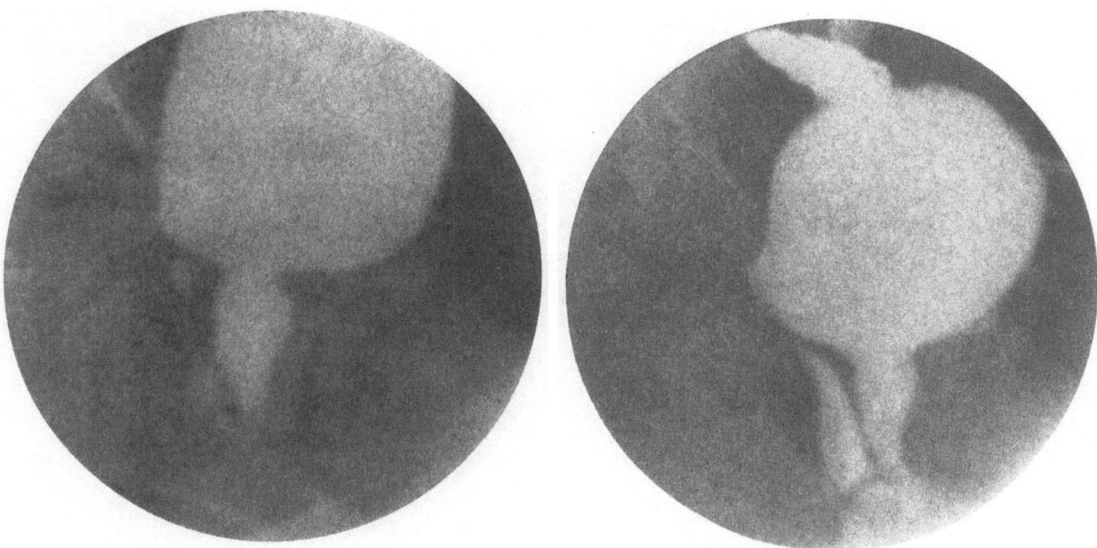

Fig. 12   Fig. 13

Fig. 12. Minor degree of bladder neck dysfunction resulting in "carrot" appearance of urethra

Fig. 13. Mild dysfunction of neck with moderate urethral dilatation

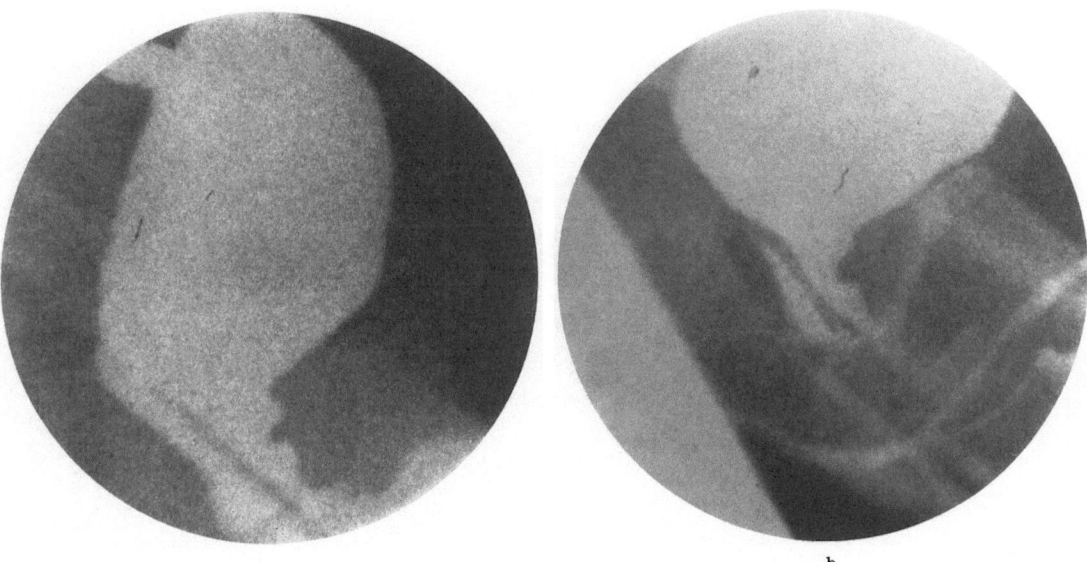

a   b

Fig. 14. a) Anterior protrusion of the bladder wall proximal to bladder neck with some dilatation of the urethra due to minor degree of bladder neck dysfunction. b) Same child studied 14 months later

no clinical significance, but the appearance of the bladder neck and urethra in such instances causes great confusion in interpretation of voiding cystourethrograms. Examples of this so-called "carrot" appearance are shown in Fig. 12 and 13. It is most important to note that the bladder neck is not fixed but varies in size during voiding in these patients so that functional obstruction does not exist but there is enough disturbance of urine flow to cause dilatation of the urethra distally. Distal urethral stenosis or meatal obstruction was not present in these

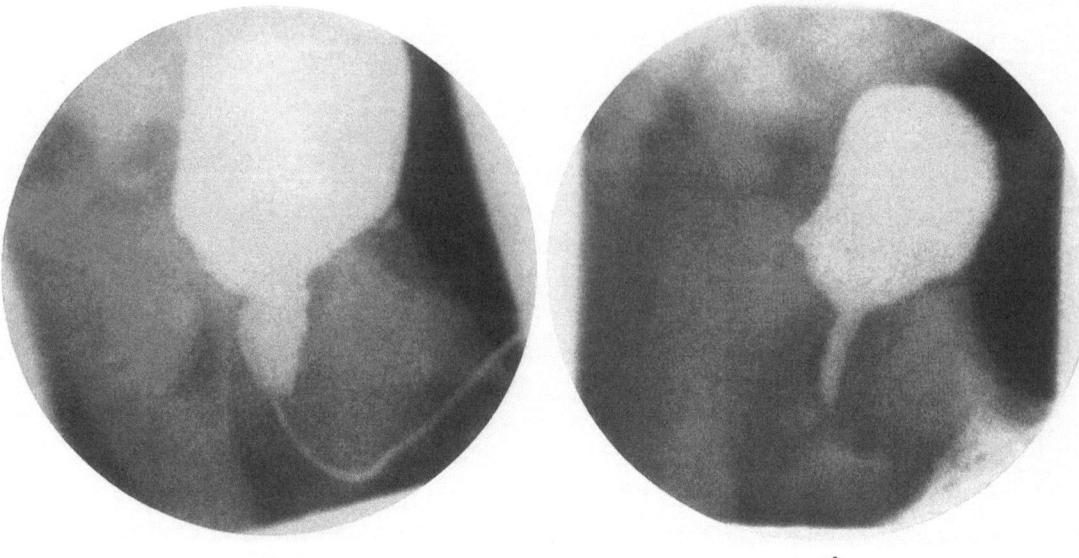

a   b

Fig. 15. a) Misleading abnormal appearance of bladder neck due to presence of urethral catheter at onset of voiding. b) Normal appearance following removal of catheter

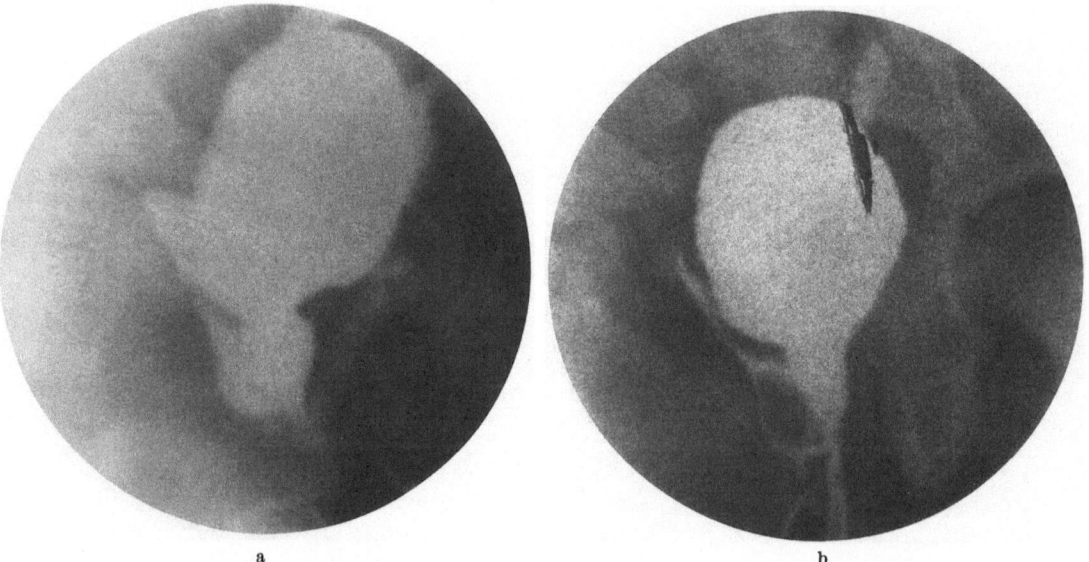

a   b

Fig. 16. a) Appearance of bladder neck of 4 year old female treated for over one year for urinary tract infection without success. b) Appearance of bladder neck if same child 27 months after surgical revision of neck

patients as determined by urethral calibration using sounds, bougie a boule and endoscopy.

Another variation of the appearance of the bladder neck and urethra due to minor degree of dysfunction is demonstrated in Fig. 14a and b. These are voiding cystourethrograms made on the same patient at an interval of 14 months. In A the appearance of the "double dimple" is due to an apparent protrusion of the anterior bladder wall just proximal to the neck itself with some dilatation of the urethra distal to the bladder neck proper. Essentially the same picture is shown in b, 14 months later. This child had no demonstrable obstruction in the distal urethra.

The presence of a catheter in the urethra at the beginning of voiding may lead to artefacts in appearance which are confusing unless the entire voiding act is recorded. Fig. 15a and b demonstrates such an artefact in a four year old white female. In a the child is beginning to void while the catheter is still in place, while in b the normal appearance of the bladder neck and urethra is apparent after the removal of the catheter.

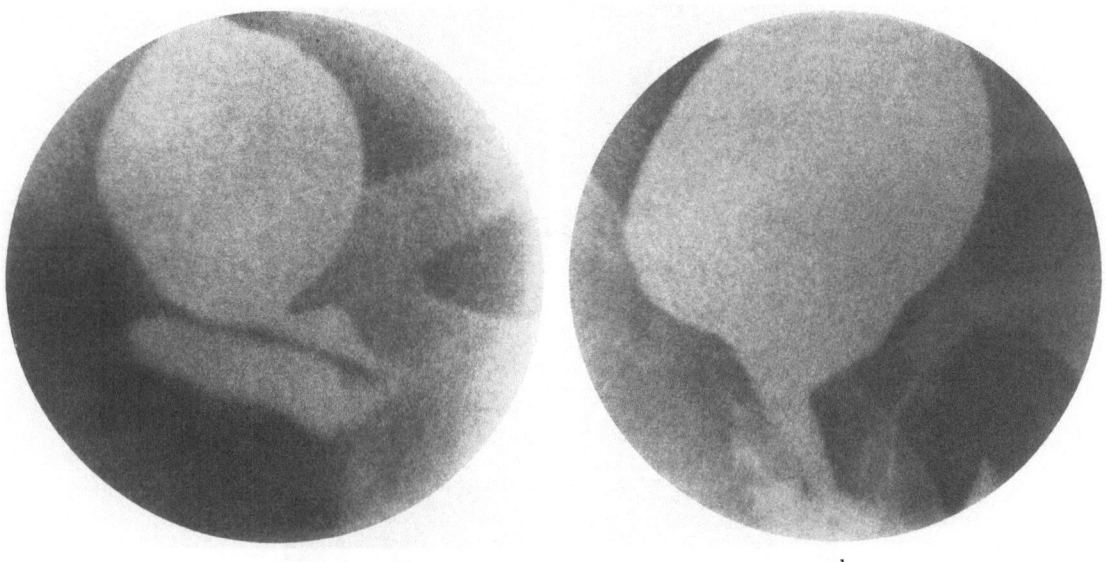

a   b

Fig. 17. a) Bladder neck dysfunction in 9 year old colored female with persistent urinary tract infection. b) Normal appearance of neck in same child 15 months after surgical revision of neck

Bladder Neck Dysfunction

Careful study of large numbers of cinefluorographic voiding cystourethrograms combined with detailed analysis of clinical history and laboratory findings convince us that in some children persistence or recurrence of urinary tract infection are due to functional obstructions at the vesical neck produced by dysfunction of the smooth muscle in this area. Whether this dysfunction is a congenital defect in the smooth muscle itself, neurogenic, induced by distal urethral obstruction or infection is often difficult to determine. It seems logical, however, to consider this dysfunction as a distinct entity requiring treatment if it persists despite a reasonable trial of more conservative measures. If urethral meatotomy, dilation and prolonged antibacterial therapy do not change the appearance of a bladder neck which has been labeled dysfunction by repeated voiding cystourethrograms and urinary tract infection persists it seems reasonable to revise the muscle in this area to permit better coordination of vesical neck activity with the detrusor contraction. It must be emphasized that the decision can not be made on the appearance of a single voiding cystourethrogram alone and that a trial of conservative management is justifiable in almost every instance.

Examples of persistent bladder neck dysfunction despite conservative therapy and their appearance after successful operative management are demonstrated

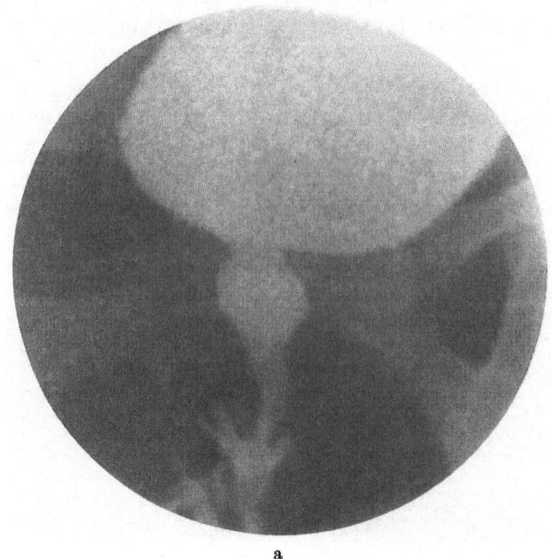

a

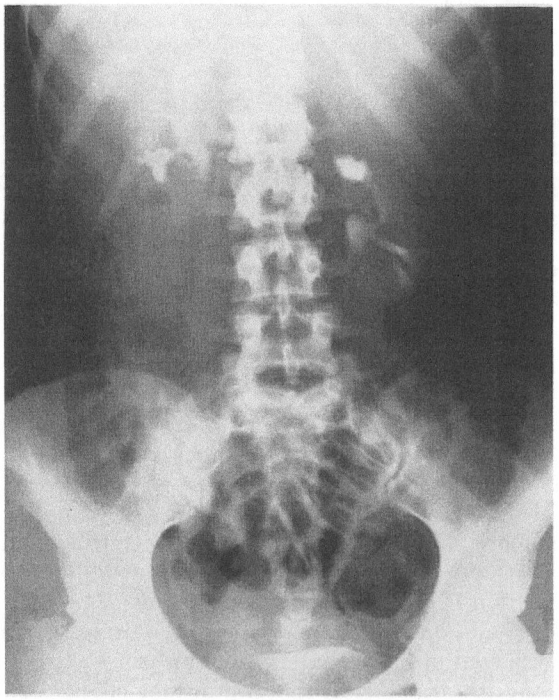

b

Fig. 18. a) Appearance of bladder neck in 22 year old female with chronic pyelonephritis and history of recurrent urinary tract infection since age 10 years. b) Excretory urogram showing scarring of upper collecting system of both kidneys in the same patient

in Fig. 16a and b, and 17a and b. The initial study, Fig. 16a, was made when the child was four year old. She had been treated with antibacterial therapy and urethral dilations for more than a year without success. Simple surgical revision of the bladder neck was recommended. She was free of infection nine months

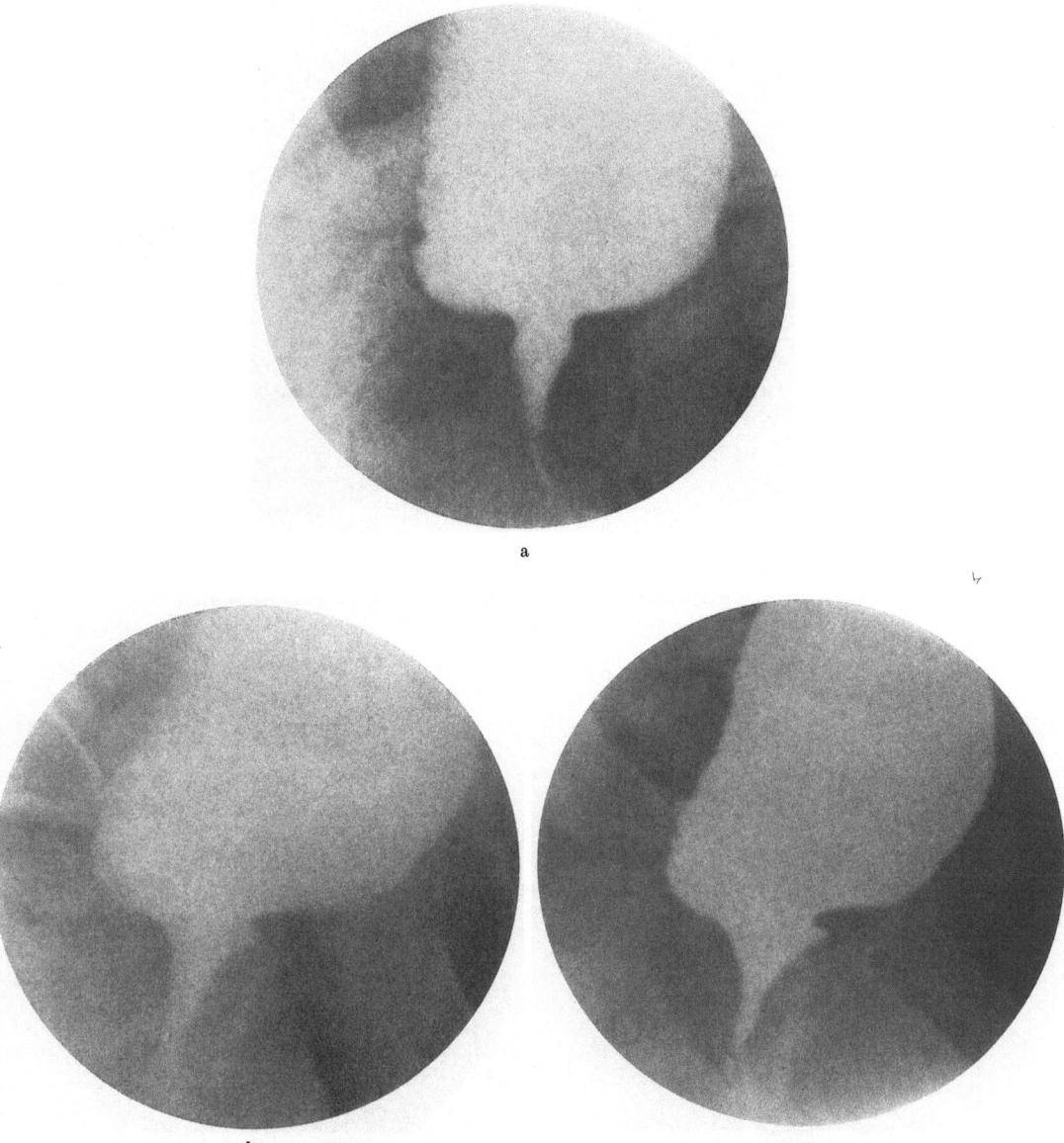

Fig. 19. a) Voiding cystourethrogram in a female with persistent urinary tract infection — misinterpreted as bladder neck dysfunction. b) Appearance of bladder neck after surgical revision. c) Appearance of bladder neck after several revisions

after surgery and the appearance of her bladder neck (27 months postoperatively) is that of a normal child. Fig. 17a is unique in our experience in that it demonstrates an apparent bladder neck dysfunction in a 9 year old colored female. This child was treated for more than two years with antibacterial therapy and repeated urethral dilations without success. Fig. 17a is an example of the appearance of her bladder neck during voiding upon repeated studies while Fig. 17b shows the appearance of her bladder neck 15 months after surgical revision. The child is now without symptoms referable to her urinary tract and free of bacilluria without antibacterial therapy.

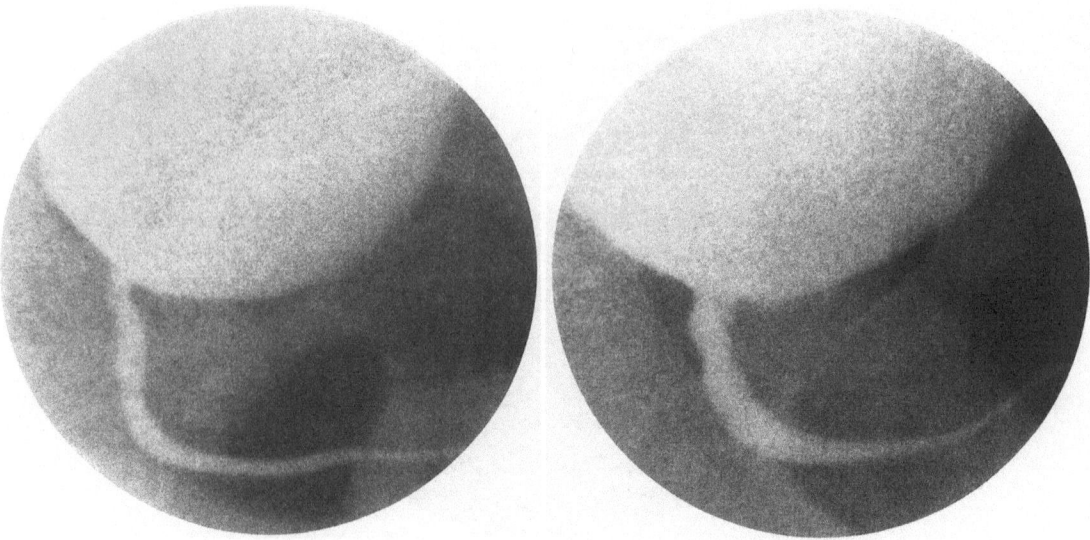

Fig. 20   Fig. 21

Fig. 20. Voiding cystourethrogram of normal bladder neck in 9 year old white male

Fig. 21. Voiding cystourethrogram of normal bladder neck in 11 year old white male

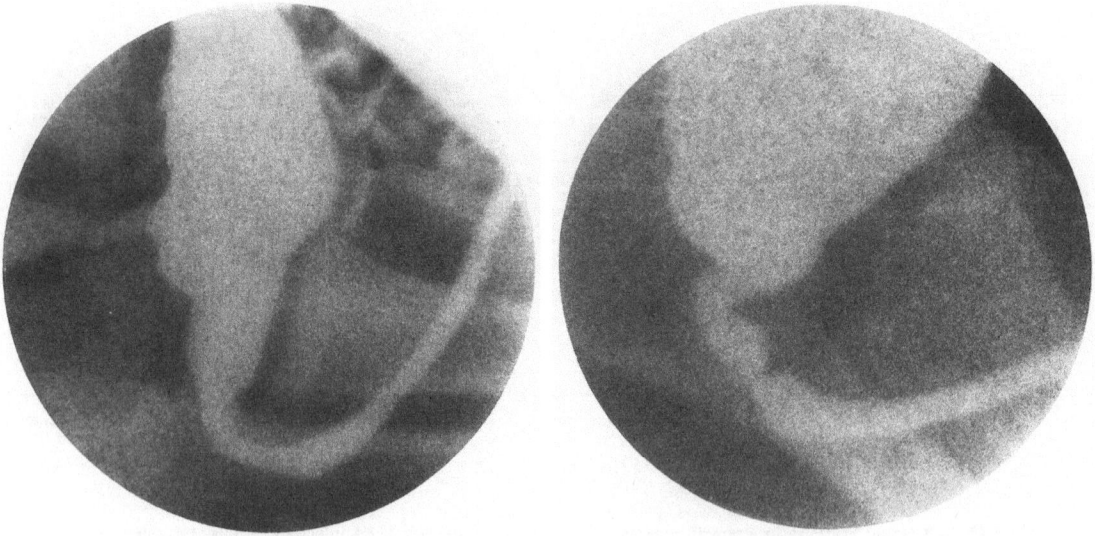

Fig. 22   Fig. 23

Fig. 22. Voiding cystourethrogram of wide neck in a normal 12 year old male

Fig. 23. Artefact — compressor nudae in a 3 year old white male

It is apparent that after careful selection of patients through the combined analysis of clinical picture and cinefluorographic voiding cystourethrography properly accomplished surgical revision of the bladder neck can result in restoration of normal appearance and function with resultant clinical improvement.

Typical bladder neck dysfunction is encountered less frequently in adult females than in children, but is found occasionally and usually with evidence of upper urinary tract damage. An example of this is shown in Fig. 18a and b. The patient, a 22 year old white female, had recurrent signs of lower urinary tract infection from age 10 years, but was not studied until she complained of back pain and was found to have infection due to a coliform organism at age 22. The voiding cystourethrogram, Fig. 18a, demonstrates a characteristic bladder neck dysfunction, while 18b shows pyelographic evidence of bilateral changes due to chronic inflammation.

Errors in Diagnosis and Management

Misinterpretation of the voiding cystourethrogram usually arises because of failure to completely evaluate the patient. Especially important is the consideration of neurological deficiency of the detrusor. This is well demonstrated in the

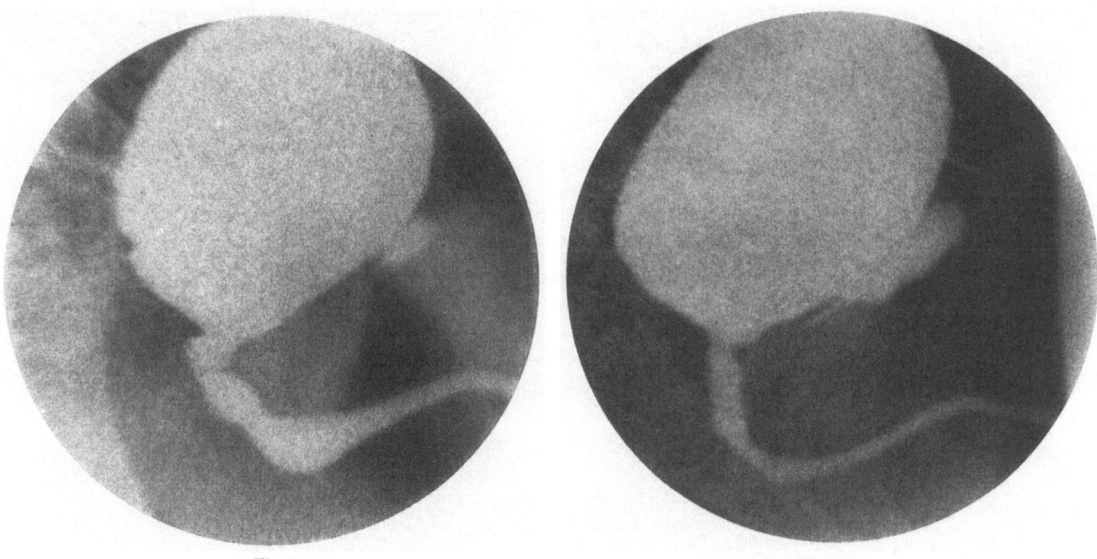

Fig. 24   Fig. 25

Fig. 24. Defect opposite verumontanum in a normal 8 year old male

Fig. 25. Incomplete funneling of bladder neck in an normal 3 year old white male

case of the patient depicted in Fig. 19a, b and c. Persistent urinary tract infection accompanied by fevers and failure to develop normally were thought to be due to bladder neck dysfunction because of the first voiding cystourethrogram, Fig. 19a. Surgical revision of the bladder neck was recommended and carried out with resultant alteration and appearance of the voiding cystourethrogram as shown in Fig. 19b. Persistent difficulties with infection led to another revision and the voiding cystourethrogram demonstrated in Fig. 19c. Failure of these surgical procedures to correct the persistent infection resulted in complete evaluation including neurological studies which demonstrated inadequate detrusor contraction because of a neurological deficiency which was undoubtedly the primary defect. It is extremely important that part of the initial evaluation of such patients include careful cystometric studies including the denervation-hypersensitivity test described by LAPIDES [13].

The Roentgen Appearance of the Normal Male Bladder Neck

There is less variation apparent on voiding cystourethrography in the normal male as compared with the female. Certain variants and artifacts most be recognized and should be familiar to the radiologist and urologist who must interpret

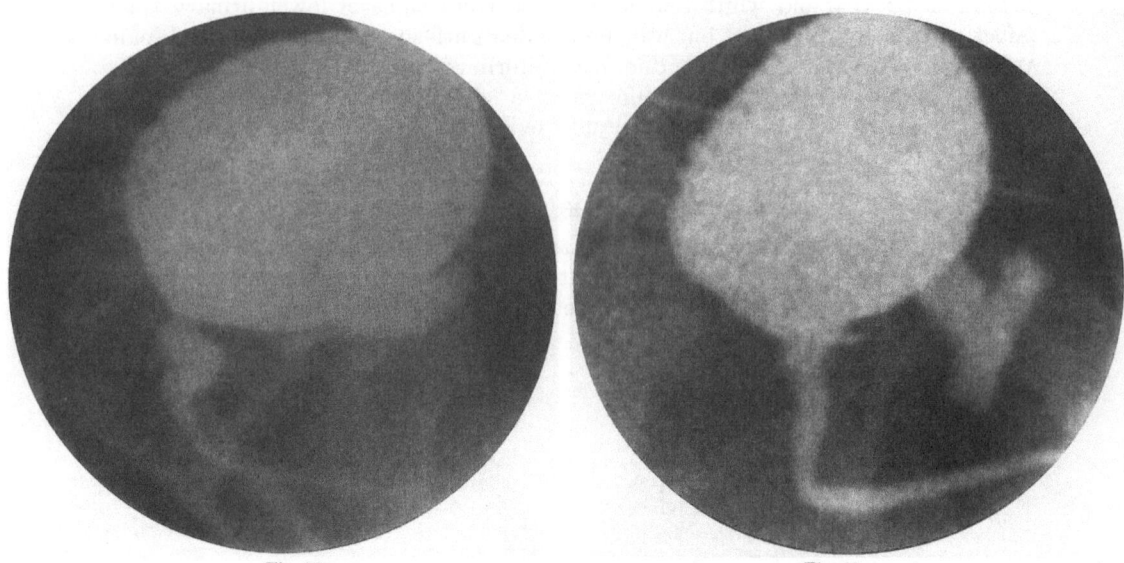

Fig. 26   Fig. 27

Fig. 26. Typical bladder neck dysfunction in a 5 year old male

Fig. 27. Postoperative cystourethrogram — same patient

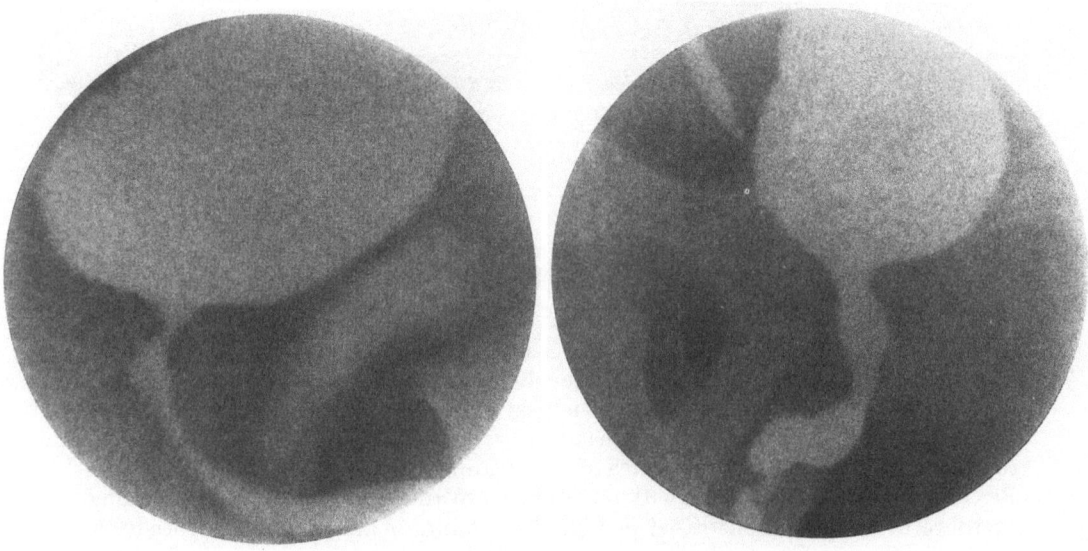

Fig. 28   Fig. 29

Fig. 28. Bladder neck dysfunction in a 21 year old white male

Fig. 29. Postoperative appearance — same patient

these studies. Typical normal appearing bladder necks are shown in Fig. 20, the voiding cystourethrogram of a nine year old white male, and Fig. 21, a similar study in an 11 year old white male. An unusually wide funnel-like appearance of the bladder and bladder neck is shown in Fig. 22, a voiding cystourethrogram in the middle of voiding on a 12 year old white male. A sometimes confusing artefact produced by the muscle compressor nudae in the voiding cystourethrogram

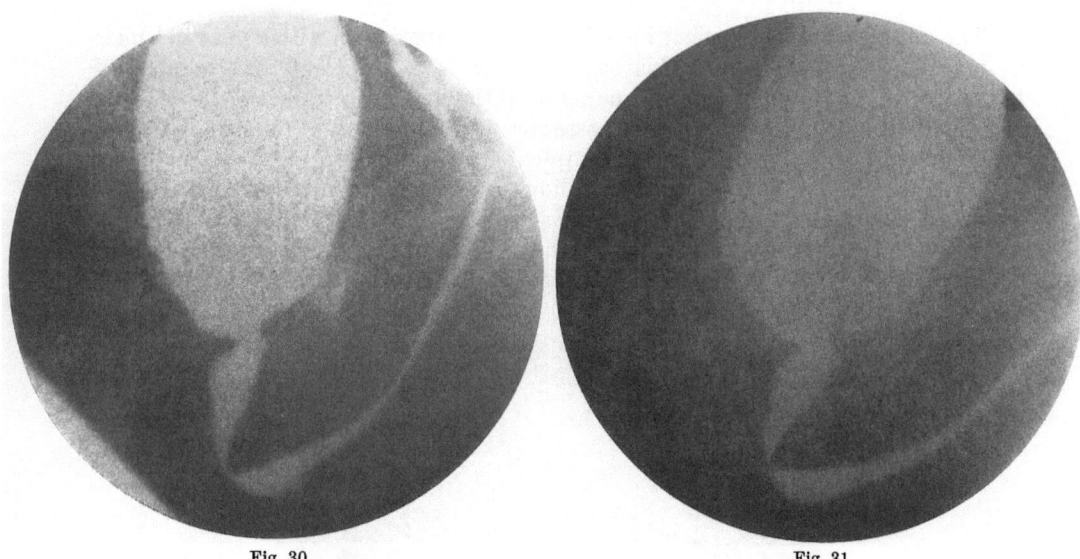

Fig. 30   Fig. 31

Fig. 30. Bladder neck hypertrophy and posterior urethral valve in a 4 year old male

Fig. 31. Persistent bladder neck dysfunction after resection of valves

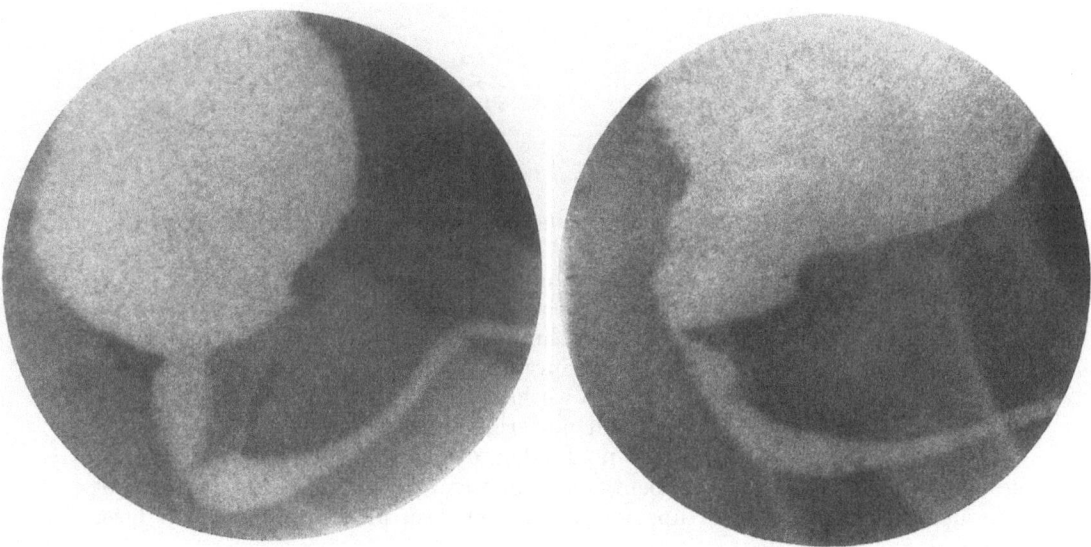

Fig. 32   Fig. 33

Fig. 32. Persistent dilatation of posterior urethra

Fig. 33. Fibroelastosis of posterior urethra in a 9 year old white male

of a three year old white male is shown in Fig. 23. Another compression defect, in this instance opposite to the verumontanum, is demonstrated in Fig. 24, the voiding cystourethrogram of a normal eight year old white male. Incomplete funneling of the bladder neck which might be confused with bladder neck dysfunction or contracture is demonstrated in Fig. 25, the voiding cystourethrogram of a normal three year old white male.

Bladder Neck Dysfunction in the Male

The appearance of a typical bladder neck dysfunction in a five year old male is demonstrated in Fig. 26. This study, made in a child referred because of frequency, urgency and enuresis demonstrated the fixed appearance of the bladder neck with some dilatation of the posterior urethra distal to this on repeated occasions and throughout voiding. Surgical revision by simple incision of the anterior aspect of the bladder neck followed by transverse closure resulted in complete disappearance of the clinical symptoms and the bacilluria. A post-operative voiding cystourethrogram is shown in Fig. 27. The appearance of the outlet is that of a wide funnel. There is a small diverticulum anteriorly at the site

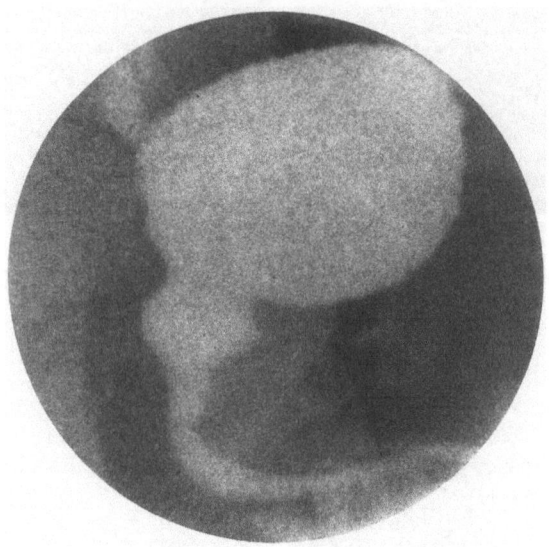

Fig. 34. Cystourethrogram after transurethral resection

of the suture line. This is apparently of no clinical significance. A similar lesion in a 21 year old white male studied because of urinary retention is shown in Fig. 28. The bladder neck-prostatic urethra junction is quite narrow and never filled out during the course of voiding. Surgical revision resulted in the improved appearance noted in Fig. 29, made two weeks postoperatively. This demonstrates a normal appearing bladder neck and posterior urethra. The patient voids with an excellent stream and has had no further difficulty. The unilateral vesico-ureteral reflux shown on this roentgenogram was attributed to the presence of the suprapubic cystostomy tube for it was not seen after this tube had been removed.

Bladder Neck Dysfunction Accompanying Posterior Urethral Valves

The apparent hypertrophy of the bladder neck which is frequently seen with posterior urethral valves may, in some instances, regress and produce no clinical symptoms when the valves have been removed. Not infrequently the dysfunction occasioned by this hypertrophy persists and requires surgical revision. An example of this is demonstrated in Fig. 30, 31, and 32. The patient was a four year old boy referred for evaluation because of persistent urinary tract infection with frequency, urgency and enuresis. The first study, Fig. 30, was thought to suggest a posterior urethral valve with secondary bladder neck hypertrophy. Urethrocysto-scopy confirmed the presence of a valve and this was resected transurethrally. Postoperatively the patient continued to have symptoms, voided poorly and had persistent bacilluria. Re-evaluation, Fig. 31, failed to visualize any posterior urethral valves, but the bladder neck dysfunction was apparent in all phases of voiding. Surgical revision of the vesical neck was carried out by simple anterior incision of the bladder neck with transverse closure. Following this, the patient voided with an excellent stream and his bacilluria quickly cleared. He was asymptomatic when studied two months postoperatively, Fig. 32, and this cysto-urethrogram demonstrates a more funnel-like appearance of the bladder neck though some posterior urethral dilation persists.

The appearance of the bladder neck in a nine year old white male with fibro-elastosis of the bladder neck and posterior urethra is shown in Fig. 33. This boy was studied because of persistent urinary tract infection. Endoscopy confirmed the presence of posterior urethral obstruction and transurethral resection of this tissue was carried out. The patient's symptoms disappeared and his urine became sterile. The greatly improved appearance of this bladder outlet on voiding cysto-urethrography performed two and one-half months after surgery is demonstrated in Fig. 34.

Summary

Experience with cinefluorographic voiding cystourethrography has demon-strated a large number of normal variations in the appearance of the bladder neck. Appreciation of these and correlation of abnormal appearances with clinical problems indicates the value of this diagnostic approach to anomalies in this area and provides a useful method for selection of the proper therapeutic approach as well as determination of the results of such treatment.

References

1. LAPIDES, J.: Structure and function of the internal vesical sphincter. J. Urol. Balti-more 80, 391 (1958).
2. WOODBURNE, R. T.: Structure and function of the urinary bladder. J. Urol. (Balti-more) 84, 79—85 (1960).
3. NANSON, E. M.: Marions' disease or bladder neck stenosis. Aust. N. Z. J. Surg. 20, 215 (1951).
4. MARION, G.: Surgery of the neck of the bladder. Brit. J. Urol. 5, 351 (1933).
5. BODIAN, M.: Some observations on pathology of congenital "Idiopathic bladder neck obstructions (Marions's disease)". Brit. J. Urol. 29, 393 (1957).
6. LEADBETTER, G., and W. F. LEADBETTER: Diagnosis and treatment of congenital bladder neck obstruction in children. New Engl. J. Med. 260, 633 (1959).
7. HURST, A. F., and J. G. JONES: Case of megaloureter due to achalasia of the uretero-vesical sphincter. Brit. J. Urol. 3, 43 (1931).
8. MURPHY, J. J., H. W. SCHOENBERG, and T. A. TRISTAN: The prevention of chronic pyelonephritis. Brit. J. Urol. 37, 58 (1965).
9. MURPHY, J. J., and H. W. SCHOENBERG: Observations on intravesical pressure changes during micturition. J. Urol. (Baltimore) 84, 106 (1960).
10. GLEASON, D. M., and J. K. LATTIMER: The pressure-flow study: A method for measuring bladder neck resistance. J. Urol. (Baltimore) 87, 844 (1962).
11. ZINNER, N. R., and A. J. PAQUIN jr.: Clinical urodynamics: I. Studies of intravesical pressures in normal human female subjects. J. Urol. (Baltimore) 90, 719 (1963).
12. SCOTT, F. B., E. M. QUESADA, and D. CARDUS: Studies on the dynamics of micturition: Observation of healthy men. J. Urol. (Baltimore) 92, 455 (1964).
13. LAPIDES, J., C. R. FRIEND, E. P. AJEMIAN, and W. F. REUS: A new test for neurogenic bladder. J. Urol. (Baltimore) 88, 245 (1962).

Anomalies of the Urethra

KEITH WATERHOUSE

With 34 Figures

I. Introduction

This chapter will consider congenital anomalies of the urethra, but will not include epispadias (Chapter 5), hypospadias (Chapter 10), or ectopic ureters draining into the urethra (Chapter 4). Bladder neck obstruction is the subject of a separate communication (Chapter 6) and will be considered only in so far as it arises secondarily to diseases originating more distally in the urethra.

The clinical material represents mainly that drawn from the State University of New York, Downstate Medical Center — Kings County Hospital Center and consists of some 2,500 children examined between 1958 and 1965. This series contains relatively few referred cases. Cases from other sources that are mentioned in the chapter will be identified as such. Figures for incidence relate to the State University of New York, Downstate Medical Center — Kings County Hospital Center alone.

In considering the incidence of any disease, many factors must be taken into account if the figures are to be compared with those of other groups. Of importance is the rate of referral of cases already diagnosed, as is the ethnic group of the sample involved. We have noticed that major congenital abnormalities are much less common in negro children than in white children and some otherwise common diseases have not been seen by us in a coloured child. The best example of this is the megacystis syndrome (PAQUIN, MARSHALL and McGOVERN, 1960) which is seen quite commonly in white children but neither in our clinic nor in PAQUIN's (1965) has a case of this disease been seen in a coloured child.

This difference in rate of congenital abnormalities in various ethnic groups is not limited to genitourinary abnormalities but has also been noted in cardiac and other abnormalities. Table 1 is reproduced from the figures of HELLMAN and KOHL and shows the rates of congenital malformations at the State University of New York, Downstate Medical Center — Kings County Hospital Center for the years 1956—1958 (15,000 births). Table 2 shows the figures taken from community hospital for the years 1957—1959 (28,000 births). As can be seen the total rate of abnormality in white and non-white groups is equal but significant abnormalities are considerably less common in non-white infants.

A most important advance in the clinical recognition of urethral anomalies has been the widespread adoption of voiding cystourethrography in the investigation of children suspected of having lower urinary tract obstruction. This technique advocated by BRODNY and ROBINS (1948) in the United States, by STEPHENS (1955) in Australia and by KJELLBERG, ERICSSON and RUDHE (1957) in Sweden has now been generally recognised as being the most satisfactory method of demonstrating anomalies in the male urethra. However, as WILLIAMS (1965) has pointed out, in children with intersex the use of voiding urethrography alone may lead to errors as in many instances the urethra will appear normal and there will be no reflux into the aberrant vagina. This structure will only be shown on injection urethrography.

Voiding urethrography is much less satisfactory, although still useful, in studying the female urethra. Occasionally valvular obstruction will be clearly seen (Fig. 6) but in general the shorter length and the marked variations in normal anatomy make interpretation difficult.

Table 1. *Rate of congenital malformations*

State University of New York, Downstate Medical
Center — Kings County Hospital, 1956—1958, 15,000 births

		White	Non-White
1.	Incompatible with life	0.6%	0.5%
2.	Consequential anomalies	1.5%	1.0%
3.	Inconsequential anomalies	0.4%	1.0%
	Total 1 and 2	2.1%	2.5%
	Grand Total	2.5%	2.5%

Table 2. *Community obstetrical study*
1957—1959, 28,000 births

		White	Non-White
1.	Incompatible with life	0.4%	0.3%
2.	Consequential anomalies	0.9%	0.5%
3.	Inconsequential anomalies	1.4%	2.1%
	Total 1 and 2	1.3%	0.8%
	Grand Total	2.7%	2.9%

II. Congenital Absence of the Urethra

Absence of the urethra may be total, in which case it is associated with the absence of the bladder, or partial. Partial obliterations may occur at the level of the bladder neck, in the membranous urethra or in the penile urethra.

1. Complete Absence of the Urethra

Complete absence of the urethra has been reported in detail by MILLER (1948). It was associated with absence of the bladder, the ureters opening externally. The patient was a 27 year old female with surprisingly little disability. The author thought that the two ureters entered the vagina at a common point; the left coming in at an abrupt angle. When first seen the patient had hydronephrosis bilaterally and cutaneous ureterostomies were advised. This was refused by the patient who was well acclimated to her disability, but performed later because of recurring infections.

2. Obliteration at the Level of the Bladder Neck

This anomaly is discussed in detail by KRUGER (1931). It is associated with gross enlargement of the bladder; KRUGER described his case with the term "vesica gigantea". In all cases there is serious hydronephrotic atrophy of the kidneys and the process has been uniformly fatal. The external genitalia are often rudimentary.

3. Obliteration at the Level of the Membranous Urethra

May (1949) has presented the case of a 9 year old boy with an obliteration of the membranous urethra. This was relatively short and urethrography showed an adequate anterior urethra. The posterior urethra proximal to the obliteration was also demonstrated. There was a fistula between the rectum and the posterior urethra. Surgical repair was performed.

4. Obliteration of the Penile Urethra

A very complete study of this problem has been presented by Menegaux and Boidot (1934). They collected 41 cases from the literature and added one of their own. Eight were in female patients and thirty-four in male patients. In the female the condition was commonly associated with other anomalies and four of the eight patients had a patent urachus. Only one of eight females reported survived.

In males the condition was rarely associated with other anomalies, only three of thirty-four infants having a patent urachus. It is also of great interest that in at least twenty-three of these thirty-four cases there was a successful surgical outcome. In only three patients was the large bladder suggestive of chronic obstruction seen. It would seem therefore, that in many of these patients there was not true absence of the urethra but that there had been temporary occlusion by epithelial debris. It can only be assumed from the generally excellent state of these children that this occlusion was of recent origin.

III. Meatal Stenosis

Stenosis of the external urinary meatus may be congenital or acquired. Congenital stenoses are usually associated with hypospadias and are part of this deformity. There has been an occasional case report of congenital stenosis unassociated with hypospadias and causing severe damage to the urinary tract (Nesbit and Baum, 1954; Higgins, Williams and Nash, 1951).

In most instances urethral meatal stenosis is secondary to meatal ulceration, the meatal ulceration being caused by irritation due to urine soaked diapers. It is seen much more commonly in circumcised than uncircumcised babies. The radiographic findings are well demonstrated in the voiding cystourethrogram shown in Fig. 1. There is dilatation of the posterior urethra and the anterior urethra with a narrowing at the level of the urogenital diaphragm. The urinary stream is very fine.

Correction of this condition by meatotomy and periodic dilatations is usually a satisfactory procedure but in rare instances the damage to the upper urinary tracts proves to be fatal.

IV. Distal Urethral Stenosis

In 1962 Lyon and Smith suggested that many recurrent urinary tract infections in girls may be due to wide caliber urethral stenosis. The areas of narrowing are not at the external meatus and the condition is to be distinguished from meatal stenosis. The most satisfactory method of demonstrating the disease is by calibration with the bougie à boule when it is found that the narrow portion of the urethra is the distal margin of the urethral musculature. By studying recorded urine flow patterns before and after rupture of this wide caliber stricture, Lyon and Smith advanced further evidence that what they were presenting was a relative obstruction.

In another study of this problem, Lyon and Tanagho (1965) showed that in 137 of 152 girls with recurrent infections of the urinary tract a distal urethral stenotic ring could be found. The ring is composed of collagenous tissue and lies at the distal extremity of the urethral musculature. It becomes a problem when it encroaches on the urethral lumen and acts functionally as an obstructive lesion although there is only slight evidence that serious obstructive changes occur other than those connected with infection.

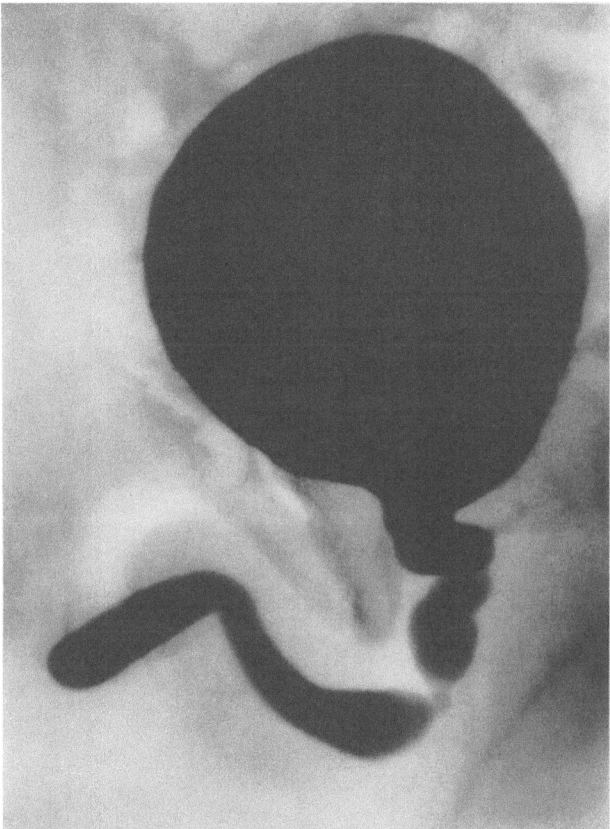

Fig. 1. Urethral meatal stenosis. Voiding cystourethrogram. There is dilatation of the posterior and the anterior urethra. A fine jet of urine can be seen spurting vertically from the meatus

Lyon has suggested that the ring be ruptured by dilatation of the urethra by passage of sounds up to 32 F in children over 3 years of age and up to 28 F in those under 3 years. In a 2 year follow-up study 70 per cent were cured by this treatment. If children in whom reflux had been demonstrated by cystography were excluded the cure rate was 90 per cent.

Marked improvement in personality was also noted in these children following dilatation. In this regard attention is drawn to the "Sham syndrome" of Stephens (1963).

Our experience has been similar to that of Lyon. In examining children in our clinic with recurrent urinary tract infection calibration of the urethra with the bougie à boule has proved much more satisfactory in the detection of this disease than has voiding cystourethrography. Although we perform a voiding cystourethrogram routinely, we are more interested in the presence or absence of reflux than in the roentgen appearance of the urethra. Fig. 2 shows the voiding cystourethrogram in a 6 year old female child with distal urethral stenosis. In this instance there is bilateral vesico-ureteral reflux. The urethra calibrated at 16F with the narrow point just inside the external meatus. The urethra was dilated to 32F with a marked improvement in symptoms. At follow-up the reflux on the right side had disappeared. The reflux on the left was still present but much less.

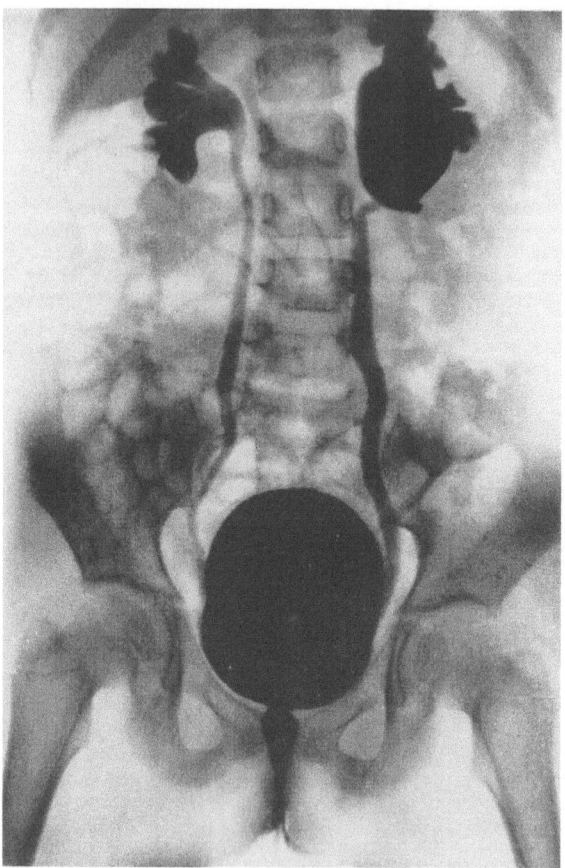

Fig. 2. Distal meatal stenosis. Bilateral pyelogram by reflux. The patient is a 6 year old female whose urethra calibrated at 16F

It would appear to us that this concept of Lyon and Smith is an important one, that the condition they are describing is real and that it is common.

V. Congenital Urethral Stricture

If meatal stenosis is excluded congenital strictures of the urethra would appear to be limited to male children. These strictures usually are short and it is difficult to distinguish them from the diaphragmatic urethral obstructions described by Young, Frontz and Baldwin (1919) as Type 3 urethral valves. There are a few reports in the literature of longer strictures presumably congenital in origin. Leadbetter (1962) reports two congenital strictures 0.5 to 1.5 cms in length

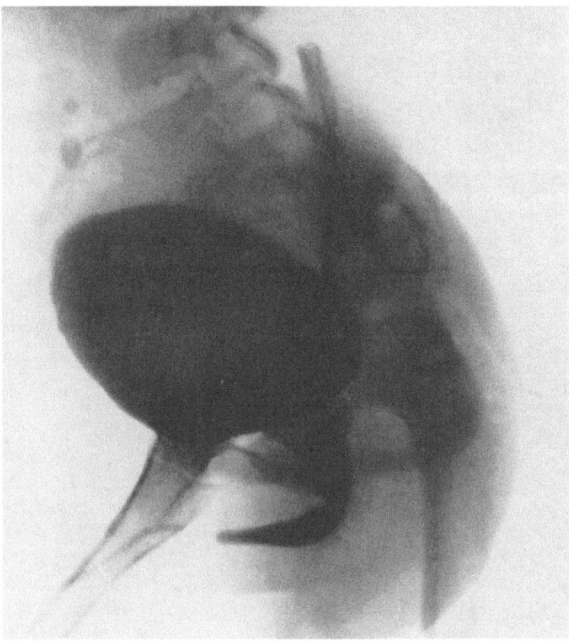

Fig. 3a. Urethral stricture. Voiding cystourethrogram. Patient is a six year old boy who during infancy had required an indwelling urethral catheter. Penoscrotal stricture is probably secondary to trauma at this time

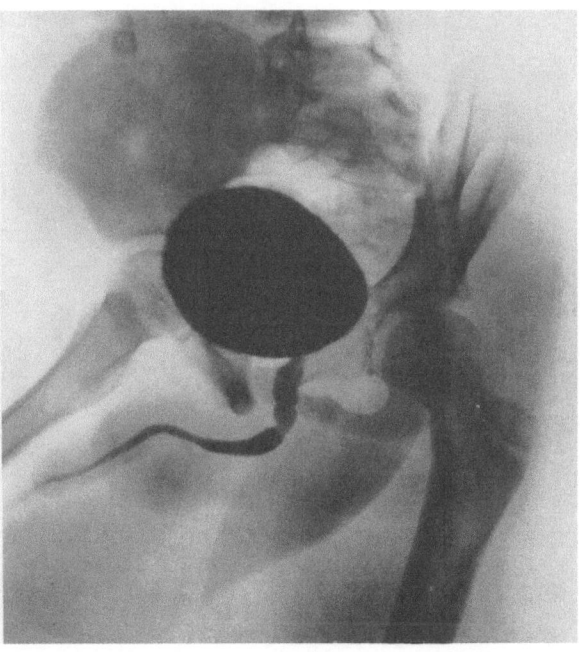

Fig. 3b. Urethral stricture. Voiding cystourethrogram. Postoperative film after dilatation of the stricture with sounds

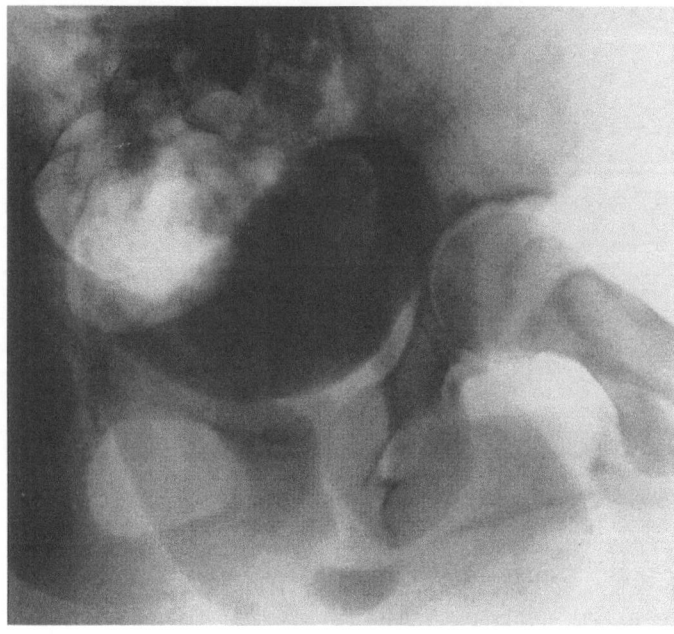

Fig. 4a. Urethral stricture. Voiding cystourethrogram. There is considerable dilatation of the urethra proximal to the stricture. Patient was kicked in the perineum eight years prior to examination

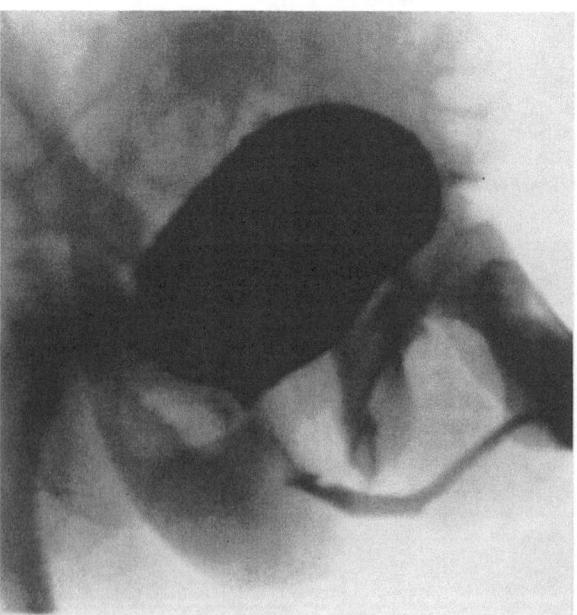

Fig. 4b. Urethral stricture. Voiding cystourethrogram. Postoperative after excision of the stricture and repair by Hamilton Russell urethroplasty

and located proximal to the fossa navicularis. It is important that in considering the origin of urethral strictures in child the possibility of injury many years previously be considered. Fig. 3 is the voiding cystourethrogram in a six year old male who presented with signs of lower urinary tract obstruction. There is clearly a stricture at the penoscrotal angle with evidence of dilatation proximal to the

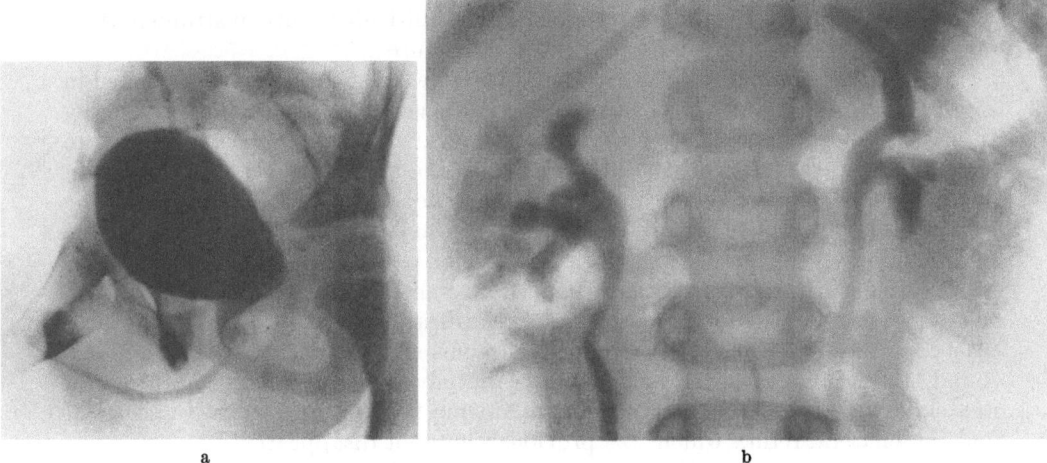

a   b

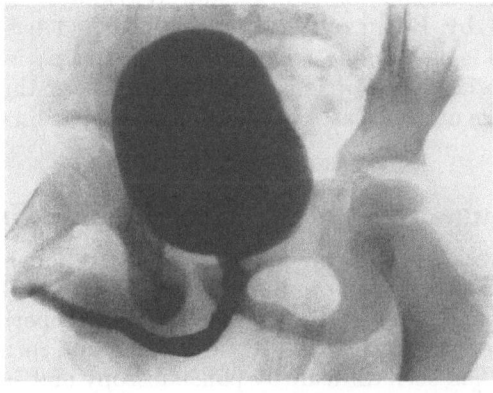

c

Fig. 5a. Urethral stricture. Diaphragmatic type (Young's type III valve). Voiding cystourethrogram. Preoperative. There is dilatation of the urethra proximal to the obstruction. Bilateral vesicoureteral reflux is demonstrated

Fig. 5b. Urethral stricture. Diaphragmatic type (Young's type III valve). Pyelogram by reflux showing pyelo-tubular back flow on the right side

Fig. 5c. Urethral stricture. Diaphragmatic type (Young's type III valve). Voiding cystourethrogram. Post-operative study following rupture of the diaphragm by passage of a sound

site of obstruction. This could have been thought to be congenital in origin if it had not been discovered that during infancy the patient had been in coma for 72 hours because of lead encephalopathy. During this time the bladder was drained by an indwelling Foley catheter. It would appear that trauma from the catheter at this time was the reason for the subsequent development of the stricture. Fig. 4 shows the voiding urethrogram in a similair patient who presented with a severe stricture at age sixteen. Again on careful questioning he had been kicked in the perineum eight years previously.

The presenting symptoms and signs in congential urethral strictures do not differ from other patients with urinary obstruction at or below the level of the bladder neck. The diagnosis which may be suspected because of difficulty with instrumentation should be confirmed by injection and voiding urethrography prior to urethral dilatation. If dilatation is performed first it may be difficult to assess the length and the severity of the stricture. It would appear that with the known rarity of congenital urethral stricture the diagnosis should be made with caution and it should be recognised that difficultiy with instrumentation in a male child does not necessarily imply a stricture.

The degree of damage to the kidneys is variable. In some strictures reported in the literature the upper tracts have been very severely damaged but on reading these case reports it would seem that the authors are reporting on conditions more accurately called atresia of the urethra rather than stricture (ENGEL, 1938). In our own cases damage to the upper tracts has not been severe although in one case of a diaphragmatic type of stricture there was bilateral reflux which, on the right side, was so severe as to cause pyelotubular back flow.

Treatment of urethral strictures may be by urethral dilatation, excision and reanastomosis of the urethra or by some modification of the Johanson urethroplasty. In most instances of urethral diaphragms simple rupture by passing of a sound is all that is required. Fig. 5 shows the preoperative voiding urethrogram in a child with a diaphragm. Following dilatation the urethra returned to normal and the bilateral reflux which was previously present disappeared.

In strictures between 0.5 cm and 1.0 cm in length we have used excision and reanastomosis of the urethra (McGOWAN and WATERHOUSE, 1964). This is the technique described by HAMILTON RUSSELL (1923) and has proven extremely satisfactory. Fig. 4b is the voiding cystourethrogram in a patient after a successful Hamilton Russell procedure. In longer strictures where this procedure is not applicable a two stage urethroplasty has been performed (JOHANSON, 1953; LEADBETTER, 1962).

VI. Congenital Valves of the Female Urethra

Obstructing mucosal folds of the female urethra are much less common than in the male but have been reported by STEVENS (1936), BAKKER (1958) and NESBIT, McDONALD and BUSBY (1963). They have been reported in both children and adults who have presented with difficult voiding and recurrent urinary tract infection. Diagnosis has been difficult by panendoscopy and suggested on voiding cystourethrography. Fig. 6 shows a female patient with a clear valve in the urethra causing damage to the bladder and reflux. Following removal of the valve the voiding urethrogram returned to normal.

The valves may be removed either by pulling them into view with a hookshaped probe and then removing them with scissors or by the cautious use of the resectoscope (BRACK and GUILD, 1958).

VII. Congenital Hypertrophy of the Verumontanum

The first description of this condition is by BUGBEE and WOLLSTEIN (1923). They reported one clinical case and eight post mortem cases. The patient was aged $2^1/_2$ years and the ages of the post mortem group varied from three weeks to three and one-half years although all but two were under one year.

Further cases of this rare anomaly were added to the literature by ROBINSON (1927), PILCHER and PRICE (1940) and EMMETT (1940).

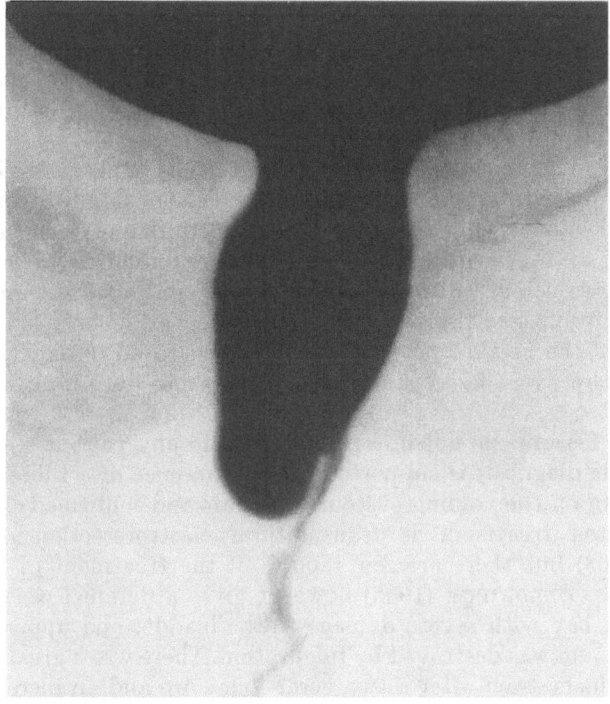

a

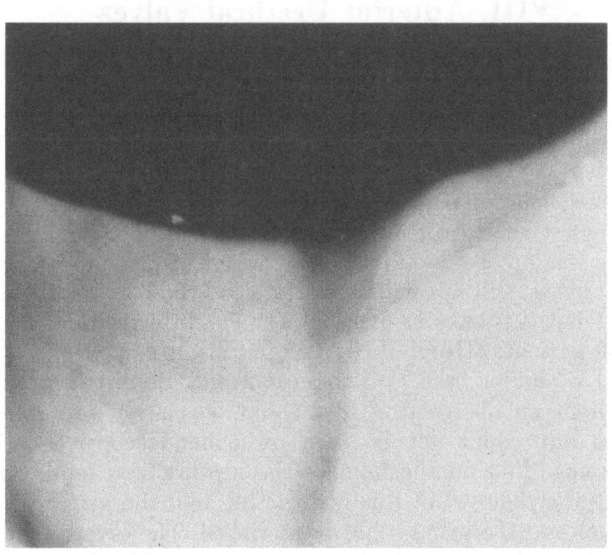

b

Fig. 6a and b. Valve in the female urethra. a Voiding cystourethrogram. Preoperative. b Voiding cystourethrogram. Postoperative

BALDRIDGE (1935) reported on the succesful treatment of this disease.

Large series of children have been examined without this anomaly being noted. KJELLBERG, ERICSSON and RUDHE (1957) have three patients in their series with enlargement on cystourethrography but no evidence of obstruction. In STEPHENS

(1963) series of children with urethral obstruction no case was recorded and in our series no child has been seen with obstruction due to congenital hypertrophy of the verumontanum although one instance has been noted of enlargement by cystourethrography. The significance of enlargement on cystourethrography alone without evidence of obstruction would seem to be negligible.

The etiology of this condition is obscure. In the few publications in which there is comment on the histology of the tissue removed either surgically or post mortem there is general opinion that the condition is not inflammatory but is a simple hypertrophy of an otherwise normal verumontanum. This hypertrophy may be so great as to completely fill the posterior urethra and for the cranial end of the verumontanum to project through the internal meatus.

Dilatation of the urethra proximal to the disease and obstructive changes in the bladder, ureters and kidneys are similair to those due to any lower urinary obstruction.

The clinical features do not differ from those in any patient with infravesical obstruction. The diagnosis is supported by the presence of a filling defect in the posterior urethra on the voiding cystourethrogram and confirmed cystoscopically.

The suggested treatment is transurethral electroresection or fulguration (Campbell, 1963) but there are few reports of the treatment of this condition in the literature. Baldridge (1935) however gives a detailed case report on an eleven year old boy with severe damage to the bladder and upper urinary tract The verumontanum was destroyed by fulguration. There was a great improvement in the upper urinary tract after a two yerar follow-up and an increase in phenol-sulphaphthalein output from 0% in 30 mins. to 38% in 30 mins.

VIII. Anterior Urethral Valves

The distinction between anterior urethral valves and anterior urethral diverticula is a fine one and it is probable that they are variations of the same pathological process. In some instances patients reported as having valves have had what was probably a diverticulum with a distal obstructing fold of mucosa (Boisonnat, 1954).

Williams (1958) has reported three cases in children, aged 3, 4 and 9 years. Each child presented with difficult micturition, distended bladder and hydroureters, but there was no obstruction to the passage of instruments. As Williams points out the diagnosis will not be missed if a good quality voiding cystourethrogram is obtained but it is easy to overlook the condition on urethroscopy. Hope has had two such patients (Hope, Jameson and Michie, 1960).

We have had a similar case (Waterhouse and Scordamaglia, 1963) in a male child in whom an obstructing fold could be clearly seen on the voiding cystourethrogram but could not be seen on panendoscopy (Fig. 7). At open operation a crescentic fold on the floor of the urethra was found and easily removed. Following the removal of this obstructing fold the upper tract dilatation returned to normal and the urine could be sterilized. The severe reflux which was present preoperatively disappeared after the operation (Fig. 7).

IX. Diverticula of the Anterior Urethra

Congenital diverticula of the anterior urethra occur in two forms. In one there is a discrete opening in the ventral surface of the urethra leading into a diverticulum lying below the urethra; in the other a whole segment of the urethra is dilated. This diffuse type of diverticulum has been described by Nesbitt (1955) as megalourethra.

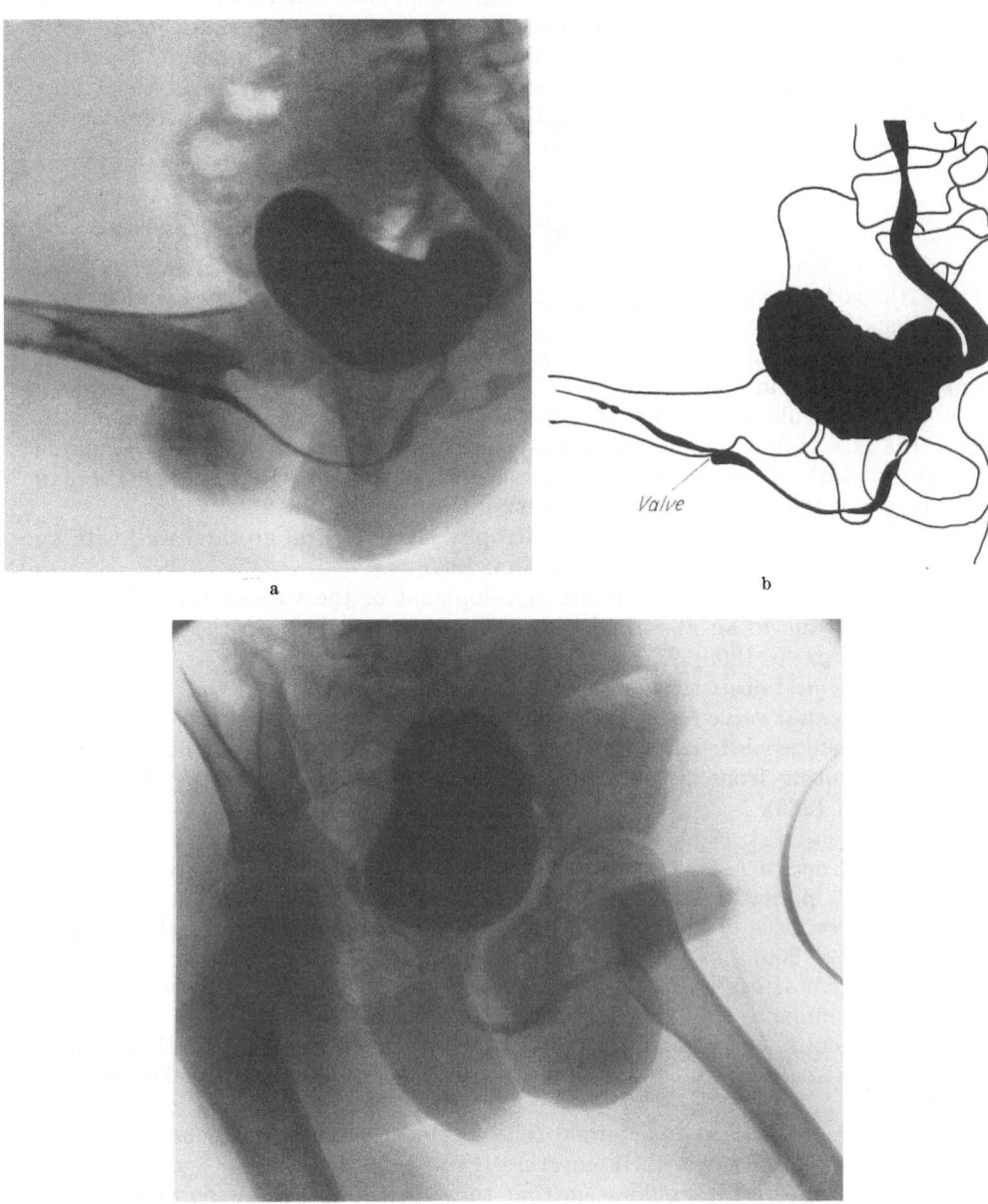

Fig. 7a—c. Anterior urethral valve. a Voiding cystourethrogram. Preoperative. The valve can be seen in the pendulous urethra. There is free reflux to the left kidney and a diverticulum in association with the left ureteral orifice. b Schematic drawing of preoperative voiding cystourethrogram. c Voiding cystourethrogram. Postoperative. Note the absence of reflux

1. Incidence

In an extensive review of the literature ABESHOUSE (1951) collected 93 cases of congenital anterior urethral diverticula and added one of his own. Since this review a number of additional cases have been added to the literature. KJELL-BERG, ERICSSON and RUDHE (1957) reported eight cases in their series, only four of which were the cause of symptoms. Stephens (1963) reported seven cases, four of the saccular type and three of the diffuse type. Other reports of interest are by Mills (1955), Demos, Gillis and Barber (1962), and Forshall and Rickham (1953). Additional examples, not specifically cited, may be found in the bibliography. From the literature it would appear that megalourethra is considerably less common that the saccular form of the disease.

2. Etiology and Pathology

The various theories as to the embryological origin of anterior urethral diverticula have been reviewed by Abeshouse (1951) and grouped under three headings.

a) Disturbances in the development of one or more of the tissue components of the primitive urethral plate.

α) Diverticula are due to the faulty development of the spongy tissue of the urethra in a manner similar to hypospadias. In the former, the defect is limited to the ventral wall of the urethra whereas in the latter the defect extends through all layers of the urethra including the skin (Vollemier, 1868).

β) Diverticula are due to primary atrophy of the central urethral wall with an unresisting corpus spongiosum (de Paoli, 1885).

γ) Diverticula are due to faulty development of the various tissue layers of the urethra due to an abnormal blood supply, particularly in the corpus spongiosum (Durand, 1900).

b) Congenital obstructive lesions in the anterior urethra.

α) Congenital valve formation (Hueter, 1869).

β) Congenital obstructions of a partial or complete, temporary or permanent nature resulting from faulty union between the glandular and penile urethra (Kaufman, 1886).

γ) Congenital stricture formation (Petz, 1900).

δ) Obstructive lesions of the urethra manifested in the intrauterine life or infancy, i.e. preputial adhesions, phimosis with narrowing of preputial orifice, congenital stenosis of the external meatus, congential stricture of the urethra, etc. (Watts, 1906).

c) Congenital cystic dilations developing in or close to the ventral urethral wall and communicating with the urethral channel.

α) Congenital cystic dilatation of the normal or accessory periurethral glands or ducts with subsequent communication with the ventral urethral wall (Johnson, 1923).

β) Cystic dilatation of epidermal cell nests near the urethral floor which subsequently communicate with the urethra (Suter, 1908).

Of these theories it would seem that those of Group b may be dismissed as many diverticula have been reported in the absence of obstruction. It would appear that a single explanation could cover both the saccular and the fusiform type of diverticulum if it could be demonstrated that there are changes in the corpus spongiosum. Dorairajan (1963) has presented evidence which seems to suggest that this is so. He examined at post mortem a child dying from overwhelming infection and toxaemia due to obstruction caused by a saccular diverticulum. The wall of the diverticulum was composed of an epithelial lining covered by a thin layer of tissue which was mainly fibrous and rudimentary spongy tissue. There was also an abnormality in the spongy tissue in the roof of the urethra opposite the orifice of the diverticulum. Furthermore the corpus spongiosum proximal to the diverticulum also exhibited a greater fibrous component than

normal but the most obvious abnormality occurred in the anatomical conformation of the spongy tissue relative to the urethra in this region. The floor of the
urethra was embedded in spongy tissue but the roof and side walls lay uncovered.
DORAIRAJAN (1963) also had the opportunity to examine a stillborn baby with
fusiform megalourethra. The penis was long and flabby and was clinically devoid
of corpora cavernosa. Histological examination revealed extensive deficiency of
the erectile tissue. There was absence of both corpora cavernosa and the corpus
spongiosum in the region of the dilated segment of the penile urethra. The vascular
spongy tissue was represented by a thin fibrous layer which was presumably the
tunica albuginea. Proximal to the dilatation the corpus spongiosum was found
to be an attenuated bed of spongy tissue grooved to accommodate the floor of the

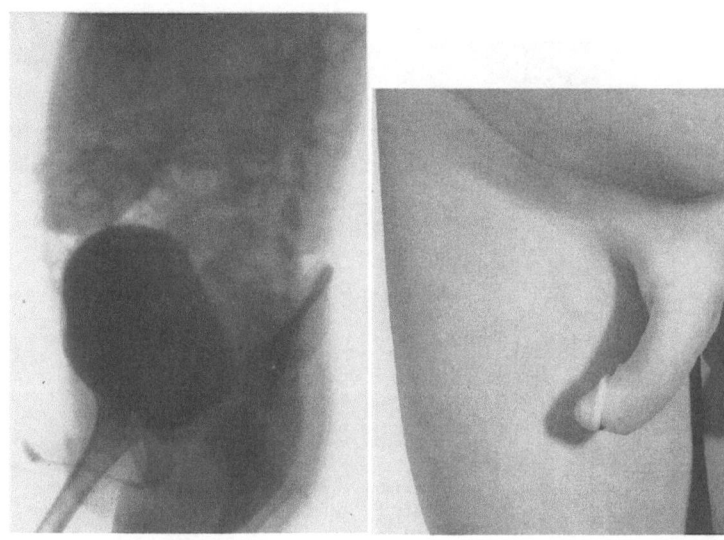

Fig. 8   Fig. 9

Fig. 8. Anterior urethral diverticulum. Diffuse type. Triad syndrome. Voiding cystourethrogram. There is a
scaphoid type of megalourethra. Note the dilatation of the posterior urethra. This is usually seen in patients
with the "triad syndrome"

Fig. 9. Patient with the "triad syndrome" and with a defect of the corpora cavernosa causing deformity of the
penis. The anterior urethra was normal on voiding cystourethrography

urethra. It did not encircle the urethra. The two corpora cavernosa formed narrow
cylinders which tapered distally and disappeared entirely when they reached the
megalourethra. Distal to the dilatation, the corpora cavernosa reappeared as a
fused mass, then separated into two short cylinders of spongy tissue which
terminated as the glans. The glans penis was fully developed although somewhat
fibrotic. DORAIRAJAN concluded that this studies revealed that congenital urethral
expansions are associated with fibrosis, deficiency or absence of the spongy tissue
of the penis. The defect is local in the saccular type but more extensive in the
diffuse type, extending in some beyond the corpus spongiosum to the corpora
cavernosa. These defects of spongy tissue may be explained embryologically on
the basis of failure of differentiation of the mesenchyme into specialised erectile
tissue and of failure of migration of this tissue to surround the urethra. In considering further the validity of this theory it may be pointed out that in one of
DORAIRAJAN's patients there were other associated anomalies usually described
under the descriptive term of "absent abdominal muscle syndrome" (LATTIMER,
1958), or "triad syndrome" (NUNN and STEPHENS, 1961). BOISSONAT (1962) has

described a patient with the "triad syndrome" and a scaphoid megalourethra and we have had a similar patient in our series (Fig. 8). In addition we have had a patient with the absent abdominal muscles syndrome and with a normal urethra but with a defect of the corpora cavernosa causing deformity of the penis (Fig. 9).

This evidence leads me to favor the theory that diverticula of the urethra are associated with, and are probably caused by, defects in the development of the corpus spongiosum.

3. Clinical Presentation and Diagnosis

In some patients the diagnosis is obvious on inspection of the penis particularly during voiding; in others there are no physical findings.

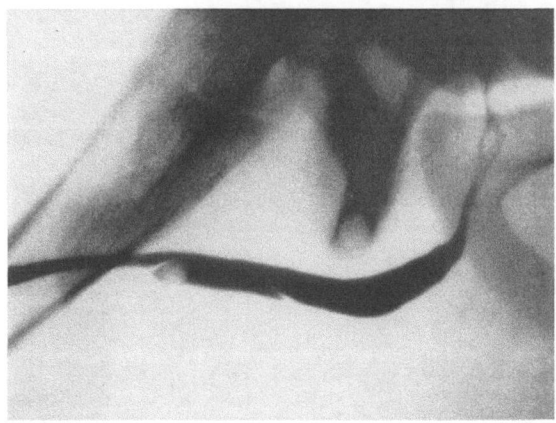

Fig. 10. Diverticulum of the anterior urethra. Saccular type. Voiding cystourethrogram. The proximal flap of mucosa is obstructive and may lead to serious changes in the upper urinary tracts

Patients with the saccular form of diverticulum usually have obstructive symptoms due to the flap-like action of the proximal mucosal fold. These symptoms are not distinguishable from those of any patient with obstruction at or below the neck of the bladder. As discussed by GROSS and BILL (1948) small saccular diverticula of the type shown in Fig. 10 are particularly dangerous and are difficult to diagnose. There are no physical findings as the diverticulum is not large enough to be seen or felt at the time of voiding and catheters and endoscopic instruments will slip past the obstructing fold with great ease.

Patients with megalourethra do not have any urethral obstruction although as mentioned previously in a number of patients there have been changes in the upper urinary tracts associated with the absent abdominal muscle syndrome. We believe these changes to be independent of the urethral diverticulum.

The diagnosis may be made or confirmed by voiding cystourethrography. Panendoscopy will allow the position of the opening into the saccular type of diverticulum to be seen and the obstructive flap of mucosa may be identified.

4. Treatment and Prognosis

The best treatment of megalourethra is that described by NESBITT (1955). In this method the urethra is exposed by incising the penile skin around the corona and allowing it to drop back towards the root of the penis. The excess urethra is then trimmed off and resutured around an indwelling urethral catheter. The skin is then replaced with a few interrupted sutures. In this method there are no overlapping suture lines thus minimising the likelihood of fistula formation.

Patients with saccular diverticula have been treated in some instances quite successfully by endoscopic methods (BOISSONAT, 1954). The obstructing flap of mucosa being destroyed by diathermy and the diverticulum sclerosed with the fulgurating electrode. This method is applicable only to relatively small diverticula.

In other cases open exposure and excision of the diverticulum has been performed. Urinary diversion by proximal external urethrotomy is generally advised. Following excision the urethra may either be closed primarily or, and probably more satisfactory, it may be allowed to reform from a buried strip following the principles of urethral repair outlined by RUSSELL (1923) and as described for repair of hypospadias by BROWNE (1950).

The prognosis in megalourethra is excellent, many patients not requiring surgical treatment. The prognosis in the saccular type is more serious as in many patients there has been considerable damage to the upper urinary tracts. In our experience however, this damage has usually been reversible.

X. Double Urethra and Accessory Urethra

Although relatively rare this condition has attracted a good deal of surgical interest and there have been reviews of the subject by BOISSONNAT (1954), CHAUVIN (1927) and LOWSLEY (1939). An extremely comprehensive review and a simple classification is that of GROSS and MOORE (1950) who found 81 cases reported in the literature and added 2 of their own. They classified the condition in the male as follows:

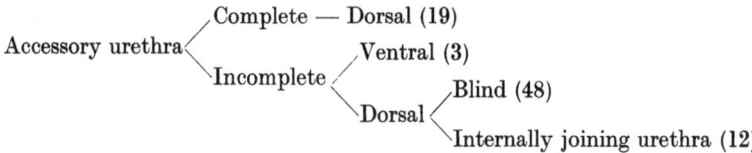

The figures in brackets after the minor classifications are the number of cases found in each group of the eighty-three cases studied.

1. Complete Double Urethra

Since the report of GROSS and MOORE (1951) there have been six cases added to the literature, ARNOLD and KAYLOR (1953), WRENN and MICHIE (1957), BOISSONNAT (1961), BROWN (1956), FUNFACK (1953) and THEVATHASAN (1961).

In all the cases described except that of BOISSONNAT (1961) the accessory urethra belongs to the epispadiac type with the upper opening in a groove on the glans or on the dorsum of the penis. BOISSONNAT presented the only case of the upper urethra opening at the normal site on the apex of the glans and the second urethra being hypospadic opening at the penoscrotal angle.

Fig. 11 shows the voiding cystourethrogram in a previously unreported case of double urethra. The case is in almost all respects similar to that of BOISSONNAT. The patient was a 14 year old male who had had a previous repair of the urethra for hypospadias. At that time it was not appreciated that another urethra was associated with the meatal dimple. The patient represented some time after an apparently succesful repair because of enuresis. Voiding cystourethrography showed the presence of a complete double urethra. The nature of the problem was explained to the patient but he declined surgical intervention. Fig. 11b shows a drawing of our interpretation of the urethrogram.

The presenting symptoms in patients with complete double urethra have included double urinary stream, incontinence, urinary tract infection, dorsal chordee and the abnormal appearance of the penis. Eleven of twenty-one patients were continent. In five patients there was minor incontinence and in five more severe incontinence. Dorsal chordee of a serious degree occurred in five patients. Three patients had gonococcal infection of the accessory urethra.

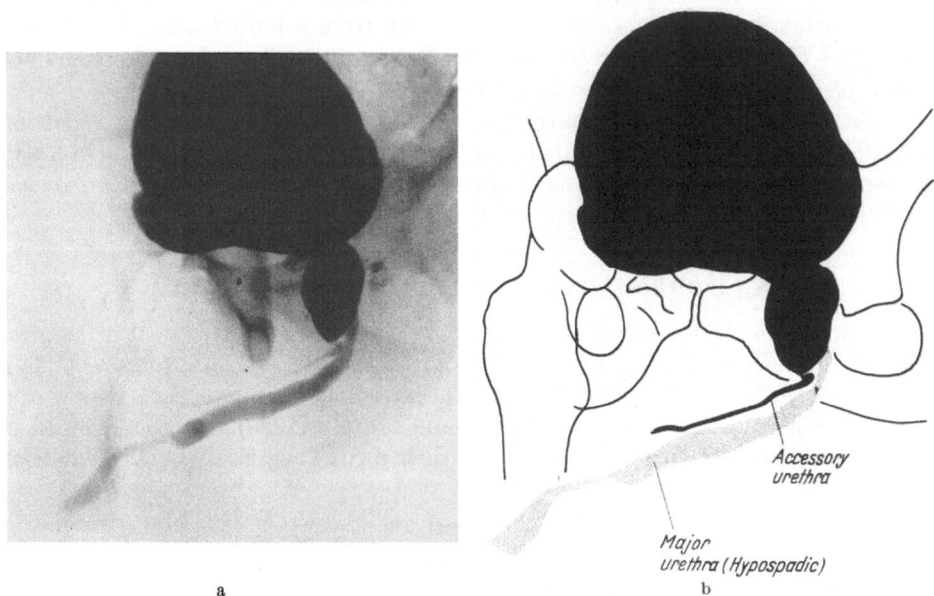

a   b

Fig. 11a. Accessory urethra, complete. Voiding cystourethrogram. The major urethra was hypospadic. The accessory urethra opened at the meatal dimple

Fig. 11b. Accessory urethra, complete. Sketch of our interpretation of the abnormal anatomy seen in Fig. 11a

Diagnosis is best confirmed by X-ray study. Either voiding cystourethrography or injection urethrography may provide a diagnostic study. In general most authors who have studied this problem have preferred injection urethrography.

Not all cases have required surgery. In only about half of the reported cases is there a note of surgical therapy. This had varied from a well described surgical excision and plastic repair of the penis as reported by GROSS and MOORE (1950) to fulguration of the accessory channel (FUNFACK, 1953).

2. Incomplete Dorsal Accessory Urethra — Blind

A large number of such cases are available in the literature described in varying degrees of detail. The best review is that of LOWSLEY (1939) who collected 39 cases of this type. Fifteen of the accessory canals opened just above the normal meatus, ten opened between the corona and the glans and six opened at the balanopreputial fold on the dorsal aspect of the penis. In three the urethra began on the dorsum of the penis at the edge of the pubic hair. This author also includes in his study three examples of incomplete ventral accessory urethrae. Symptoms were minimal in most instances except when the urethra became infected with the gonococcus. It was this which brought twenty-six patients for medical care.

There are reports of twelve patients being operated on. In six the urethra was excised and six an incision into the canal was made and the epithelium destroyed by currettage.

3. Incomplete Dorsal Accessory Urethra — Communicating with the Urethra

GROSS and MOORE (1950) discovered twelve examples of this anomaly in the literature. In this variation of the anomaly the terminal urethra is Y shaped and the patient voids through two orifices. Apart from the minor disadvantage of having to control two urethral streams these patients are asymptomatic and are usually discovered in the course of examination for some unrelated problem. Fig. 12 shows the urethrogram in a child with a Y shaped terminal urethra. Surgical treatment was not thought to be necessary. Fig. 13 shows an excretory voiding urethrogram of incomplete dorsal and ventral accessory urethrae.

4. Incomplete Ventral Accessory Urethra — Blind

This appears to be the rarest of all the urethral anomalies of number LOWSLEY (1939) finding only three examples of this anomaly.

5. Double Urethra in the Female

BOISSONNAT (1961) in a review of the literature concluded that there were two examples of complete double urethra with a single bladder in the female. They were one case of his own and that of DeNICOLA and McCARTNEY (1949). There is possibly a third authentic case (DANNREUTHER, 1923). Since this paper additional cases have been reported by BROWN (1956) and ANSELMO (1964).

XI. Posterior Urethral Valves

These are mucosal folds in the posterior urethra which are by definition obstructive. We are not here concerned with those ridges normally seen in the urethra and usually attached to the verumontanum, nor with what might be termed nonobstructive folds of the mucosa. It is important that the concept of obstructiveness be adhered to in the diagnosis of posterior urethral valves as otherwise many mild exaggerations of the normal mucosal redundancy could be so described.

PRESMAN (1961) has suggested that the diaphragm type of valve (Type III in YOUNG's classification) should be referred to as "congenital stenosis of the urethra". We prefer to continue to classify this type of obstruction as a posterior urethral valve.

1. Incidence

LANGENBECK is generally credited with the original description of posterior urethral valves in 1802 (LANGENBECK, 1802). The condition is mentioned again by VELPEAU in 1832 but the description is unclear (VELPEAU, 1832). TOLMATSCHEW published in 1870 a description and drawing of an undoubted case (TOLMATSCHEW, 1870). This is reproduced by YOUNG, FRONTZ and BALDWIN (1919). The early cases were all post mortem reports and it was not until 1912 that HUGH YOUNG recognised a case during life and treated it surgically. In 1913 YOUNG reported this first case and three others before the John Hopkins Medical Society. In 1919 YOUNG, FRONTZ and BALDWIN reported twelve cases, six of whom had been operated on. They also found twenty-four additional cases in a search of the literature.

17*

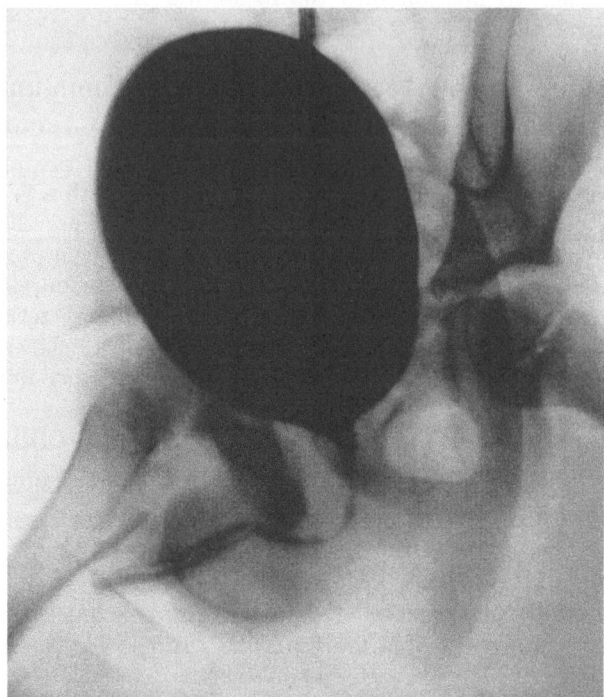

Fig. 12. Accessory urethra, incomplete, dorsal. Voiding cystourethrogram

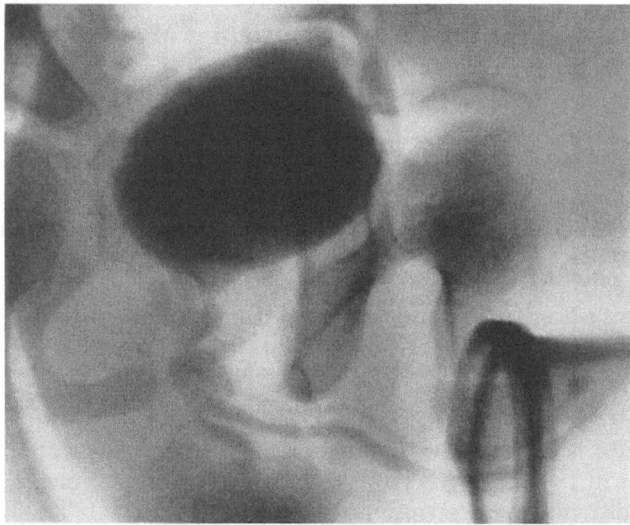

Fig. 13. Accessory urethra, incomplete, dorsal and ventral. Excretory voiding cystourethrogram

By 1934 LOWSLEY and KIRWIN could report 130 cases from the literature and 3 of their own and in 1937 in his "Pediatric Urology" CAMPBELL gave the details of his personal series of 63 cases. Since this time several large series have been presented. KJELLBERG, ERICSSON and RUDHE (1957) analysed the cystourethrograms of 1,461 children examined in their clinic at the Karolinska Institute,

Stockholm. They note that in their material several congential malformations are unquestionably over-represented because, as well as the material from the City and County of Stockholm, some of the cases were referred from clinics in various parts of Sweden. They found however, 52 valves in 1,461 examinations. STEPHENS (1963) in Australia has discussed 47 cases of posterior uretrhal valves seen by him, and NESBIT, THIRLBY and RAPER (1951) presented 22 cases from Michigan. Among other large series must be mentioned that of WILLIAMS (1965) from London, which now numbers 104, that of FORSYTHE and McFADDEN (1959) of Belfast, 35 cases, and BOISSONNAT (1953) of France, 12 cases.

In a relatively pure series we have seen 27 examples in the years 1958—1965 in examinations of 2,500 children. The annual rate is rather constant, being about 5 cases per year.

The total number of cases reported in the literature is now large — and mention must be made of those series of children in which the diagnosis has only rarely been made. Prominent amongst such series is an analysis by LATTIMER (1961) of 2,063 admissions during the 11 year period 1950—1961 to the Childrens Service of the Squier Urologic Clinic, Columbia-Presbyterian Hospital. The diagnosis was made in only 3 cases. Similarly in the large series of bladder neck obstructions reported by BURNS (1955) there were 78 instances of bladder neck obstruction but only three of posterior urethral valves. Even in later reports, when awareness of the relatively common occurrence of valves is current, large series of children have been reported without a single instance of this diagnosis (REISMAN, 1965). In these series voiding cystourethrography was not routinely employed in the study of lower urinary tracts of children. We believe that this leads to posterior urethral valves being overlooked in many instances.

2. Etiology and Pathology

YOUNG, FRONTZ and BALDWIN (1919) classified posterior urethral valves into three types. Type I in which the valve leaflets take origin from the verumontanum and pass inferiorly and laterally to become attached to the walls of the urethra. The stream of urine catches in these folds which balloon out and become obstructive when filled with urine. Fig. 14 shows the urinary tract of a stillborn male infant with Type I valves. Fig. 15 is a close-up of the posterior urethra showing the valves attached to the verumontanum. Type II in which the valve leaflets take origin from the verumontanum and pass superiorly and laterally towards the bladder neck, and Type III which are diaphragms, unattached to the verumontanum and which are situated either above or below the verumontanum. A diagram of these types, redrawn from YOUNG's original article, is shown in Fig. 16.

Most authors have found this classification useful with the exception of KJELLBERG, ERICSSON and RUDHE (1957) who felt that indefinite transitions between the forms occurred and that the classification was of little practical value as neither the degree of obstruction nor the treatment is influenced by the type. They did however, point out that no valve of Type II was demonstrated in their series. Doubts have been previously expressed about the occurrence of Type II valves by WILLIAMS (1958), who has not encountered an example of Type II valves clinically nor has he found a specimen of Type II valves in the Museum of the Hospital for Sick Children, Great Ormond Street, London. STEPHENS (1963) also says that Type II valves do not exist and we have not seen such a case. It would appear that the earlier investigators mistook the multiple mucosal folds which are commonly seen radiating upward from the verumontanum toward the bladder in patients with a dilated urethra due to distal obstruction, for the cause

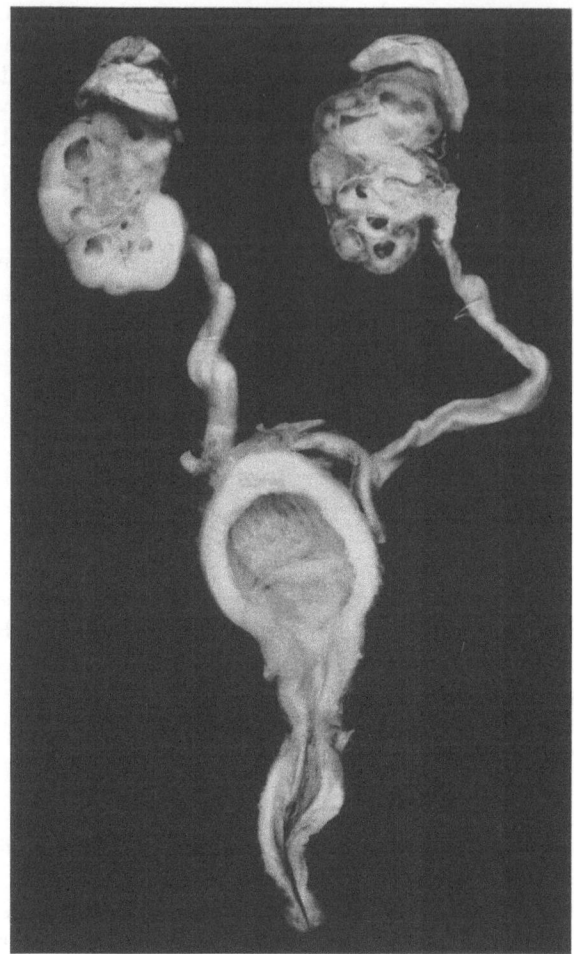

Fig. 14. Valves of the urethra, Young's type I. Kidneys, ureters, bladder and urethra of a stillborn male. The posterior urethra is dilated, the bladder is thickened and the ureters dilated and tortuous. There is a bilateral hydronephrosis

of the obstruction. In Fig. 13 four such folds can be clearly seen. Fig. 17 is reproduced from FRONTZ (1932) and shows examples of Type II valves in association with obstruction due to an anterior urethral diverticulum. Most authors now agree that this type of fold is not obstructive. For practical purposes then valves are either of YOUNG's Type I or YOUNG's Type III. In all series Type I are much commoner than Type III. STEPHENS (1963) has 37 examples of Type I valves and 10 of Type III. We have twenty-four of Type I and three of Type III.

The congenital origin of these anomalies is not disputed but there is considerable controversy as to the mode of their production.

BAZY (1903) put forward the suggestion that valves represented a persistence of the urogenital membrane. His theory was based on the fact that this structure, in its later development, occupies the site corresponding to the common location of valves.

TOLMATSCHEW (1870) explained the occurrence of valves by stating that they were simply enlargements of the folds commonly occurring in the normal urethra

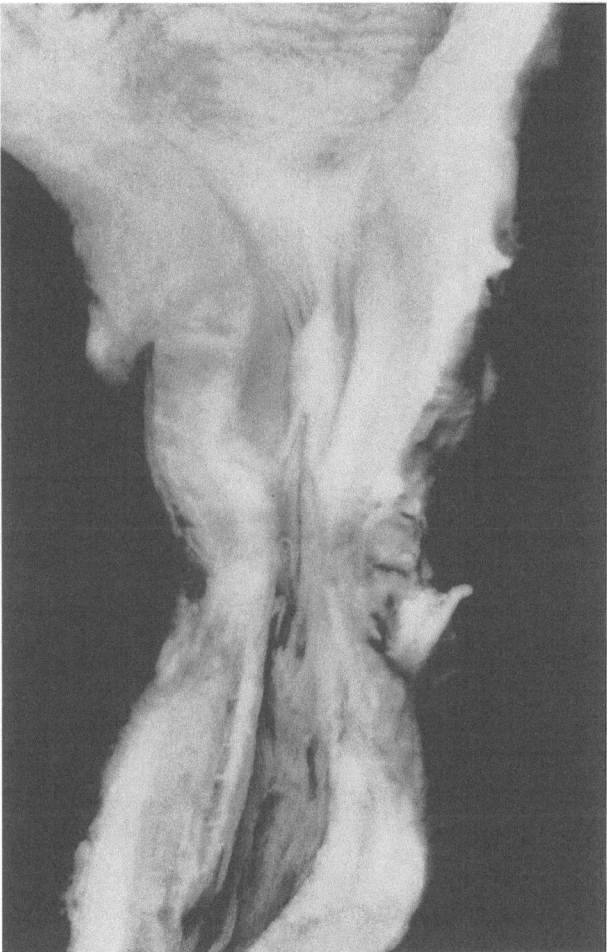

Fig. 15. Valves of the urethra, Young's type I. Close up of posterior urethra in same specimen as Fig. 14. Note the absent inferior urethral crest and the presence of additional but non-obstructive folds running upwards from the verumontanum towards the bladder neck

and stating that if these became hypertrophied, obstruction would result. WILLIAMS (1958) has recently supported this view.

WATSON (1918), who worked on the embryological development of the verumontanum found in examining cross sections of the fetal urethra a specimen with three fibrous bands. These bands appeared to represent attachment of the tip of the colliculus to the urethra and he concluded that congential valves were the result of the fusion of the colliculus, at an early stage of its development, with the epithelium on the roof of the posterior urethra.

LOWSLEY (1914) as a result of histological studies believed that valves represented anomalous development of the Wolffian or Mullerian system.

STEPHENS (1963) in a post mortem study of nineteen infants with posterior urethral valves pointed out that in all patients the verumontanum, the inferior urethral crest and the fins were larger and tougher than normal. He also showed that in sixteen of the nineteen cases the inferior urethral crest was either absent or very short. A control examination of thirty stillborn males with normal urinary

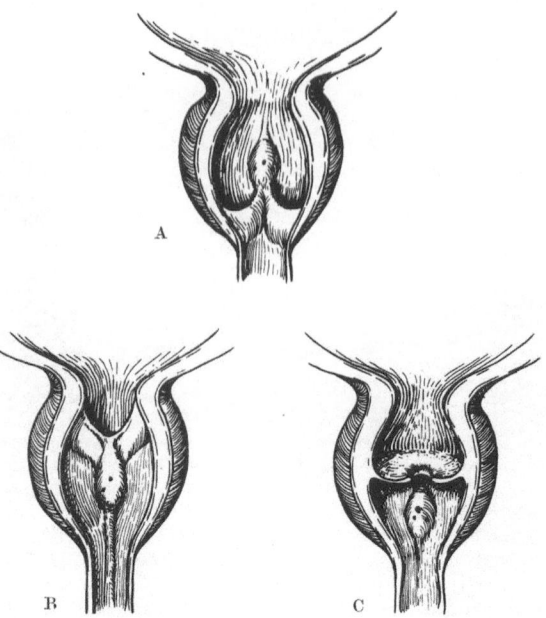

Fig. 16. Drawing of Young's classification of posterior urethral valves, A Type I, B Type II, C Type III

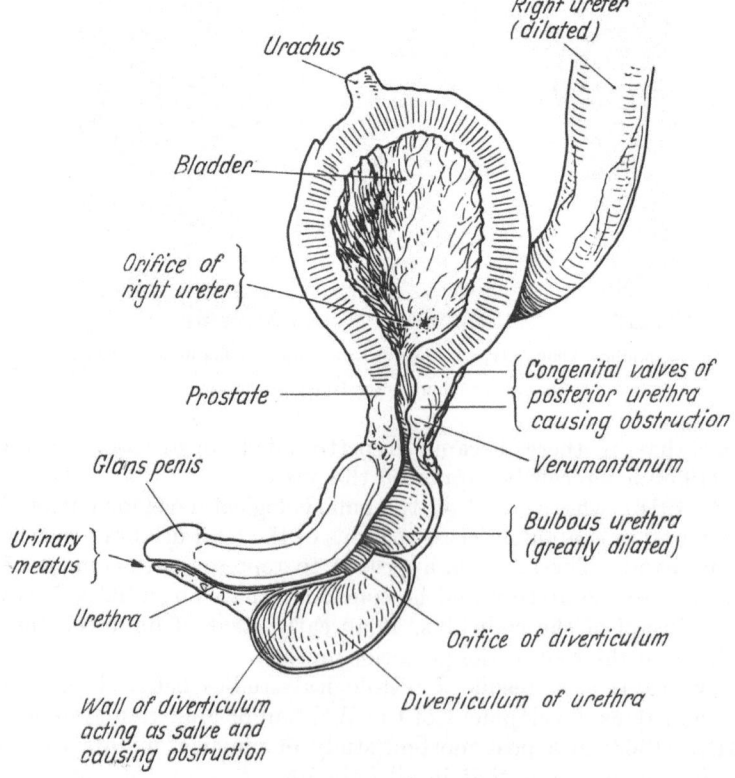

Fig. 17. Drawing of specimen demonstrating an anterior urethral diverticulum and Type II posterior urethra valves. Reproduced from FRONTZ, J., Urol. 27, 489 (1932). These Type II valves are commonly associated with distal obstruction and are probably non-obstructive

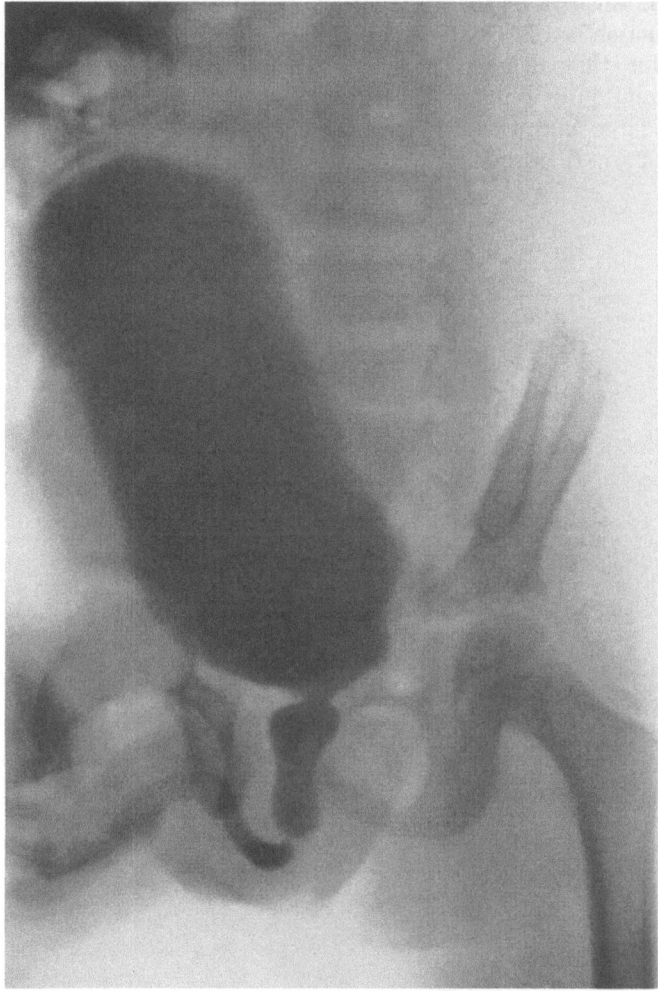

Fig. 18. Posterior urethral valves. Voiding cystourethrogram. Patient has responded to obstruction by the development of a large capacity bladder

tracts showed the crest to be absent in not a single case and short in only two cases. In the pathological cases the fins were thick, unbroken, mucosal structures whereas in the normals the fins were brief, delicate, tapering ridges.

He further reports the findings in a stillborn infant with congenitally absent Wolffian and Mullerian systems. The urethra was of normal caliber but the verumontanum more elongated than usual. The inferior urethral crest was absent and there were no finlike folds radiating from any part of the midline below the verumontanum. STEPHENS concludes that "Valves of the urethra appear to be related to deficient integration of the Wolffian ducts into the walls of the urethra, to abnormal locations of their original orifices into the cloaca and to an abnormal course of their distal ends".

It would appear that STEPHEN's theory would explain the occurrence of YOUNG's Type I valves and that BAZY's theory would explain the occurrence of YOUNG's Type III valves.

Obstruction in the urethra causes changes in the kidneys, ureters and bladder, and in the urethra proximal to the obstruction. The degree of change is directly related to the severity of the obstruction. In all instances the urethra proximal to the obstruction is dilated and the bladder shows signs of obstruction. We have noted however, that although in most instances the bladder becomes thickened with the formation of cellules and diverticula, in some patients the bladder becomes very large and there is much less evidence of muscular hypertrophy (Fig. 18). YOUNG (1919) mentioned that in no case in his series nor in any cases

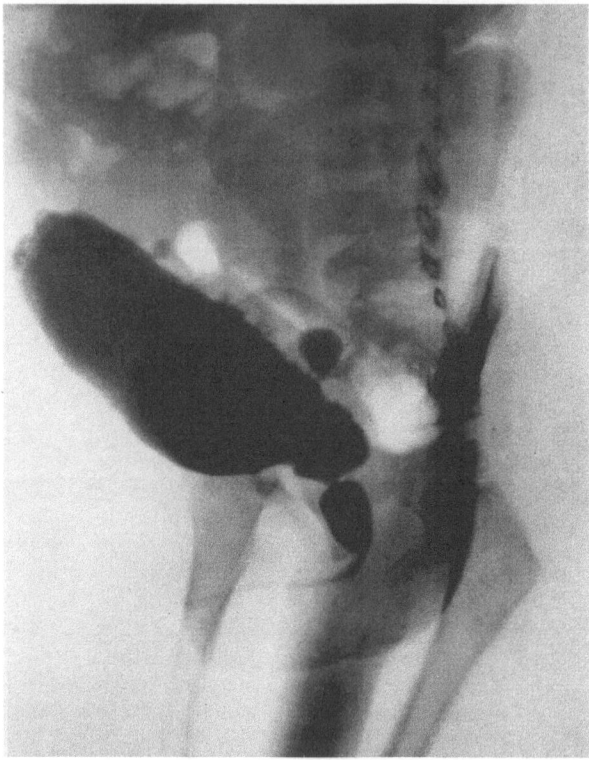

Fig. 19. Posterior urethral valve. Voiding cystourethrogram. Patient has a number of diverticula of the bladder due to the obstruction

coming to his notice were there diverticula in the bladder. We have seen diverticula associated with this disease (Fig. 19) on many occasions as have other authors (KJELLBERG, ERICSSON and RUDHE, 1957).

Changes at the level of the internal sphincter are of considerable importance and must be commented on. Some authors, WOLGIN, ROSENBERG and MUSCHAT (1952) have reported on the concommitant occurrence of valvular obstruction and bladder neck stenosis, others, MITCHELL (1963) believe that valves may be caused by bladder neck stenosis. We believe that neither of these views is correct, but that the changes at the internal sphincter are secondary to the obstruction caused by the valve. The internal sphincter is part of the detrusor muscle and hypertrophies along with the detrusor muscle as a response to obstruction (Fig. 20). MITCHELL (1963) has stated that obstruction due to causes other than valves is not associated with bladder neck hypertrophy. This has not been our experience. Fig. 4 show bladder neck hypertrophy associated with obstruction due to stricture

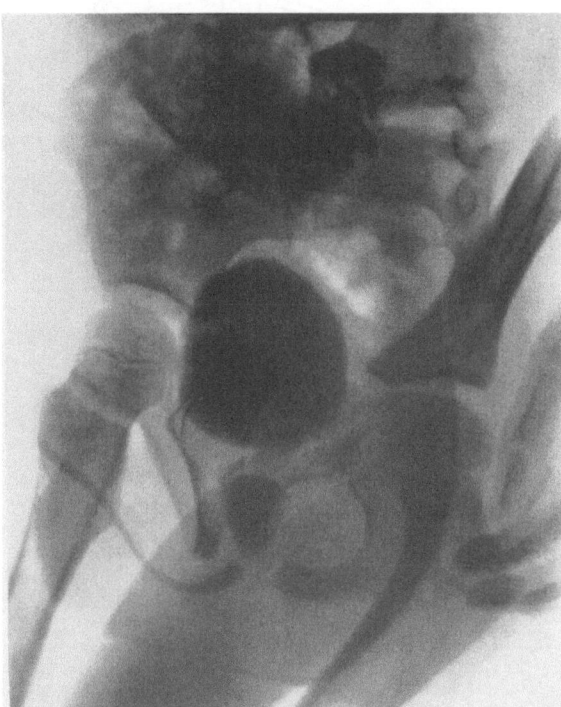

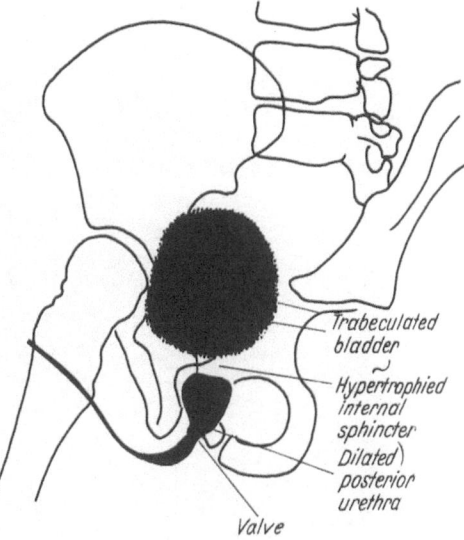

Fig. 20. Posterior urethral valve. Voiding cystourethrogram. The patient aged $4^1/_2$ years at the
time of this study had had a V-Y plasty performed
on the neck of the bladder at age $2^1/_2$ years

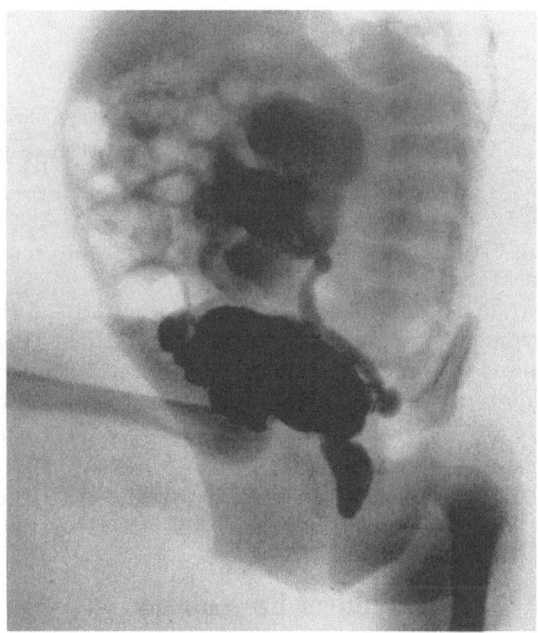

Fig. 21. Posterior urethral valves. Newborn male. Voiding cystourethrogram. There is vesico-ureteral reflux
and gross hydronephrosis

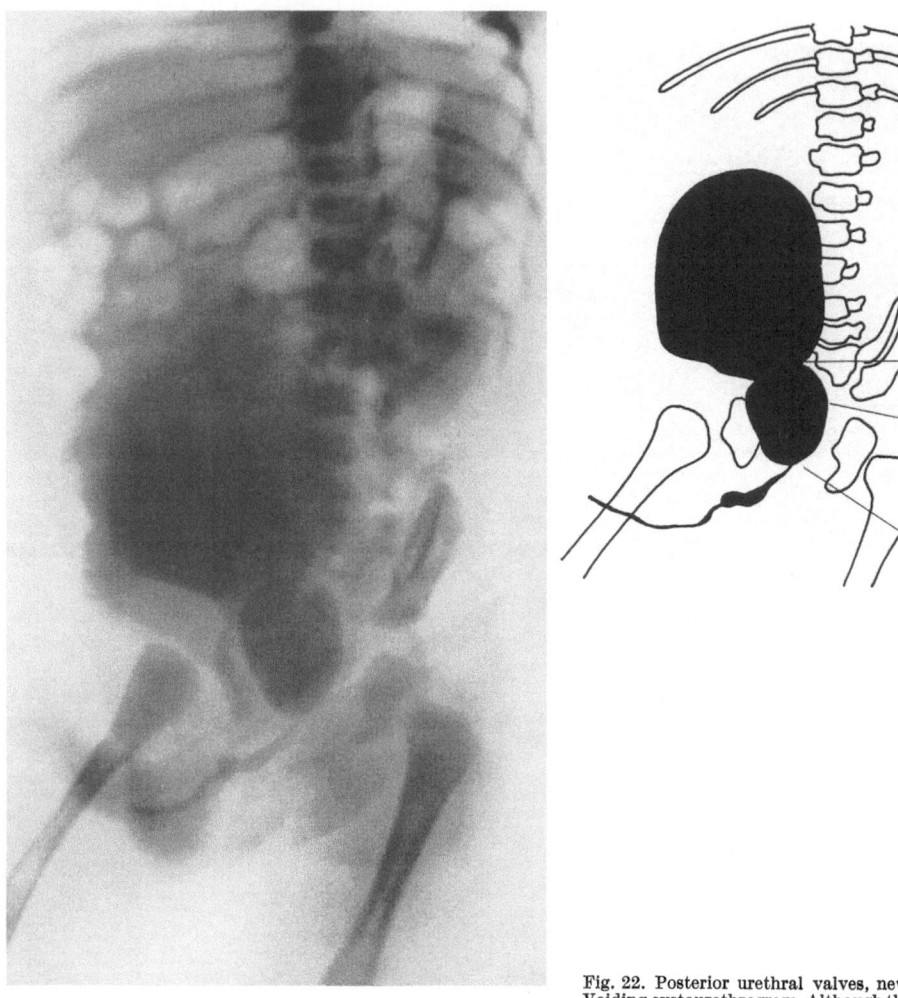

Fig. 22. Posterior urethral valves, newborn male. Voiding cystourethrogram. Although there is severe obstruction and the urogram showed severe hydronephrosis there is no evidence of reflux

in the urethra and Fig. 1 shows bladder neck hypertrophy associated with obstruction due to meatal stenosis.

Vesicoureteral reflux occurs in approximately one third of the patients and is commoner in the younger more severely obstructed patient (Fig. 21). The reflux is due to incompentence of the ureterovesical valves secondary to the obstruction. Although most children diagnosed in the newborn period reflux we have one patient with severe obstruction diagnosed at aged ten days who does not do so (Fig. 22).

The changes in the ureters and the kidneys do not differ from those resulting from any lower urinary tract obstruction of long standing and will not be discussed further at this time.

3. Clinical Presentation

Obstruction due to valves in the posterior urethra is very variable in degree. This wide variation causes children born with this anomaly to present clinically

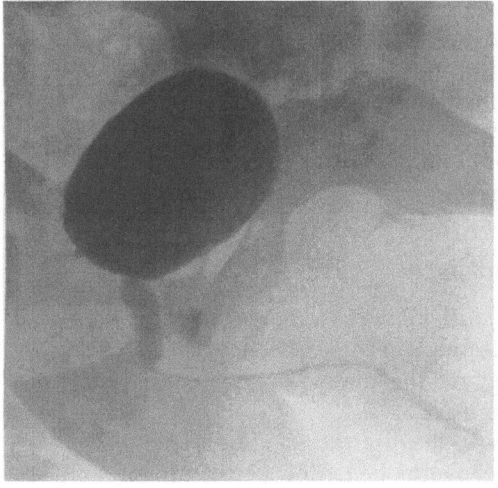

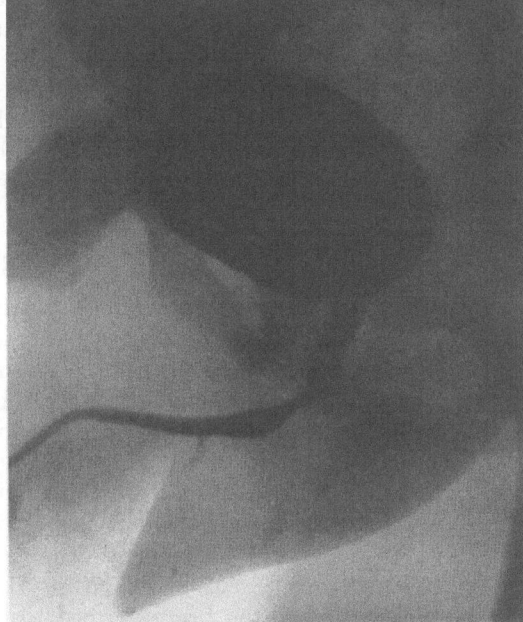

Fig. 23a and b. Posterior urethral valve. 13 year old boy presenting with enuresis. a Voiding cysto-urethrogram. Preoperative. b Voiding cystoure-throgram. Following transurethral removal of valve. The external urethrotomy through which the operation was performed is still not comletely healed

Table 3. *Presenting complaint in 27 cases of posterior urethral valves*

Obstructive complaints	
(poor stream, dysuria, frequency)	7
Fever	8
Palpable bladder	3
Failure to thrive	4
Enuresis	3
Examined because of other congenital anomalies	2
	27

at any age between the newborn period and adolescence. The earliest diagnosed case in that of FUCHS (1900) who described the anomaly in a five month old fetus. There are reports in the literature of a number of cases in adults, even one first diagnosed by IVERSEN (1914) in an eighty-five year old man. Careful reading of most of the adult reports suggest however, that they are largely spurious and that in many instances there has been considerable dilatation of the posterior urethra secondary to anterior urethral strictures. Such patients often develop mucosal redundancy in the posterior urethra which appears to be valvular and may be misinterpreted as a posterior urethal valve. In our series the youngest patient seen was aged ten days (Fig. 22) and the oldest was aged thirteen years (Fig. 23).

The presenting sign in the infant is usually a mass palpated during routine examination in the newborn nursery. Even with very severe obstruction and serious destruction of the kidneys these children look quite healthy as the placenta has been acting as an artificial kidney for them. If the diagnosis is not made at this time these children will be admitted later with failure to thrive, diarrhea or vomiting. By this time they will look much worse than when they were born and will have a raised blood urea nitrogen. Children who are less severely obstructed present a little later either because of recurrent urinary tract infection or hematuria. Those children with the least degree of obstruction present usually at aged 5—6 as enuretics (HAMM and WATERHOUSE, 1961).

On physical examination in all age groups a significant finding has been the presence of a palpable bladder. NESBIT (1951) pointed out that in their twenty-two cases the bladder was palpable in all but one patient. This has also been very common in our series. Table 3 gives the presenting clinical features in our cases.

4. Diagnosis

There are no physical findings which will allow the diagnosis of posterior urethral valves to be made with certainty although in some instances the palpation of a smoothly enlarged prostate may suggest the diagnosis. For confirmation radiologic and endoscopic methods must be used.

In the past reliance upon panendoscopy to localise the site of obstruction has often led to posterior urethral valves being overlooked. The valves are situated low in the urethra in an area difficult to visualize even in the adult and especially so in the child and they are flattened against the side of the urethra by the flow of water down the instrument. The secondary hypertrophy of the internal sphincter already discussed is, on the contrary, easily seen and the obstructive complaints have often been ascribed to its presence. Fig. 11 shows the voiding urethrogram on a child who had already had a V-Y plasty performed on him at aged two and a half before presenting to our clinic aged four and one-half. As can be seen the hypertrophy of the detrusor had continued and the internal sphincter became so muscular that it is difficult to seen the communication between the bladder and the posterior urethra.

Simple cystography and injection urethrography is also unsatisfactory for demonstrating valves but they are clearly seen on a voiding cystourethrogram (HAMM and WATERHOUSE, 1961; WATERHOUSE, 1961). We feel that it is important that a voiding cystourethrogram be performed on all children prior to cystoscopy (WATERHOUSE, 1963), so that the attention of the surgeon may be directed to those areas which require particularly careful evaluation.

The characteristic roentgen findings of posterior urethral valves as seen on the voiding cystourethrogram are dilatation of the urethra proximal to the valve and poor filling of the urethra distally. The internal meatus is, in most cases prominent, but in some patients who are very severely obstructed the dilatation of the posterior urethra is so great that the posterior urethra and bladder become one indistinguishable cavity. Reflux of contrast material into the ureters is common in young severely obstructed patients but is unusual in the older age group. The bladder itself shows signs of damage, usually hypertrophy with the formation of cellules and diverticula, but occasionally a relatively large smooth walled bladder may develop.

Although the commonest cause of dilatation of the posterior urethra is valvular obstruction, other obstructions either organic or functional may cause somewhat similar findings and care must be taken to distinguish between anterior urethral valves, anterior urethral strictures, absent abdominal muscles and cases of neurogenic bladder. Diseases in the anterior urethra may be separated from posterior urethral valves on the basis of the dilatation of the urethra ending much further down the urethra, the bulb being well filled with contrast material. This is rare finding in valves. Fig. 7 is the voiding cystourethrogram of a male child with an anterior urethral stricture. As can be seen the dilatation extends through the urogenital diaphragm with distal point of the dilatation at the site of the obstruction (WATERHOUSE, 1964).

In neurogenic bladder there is usually relaxation of the internal sphincter so that the form of the dilated posterior urethra differs from that seen in valves. Changes in the mechanism of voiding cause a rather acute angulation between the anterior urethra and the membranous urethra. This has been seen in many of our cases (Fig. 24). This angulation does not occur in patients with valvular obstruction. The urethral anomaly associated with absent abdominal muscles may at times be similar to that occurring in valvular obstruction but the bladder is always, in our experience, of very large capacity and smooth walled (Fig. 25).

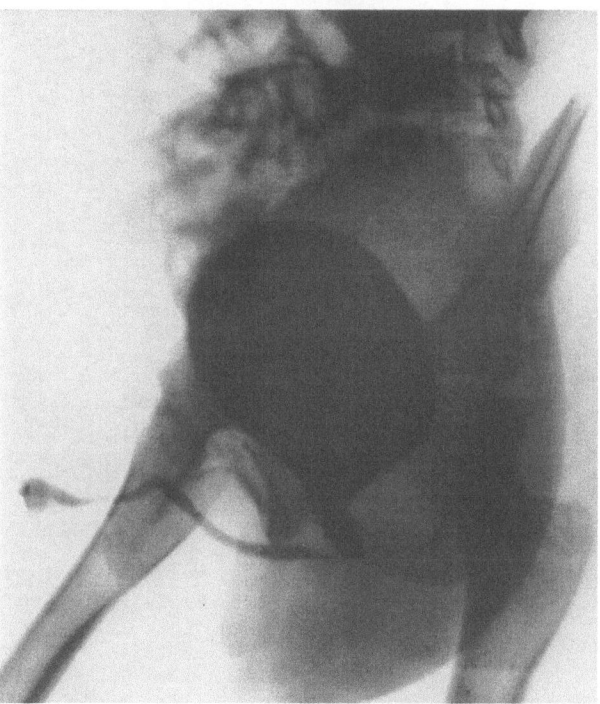

Fig. 24. Neurogenic bladder (Pott's disease). Voiding cystourethrogram. The internal sphincter is wide open. There is an acute angle at the junction of the anterior urethra and membranous urethra. This is commonly seen in patients with neurogenic bladders

Free reflux up both ureters is common and narrowing at the level of the internal sphincter is not seen. These features will usually distinguish the two anomalies.

Once the diagnosis of posterior urethral valves has been confirmed by voiding cystourethrography it is our practice to perform the endoscopic study with an infant resectoscope so that the valves may be destroyed at the same sitting.

5. Treatment

As WILLIAMS and ECKSTEIN (1965) have so aptly pointed out diagnosis and treatment cannot be clearly distinguished in practice as they can on paper. In many instances these children are so ill that although the diagnosis of posterior urethral valves is suspected the uremia must be treated prior to attempts at diagnosis.

In attempting to prepare seriously ill infants for operation it has been customary to divert the urine in some way. Urethral catheter drainage is both unsatisfactory and dangerous. Catheters of adequate size to drain the bladder cannot be
introduced per urethram in the newborn and even small catheters often do serious
damage of the urethra. Suprapubic cystostomy drainage has also proven unsatisfactory as the hypertrophied bladder contracts down on the tube and following this the drainage from the dilated upper tracts is extremely poor. We have
seen two instances of complete anuria following suprapubic cystostomy. Drainage
by pyelostomy and nephrostomy whilst easy to institute is difficult to maintain

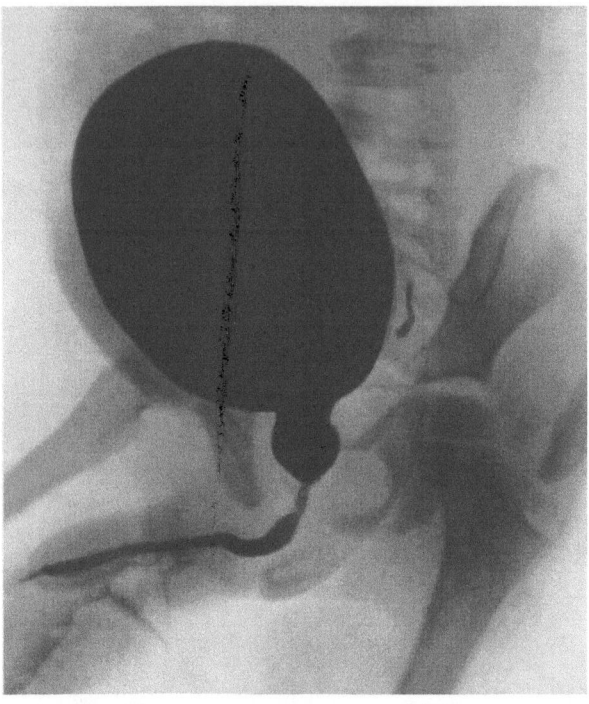

Fig. 25. Absent abdominal muscles syndrome. Voiding cystourethrogram. The posterior urethra is dilated and
has many of the characteristics associated with the changes due to posterior urethral valves. Note however
that the bladder is smooth walled and large and there is reflux into the left ureter

and also leads inevitably to infection in the already damaged kidneys. It is
probably the most satisfactory of the diversionary procedures but leaves much
to be desired. We have no experience with the loop ureterostomy described by
WILLIAMS and ECKSTEIN (1965).

Because of the unsatisfactory nature of urinary diversion in these infants and
the hazards of infection subsequent to indwelling catheters we have recently
treated two cases by peritoneal dialysis prior to surgery. This allows the general
condition of the infant to be rapidly improved and operation be performed. It is
as yet too early to say whether this can be adopted as a general method of
management but in the two cases in which it has been tried it was satisfactory.
As the technique of peritoneal dialysis in the newborn has not been described
previously a brief note on the method will be presented.

The abdomen is puntured with a trocar (McDONALD, 1963) and a number
11 Fr. nylon catheter is introcuced. As the pelvis is undeveloped in the infant we

have made the puncture above the umbilicus and to the left of the midline to avoid the liver (Fig. 26). The infant is then placed in a crib and the both are placed on a sensitive balance (Fig. 27). Dialysis is begun by running in between 150 and 200 ccs of dialysis fluid. The weight changes in the cycles are measured and recorded and in this way the dangers of overhydration avoided. After allowing thirty minutes for equilibration to occur the direction of flow is reversed and the fluid emptied from the peritoneal cavity (MAXWELL, 1959).

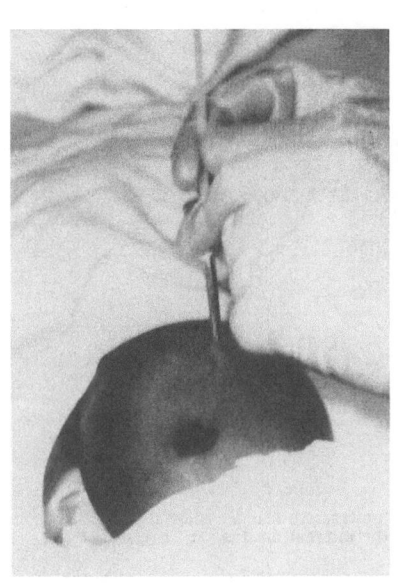 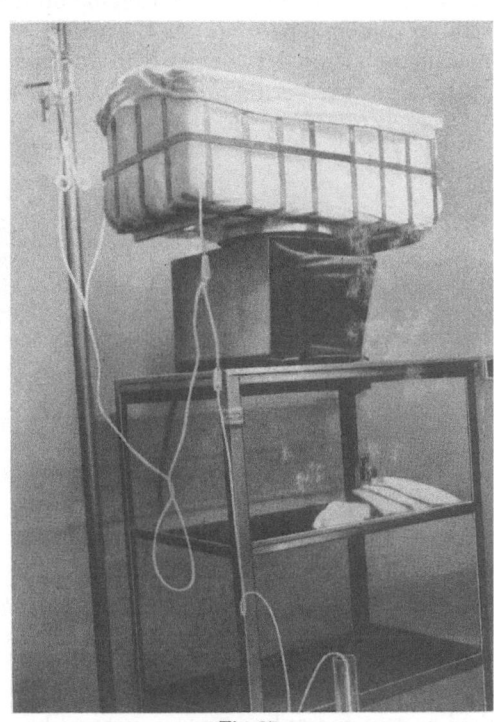

Fig. 26   Fig. 27

Fig. 26. Peritoneal dialysis in the newborn. Puncture of the abdomen with the McDonald trocar and insertion of the dialysis catheter

Fig. 27. Peritoneal dialysis in the newborn. A modification of the technique of MAXWELL is used. The baby and crib are supported on an accurate balance to avoid the dangers of overhydration

We have found that twelve one hour cycles; fifteen minutes for the fluid to run in, 30 minutes for equilibration and 15 minutes for the fluid to run out, have returned the serum electrolytes to near normal.

Operative removal of the valves has been performed in four ways

(i) Open removal by the suprapubic route, either with or without splitting the pubic symphysis (GROSS, 1953).

(ii) Retrograde passage of urethral sounds through a suprapubic cystostomy (YOUNG, 1919).

(iii) Perineal exposure and excision of the valves (YOUNG, 1919; HIGGINS, WILLIAMS and NASH, 1951).

(iv) Transurethral excision of the valves (CAMPBELL, 1963; KJELLBERG, ERICSSON and RUDHE, 1957; FORSYTHE and McFADDEN, 1959).

We favour, as do all recent authors, the endoscopic removal of the valves. We use a 12 Fr. McCarthy Infant Resectoscope introduced either directly through the urethra in the older child or through a perineal urethrostomy. This approach has

been used in all but one of our cases. It has been suggested by some authors that
the valves will be more easily seen if the water flow through the instrument is
turned off. We have not found this to be true but have found that if the instrument
is turned a little so that the side wall of the urethra and the verumontanum are
in view and that if the instrument is then withdrawn so that the beak of the
instrument is only just holding open the external sphincter the valves will be
seen as semi lunar folds running laterally from the lower end of the verumontanum.
They may be trapped with the loop of the resectoscope which has not yet been
activated and then by gently tugging on them their valvular nature can be demonstrated. We prefer to cut off the valve at its attachment ot the verumontanum
making no attempt to excise it in its entireity. We have observed that in children
with valves the verumontanum does not pop up into and occlude the endoscopic

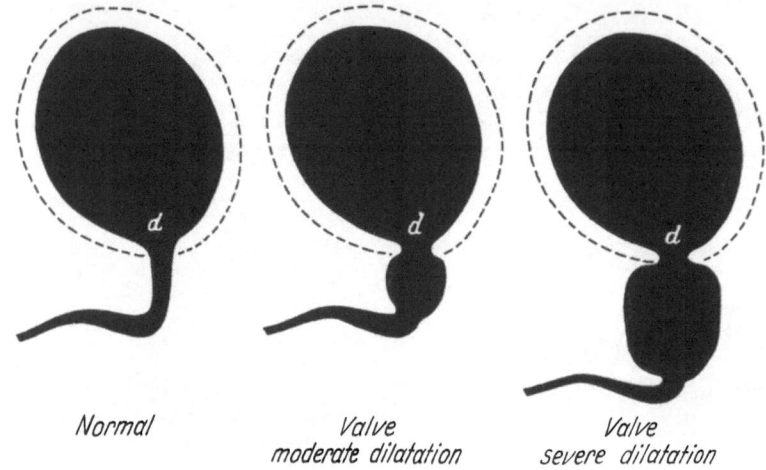

Fig. 28. Schematic drawing showing apparent development of an "obstructive bladder neck" by the dilatation
of the posterior urethra. In this drawing the diameter of the internal meatus "d" is constant

field as so commonly occurs in normal boys. When however the valves have been
cut off the verumontanum this phenomenon of "popping-up" comes back. We
have found in the past few years this to be a valuable sign that the obstruction
has been relieved.

Following the valvulotomy no catheters are put in as we wish to limit as much
as possible the introduction of infection. In most patients voiding occurs spontaneously but occasionally catheter drainage is needed because of obvious retention.
This is used for a short a period as possible.

With regard to the internal sphincter we originally believed this to be hypertrophied and obstructive and removed it either endoscopically or suprapubically.
However we now believe this to be unnecessary as much of the appearance both
on the X-ray and at cystoscopy is due to a hollowing out below the base of the
bladder and it to some extent artifactual. Fig. 28 shows a series of drawings in
which the diameter of the internal meatus is kept constant but a hollowed out
posterior urethra has been added below the bladder. As can be seen the appearance
is that of a bladder neck constriction. In 1951 HIGGINS, WILLIAMS and NASH
commenting on the open surgery of bladder neck disease wrote that if on ex-
ploring the bladder the internal meatus was rigid and difficult to dilate the
diagnosis was primary bladder neck contracture. Whereas if the bladder neck was
soft and dilatable the obstruction was more distally placed and the bladder neck

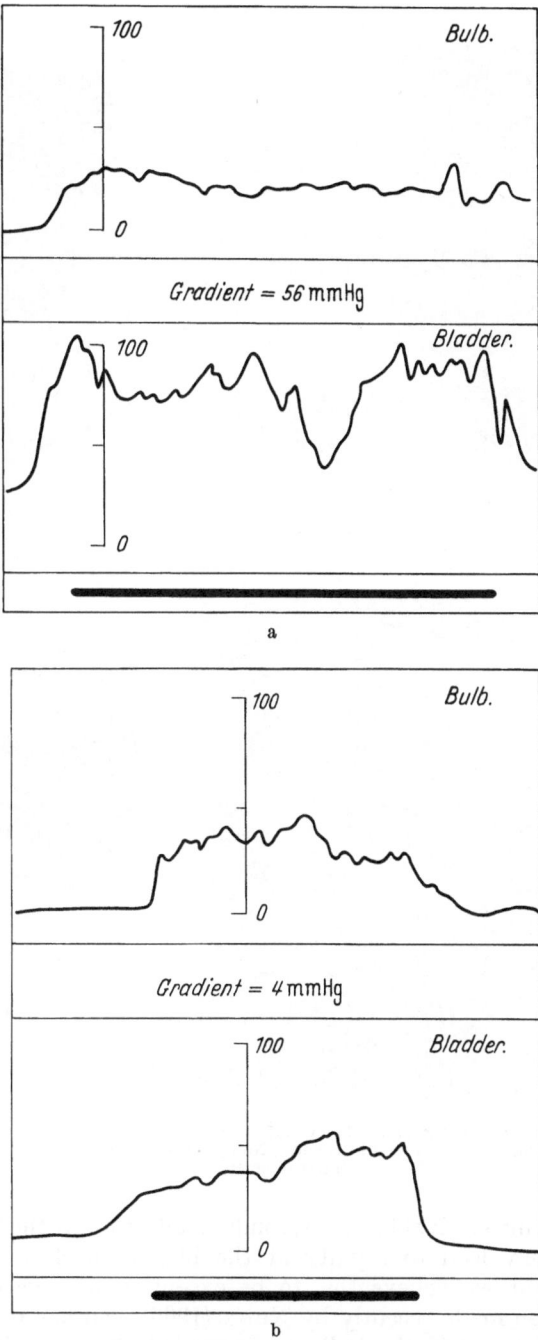

Fig. 29a and b. Urethral obstruction from congenital valves. a P.C. aged 11 years. Preoperative curves. Difefrence between bladder and urethral voiding pressures is marked, gradient being 56 mm. mercury. b P.C aged 14 years. Postoperative curves in same patient. Clinically, urethral obstruction has been cured and voiding pressures are now normal. Gradient is now well within normal limits.
[From Nunn, J. Urol. **93**, 693 (1965)]

changes were secondary. STEPHENS (1963) is also of the opinion that it is not necessary to remove the bladder neck in children with valvular obstruction. We have not removed the bladder neck in a child with valves in the last four years.

18*

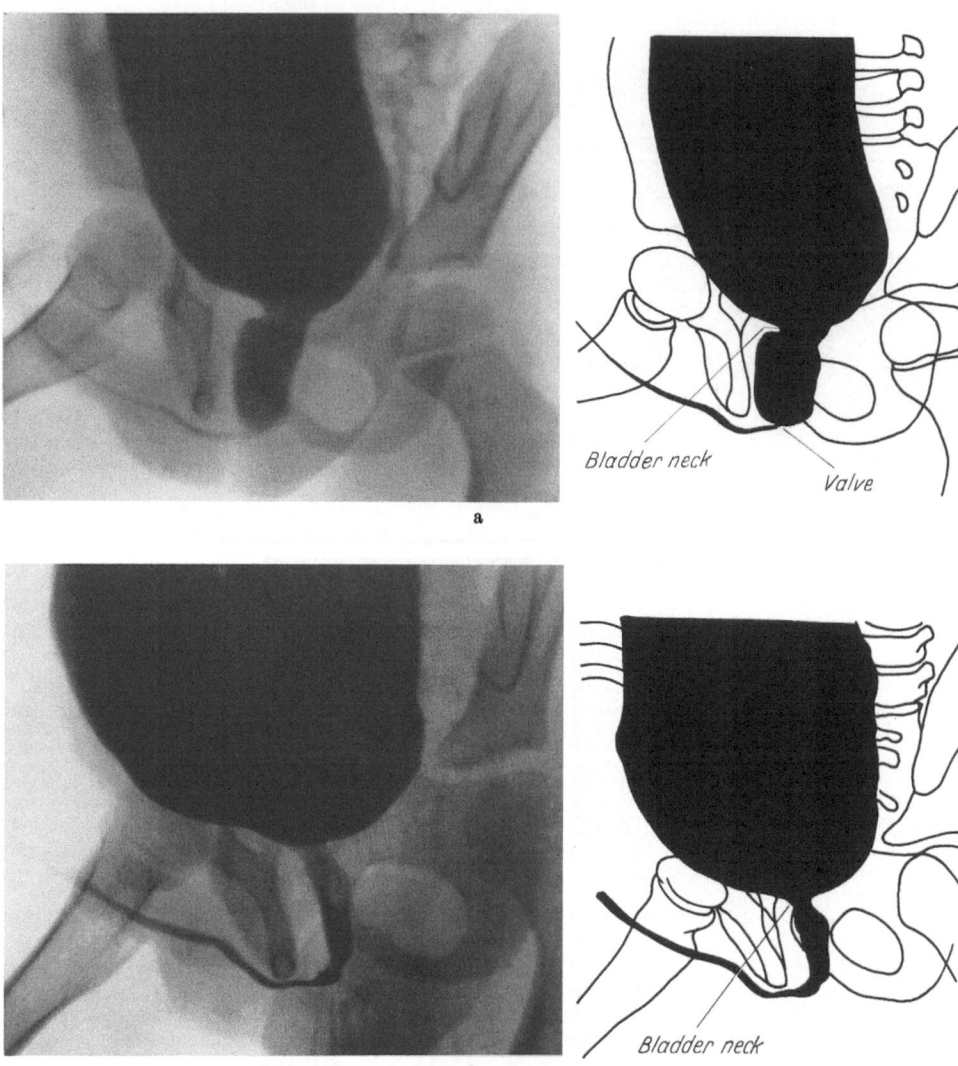

Fig. 30a and b. a Voiding cystourethrogram. Preoperative. b Voiding cystourethrogram. Postoperative. Note that the only procedure was excision of the valve. No operation was performed at the level of the internal meatus

It is of course not inconceivable that secondary infection in the bladder and subsequent fibrosis may lead to rigidity at the bladder neck which will require surgical therapy but we believe this to be exceptional. A very significant investigation has been made recently by NUNN (1965) who has recorded the pressures in the bladder and in the bulbous urethra of children with urinary tract obstruction. Fig. 29a and Fig. 29b are reproduced from NUNN's paper and show the pressures found in a child with a posterior valve. Preoperatively the bladder voiding pressure was extremely high, about 100 mms/Hg and there was a 56 mms/Hg gradient between the pressure in the bladder and that in the bulb. After removal of the valve, with no surgery at the internal spincter, the bladder voiding pressure fell to less than 50 mms/Hg and there was a gradient of only 4 mms/Hg between the bladder and the bulb.

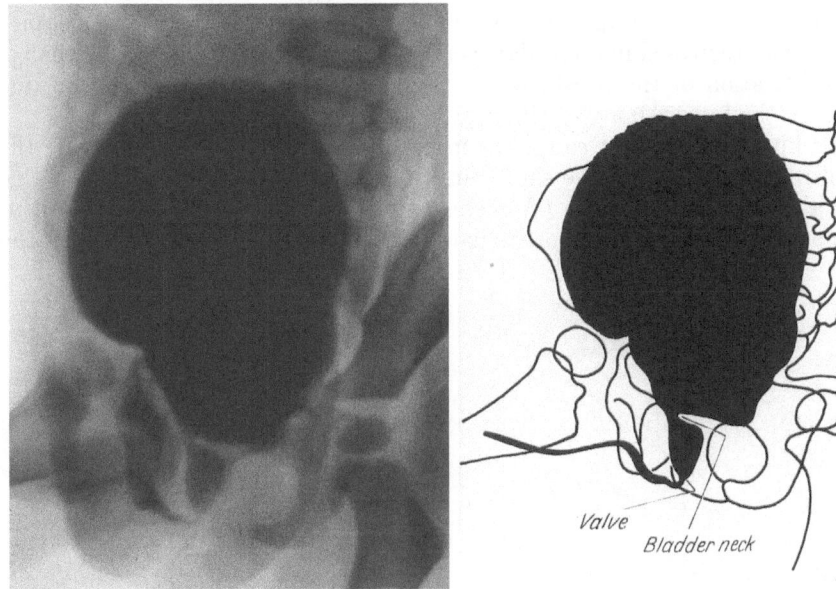

a

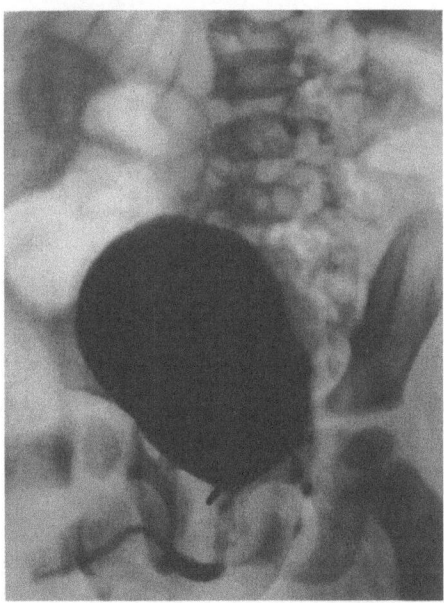
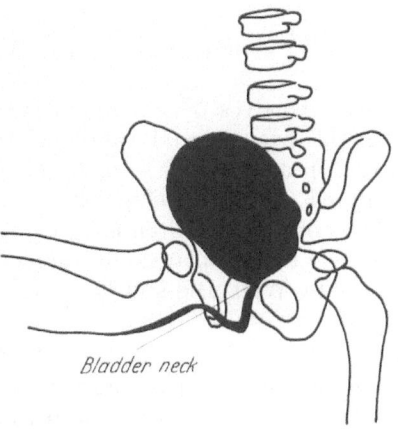

b

Fig. 31a and b. Posterior urethral valve. a Voiding cystourethrogram. Preoperative. b Voiding cystourethrogram. Postoperative. Note that the only procedure was excision of the valve. No operation was performed at the level of the internal meatus

Patients treated simply by removal of the valvular obstruction are shown in Figs. 30 and 31. In these patients the marked changes seen at the level of the internal sphincter have disappeared after removal of the valve. In one instance the changes had regressed in a ten day period suggesting that much of the change is due to dilatation of the prostatic urethra and that when this dilatation disappears then the hypertrophy of the bladder neck is no longer as evident.

The striking changes that can occur in the upper urinary tracts following the successful removal of valves are shown in Fig. 32. We have seen changes of this order on a number of occasions (WATERHOUSE, 1963) and have also noted considerable straigthening in previously dilated and tortuous ureters. Following the

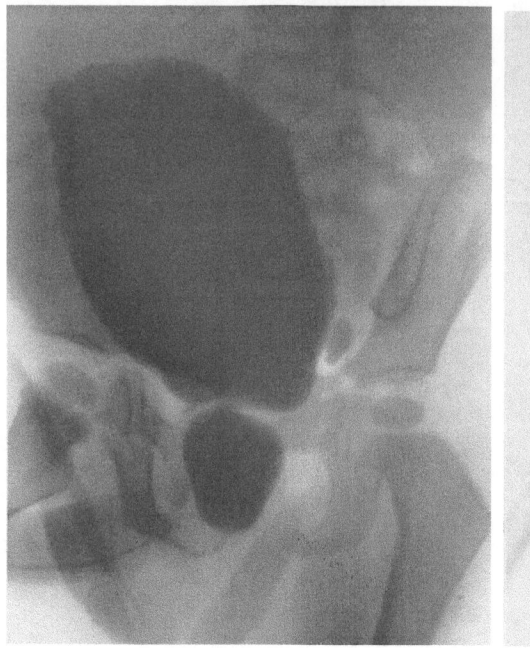 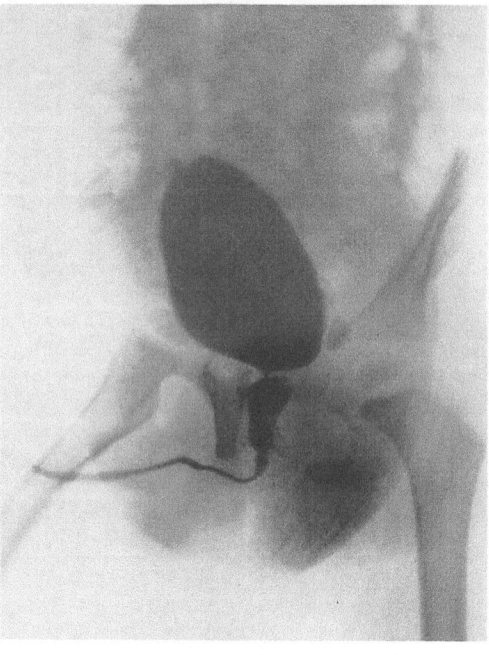

a   b

Fig. 32 a and b. Posterior urethral valves. a Voiding cystourethrogram. Preoperative. b Voiding cystourethrogram. Postoperative. c Urogram. Preoperative. d) Urogram. Postoperative

removal of valves it is common for vesicoureteral reflux to disappear. Cases in which this has occurred are illustrated in Fig. 33 and Fig. 34. If the reflux does not disappear it would seem that an operation of the antireflux type should be attempted (PAQUIN, 1959). I have been hesitant however to attempt this as the bladders of these children are so thick and fibrotic and would appear to be very unsuitable for reimplantation surgery. A similar view to mine has been expressed recently by WILLIAMS and ECKSTEIN (1965).

6. Prognosis

The prognosis in patients with posterior urethral valves is dependent on the severity of the damage to the upper urinary tracts. As has been mentioned already severely obstructed children present early in life and in those presenting before the age of one, of which we have twelve, six are already dead. The remainder are alive with varying degrees of abnormality in the upper urinary tracts. Of those presenting over the age of one all fifteen patients are alive and eight have upper

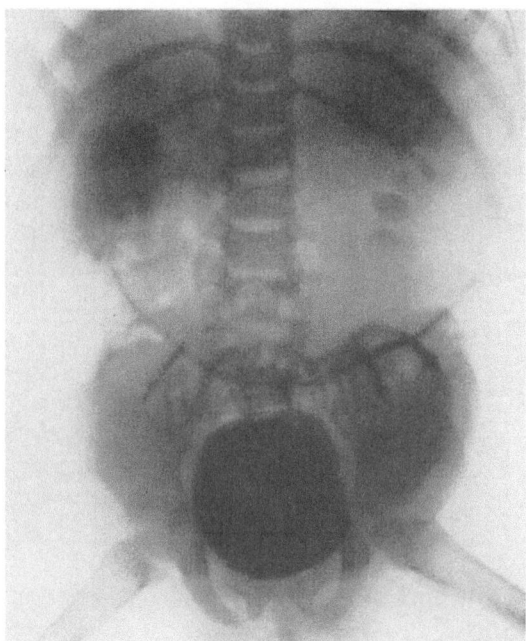

Fig. 32 c

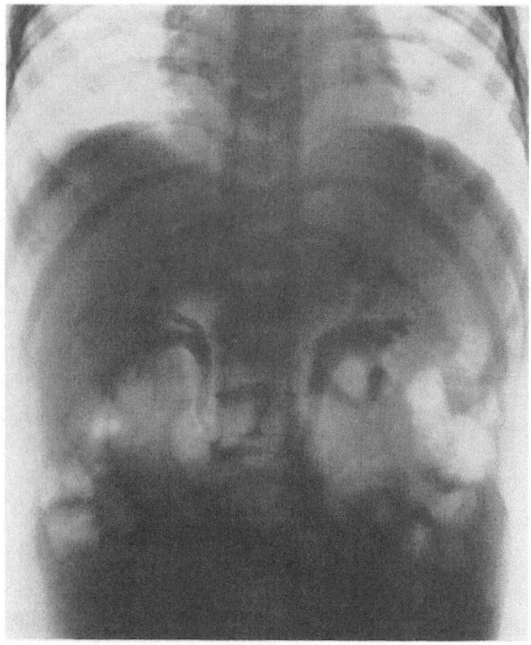

Fig. 32 d

tracts showing little or no signs of hydronephrosis. It would appear that the prognosis in these eight out of a total of twenty-seven patients is good. The re- maining seven must have a guarded prognosis as we have seen recently two

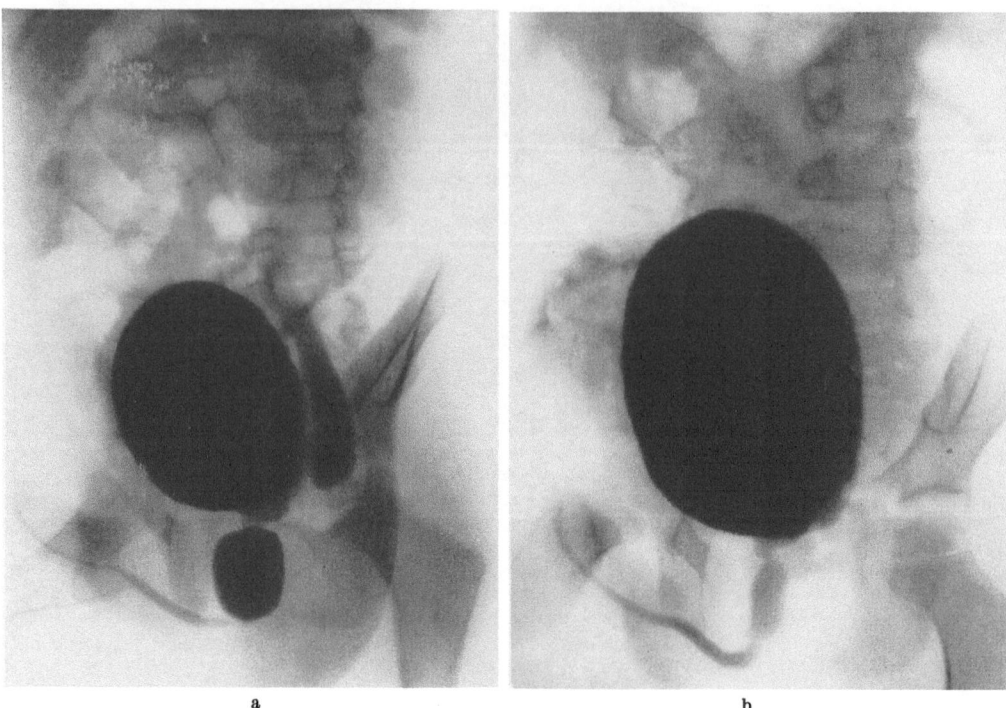

a   b

Fig. 33a. Posterior urethral valves. Voiding cystourethrogram. Preoperative. There is free reflux up a dilated ureter

Fig. 33b. Voiding cystourethrogram. Postoperative. The reflux present preoperatively is no longer evident

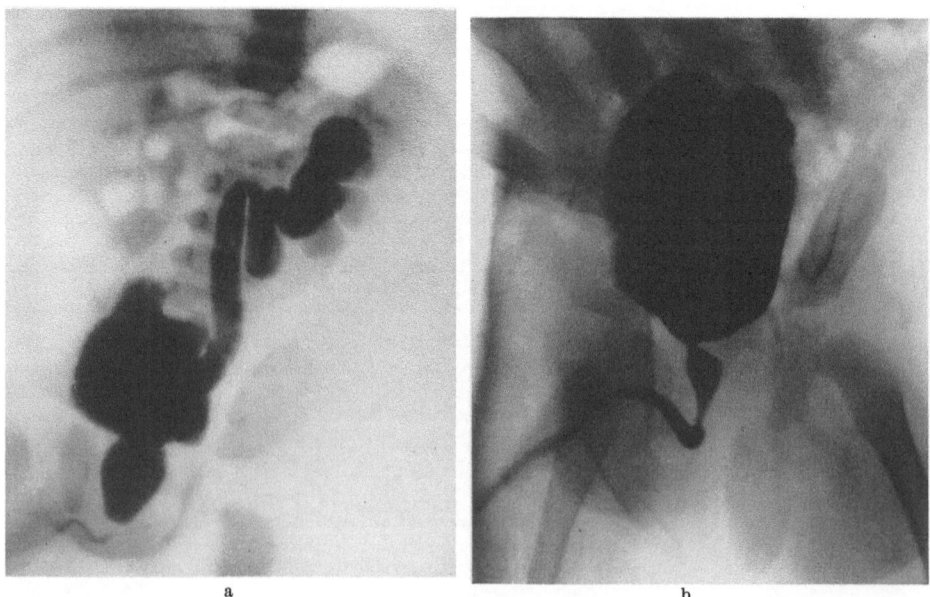

a   b

Fig. 34a. Posterior urethral valve. Voiding cystourethrogram. Preoperative. There is free reflux up a dilated ureter

Fig. 34b. Posterior urethral valve. Voiding cystourethrogram. Postoperative. The reflux present preoperatively is no longer evident

patients who had had valves removed many years previously in other institutions and now presented in their early twenties with terminal uremia. Examination of the lower urinary tracts of these patients showed that the obstruction appeared to have been adequately removed.

Acknowledgements

The author wishes to thank the following colleagues for permission to include their cases: Dr. R. M. NESBIT and Dr. S. BUSBY, Fig. 6; Dr. H. McDONALD, Jr., Fig. 6 and Fig. 11; Dr. T. MALETTA, Fig. 5 and Fig. 12; Dr. J. A. KIRKPATRICK, Fig. 13; Dr. I. N. NUNN, Fig. 29.

The author also thanks the following editors and publishers for permission to republish: The Editors of the Journal of Urology and Messrs. WILLIAMS and WILKINS, Fig. 4, Fig. 6, Fig. 7, Fig. 16, Fig. 17, Fig. 24, Fig. 28, Fig. 29, Fig. 30 and Fig. 31. The Editors of Biochemical Clinics and The Reuben H. Donnelley Corp., Fig. 18. The Editors of the American Journal of Roentgenology, Radium Therapy and Nuclear Medicine and Messrs. CHARLES C. THOMAS for Fig. 19, Fig. 20, Fig. 22 and Fig. 24.

References

Introduction

BRODNY, M. L., and S. A. ROBINS: Urethrocystography in the male child. J. Amer. med. Ass. **137**, 1511 (1948).

HELLMAN, L. M., and S. G. KOHL: Personal communication.

KJELLBERG, S. R., N. O. ERICSSON, and U. RUDHE: Lower urinary tract in childhood: Some correlated clinical and roentgenologic observations. Chicago: Yearbook Publ. Inc. 1957.

PAQUIN jr., A. J.: Personal communication 1965.

— MARSHALL, V. F., and J. McGOVERN: The megacystis syndrome. J. Urol. (Baltimore) **83**, 634 (1960).

STEPHENS, F. D.: Urethral obstruction in childhood. The use of urethrocystography in diagnosis. Aust. N. Z. J. Surg. **25**, 89 (1955).

WILLIAMS, D. I.: Urologist letter club 1965.

Congenital Absence of Urethra

KRUGER, R.: Über Vesica gigantea. Z. urol. Chir. **32**, 330 (1931).

MAY, F.: Ein Fall von kongenitalem Verschluß der Urethra membranacea. Z. Urol. **42**, 245 (1949).

MENEGAUX, G., et M. BOIDOT: Des obliterations congenitales du meat et de la portion balanique de l'urètre (hypospadias excepte). J. Chir. (Paris) **43**, 641 (1934).

MILLER, H. L.: Agenesia of the urinary bladder and urethra. J. Urol. (Baltimore) **59**, 1156—1163 (1948).

Meatal Stenosis

HIGGINS, T. T., D. I. WILLIAMS, and D. F. ELLISON NASH: Urology of childhood, p. 113. London: Butterworth & Co. 1951.

NESBIT, R. M., and W. C. BAUM: Diagnosis and surgical management of obstructive uropathy in childhood. Amer. J. Dis. of Child. **88**, 239 (1954).

Distal Urethral Stenosis

LYON, R. P., and D. R. SMITH: Distal urethral stenosis. J. Urol. (Baltimore) **89**, 414—421 (1963).

—, and E. A. TANAGHO: Distal urethral stenosis in little girls. J. Urol. (Baltimore) **93**, 379 (1965).

STEPHENS, F. D.: Congenital malformations of the rectum, anus and genito-urinary tracts, p. 243. Edinburgh and London: Livingstone 1963.

Congenital Urethral Stricture

ENGEL, W. J, and F. C. SCHLUMBERGER: Urinary extravasation in newborn infant associated with congenital stenosis of the urethra; report of case. Cleveland Clin. Quart. **5**, 278—283 (1938).

JOHANSON, B.: Reconstruction of the male urethra in strictures. Acta chir. scand. Suppl. 176
  (1953).
LEADBETTER, G. W., and G. F. LEADBETTER: Urethral strictures in male children. J. Urol.
  (Baltimore) 87, 409—415 (1962).
McGOWAN jr., A. J., and K. WATERHOUSE: Mobilisation of anterior urethra. Bull. N.Y.
  Acad. Med. 40, 10 776 (1964).
RUSSELL, R. H.: The treatment of urethral stricture by excision. Med. J. Aust. 1, 231—234
  (1919).
YOUNG, H. H., W. A. FRONTZ, and BALDWIN: Congenital obstruction of the posterior urethra.
  J. Urol. (Baltimore) 3, 289 (1919).

Congenital Valves of the Female Urethra
BAKKER, N. J.: Valves in the female urethra. Urol. int. (Basel) 6 (1—2), 187 (1958).
BRACK, C. B., and H. G. GUILD: Urethral obstructions in the female child. Amer. J. Obstet.
  Gynec. 76, 1105 (1958).
EVERETT, H. S., and C. B. BRACK: Unusual lesions of the female urethra. Obstet. and
  Gynec. 1, 571 (1953).
NESBIT, R. M., H. P. McDONALD, and S. BUSBY: Obstructing valves in the female urethra.
  Trans Amer. Ass. gen.-urin. Surg. 55, 21 (1963).
STEVENS, W. E.: Congenital obstructions of female urethra. J. Amer. med. Ass. 106, 89 (1936).

Congenital Hypertrophy of the Verumontanum
BALDRIDGE, R. R.: Congenital hypertrophy of verumontanum; case. New Engl. J. Med. 213,
  46 (1935).
BUGBEE, H. G., and M. WOLLSTEIN: Retention of urine due to congential hypertrophy of
  verumontanum. J. Urol. (Baltimore) 10, 477 (1923).
CAMPBELL, M. F.: Urology, vol. 2, p. 1765. Philadelphia and London: W. B. Saunders Co.
  1963.
EMMETT, J. L.: Obstruction of the vesical neck of a male infant produced by hypertrophy
  of the verumontanum. Proc. Mayo Clin. 15, 364 (1940).
KJELLBERG, S. R., N. O. ERICSSON, and U. RUDHE: Lower urinary tract in childhood: Some
  correlated clinical and roentgenologic observations. Chicago: Yearbook Publ. Inc. 1957.
PILCHER jr., F., and H. W. PRICE: Congenital hypertrophy of verumontanum. J. Amer.
  med. Ass. 115, 2072 (1940).
ROBINSON, W. W.: Congential hypertrophy of verumontanum as a cause of urinary retention.
  J. Urol. (Baltimore) 17, 381 (1927).
STEPHENS, F. D.: Congenital malformations of the rectum, anus and genito-urinary tracts.
  Edinburgh and London: Livingstone 1963.

Anterior Urethral Valves
BOISSONNAT, P., and P. BOUTEAU: Valve of the anterior urethra, accessory diverticular canal
  and disease of the vesical neck in boy 10 years old. J. Urol. (Paris) 60, 949 (1954).
HOPE, J. W., P. J. JAMESON, and A. J. MICHIE: Diagnosis of anterior urethral valve by
  voiding urethrography: Report of two cases. Radiology 74, 798 (1960).
WATERHOUSE, K., and L. J. SCORDAMAGLIA: Anterior urethral valve: A rare cause of bilateral
  hydronephrosis. J. Urol. (Baltimore) 87, 556 (1962).
WILLIAMS, D. I.: Urology in childhood. Encyclopedia of urology, vol. 15. Berlin-Göttingen-
  Heidelberg: Springer 1958.

Diverticula of the Anterior Urethra
ABESHOUSE, B. S.: Diverticula of anterior urethra in male: report of 4 cases and review of
  literature. Urol. cutan. Rev. 55, 690 (1951).
BOISSONNAT, P., and P. BOUTEAU: Valve of the anterior urethra, accessory diverticular
  canal and disease of vesical neck in boy 10 years old. J. Urol. (Paris) 60, 949 (1954).
—, and B. DUHAMEL: Congenital diverticulum of the anterior urethra associated with aplasia
  of the abdominal muscles in a male infant. Brit. J. Urol. 34, 59 (1962).
BOURNE, W. I.: Congenital diverticulum of the urethra. Brit. J. Radiol. 30, 327 (1957).
BROWNE, D.: Techniques in British surgery: Hypospadias, p. 412. London and Philadelphia:
  W. B. Saunders Co. 1950.
DAHL-IVERSEN, E., and H. JOERGENSEN: Congenital diverticula in male; case. Lyon chir. 35,
  37 (1938).
DEES, J. E.: Congenital diverticulum of anterior male urethra. Urol. cutan. Rev. 54, 480
  (1950).

DEMOS, N. J., D. A. GILLIS, and K. E. BARBER: Congenital diverticula of the anterior urethra in male infants: Report of two cases. J. Urol. (Baltimore) 88, 252 (1962).

DE PAOLI, E.: Delle borse urinose uretroli. Abstract from Gazz. med. di Torrino 1885. Zbl. Chir. 12, 905 (1885).

DORAIRAJAN, T.: Defects of spongy tissue and congenital diverticula of the penile urethra. Aust. N.Z. J. Surg. 32, 209 (1963).

DURAND, M.: Un cas de poche diverticulaire congénitale de l'urètre pénien. Bull. Soc. Chir. de Lyon 4, 23 (1900).

FORSHALL, I., and P. P. RICKHAM: Case of congenital diverticulum of anterior urethra in male infant. Brit. J. Urol. 25, 142 (1953).

GEIRINGER, D., and M. O. ZUCKER: Diverticulum of anterior urethra in male child. Amer. J. Surg. 44, 463 (1939).

GROSS, E. E., and A. H. BILL: Concealed diverticulum of the male urethra as a cause of urinary obstruction. Pediatrics 1, 44 (1948).

HALPERSTEIN, J. E.: Congenital diverticula of urethra. Z. urol. Chir. 19, 79 (1926).

HUETER, F.: Großes angeborenes Divertikel der Urethra mit klappengoringen Verschluß der Urethra und Incontinentia urinae. Operation. Heilung. Virchows Arch. path. Anat. 46, 32 (1896).

JOHNSON, F. P.: Diverticula and cysts of the urethra. J. Urol. (Baltimore) 10, 295 (1923).

KAUFMAN, C.: Verletzungen und Krankheiten der maennlichen Harnröhre und des Penis. Stuttgart: Ferdinand Enke 1886.

KHOURY, E. N.: Diverticula of male urethra. J. Urol. (Baltimore) 69, 291 (1953).

KNOX, W. G.: Congenital diverticulum of male urethra. J. Urol. (Baltimore) 58, 344 (1947).

KOTT, B.: Case of congential diverticulum of urethra. Z. Urol. 20, 499 (1926).

KRETSCHMER, H. L.: Diverticula in the anterior urethra in male children. Surg. Gynec. Obstet. 62, 634 (1936).

LATTIMER, J. K.: Congenital deficiency of the abdominal musculature and associated genito-urinary abnormalities. J. Urol. (Baltimore) 79, 343 (1958).

LOWSLEY, O. S., and R. GUTIERREZ: Congenital diverticula of the male urethra. Verh. dtsch. Ges. Urol. 8, 312 (1928).

McGAURAN, H. G.: Diverticula of male; 2 cases. J. Amer. med. Ass. 120, 1381 (1942).

MILLS, W. G. Q.: Chronic retention in boys caused by diverticula in anterior urethra. Brit. J. Urol. 27, 292 (1955).

MOUCHA, D.: Voluminous urethral scrotal diverticulum in a 2-year-old boy. Review of the literature appropos of an operated case. J. Chir. (Paris) 81, 581 (1961).

NESBITT, T. E.: Congenital megalourethra (variation of diverticulum). J. Urol. (Baltimore) 73, 839 (1955).

NUNN, I. N., and F. D. STEPHENS: The triad syndrome: A composite anomaly of the abdominal wall, urinary system and testes. J. Urol. (Baltimore) 86, 782 (1961).

PETZ: Quoted by BOKAY. Jahrbuch für Kinderheilkunde, Bd. 52, No 2. (1900); — Derm. Z. 7, 741 (1900).

RIOSECO, E., and J. VARGAS: Congenital diverticulum. Urol. cutan. Rev. 48, 209 (1944).

RUSSELL, R. H.: Papers and addresses in surgery, p. 298. Melbourne: Grant 1923.

STEPHENS, F. D.: Congenital malformations of the rectum, anus and genitourinary tracts. Edinburgh and London: Livingstone 1963.

SUTER, F.: Ein Beitrag zur Histologie und Genese der congenitalen Divertikel der männlichen Harnröhre. Langenbecks Arch. klin. Chir. 87, 225 (1908).

TERNOUSKY, S.: Congential diverticula. Urol. cutan. Rev. 34, 578 (1930).

VINCENT, E.: Contribution to the study of diverticula of the male posterior urethra. J. Urol. méd. chir. 66, 475 (1960).

VOLLEMIER: Traite des maladies des voies urinaires, p. 382. 1868.

WATTEN jr., J. W.: Congenital diverticulum. Amer. Surg. 21, 385 (1955).

WATTS, S. H.: Urethral diverticula in the male with report of a case. Johns Hopk. Hosp. Rep. 13, 49 (1906).

Double Urethra and Accessory Urethra

ANSELMO, G.: On a case of duplication of the female urethra. Arch. ital. Urol. 36, 403 (1964).

ARNOLD, M. W., and W. M. KAYLOR: Double urethra. J. Urol. (Baltimore) 70, 746 (1953).

BOISSONNAT, P.: Accessory urethral canals; anatomic classification in connection with 5 cases. J. Urol. méd. chir. 60, 954 (1954).

— Two cases of complete double functional urethra with a single bladder. Brit. J. Urol. 33, 453 (1961).

BONANOME, L.: Bifid urethra; case. Arch. ital. Urol. 19, 361 (1942).

284 K. WATERHOUSE:

BROWN, J. J. M.: Lesions of the anterior urethra in infancy and childhood. Proc. roy. Soc. Med. **49**, 891 (1956).

CHAUVIN, E.: A propos des urètres doubles, en particulier de leurs variétés postérieures. J. Urol. méd. chir. **23**, 293 (1927).

COUVELAIRE, R.: Roentgenography of rare form of duplication in infant 6 months old; case. J. Urol. méd. chir. **52**, 208 (1944/45).

DANNREUTHER, W. T.: Complete double urethra in a female. J. Amer. med. Ass. **81**, 1016 (1923).

DE NICOLA, R. R., and R. C. McCARTNEY: Duplication in female child treated with sclerosing solution. J. Urol. (Baltimore) **61**, 1065 (1949).

FUNFACK, M.: Complete double urethra in male; case. Z. Urol. **46**, 391 (1953).

GROSS, R. E., and T. G. MOORE: Duplication; report of 2 cases and summary of literature. Arch. Surg. **60**, 749 (1950).

IRMISCH, G. W., and E. W. COOK: Double and accessory urethra. Minn. Med. **29**, 999 (1946).

LOWSLEY, O. S.: Accessory urethra; Report of 2 cases with review of literature. N.Y. St. J. Med. **39**, 1022 (1939).

MAY, F.: Double urethra in male and its surgical therapy. Urologia **21**, 319 (1954).

MOORE, C. B.: Reduplication of urethra. J. Urol. (Baltimore) **56**, 130 (1946).

RINKER, J. R.: Accessory urethra in boy. J. Urol. (Baltimore) **50**, 331 (1943).

SCHURR, P. H.: Accessory urethral canal in male. Brit. J. Surg. **36**, 181 (1948).

SLOTKIN, E. A., and A. MERCER: Case of epispadis with double urethra. J. Urol. (Baltimore) **70**, 743 (1953).

THEVATHASAN, C.: Accessory urethra in the male child. Report of 2 cases. Aust. N.Z. J. Surg. **31**, 134 (1961).

WRENN, T., and A. J. MICHIE: Ann. Surg. **145**, 119 (1951).

Posterior Urethral Valves

ADDISON, O.: Congenital valvular obstruction. Arch. Dis. Childh. **4**, 255 (1929).

BAZY, P.: A propos du diagnostic des lesions renale unilaterales. Bull. Soc. Chir. Paris **29**, 32 (1903).

BOISSONNAT, P.: Congenital dysurias not due to bladder neck obstruction; 12 cases of urethral valves. Mém. Acad. Chir. **79**, 242 (1953).

BURNELL, G. H.: Congenital valvular obstruction of posterior urethra. Aust. N.Z. J. Surg. **4**, 322 (1935).

BURNS, E., A. M. PRATT, and R. G. HENDON: Management of bladder neck obstruction in children. J. Amer. med. Ass. **157**, 570 (1955).

CAMPBELL, M. F.: Posterior urethral valve obstruction in infancy and childhood; study of 18 cases. J. Amer. med. Ass. **96**, 592 (1931).

— Pediatric urology, vol. 2. New York: Macmillan Co. 1937.

CHADWICK, jR. D., and S. P. MEADOWS: Congenital obstruction of posterior urethra. Brit. med. J. **1930 I**, 443.

COUNSELLER, R. S., and J. G. MENVILLE: Congenital valves of posterior urethra. J. Urol. (Baltimore) **34**, 268 (1935).

DAVIDSON, I., and C. NEWBERGER: Congenital valves of posterior urethra in twins. Arch. Path. **16**, 57 (1933).

DEROW, H. A., and M. L. BRODNY: Congenital posterior urethral valves causing renal rickets; case. New Engl. J. Med. **221**, 685 (1939).

DODSON, A. I., and H. LORRAINE: Congential obstruction of posterior urethra. Virginia med. Mth. **58**, 102 (1931).

FAGERSTROM, D. P.: Congenital obstruction of lower urinay tract in male with particular reference to valve formations. J. Urol. (Baltimore) **37**, 166 (1937).

FORSYTHE, W. F., and D. F. McFADDEN: Congenital posterior urethral valves; a study of 35 cases. Brit. J. Urol. **31**, 63 (1959).

FRONTZ, W. A.: Congenital urinary obstructions in male children, with reports of cases presenting unusual anomalies. J. Urol. (Baltimore) **27**, 489 (1932).

FUCHS, N.: Zwei Fälle von kongenitaler Hydronephrose. Inaug.-Diss. Zürich 1900.

GRIESBACH, W. A., K. WATERHOUSE, and H. Z. MELLINS: Voiding cystourethrography in the diagnosis of congential posterior urethral valves. Amer. J. Roentgenol. **82**, 521 (1959).

GROSS, R. E.: Surgery of infancy and childhood. Philadelphia: W. B. Saunders Co. 1953.

HANSMANN, G. H.: Obstructions of posterior urethra by congential valves; report of case. Boston med. surg. J. **190**, 12 (1924).

HASEN, H. B., and Y. S. SONG: Congenital valvular obstruction of posterior urethra in 2 brothers. J. Pediat. **47**, 207 (1955).

HESS, E., and C. O. PETERS: Congenital valves of posterior urethra; 2 cases. Penn. med. J. **35**, 460 (1932).

HIGGINS, T. T., D. I. WILLIAMS, and D. F. E. NASH: The urology of childhood. London: Butterworth & Co. 1951.

HINMAN, F., and A. A. KUTZMANN: Congenital valvular obstruction of posterior urethra. J. Urol. (Baltimore) **14**, 71 (1925).

IVERSEN, T.: A peculiar valve formation in the prostatic urethra. Hospitalstidende **7**, 1367 (1914).

JORUPS, S., and S. R. KJELLBERG: Congenital valvular formations in the urethra. Acta radiol. (Stockh.) **30**, 197 (1948).

KJELLBERG, S. R., N. O. ERICSSON, and U. RUDHE: Lower urinary tract in childhood: Some correlated clinical and roentgenologic observations. Chicago: Yearbook Publ. Inc. 1957.

KRETSCHMER, H. L., and L. E. PIERSON: Congenital valves of posterior urethra. Amer. J. Dis. Child. **38**, 804 (1929).

LANDES, H. E., and R. RALL: Congenital valvular obstruction of posterior urethra. J. Urol. (Baltimore) **34**, 254 (1935).

LANGENBECK: Memoire sur la lithotomie 1802.

LATTIMER, J. K., and M. HUBBART: Relative incidence of pediatric urological conditions. J. Urol. (Baltimore) **71**, 759 (1954).

LOWSLEY, O. S., and T. J. KIRWIN: A clinical and pathological study of congenital obstruction of the urethra; report of 4 cases. J. Urol. (Baltimore) **31**, 497 (1934).

MAXWELL, M. H., C. R. KLEEMAN, and R. ROCKNEY: Peritoneal dialysis-technique and applications. J. Amer. med. Ass. **170**, 917 (1959).

McCREA, L. E.: Congenital valves of posterior urethra. J. int. Coll. Surg. **12**, 342 (1949).

McDONALD, H. P.: A peritoneal dialysis trocar. J. Urol. (Baltimore) **89**, 946 (1963).

McKAY, R. W.: Obstructions of vesical neck (due to congenital valves of the urethra) in children. Sth. med. J. (Bgham, Ala.) **33**, 377 (1940).

MINKOWSKI, A., and P. CLAISSE: Congenital urethral valves as a cause of giant dilation of urinary tract. Arch. franç. Pediat. **8**, 840 (1951).

MITCHELL, J. P.: Association of valves in the posterior urethra with bladder neck obstruction. Acta urol. belg. **31**, 507 (1963).

NESBIT, R. M.: Congenital valvular obstruction of prostatic urethra; surgical procedure. J. Urol. (Baltimore) **51**, 167 (1944).

—, R. L. THIRLBY, and F. P. RAPER: Diagnosis and treatment of congenital urethral valves. J. Mich. med. Soc. **50**, 1244 (1951).

NUNN, I. N.: Bladder neck obstruction in children. J. Urol. (Baltimore) **93**, 693 (1965).

PAQUIN Jr., A. J.: Ureterovesical anastomosis: The description and evaluation of a technique. J. Urol. (Baltimore) **82**, 573 (1959).

PRESMAN, D.: Congenital valves of the posterior urethra. J. Urol. (Baltimore) **86**, 602 (1961).

RANDALL, A.: Congenital valves of posterior urethra with report of case. J. Urol. (Baltimore) **23**, 57(1930).

RAPER, F. P.: Recognition and treatment of congenital urethral valves. Brit. J. Urol. **25**, 136 (1953).

RATTNER, W. H., U MEYER, and J. BERNSTEIN: Congenital abnormalities of the urinary system. IV. Valvular obstruction of the posterior urethra. J. Pediat. **63**, 84 (1963).

REISMAN, D. D.: Bladder neck obstructions in children. J. Amer. med. Ass. **188**, 1057 (1964).

RUTHERFORD, R., and J. H. FOLLOWS: Congenital valvular obstruction in prostatic urethra. Brit. J. Child. Dis. **31**, 297 (1934).

SCHACHT, F. W.: Congenital valvular obstruction with vesical diverticulum. J. Urol. (Baltimore) **24**, 83 (1930).

SHELDON, W.: Posterior urethral obstruction in childhood (due to valve formations). Proc. roy. Soc. Med. **31**, 1366 (1938).

STEPHENS, F. D.: Congenital malformations of the rectum, anus and genito-urinary tracts. Edinburgh and London: Livingstone 1963.

THOMPSON, G. J.: Urinary obstruction of vesical neck and posterior urethra of congenital origin. J. Urol. (Baltimore) **47**, 591 (1942).

TOLMATSCHEW, N. v.: Ein Fall von Semilumarron Klappen der Harnrohre und von Vergrossater Vesicula Prostatica. Virchows Arch. path. Anat. **49**, 348 (1870).

TORP, K. H.: Congenital valvular formation of posterior urethra. Acta paediat. (Uppsala) **43**, 192 (1954).

VELPEAU: Nouveaux elements de medicine operatoire. vol. 111, p. 907. Paris 1832.

WATERHOUSE, K.: Voiding cystourethrography: A simple technique. J. Urol. (Baltimore) **85**, 103 (1961).

— Lower urinary tract obstruction in children. Biochem. Clin. **2**, 245 (1963).

WATERHOUSE, K.: The dilated posterior urethra: I. Male. J. Urol. (Baltimore) **91**, 71 (1964).
—, and F. C. HAMM: The importance of urethral valves as a cause of vesical neck obstruction in children. Trans. Amer. Ass. gen.-urin. Surg. **53**, 138 (1961).
WATSON, E. M.: The structure of the verumontanum — A study of the origin and development of its inherent glandular elements. J. Urol. (Baltimore) **2**, 337 (1918).
WILLIAMS, D. I.: Congenital valves in posterior urethra (2 cases). Proc. roy. Soc. Med. **46**, 427 (1953).
— Urology in childhood. Encyclopedia of urology, vol. 15. Berlin-Göttingen-Heidelberg: Springer 1958.
— and H. B. ECKSTEIN: Obstructive valves in the posterior urethra. J. Urol. (Baltimore) **93**, 236 (1965).
WISIOL, E.: Congenital valvular obstruction of posterior urethra; case. J. Urol. (Baltimore) **35**, 524 (1936).
WOLGIN, W., M. ROSENBERG, and M. MUSCHAT: Co-existence of congenital median bar and urethral valves. J. Urol. (Baltimore) **68**, 506 (1952).
YOUNG, H. H., W. A. FRONTZ, and BALDWIN: Congenital obstruction of the posterior urethra. J. Urol. (Baltimore) **3**, 289 (1919).

Anomalies of the Male Genitalia

R. J. Prentiss

With 15 Figures

Introduction

This section is devoted to the testicle, its ducts and appendages, and to the prostate and seminal vesicles. Anomaliès of the urethra are presented in Chapter 6, and those of the penis in Chapter 8. The undescended testicle is the commonest of all urogenital anomalies. I intend to present my own experience in these problems. Those who wish to read extensive reviews of the literature should seek other sources.

A. The Undescended Testicle

I. Causes

Of many theories to account for interruption in descent of the testicle, lack of hormonal influence, either in utero, or at the time of puberty seems most rational. Maldevelopment, often found in the undescended male gonad, lends support ot this view. Congenital absence, of course, would account for failure of appearance.

II. Associated Conditions

Inguinal hernia is present in all cases. Finding the epididymis in tandem with the testicle is not unusual. Hydrocele of the tunica vaginalis and hydrocele of the cord occur fairly often. Redundancy of the vas deferens occurs sufficiently often to constitute a surgical problem. Failure to recognize may lead to injury of the vas deferens during transplantation. Torsion of the testicle either in the canal or at the external ring, or intraperitoneally, has been reported. Torsion of the appendages of the testicle occurs in the descended testicle; it rarely occurs in the undescended testicle. Anomalies of the urinary tract are sometimes associated with genital anomalies. This suggests the need for taking an excretory urogram on every child with undescended testicle even though he presents no symptoms or signs of a urinary disorder.

III. Differential Diagnosis

Normal, retractile testicles may be mistaken for undescended testicles. These are often reported during school examinations when the child is frightened. The condition is differentiated by the history of competent examiners finding the testicle in the past, by having the parents examine the child under normal conditions at home, and by repeated examination of the child by a physician. Ectopic testicles may be located in the perineum, thigh and at the base of the penis. This condition must be considered if the testicle is not in the scrotum. Absence of the testicle can only be determined by operation. One must not forget pseudo- and true hermaphrodites. This is covered in Chapter 10.

IV. Method of Examination

A careful history, general physical examination and examination of the urine are important steps. Excretory urography is also indicated.

Local examination should be careful and complete to exclude ectopia. Repeated examinations will differentiate the migratory testicle of physiologic cause. If the testicle is not readily palpated at the external ring, at the entrance to the scrotum or under the skin of the abdomen, it is helpful to have the child cough or strain, at which time the palpating hand slides downward over the internal inguinal ring, in an attempt to trap the testicle as it is forced out of the abdomen. At the same time, the index finger of the other hand is at or in the external ring. This maneuver will lead to the detection of most undescended testicles. The same procedure should be repeated standing, with proper support for the child's buttocks. Having the child in the supine position, grasping his knees and abducting and flexing both thighs to extreme degree will reveal a few testicles otherwise undiscovered. In the course of the examination, the presence of a clinical hernia must be ascertained. In normal retractile testicles the condition is usually bilateral. The parents then are instructed in the methods of examination at home, under normal contitions and during or after a warm bath. Careful diagnosis eliminates unnecessary medical and surgical treatment.

V. Medical Treatment

Medical treatment revolves around the proper use of chorionic gonadotropins or anterior-pituitary-like hormones. This form of endocrine therapy can be used whether the undescended testicle is unilateral or bilateral. It is best to start at the age of 3 to 5 years. Although it can be used at any age. I do not feel it is of any avail at or after the onset of puberty.

For children from 3—6 years of age, 300 I.U. of Antuitrin-S, or similar gonadotropic hormone, twice weekly up to a total of 6000 units given by injection subcutaneously or intramuscularly, is the appropriate dose. Above six years of age, depending on the size of the patient, 500 I.U. are given at the same interval up to a total of 9000 I.U. If the child is young and far from puberty, if no result is obtained, the course may be repeated at a later date.

Some urologists use different dosages. Some give as much als 4000 I.U. intramuscularly every second or third day for one or two weeks, and operate promptly if descent does not occur. Some combine this high dosage followed by smaller doses over longer periods of time. Extremely large doses are sometimes given in one or two doses and, if descent fails to occur within a week, operation is carried out.

If treatment is continued over long periods of time, one must watch for signs of over dosage as penile enlargement, frequent erections and appearance of pubic hair in young children. If this occurs, injections should be interrupted and resumed after a rest period. In any situation, prolonged treatment should be avoided because of the danger of closure of the epiphyses.

The use of negative mechanical suction over the inguinal canal, external ring and scrotum has been tried by many and also by the author. Children dislike it and parents object to it. I have seen no good come of it, and once, when the testicle did finally descend, a hydrocele had to be repaired later. I feel this method of management should be discarded.

VI. Surgical Treatment
1. Indications

The indications for surgical transplantation of the testicle to the scrotum are present when medical treatment fails. The best time to transplant the testicle surgically is between the fourth and sixth year of age. The primary aim in the surgical treatment of undescended testicle is to place a normal functioning testicle

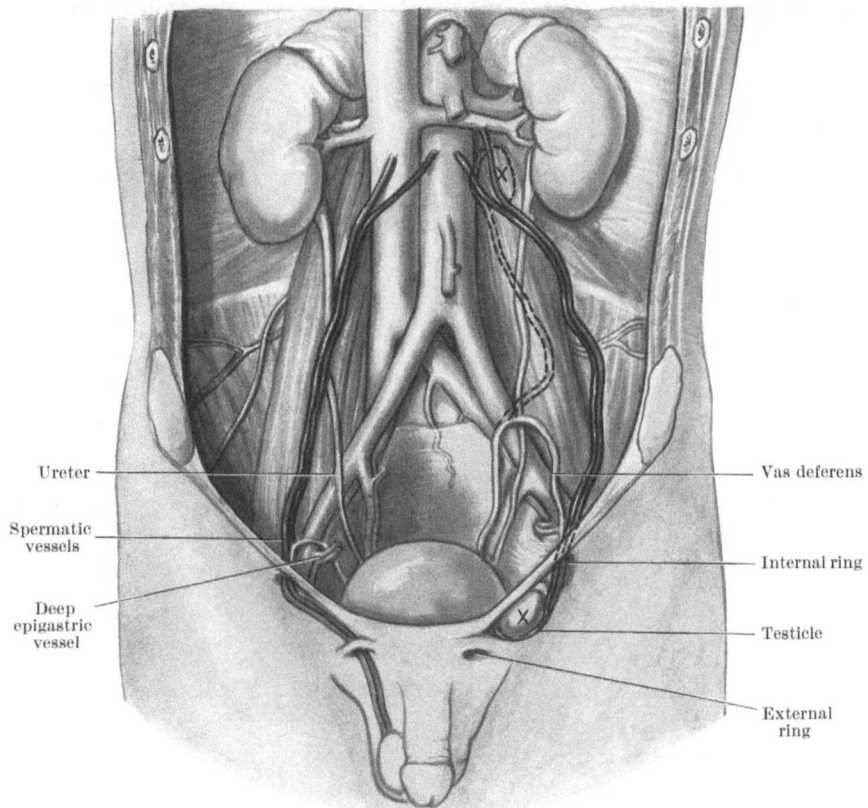

Ureter

Spermatic vessels

Deep epigastric vessel

Vas deferens

Internal ring

Testicle

External ring

Fig. 1. Abdominal-inguinal course of spermatic vessels. Frontal spermatic triangle

with proper blood supply and intact ductile system in the scrotum where it will have the best chance to perform its spermatogenic as well as its androgenic function. As in all congenital malformations, it is best to correct the condition as early as possible in life to permit the child to develop normally. However, I do not believe that transplantation of the testicle is ever indicated before the age of four years unless the primary indication is to repair a hernia which is clinically bothersome. Other indications for surgical repair are to settle any doubt, to get a better cosmetic result and for psychologic reasons. If the child has an atrophic testicle or lacks a testicle, implantation of a suitable prosthesis is advisable for psychologic and cosmetic reasons.

2. Surgical Anatomy

The chief points in the surgical anatomy of the treatment of undescended testicle are:

a) The frontal and sagittal spermatic surgical triangles (Figs. 1, 2 ,3 and 5).

b) The retroperitoneal space.

c) The lateral spermatic ligament (Figs. 3 and 9).

d) The floor of the inguinal canal (Fig. 9).

e) The deep epigastric vessels (Fig. 9).

f) The dartos muscle of the scrotum.

These points in surgical anatomy have been developed by careful dissections of many fresh cadavers, operations on the living, visualization of the vascular

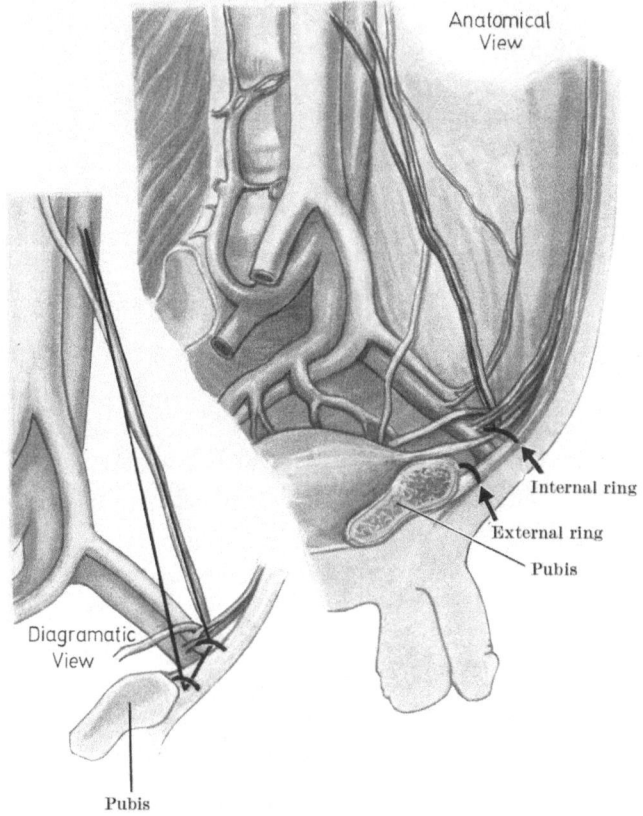

Fig. 2. Abdominal course of spermatic vessels. Sagittal spermatic triangle

supply of the spermatic system by arteriography and venography, visualization of the lymphatic drainage of the testicle by lymphangiography, and by radiography in the anterior-posterior and lateral view in the living after attaching opaque ureteral catheters to the spermatic vascular system (Figs. 6, 7, and 8).

These studies have shown that in the frontal view, the spermatic vessels course obliquely downward and laterally from the renal area in the retroperitoneal space to exit through the internal inguinal ring. From that point they course medially and inferiorly to the external ring (Figs. 1 and 4). When one views the vessels from the sagittal plane, the spermatic vessels course anteriorly, as well, to the internal ring, and then postero-inferiorly in the inguinal canal (Figs. 2 and 5). Throughout the retroperitoneal course, the vessels are enveloped by the transversalis fascia behind the peritoneum which, thus, fixes the vessels to the lateral abdominal wall (Figs. 3 and 9). This is GEROTA's fascia which encloses the kidney above and the ureter farther down. Thus, severance of the lateral spermatic ligament will allow medial displacement of the spermatic vessels, allowing them to traverse a shorter distance to the scrotum.

However, this is not possible unless the floor of the inguinal canal and deep epigastric vessels are divided to permit the spermatic cord to exit from the external-inguinal ring only. Likewise, this division permits wide retroperitoneal exposure and excellent visualization to the lower pole of the kidney from the inguinal

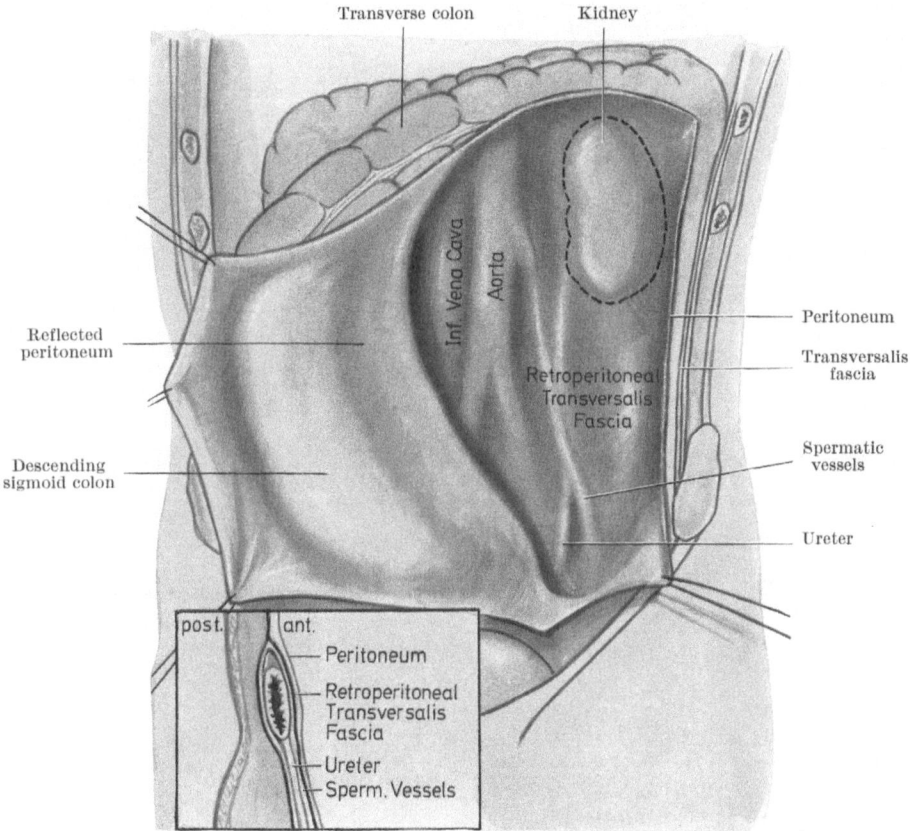

Fig. 3. Spermatic vessels enveloped in posterior transversalis fascia (lateral spermatic ligament)

incision (Fig. 9). It also allows for wide incision of the peritoneum when this becomes necessary because the testicle cannot be found in the pelvis of adjacent retroperitoneal area.

From the above description, then, the lateral spermatic ligament has been defined as that portion of the transversalis fascia behind the peritoneum which fixes the spermatic vessels to the lateral abdominal wall. The frontal spermatic triangle then can be defined as the normal course of the spermatic vessels in the retroperitoneum from origin to the internal ring with the lower side of the triangle being formed by the inguinal course of the vessels. The medial side of the triangle is an imaginary line from the origin of the vessels to the external ring. Likewise, the sagittal spermatic surgical triangle is defined by the course of the vessels when viewed from the lateral aspect. In this view the spermatic vessels in the retroperitoneal space traverse in an anterior course to the internal ring, and then in a postero-inferior course in the inguinal canal. The third side of the sagittal triangle is formed by an imaginary line from the origin of the vessels to the external ring. The division of the deep epigastric vessels and the floor of the inguinal canal cause elimination of two sides of both triangles, giving relative increase in length through allowing the vessels to traverse a straight line from origin to the external ring and scrotum.

The dartos muscle of the scrotum is an active involuntary muscle which is best dissected away from the skin of the scrotum to allow the testicle to be placed in a non-retractile area of the scrotum.

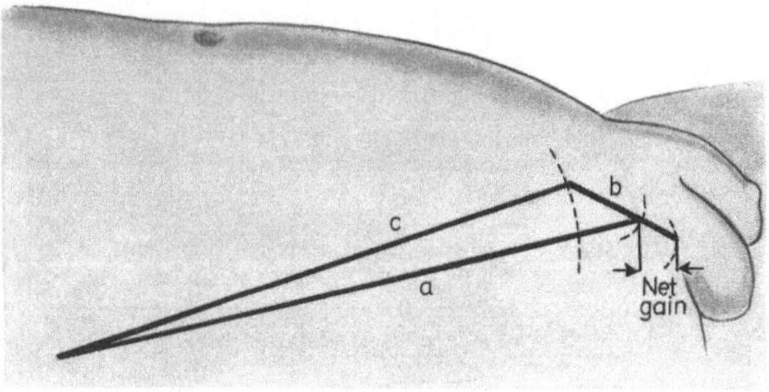

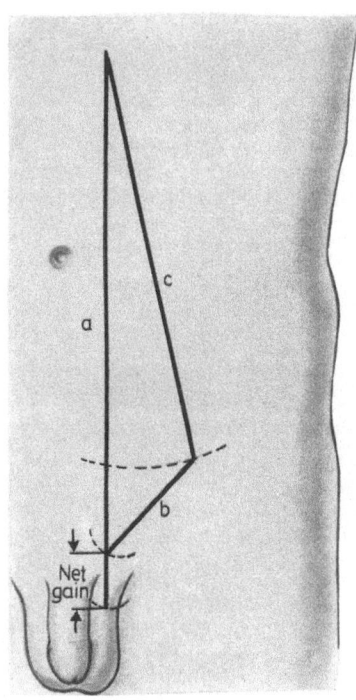

Fig. 4

Fig. 5

Fig. 4. Frontal spermatic surgical triangle superimposed on body

Fig. 5. Sagittal spermatic surgical triangle superimposed on body

3. Surgical Principles

The principles of operation for undescended testicle evolve from the surgical anatomy just described. The anatomy is the same whether or not one uses an inguinal approach or midline approach. There should be adequate exposure. The fact that the shortest distance between two points is a straight line allows for the relative increase in length of the spermatic vascular system through elimination of two sides of both spermatic surgical triangles. The third principle involves avoidance of injury to the blood vessels and ductile system of the testicle either by dissection or tension. This is best achieved by the complete dissection in the retroperitoneum displacing the vessels as suggested. Obviously, the scrotum must have adequate space for the testicle; incision of the dartos insures this. Despite the lack of tension on the vessels after this dissection, temporary fixation of the transplanted testicle by suturing the testicle to the skin of the thigh may prove helpful.

4. Technique

The technique of the operation develops logically and easily of one follows the surgical anatomy and principles described. Complete exposure and identification of the spermatic vascular and ductile systems is assured by a complete inguinal incision, division of the floor of the inguinal canal and the deep epigastric vessels, and development of the retroperitoneal space. The chief points in the technique are:

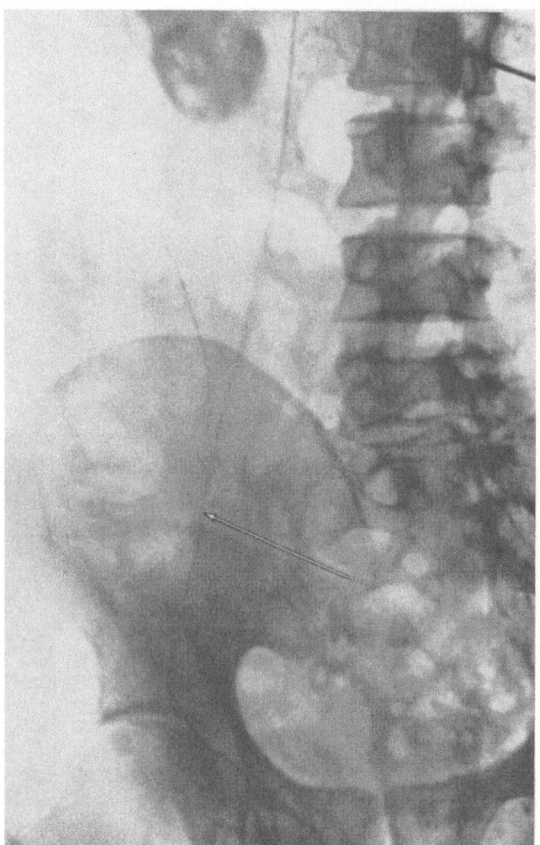

Fig. 6. Aortogram demonstrating two sides of frontal triangle (course of spermatic artery)

a) Skin incision from the anterior-superior spine of the ilium to the entrance of the scrotum. The external oblique fascia is divided throughout the length of the incision.

b) The internal oblique muscle must be divided lateral to the internal ring to the upper limit of the incision or as indicated.

c) If the cord is found in the canal, the dissection should be from above downward with preservation of the ilio-inguinal nerve. Fine plastic forceps and iris scissors are best for this dissection.

d) The internal ring is developed posterior to the spermatic cord and the peritonium is reflected from the posterior surface or floor of the inguinal canal.

e) The transversalis fascia and inferior epigastric vessels in the floor of the inguinal canal are divided with suitable clamping and ligature (Fig. 9).

f) At this point, if one is sure the testicle will be transplanted, or if one plans
a prosthesis, the scrotum is developed digitally from under the superficial fascia.
It is spread down to the anterior perineum. The scrotum is everted and the dartos
muscle incised and then bluntly dissected away from the skin. Blood vessels are
controlled with ligature of fulguration, being careful of the skin.

g) The retroperitoneal space is developed to the kidneys if necessary.

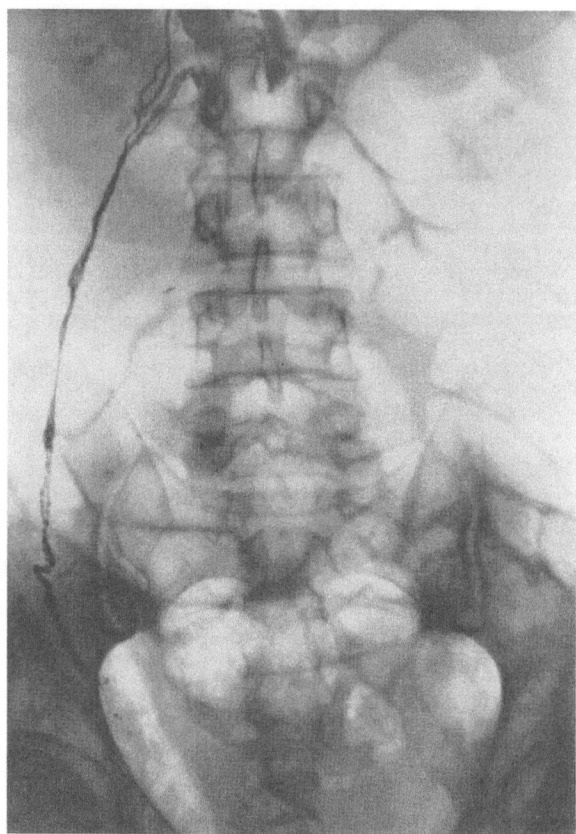

Fig. 7. Spermatic lymphangiogram demonstrating two sides of frontal triangle (after BUSH)

h) The hernia sac is identified and elevated with saline, facilitating the dissection of a hernia sac from the spermatic vessels and separation of the posterior
peritoneum from the spermatic vessels and the lateral spermatic ligament and
vas deferens.

i) Hernia sac is closed.

j) The lateral spermatic ligament is developed further and divided permitting
progressive medial and posterior displacement of the vessels and eliminating two
sides of the surgical triangles (Fig. 9).

k) Occasionally the vas deferens must be dissected as far down as the seminal
vesicle.

l) The sponge, placed in the scrotum after development of the scrotal space
and division of the dartos, is removed and the scrotum inspected for any bleeding.

m) The tunica albuginea of the testicle at its most inferior point is pierced
with a small needle with O chromic catgut. Another straight needle is on the

other end of this suture. These two needles are placed through the bottom of the scrotum and the testicle is drawn into the depth of the scrotum. The suture is then tied after piercing the skin of the thigh to hold the testicle, until fixed, for two or three days (Fig. 10).

n) Harmful dissection within the cord is avoided, although at times peritoneum on the cord is eliminated. The cremaster muscle is always removed if present. If

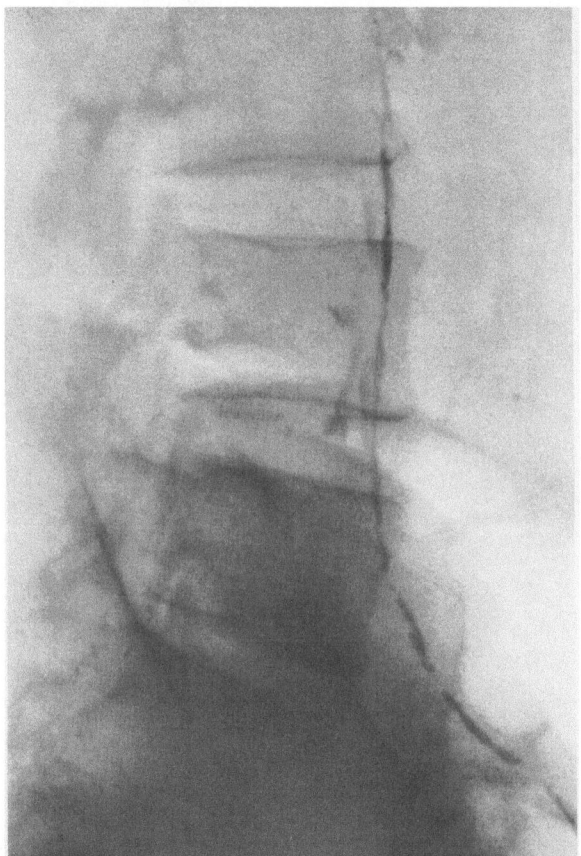

Fig. 8. Lateral spermatic lymphangiogram demonstrating sagittal triangle (after BUSH)

the tunica vaginalis is present, it is always everted and trimmed, being careful to avoid injuring a redundant vas deferens.

o) Long hemostats and long retractors for high retroperitoneal visualization or peritoneal visualization are a sine qua non of this operation.

p) The abdominal wall is repaired in layers eliminating the internal ring completely.

If one adheres to these principles and technique, the testicle will be in the bottom of the scrotum without injury to blood vessels or ducts. Only congenitally short vessels or vas deferens could prevent this. This problem is not amenable to surgery. However, one can try a second stage if such is the case.

If one elects to do a two-stage orchiopexy because of short vessels or vas deferens despite adequate exposure and dissection, a period of two or three years should elapse. However, it could be done within one year if the tissues had

softened properly. In the second stage the same principle of adequate exposure
obtains. A complete incision should be made. Since there will be some scarring,
great care must be exerted to avoid injury to the vessels and vas deferens until
they are completely visualized. The same principles of freeing the vessels is
followed and there is a reasonable chance that the testicle can be placed in the
scrotum at this second stage.

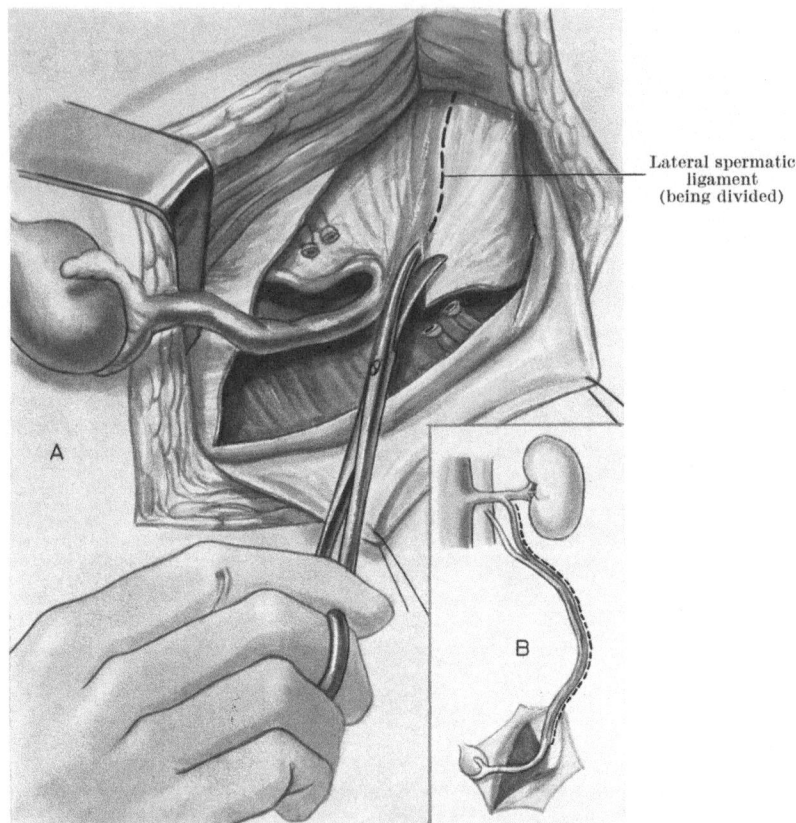

Fig. 9. Incision of lateral spermatic ligament

5. Special Situations

If the testicle has never been palpated, the family should be advised that it
may be absent, lie at the renal area, may be impossible to find, or if found, the
vessels may be too short to permit transplantation. Conversely, a testicle which
has been palpable at one time or another can almost certainly be transplanted.

If the testicle is not found in the canal, the retroperitoneum should be explored thoroughly to above the pelvic brim. One should look particularly for the
vas deferens since it must traverse the true pelvis regardless of the location of
the testicle. If the vas is found and the testicle is not close by, it is probably
non-existent or hypoplastic. Rarely, the vas may be present and fail to attach
to the epididymis and the epididymis may fail to attach to the testicle. If one
has found no vas or evidence of a testicle in the retroperitoneum, the peritoneum
must be opened widely and inspected on both sides of the colon to the renal level.

Under no circumstances do I feel the incision should be extended or a new one be made to explore the renal pedicle to find an organ which may not be there because of the questionable future danger of malignancy. In other words, the surgical procedure outweighs the danger of future malignancy in an organ that is probably not present.

If the vas is connected to the epididymis and the epididymis is in tandem with the testicle, no specific treatment should be carried out. If the ductuli efferentes

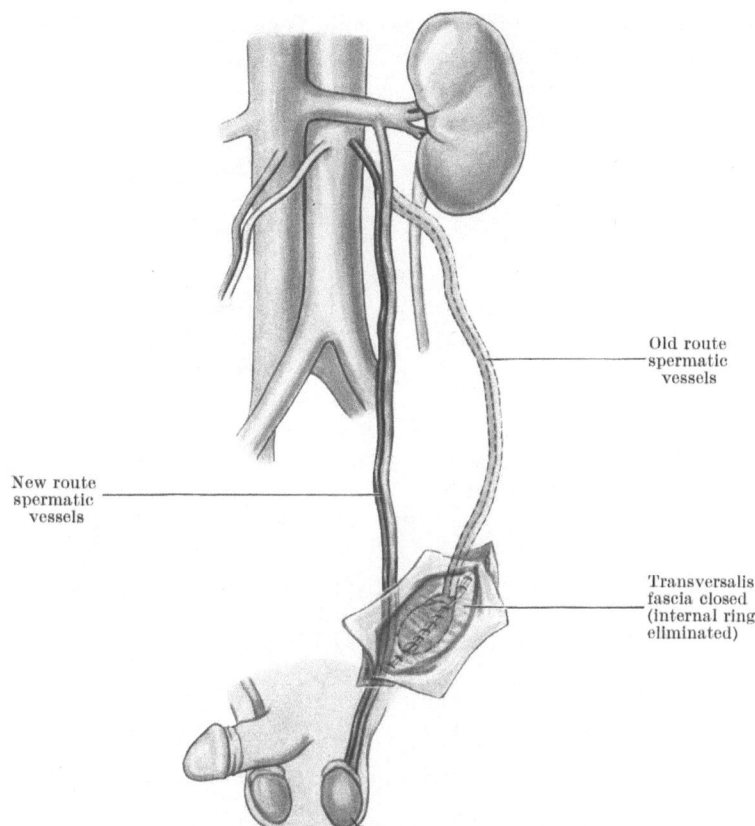

Old route
spermatic
vessels

New route
spermatic
vessels

Transversalis
fascia closed
(internal ring
eliminated)

Fig. 10. Direct route of spermatic vessesls to scrotum after dissection

are present, spermatozoa can get out. If they are not present in this tandem situation, nothing can be done about it. If the vas does not connect to the epididymis one can consider epididymovasostomy. Although my experience with this would indicate not much hope of success would exist.

All of the appendages of the epididymis or testicle such as the appendix epididymis, the appendix testis and the organ of Giraldes, if present, should be removed at the time of transplantation of the testicle.

If no testicle has been palpated before operation, a prosthesis should be at hand. We have found silicone rubber prostheses (Silastic S-6508 of the Dow Corning Corporation) to be the most suitable for use because its specific gravity is about that of a normal testicle, it is easily shaped, and can be sterilized by auto-claving without change. It is also non-reactive in tissues. In small children, we use the juvenile size ($2.5 \times 2 \times 2$ cm.) and in children at, or close to, puberty, we use the adult size (3.5×2.5×2.5 cm.). In the past we have attempted to place these prostheses within a dartos pocket. We have found it much simpler to place the prosthesis in the depth of the scrotum and purse-string the dartos and scrotal layers above it to hold it in proper place.

6. Other Methods

The Bevan technique is mentioned only to condemn it since it depends on dissection within the inguinal canal only, with dissection within the cord releasing so-called "adhesions" to gain length. This usually destroys the blood supply of the testicle and so no advantage to the child is gained by this technique. It fails to use the "shortest distance principle".

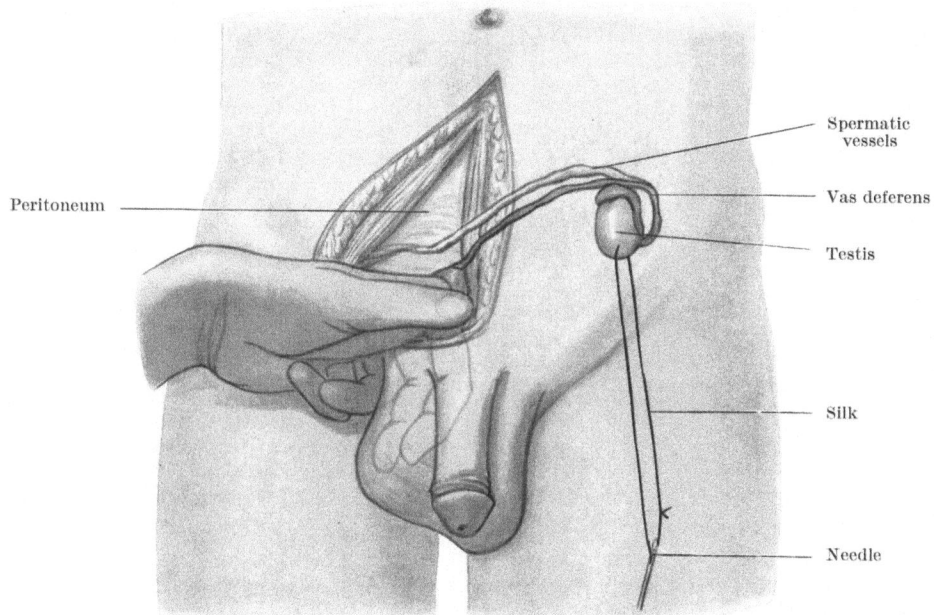

Fig. 11. Midline orchiopexy. Dissection complete. Testicle ready for scrotal placement (after JACOBSON)

The Torek method is likewise mentioned only to state that it is unnecessary if the principles and techniques as described above have been followed. There is no point in fixing the testicle to the thigh fascia under tension and hope that it will survive when one can elongate the vessels by following the surgical anatomy and principles outlined above. If one has to fix the testicle to the high because of inadequate dissection and freeing of the cord, ischemia will result in such a large number to make this not only undesirable but a dangerous technique. The Torek and the Bevan methods I believe are the source of bad reputation that orchiopexy has enjoyed among pediatricians for a great many years.

The midline abdominal incisional approach to the retroperitoneal space follows all of the surgical anatomy and surgical principles outlined (THOMAS FLORENCE, 1957; I. PONCE DE LEON, 1964; CHARLES E. JACOBSEN, Jr., 1966). Dr. JACOBSEN reports that Dr. JEAN-PAUL BOURQUE of Montreal inspired him in this approach many years ago. However, Dr. BOURQUE apparently did not publish this. I am sure others have used it but I do not find their work published. Among the latter is Dr. O. C. BERG of Witchia Falls, Texas.

Because of stimulation by Dr. JACOBSEN, I have personally operated on four boys and transplanted five testicles by this method. This is a good way to transplant the undescended testicle. After the midline incision and reflection of the peritoneum, the technique is exactly as described by me. The testicle is brought out through the external ring only. There is no difficulty in developing the scrotum by the same method, under the superficial fascia. Fixation is the same. The advantages of the midline approach include allowing bilateral operation in one stage through one incision. The exposure is excellent and allows all the anatomic surgical points and principles to be followed (Fig. 11).

However, there are some disadvantages to the midline approach. In operating on a child of 4 or 5 years of age, complete dissection from the kidney level to the pelvic diaphragm would lay open such a wide space that there would certainly

Table 1. *Postmortem results. Spermatic cord length increase*

	Cord dissection only	Retroperitoneal dissection added	Vas dissection added
Subject 1	Not done	5.00	Not done
Subject 2	Not done	8.00	Not done
Subject 3	Not done	7.50	Not done
Subject 4	Not done	6.50	Not done
Subject 5	Not done	7.00	Not done
Subject 6	Not done	3.25	Not done
Subject 7	Not done	8.00	Not done
Subject 8	Not done	7.00	Not done
Subject 9	Not done	4.00	Not done
Subject 10	1.25 cm	7.50	3.25 cm
Subject 11	2.00 cm	8.00	2.00 cm
Subject 12	2.50 cm	7.50	2.50 cm
Subject 13	1.50 cm	6.50	3.50 cm
Subject 14	2.00 cm	8.00	2.25 cm
Subject 15	1.50 cm	5.00	3.75 cm
Subject 16	1.75 cm	6.50	1.75 cm
	1.78 cm, av.	6.58 cm, av.	2.71 cm, av.

be extra risks for the child, besides increasing the time of the operation for the surgeon. This would occur only in those testicles which are not easily found. In the situation of the testicles being palpable, bilaterally, in the area of the inguinal canal, the chances are that the operation could be done promptly without laying open widely the whole retroperitoneal space bilaterally. Nevertheless, sometime, even with palpable testicles, this is necessary. The advantage claimed for the midline approach in making it easier to explore the peritoneum does not obtain; one can explore the peritoneum just as well through the incision that I have described. Likewise, there is no advantage in this approach in transplanting the palpable unilateral undescended testicle.

Of more serious nature is the problem of the vestigial testicle. If one performs a midline incision and finds the spermatic vessels and vas deferens exiting from the internal ring, traction dissection on this might produce considerable harm to the previously unpalpated vestigial testicle. If one placed traction on this cord and vessels and either with sharp or blunt dissection, in pulling up the cord, certainly this vestigial remnant, which may have interstitial cells, would be injured or left behind. Even if one cut the deep epigastric vessels and opened the inguinal canal from within by incising the transversalis fascia, dissection is not as easy as it would be looking directly at it from the exterior inguinal approach.

Therefore, unless the surgeon is willing to stop and make an inguinal approach in this situation, there is grave danger of injuring a potentially useful vestigial organ.

If one should encounter an undescended testicle in the course of repairing an acute inguinal hernia in the presence of a vestigial or undeveloped scrotum, the Torek principle can be used to enlarge this tiny scrotum with subsequent use of thigh skin to complete the scrotum later.

7. Results

If one follows the surgical anatomy and surgical principles described, one can gain an average relative lengthening of the spermatic cord of 6.58 cm. If the vas deferens limits, complete dissection of the vas to the seminal vesicles will produce an additional 2.7 cm. of length. It is calculated that this dissection will add about 26% increase in length of the cord whether it is relative or actual. This should allow the placement of the testicle in the depths of the scrotum with a good ductile and vascular system (Table 1).

B. Other Abnormalities of the Testicle
Anorchism and Hypoplasia

Bilateral anorchism has occurred clinically twice in my experience. Both were proven by extensive urologic studies and abdominal exploration. Both were treated with hormonal substitution either by injection or by implantation of testosterone pellets. One had testicular prostheses inserted because, after his artificial puberty, he wanted to fight. The armed forces rejected him until the prostheses were placed. Generally speaking, however, anorchism may well be associated with severe other abnormalities of the urogenital system or be involved with hermaphroditism.

Unilateral anorchism or unilateral hypoplasia or bilateral hypoplasia are more common. The former has occurred three times in my experience and proven by operation. Hypoplasia is the most common. It is my feeling through clinical experience in bilateral hypoplasia followed by normal puberty when preserved, that these remnants should be placed in the scrotum where possible for their probable future hormonal value. If they are so transplanted, should tumor develop, early detection is possible.

I have never seen duplication of the testicle. Synorchism is a curiosity. Testicles may have other organ rests such as the adrenal or spleen, but these are also extremely rare and I have never seen them.

C. Appendages of the Testicle, Epididymis and Cord

The appendix testis is quite commonly found, while the appendix epididymis is much more rare. I have seen only one appendix epididymis. Occasional torsion of these vestigial remnants will occur and simulate epididymitis, orchitis or testicular torsion. If one can be sure, watchful waiting with antiseptics will be enough. If not, operation is indicated. Such a scrotal exploration is routine with eversion of the tunica vaginalis, amputation of the torsed vestige, followed by routine closure and drainage for one or two days.

The organ of Giraldes is superior to the epididymis and usually on the spermatic cord. I have had one of these which develop torsion and hemorrhage and required routine surgical treatment. It is rare.

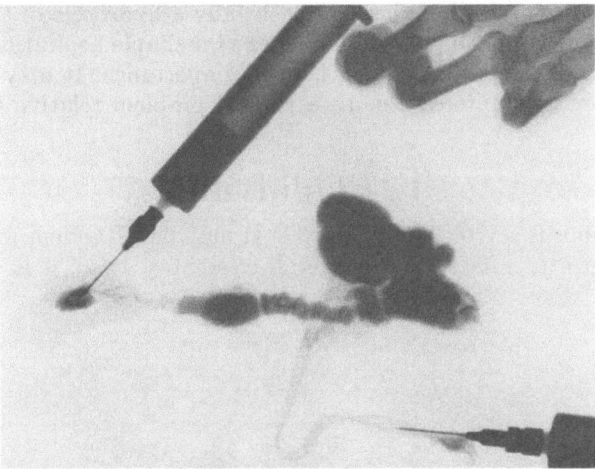

Fig. 12. Duplication of vas with union and insertion into ureter from pelvis of aplastic kidney. Ureter drained into seminal vesicle

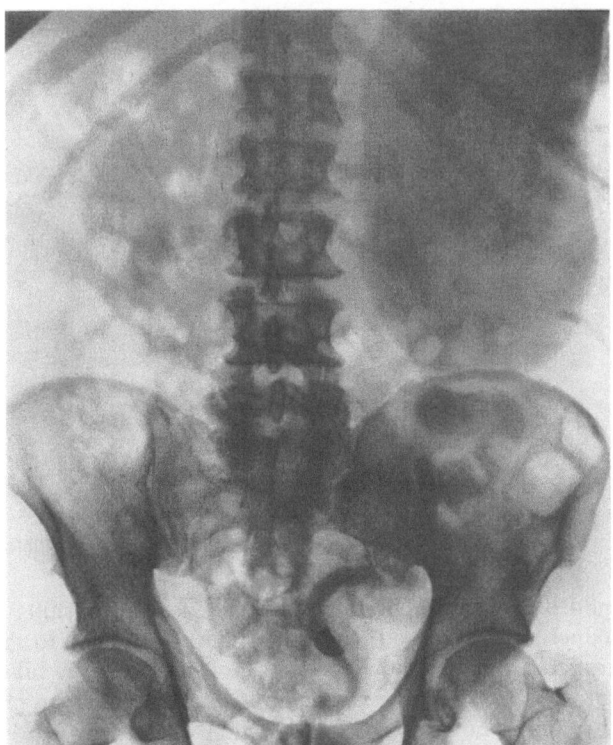

Fig. 13. Dilated aberrant ureter containing a stone entering prostatic urethra

D. Anomalies of the Tunics of the Testicle and Scrotum

Improper closure of the processus vaginalis of the cord may produce hernia, communicating hydrocele, hydrocele of the tunica vaginalis, hydrocele of the cord alone or in combination. They may be associated with other anomalies. Routine surgical treatment cures all of them. Occasionally a hydrocele of the tunica vaginalis in an infant will disappear after one or two simple aspirations of the fluid.

The scrotum may be bifid, which is of no importance. It may be very small, but I have never encountered this as a clinical problem relative to smallness or absence.

E. Epididymis

The epididymis is a Wolffian derivative. It may be in tandem with the testicle, but with proper attachment of the ductuli efferentes. If so, it is of no problem.

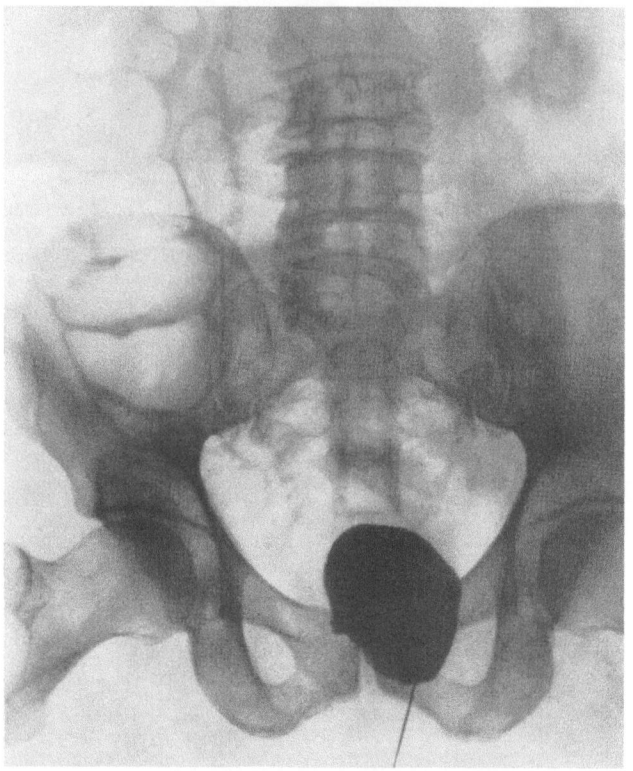

Fig. 14. (Case 1 of MOORE.) Cyst of prostate visualized by injection. Removed by perineal dissection

If not so, it would be nearly impossible to determine, and nothing could be done if there were no ductuli efferentes. If the epididymis is attached to the vas deferens but not to the testicle, the patient will be sterile if the lesion is bilateral. I doubt that either vasoorchiostomy or epididymoorchiostomy would be of any avail. It is possible for the epididymis to be absent. If the vas deferens is present and close enough to the testicle, possibly it could be implanted in the testicle, but the results would be poor.

F. Vas Deferens

This also is a Wolffian derivative. Duplication of the vas deferens occurs at times. I have seen two proven cases. One was during routine vasectomy before prostatectomy, when both vasa deferentia were within the same sheath. This was proven histologically. There was no other abnormality. I feel that this is a rare problem, but it is a convenient excuse for failure of vasectomy for sterilization purposes.

Another duplication of the vas deferens occurred in my experience associated with a vestigial right kidney draining into the right seminal vesicle. The proximal end of the second vas ended in a cyst at the external ring, but joined its mate within the pelvis. The common vas then joined the ureter as it entered the seminal

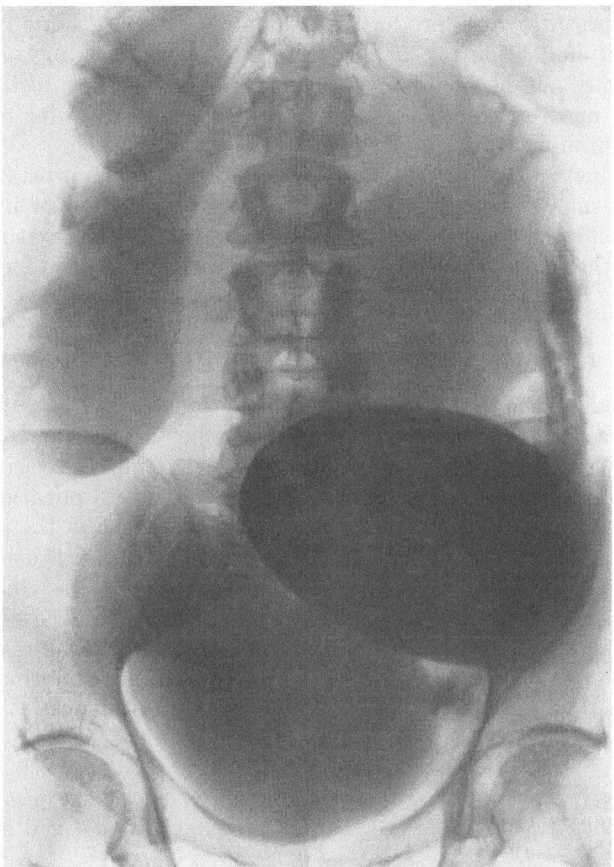

Fig. 15. Mullerian remnant attached to prostate and utricle visualized by abdominal injection. Removed abdominally

vesicle. In this situation treatment was removal of the vas deferens, seminal vesicle, ureter and vestigial pelvic kidney (Fig. 12).

Absence of the vas deferens or failure to attach to the epididymis occurs. I have not seen the latter. One personal case proven by exploration, had no vas on either side, and of course was sterile. There was also deterioration of the testicles. Absence of the vas usually indicates that there may be other congenital anomalies of the urinary or genital system. One personal case presented no vas on the right side associated with absence of the right kidney and ureter, but with a tremendous Mullerian remnant cyst. A third patient had no vas deferens on the right and exactly the same problem as cited above. However, this same patient had a left vas deferens, but with evidence also of a Mullerian remnant on the left as shown by extra-genital and urinary stones associated with persistent pyuria after cure of the right congenital problem. Therefore, one must be on guard that there will be serious other associated and probable symptomatic disease in the absence of one or both vasa deferentia.

G. Prostate

Congenital abnormalities of the prostate are rare. It may be absent but I have never seen this. One of the ureters of a duplicated kidney may enter the urethra through the prostate and produce considerable trouble. Usually there is obstruction with indication for nephroureterectomy. Depending on the level of insertion, it may not be possible to close the defect in the prostatic urethra. A personal case needed removal of his enlarged prostate in order to close the defect after nephroureterectomy (Fig. 13).

Mullerian duct remnant cysts may occur in or about the prostate and produce obstruction. Usually the symptoms are those of obstruction or infection, and a mass can be felt either abdominally, perineally, or both. The treatment is excision of the cyst, either abdominally or perineally, depending on the location (Figs. 14 and 15).

H. Seminal Vesicles

The seminal vesicles may be absent on one or both sides, duplicated, hypoplastic, produce cysts, or have the ureter attached to one or both.

Careful investigations cystoscopically and with seminal vesiculography, and urography, will produce a diagnosis. It is doubtful that duplication is of clinical note, absence may be associated with sterility. The cysts and the aberrant ureter entering the seminal vesicles will produce problems. The latter is usually associated with hypoplastic pelvic kidneys and abnormalities of the vas deferens (Fig. 12). The treatment is excision of the enlarged abdominal seminal vesicle, part of the vas deferens and the aplastic kidney with its attached ureter. In one of our patients with this problem, the vas deferens entered the abnormal ureter.

References

Abstract of Literature of Undescended Testicle. Therap. Notes **60**, 11 (1953).
Anson, B. J., and W. G. Maddock: Callander's surgical anatomy, 3rd ed. Philadelphia: W. B. Saunders Co. 1952.
Barzilay, B.: Some new observations on the problem of cryptorchidism. Harefuah **51**, 173—180 (1957).
Baumrucker, G. O.: Testicular prosthesis for an intracapsular orchiectomy. J. Urol. (Baltimore) **77**, 756 (1957).
Beach, E. W.: Undescended testes: Cause and treatment. J. Urol. (Baltimore) **60**, 623—630 (1948).
Berg, O. C., Wichita Falls, Texas: Personal communication 1964.
Bevan, A. D.: Surgical treatment of undescended testicle: Further contribution. J. Amer. med. Ass. **41**, 7-8-724 (1903).
Bourque, J. P., Montreal, Canada: Quoted in personal communication from C. E. Jacobson, Manchester, Conn.
Bunce, P. L.: Diagnosis of undescended testes: Letter to editor. Pediatrics **27**, 165—166 (1961).
Cabot, H.: Modern urology, p. 397—400. Philadelphia: Lea & Febiger 1936.
Campbell, M.: Pediatric urology, vol. 1, p. 364; vol. 2, p. 496. New York: Macmillan Co, 1937.
— Cryptorchidism and hypospadias. Amer. J. Surg. **82**, 3 (1948).
Clatworthy, H. W., and associates: The inguinal hernia, hydrocele and undescended testleic problem in infants and children. Postgrad. Med. **22**, 122—131 (1957).

CRAWFORD, E. S., M. E. DE BAKEY, and D. A. COOLEY: Clinical use of synthetic arterial substitutes in three hundred seventeen patients. Arch. Surg. 76, 261 (1958).

DEMING, C. L.: Evaluation of therapy in cryptorchidism. J. Urol. (Baltimore) 30, 141—150 (1950); 68, 354—357 (1952).

DODSON, A. I.: Urologic surgery, p. 661. St. Louis: C. V. Mosby Co. 1944.

— Urologic surgery, 3rd ed. St. Louis: C. V. Mosby Co. 1956.

EISENSTAEDT, J. S.: Imperfect descent of the testis and its management. Surg. Clin. N. Amer. 30, 141—150 (1950).

ELLIK, M.: Personal communication 1962.

FLORENCE, T. J.: Surgical approach to cryptorchidism. Trans. s.-east. Sec. Amer. urol. Ass. 89—98 (1957).

FOWLER jr., R., and F. D. STEPHENS: The role of testicular vascular anatomy in the salvage of high undescended testes. Aust. N.Z. J. Surg. 29, 92—106 (1959); abstracted in Surg. Gynec. Obstet. 110, 164—172 (1960).

GIRDANSKY, J., and H. F. NEWMAN: Use of a vitallium testicular implant. Amer. J. Surg. 53, 514 (1941).

GOSS, C. M.: Gray's anatomy, 26th ed. Philadelphia: Lea & Febiger 1954.

GRANT, J. C. B.: Atlas of anatomy. Baltimore: Williams & Wilkins Co. 1951.

GREATBATCH, W., W. M. CHARDACK, and A. A. GAGE: Implantable pacemaker. Bull. Dow-Corning Center Aid Med. Res. 3, 1 (1961).

GROSS, R. E.: Surgery of infancy and childhood, pp. 467—481. Philadelphia: W. B. Saunders Co. 1953.

—, and T. C. JEWETT jr.: Undescended testis. J. Amer. med. Ass. 160, 634—642 (1957).

HALLMAN, N., and associates: Undescended testis, indication and results of operation. Ann. Chir. Gynaec. Fenn. 46, 22—35 (1957).

HAMM, F. C.: Management of undescended testicle. N. Y. St. J. Med. 53, 295—298 (1953).

HAND, J. R.: Treatment of undescended testis and its complications. J. Amer. med. Ass. 164, 1185—1191 (1957).

HAZZARD, C. T.: The development of a new testicular prosthesis. J. Urol. (Baltimore) 70, 959 (1953).

JACOBSON, C. E., Manchester, Conn.: Personal communication 1964.

— Midline approach to orchiopexy. J. Urol. (Baltimore) 95, 74—76 (1966).

JONES jr., H. W., and W. W. SCOTT: Hermaphoditism, genital anomalies and related endocrine disorders. p. 286—287. Baltimore: Williams & Wilkins Co. 1958.

KEELEY, J. L.: Orchiopexy. Arch. Surg. 79, 994—998 (1959).

KEYES, E. L.: Urology, p. 695—697. New York: D. Appleton-Century Co. 1936.

KIEFER, J. H.: Surgical treatment of cryptorchidism. J. Urol. (Baltimore) 68, 358—365 (1952).

KIESWELTER, W. B.: Undescended testicles. Gen. Practit. Aust. 19, 95—100 (1959).

KIMBROUGH, J. C., and J. F. REED: Treatment of undescended testis. J. Amer. med. Ass. 163, 621—625 (1957).

— — Treatment of undescended testis. Arch. Surg. 75, 898—905 (1957).

KOHLER, F. P., and J. J. MURPHY: A mechanical ureteral valve. Surg. Gynec. Obstet. 109, 703 (1959).

KOOP, E. C.: Undescended testicle, differential diagnosis and management. Med. Clin. N. Amer. 36, 1779—1785 (1952).

—, and C. L. MINOR: Observations on undescended testis. Arch. Surg. 75, 898—905 (1957).

LATTIMER, J. K.: Scrotal pouch technique for orchiopexy. J. Urol. (Baltimore) 78, 628—632 (1957).

LAUGHLIN, V. C.: Orchidofunicolysis. J. Urol. (Baltimore) 77, 39—46 (1957).

LEMEH, C. N.: A study of the development and structure relationships of the testis and gubernaculum. Surg. Gynec. Obstet. 110, 164—172 (1960).

LEWIS, L.: Cryptorchism. J. Urol. (Baltimore) 60, 345—356 (1948).

LINKE, C. A., and J. H. KIEFER: Occurrence of testis tumor in undescended testes. J. Urol. (Baltimore) 82, 347—351 (1959).

LOWSLEY, O. S.: The sexual glands of the male, p. 213—216: Oxford University Press 1942.

McCREA, L. E.: Lucite: A new synthetic material suitable for testicular prosthesis. Urol. cutan. Rev. 42, 732 (1938).

MINOR, C. L.: The empty scrotum. Öediat. Clin. N. Amer. 6, 1137—1146 (1959).

MOORE, V., and G. E. HOWE: Mullerian duct remnants in the male. J. Urol. (Baltimore) 70, 781—788 (1952).

MOOREHEAD, S. W.: Treatment of undescended testicle. Surg. Clin. N. Amer. 27, 1541—1549 (1947).

Patton, J. F., D. N. Seitzman, and R. A. Zone: Diagnosis and treatment of testicular tumors. Amer. J. Surg. **99**, 525—532 (1960).

Ponce de Leon, I.: Tratamineto funcional de la criptorquidia resolucion quirurgica por via abdominal. Arch. esp. Urol. **17**, 69 (1964).

Prentiss, R. J., and associates: Undescended testes in a young boy. J. Amer. med. Ass. **150**, 13 (1952).

—, and associates: Medical and surgical treatment of cryptorchidism. Arch. Surg. **70**, 283—290 (1955).

—, and associates: Medical and surgical treatment of cryptorchidism. A. M. A. Scientific Exhibits Nos 766—770 (1955).

—, and associates: Undescended testis: Surgical anatomy of spermatic vessels, spermatic surgical triangles and lateral spermatic ligament. J. Urol. (Baltimore) **83**, 686—691 (1960).

—, and associates: Medical and surgical treatment on the undescended testicle. J. Okla. med. Ass. 952—597 (1961).

—, and associates: Surgical repair of undescended testicle. Calif. Med. **96**, 401—405 (1962).

—, and associates: Testicular prosthesis: Materials, methods and results. J. Urol. (Baltimore) **90**, 208—210 (1963).

Rea, C. E.: The use of a testicular prosthesis made of lucite with a note concerning the size of the testis at different ages. J. Urol. (Baltimore) **49**, 727 (1943).

Schaffer, J. P.: Morris' human anatomy, 10th ed. New York: Blakiston Co. 1942.

Snyder jr., W. H.: Undescended testis. Calif. Med. **72**, 239—242 (1950).

Spalteholz, W.: Hand atlas of human anatomy. Philadelphia: J. B. Lippincott Co. 1937.

Torek, F.: Orchiopexy for undescended testicle. Ann. Surg. **94**, 97—110 (1931).

Utz, D. C.: Surgical management of cryptorchidism. Surg. Clin. N. Amer. **39**, 995—1005 (1959).

Warren, J., and R. M. Green: Warren's handbook of anatomy. Boston: Harvard University Press 1937.

Wershub, L. P.: Orchiopexy. J. Urol. (Baltimore) **60**, 631—635 (1948).

Hypospadias

ORMOND S. CULP and J. WILLIAM McROBERTS

With 28 Figures

Hypospadias is characterized by incomplete development of the urethra and usually is accompanied by other anomalies of the genitalia. The clinical composite has been challenging surgical imagination and ingenuity for several generations.

According to WAIN, the Roman physician, CLAUDIUS GALEN, first described and named this anomaly during the second century. The term is composed of the prefix "hypo," meaning "under," and the Greek word "spadizo" meaning "to pull" or "tear off." Thus, it literally means that the urethra was "torn off and ends under" the penis.

Embryology and Morphology

The embryogenesis of hypospadias is relatively uncomplicated. It involves the failure of the margins of the urethral groove to meet and fuse in the midline to form the urethral floor. As pointed out in preceding chapters, the genital tubercle is the primordium of the glans penis in the male and the glans clitoridis in the female. On the caudal aspect of the genital tubercle are slender structures, known as genital or urethral folds, that are separated by the urethral groove. In the male embryo, at about 8 to 10 weeks (30 to 50 mm stage), the urethral folds close over the urethral groove and become progressively fused from the basal to the distal aspect to form the urethra. Premature arrest may occur at any point during this closure, producing hypospadias in its various forms (Fig. 1).

In the normal closing process the urethra penetrates the glans and joins the preformed fossa navicularis. The fossa is formed separately by an ingrowth of epithelium from the surface of the glans. Thus the male urethra has a dual embryologic origin. This is reflected by the presence of two meatuses in many patients with hypospadias. One is located on the glans and opens into a cul-de-sac representing the fossa. The other is situated posteriorly and is the opening through which the urine passes (Fig. 2).

In the most common types of hypospadias the undeveloped urethra distal to the displaced urinary meatus consists of fibrous elements, reputed to be remnants of urethral mucosa and arrested corpus spongiosum. These fibrous bands are shorter than the corresponding length of the penis and produce the characteristic ventral curvature of the penis (chordee) usually seen with hypospadias (Fig. 1 B, C, and D).

SMITH and BLACKFIELD have expressed the view that the presence of rudimentary corpus spongiosum is a "myth" kept alive by the homage of authors and the dutiful depiction of medical illustrators. Of 46 patients who underwent correction of chordee by these authors, 43 had "perfectly normal corpora spongiosum." In the three cases in which a rudimentary corpus was seen, the patients had perineal hypospadias. VAN DER MEULEN has said that the rudimentary corpus spongiosum theory is untenable because the corpus "is formed in the periurethral tissue only at the end of the fourth month." PAUL and KANA-GASUNTHERAM have expressed the view that the fibrous band is a remnant of the urethral plate.

We infrequently have seen normal corpora spongiosa in these patients and are inclined to agree with CREEVY that "this is a matter of semantics." Confusion arises from the difficulty of differentiating fibrous, arrested corpus spongiosum from Buck's fascia. Occasionally congenital chordee occurs without hypospadias (see Fig. 19). We have encountered 14 patients with marked chordee and ab-

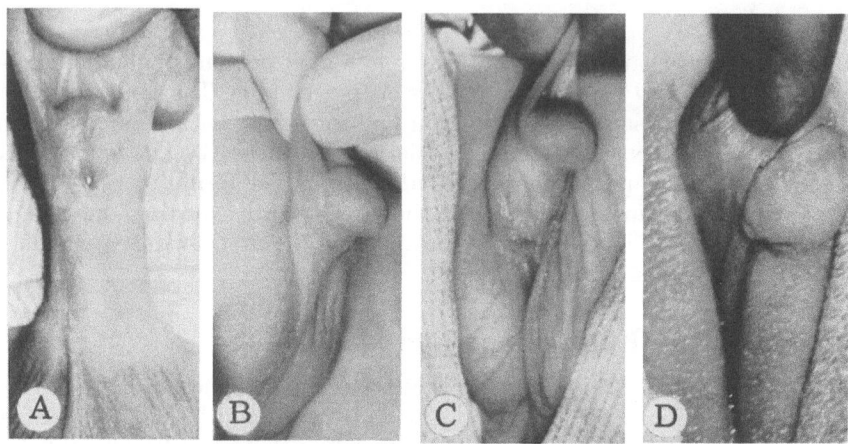

Fig. 1. Examples of untreated hypospadias with varying degree of deformity. *A*, Coronal without chordee *B*, Penile, *C*, scrotal, and *D*, perineal, each with chordee

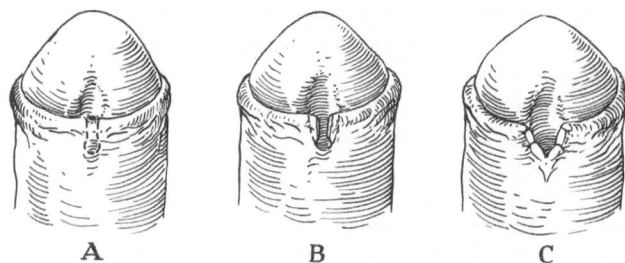

Fig. 2. Denis Browne anterior meatotomy. *A*, Dotted lines identify bridge of tissue between hypospadiac urethral meatus and smaller distal opening. *B*, Intervening tissue divided and (*C*) edges sutured

normally short urethra with the meatus in the normal position at the tip of the glans. Each of these had a short but otherwise normal corpus spongiosum. The converse also has been observed: hypospadias severe enough to require urethroplasty in the absence of any chordee (four cases). The corpus spongiosum was not recognizable in the terminal segments of these urethras.

Concurrently with the formation of the urethra (at about the 55 mm stage) the prepuce is formed over the glans penis. That the process is intimately related to the fusion of the urethral folds is reflected in the failure of the prepuce to develop ventrally in the presence of hypospadias. Accordingly, its normal redundancy on its dorsal and lateral aspects gives it a typically hooded appearance. The glans penis is well developed, but appears rolled out in its ventral surface by the open, undeveloped urethral strip, thereby often creating a cobra-head deformity, characteristic of many of the situations.

The genital folds develop in their caudal aspect by expanding and merging in the midline to form the scrotum. The scrotal swellings, by the seventh month, are pouched out as extensions of the peritoneal cavity when the testes begin to pass through the inguinal canal. By the eighth month the testes are generally in the scrotum. However, hypospadias is associated with cryptorchism in about 15% of cases, the cryptorchism being bilateral in two thirds of the cases (CAMPBELL, SORENSON, ROSS and colleagues, KENNEDY). SORENSON also pointed out that the incidence of cryptorchism increases with the severity of the anomaly, being most frequent in the scrotal and perineal forms.

When the urethral meatus is proximal to the penoscrotal angle, the scrotum is bifid. This is probably a result of a disturbance of the mesenchymal proliferation of the genital tubercle in its normal cephalic development relative to the caudal development of the genital swellings as they form the scrotum.

Most embryologists conclude that up until the ninth week (40 to 45 mm stage) the development of the external genitalia is essentially the same in both male and female embryos. However, SPAULDING found that he could reliably determine the sex of the human embryo "practically from the first appearance of the urethral groove," which would be about the sixth week, by noting the "marked difference" in its length in the two sexes.

According to the monohormonal thesis of WIESNER, the male embryo, from a common starting point (40 to 45 mm stage), differentiates to its definitive form under the influence of "morphogenic hormone" from the embryonic testes. Castration while the embryonic gonad is still bipotential leads to a complete female sex differentiation (JOST). This process can be reversed by giving testosterone. Conversely, ovarian hormones play no role in the differentiation of feminine sex characteristics, as castration does not affect this process.

Thus we can appreciate hypospadias as being a hermaphroditic phenomenon in which the urogenital primordia of the male develop toward those of the female. The greater the degree of hypospadias, the greater is the tendency to feminization. In extreme forms, as in the perineal variety, it may be impossible to decide the baby's sex. The penis may be small enough, with ventral curvature, to resemble an enlarged clitoris, and the empty, bifid scrotum may be mistaken for labia. In such cases, by spreading the labial-like edges of the perineal meatus, the enlarged prostatic utricle can sometimes be visualized. HOWARD, on the basis of urethroscopic examinations made on 14 patients with hypospadias, demonstrated a correlation between the size of the utricular dilatation and the severity of the anomaly. It is in these cases that it is particularly important to determine the patient's true sex. In addition to thorough physical examination, necessary investigations usually include urethrography, panendoscopy, buccal smears, and quantitative determination of 17-ketosteroids in the urine. Even so, complicated intersex problems may exist, as described in chapter 10.

Incidence

Male hypospadias is generally classified in the literature according to the location of the urethral meatus: glandular (juxtaglandular or coronal), penile, penoscrotal, scrotal, and perineal. Unfortunately there is no classification that encompasses all the ramifications of this deformity. SORENSON estimated that in three quarters of all cases the hypospadias is glandular or juxtaglandular, while in about one eighth it is of the penile or perineal variety. CAMPBELL found more of the penile (25 to 30%) and penoscrotal-perineal varieties (10 to 15%) and less of the glandular variety (40 to 50%).

With surgical correction of the chordee the meatus is shifted more proximally on the ventral surface of the penis, paradoxically increasing the deformity of the urethra to more accurately reflect the degree of its underdevelopment. We have found that once the associated chordee was corrected, approximately 28% of the meatuses were penile in location, 45% were penoscrotal, 16% were scrotal, and 10% were perineal (Fig. 3). It is advisable to forewarn parents of this further displacement before correcting the chordee.

Sorenson found the incidence of hypospadias to be 1 per 300 live births in Denmark. Other authors (for example, Crawford in England and Campbell in the United States) give a somewhat lesser incidence. In the United States with

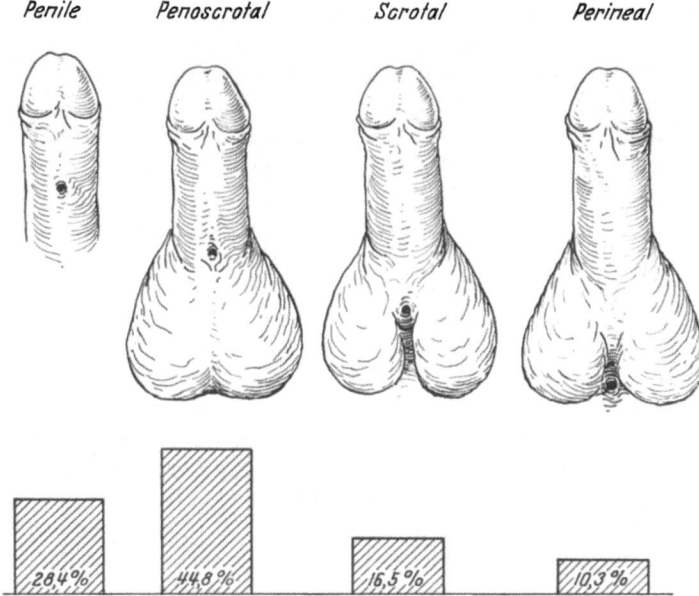

Fig. 3. Location of urethral meatus after correction of chordee

current births numbering nearly 2,500,000 males yearly, approximately 8,000 hypospadiacs are estimated to be born each year.

Heredity seems to be a factor, as there is a tendency for the anomaly to run in families. Ross and colleagues found that 15 of 108 patients (14%) gave a history of more than 1 case of hypospadias occurring in the same family. Sorenson also found a significantly high incidence of hypospadias occurring in the siblings of patients with the anomaly. He pointed out that the anomaly was frequently found in both monozygotic twins, but not in both twins in all instances and hence that there must be exogenous factors in play as well as endogenous ones.

Sorenson had no statistical evidence that maternal disease per se during the pregnancy contributed to the production of the anomaly. He expressed the view that, on the basis of his statistics, hypospadias is transmitted by a recessive gene, and that, if parents already have a hypospadiac child, the chance that the next child would also have hypospadias is about 10%.

Female Hypospadias

Although male hypospadias has an overwhelming statistical predominance, the female variety cannot be ignored completely. A few writers (for example, Whar-

TON, GARSKE) have regarded female hypospadias as being fairly common but according to CAMPBELL only 46 cases have been reported.

The external urethral meatus usually is situated obliquely on the anterior vaginal wall proximal to its normal site in the vestibule. In extreme situations the meatus is located just outside the vesical neck, thereby producing, in essence, a vesicourethrovaginal fistula with incontinence. This degree of female hypospadias is usually associated with other conspicuous genital anomalies, notably incomplete separation of the urethra and vagina and posterior displacement of both toward the anus. This must not be confused with congenital absence of the entire female urethra or with female epispadias.

The clinical importance of female hypospadias varies with the degree of the deformity. The mild forms may cause little or no symptoms beyond the greater susceptibility of the patient to ascending infections from the vagina. Therapy is directed toward meticulous hygiene and especially vaginal cleanliness. Meatal stenosis is not uncommon in the milder forms and is best treated by dilatation, since the urethra already is short and meatotomy could produce incontinence.

In the more severe forms of female hypospadias, characterized by leakage of urine, the primary therapeutic objective is establishment of urinary control. Treatment is surgical but construction of *additional* urethra is not necessary in all instances. Other successful operative procedures have been the Kelly, the Kennedy, and the Marshall-Marchetti operations.

In extreme and refractive cases, operations similar to those employed in female epispadias may be necessary. Indeed, urinary diversion may be the only practical solution, especially when efforts to construct new urethra have failed.

Therapeutic Principles

The treatment of male hypospadias also is strictly surgical. However, in glandular or coronal hypospadias without chordee (Fig. 1 A) many of us are convinced that patients who have a well-directed urinary stream should be left alone. However, in about 50% of patients with glandular hypospadias and no chordee there is a small, stenotic meatus (SMITH and FORSYTHE, and BARCAT and STEPHAN). These individuals have a poor urinary stream and require meatotomy. If this is performed in the customary manner, the degree of hypospadias will be aggravated and probably sufficiently so to necessitate subsequent construction of additional urethra.

DENIS BROWNE pointed out that patients in this category usually have a more distal vestigial channel parallel to the one through which the urine passes. By dividing the intervening septum and suturing the edges he enlarged the meatus without sacrificing any of the useful urethra (Fig. 2).

Despite the many controversial facets of hypospadias, it is generally agreed that any significant degree of chordee (Fig. 1 B, C, and D) makes surgical treatment mandatory. As noted previously, straightening the phallus usually increases the degree of hypospadias (unless it already is scrotal or perineal), and some type of urethroplasty becomes obligatory.

Surgery has several goals. The corrected chordee releases the angulated penis and permits normal sexual relations in adulthood. The reconstructed urethra gives the youngster a manageable urinary stream, permitting normal voiding in a standing position without soiling his clothing. Fertility is enhanced as the additional urethra permits the semen to be deposited properly within the inner one third of the vagina with easy access to the external cervical os.

It also is conceded by all authorities that new urethra should not be constructed until the penis is truly straight. Since no operation is entirely foolproof, most writers have favored separate surgical procedures for straightening and for construction of new urethra.

There have been frantic pleas throughout the voluminous literature to avoid circumcision of the hypospadiac child. The original preputial skin is desirable and useful in all types of straightening operations.

Creevy maintains that, from the standpoint of surgical technique, the age at which the penis is straightened is not important. However, Cecil believes that if the patient has reached any considerable age, potential atrophy of the blood spaces on the ventral surface of the corpora can complicate the chordee. He, therefore, does the straightening operation at about 2 years of age. Van der Meulen also found that the curvature of the penis can increase as the patient grows older, and he attributed this to the growth discrepancy between the corpora cavernosa and the fibrous elements of the chordee.

Once the penis is straight, if it has grown substantially before urethroplasty is undertaken, probable chances of success are enhanced. Nevertheless, we believe strongly that it is the inalienable right of every boy to be a "pointer" instead of a "sitter" before he starts to school.

Higgins and Creevy, among others, regard postoperative erections as being detrimental to wound healing, and accordingly give their patients estrogens starting a few days before operation and continuing this medication until the wound is fairly well healed (for example, 10 mg of diethylstilbestrol daily). In our experience estrogens, even in larger doses, have been disappointing in teenagers and young adults. By far the best way to avoid troublesome postoperative erections is to perform all surgical stages during the preschool period.

Techniques for Correction of Chordee

Almost 50 allegedly different procedures have been proposed for correction of chordee. Obviously all of these cannot be acknowledged in this presentation. Most of the methods of straightening the penis vary only in the manner of deploying skin from the hooded prepuce to the denuded ventral surface of the penis. Before discussing specific techniques, a few basic tenets warrant special consideration.

Most surgeons advocate an indwelling urethral catheter during the dissection. This also facilitates preservation of the original meatus when it is of adequate caliber.

Some use a tourniquet at the base of the penis to control bleeding during the operation. But this is seldom practical or necessary during infancy.

Too frequently the straightening operation is too conservative. It is imperative to incise or excise all tissue contributing to the angulation. This usually demands extensive freeing of the intact urethra. It also usually entails removal of some of Buck's fascia and occasionally some of the ventral wall of the corpora cavernosa.

Some surgeons unhesitatingly attempt to correct the chordee when they have no intention of ever constructing the additional urethra. It is possible to divert too much skin to the under surface of the penis for some types of urethroplasty. For example, excess ventral skin, ideal for free tube grafts, presents troublesome redundancy during any procedure based on the Thiersch-Duplay or Denis Browne principles. Best results will be achieved when the same surgeon performs each operation so it will assure proper preparation for the next stage.

The first practical method of straightening the penis was that of DUPLAY published in 1874 (Fig. 4). He transfixed the penis with a traction suture through the glans, which not only helped to immobilize the penis but it also accentuated the constricting bands of tissue. A transverse incision was then made distal to the meatus and extended upward and laterally into the hooded prepuce to aid in diverting the dorsal skin to the ventrum. DUPLAY was also one of the first to emphasize removal of all of the fibrous tissue rather than simply dividing it. The original transverse skin incision was closed longitudinally with interrupted sutures which served to divert the ample dorsal skin ventrally. In this operation the original urethral meatus was displaced posteriorly through a circular aperture in the skin near or back of the penoscrotal angle.

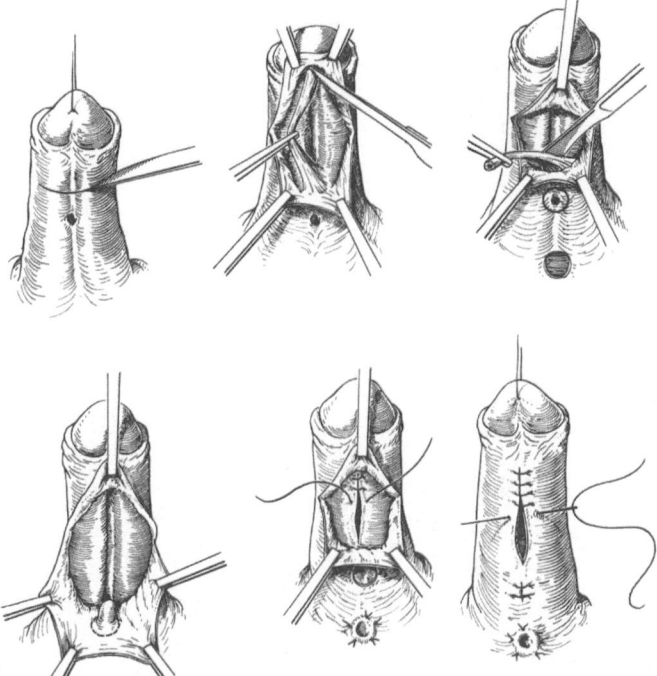

Fig. 4. DUPLAY'S operation for correction of chordee, based on transverse skin incision, excision of fibrous bands, posterior displacement of urethral meatus, and longitudinal closure of fascia and skin

Modification of the Duplay procedure was exceedingly difficult, so the hooded prepuce was attacked from other angles.

EDMUNDS' innovation was a two-stage procedure in which the preliminary step consisted of a bipedicle tubing of the preputial hood with formation of a buttonhole in the skin (Fig. 5 A). EDMUNDS stated that "the object of this incision is to divide the dorsal vessels of the prepuce, and so lead to the formation of a number of smaller vessels which will vascularize the prepuce from the sides." After a 3-month healing period the second stage was performed in which the tube was divided in its center and unfolded so as to form two proximal pedicle flaps, which in turn were applied to the denuded shaft of the penis after elimination of the chordee.

BLAIR, by omitting tubulation of the prepuce (Fig. 5 B), accomplished with only one operation the same end result as did the Edmunds method. He made a long dorsal slit in the redundant foreskin, thereby developing two lateral flaps

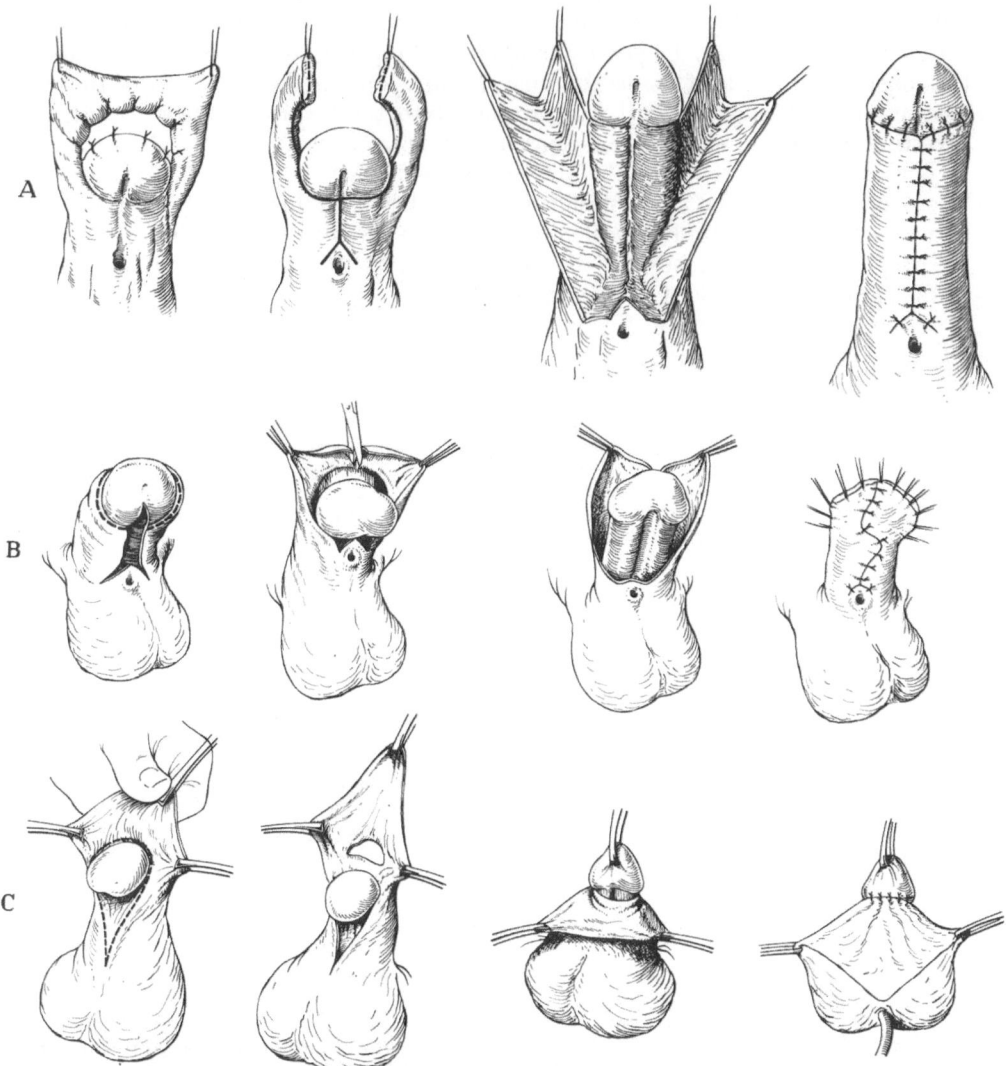

Fig. 5. Straightening operations. *A*, Two-stage Edmunds technique. *B*, One-stage Blair procedure. *C*, Nesbit operation, which is completed in a single stage

which were swung ventrally and tailored to cover the denuded ventral surface after the constricting fibrous tissue had been removed.

BECK (1917) was the first to divert the dorsal skin ventrally by means of a buttonhole flap in the prepuce. The preputial skin was unfolded. The buttonhole was made by incising the lateral free edges and making a complete circumferential incision just proximal to the glans. After dissecting away the constricting fibrous tissue, the glans was passed through the buttonhole, thus employing the dorsal foreskin to close the defect on the ventral side of the penis.

Little attention was given to BECK's original contribution until NESBIT published his modification in 1941 (Fig. 5 *C*). This permitted even more skin to be transferred ventrally. NESBIT began his operation with a pericoronal circumferential incision. He then mobilized the foreskin completely around the penis distal to the meatus, removed the constricting tissue, drew the prepuce distally, and made a transverse incision in it on the dorsal aspect just over the corona. The glans was then passed through the buttonhole, transferring the preputial skin ventrally. The skin was then carefully tailored to cover the denuded area on the ventral aspect of the penis. If this was not done meticulously, troublesome redundancy of the skin could be created.

The transverse closure of skin at the base of the penis was considered a cardinal virtue in the prevention of persistent or recurrent chordee. Another desirable feature of this technique was removal of a wedge of tissue from the ventral aspect of the glans. This defect was covered with skin in hope that when the urethra was constructed into this area the end of the penis would appear more normal and a well-directed nonspraying stream could be created more readily.

Grooving the glans was not new. BLAIR and colleagues (1933) had advocated doing so by drawing a quadrangular flap of skin from each side to cover the denuded portion. BRENDLER (1948) refocused attention on this aspect by excising a wedge of glandular tissue and letting it epithelize spontaneously. Healing usually required 2 weeks. During this period the edges of the denuded portion were held open by sutures anchored on the abdominal wall.

Efforts directed to revision of the glans during correction of chordee obviously have been motivated by attempts to reconcile functional and cosmetic features of the definitive location of the new urethral meatus.

Position of New Meatus

Not all types of urethroplasty are adaptable to creation of a new urethral meatus in the same relative anatomic site. Surgeons, therefore, must weigh the pros and cons of sundry techniques before settling on a method which they hope to master.

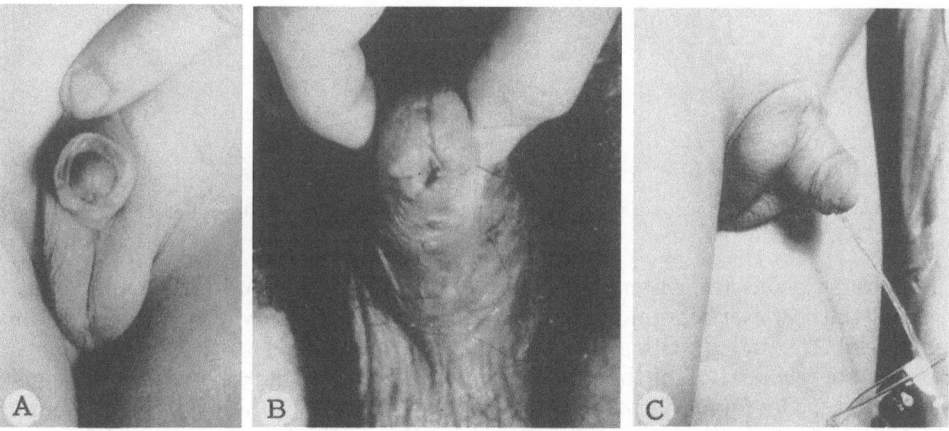

Fig. 6. Urethral meatus on base of glans. *A*, Child. *B*, Young adult. *C*, Urinary stream directed from meatus in this location

We are convinced that functional virtues should supersede stubborn dedication to anatomic perfectionism. Some of the most commendable cosmetic results have left much to be desired from the practical aspect. Best interests of the patient should take precedence over the aspirations of the surgeon. It is the basic right and privilege of every hypospadiac to be able to write his name in the snow and have it legible. Anything short of this is undesirable.

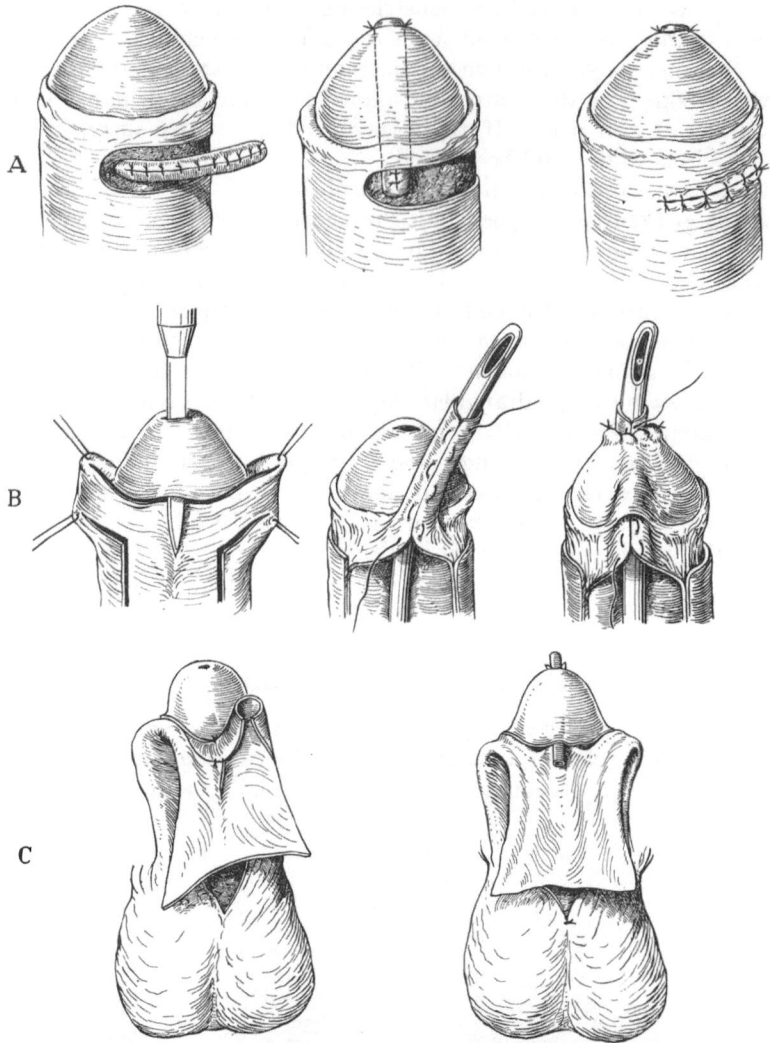

Fig. 7. Construction of new glandular urethra from ventral surface of penis. *A*, Russell operation. *B*, Young method. *C*, Mays technique

Most surgeons are content with having the external meatus of the new urethra end at the posterior margin of the preformed groove on the ventral surface of the glans (Fig. 6 *A* and *B*). Others insist that it must be situated at the tip of the penis. Deliberate termination of the urethra on the base of the glans poses no functional limitations. The urinary stream can be directed normally (Fig. 6 *C*). Furthermore, several patients in the series represented by this patient are now proud fathers. We feel that extending the urethra to the tip of the glans is fraught with hazards that more than offset any added cosmetic effect. Some glandular segments created by other surgeons had to be sacrificed because of refractive complications. Some of the successful efforts in this region have been only technical triumphs rather than justifiable grounds for cosmetic or functional pride.

Aside from previously mentioned grooving operations on the glans, procedures designed to provide a glandular segment of new urethra have entailed some type of free or pedicular tube graft.

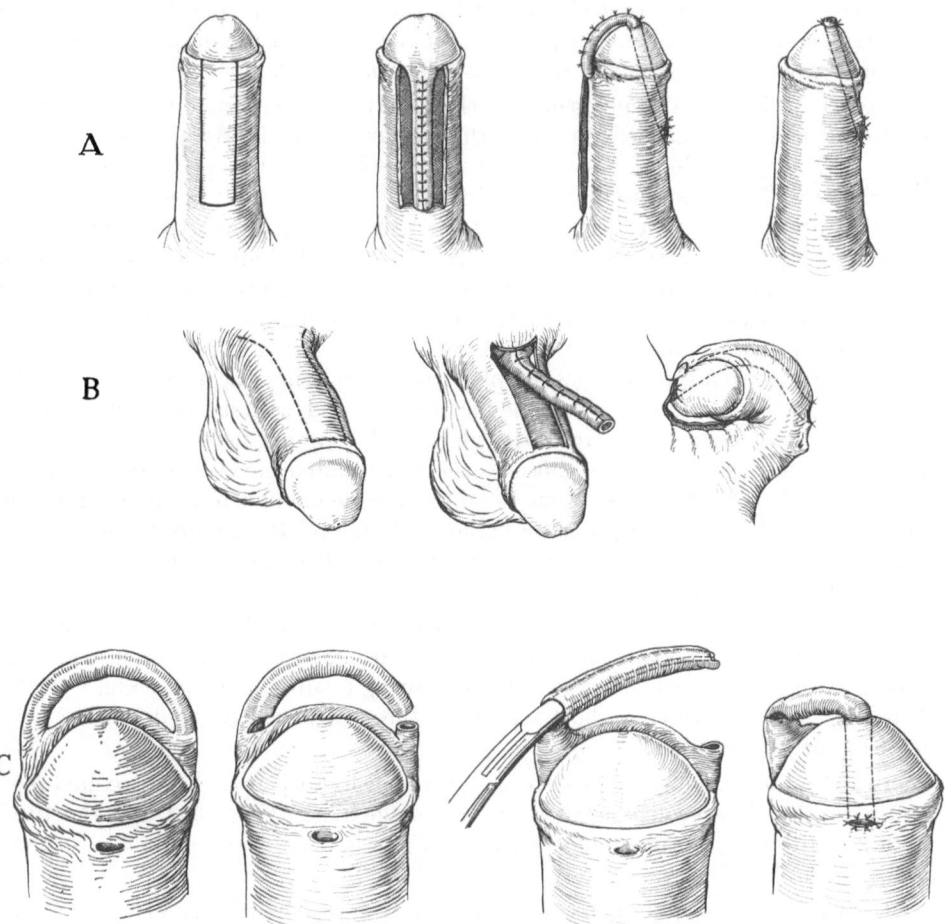

Fig. 8. Construction of new glandular urethra from dorsal surface of penis. *A*, Mayo operation. *B*, Davis method. *C*, Levi technique

RUSSELL pioneered the latter group and his operation was a lateral type of maneuver (Fig. 7*A*) with the pedicle attached ventrally. He began his operation with an incision just proximal to the corona, starting to one side of the midline and continuing across it circumferentially three quarters of the way around the penis. A second incision paralleled the first, 1 cm below it, and joined it on the dorsal aspect of the penis. This skin flap was dissected free except for its ventral attachment at the meatus and fashioned into a skin-lined tube which was brought out through a tunnel to the tip of the glans where it was secured.

YOUNG placed one half of the transverse flap on either side of the midline, turning both ends up and around a catheter, suturing them, and then pulling the tube through the glans in the manner of RUSSELL (Fig. 7 *B*). Both of these operations tried to utilize remaining portions of the hooded prepuce *after* straightening had been completed. This would be feasible only in relatively mild degrees of chordee.

MAYS constructed a glandular segment of urethra as a modification of the Nesbit straightening operation and based this new tube on the ventrum, as shown in Fig. 7 *C*. He squared the buttonhole in the prepuce and made the resulting flap into a tube over a catheter which was tunneled through the glans.

MAYO was one of the first to construct a tube of *dorsal* penile skin and bring it through the glans to a convenient point on the ventrum of the penis (Fig. 8 *A*). In this operation the tube was based distally near the glans.

DAVIS utilized the same principle (Fig. 8 *B*) but constructed a dorsal tube that remained attached proximally at the base of the penis. By angulation of the penis dorsally, additional *useful* urethra was provided. The procedure required three stages. The straightening operation diverted only the minimal amount of required dorsal skin. The first stage of the urethroplasty involved tunneling of the dorsal skin tube through the glans. It was begun by outlining, on the dorsal aspect of the penis, a rectangular skin flap of remaining prepuce which included "the subcutaneous tissue external to the tunica albuginea." From this flap a skin-lined tube was constructed with its raw surface outward and its base placed to include the preputial arterial supply, thus permitting the tube graft to "be made as long as the length of the penile skin allowed without sloughing." Then an incision was made at the tip of the glans and a tunnel fashioned through the ventral subcutaneous tissue as close to the hypospadiac meatus as the length of the tube graft permitted. The tunnel had to be wide enough to accommodate the tube graft readily and prevent later stenosis of the meatus or any portion of the new urethra. The penis was bent backward to utilize the entire length of the tube graft.

If the tube was long enough it was anastomosed directly to the hypospadiac meatus. If not long enough, the tube was brought out distal to the original meatus and the two urethral openings were joined during a subsequent procedure by the Thiersch-Duplay method. In both situations the pedicle was divided from its base after an interval of 2 or 3 weeks.

In the more severe cases of scrotal or perineal hypospadias, the proximal new scrotal urethra was reconstructed at an early stage by the Thiersch-Duplay method so that the dorsal skin tube might reach it. Urine was diverted by perineal urethrostomy during each stage.

LEVI constructed a similar tube transversely from the residue of hooded prepuce and tunneled it through the glans (Fig. 8 *C*). The tube graft was severed at the tip of the glans at a later stage.

Basic Types of Urethroplasty

Space precludes discussion of all types of urethroplasty. It has been estimated that more than 150 supposedly original operations have been proposed. CREEVY devoted more than 40 printed pages to consideration of the procedures that he considered most significant. This plethora of technical variations is proof positive that the ideal operation continues to be most elusive.

Urethroplasty is most readily classified on the basis of the source of the new urethra and its type of covering after the reconstruction. The surgical possibilities available to the surgeon depend in large part on the location of the external meatus after the straightening operation, notably whether it is at, distal to, or proximal to the penoscrotal angle.

The following general classification (modified from BARSKY) incorporates those procedures that have fundamental historical interest, those that have contributed notable innovations, and those that are enjoying significant vogue today. Personal variants of the senior author will be forthcoming in a later section of this presentation.

I. Repairs employing penile skin only
   A. Ventral skin tube covered by adjacent skin flaps (THIERSCH, ANGER, DUPLAY, BYARS)
   B. Ventral skin strip, not tubed, covered by adjacent skin flaps, using special tension sutures (DUPLAY-MARION, DENIS BROWNE)
   C. Ventral skin made into a pouch (OMBRÉDANNE, FARMER)
   D. Dorsal skin tube tunneled through glans (DAVIS)
II. Repairs employing penile and scrotal skin
   A. Ventral penile skin tube buried in scrotum (BUCKNALL, BECK, BLAIR, CECIL, CULP)
   B. Ventral penile skin strip, not tubed, buried in scrotum (MICHALOWSKI and MODELSKI)
   C. Ventral penile skin tube covered by previously constructed scrotal tube pedicle (WEHRBEIN, SMITH and BLACKFIELD)
III. Repairs employing free skin grafts
   A. Skin graft on a catheter tunneled through glans and penile shaft (NOVÉ-JOSSERAND, MCINDOE, HAVENS and BLACK, YOUNG and BENJAMIN)
   B. Free graft at time of release of chordee
      1. Split-thickness skin graft forming urethra (MCCORMACK)
      2. Full-thickness graft of prepuce (CLOUTIER)
IV. One-stage repairs combining release of chordee and urethroplasty (HUMBY, DEVINE and HORTON, BROADBENT, DES PREZ, ARONOFF).

Irrespective of techniques of preference, it cannot be overemphasized that no type of urethral reconstruction should be attempted until (1) all chordee has been eliminated, (2) the resultant urethral meatus is of adequate caliber, and (3) sufficient time has elapsed between operations to assure adequate blood supply to all tissues about to be deployed.

Evolution of the current status of hypospadiac surgery is a fascinating but frustrating tale of diversified failures punctuated by occasional ingenious and productive innovations.

DIEFFENBACH in 1837 was credited with the first attempt to correct minor degrees of hypospadias by freshening the edges of the tissues and "uniting them with needles," as well as tunnelization of the glans back to the hypospadiac opening. But his pioneer efforts failed. The distinction of the first clinical cure fell to ANGER in 1874 who developed a technique described by THIERSCH in 1869.

THIERSCH made a U-shaped incision on the ventral aspect of the penis with one limb of the U a few millimeters from the midline, and its parallel limb on the other side far enough away to outline a flap as wide as the desired new urethral circumference (Fig. 9). The bottom of the U extended a few millimeters below the meatus. The limbs extended to the posterior margin of the glans. The lateral margin of the strip of ventral skin was freed to include as much subcutaneous tissue as possible. This dissection was carried far enough toward the midline to permit its being wrapped around a temporarily placed sound or catheter. It was sutured to the contralateral skin edge.

The skin and underlying connective tissue lateral to the new urethra were undermined on each side to permit closure over the urethra without tension. This maneuver was designed to avoid overlapping suture lines in the hope that fistulas would be prevented. Nevertheless, they occurred. Even so, THIERSCH had a lasting influence on all subsequent operations for this type of anomaly.

The next era in the treatment of hypospadias involved "the big three" of French contributors — DUPLAY, NOVÉ-JOSSERAND, and OMBRÉDANNE — each of whom had his own particularly enthusiastic following.

DUPLAY made the most significant contributions. His first method of repair (1874) was similar to THIERSCH's except that the parallel incisions which outlined the new urethra were equidistant from the midline. It had all the disadvantages of the Thiersch method plus the additional one of superimposed suture lines. Six years later, DUPLAY published his second method which was based on the principle that a subcutaneously buried strip of skin grows to form a tube. To secure as much raw surface for healing as possible, skin edges were doubly locked on both sides with malleable silver clips. DUPLAY specifically noted that the buried strip of skin should have a width equal to one-half the circumference of the anticipated new urethra. He reported five cures with his second method. But DUPLAY again was plagued by fistulas.

BYARS advocated the basic Duplay principle in 1951 but urged that the lateral skin be undermined sufficiently to permit several layers of closure over the urethra. He also terminated the new urethra short of the original meatus and joined the two urethral segments at a later stage.

The idea of reconstructing the urethra with a tube of free skin was devised by NOVÉ-JOSSERAND in 1897. He employed a split-thickness graft from the lateral

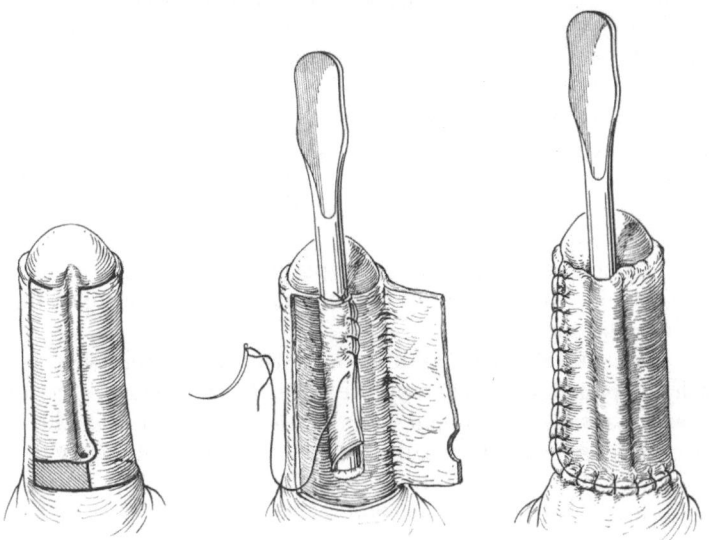

Fig. 9. Original Thiersch urethroplasty

aspect of the thigh which was rolled, with the raw side out, over a metal probe. The tube was inserted by means of a trocar through a subcutaneous tunnel on the ventral side of the penis. However, the results did not match the ingeniousness of the concept, and frequent fistula and stricture formation dampened early enthusiasm for the procedure.

Nevertheless, ardor for the free grafts did not disappear. In some respects this merely reflected the despair with which many surgeons viewed the treatment of hypospadias. Materials used for free grafts included veins, ureter, appendix, and, more recently and more rationally, vesical mucosa.

Results were not encouraging until McINDOE in 1937 introduced his original technique of using free inlay skin grafts (Fig. 10 A). HAVENS and BLACK carried on his work. Many of their cases were inherited by the authors. YOUNG and BENJAMIN (1949) championed further use of free grafts of split-thickness skin during the ensuing period.

McINDOE's procedure duplicated NOVÉ-JOSSERAND's except for utilization of a special trocar and bougie. He made a tunnel in the penis with the trocar. This extended from the tip of the glans to a point just distal to the hypospadiac meatus. The obturator of the trocar was then withdrawn. In its place was inserted a bougie around which a free split-thickness graft from a non-hair-bearing area had been sutured. With traction on the bougie, the trocar was removed from the tunnel in the penis, leaving the graft and bougie in place. The proximal and distal ends of the graft were sutured to the adjacent skin edges. At the end of 10 days, the bougie was removed, the urethra was irrigated and a special plastic stent was inserted for 6 months.

The grafted urethra was then anastomosed to the original meatus and the urine was diverted by perineal urethrostomy.

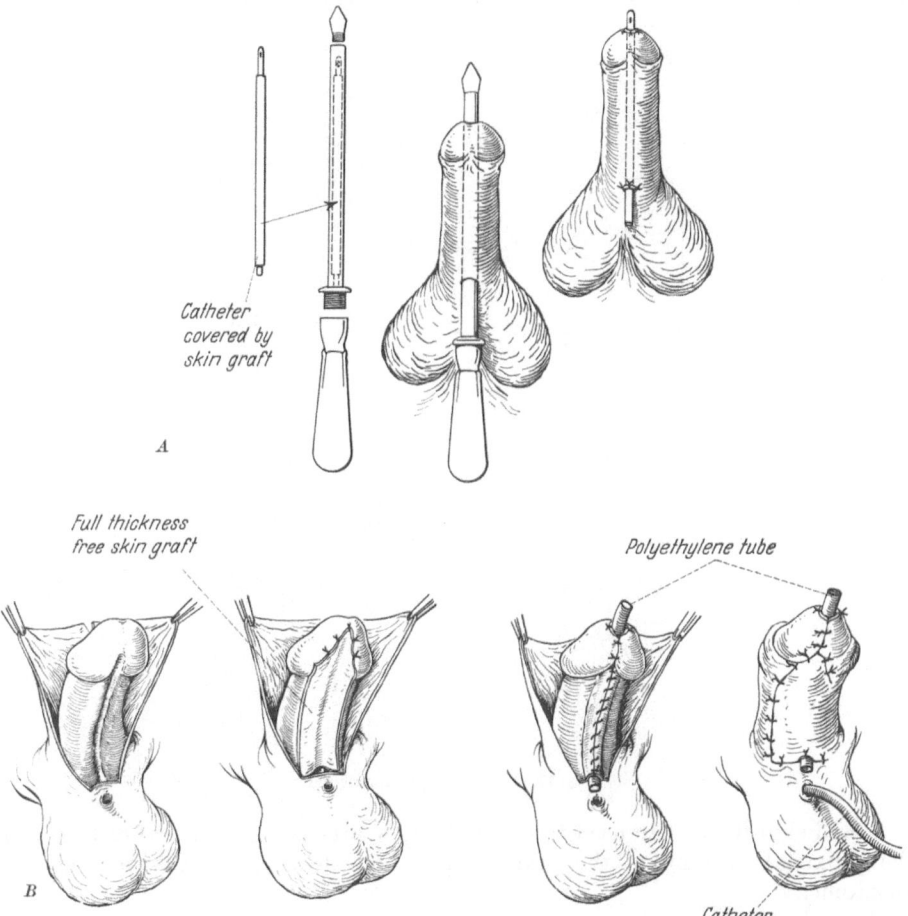

Fig. 10. Free skin grafts. *A*, McIndoe method. *B*, Technique of McCormack

Despite prolonged splinting, one of the serious disadvantages of the split-thickness skin graft proved to be a tendency for the new urethra to contract both longitudinally and transversely. Failure of the graft to grow and thereby keep pace with development of the penis predisposed recurrence of chordee. For this reason, HAVENS advised that this type of operation be deferred until the age of 16.

YOUNG and BENJAMIN, on the other hand, recommended that their similar urethroplasty be done at 2 to 3 years of age. They modified McINDOE's procedure by cementing the free graft to a catheter and made their subcutaneous tunnel with scissors and a hemostat. Leaving the catheter in as a splint, the urethral graft was anastomosed to the hypospadiac meatus and the urine was diverted by perineal urethrostomy. The splinting catheter was removed after 1 week. The perineal catheter was taken out after 8 days. The urethral stent advocated by McINDOE and HAVENS was omitted. Although no contractures or strictures were reported, YOUNG and BENJAMIN acknowledged fistulas in 11 of their 16 patients.

McCORMACK (Fig. 10 B) inserted a free graft at the time of the straightening operation. He corrected the chordee after the manner of BLAIR and constructed the new urethra from a split-thickness graft of hairless skin from the inner surface of the upper arm. The tube was splinted over polyethylene and the previously

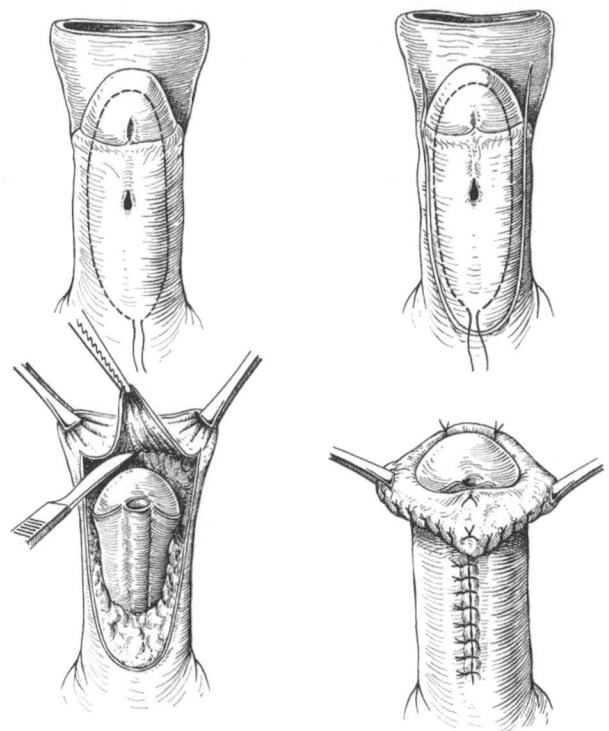

Fig. 11. Operation of Ombrédanne

dissected skin flaps were used to cover the new urethra. At the end of 3 months, the proximal opening of the grafted tube and the hypospadiac meatus were anastomosed.

Like McCORMACK, CLOUTIER (1962) performed a two-stage procedure, but used a full-thickness preputial skin graft at the time of the straightening operation. Three months later a modified Denis Browne procedure was employed to join the urethral segments.

The now infamous Ombrédanne operation (Fig. 11) once enjoyed popular acclaim. Unfortunately it created poorly draining, sacculated segments of new urethra characterized by stagnation of urine, infection, and postvoiding dribbling. We have encountered many of these that had to be sacrificed and reconstructed later by more acceptable methods.

Basically the Ombrédanne technique consisted of advancing the meatus with as many pouches as might be necessary to reach the ultimate destination. It was practical only in minor degrees of hypospadias but it was employed promiscuously with disastrous results, even in instances of scrotal hypospadias in which hair-bearing skin was employed with regrets.

OMBRÉDANNE extended the meatus by placing a purse-string suture around the periphery of an elliptical flap with the meatus in its center. When the purse-string suture was pulled taut, a pouch enclosing the meatus was created. The buttonholed preputial hood was drawn over the glans to cover the ventral raw surface.

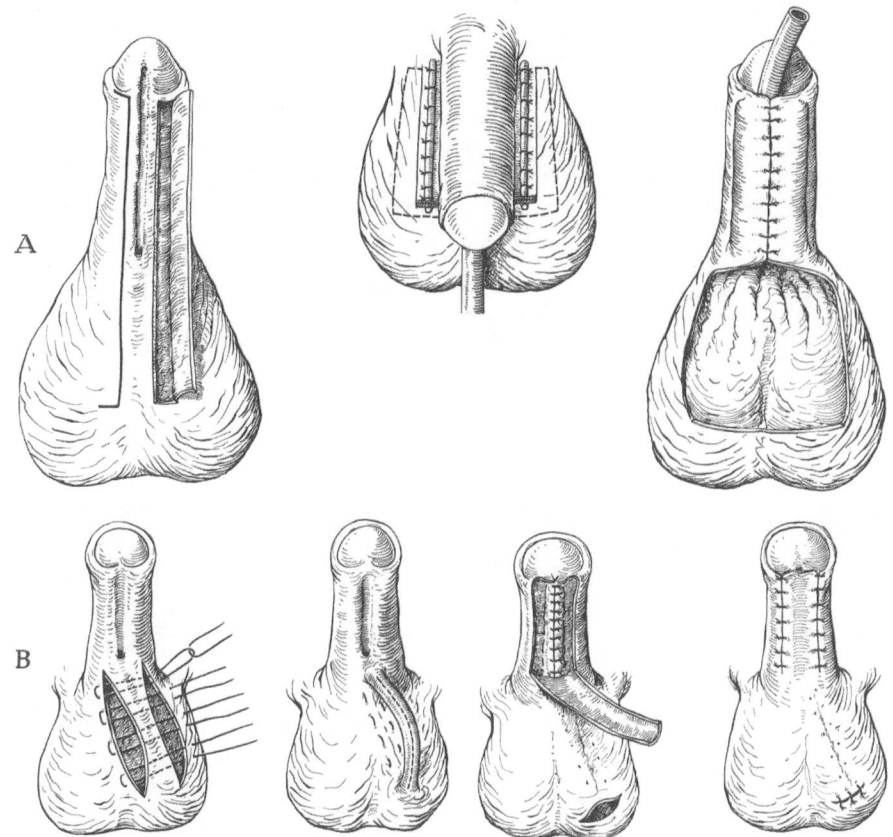

Fig. 12. Types of urethroplasty. *A*, Bucknall. *B*, Wehrbein

FARMER described a modification of the Ombrédanne operation in which a wrap-around flap of skin was used to cover the resultant ventral penile skin defect, rather than by OMBRÉDANNE's buttonhole incision. Although this modification did give a better cosmetic effect with more rapid healing, it, like the Ombrédanne operation, was limited in application.

BUCKNALL (1907) first attached the pendulous portion of the penis to the scrotum to avoid fistulas (Fig. 12 *A*). But he used scrotal skin for the bottom half of the new urethra and this hair-bearing tissue led to difficulties.

WEHRBEIN utilized the auxiliary scrotal concept 36 years later by forming a tube of scrotal skin during the straightening operation (Fig. 12 *B*). He opened this after building new urethra by the Thiersch-Duplay method and used the flap of opened scrotal skin to cover the new urethra and adjacent tissues. SMITH has found the method to be especially desirable and the results to be most gratifying. Nevertheless, fistulas have occurred.

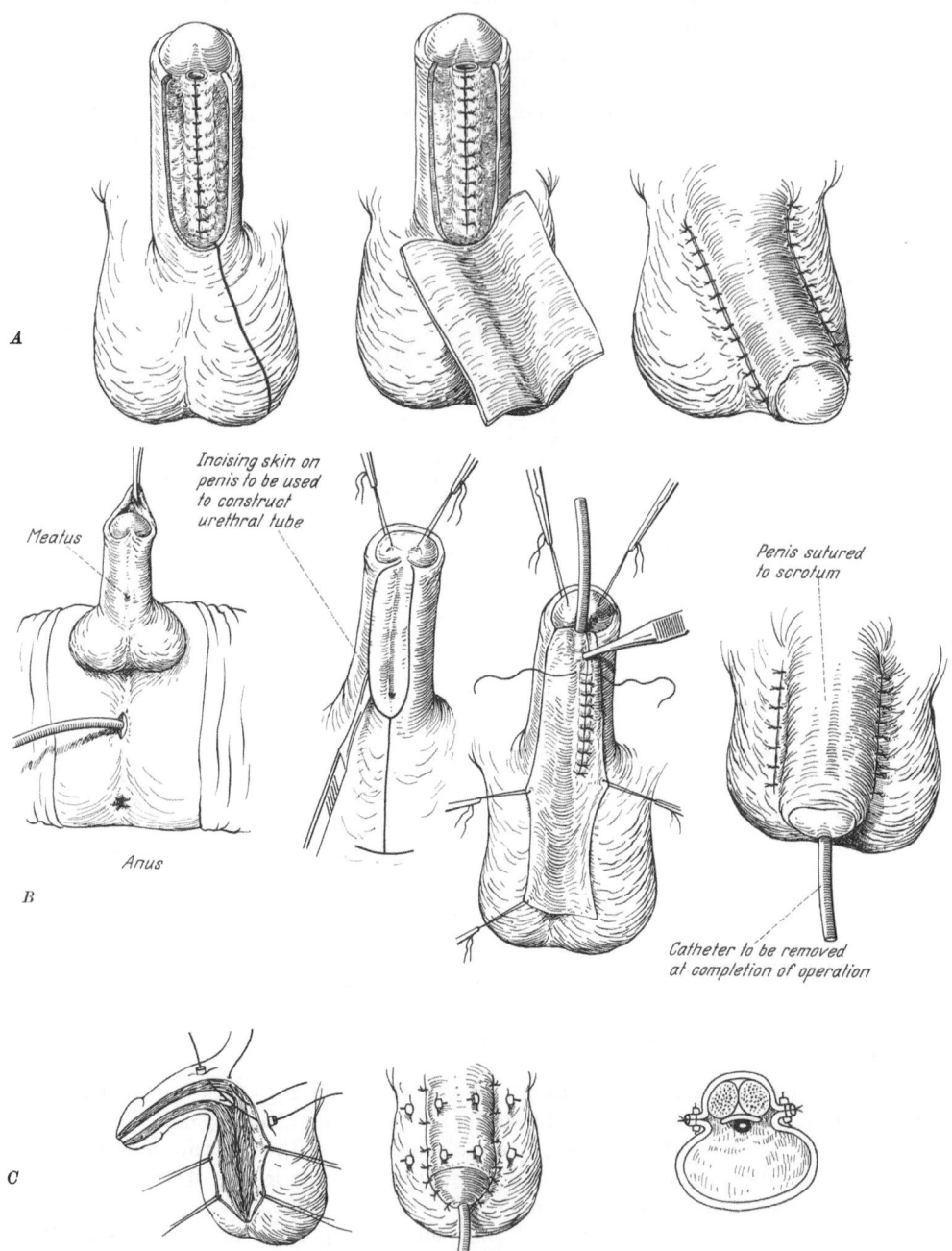

Fig. 13. Types of urethroplasty. *A*, Beck. *B*, Cecil. *C*, Michalowski and Modelski

BECK (1917) buried the entire pendulous urethra, constructed solely of penile skin, in the scrotum to prevent persistent fistulas (Fig. 13 *A*). But this maneuver gained little recognition and even less popularity.

Beginning in 1936, CECIL truly championed scrotal burial of new pendulous urethra (Fig. 13 *B*). Urethra was constructed by the Thiersch-Duplay technique

Table 1. *Complications after various types of urethroplasty*

Author	Method	Cases	Fistulas		Strictures	Dehiscences
			No.	%		
	Penile skin only					
BYARS	Duplay	77	13	17	1	
CREEVY	Thiersch-Duplay	43	18	42		
CULP	Thiersch-Duplay	10	5	50	2	
VAN DER MEULEN	Duplay-Raadsveld	9	2	22		1
VAN DER MEULEN	Duplay Y-V-plasty	33	6	18		1
COMPILED	Denis Browne	318	44	14		
WALTERS	Ombrédanne	32	4	13		
LOUGHRAN	Ombrédanne	29	4	14		
DAVIS	Davis	46	24	52	3	
	Penile and scrotal skin					
CREEVY	Cecil	9	0	0		
CULP	Cecil	121	7	6	18	6
WEHRBEIN	Wehrbein	11	3	27		
SMITH-BLACKFIELD	Blair	21	8	38		
SMITH-BLACKFIELD	Blair-Wehrbein	44	6	14		
	Free grafts					
McINDOE	Nové-Josserand	35	?	?		
HAVENS-BLACK	McIndoe	29	?	?		
YOUNG-BENJAMIN	McIndoe	16	11	69		
McCORMACK	Own variation	3	?	?		
CLOUTIER	Own variation	22	7	32		
	One-stage					
HUMBY	Own technique	12	2	17		1
DEVINE-HORTON	Own technique	23	7	30		
DES PREZ et al.	Own technique	10	3	30	3	
BROADBENT et al.	Own technique	8	3	38		
ARONOFF	Own technique	29	4	14		

and urine was diverted by perineal urethrostomy. The catheter over which the new urethra was formed was removed at the termination of the operation. Scrotal and penile skin were closed with silk sutures.

CECIL's operation carried the least risk of fistula formation of any urethroplasty devised up to that time. It had the additional advantages of construction of a urethra lined completely by epithelium that was non-hair-bearing, of providing sufficient tissue for later skin closure without tension, and of being accompanied by little or no edema of the involved tissues. CREEVY was convinced, after his comprehensive review, that edema is the nemesis of most types of urethroplasty.

Despite the fact that the Cecil technique requires two stages, the enthusiasm of the authors for these principles will be manifest in a later section of this discussion.

Early in our experience with the Cecil operation we found that six gaping wound separations (Table 1) due to virulent infection required only simple closure of the skin even though the urethra had a sizable associated defect. It was postulated at that time that burial of only a strip of skin no doubt would produce a satisfactory urethra.

MICHALOWSKI and MODELSKI proposed the combination of Cecil and Denis Browne principles in 1963 (Fig. 13 C). Double-stop tension sutures were employed for the approximation of penile and scrotal tissues.

It is unlikely that any single surgical procedure has ever attained more acclaim than the Denis Browne operation which was published in 1949. Despite the fact that Duplay showed that a strip of buried penile skin would create a tubular channel, useful as new urethra, virtually no attention was paid to this principle until Browne's announcement. Regardless of alleged and postulated priorities and the ensuing controversies and innuendoes, Denis Browne offered a unique solution to this basic problem.

By wide separation between the buried skin strip and the outer skin flaps, Browne thought, it was more difficult for the new epithelium to grow across and form a fistula as compared with the Duplay technique, in which both surfaces were left to heal close together. Additionally, Duplay's operation involved

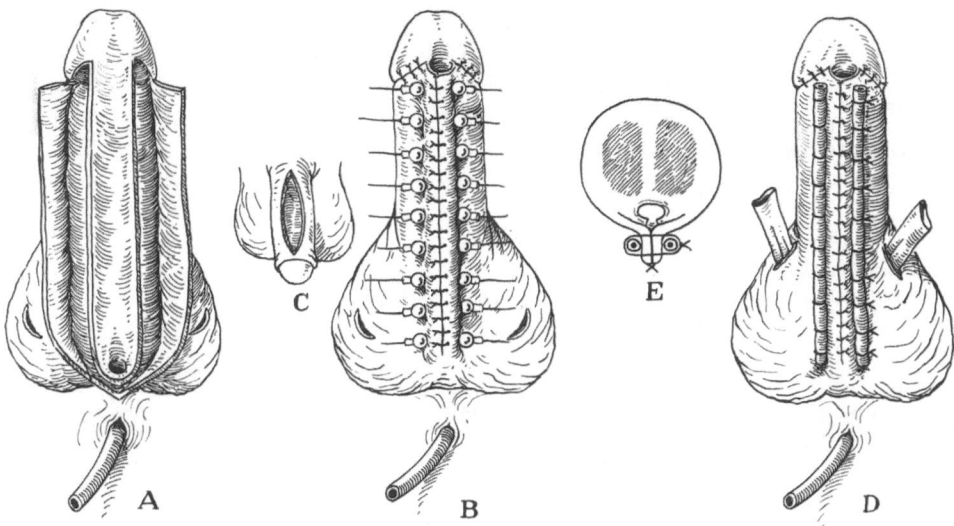

Fig. 14. Denis Browne urethroplasty. *A*, Basic incisions. *B*, Original type of closure with double-stop sutures. *C*, Relaxing dorsal slit. *D*, Modified closure with mattress sutures over strips of rubber tubing. Note through-and-through drain at penoscrotal juncture. *E*, Diagrammatic relationship of nylon mattress sutures

two stages; Denis Browne's was completed in one stage. Browne also omitted the tunnel through the glans, stating that it was not worth the added difficulties.

Urine was diverted by means of perineal urethrostomy. The urethroplasty began with a U-shaped incision on the vertical surface of the penis which extended from the ventral aspect of the glans to a point just behind the external urinary meatus (Fig. 14 *A*). A small triangle of glans was denuded on each side and the lateral skin margins of the U were widely undermined to form thick flaps of skin and subcutaneous tissue. Wide approximation of skin edges was employed so that they overlapped in the midline without tension. Stab wounds for drainage were made through the flaps lateral to the original meatus.

Skin flaps were approximated with double-stop sutures to keep all tension off the tiny catgut sutures which joined the skin edges (Fig. 14 *B*). The double-stop sutures were single nylon ones with a glass bead threaded on either end with an aluminium cylinder behind each bead. When these were crushed they locked the sutures in place. These sutures were so placed that at least 0.5 cm of slack was left between beads and skin on each side.

Browne advocated making a dorsal relaxation incision (Fig. 14 *C*) in all cases to reduce further any possibility of tension on suture lines. No dressing was

applied. The patient was confined to bed for the most part during the first 7
postoperative days, after which the double-stop sutures were removed. The
perineal urethral catheter was removed the following day.

The most serious drawback to the Denis Browne operation has been fistula
formation. In fact, from 10 to 30% of patients develop fistulas as the operation
is generally performed. However, the advantages of the procedure substantially
outweigh the drawbacks. This operation requires less skin than most other one-
stage urethroplasties and is adaptable to all degrees of hypospadias.

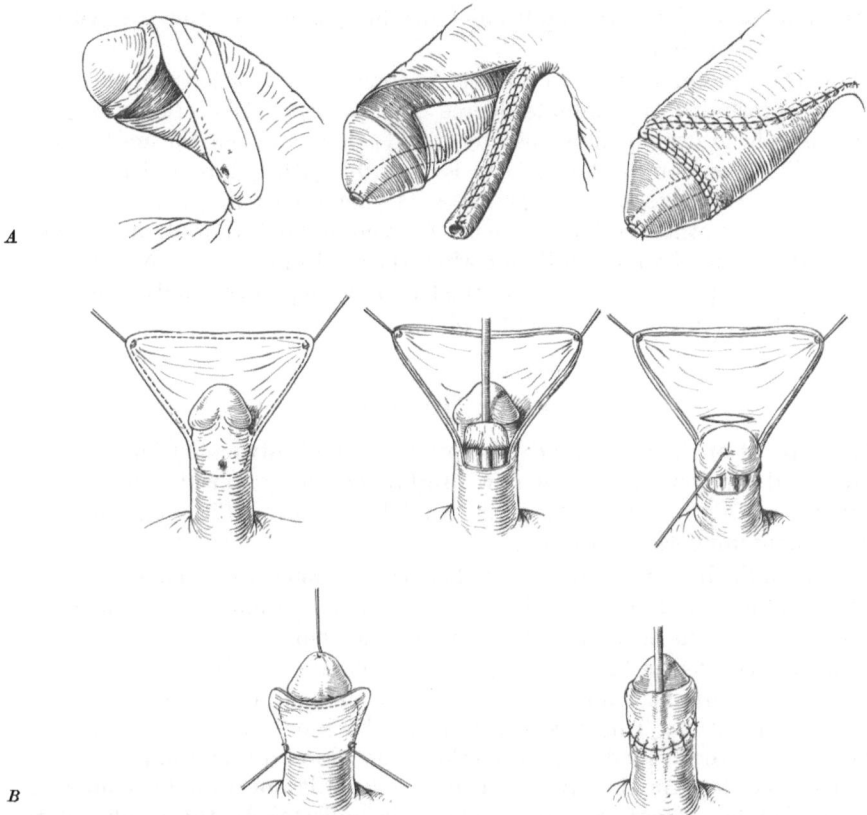

Fig. 15. One-stage operations. *A*. Method of Broadbent and colleagues. *B*, Aronoff techniques

As will be pointed out later, the technique has been modified by the authors
with a substantial reduction in the incidence of postoperative fistulas.

Admittedly, a dependable one-stage simultaneous correction of chordee and
construction of additional new urethra would be the ideal solution to this problem.
In 1941, HUMBY reported a fairly effective one-stage procedure. After the constricting chordee was removed, new urethra was constructed by using a full-
thickness tubed skin graft from the upper part of the arm. The graft was glued
around a rubber catheter and joined to the old meatus. Interestingly, only silk
was used as suture material. The procedure never became popular but it undoubtedly stimulated subsequent efforts by other enterprising surgeons.

Interest in one-stage operations was revived in 1961 by DEVINE and HORTON
who used a full-thickness tube graft of prepuce for the new urethra. This was
attached to the original meatus and to a denuded groove in the glans. It was then

covered with scrotal flaps and the remains of the perforated hooded prepuce. Since it is difficult to summarize all of the crucial technical aspects of this method diagrammatically, the reader is referred to the original publication.

Des Prez, Persky, and Kiehn also reported a one-stage operation during 1961. Broadbent, Woolf, and Toksu added another (Fig. 15 A) during the same year.

Aronoff digressed from the customary tube techniques in 1963 (Fig. 15 B). The hooded prepuce was separated into mucosal and skin layers. Ventral skin adjacent to the original meatus was freed as a flap and sutured to the buttonholed preputial mucosa to form the additional urethra. This was covered with the ventrally deployed preputial skin.

Despite the ingenuity of these and other one-stage operations, obvious limitations have precluded any notable degree of popularity. Dissection during the straightening phase usually must be restrained, whereas even the most radical techniques for correction of chordee can be fraught with failure and have to be repeated. Not all degrees of hypospadias lend themselves to one-stage repairs. Hazards inherent in other methods have not been eliminated. Indeed, management of complications may be more difficult after the "all out" effort. Nevertheless, these consolidated procedures have worked and perhaps eventually the refinements will displace all other therapeutic plans.

General Results

Glowing proof of the advantages of most operations advocated for correction of chordee and for construction of new urethra has been conspicuously absent in the available literature. There has been a plethora of generalities and relatively few truly enlightening series of cases.

In the main, the literature on hypospadias before 1950 consisted of descriptions of novel techniques, modifications of other operations, meager discussions of few cases, and the problems encountered in isolated incidents.

Only in the recent past has it been possible to critically appraise and compare the various procedures. Scant attention has been paid to the results of the straightening operation. This possibly reflects the more spectacular nature of the urethroplasty and a tendency to overlook the more subtle complications of the repair of chordee. As already noted, if the penis is not straight, completely healed, and flexible before the urethroplasty is undertaken, the results of the latter reflect this "shortcoming." The common complications of straightening operations include residual chordee, meatal stricture, and wound dehiscence. These may be caused by insufficient removal of the fibrous elements producing the curvature, by initial compromising of the external meatus, or by postoperative fibrosis.

Creevy reported that he had 1 instance of residual chordee, 2 of meatal stricture, and 1 of wound dehiscence after 78 straightening operations. In an earlier report, one of us (Culp) acknowledged 3 persistent chordees, 6 meatal strictures, and 4 wound separations as complications of 124 corrections of chordee.

Factual results of the various types of urethroplasty have been equally elusive. There is overwhelming evidence that surgeons do not appraise their products as critically as do the subjects. There is virtually nothing in the literature regarding the "spraying meatus" after sundry operations. Fistulas have been more dramatic and less nebulous. Yet, for practical purposes, a gaping or retracted new meatus is a fistula in the terminal segment.

A concerted attempt was made to review the most significant series of all types of urethroplasty reported to date. Comparably small groups were included under unique circumstances because of their academic value. The incidence of major complications following these varied methods is shown in Table 1.

There is no type of urethral construction that is ideal for all degrees of hypospadias. The most dependable varieties seemingly utilize existing penile skin in some manner, and coverage of the new urethra is accomplished in various ways. Tube grafts (free or attached) alone assure a meatus at the tip of the glans. But the desirability of such is debatable. Fistulas have been the major pitfall in most therapeutic schemes. Small ones can be just as annoying and refractive as large ones.

Anyone who elects to treat hypospadias not only must choose from a variety of methods but also should strive to perfect his techniques of choice. Even so, correction of hypospadias is not always as simple as some reports have implied. Associated problems often compound the fundamental one. Ill-advised initial treatment frequently jeopardizes all further therapeutic efforts. Some complications seem to be inevitable regardless of surgical experience.

Personal Experiences

From August 1950 to September 1965, more than 70 linear feet of new, watertight urethra were constructed for the hypospadiacs included in this series. These 422 individuals required 1,071 operations in addition to 383 earlier surgical procedures by other surgeons on 134 of the patients. From this plethora of tedious dissecting and meticulous sewing came many headaches, heartaches, and poignant therapeutic lessons.

Not all patients presented the same problem when first seen. They conformed to 10 basic categories (Fig. 16) each of which required some special consideration. Chordee had to be corrected in six of these groups, or 68% of all cases. From one to six previous unsuccessful straightening operations had been tried on 10% of patients.

When the customary hooded prepuce is still present, the simple method of correcting chordee shown in Fig. 17 is preferred. If the patient has been circumcised, some form of Z-plasty (see Fig. 19 C) is employed. Thorough dissection is imperative. If the original urethral meatus is of adequate caliber, it should be preserved to avoid postoperative meatal stenosis. Fine wire sutures (6—0) in the skin minimize scarring. (More recently, fine Mersilene sutures have been equally effective).

The type of dressing after this type of operation is important (Fig. 18). A layer of Telfa (to expedite removal) held in place by conforming gauze and covered with Elastoplast prevents edema and controls oozing. If the Elastoplast is applied too tightly, healing of the skin edges may be impaired. Secondary closure was necessary in 5% of the early cases.

The catheter is left indwelling for only a day or so unless the original meatus has been revised or compromised; then, it remains in situ until the dressing and the skin sutures are removed on or about the tenth postoperative day. Rectal anesthesia is usually employed at that time for very young patients. This also is an excellent opportunity for routine excretory urography.

This type of straightening has been used successfully on patients from 1 to 45 years of age. The ideal time seems to be around the age of 18 months. Persistent chordee required repetition, 6 to 12 months later, in 4% of 285 cases managed in this fashion. Most of these failures were in individuals beyond

puberty. Five of these patients had ventral clefts or contractures of the corpora
cavernosa that required partial excision of the undersurface of the corpora before
straightening could be accomplished. Only skin and the pressure dressing were
used to cover the exposed vascular channels in these cases. Subsequent erections
were normal.

Fourteen untreated patients had typical chordee but the urethra was intact
and the meatus was at the tip of the glans (Fig. 19). Simple freeing of the pendulous

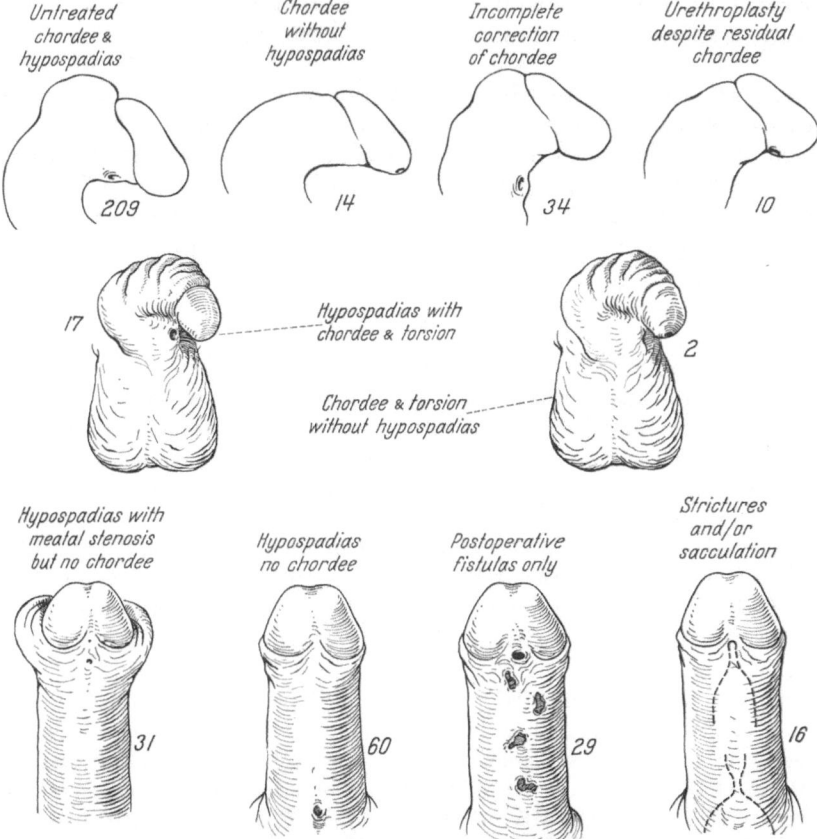

Fig. 16. Status of 422 patients when first seen

urethra did not suffice. In each instance the shortened urethra was divided. It is
best to transect the urethra just beyond the penoscrotal angle, free the distal
segment, excise any constricting tissue, close the skin by Z-plasty, keep a catheter
in both urethral segments for at least a week, calibrate all openings frequently,
revise the unused distal one if it constricts, and bridge the gap with new urethra
6 to 12 months later.

Essentially the same technique was employed for the 10 unfortunates who
had new urethra constructed elsewhere before all chordee had been corrected.

Nineteen untreated patients had chordee and 90° counterclockwise penile
rotation (Fig. 20). The urethral meatus was normally situated at the tip of the
of the glans in two of these. The other 17 patients had typical hypospadias.

Torsion may occur without chordee, may be clockwise or counterclockwise,
and may be as much as 180°. It has been a revelation how frequently various

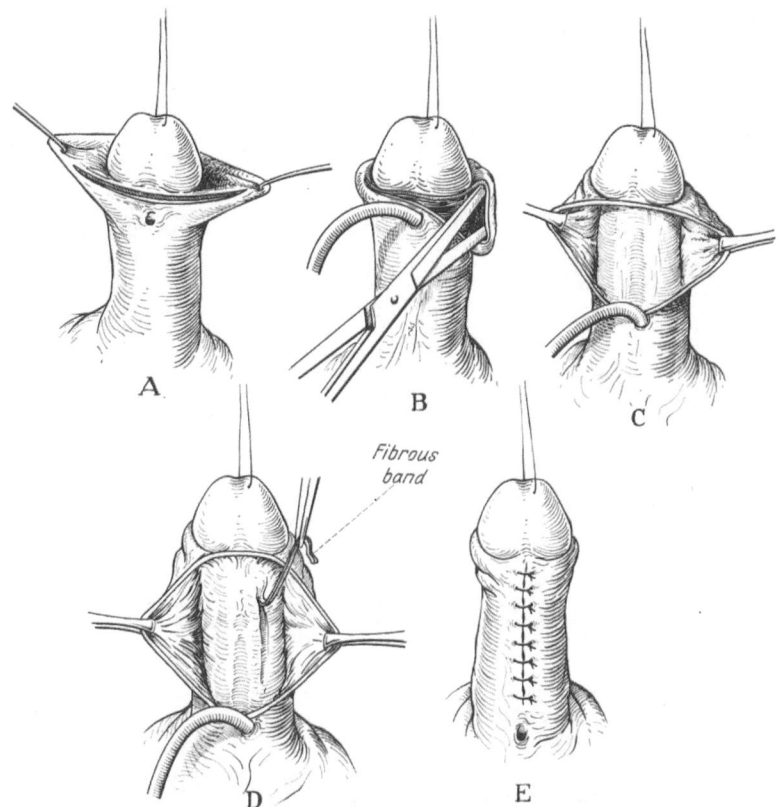

Fig. 17. Correction of chordee. *A*, Transverse incision distal to meatus is extended into hooded prepuce and (*B*) latter is freed with blunt scissors. *C*, Urethra is freed from corpora with indwelling catheter as guide. *D*, All constricting tissue is excised after freeing penile skin laterally. Meatus gravitates to most convenient position. *E*, Penile skin is closed longitudinally with interrupted sutures

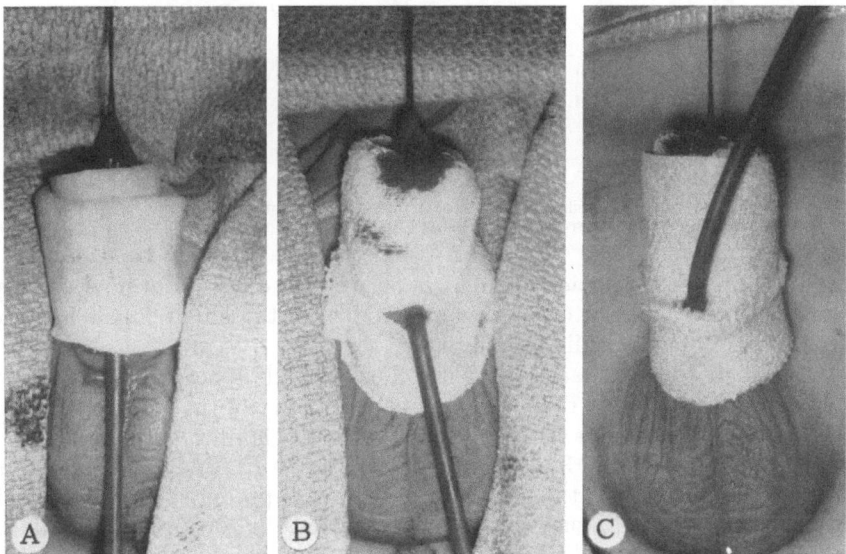

Fig. 18. Dressing after correction of chordee. *A*, Layer of Telfa. *B*, Conforming gauze. *C*, Covering of Elastoplast

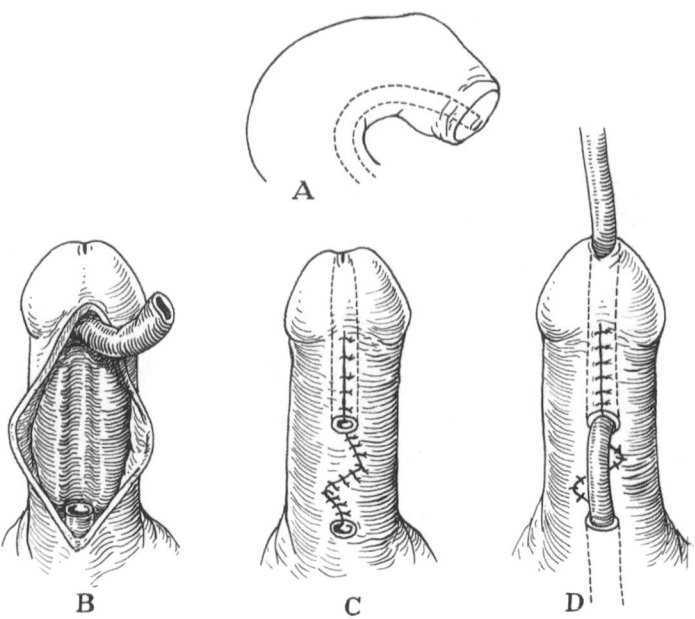

Fig. 19. *A*, Chordee without hypospadias. *B*, Urethra transected near penoscrotal juncture, distal segment freed, and constricting tissue excised. *C*, Skin diverted by Z-plasty and severed urethra attached to adjacent skin. *D*, Catheter left in both urethral segments

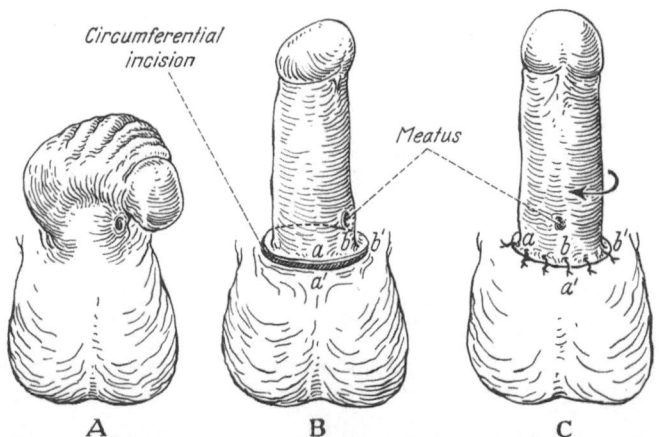

Fig. 20. *A*, Chordee, hypospadias, and counterclockwise torsion. *B*, Deep circular incision after correction of chordee. *C*, Counterrotation of penis and reapproximation of adjacent tissues

forms of asymptomatic torsion alone have been noted during routine cystoscopies once one became cognizant of this anomaly. Most of these individuals have been totally unaware of any penile peculiarity.

In the past, torsion associated with chordee and hypospadias usually was ignored and was attributed to developmental inequality of the corpora cavernosa. Neither of these concepts is tenable today. Once all chordee was eliminated, the glans continued to face the 3 o'clock position (Fig. 20 *B*). Simple circumferential incision and deep freeing at the penoscrotal juncture, appropriate counterrotation, and reapproximation of tissues completely corrected the torsion in these 19 cases (Fig. 21). Treatment should constitute a separate stage and should be deferred until at least 6 months after successful straightening.

After all chordee had been corrected the urethral meatus was at or distal to the penoscrotal juncture in almost three fourths of the patients (Fig. 3). Additional urethra was not constructed until the penis seemed to be truly straight, the displaced meatus appeared to be of adequate caliber, and at least 6 months had elapsed since the last operation. Even so, personal judgments were not always infallible despite more than average clinical experience.

The ideal age for urethroplasty seems to be about 5 years, but of late more patients have undergone the procedure during the preceding 12 months without difficulty or regrets. The degree of genital growth has been a potent motivating factor.

When the meatus is at or just distal to the penoscrotal juncture, the modification of the Cecil operation shown in Fig. 22 A is employed. The size of the

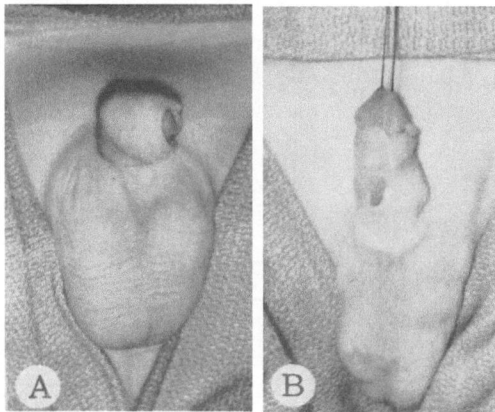

Fig. 21. Typical example of counterclockwise penile torsion and hypospadias with chordee. A, Before treatment. B, After correction of chordee and later of torsion. Patient ready for urethroplasty

indwelling catheter is dictated by the amount of tissue available at the contemplated site of the new meatus and not by the caliber of the remainder of the new urethra. Usually a 10 or 12 F catheter is used for 1 week in children and a 16 or 18 F is employed in older patients. Chromic catgut (4—0) is used throughout the operation. It is important to wrap the deep scrotal tissue snugly over the new urethra and attach it high on the corpora (Fig. 22 A inset) to assure having a long tract if a fistula develops. Reinforcing sutures around the new meatus help prevent meatal retraction. Deviations from CECIL's original technique include use of the indwelling urethral catheter, elimination of perineal urethrostomy, special emphasis on high attachment of the scrotal base on the corpora, and use of slightly different types of sutures.

Penis and new urethra are freed from the scrotum after an interval of at least 2 months (Fig. 22 B). A longer period of attachment is advisable if there were earlier unsuccessful urethroplasties. The catheter or sound is used only during the dissection. It is prudent to provide an ample strip of scrotal skin on each side. These can be trimmed or sacrificed entirely if redundant. Meticulous hemostasis in the scrotum is essential. Since this may be deceptive, a dependent scrotal drain is employed for 2 or 3 days. This has prevented hematomas, which occurred in 6% of early cases. Silk skin sutures are removed after 1 week.

If the meatus is on the penile shaft and well removed from the penoscrotal juncture (28% of all cases), only the distal portion of the penis is attached to the scrotum (Fig. 22 C). It is imperative also to bury at least 1 cm of normal urethra

to avoid a possible fistula in the unattached recess at the base. The free penoscrotal angle greatly simplifies the second stage. This innovation has been employed with increasing frequency in our therapeutic plan.

The two-stage Cecil-type operations have been preferred because they have
eliminated persistent postoperative fistulas. If leakage occurs after removal of the
catheter, the long tract heals spontaneously. The second stage also affords an
opportunity to correct inadequacies that may follow the first stage. Many of
these otherwise might be ignored or discredited.

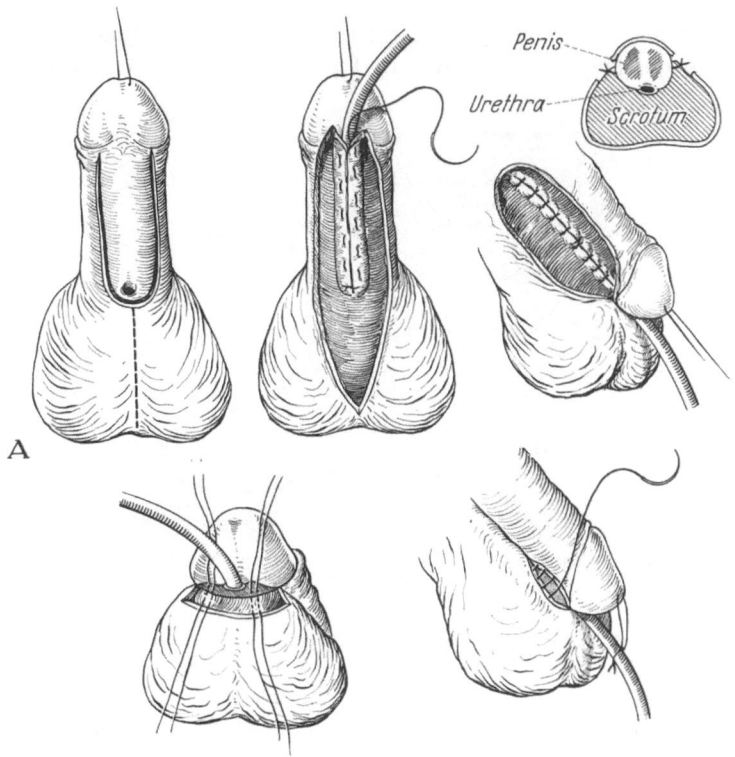

Fig. 22. Cecil-type urethroplasty. *A*, Modification I for penoscrotal hypospadias. *B*, Second-stage freeing of
penis and new urethra from scrotum. *C*, Modification II for penile hypospadias

But the Cecil techniques should be confined to penile and penoscrotal degrees
of hypospadias. Early in this study the same method was applied to scrotal hypospadias by interposing deep scrotal tissue between scrotal and pendulous segments
of new urethra. This maneuver was effective in several cases. But eventually
urethrourethral fistulas occurred and the scheme was abandoned. It is impractical
to bury new pendulous urethra in apposition to new scrotal urethra. Even if the
integrity of both components is maintained, freeing is notably difficult.

Many years ago attempts were made to construct new scrotal urethra by the
Thiersch-Duplay method at the same time that chordee was corrected. It was
hoped that the situation would then be adaptable to the Cecil operation. Unfortunately the meatus never quite reached the penoscrotal juncture and the
procedure was abandoned. A few attempts also were made to construct independent scrotal and pendulous segments in the hope of joining them later. Strictures
prevailed in the unused portion, so this plan also was discarded.

The Denis Browne principle has been employed when the meatus is scrotal or perineal (Fig. 14). By replacement of the conventional double-stop sutures with mattress ones of fine nylon placed around strips of 8 F rubber tubing (Fig. 14 *D*), the incidence of persistent fistula was reduced from 30% to 15%. The through-and-through drain near the penoscrotal juncture provides a better safety valve than the simple stab wounds of the original technique. This is removed on the third day and the nylon sutures come out routinely on the seventh postoperative day. Experience has shown that leaving them in place longer can create fistulas.

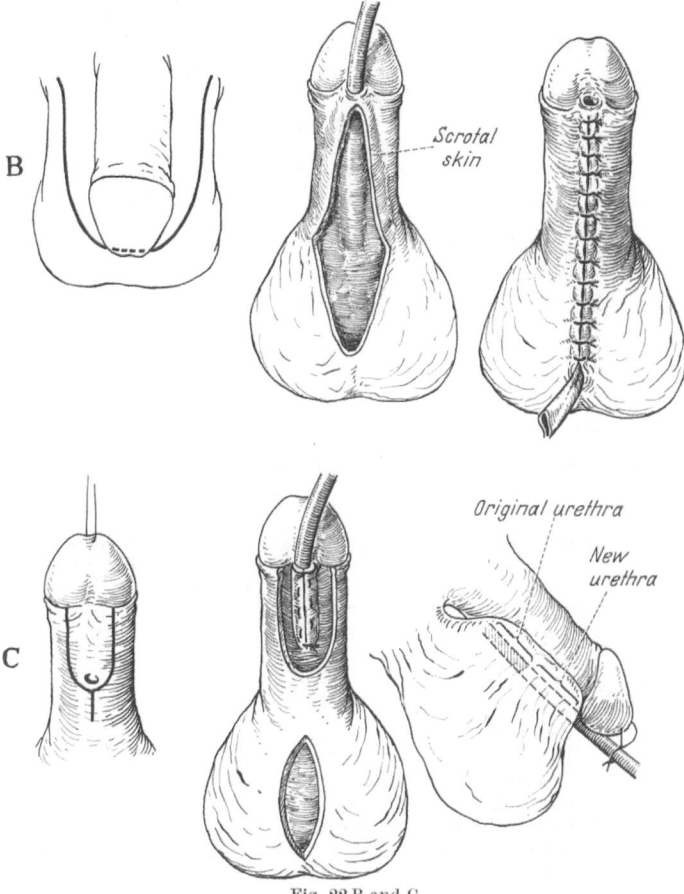

Fig. 22 B and C

The perineal catheter is removed 10 to 12 days after operation. If a pendulous defect has developed, prolonged diversion of urine alone will not close it. The scrotal segment always heals. Fistulas in the pendulous portion can be managed in a variety of ways.

Thirty-one patients required only meatotomy. These had meatal stenosis, coronal hypospadias, and no chordee, as described earlier by DENIS BROWNE (Fig. 23 *A*). Each had a distal opening or indentation which on probing communicated with a separate, vestigial terminal urethra parallel to the one through which the urine passed (Fig. 23 *B*). Frequently this dual passageway extended for at least 2 cm along the distal urethra.

Instead of slitting the terminal portion of the intervening barrier and suturing the respective mucosal margins (Fig. 2), we have eliminated the entire septum

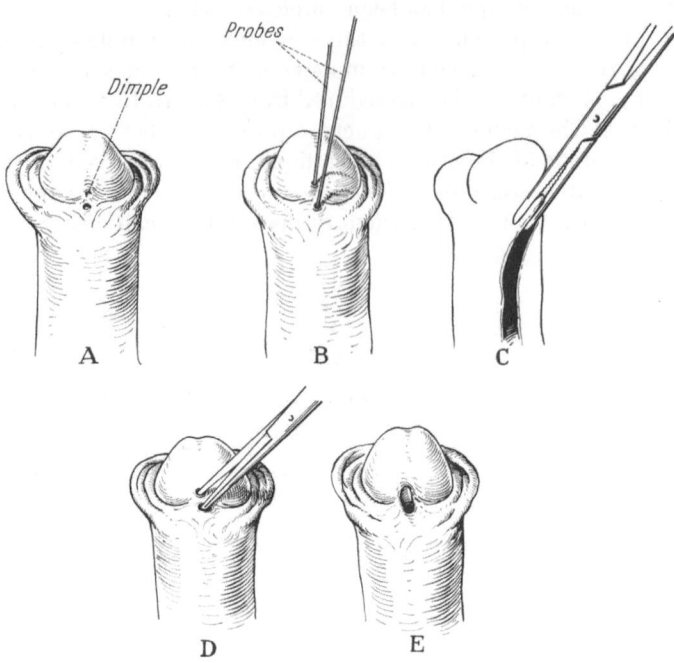

Fig. 23. *A*, Mild hypospadias with meatal stenosis but no chordee. *B*, Dimple on glans communicates with separate short channel parallel to normal urethra. *C*, Intervening septum crushed with mosquito clamp and (*D*) divided with scissors, thereby enlarging meatus (*E*) without sacrificing any urethra

Table 2. *Status of 422 hypospadiacs as of Sept. 1, 1965*

Status		Patients
Treatment completed		345
Successful urethroplasty	314	
Required only meatotomy	31	
Treatment incomplete		77
Only chordee corrected	58	
Too young for urethroplasty	17	
Too soon for urethroplasty	8	
Ready for urethroplasty	33	
Urethra still buried in scrotum	11	
Persistent complications	8	

Table 3. *How new urethra was constructed in 314 cases*

	Total cases	Initial operation				
		Cecil		Browne		Thiersch-Duplay
		Mod. I	Mod. II	Usual	Modified	
Number tried	314	144	80	27	60	3
No more treatment	175 (56%)	81 (56%)	45 (56%)	13 (48%)	36 (60%)	0
Minor trouble — no additional operations	65	32	20	2	11	0
Complications: additional operations	74 (24%)	31 (22%)	15 (19%)	12 (44%)	13 (22%)	3 (100%)

(including the deepest extensions) by crushing it with a mosquito clamp (Fig. 23 *C*) and dividing it with small scissors (Fig. 23 *D*). Then the enlarged meatus (Fig. 23 *E*) does not tend to become stenotic again, even though none of the normal urethra has been sacrificed.

Since there is no chordee and the meatus is in satisfactory functional position, the only further treatment necessary in most of these cases is circumcision. It is best to defer this until one is absolutely certain that there is no chordee.

The present status of the 422 patients included in this review is summarized in Table 2. Seventy-seven require further treatment for the reasons that are listed. Although 314 have had successful urethroplasties, only 56% had per primam healing, uneventful convalescence, and no further trouble of any sort (Table 3).

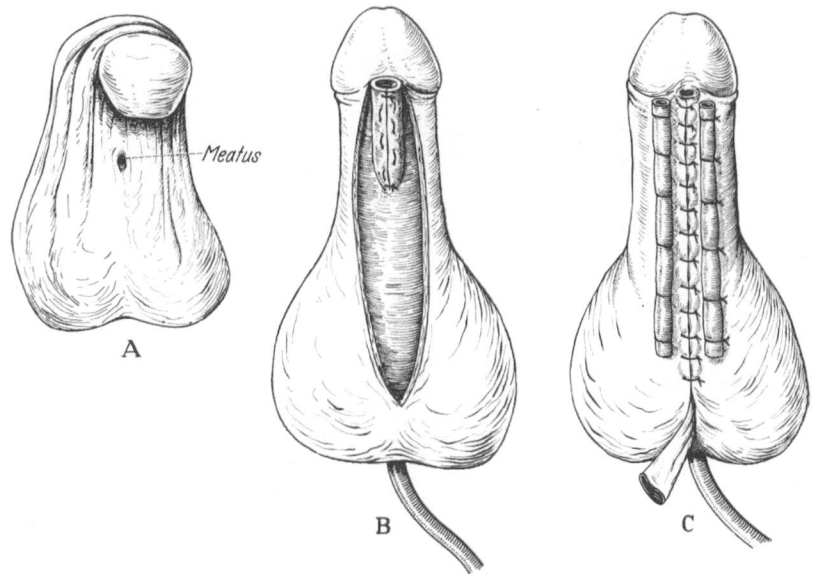

Fig. 24. Revision of new meatus. When meatus retracts into scrotum (*A*) following first stage of Cecil urethroplasty, new terminal segment is constructed (*B*) when penis is freed from scrotum. Urine is diverted by perineal urethrostomy. *C*, Skin is closed and scrotum is drained in usual manner

Complications

The incidence of complications varies with the degree of critical postoperative appraisal employed by each surgeon. Too frequently surgeons have managed to ignore postoperative shortcomings much more easily than patients have been able to tolerate them.

Over 20% of the individuals in this series had minor annoyances that were corrected without extra operations. Most of these were urinary fistulas that healed spontaneously or mild meatal strictures that were eliminated by a few dilatations.

Major complications necessitated additional surgical procedures in almost one fourth of all cases. It is noteworthy that the three techniques used in recent years (both modifications of CECIL's operation and the modified Browne method) had essentially the same degree of initial therapeutic reliability. But the magnitude of undesirable sequelae varied considerably.

The new urethral meatus slipped posteriorly after 10% of the Cecil operations and after 17% of the Browne procedures. Any time the patient is unable to direct his urinary stream satisfactorily and sprays urine instead, revision of the new meatus is mandatory. This can be accomplished during the second stage of the Cecil plan as shown in Fig. 24. Urine usually is diverted by perineal urethrostomy. The same type of advancement has been effective after deficient Browne operations. But most of the gaping post-Browne meatuses have required reconstructions similar to the modified Cecil operation shown in Fig. 22 C. During many of the second-stage Cecil procedures the urethra has been advanced successfully for $^1/_4$ inch or so without diversion of the urine. This enhances the final result.

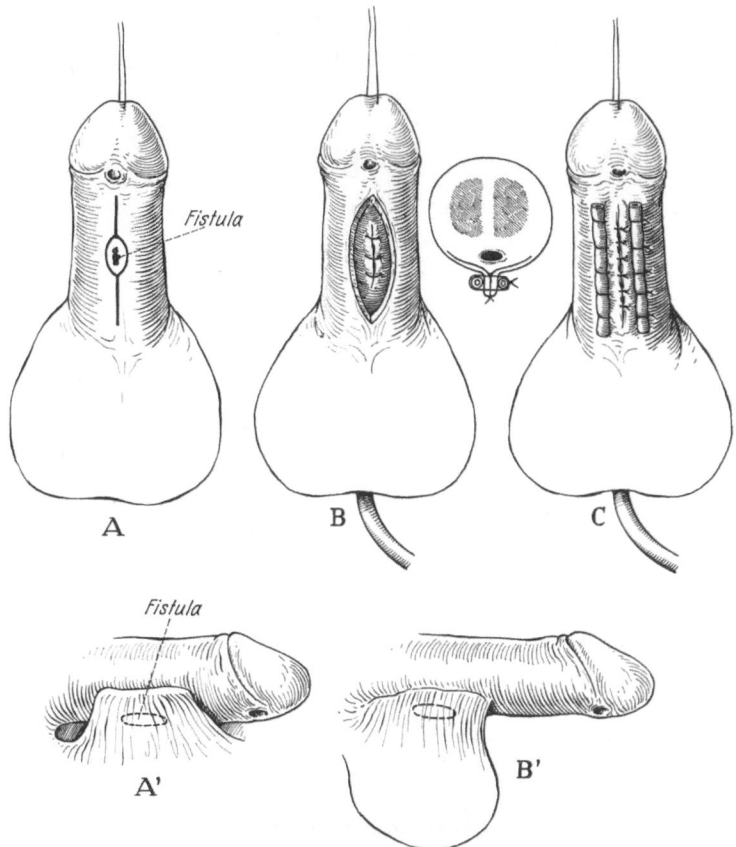

Fig. 25. Closure of fistula. *A*, Defect is circumcised and adjacent skin freed. Incision is extended in both directions if necessary for adequate exposure. *B*, After perineal urethrostomy, deep tissues are closed with fine catgut sutures and (*C*) skin is closed by the modified Browne method. Larger fistulas are buried in the scrotum. This may require attachment of mid urethra (*A'*) or the proximal segment (*B'*). Indwelling catheter drainage usually is included

Just about every conceivable type of fistula was created or was included in the 29 cases that were inherited in this condition. They varied from minute openings to multiple defects to loss of most or all of the pendulous segment. Practically all of the persistent fistulas produced during personal management of these patients followed Browne procedures. "Homemade" as well as "bequeathed" openings required variable numbers of subsequent operations.

Variations and combinations of the modified Browne and Cecil methods were the most dependable for closure of fistulas (Fig. 25). With sizable or multiple openings the modified Cecil method usually was employed. Any portion of the pendulous urethra can be attached to the scrotum effectively, provided some

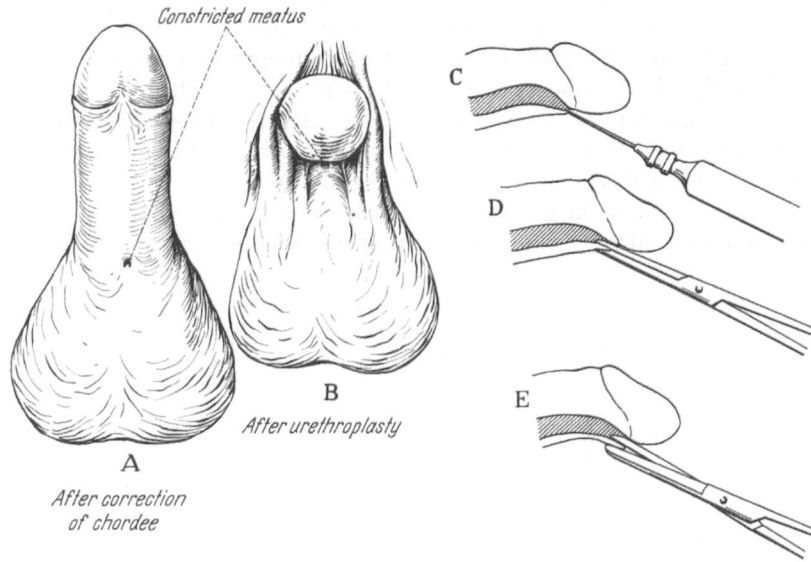

Fig. 26. Usual type of meatotomy. This is effective after correction of chordee (*A*) or urethroplasty (*B*). *C*, Posterior lip is infiltrated with anesthetic agent. *D*, Constricted portion is compressed with mosquito hemostat. *E*, Avascular strip is divided with scissors

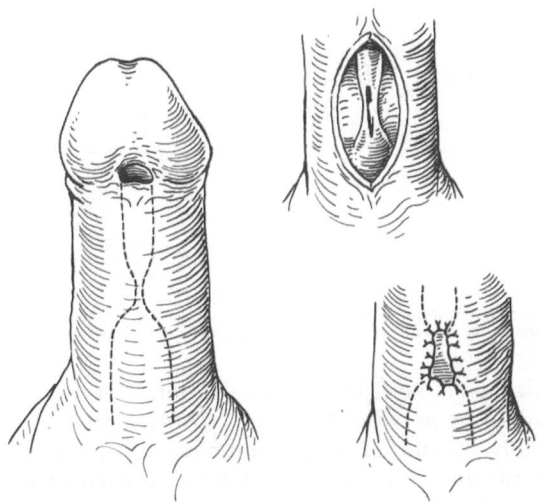

Fig. 27. Stricture at original hypospadiac meatus. Constricted portion of urethra incised and mucosa sutured to adjacent skin, pending later closure of the defect

normal urethra also is buried. Huge fistulas usually required burial of the entire pendulous segment as shown in Fig. 22 *A*.

Although fistulas became extinct after properly executed Cecil operations and more meticulous suturing around the new urethral meatus reduced the incidence of meatal retraction, meatal strictures became the counterpart and occurred in 15% of the patients. The meatus did not constrict after Browne operations. It tended to gape instead. Enlarging a meatus proved to be much simpler than trying to reduce the caliber of one that sprayed. The type of meatotomy shown in Fig. 26 requires only local anesthesia. A few post-Cecil stenoses required more extensive revision during the second stage.

22*

The most inexcusable and refractive strictures were at the original meatus. Fortunately they were rare. They were unrelated to the type of urethroplasty. Dilatations and internal urethrotomy were disappointing. Usually the stenotic portion of urethra had to be opened widely (Fig. 27), attached to the skin, left opened as a fistula for several weeks, and closed later. These disasters can be prevented by careful calibration of all urethral openings and appropriate revision of the constricted ones before new urethra is constructed.

Sixteen patients had chronic strictures near the original meatus or at the new one when first seen. Most of these strictures were complicated further by fistulas,

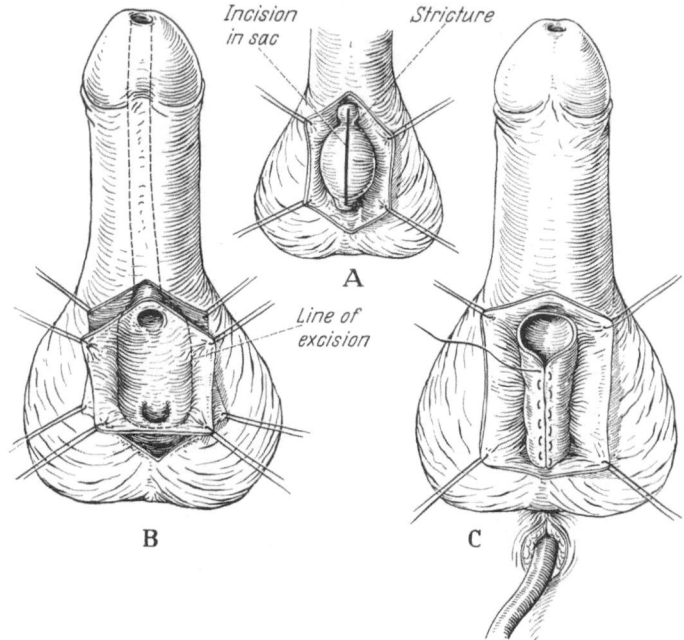

Fig. 28. Urethral sacculation. *A*, Pouch and adjacent stricture incised. *B*, Excess sac wall excised. *C*,Remainder used to construct urethra of proper caliber. Urine diverted by perineal urethrostomy

sacculation, poor drainage, and persistent infection. Several pendulous segments of new urethra had to be sacrificed entirely. Construction of disproportionately large urethra, notably by the infamous Ombrédanne operation, required similar management. New urethra was constructed later in each of these cases.

Pouches deeper in the urethra were more amenable to simple revision (Fig. 28). Most of these pseudodiverticula had followed use of free tube grafts and strictures at the site of the anastomosis. Obviously any associated constriction must be corrected simultaneously or the sacculation will recur.

The generous dorsal slit inherent in the Denis Browne technique usually heals uneventfully; but in three cases painful scars had to be excised.

One young boy was provided with an excellent new urethra but through some peculiarity of healing the meatus moved to the left lateral aspect of the corona instead of remaining in the midline. Six months later the new urethra was freed and moved to the midline without compromising its integrity.

Space precludes discussion of the numerous odd balls, ball-bearing females, and latest models of convertibles that were included in this series. Suffice it to say that 13% of the patients had at least one side of the "scrotum" empty for the

Table 4. *Concomitant operations*

Type	Cases		
	Total	Straighten-ing	Urethro-plasty
Orchiopexy	30	17	13
Hernia only	6	4	2
Hydrocele	5	2	3
Orchiectomy	3	3	
Testicular biopsy	4		4
Exploration, testis	3		3
Hysterectomy, etc.	5	5	
Vaginectomy	4		4
Laparotomy	3	3	
Total	63	34	29

Table 5. *Age of 422 hypospadiacs*

Age, years	When first seen		Treatment completed	
	Patients	% of 422	Patients	% of 345
1—3	113	27⎫47⎫71⎫90	4	28⎫64⎫89
4—5	85	⎭	94	⎭
6—8	53	⎭	75	⎭
9—11	50	⎭	47	⎭
12—15	51	⎭	59	⎭
16—19	29	⎭	28	⎭
20—29	28	⎫10	29	⎫11
30—64	13	⎭	9	⎭

following reasons: cryptorchism in 44 cases (bilateral in 23 and unilateral in 21), testicular agenesia in 2, adrenogenital syndrome in 6, and hermaphroditism in 2.

Excretory urograms were obtained routinely. Retrograde urethrography and cystoscopy were employed frequently. Surprisingly few other urologic anomalies were found. They were limited to rectourethral fistula in three cases, megalo-ureter in two, vesical diverticulum in one, ureteropelvic obstruction in two, duplication of pelvis and ureter in one, and renal malrotation in one.

Deformities of other systems were more common. These deformities and systems included the feet in seven cases, heart in five, ears in five, strabismus in three, cleft palate in two, speech (mental?) in three, gynecomastia in two, hemangioma in two, tracheoesophageal fistula in one, and hydrocephalus in one.

Numerous investigative and definitive surgical procedures were combined effectively and uneventfully with correction of chordee and with urethroplasty, especially the second-stage Cecil operation (Table 4). The embarrassing and tragic predicaments of yesteryear can be avoided today by buccal smears and allied studies of all genital enigmas soon after birth.

But one pathetic and disheartening group will persist as long as surgeons digress from sound surgical principles and well-established procedures. Grotesque postoperative deformities produced by misconceived and maldirected efforts can be resolved only by excising offending tissues, trying to convert the hypospadias to some semblance of its original status, and starting anew with the type of repair that promises the most.

Only 28% of the patients were seen early enough to complete their treatment during the preschool period (Table 5). Indeed, only 64% were cured before 12 years of age. There is urgent need of widespread missionary work on behalf of hypospadiacs.

Many of the patients were teenagers when first seen. They were victims of the now outmoded philosophy that urethroplasty should be deferred until full growth had been attained.

Adults in this series invariably had chronic complications. Many had become discouraged after multiple surgical failures. The oldest patient finally resolved at the age of 64 years to try again to have his persistent fistula closed. He now has an intact urethra.

Admittedly, the treatment of hypospadias and the associated anomalies is challenging and demanding. At times it can be frustrating and soul-searching. But it also can be incomparably gratifying and rewarding if sound and proved operative techniques are employed.

References

ANGER. T.: Quoted by C. D. CREEVY 1958.

ARONOFF, M.: A one-stage operative technique for the treatment of sub-glandular hypospadias in children. Brit. J. plast. Surg. 16, 59—62 (1963).

BACKUS, L. H., and C. A. DE FELICE: Hypospadias — then and now. Plast reconstr. Surg. 25, 146—160 (1960).

BARCAT, J., et J. C. STEPHAN: L'hypospadias: Étude clinique et thérapeutique; à propos de 140 nouveaux cas. J. Chir. 81, 551—580 (1961).

BARSKY, A. J., S. KAHN, and B. E. SIMON: Principles and practice of plastic surgery, 2. ed. New York: Blakiston Co. — Division of McGraw-Hill Book Co. Inc. 1964.

BECK, C.: Hypospadias and its treatment. Surg. Gynec. Obstet. 24, 511—532 (1917).

BLAIR, V. P., J. B. BROWN, and W. G. HAMM: The correction of scrotal hypospadias and of epispadias. Surg. Gynec. Obstet. 57, 646—653 (1933).

BRENDLER, H.: A new method for the construction of a glandular urethra in hypospadias repair. J. Urol. (Baltimore) 59, 1164—1168 (1948).

BROADBENT, T. R., R. M. WOOLF, and E. TOKSU: Hypospadias: One-stage repair. Plast. reconstr. Surg. 27, 154—159 (1961).

BROWNE, D.: Hypospadias. Postgrad. Med. 25, 367—372 (1949).

— A comparison of the Duplay and Denis Browne techniques for hypospadias operations. Surgery 34, 787—798 (1953).

BUCKNALL, R. T. H.: A new operation for penile hypospadias. Lancet 1907 II, 887—890.

BYARS, L. T.: Surgical repair of hypospadias. Surg. Clin. N. Amer. 30, 1371—1378 (1950).

— Functional restoration of hypospadias deformities: With a report of 60 completed cases. Surg. Gynec. Obstet. 92, 149—154 (1951).

— A technique for consistently satisfactory repair of hypospadias. Surg. Gynec. Obstet. 100, 184—190 (1955).

CAMPBELL, M. F.: Undescended testicle and hypospadias. Amer. J. Surg. 82, 8—17 (1951).

— Clinical pediatric urology. Philadelphia: W. B. Saunders Co. 1951.

— Anomalies of the genital tract. In: Urology, 2. ed., vol. 2, p. 1713. Philadelphia: W. B. Saunders Co. 1963.

CECIL, A. B.: Surgery of hypospadias and epispadias in the male. J. Urol. (Baltimore) 27, 507—537 (1932).

— A further report on the cure of hypospadias and epispadias. J. Urol. (Baltimore) 34, 278—283 (1935).

— Repair of hypospadias and urethral fistula. J. Urol. (Baltimore) 56, 237—242 (1946).

— Modern treatment of hypospadias. J. Urol. (Baltimore) 67, 1006—1011 (1952).

— Symposium on pediatric urology. Hypospadias and epispadias: Diagnosis and treatment. Pediat. Clin. N. Amer. 2, 711—728 (1955).

CLOUTIER, A. M.: A method for hypospadias repair. Plast. reconstr. Surg. 30, 368—373 (1962).

CONNOLLY, N. K.: Results of Denis Browne's operation for hypospadias. Brit. J. Surg. 41, 615—618 (1954).

CRAWFORD, B. S.: The management of hypospadias. Brit. J. clin. Pract. 17, 273—280 (1963).

CREEVY, C. D.: The operative treatment of hypospadias: With a report of 13 cases. Surgery 3, 719—731 (1938).

— The correction of hypospadias: A review. Urol. Surv. 8, 2—47 (1958).

CULP, O. S.: Early correction of congenital chordee and hypospadias. J. Urol. (Baltimore) 65, 264—274 (1951).

— Experiences with 200 hypospadias: Evolution of a therapeutic plan. Surg. Clin. N. Amer. 39, 1007—1023 (1959).

DAVIS, D. M.: The pedicle tube-graft in the surgical treatment of hypospadias in the male: With a new method of closing small urethral fistulas. Surg. Gynec. Obstet. **71**, 790—796 (1940).
— Results of pedicle tube-flap method in hypospadias. J. Urol. (Baltimore) **73**, 343—348 (1955).
DES PREZ, J. D., L. PERSKY, and C. L. KIEHN: One-stage repair of hypospadias by island flap technique. Plast. reconstr. Surg. **28**, 405—411 (1961).
DEVINE jr., C. J., and C. E. HORTON: One stage hypospadias repair. J. Urol. (Baltimore) **85**, 166—172 (1961).
DUPLAY, S.: De l'hypospadias périnéo-scrotal et de son traitement chirurgical. Arch. gen. méd. **23**, 513—530 (1874).
— Sur le traitement chirurgical de l'hypospadias et de l'épispadias. Arch. gén. Méd. **145**, 257—274 (1880).
EDMUNDS, A.: An operation for hypospadias. Lancet **1913 I**, 447—449.
— Pseudo-hermaphroditism and hypospadias: Their surgical treatment. Lancet **1926 I**, 323—327.
FARMER, A. W.: Hypospadias. Surgery **12**, 462—470 (1942).
FEVRÉ, M.: Généralités sur le traitement de l'hypospadias. Sem. Hôp. Paris **23**, 893—894 (1947).
FOGH-ANDERSEN, P.: Hypospadias: Thirty-four completed cases operated on according to Denis Browne. Acta chir. scand. **105**, 414—423 (1953).
GARSKE, G. L.: The female urethra. J. Lancet **83**, 45—52 (1963).
GELBKE, H.: Harnröhrenplastiken bei Hypospadien, Defekten, Stenosen und Fisteln. Z. Urol. **48**, 65—83 (1955).
GLENISTER, T. W.: The origin and fate of the urethral plate in man. J. Anat. (Lond.) **88**, 413—425 (1954).
— A consideration of the processes involved in the development of the prepuce in man. Brit. J. Urol. **28**, 243—249 (1956).
GREENE, R. R.: Embryology of sexual structure and hermaphroditism. J. clin. Endocr. **4**, 335—348 (1944).
— M. W. BURRILL, and A. C. IVY: Experimental intersexuality: Effects of estrogens on antenatal sexual development of rat. Amer. J. Anat. **67**, 305—345 (1940).
GYARMATHY, F.: Unsere Erfahrungen mit der Hypospadieoperation nach DENIS BROWNE. Z. Urol. **56**, 223—227 (1963).
HAND, J. R.: Surgery of the penis and urethra. In: M. F. CAMPBELL, Urology, ed. 2, vol. 3, p. 2690. Philadelphia: W. B. Saunders Co. 1963.
HAVENS, F. Z., and A. S. BLACK: The treatment of hypospadias. J. Urol. (Baltimore) **61**, 1053—1064 (1949).
—, and T. J. LITZOW: Treatment of hypospadias: Ten years' experience with tunnel-graft urethroplasty. J. Urol. (Baltimore) **72**, 677—680 (1954).
HIGGINS, C. C.: Hypospadias. Cleveland Clin. Quart. **14**, 126—127 (1947).
HOWARD, F. S.: Hypospadias with enlargement of the prostatic utricle. Surg. Gynec. Obstet. **86**, 307—316 (1948).
— The surgery of intersexuals. J. Urol. (Baltimore) **65**, 636—649 (1951).
HUMBY, G.: A one-stage operation for hypospadias. Brit. J. Surg. **29**, 84—92 (1941).
Inter-American Conference on Congenital Defects: Papers and discussion of the first conference, 1962, compiled and edited for the Internat. Medical Congr., Ltd. Philadelphia: J. B. Lippincott Co. 1963.
JOHNSON, F. P.: The later development of the urethra in the male. J. Urol. (Baltimore) **4**, 447—492 (1920).
JOST, A.: Recherches sur la différenciation sexuelle de l'embryon de lapin. Arch. Anat. micr. Morph. exp. **36**, 242—270 (1946/47).
KENNEDY, P. A.: Hypospadias: A twenty year review of 489 cases. J. Urol. (Baltimore) **85**, 814—817 (1961).
LEVI, D.: Quoted by C. D. CREEVY, The correction of hypospadias. A review. Urol. Surv. **8**, 2—47 (1958).
LOUGHRAN, A. M.: Observations on hypospadias: Including late results of Ombrédanne's urethroplastic operation. Brit. J. plast. Surg. **1**, 147—158 (1948).
MARION, G.: Traité d'urologie, 4. ed. Paris: Masson & Cie. 1940.
—, et J. PERARD: Technique des opérations plastiques sur la vessie et sur l'urètre. Paris: Masson & Cie. 1942.
MARSHALL, V. F., and R. M. SPELLMAN: Construction of the urethra in hypospadias using vesical mucosal grafts. J. Urol. (Baltimore) **73**, 335—342 (1955).
MAYO, C. H.: Quoted by C. D. CREEVY, The correction of hypospadias: A review. Urol. Surv. **8**, 2—47 (1958).

Mays, H. B.: Quoted by C. D. Creevy, The correction of hypospadias: A review. Urol. Surv. 8, 2—47 (1958).

McCormack, R. M.: Simultaneous chordee repair and urethral reconstruction for hypospadias. Plast. reconstr. Surg. 13, 257—274 (1954).

McIndoe, A. H.: The treatment of hypospadias. Amer. J. Surg. 38, 176—185 (1937).

Memmelaar, J.: Use of bladder mucosa in a one-stage repair of hypospadias. J. Urol. (Baltimore) 58, 68—73 (1947).

Meulen, J. C. H. M. van der: Hypospadias. Springfield (Ill.): Ch. C. Thomas 1964.

Michalowsky, E., and W. Modelski: Operative treatment of hypospadias. J. Urol. (Baltimore) 89, 698—701 (1963).

Nesbit, R. M.: Plastic procedure for correction of hypospadias. J. Urol. (Baltimore) 45, 699—702 (1941).

— The surgical treatment of congenital chordee without hypospadias. J. Urol. (Baltimore) 72, 1178—1180 (1954).

— W. J. Butler, and W. Whitaker: Production of epithelial lined tubes from buried strips of intact skin. J. Urol. (Baltimore) 64, 387—392 (1950).

Nové-Josserand, G.: Nouvelle technique pour la restauration en une séance des hypospadias étendus par la tunnelisation avec greffe dermo-épidermique. J. Urol. 8, 449—456 (1919).

Patten, B. M.: Human embryology, 2. ed. New York: McGraw-Hill Book Co. Inc. 1953.

Paul, M., and R. Kanagasuntheram: The congenital anomalies of the lower urinary tract. Brit. J. Urol. 28, 118—125 (1956).

Ross, J. F., A. W. Farmer, and W. K. Lindsay: Hypospadias: A review of 230 cases. Plast. reconstr. Surg. 24, 357—368 (1959).

Russell, R. H.: Operation for severe hypospadias. Brit. med. J. 1900 II, 1432—1435.

Smith, B. T., and I. W. Forsythe: Quoted by J. C. H. M. van der Meulen.

Smith, D. R.: Surgical treatment of hypospadias. J. Urol. (Baltimore) 73, 329—334 (1955).

—, and H. M. Blackfield: A modification of the Blair procedure for the repair of hypospadias. J. Urol. (Baltimore) 59, 404—413 (1948).

— — A critique on the repair of hypospadias. Surgery 31, 885—899 (1952).

Sorenson, R.: Quoted by J. C. H. M. van der Meulen.

Spaulding, M. H.: Quoted by J. C. H. M. van der Meulen.

Thiersch, C.: Quoted by C. D. Creevy: The operative treatment of hypospadias: With a report of 13 cases. Surgery 3, 719—731 (1938).

Wain, H.: The story behind the word. Springfield (Ill.): Ch. C. Thomas 1958.

Walters, W.: Successful operations for hypospadias. Amer. Surg. 103, 949—958 (1936).

Wehrbein, H. L.: Hypospadias. J. Urol. (Baltimore) 50, 335—340 (1943).

Wharton, L. R.: Gynecology: With a section on female urology, 2. ed. Philadelphia: W. B. Saunders Co. 1947.

Wiesner, B. P.: The post-natal development of the genital organs in the albino rat: With a discussion of a new theory of sexual differentiation. J. Obstet. Gynaec. Brit. Emp. 41, 867—922 (1934); 42, 8—78 (1935).

Wilson, K. M.: Quoted by J. C. H. M. van der Meulen.

Young, F., and J. A. Benjamin: Preschool age repair of hypospadias with free inlay skin graft. Surgery 26, 384—404 (1949).

Young, H. H.: The abnormalities and plastic surgery of the lower urogenital tract. J. Urol. (Baltimore) 35, 436—440 (1936).

Anomalies of the Female Genitalia *

Howard W. Jones Jr.

With 40 Figures

I. Anomalies of the Vulva and Vagina

1. Reduplication of the Vulva

Reduplication of the vulva in whole or in part is an extremely rare but striking anomaly. The few duplications which have been reported are uniformly lateral and in the most extreme form have consisted of complete vulvar duplication with two clitorides, four labia majora, and four labia minora. Such duplications are associated with other anomalies, such as reduplication of the internal structures, i.e. vagina and uterus, bladder, rectum and colon, as well as an occasional anomaly of the bony structures where this has been investigated.

This anomaly seems much more profound than the problems of lateral fusion of the müllerian ducts, in spite of the fact that failure of fusion is often a part of the anomaly of reduplication. Embryologically, as is well known, the external genitalia arise from the so-called genital tubercle and scrotolabial folds, external structures which are quite distinct from the internal müllerian ducts which form the uterus and part of the vagina. Thus, the disorder which causes reduplication involves much more than a single organ system.

Breen and Weinberg (1965) reported a remarkable example of reduplication and reviewed the literature up until that time. They reported a 20 year old Mexican indian female whose history revealed that puberty had occurred at the normal age of 12 and that her menstrual periods were quite regular with a 30 day interval and a duration of 8 days. Her first pregnancy terminated prematurely at 28 weeks and her second pregnancy resulted in a stillborn delivery at home. She was seen when her third pregnancy was at term. All pregnancies had occurred through the right vagina and the patient apparently felt that psychologically she would be more normal if only one of her genital tracts was utilized for coitus, conception and delivery. The third pregnancy terminated normally after a 10 hour spontaneous delivery.

On further examination before discharge from the hospital, the complete double vulva was noted with two vaginas, each of which had a uterus at the vault. Catheters in each urethra produced independent drainage from two separate bladders (Fig. 1). There were two anal orifices. An intravenous pyelogram revealed normal kidneys but each ureter opened into a separate bladder (Fig. 2). A barium enema revealed two normal colons which ran parallel from the anus to the cecum (Fig. 3).

Unfortunately, further x-ray studies of the gastro-intestinal tract and of the generative tract were not performed.

---

* Based in part on material from *Pediatric and Adolescent Gynecology* by Howard W. Jones, Jr., and Richard H. Heller, Williams and Wilkins Company, Baltimore, 1966.

Eight other cases, very similar to this, all of which have been reported previously are shown in Table 1 which is modified from the paper by BREEN and
WEINBERG.

Our experience with reduplication of the external genitalia is limited to a
single patient who had a single clitoris but reduplication of the urethra with two
ani, double vagina, double uterus and partial reduplication of the colon which
branched in the region of the sigmoid (Fig. 4). An intravenous urogram disclosed

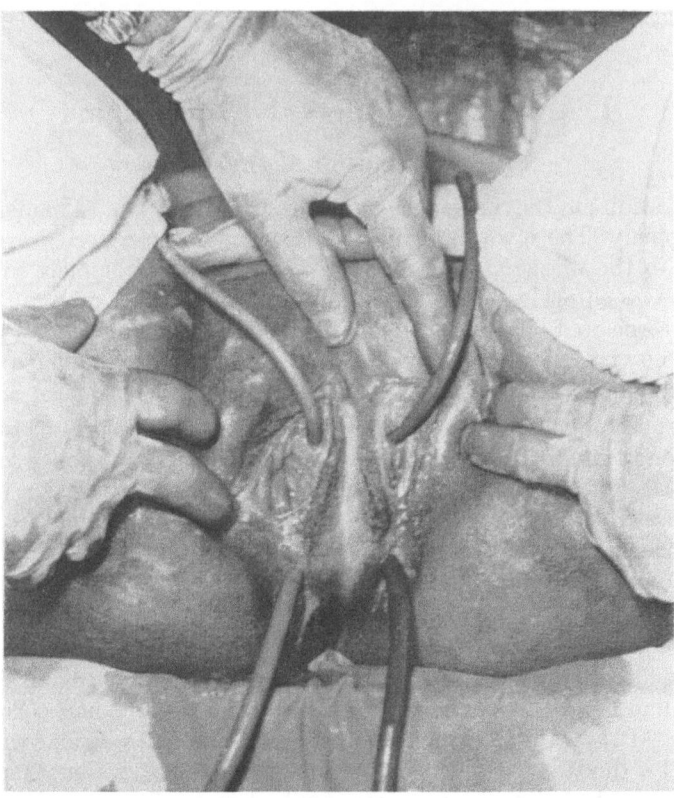

Fig. 1. Photograph showing the external genitalia of a 20 year old female with reduplication of the vulva. There
is a catheter in each of the urethras and a catheter in each of the ani. (From: J. L. BREEN and C. R. WEINBERG,
[1965)

normal kidneys which entered a single bladder which could be emptied through
either urethra. However, there was a peculiar duplication of the sacrum (Fig. 5).
This patient was severely handicapped by a large extra buttock and in this
respect differed from most patients with reduplication of the vulva who seem to
have no large bulky masses of tissue.

2. Masculinization of the Vulva

See chapter on the intersex states.

3. Imperforate Hymen

The hymen is located at the site where the embryonic vagina buds from the
urogenital sinus. The hymenal area is, therefore, composed entirely of urogenital

sinus epithelium. If a lumen fails to develop at the point where the budding vagina arises, the result is an imperforate hymen.

It is unusual for an imperforate hymen to be discovered prior to the onset of puberty. However, if the condition is discovered, the hymen nevertheless should be excised as it may give rise to obstructive symptoms during infancy (Fig. 6). Sometimes an accumulation of a clear mucoid fluid collects above the obstruction apparently as a result of excess secretion of cervical mucus. This is referred to as

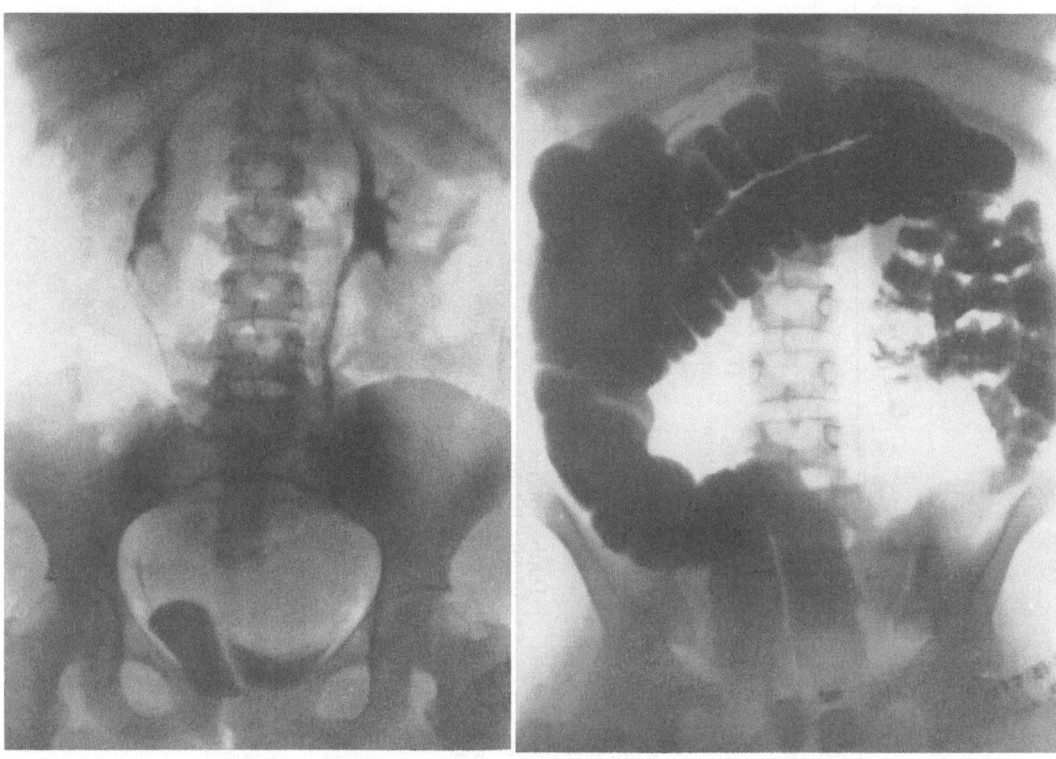

Fig. 2   Fig. 3

Fig. 2. An intravenous pyelogram of the same patient shown in Fig. 1. This shows bilateral relatively normal kidneys with ureter which enter in separate bladders (From: BREEN and WEINBERG, 1965)

Fig. 3. Barium enema in the same patient shown in Fig. 1. This reveals two normal and parallel colons (From: BREEN and WEINBERG, 1965)

a hydrocolpos or mucocolpos and it may reach considerable proportions. The collection of mucus may be sufficient to involve the uterus which becomes enlarged and may even obstruct the cervical tract. Fatalities in infants from an unrecognized hydrometrocolpos have been reported. Most often a large unrecognized hydrometrocolpos is due to a transverse vaginal septum rather than an imperforate hymen and further discussion of this condition will be found under that heading.

If the imperforate hymen is not discovered until after the onset of the menarche, symptoms may arise from the accumulation of menstrual blood. Although cyclic abdominal pain is a common symptom, it is remarkable that a large amount of blood can sometimes accummulate in the vagina, tubes and uterus and cause little discomfort (Fig. 7, 8). In rare instances flank pain due to urinary retention has been a presenting symptom.

Table 1. *Associated Genital-Urinary and Intestinal Tract Duplication in Patient with Reduplication of the External Genitalia*

Author	Year	Patient Age	External Genitalia	Internal Genitalia	Intestinal Tract	Urinary System
GEDDA, LUIGI	1860	17 yrs.	Double external genitalia	Not recorded	Not recorded	Double urinary system (no details)
SUPPIGER, J.	1876	1½ yrs.	Two complete and parallel vulvae	Double vagina; bicornuate uterus	Double colon; 1 ileocecal valve and appendix; 2 anal dimples; 1 vestibular anus	Double urethra and single bladder
CHIARLEONI, G.	1894	—	Two complete and parallel vulvae	Double vagina; bicornuate uterus	Double rectum	Not recorded
GEMMELL and PATERSON	1913	Adult	Two complete and parallel vulvae	Double vagina; uterus didelphys	Single anus	One urethra
LESBRE, R. X.	1927	Adult	Double external genitalia	Double vagina; uterus didelphys	Double colon; double anus	Double bladder
OMBREDANNE, L.	1936	2 yrs.	Two complete and parallel vulvae	Not recorded	Double colon and ceca; right anal dimple; left anal dimple; left vestibular anus	Urethra, right; blind urethral dimple left (suggestive of double bladder)
PATTERSON, N. G. and J. P. MAXWELL	1939	49 yrs.	Double external genitalia (rudimentary clitoris, left)	Double vagina; uterus didelphys	Not recorded	Single, bladder, normal right urethra, left urethra, outer portion left vagina
AITKEN, J.	1950	1 day	Double external genitalia (1 bifid clitoris)	Double vagina; bicornuate uterus	Double ileum beyond Meckel's diverticulum; 2 ceca and appendices; 2 complete double colons	One urethra at root of bifid clitoris

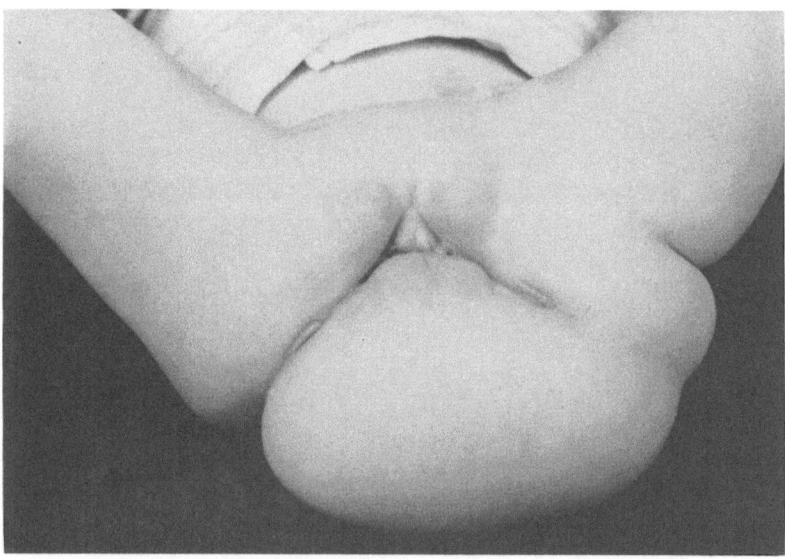

Fig. 4. Photograph of a child with reduplication of the external genitalia. As can be seen there is fusion in the region of the clitoris, but there are two separate vaginas, two urethras, a double uterus and partial duplication of the colon with 2 ani

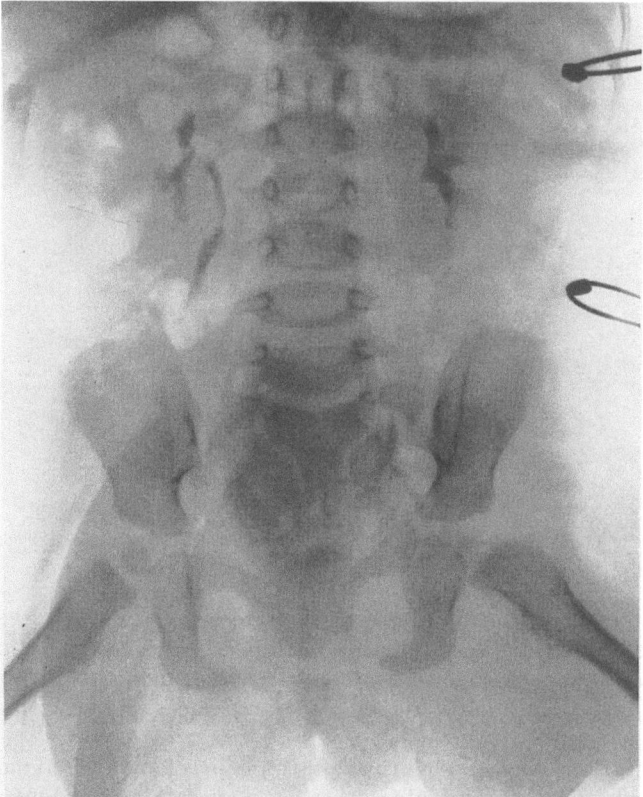

Fig. 5. Intravenous pyelogram of the same patient shown in Fig. 4. The kidneys, ureters and bladder seem quite normal but an examination of the bony structure reveals a peculiar duplication of the sacrum

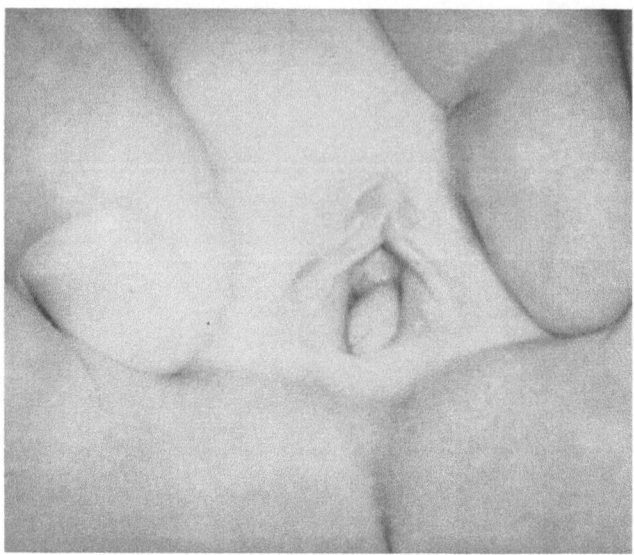

Fig. 6. Photograph of imperforate hymen in an infant

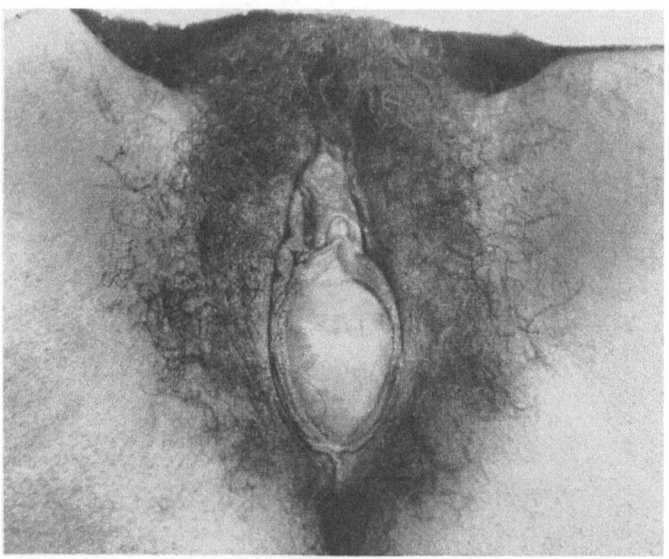

Fig. 7. Photograph of an imperforate hymen showing bulging due to a large accumulation of contained menstrual blood

It should be noted that on occasion the hymen may appear to be imperforate but will in reality have a very few tiny openings. A most extraordinary example of this appeared in a private patient who became pregnant and in whom tiny perforations of the imperforate hymen were discovered with only the most careful examination (Fig. 9).

Treatment of the condition consists of simple incision of the hymen or excision of the triangular flap. In patients with an accumulation of mucus or blood, the administration of antibiotics to prevent infection may be considered routine. It is remarkable that even large accummulations of menstrual blood may be tolerated

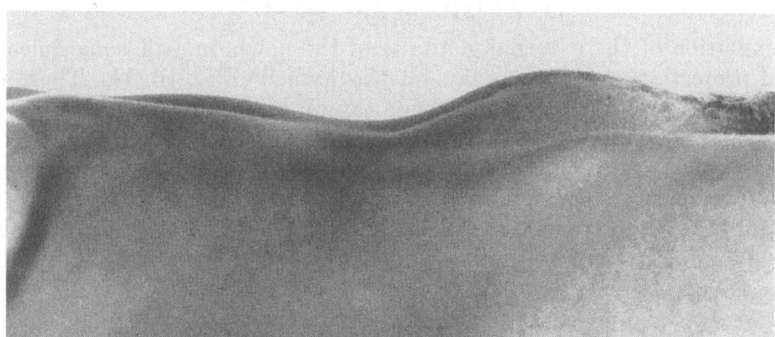

Fig. 8. Photograph of a patient with a large supra-pubic mass due to hematometra and a hematocolpos

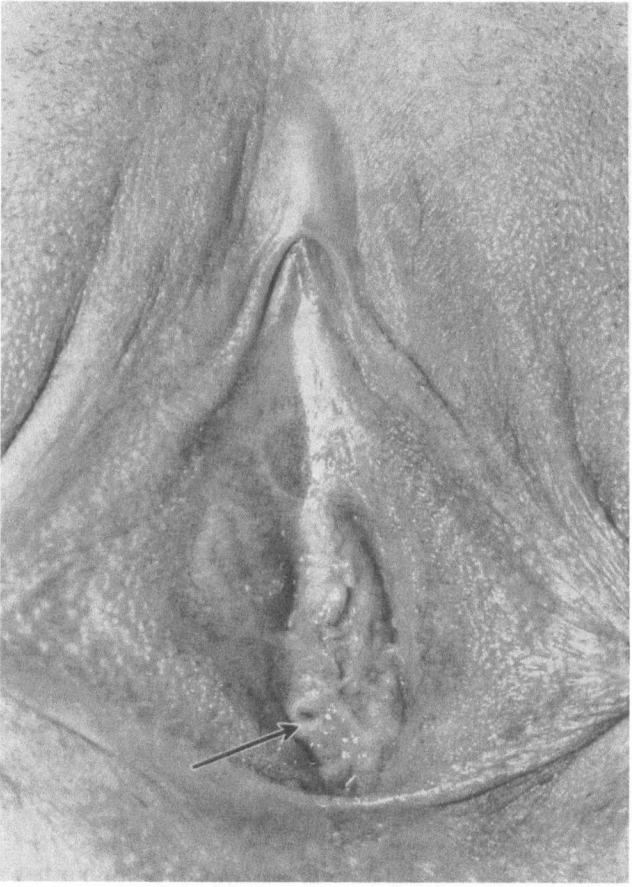

Fig. 9. Photograph of a hymen which is almost imperforate. The pinhead sized opening was sufficient to permit pregnancy. (From: R. W. TeLinde, 1962)

without permanent damage to the generative tract. A patient who had over 1,000 cc. of menstrual blood when her lesion was released in her early teens, some ten years after this operation became married, promptly became pregnant, and delivered a normal child.

4. Polyps of the Hymen

Examination of the external genitalia of the newborn will sometimes reveal polypoid projections from the region of the hymen (Figs. 10, 11). These usually are of no clinical significance but routine examination shows that they are more frequent than generally realized. BORGLIN and SELANDER (1962) examined 1,000 consecutive newborn girls and found such polyps in 63 or 6%. Such polyps are usually prominent in the newborn because of the hypertrophy and edema associated with the estrogen milieu of the first few days of life. After the first weeks they reduce in size and may disappear, but an occasionally prominent one is best removed.

Histologic examination shows that they are lined by a regular, highly differentiated squamous epithelium with cornification. Such epithelium is indistinguishable from that of the normal vagina. The center of such polyps is composed of loose edematous connective tissue.

5. Transverse Vaginal Septum

A transverse vaginal septum is the result of failure of lumen to appear in the embryonic vaginal plate. In view of the debate concerning the origin of the vaginal plate, the exact embryological interpretation and significance of the transverse vaginal septum is likewise doubtful. However, the location of the transverse vaginal septum may be given some weight as an index of the location of the junction between the down-growing müllerian epithelium and the up-growing epithelium of the urogenital sinus. By such reasoning, the fact that the most common location of the transverse vaginal septum is the junction of the upper and middle third of the vagina can be used as supporting the views of KOFF (1933) who from morphological studies on embryos concluded that the two epithelia, the müllerian and the urogenital sinus, met most often at about the junction of the upper and middle third of the vagina. KANAGASUNTHERAM and DASSANAYAKAE (1958) pointed out that histological examination of the excised septum might be decisive in some instances in settling the origin of the septum. They observed that an imperforate hymen would be expected to have squamous epithelium on both sides, whereas the septum in some of their cases had columnar epithelium on one side and squamous epithelium on the other. The presumption in this circumstance was that the columnar epithelium was müllerian in origin whereas the squamous epithelium was from the urogenital sinus. This arresting observation and its interpretation of the difference of the epithelium on the two sides of the septum was apparently first made by BLAIR-BELL (1912) who was able to collect 12 patients from whom the septum was removed and in 6 of whom the material was satisfactory for pathologic examination. The upper surface of each was covered by columnar epithelium, whereas the lower surface was covered with squamous epithelium. BLAIR-BELL used this as an argument that the septum represented the intended site of fusion of the approaching epithelial buds. It is a pity that, in so far as I am aware, no one has biopsied the remaining vaginal epithelium or even exocervix, for it would be most informative to know if the entire isolated upper vagina would be covered by columnar epithelium. Presumably it would be, for it would be most extraordinary to have an island of columnar epithelium on the upper surface of the vaginal septum and nowhere else. It might be inquired if this observation means that the müllerian epithelium in the normal formation of the vagina adopts itself when exposed to the influences of the urogenital sinus epithelium.

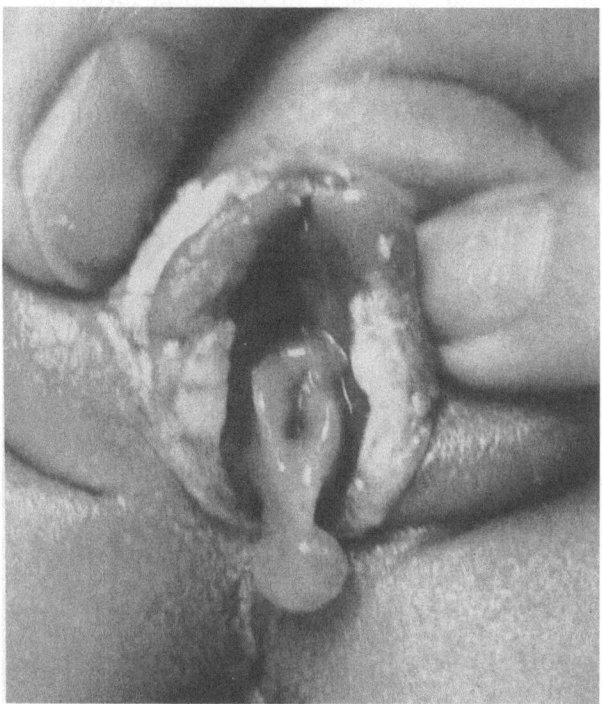

Fig. 10. Photograph of a hymenal polyp from the dorsal portion of the hymen. (From: N. E. BORGLIN, and P. SELANDER, 1962)

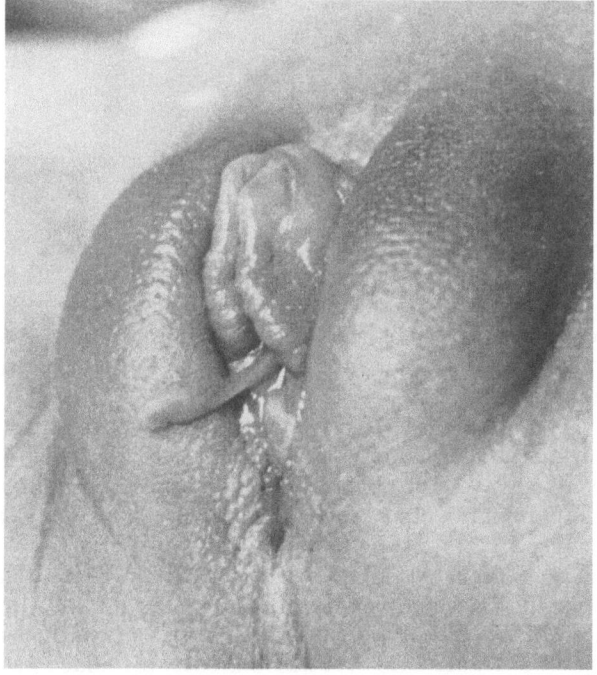

Fig. 11. Photograph of an elongated hymenal polyp. (From: BORGLIN and SELANDER, 1962)

It is extremely difficult to assign an etiological factor to most congenital anomalies, but it is interesting that McKusick et al. (1964) have noted an apparent genetic background to patients with transverse vaginal septum. These authors noted that among the Amish in Pennsylvania two distant cousins who were the offspring of consanguineous marriage were found to have identical examples of the transverse vaginal septum.

The number of such patients described in the literature is relatively modest although the number has increased rapidly in the last few years. There are probably about 200 such cases on record, mostly represented by reports of 2 or 3 cases in a series. The largest group of patients reported is that of Angelo Lodi (1951) who

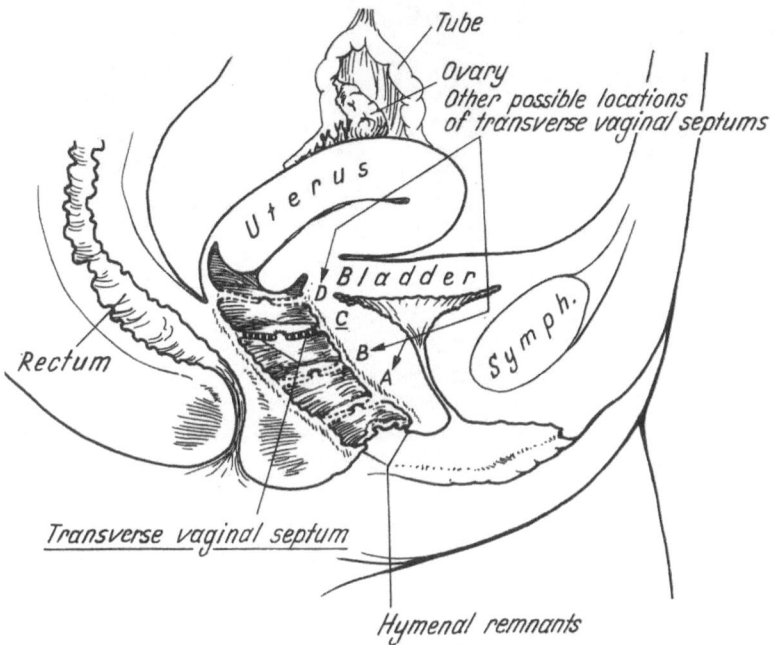

Fig. 12. A drawing of a congenital transverse vaginal septum with other possible locations in the vagina. Note that these locations are not related to the position of the hymenal remnants. (J. A. Bowman and R. B. Scott, 1954)

reported 110 cases of various vaginal abnormalities which had been studied over a 44 year period in Milan, Italy. Among these cases were 42 individuals with transverse vaginal septa, the most common of all vaginal abnormalities. Bowman and Scott recorded 4 individuals with this abnormality and presented a drawing (Fig. 12) showing the various possible locations of the transverse vaginal septa within the vagina.

The symptoms caused by the transverse vaginal septum are due to the obstruction that it causes and thus varies according to the time of its discovery. During infancy it may be unrecognized except for the accumulation of mucus where a large hydrometrocolpos may accumulate. This may be particularly difficult to recognize in view of the fact that there is no bulging at the introitus. Fatalities have been reported under such circumstances as illustrated in the patient shown in Fig. 13, which is an unpublished case of Dr. Roger Scott. Inaccurate diagnosis has resulted in undesirable laparotomy and even hysterectomy with the misdiagnosis of a urethral cyst with a greatly distended vagina (Dennison and Bacsich, 1961).

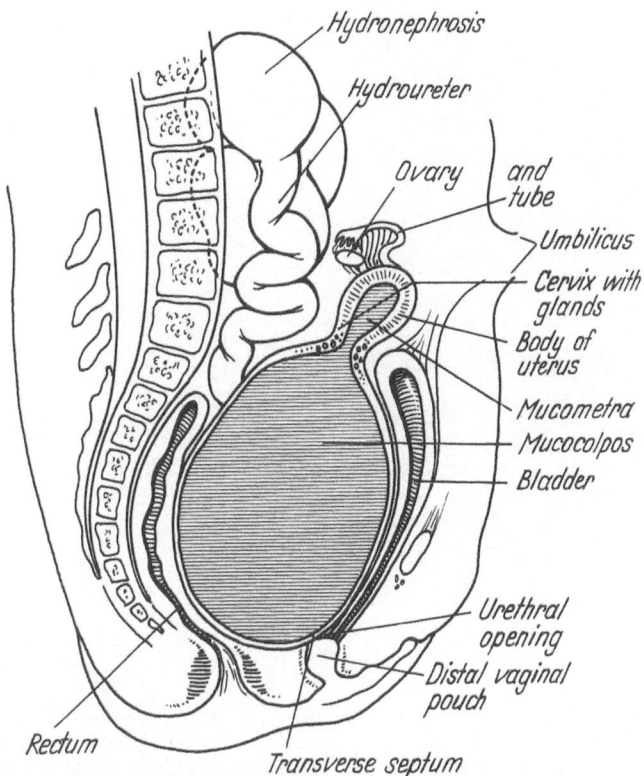

Fig. 13. Sagittal drawing of newborn with huge mucocolpos due to transverse vaginal septum. This was an un-published case observed by Dr. ROGER SCOTT. The child had marked urinary obstruction due to the extrinsic pressure of the mass. The patient did not present any bulging at the outlet as would be expected from a muco-colpos due to an imperforate hymen. The source of the mucus is apparently the cervical glands but why some newborns with complete obstruction of the vagina collect great quantities of mucus and others is completely unknown

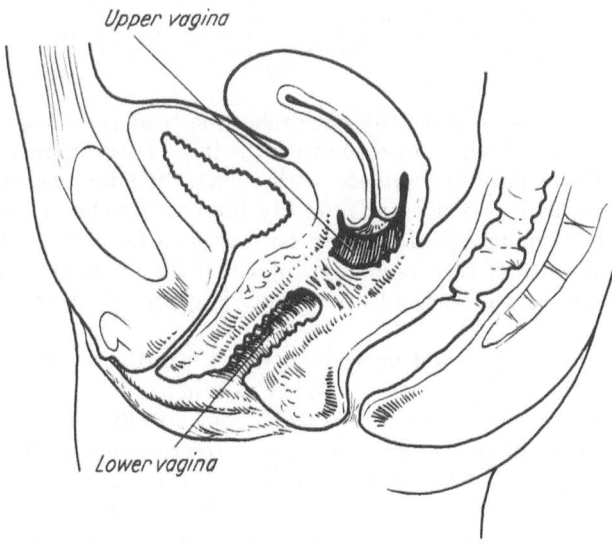

Fig. 14. A transverse vaginal septum. (From: TeLinde, 1962)

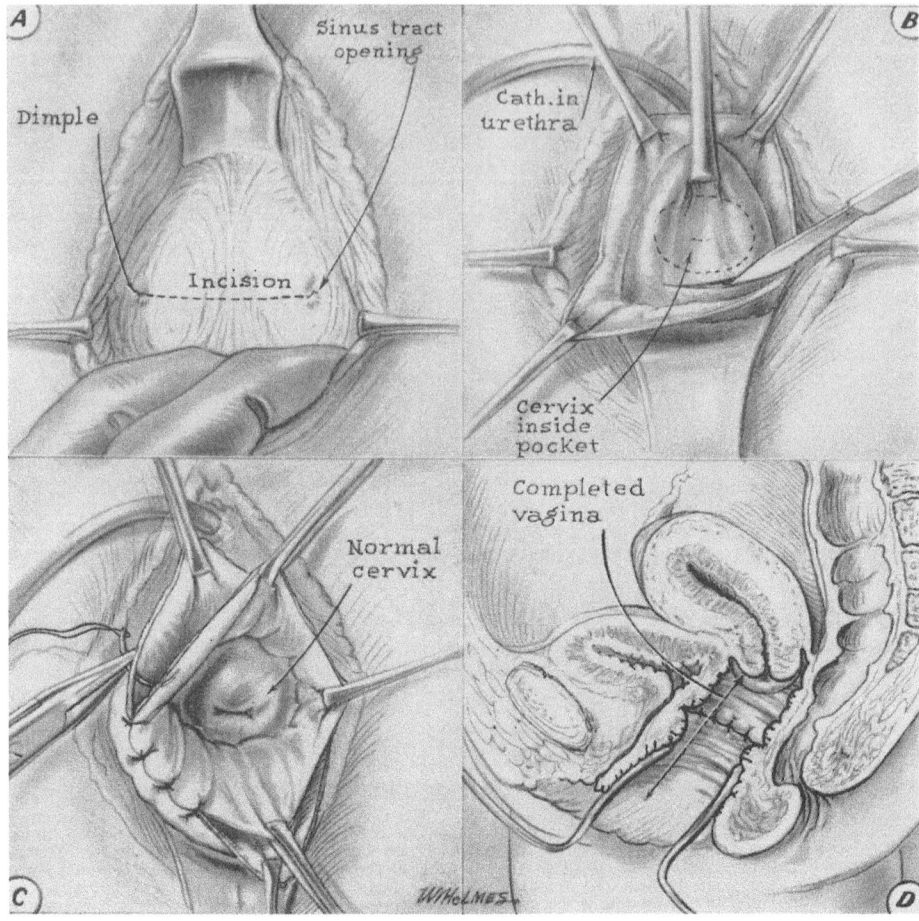

Fig. 15. A. Upper end of short vagina showing incision through mucus membrane; B. Areolar tissue has been tunneled to the pocket of the mucus membrane which enclosed the cervix; C. The anastomosis between the lower and upper vaginas; D. The completed vagina. (From: TELINDE, 1962)

If no symptoms occur until the menarche, the transverse vaginal septum will cause retention of menstrual blood. As with the imperforate hymen, large amounts of menstrual blood may accumulate with relatively minor symptoms although lower abdominal pain recurring at monthly intervals is characteristic. A patient with a transverse vaginal septum may be mistaken for a case of intersexuality or congenital absence of the vagina and uterus if the examiner mistakenly interprets the septum as the vaginal apex. This is especially true if the septum is imperforate.

In most instances the septum is less than a centimeter in thickness, but in other examples the vaginal plate has failed to canalize for a much longer distance so that a portion of the vagina seems to be congenitally absent (Fig. 14).

Various methods have been proposed for correction of the abnormality. If an opening in the septum is large enough to allow it, the simplest measure is manual dilatation. On the other hand, if there is complete obstruction, surgical incision would appear to be the most satisfactory measure in most cases. If a considerable portion of the vagina is absent, the anastomosis of the upper portion of the vagina with the lower may present technical difficulty even to the extent of the use of a split thickness graft according to the method of McINDOE (1950). However, in most examples of the transverse vaginal septum, simple excision of the septum is all that is necessary (Fig. 15).

II. Anomalies of the Müllerian Ducts

1. Congenital Absence of the Vagina and Uterus

Patients with congenital absence of the vagina usually also have absence of the uterus. Therefore, a more accurate term might be aplasia or dysplasia of the müllerian ducts. However, by common usage the term congenital absence of the vagina is used to describe the condition. Such patients may have a normally developed lower vagina a few centimeters in length. The usual lesion includes absence of the middle and upper third of the vagina and the uterus.

Some reported series of congenital absence of the vagina have included patients with a uterus and absence of various lengths of the vagina. In view of the homogeneity and specificity of the syndrome of congenital absence of the vagina, i.e. no uterus and high incidence of urinary tract anomalies, we have preferred to consider patients with the uterus and some defects of the vagina under the category of transverse vaginal septum.

It is uncommon for a diagnosis of this condition to be made in the newborn child. Indeed it cannot be made except by a very careful examination, which includes some procedure to determine the vaginal length by sounding and rectal examination to identify the presence or absence of the uterus. Such patients usually seek the physician at puberty or later because of failure of menstruation to appear. The general growth and development of such patients is quite normal, including the secondary sex characteristics, and there are ordinarily no associated anomalies except urinary tract and bone as will be mentioned below. The external genitalia are quite normal. The vagina may be shallow.

Three patients in the Hopkins series have had lower abdominal exploratory operations. One at the age of 21 because of doubt as to her sex, another at the age of 15 because of vague abdominal pain of one year's duration, and a recent patient because of multiple congenital anomalies. In each instance the findings were similar: the tubes and ovaries were entirely normal, but the uterus was represented by a small bicornuate rudimentary structure (Fig. 16). Microscopic examination of this rudimentary structure has shown muscle whorls which appear to be typical of uterine myometrium and at the center a few scraps of endometrium entirely of basalis as they exhibited not the slightest response to the ovarian hormones (Fig. 17). On rectal examination a normal uterus can, of course, not be felt; although as might be suspected from the description of the findings of the laparotomy, one can usually feel a thickening at the vaginal vault. It should be emphasized that in the average case exploratory laparotomy is not indicated.

The only possible difficulty in a differential diagnosis would be the testicular feminization syndrome in which situation there may be a shallow vagina with no uterus on rectal examination. The differential diagnosis is easily made by a buccal smear which in the congenital absence of vagina cases is entirely normal and in the cases of the testicular feminization syndrome is always negative. Chromosomes in 12 patients with congenital absence of the vagina have been investigated by AZOURY and JONES and in each instance an entirely normal female complement was discovered (1965).

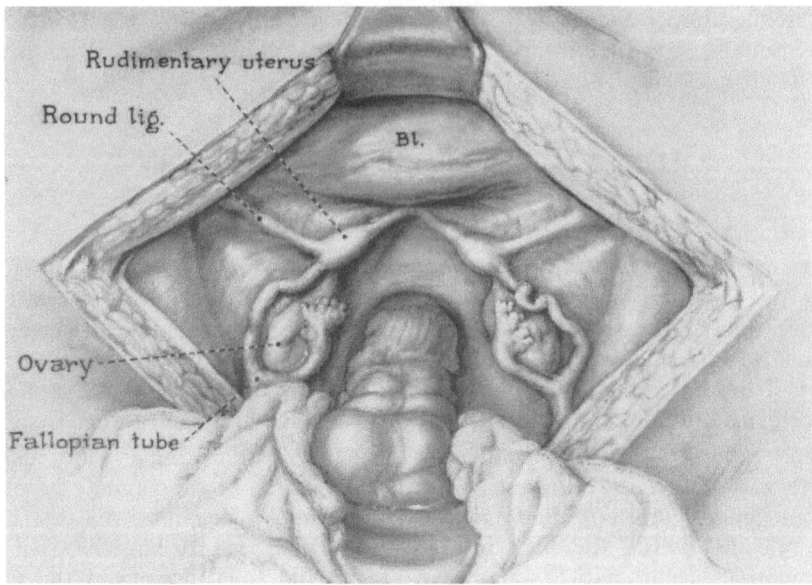

Fig. 16. View at laparotomy of patient with congenital absence of the vagina. This is the usual finding. The rudimentary uterus consists of scarcely more than a muscle bundle and a bit of endometrial tissue. The tubes are normally present as are the ovaries. (From: Jones and Scott, 1958)

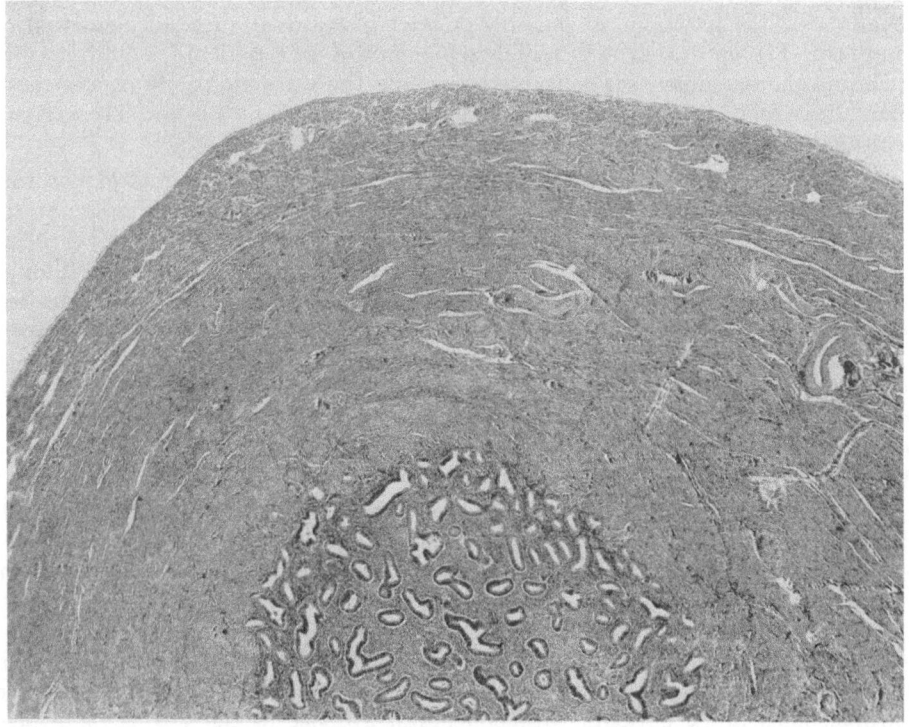

Fig. 17. Cross section of the rudimentary uterine tissue from the same patient shown in Fig. 16. The endometrial tissue is quite identifiable as such but obviously does not respond to the ovarian hormones (H & E × 20). (From: Jones and Scott, 1958)

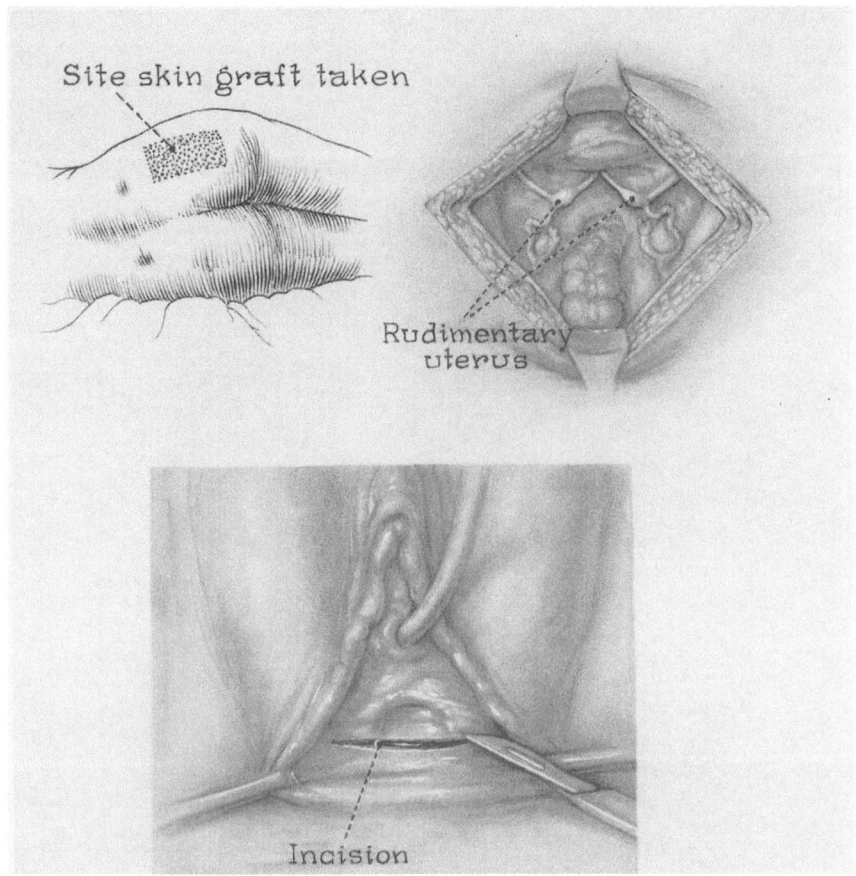

Fig. 18. The McIndoe procedure. Left upper: Although the skin graft may be taken from various areas, the sites shown in the illustration are in a concealed area which most patients find very satisfactory. Right upper: The usual appearance of the internal organs in patients with congenital absence of the vagina and uterus. Lower: A transverse incision is generally desirable as it assures maximum protection to the urethra with the available vaginal epithelium

Absence of the uterus and vagina is associated with anomalies of the urinary tract in a significant number of patients. In 17 patients investigated urologically by THOMPSON and his associates (1957), eight showed abnormalities—pelvic kidney, etc. Among 72 cases examined by PHELAN et al. (1953), twenty-six had anomalies of various sorts: 18 of these were major whereas 8 were considered more or less minor such as obstruction at the pelvoureteral junction. Roentgen visualization of the urinary tract is always desirable in patients with congenital absence of the vagina.

Anomalies of the spine and pelvis seem to be common. In 11 recent cases in the Hopkins series with x-rays of the lower spine, there was one case of lumbarization of the first sacral vertebra and in another patient who had x-rays of her entire spine because of discomfort of the neck, it was found that she had fusion of three of her cervical vertebrae. BRYAN et al. (1949) reported 6 spinal anomalies in their series of 100 patients.

Surgical construction of an artificial vagina may be very satisfactorily carried out by a variety of procedures. However, we have preferred the McIndoe technique,

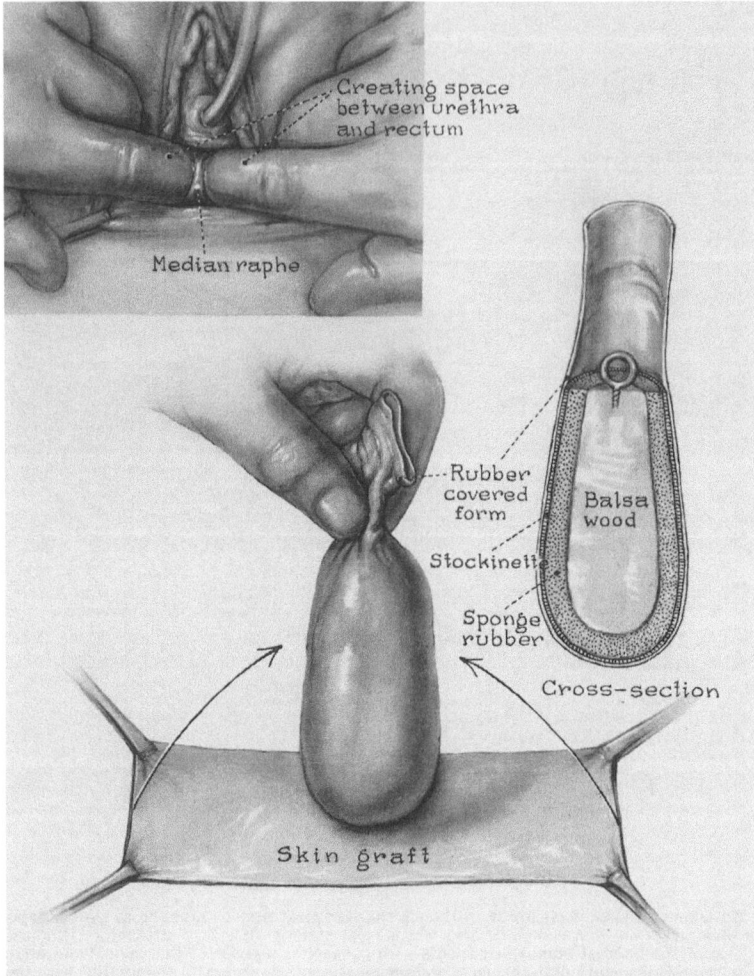

Fig. 19. The McIndoe procedure. Left upper: The blunt dissection is best carried out with the tips of the second fingers and is most easily done on either side of the median raphe which contain fibers between the rectum and the urethra which can be severed by the aid of scissors after the dissection is carried to the region of the peritoneum. Lower: After the covered balsa-wood form is selected for proper size, the skin graft is attached as shown, being sure that the external surface of the graft is against the balsa-wood form

but this is ordinarily not recommended until the child has attained her full growth. The procedure can then be carried out at any time. It is not recommended until the latter teens or marriage is intended because a good result requires cooperation and motivation on the part of the patient in wearing of the vaginal form for a few months (Figs. 18, 19, 20, 21).

2. Maldevelopment of the Vagina and Uterus

Maldevelopment of the uterus occurs in a variety of forms. Most such lesions are asymptomatic in childhood and not discovered until later in life when they may cause reproductive failure. However, if the maldevelopment has resulted in obstruction to the outflow of menstrual blood, acute abdominal pain soon after

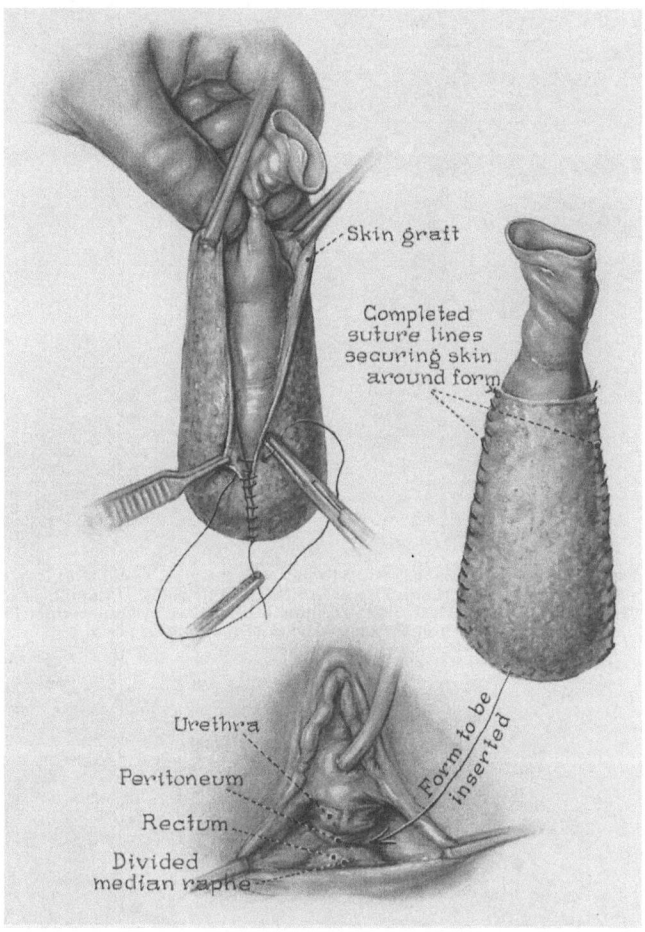

Fig. 20. McIndoe procedure. Upper: This illustration shows the graft being sewed over the balsa-wood form. It is possible to secure the same result by using dermatome cement in lieu of the sutures. Lower: Illustration of the space developed between the bladder and the rectum. The dissection should be carried until a button of peritoneum is palpable and visible at the apex of the dissection

the menarche may be expected. It is convenient, therefore, to describe maldevelopment of the vagina and uterus under two headings: obstructive and nonobstructive.

a) Obstructive Maldevelopment of the Vagina and Uterus

Obstructive maldevelopments are by and large associated with problems of lateral fusion of the müllerian ducts resulting in lateral duplication of the vagina and/or uterus in whole or in part. The lateral duplication may be symmetrical or asymmetrical in which latter situation a rudimentary maldevelopment of one side will result. Very rarely congenital obstruction may occur transversely in the cervix, in an otherwise normally formed organ (Fig. 22). Still more rarely obstruction may occur at the outlet (Figs. 23, 24).

With lateral reduplication, the obstruction may be in the vagina (Figs. 25, 26) or in the uterus (Fig. 27). Often there is considerable disproportion between the two uterine sides so that one might be termed rudimentary. Such rudimentary

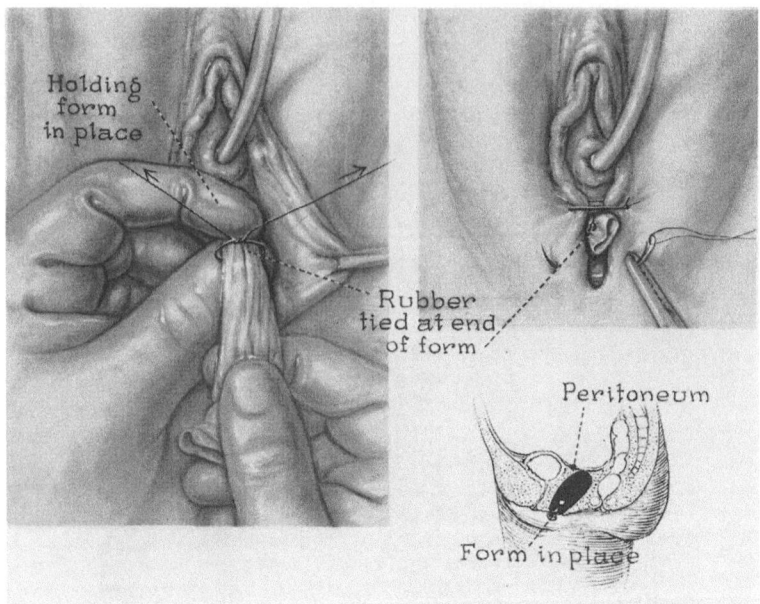

Fig. 21. The McIndoe procedure. Left: The balsa-wood form with the graft sewed over it has been inserted in the place in the developed space and the rubber sheath tied only at this point. If the rubber is tied prior to this trapped air will prove to be a problem. Upper right: The form held in place by sutures through the labia. Lower right: Sketch of the sagittal view of the form in place

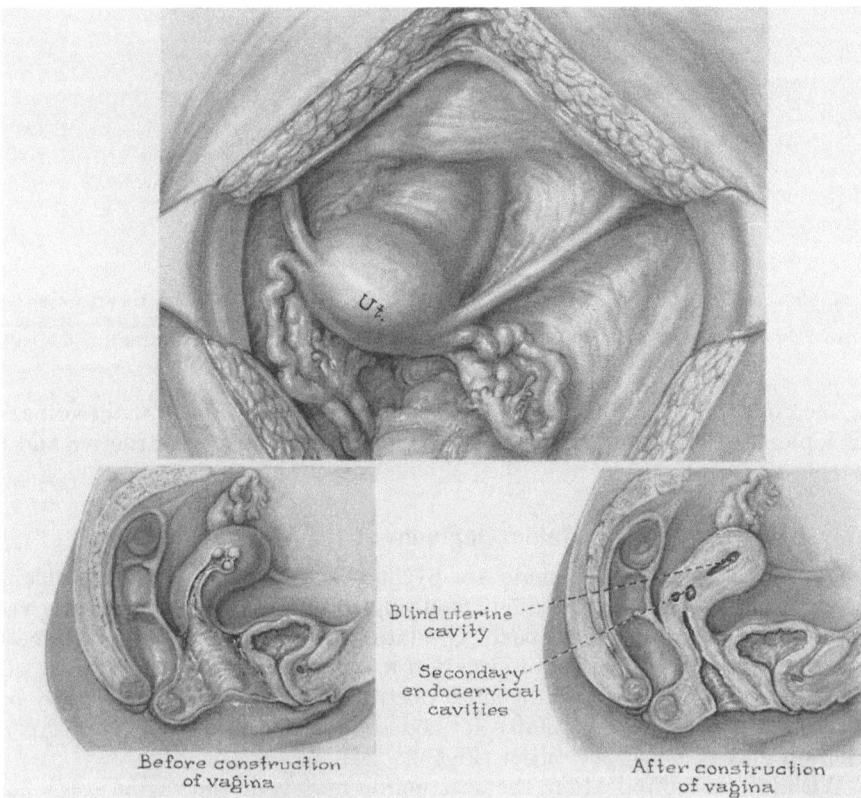

Fig. 22. Operative findings and inset diagrammatic representation of a patient who had congenital absence of the vagina and also congenital absence of the cervical canal resulting in a blind uterine cavity. Attempts to establish communication between the uterine cavity and artificially constructed vagina were unsuccesful, and it was necessary to do a hysterectomy to relieve the cyclic pain

uterine horns may be attached by only a thin thread to the more normal contralateral uterus. When there is relatively little disparity in size between the two
sides, they are often intimately associated.

Cyclic abdominal pain, which may become severe, is a common symptom.
The appearance of normal menstrual bleeding from the non-obstructive side
associated with the severe pain from the obstructive side may be misleading.
Some patients have not had impelling symptoms for an unbelievable length of
time, six, eight or more years after the onset of the menarche.

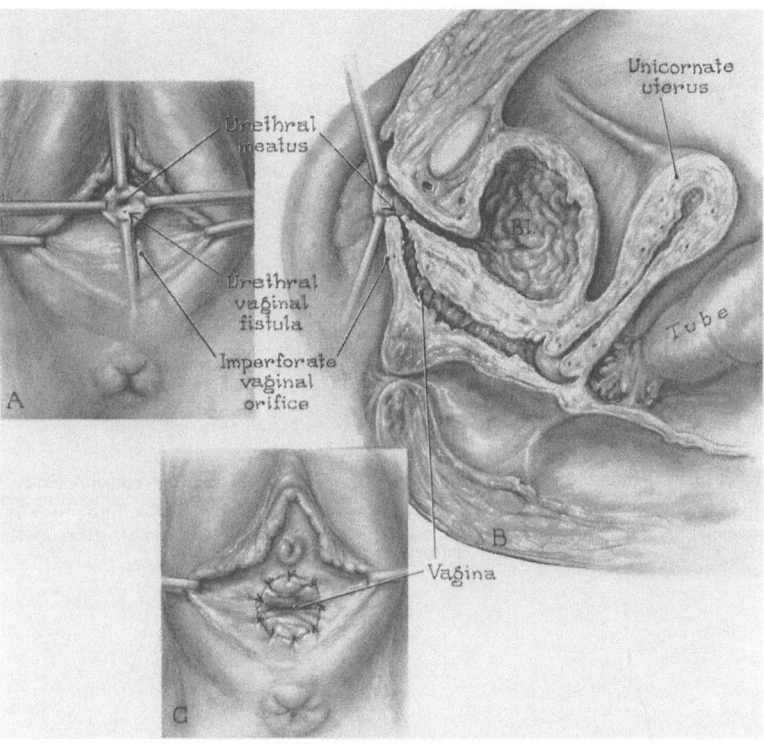

Fig. 23. A patient with a peculiar obstructive lesion of the vaginal outlet. A. A tiny communication of the vagina
to the outside just at the urethral meatus. B. A surgical view to illustrate the situation. C. The revised vaginal
outlet established at surgery

Pelvic examination may be difficult to interpret. In one patient the bulging
in the vagina of the mass of retained blood caused such distortion that it gave the
impression of a paravaginal tumor. Complete absence of one kidney is not unusual.
As was so convincingly pointed out by WOOLF and ALLEN (1953), the defect of
the urinary system is invariably on the side with the most serious müllerian
defect (Fig. 28).

If the diagnosis is promptly made, it is often possible to relieve the obstruction
without sacrificing the reproductive potentiality. With the unilateral obstructing
vaginal septum, excision is, of course, indicated. These obstructive septa are, in
my experience, quite thick and their removal is not easy. The bleeding is
vigorous and care must be exercised so that too much traction is not exerted
with the result that more than the desirable amount of tissue is removed. With
an obstruction in the bicornuate uterus, the possibility of uterine reconstruction
exists. As no two of these anomalies are exactly alike, it cannot be anticipated that

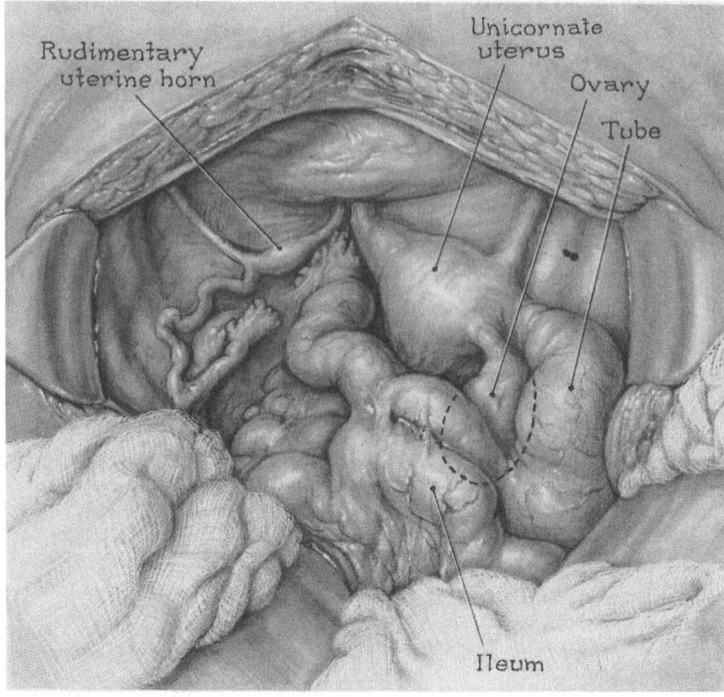

Fig. 24. The abdominal situation in the same patient shown in Fig. 23. The tiny pinpoint obstruction at the outlet which impeded the flow of menstrual blood has resulted in a hematocolpos, a hematometra and a hemato-salpinx. Although it was necessary to surgically remove these structures, it is very likely that this patient's reproductive potential was irreversibly compromised by failure to relieve the obstruction at the outlet at puberty or thereafter

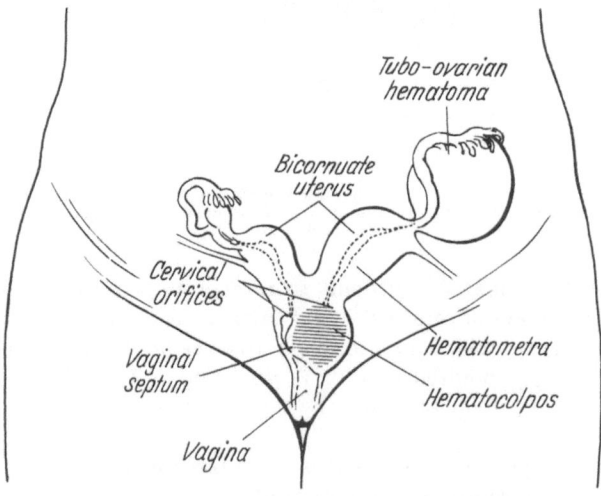

Fig. 25. Diagrammatic representation of a patient with lateral duplication of the uterus and part of the vagina but an obstructive left side. Although the patient menstruated regularly for about 2 years, she had excruciating abdominal pain with each menstrual period

a standard technique will apply in all instances. Fig. 29 and 30 illustrate the technique which was applicable to one such case.

Rudimentary obstructing noncommunicating horns can be excised.

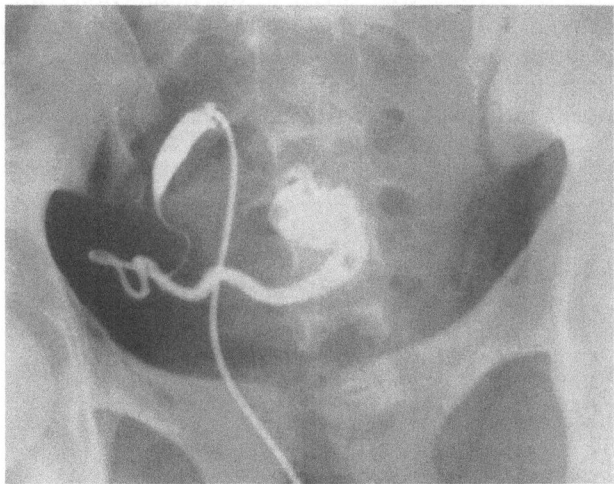

Fig. 26. Hysterogram of the same patient diagrammed in Fig. 25

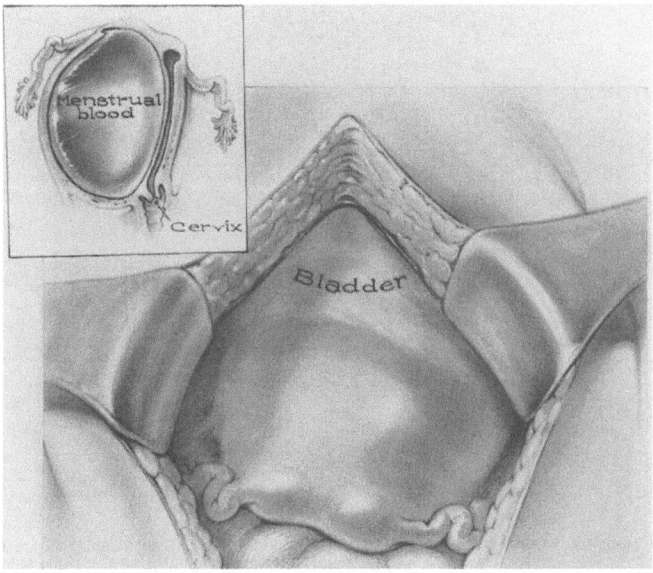

Fig. 27. Operative finding in a patient with a septate uterus and an obstructive right side

b) Non-Obstructive Maldevelopment of the Vagina and Uterus

α) *The Unicornuate Uterus.* If the development of one müllerian duct is completely arrested, the uterus and fallopian tube may be formed entirely from the other. This so-called uterus unicornis or unicorn uterus seldom causes any clinical abnormality (Fig. 31).

β) *Rudimentary Uterine Horn.* The obstructive form of a rudimentary uterine horn has already been discussed. Most rudimentary horns are indeed noncommunicating. In some cases the endometrium is non-functional, so that no clinical symptoms are present. It has been mentioned above that if the endometrium is functioning, a clinical situation may arise from the retention of menstrual blood.

In other instances the endometrial cavity of the rudimentary uterine horn may communicate through a narrow channel with the more normal opposite cavity. Under these circumstances pain, and very rarely pregnancy, may occur. If a pregnancy does occur in a rudimentary horn under such a circumstance, the patient may present the classic picture of an ectopic pregnancy including rupture (Fig. 32).

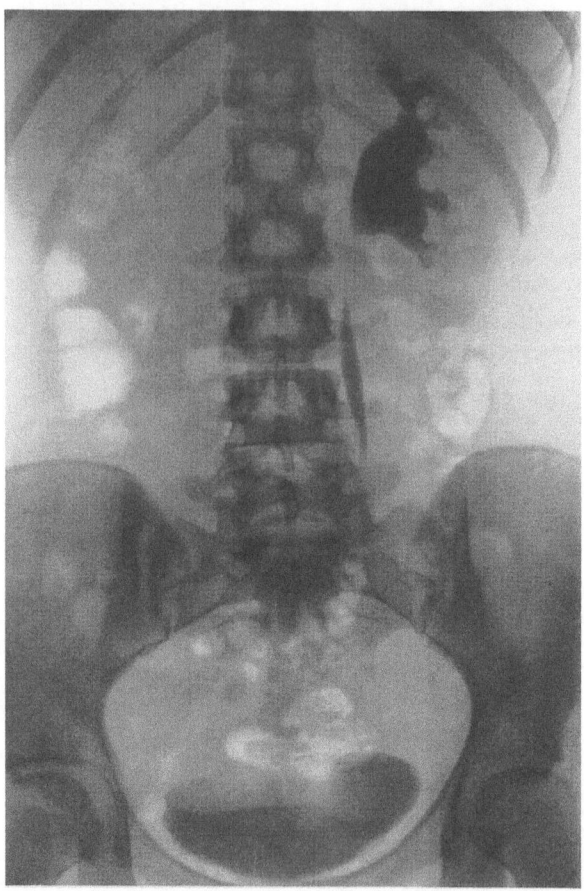

Fig. 28. Intravenous pyelogram showing a congenital absence of one kidney in a patient with rudimentary right uterine horn. This illustrates the fact that the serious renal anomaly is usually on the same side with the most serious müllerian defect

As with all examples of maldevelopments of müllerian ducts, anomalies of the wolffian ducts may also be present. With a rudimentary horn this is especially common. The anomaly of the urinary tract is, as an invariable rule, on the same side as the maldevelopment of the müllerian duct. While the kidney may be malrotated, low lying or actually within the bony pelvis complete agenesis is not uncommon.

γ) *Symmetrical Double Uterus*. When the two müllerian ducts develop side by side without communicating with each other, there is produced a complete double uterus. Each duct forms one cervix and one uterine body with a fallopian tube attached to each. The duplication may continue down into the vagina so that

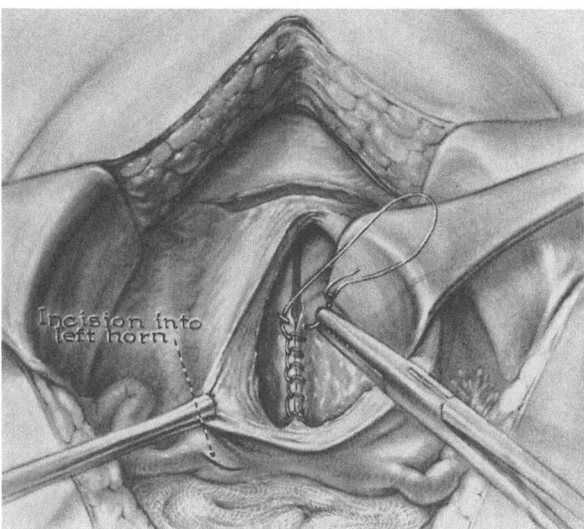

Fig. 29. Reconstruction of the uterus shown in Fig. 27. It was possible to anastomose the obstructed side to the unobstructive side

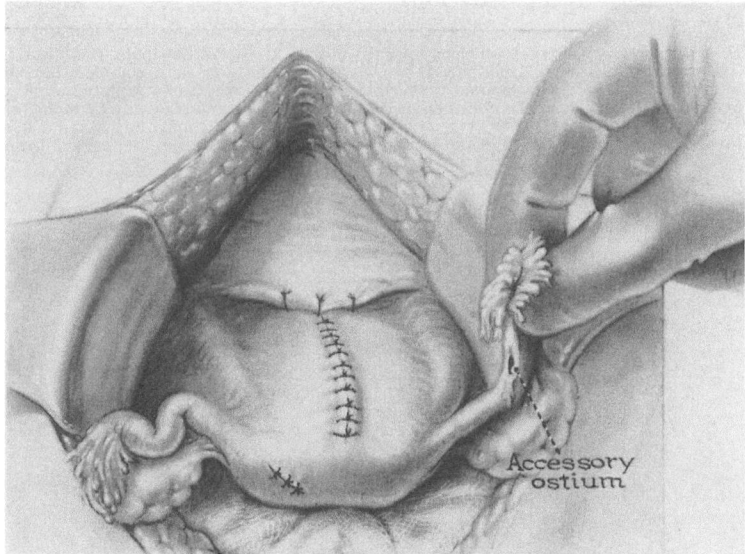

Fig. 30. A further step in the procedure illustrated in Fig. 29

that part of the vagina formed by the müllerian ducts is duplicated. Such complete duplication may be referred to as a uterus didelphys (Fig. 31).

Most reduplicated uteri, however, are not so complete and the fusion may involve only the upper portion so that the two uterine cavities communicate to a greater or a less degree. Under this circumstance there is but a single cervix and a single vagina. A rather simplified classification of such double uteri is convenient. If the two horns of such a partially fused uterus are recognizable externally, the uterus may be designated as a bicornuate uterus (Fig. 31). Some-

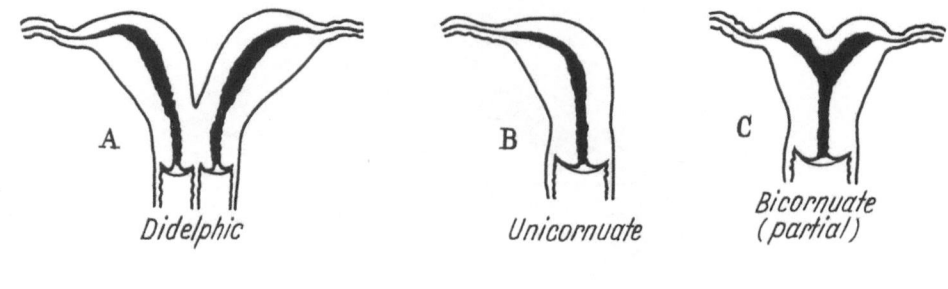

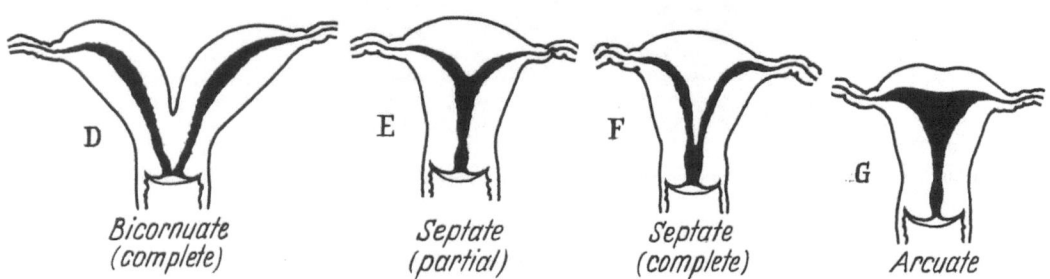

Fig. 31. Sketch of non-obstructive maldevelopments of the uterus. A. Uterus didelphis. B. Unicornuate uterus.
C. A partial bicornuate uterus as evidenced by the external configuration. D. A complete bicornuate uterus
as evidenced by the external configuration and the double cervix. E. A sketch of a septate uterus as evidenced
by a normal external configuration of the uterus and a septum which is diagnosable only by radiographic means.
F. Sketch of a complete septate uterus as evidenced by a normal external configuration and a double cervix.
G. A sketch of an arcuate uterus which is the most mild form of malformation and seldom associated with
reproductive difficulties

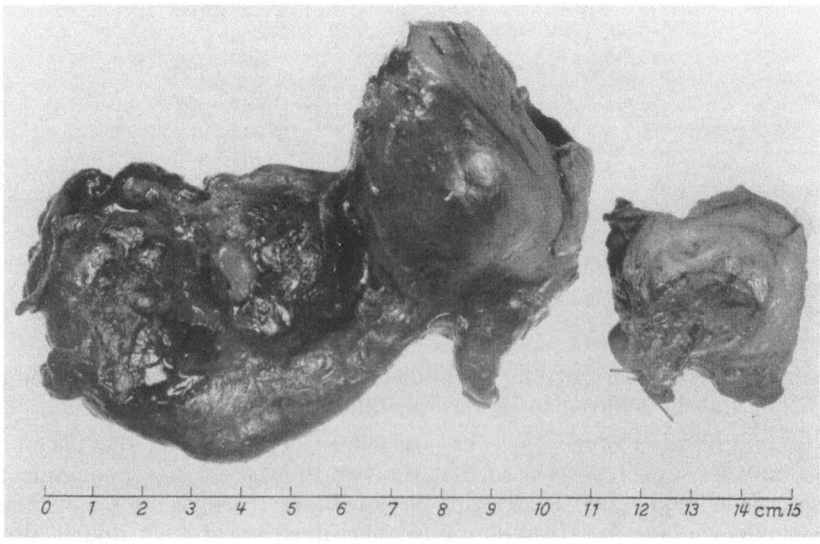

Fig. 32. Photograph of a pregnancy in a communicating rudimentary horn of a double uterus. The signs and
symptoms of this patient were those of an ectopic pregnancy

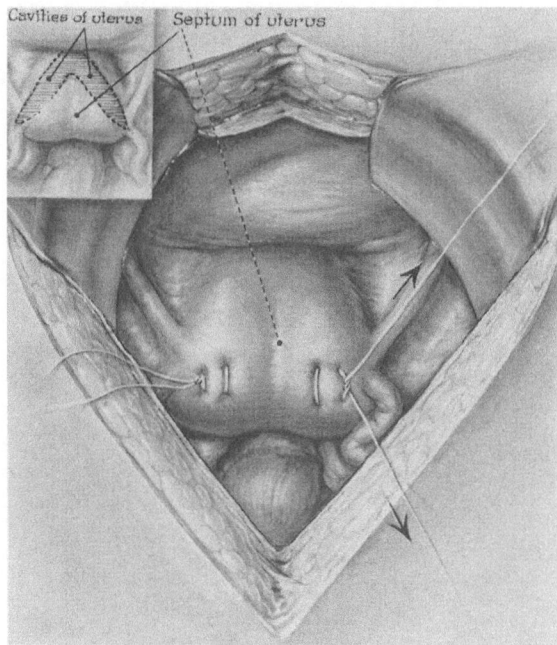

Fig. 33. The appearance of a septate uterus at operation. Inset shows a thick muscular septum. (From: H. W. JONES Jr. and G. E. S. JONES, 1953)

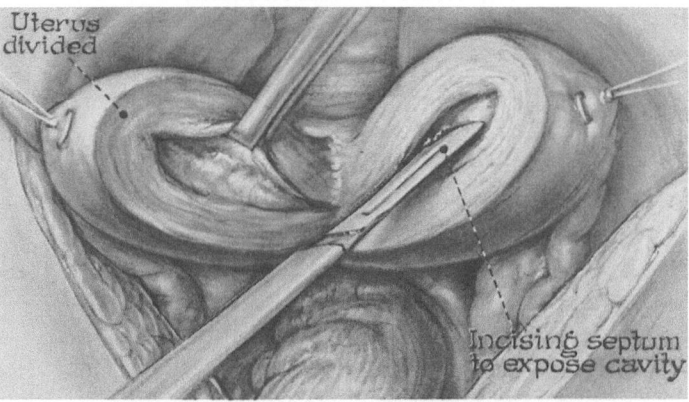

Fig. 34. The uterus bisected. After excision of wedge-shaped segment it is necessary to excise additional tissue to unroof each horn. (From: JONES and JONES, 1953)

times, however, the external configuration of the uterus is relatively normal, so that it may be almost impossible to recognize it from its external appearance. A malfusion in such a situation is represented only by a septum within the uterus. In this circumstance the uterus may be referred to as a septate uterus (Fig. 31). If the condition is minimal an arcuate uterus is said to occur (Fig. 31).

The various degrees of reduplication of the uterus may be associated with reproductive failure especially repeated miscarriages. It should be emphasized, however, that an anomalous uterus may be compatible with a normal reproductive history. Experience has shown that only about one fourth of all women with

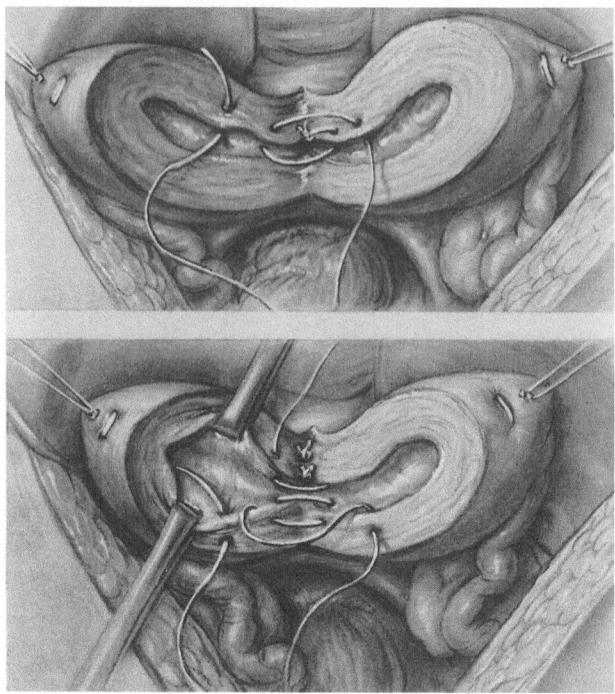

Fig. 35. Anastomosis of the two halves of the uterus. Two continuous sutures are used to join the endometrial cavity. (From: Jones and Jones, 1953)

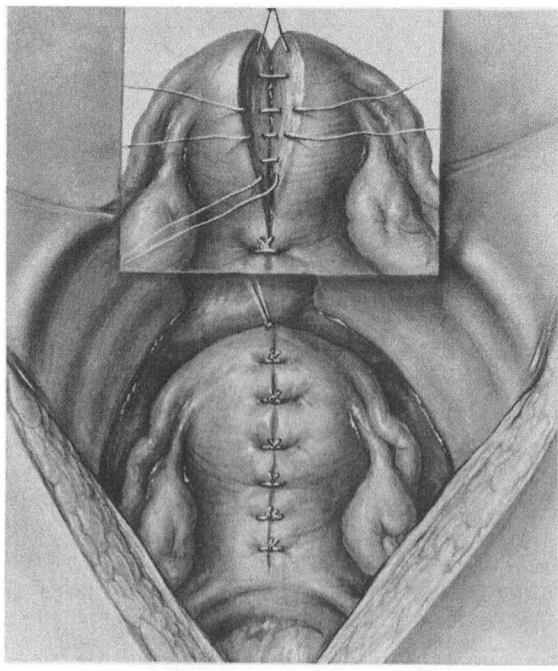

Fig. 36. The third layer unites the myometrium and serosa. The inset shows a second layer of sutures uniting the myometrium. (From: Jones and Jones, 1953)

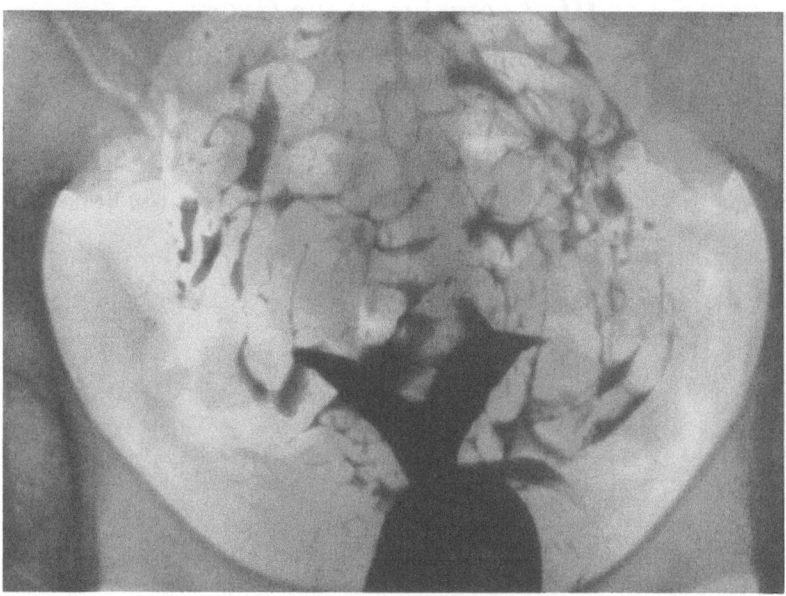

Fig. 37. Typical hysterogram of a double uterus. This particular case had a single cervix. (From: H. W. JONES Jr., E. DELFS and G. E. S. JONES, 1956)

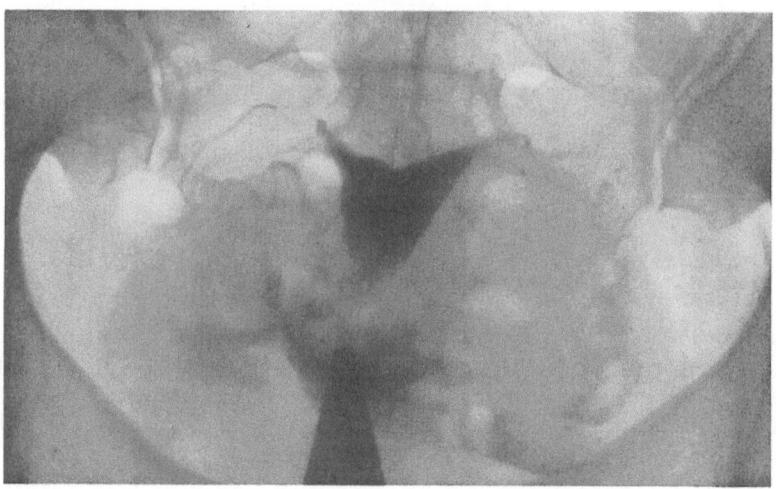

Fig. 38. Hysterogram following operative reconstruction of the uterus shown in the preceding figure. (From: JONES, DELFS and JONES, 1956)

some form of this anomaly will have reproductive problems. If it can be demonstrated by suitable tests by the exclusion of other causes that a double uterus is in fact responsible for the reproductive problem, surgical reconstruction of the uterus to form a single uterine cavity is entirely feasible. It is interesting that a septate uterus is much more apt to cause difficulty than a bicornuate uterus, although reproductive failure may be associated with either (Figs. 33, 34, 35, 36, 37, 38). In properly selected patients with repeated miscarriages reconstruction of the uterus will result in a live birth in about 3 out of 4 patients (JONES, DELFS and JONES, 1956).

24*

III. Anomalies of the Ovary

1. Congenital Absence of the Ovaries

See chapter on the Intersex States.

2. Supernumerary Ovaries

Abnormally located ovarian tissue is an extremely rare condition. Wharton (1939) thoroughly reviewed the subject.

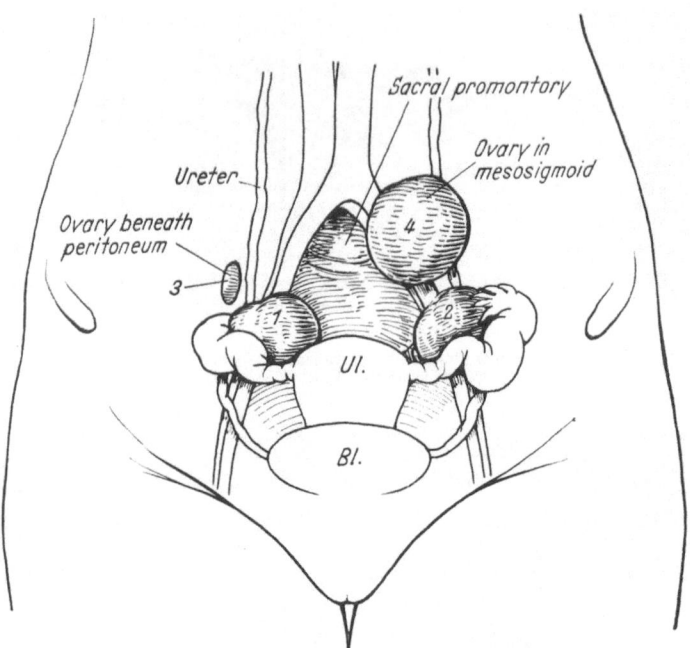

Fig. 39. A composite drawing showing the four locations of the four acceptable reported patients with supernumerary ovaries. (From: L. R. Wharton, 1959)

Supernumerary ovaries include those cases in which one or more extra ovaries are entirely separate from the normally placed ovaries (Fig. 39). According to Wharton, there are only 4 such acceptable cases in the literature. Such ovaries apparently arise through a duplication of the embryological process which is responsible for the development of normal ovaries.

3. Accessory Ovaries

Accessory ovaries include those cases in which excess ovarian tissue is situated near the normally placed ovary, may even be connected to it, and seem to have developed from it. Such accessory tissue is variably located near the normally placed ovary. Generally speaking these accessory ovaries are small and almost always measure less than a centimeter in diameter. Most frequently they have been found attached to the broad ligament near the normal ovary. However, they have been described in the cornu of the uterus and between the leaves of the broad ligament. These small masses of tissue have usually been grossly mistaken for lymph nodes and their true nature has only been revealed by microscopic examination. For the most part, they have been solitary, but a few cases of two or even three accessory ovaries have been reported. They have universally been an incidental finding. MILLER (1937) states that if carefully searched for such accessory ovaries may be found in about 4% of patients in an autopsy series.

4. Displaced Ovaries

The descent of the ovaries which normally takes place during fetal life may cease prematurely at different stages. The abnormality may be unilateral or

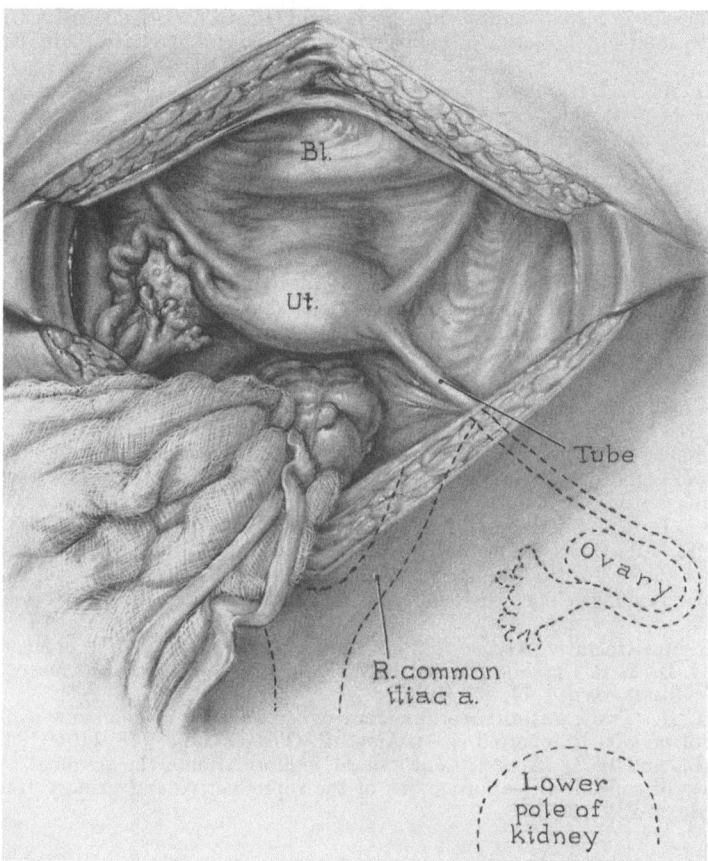

Fig. 40. Operative drawing from a patient with a displaced ovary. In this situation a defect in descent occurred in such away as to place the ovary at the lower pole of the kidney on the right side

bilateral. An extreme case was described by NICHOLS and POSTLOFF (1951) where one ovary was completely undescended and located immediately below the liver. We have encountered one such case where the right ovary was just at the lower pole of the right kidney (Fig. 40).

References

AITKEN, J.: A case of colon and ileum duplex. Brit. J. Surg. 37, 349 (1950).
AZOURY, R. S., and H. W. JONES jr.: Cytogenetic findings in patients with congenital absence of the vagina. Amer. J. Obstet. Gynec. 93, 335 (1965).
BELL, B. W.: Further investigation into the chemical composition of menstrual fluid and the secretions of the vagina, as estimated from an analysis of haematocolpos with a discussion of the clinical features associated with haematocolpos, and a description of the character of the obstructing membrane. J. Obstet. Gynaec. Brit. Emp. 21, 209 (1912).

Borglin, N. E., and P. Selander: Hymenal polyps of the newborn infant. Acta pediat., Suppl. 135, 28 (1962).

Bowman, J. A., and R. B. Scott: Transverse vaginal septum; report of four cases. Obstet. Gynec. 3, 441 (1954).

Breen, J. L., and C. R. Weinberg: Genitourinary and intestinal duplication. Case report and survey of the literature. Obstet. Gynec. 26, 804 (1965).

Bryan, A. L., J. A. Nigro, and V. S. Counseller: One hundred cases of congenital absence of the vagina. Surg. Gynec. Obstet. 88, 79 (1949).

Chiarleoni, G.: Ann. Ostet. Ginec. 16, 469 (1894).

Dennison, W. M., and P. Bacsich: Imperforate vagina in the new born. Arch. Dis. Child. 36, 156 (1961).

Gedda, L.: Twins in history and science. Springfield (Ill.): Ch. C. Thomas 1963.

Gemmell, R., and M. Paterson: Double vulvae. J. Obstet. Gynaec. Brit. Emp. 23, 139 (1913).

Jones jr., H. W., E. Delfs, and G. E. S. Jones: Reproductive difficulties in double uterus. The place of plastic reconstruction. Amer. J. Obstet. Gynec. 72, 865 (1956).

—, and G. E. S. Jones: Double uterus as an etiological factor in repeated abortion: Indications for surgical repair. Amer. J. Obstet. Gynec. 65, 325 (1953).

Kanagasuntheram, R., and A. G. S. Dassanayakae: Nature of the obstruction membrane in primary cryptoamenorrhea. J. Obstet. Gynaec. Brit. Emp. 65, 487 (1958).

Koff, A. K.: Development of the vagina in the human fetus. Contrib. Embryol. 24, 59 (1933).

Lesbre, F. X., et F. Vigot: Traité de tératologie, 1927.

Lodi, A.: Contributo clinico statistico sulle malformazioni dela vagina osservate nella clinica, Obstetrica e Ginecologica di Milano del 1906 al 1950. Ann. Ostet. Ginec. 73, 1246 (1951).

McIndoe, A.: Treatment of congenital absence of obliterative condition of the vagina. Brit. J. plast. Surg. 2, 254 (1950).

McKusick, V. A., L. Bauer, C. E. Koop, and B. Scott: Hydrometrocolpos as a simply inherited malformation. J. Amer. med. Ass. 189, 813 (1964).

Miller, J.: Die Krankheiten des Eierstockes. In: Henke/Lubarsch, Handbuch der speziellen pathologischen Anatomie und Histologie, Bd. VII/3. Berlin 1937.

Nichols, D. H., and A. V. Postloff: Congenital ectopic ovary. Amer. J. Obstet. Gynec. 62, 195 (1951).

Ombredanne, L.: Une fillete Splanchnodyme. Mém. Acad. Chir. 62, 747 (1936).

Phelan, J. T., V. S. Counseller, and L. F. Greene: Deformities of the urinary tract with congenital absence of the vagina. Surg. Gynec. Obstet. 97, 1 (1953).

Suppiger, J.: Bildungsfehler der weiblichen Beckenorgane. Korresp. Bl. schweiz. Ärzte 6, 418 (1876).

TeLinde, R. W.: Operative Gynecology, 3rd ed. Philadelphia: J. B. Lippincott Co. 1962.

Thompson, J. D., L. R. Clinton sr., and R. W. TeLinde: Congenital absence of the vagina. Amer. J. Obstet. Gynec. 74, 397 (1957).

Wharton, L. R.: Two cases of supernumerary ovary and one of accessory ovary, with an analysis of previously reported cases. Amer. J. Obstet. Gynec. 78, 1101 (1939).

Woolf, R. B., and W. M. Allen: Concomitant malformations; the frequent simultaneous occurrence of congenital malformations of the reproductive and urinary tracts. Obstet. and Gynec. 2, 234 (1953).

The Intersex States*

Howard W. Jones Jr.

With 73 Figures

Sex identification may be described in terms of at least seven characteristics. The first five of these are organic and the last two psychological: 1) sex chromatin and sex chromosomes; 2) gonadal structure; 3) morphology of the external genitalia; 4) morphology of the internal genitalia; 5) hormonal status; 6) sex of rearing and 7) gender role.

I. Criteria of Sex

1. The Sex Chromatin and Sex Chromosomes

The understanding of the genetic background of sex at the clinical level was greatly stimulated and aided by the discovery of Barr and Bertram (1949) of a simple technique for the recognition of sexual dimorphism in interphase nuclei. The imprint of sex is found in the so-called sex chromatin which is distinctive for the human female as well as many other mammals (Fig. 1).

It now seems established that the sex chromatin is one of the two X chromosomes which has become stainable and, therefore, visible during interphase. Because of the findings in certain X-linked characteristics in rodents, Lyon suggested that the visible X chromosome may be of maternal origin in a fixed portion of cells and the remainder of paternal origin. Furthermore the Lyon proposal suggested that the sex chromatin was genetically inactive and that the ratio of maternal to paternal sex chromatin remains constant in any one individual after being fixed by chance early in embryonic life (Lyon, 1962).

For many years it was thought that the human diploid chromosome number was 48. However, in 1956 Tjio and Levan using a tissue culture technique found but 46. By taking advantage of recent technical advances, it now seems to have been clearly established that 46 is the correct human chromosome number. In addition to the establishment of the chromosome number, the newer techniques allow further study of chromosome morphology. Karyotypes of human somatic chromosomes have been published by a number of workers, and the identification of most chromosomes including both X and Y is reasonably clear (Fig. 2).

In the normal female, the percentage of cells showing the sex chromatin varies with the technique of preparations and the tissue used. In clinical work cells from the buccal mucosa are readily obtained and 20 to 50% of them show a single sex chromatin body in a normal woman. From a study of pathological states, it has been found that the number of sex chromatin bodies per cell is one less than the number of X sex chromosomes in each cell. Thus, a cell with three X chromosomes has two sex chromatin bodies during interphase (Fig. 3).

---

* Based on material from *Pediatric and Adolescent Gynecology* by Howard W. Jones, Jr. and Richard H. Heller, Williams and Wilkins Company, Baltimore, 1966.

Chromosomal mosaicism is said to exist when cell lines with varying numbers of chromosomes are found in the same individual. Thus, if O is used to indicate the absence of a chromosome an example of mosaicism of the sex chromosomes may be represented in a pathologic state where one cell line has a sex chromosome complement of XO and another line of XX. This state may be represented as XO/XX. When there is dilution of the normal cell line by a strain of cells which does not have a positive sex chromatin, the quantitative count will be reduced. Therefore, in patients with a low quantitative sex chromatin count, chromosomal mosaicism of the sex chromosomes must be suspected.

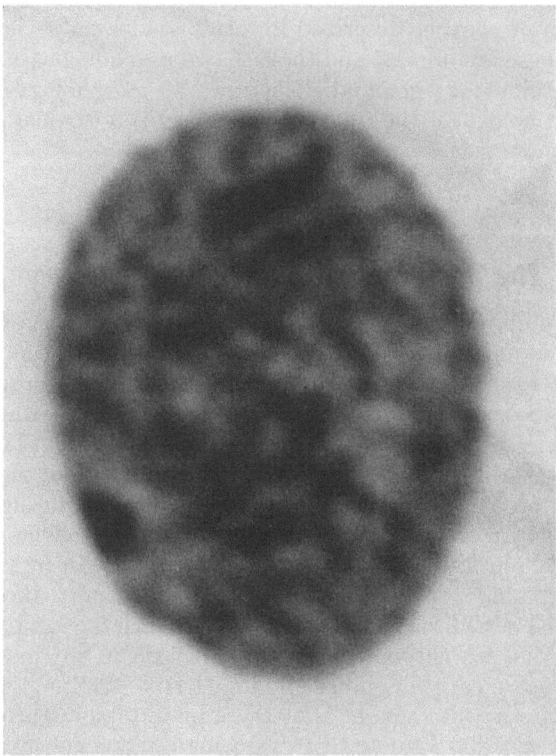

Fig. 1. Photomicrograph of cell from the buccal mucosa showing the sex chromatin just under the nuclear membrane at about 8 o'clock (acetic-orcein × 2000).

Not only have abnormal numbers of chromosomes been found, but structural abnormalities of individual chromosomes have been encountered. This may take the form of deletion of a portion of the chromosome, translocation of one part of a chromosome to another chromosome, or the formation of an isochromosome by an error in mitosis, e.g., an abnormal X chromosome may be composed of four long arms or four short arms instead of the normal configuration of two long and two short arms. In these abnormal forms, the size of the sex chromatin body is affected; an isochromosome of the long arms of an X chromosome gives a large chromatin body and an isochromosome of the short arms gives a small sex chromatin body.

"Drumsticks" of polymorphonuclear leucocytes are also an expression of the sex chromatin but are not as convenient or reliable as a study of the chromatin bodies of a smear made from the buccal mucosa membrane (Fig. 4).

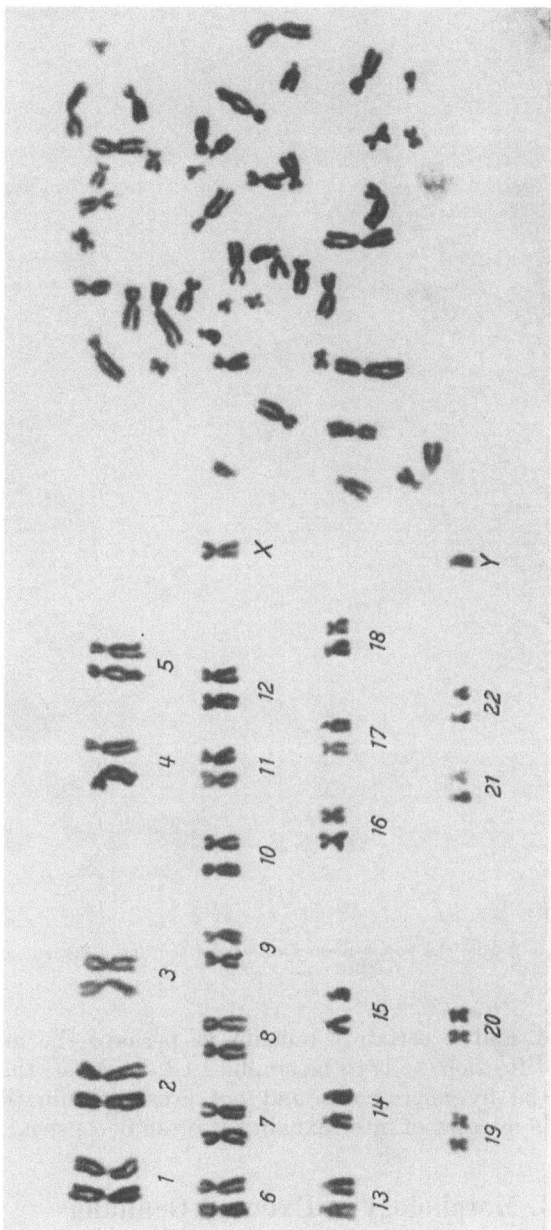

Fig. 2. Photograph of a normal male karyotype

2. Gonadal Structure

The identification of the sex of the gonads by microscopic examination is the basis for the well known Kleb's classification of hermaphrodites. It is obvious that various other criteria of sex identification are often in conflict with the gonadal sex. For example, in a study by MONEY et al. (1955a) of 76 hermaphroditic patients, there were 20 in whom a contradicition was found between gonadal sex and sex of rearing. Adequate therapy often depends upon proper identification

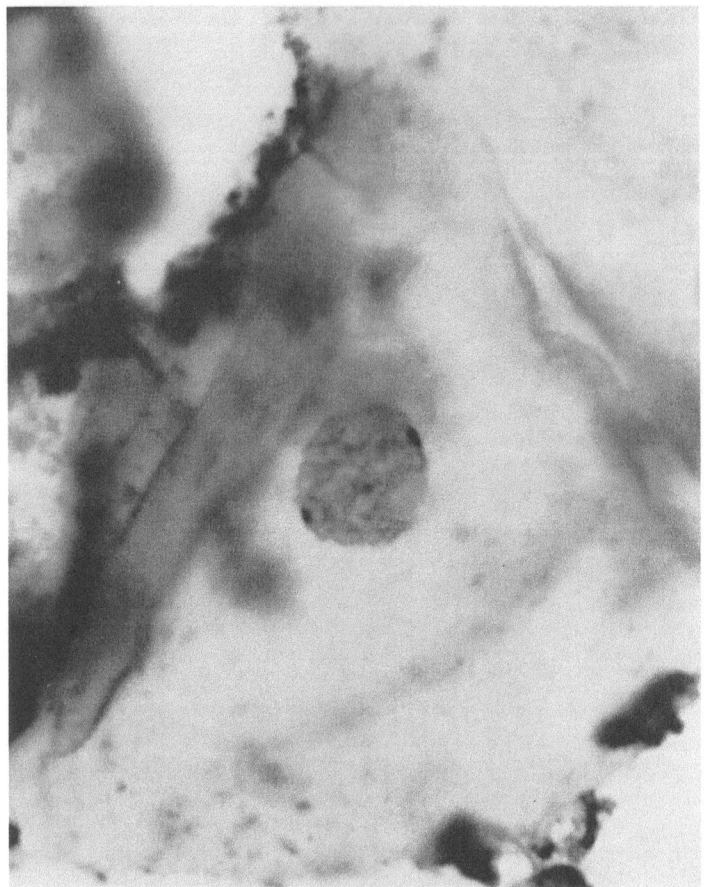

Fig. 3. Photomicrograph of cell from the buccal mucous membrane showing 2 normal sex chromatin bodies
(acetic-orcein × 2000)

of gonadal structure, and it certainly remains as perhaps the most important
criterion of sex identification. It is to be emphasized, however, that the gonadal
identification must be by microscopic and not gross examination. The gross
appearance of gonads in cases of intersexuality is often deceptive.

3. Morphology of External Genitalia

It goes without saying that the morphology of the external genitalia is the
criterion most often used by the obstetrician in assigning a sex to the newborn.
With few exceptions, therefore, the sex of rearing is dependent upon the morphology of the external genitalia. If a contradiction does exist between the predominant appearance of the genitalia and the sex of rearing, it has been shown by
MONEY et al. (1955a) that interestingly enough the patient often succeeds in
coming to terms with the morphologic anomaly and assumes a role consistent
with that of the assigned sex and rearing. This is not to say that such patients have
no psychologic difficulties—quite the contrary—and the responsibility which the
obstetrician has in assigning sex is indeed a grave one.

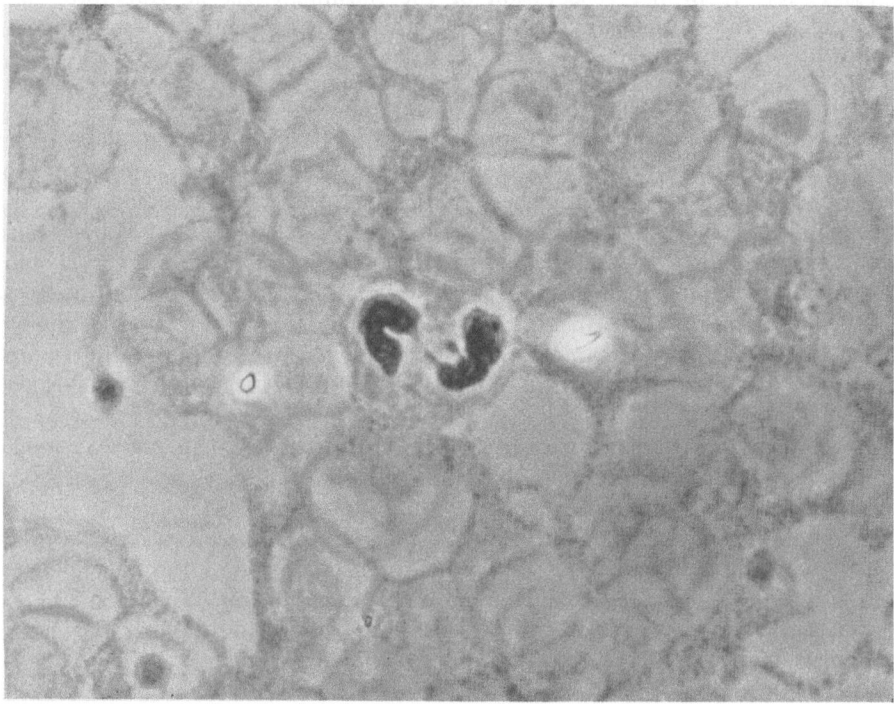

Fig. 4. Photomicrograph of leucocyte showing a "drumstick" (Wright's stain × 2000)

4. Morphology of the Internal Genitalia

In every normal embryo both the mesonephric (wolffian) and paramesonephric (müllerian) ducts are present and capable of development into perfect male or female internal genitalia depending upon the hormonal environment. Whether this development occurs in a normal manner or whether there is some developmental contradiction, as in the hermaphroditic state, is of little importance with regard to assigned sex or gender role, both of which are largely dependent upon external morphology. Nevertheless, the development or atrophy of these structures is of surpassing importance as far as the function of the genital apparatus is concerned. The presence of normal or contradictory internal genitalia is, therefore, a most important measure of sex status.

5. Hormonal Status

The hormonal milieu is of great importance as a criterion of sex and is by no means identical with gonadal structure. The endocrine environment is not only an important determining factor in the morphology of the genitalia, but is responsible for such secondary sex characteristics as breast development, hair growth, bone growth and epiphyseal fusion, fat distribution, general body habitus and the like. Not only do the testes sometimes secrete estrogens which feminize the body but glands other than the gonads exert controlling hormonal influences of a contradictory type in some instances. The adrenals in congenital adrenal hyperplasia, for example, secrete large quantities of a virilizing hormone responsible for the hermaphroditic state in one of the most common forms of female intersexuality. A proper recognition of the hormonal sex is, therefore, most necessary for the proper understanding and therapy of many cases of inter-sexuality.

6. Gender Role

The gender role has also been emphasized by Money et al. (1955a) as an important psychologic criterion of sex. It is defined as all those things a person says or does to disclose himself or herself as having the status of boy or man, girl or woman respectively. It includes, but is not restricted to sexuality in the sense of eroticism. Gender role is appraised in relation to the following: general man-nerisms, deportment and demeanor; play preferences and recreational interests; spontaneous topics of talk in unprompted conversation and casual comment; content of dreams, day dreams, and fantasies; replies to oblique inquiries and projective tests; evidence of erotic practices and, finally, the individual's replies to direct inquiry.

On the basis of experience with hermaphrodites, it was concluded that gender role more nearly followed assigned sex than any other factor. Of 86 cases studied, there were but 4 exceptions to this rule.

Although few mistakes would be made if assigned sex were used as a guide to sex orientation, the gender role represents an extension of the psychological aspect of sex and as such is an additional important criterion. When the assigned sex and gender role are at variance, an extremely careful psychiatric evaluation of the patient is necessary to assist the physician in the correct therapeutic orientation.

II. The Definition and Classification of Hermaphroditism

Although any of the several criteria of sex might be the basis of a classification of ambisexual individuals, the well established classification of Klebs (1876) is based upon the microscopic character of the gonad. According to this view the state of true hermaphroditism can be considered to exist only if both male and female elements can be identified in the gonad. Klebs's original classification of *true* hermaphroditism is complex and attempts to give consideration to the various gonadal combinations which might exist. According to Klebs, 16 sub-divisions were thought possible.

The Latin of the original classification is quite understandable when it is kept in mind that it refers only to the gonads with respect to the classification of true hermaphroditism. The adjectives *bilateralis*, *unilateralis*, or *lateralis* refer to ambi-sexual gonadal tissue on both sides, one side only or alternating (male one side—female one side) respectively. *Completus* is used to indicate gonadal tissue on both sides and incompletus to indicate that one side is devoid of all gonadal tissue. *Masculinus* or *feminus* followed by *dexter* or *sinister* is used to describe the character and side of the gonad opposite to a unilateral ovotestis.

Klebs further specified that ambisexual individuals with entirely male gonads should be identified as male *pseudo*hermaphrodites and ambisexual individuals with entirely female gonads should be identified as female *pseudo*hermaphrodites. Furthermore, according to Klebs, there are 3 forms of pseudohermaphroditism: 1. *completus*, 2. *internus* and 3. *externus*. As explained by Creevy (1933) many *pseudo*hermaphrodites can be classified and described according to the Klebs concept by two adjectives. The first (male or female) indicates the nature of the sex gland. The second (external, internal, complete) indicates whether it is the sex of the internal, external or both groups of genitalia which differs from that of the gonad.

Table 1. *Hermaphroditic Abnormalities and Criteria of Sex Identification*

Etiology	Group	Criteria of Sex						Sex of rearing		Gender role
		Chromosomal arrangement	Sex chromatin	Gonadal structure	Morphology of external genitalia	Morphology of internal genitalia	Hormonal dominance	Actual	Preferred	
Chromosomal aberration in most cases	Ovarian agenesis and dysgenesis	XO, XO/XX, etc.	Negative or positive	None	Female	Female	None	As women	As women	Women
Unknown	True Hermaphroditism	XX, etc.	Positive or negative	Testis and ovary	Mixed	Mixed	Mixed	Either	Either	Either
Chromosomal aberration	Klinefelter's Syndrome	XXY, etc.	Positive etc.	Testes	Male	Male		As men	As men	Men
Chromosomal aberration	Double X male	XX	Positive	Testes	Male	Male ?	Male	As men	As men	Men
Chromosomal aberration	Multiple X female	XXX	Double positive	Ovary	Female	Female	Female	As women	As women	Woman
Unknown (probably genetic)	Male hermaphroditism with virilization	XY, XY/XO, etc.	Negative	Testis	Mixed	Mixed	Male	Either, but mostly as women	Either, but mostly women	Either, but mostly women
Genetic	Male hermaphroditism with feminization	XY	Negative	Testis	Female	None	Female	As women	As Women	Women
Congenital adrenal hyperplasis (genetic)	Female hermaphroditism with virilization	XX	Positive	Ovary	Mixed	Female	Male	Either, but mostly as women	As women	Either, but mostly as women
Maternal androgen in some cases	Female hermaphroditism with feminization	XX	Positive	Ovary	Mixed	Female	Female	Either, but mostly as women	As women	Either, but mostly as women

The female internal variety includes outwardly normal females with ovaries, but with such male vestigial remnants as a Gartner's duct, a rudimentary prostate and vas deferens. The female external type has normal or rudimentary internal genitalia and ovaries, but with external genitalia resembling those of the male. In complete female pseudohermaphroditism, both the external and internal genitalia would tend to be masculine in an individual with ovaries.

The male internal variety of pseudohermaphrodite would possess testes and a well developed uterus, tubes and perhaps vagina. The male external variety would have testes, but feminine external genitalia. In complete male pseudo-hermaphroditism, the patient would have testes, but feminine internal and external genitalia.

It is obvious that the Klebs classification is impractical by being at the same time too complex and yet incomplete in terms of current knowledge. However, it has been described at some length both because of its historic interest and because some hermaphrodites are described by its terms in current literature. Its widespread use and acceptance, in a simplified form, has made it desirable to continue to use gonadal sex as the basic criterion of any classification.

The classification to be used recognizes 7 main subdivisions although there are numerous other subdivisions within each (Table 1). It is to be emphasized that the designation of a hermaphroditic individual by a particular name implies nothing except the morphological character of the gonad and the implication that there is a contradicition of one of the other criteria of sex.

We, therefore, define the hermaphroditic state as existing when there is a contradicition in one of the morphological criteria of sex, i.e. sex chromatin and sex chromosomes, gonadal structure, morphology of external genitalia, or morphology of the internal genitalia.

Contradiction of hormonal status, sex of rearing or gender role are *not* to be considered in defining hermaphroditism.

III. Ovarian Agenesis and Dysgenesis

In 1938 TURNER described 7 girls from 15 to 23 years of age who exemplified "a syndrome of sexual infantilism, webbing of the neck, cubitus valgus and retardation of growth."

In retrospect it seems likely that these were examples of "ovarian agenesis" but it remained for VARNEY, KENYON and KOCH (1942) and ALBRIGHT, SMITH and FRASER (1942) to demonstrate that the syndrome was associated with an elevated titer of urinary gonadotrophin thus representing a severe ovarian deficiency. WILKINS and FLEISCHMANN (1944) recorded the pathologic findings in the syndrome and subsequently such patients were well recognized under the name "ovarian agenesis" or Turner's syndrome.

However, a survey of the literature from the last few years makes it clear that the eponym Turner's syndrome means various things to various people. After the discovery of ovarian streaks as a characteristic of the clinical syndrome described by TURNER, the expression "ovarian agenesis" was applied as a synonym for Turner's syndrome (WILKINS and FLEISCHMANN, 1944). With the discovery of the absence of the specific sex chromatin in such patients, the term ovarian agenesis gave way to "gonadal dysgenesis" (GORDON et al., 1955), "gonadal agenesis" (WILKINS, 1957) or "gonadal aplasia" (WILKINS, 1957).

Meanwhile some patients were discovered to have a normally positive sex chromatin count (JONES, FERGUSON-SMITH and HELLER, 1963). Furthermore a variety of sex chromosome complements have been found with streak gonads.

As if these contradictions were not perplexing enough, it has been found that streaks are by no means confined to patients with Turner's original classical syndrome of infantilism, webbing of the neck, cubitus valgus and retardation of growth, but may be present in girls with none of the findings just mentioned save infantilism. Since TURNER's 1938 description, a whole host of additional somatic anomalies have been found associated with his original clinical picture with varied frequencies. These include shield chest, overweight, high palate, micrognathia, epicanthal folds, low set ears, hypoplasia of nails, osteoporosis, pigmented moles, hypertension, lymphedema, cutis laxa, keloid formation, coarctation of the aorta, mental retardation, intestinal teleangiectasis, deafness and others (HADDAD and WILKINS, 1959).

This is perhaps not the medium to try to resolve the nomenclature problem of these conditions. Suffice it to say that for the purpose of this discussion the eponym Turner's syndrome is used to indicate an individual with sexual infantilism due to streaks, short stature and two or more of the somatic anomalies mentioned above. In this context such terms as ovarian agenesis, gonadal agenesis, gonadal dysgenesis lose their clinical significance and become merely descriptions of the gonadal development of the individual.

With these definitions, it now becomes possible to summarize the data available on the relation between sex chromosomes, Turner's syndrome and streak gonads. There have been at least 21 and probably more different sex chromosome complements associated with streak gonads. Interestingly enough only about 9 sex chromosome complements have been associated with Turner's syndrome. In no study has there been any suspicion of any abnormalities among the autosomes. About all that can be said is that about one half of the patients who have been studied with streak gonads including those with classical Turner's syndrome have an XO sex chromosome complement. However, about two thirds of the patients with Turner's syndrome have an XO sex chromosome complement and only $1/4$ of the patients without Turner's syndrome but with streak ovaries have an XO sex chromosome complement (JONES, 1965).

At the present time there is no obvious correlation between any of the somatic anomalies and the observed sex chromosome defects. On the other hand, it is possible to at least offer a correlation between the sex chromosome defects and the failure of gonadal development in spite of the fact that the great variety of sex chromosome complements associated with streak ovaries would seem to preclude an explanation. However, if it is noted that less than 10% of patients with streaks have a normal sex chromosome complement and if it is further noted that there is serious technical difficulty in ruling out chromosome mosaicism, it may be possible that all patients with streaks have an abnormal sex chromosome complement. We may tentatively conclude that normal ovarian develoment requires 2 normal X chromosomes.

It will be recalled that only one sex chromosome is thought to be genetically active, and it will be further recalled that the sex chromatin does not appear prior to day 16 of embryonic life. Furthermore, neither gonocytes, nor the oogonia, nor, of course, the ovocytes exhibit a sex chromatin implying that the diploid germ cell requires two normal active sex chromosomes. Since the gonocytes first appear in the epithelium and mesentery of the gut just prior to the general inactivation of the X chromosome as evidenced by the appearance of the sex chromatin, it may be suggested that germ cells cannot exist unless two normal active X chromosomes are available.

It remains to be noted that an ovary cannot develop without germ cells or, to put it another way, a streak is an ovary without germ cells and their satellites, the follicular apparatus. Thus, the essence of the pathogenesis of streak ovaries is the failure of germ cell segregation which cannot occur unless there are two normal X chromosomes in that portion of the embryo from which the germ cells segregate.

1. The Pathology of Gonadal Agenesis

It has been demonstrated that the pathological findings in the streak and broad ligament area in patients with gonadal streaks (Fig. 5) is essentially the same regardless of the cytogenetic background of that patient (Jones, Ferguson-Smith and Heller, 1963).

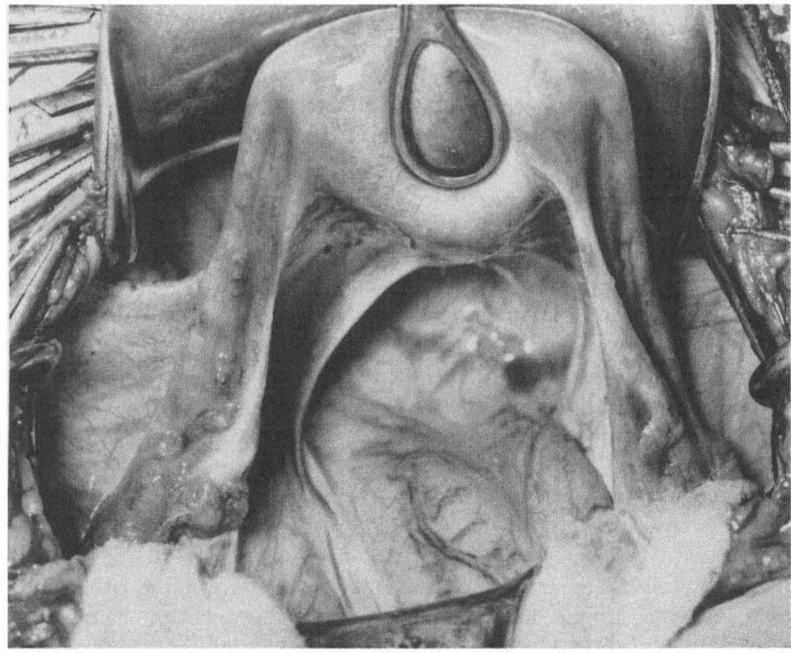

Fig. 5. Photograph of operative findings in a patient with gonadal streaks. The uterus and fallopian tubes are well developed but in place of the ovaries nothing is seen except a thin streak of white firm tissue

Fibrous tissue is the major component of the streak and is indistinguishable microscopically from that of normal ovarian stroma (Fig. 6). The so-called germinal epithelium on the surface of the structure is a layer of low cuboidal cells which appears to be completely inactive. Some ridges contained small (up to 6 mm.) surface inclusion cysts lined with this epithelium. Grossly, these suggested small follicles. In 11 cases studied by us, no patient had any structure resembling primary germ cells or follicular apparatus or structures which resembled medullary or cortical sex cords.

In blocks from about the mid portion of the streak, tubules of the ovarian rete were invariably found (Fig. 7). This is a normal finding in the hilum of all ovaries and these tubules become a normal functioning component of the appendages of the testis as the collecting tubules. The histogenesis of the ovarian rete has not been clearly established. On one hand, it has been considered to be derived from the mesonephric tubules (von Winiwarter, 1901) and on the other from the

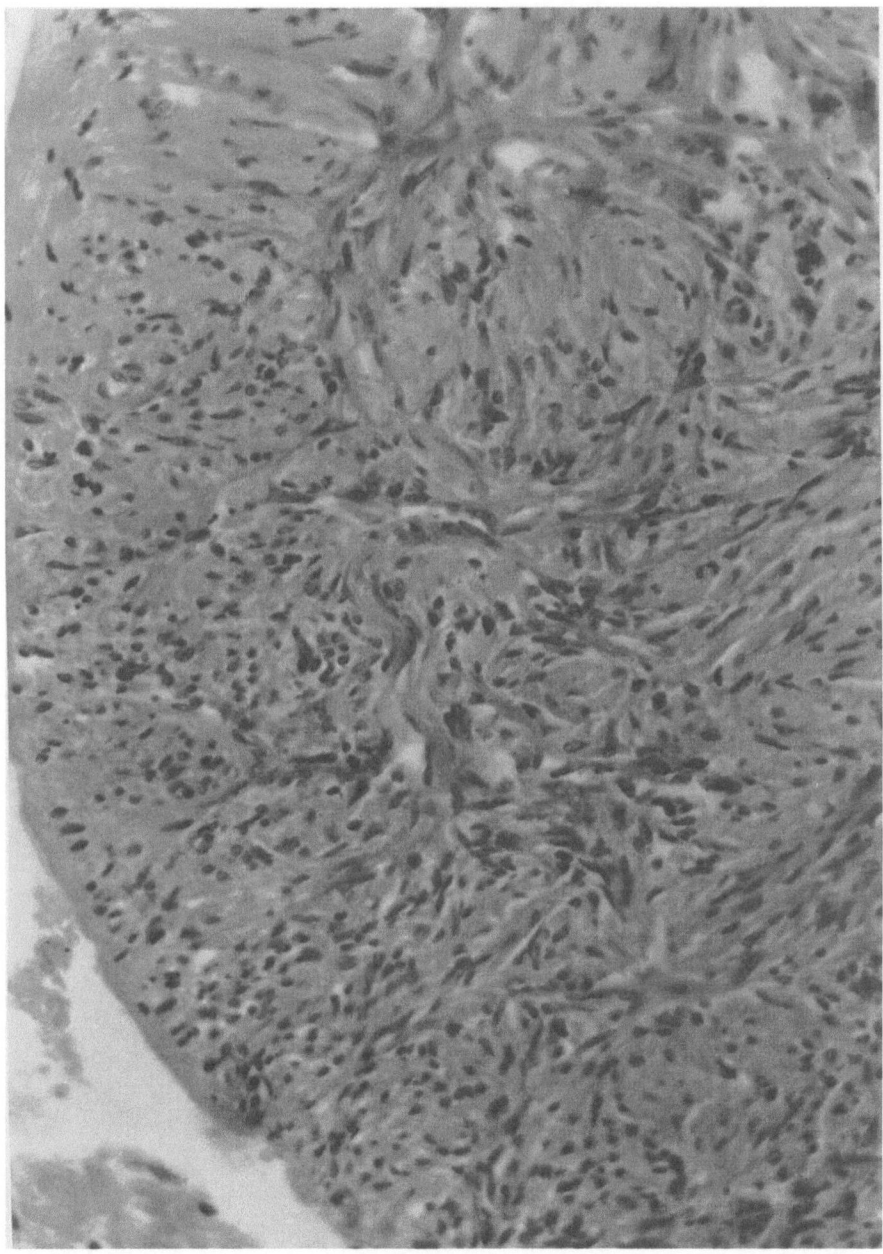

Fig. 6. Photomicrograph of section taken through the main portion of the streak gonad. In this section no structures are visible except fibrous tissue which is indistinguishable from that of a normal ovary (H & E × 300)

medullary sex cords (WILSON, K. M., 1926). Its presence in ovarian agenesis is not decisive in this regard, but of great interest, for if the medullary sex cord origin of the rete ovarii could be clearly shown, it would mean that the gonadal development in these cases had been arrested subsequent to the appearance of medullary sex cords in embryonic life and prior to the invasion of the cortical sex

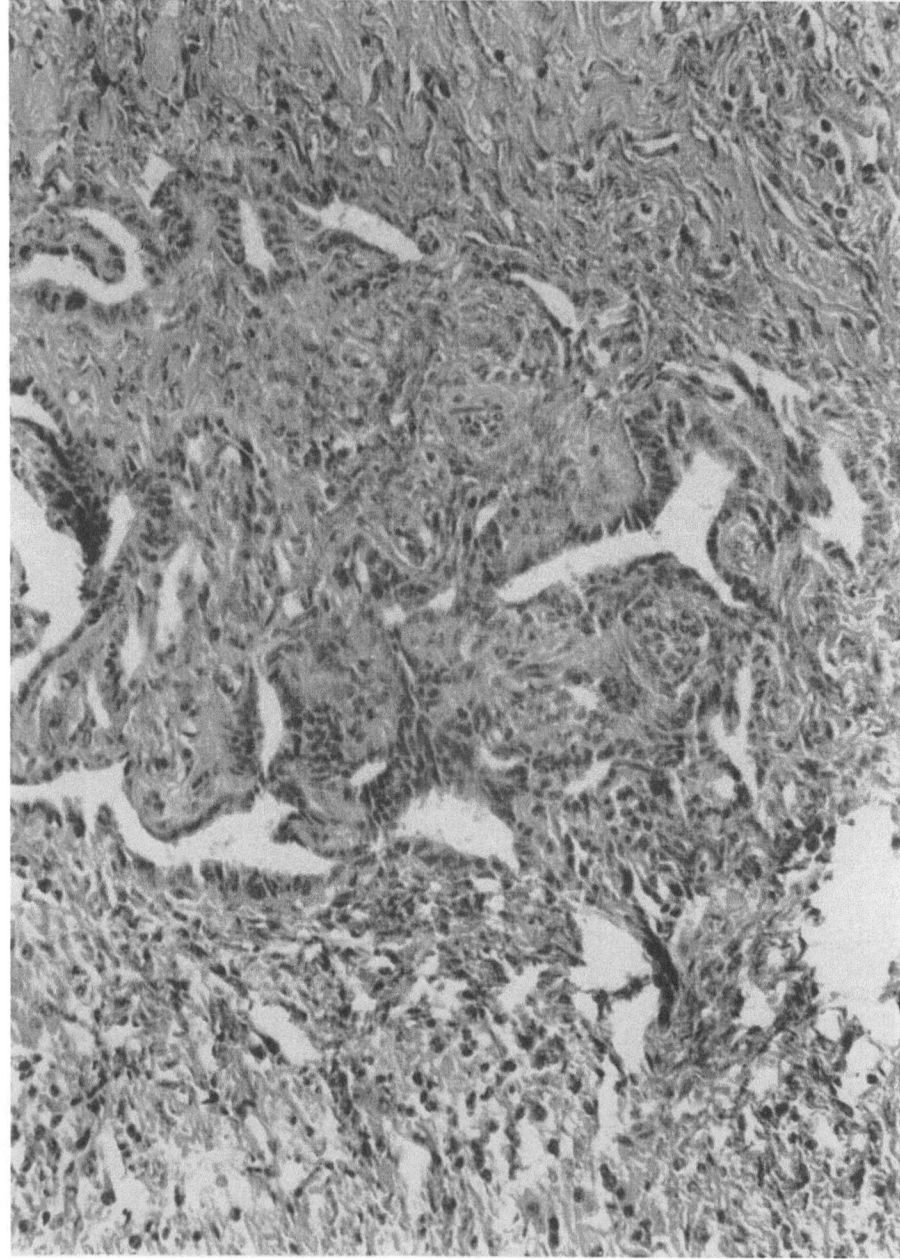

Fig. 7. Section through a gonadal streak showing rete tubules (H & E × 300)

cords. Thus, the constant presence of the ovarian rete could be used as evidence that development of the gonad proceeded as normal ovarian development until after the indifferent gonad stage (14 to 17 mm) and was arrested prior to the appearance of the cortical sex cords when the primitive oocyte can be normally identified.

In all patients above the age of normal puberty, hilar cells were also demonstrated (Fig. 8). The number of such cells seemed to vary from case to case. In patients with some enlargement of the clitoris, hilar cells were present in large numbers (Figs. 9, 10). It may be that these facts are causatively related. When the hilar cells were scarce, they were invariably found in conjunction with the rete ovarii. Hilar cells also occur in a high percentage of normal ovaries (SAURAMO, H., 1962). The origin of hilar cells, is, likewise, not clear, but they are certainly somehow associated with the development of the medullary portion of the gonad. Their presence lends further support to the concept that, in ovarian agenesis, the

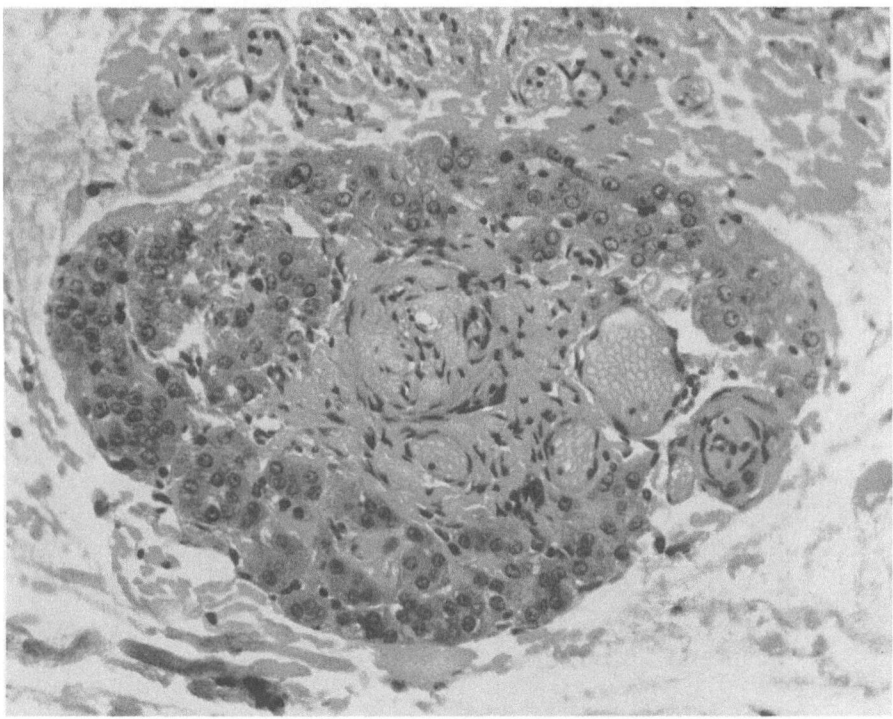

Fig. 8. Photomicrograph of hilar cells in a patient with a streak gonad (H & E × 300)

gonad seems to develop along normal lines until just prior to the expected appearance of the cortical sex cords.

In all cases where sections of the broad ligament were available for study it was possible to identify mesonephric duct and mesonephric tubules. The duct was found in the broad ligament close to the fallopian tube beginning near the fimbria and continuing to about the mid portion of the streak. The duct does not seem to be present in sections of the broad ligament adjacent to those portions of the tube most proximal to the uterus. In sections taken from the midportion of the fallopian tube, mesonephric tubules can be identified in addition to the mesonephric duct. The distinction between mesonephric duct and tubules is not always easy. However, using the criteria of GARDNER, GREENE and PECKHAM (1948), the duct can be seen to have a prominent circular muscular layer with low cuboidal epithelium. The tubules, on the other hand, have epithelium containing two types of cells, ciliated and nonciliated (Fig. 11). In this respect the mesonephric tubules

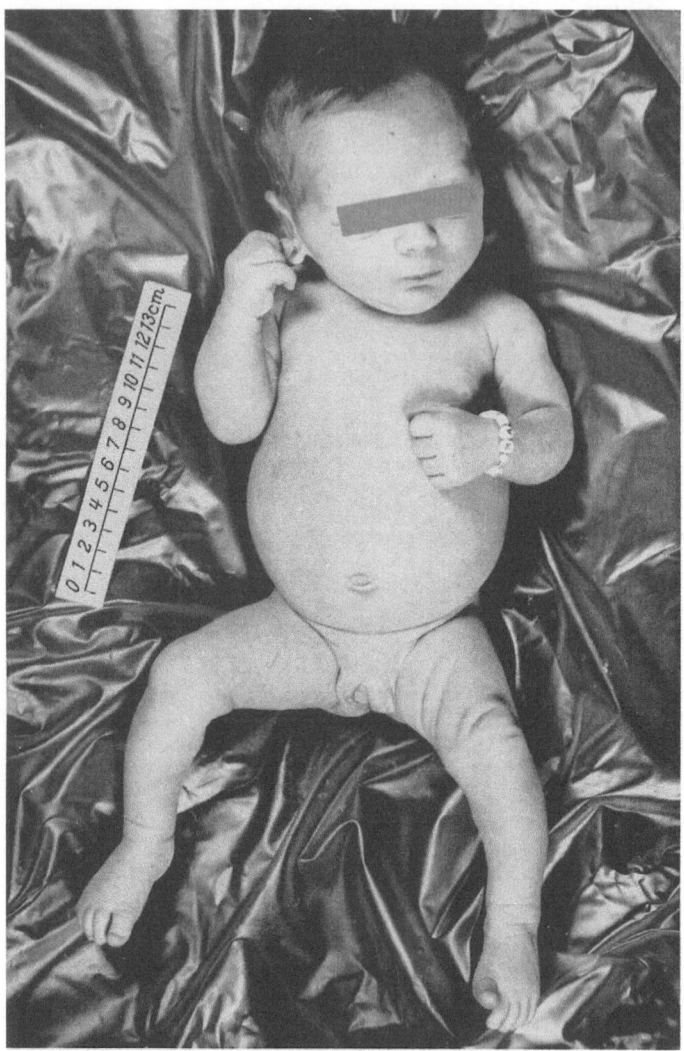

Fig. 9. Photograph of a child who on exploratory laparotomy proved to have gonadal streaks. The only
abnormality of the external genitalia was the enlarged clitoris

have epithelium very similar to that of the fallopian tube. The mesonephric
tubules may also have a thin circular layer of muscle.

These broad ligament structures just described are found in all normal females.
They, therefore, are not a specific pathological finding in patients with gonadal
agenesis. Their description is of importance only in emphasizing that the embryonic
influences concerned with the control of development of the mesonephric duct
and tubules are, apparently, similar in the normal female and in those patients

Fig. 10. Photograph of the external genitalia of a 19 year old girl with an enlarged phallus. Exploratory laparotomy
showed that typical gonadal streaks were present. The only abnormality noted on histological examination was
a large number of hilus cells in the streak

Fig. 11. Photomicrograph of wolffian tubules found in the broad ligament area of a patient with ovarian streaks
(H & E × 300)

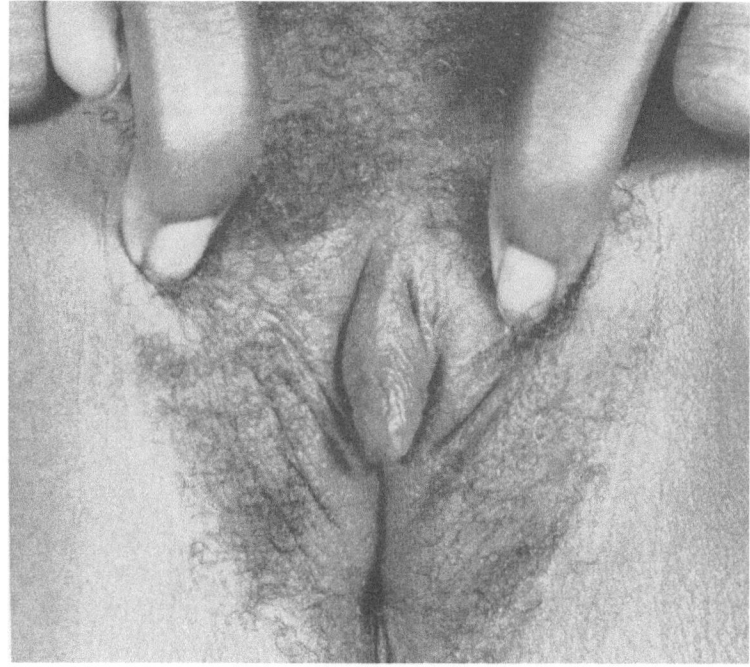

Fig. 10

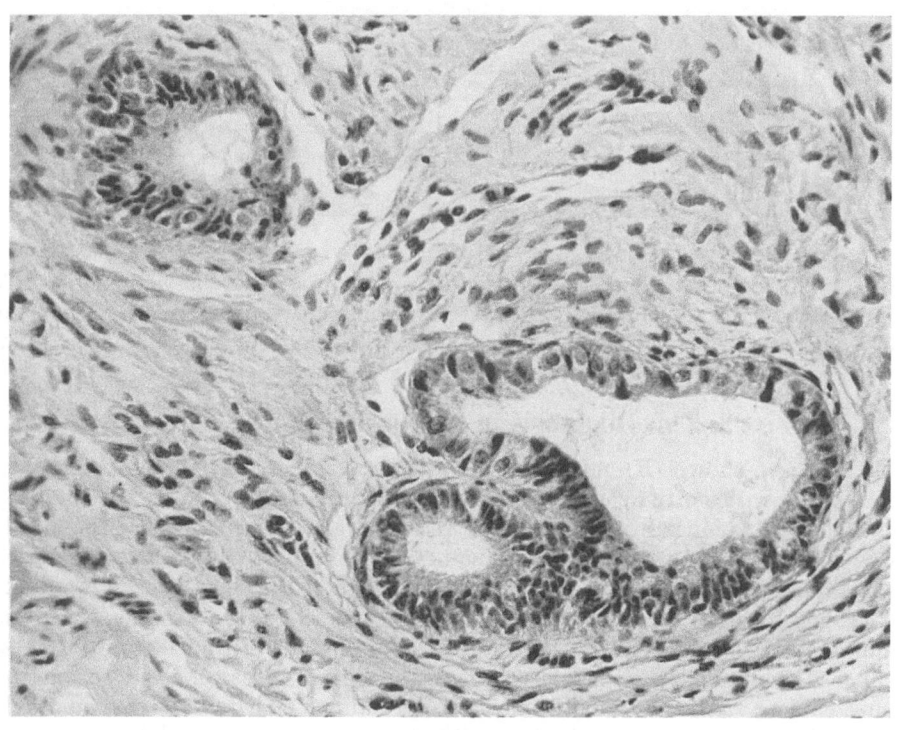

Fig. 11
Figs. 10 and 11 (for legends see p. 388)

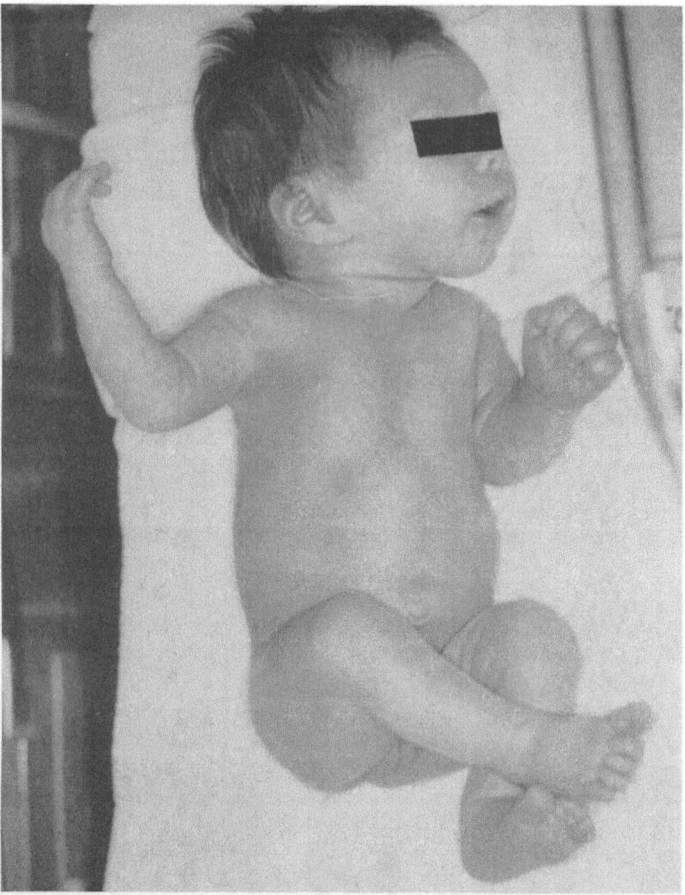

Fig. 12 A. Photograph of an infant with marked edema of the extremities often seen in newborn infants with streak gonads. (From: R. M. Richart and K. Benirschke, 1960)

with gonadal agenesis. Thus, the embryonic ovary apparently plays no active role in suppressing the development of the mesonephric (wolffian) duct such as the embryonic testis does in suppressing the development of the müllerian duct.

2. The Diagnosis in Newborn Infants

It has been shown (Richart and Benirschke, 1960) that the newborn child with streak ovaries often shows edema of the hands and feet (Fig. 12 A and B). This edema is peculiar in that it is firm and generally non-pitting. Histologically it was shown by Richart and Benirschke to be associated with large dilated vascular spaces. A second common feature in infants is the short neck. Frequently this is not a subjective observation but is generally unmistakable if careful examination is carried out. With these two findings or even with one of these findings it is obviously desirable to check the buccal smear for sex chromatin. From a clinical view, additional somatic abnormalities as listed above would help confirm the diagnosis. Some children with streak ovaries, particularly those who have few or no somatic abnormalities, cannot be recognized at birth.

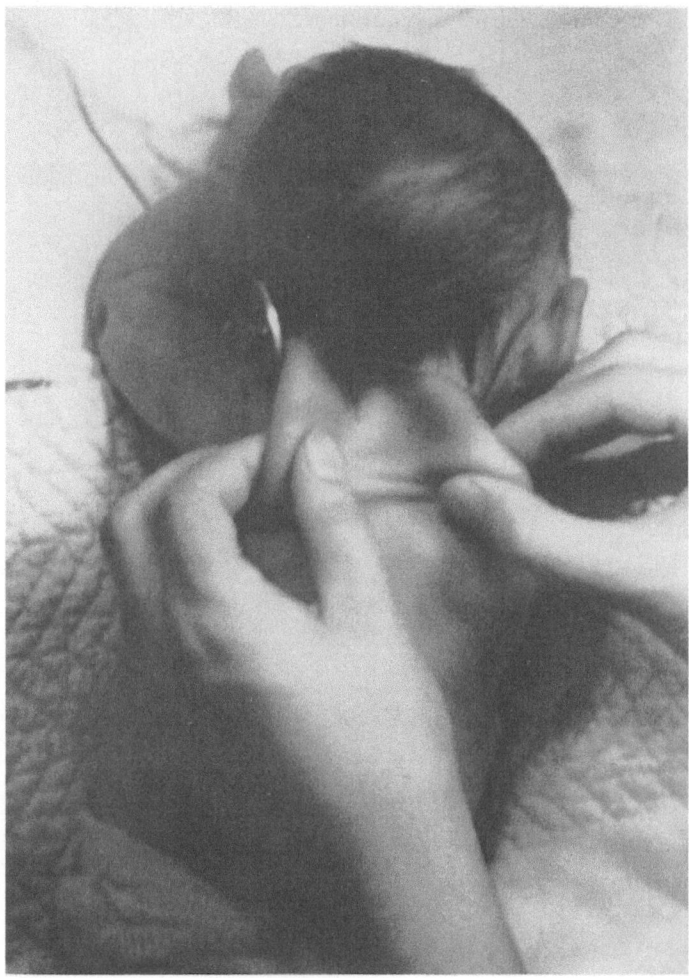

Fig. 12B. Photograph of an infant with webbing of the neck. In such children the webbing may not be clinically evident but is easily developed by palpation as shown in the photograph. (From: R. M. RICHART and K. BENIRSCHKE, 1960)

3. The Diagnosis in Adolescence

The arresting and characteristic clinical finding in many of these patients is their short stature. Typical patients seldom obtain a height of five feet (Fig. 13). In addition to this, sexual infantilism is a striking finding. As mentioned above a variety of somatic abnormalities may be present and by definition we assume that if two or more of these are noted, the patient may be considered to be suffering from Turner's syndrome. As noted above most of such patients have a negative buccal smear and $2/3$ of such patients have an XO sex chromosome complement.

It has recently been shown, however (JONES, FERGUSON-SMITH, HELLER, 1963), that tall patients without somatic abnormalities may also have gonadal streaks (Figs. 14, 15, 16). Under such circumstances, the sex chromatin is apt to be present in the cells in normal numbers or in less than normal numbers as an expression of mosaicism. However, the internal findings are exactly the same as in patients with clinical Turner's syndrome.

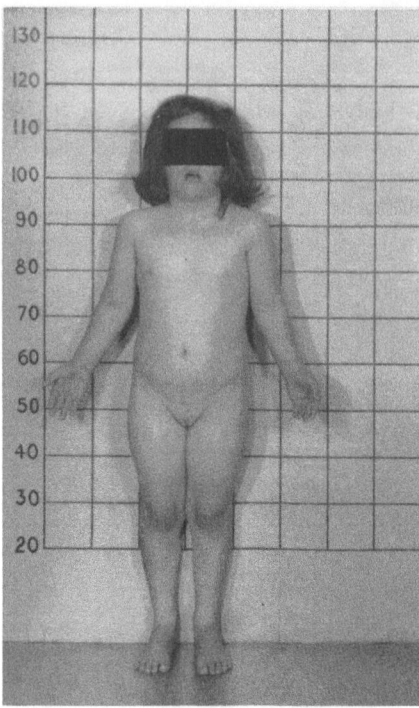

Fig. 13. Photograph of a patient with gonadal agenesis showing the typical short stature. This patient was 9 years 4 months of age and had a height age of 6 years 6 months

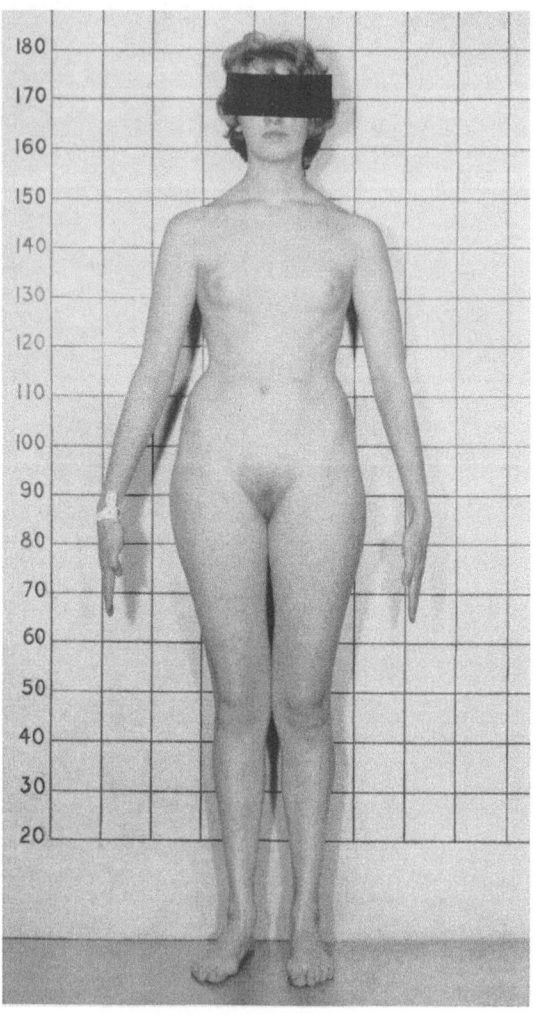

Fig. 14. Photograph of a 19 year old normally tall girl with primary amenorrhea who proved to have gonadal streaks. The breast development shown in the photograph is the result of substitution therapy which she had had at the age of 16 for 6 months. The buccal smear was normally positive (22%)

4. Endocrine Findings

An important finding in patients above the age of expected puberty, i.e. 12 years or there abouts is an elevation of the total gonadotrophin excretion. From a practical point of view above the age of 15 ovarian failure cannot be considered in the diagnosis unless the gonadotrophin value is in excess of 50 mu/24 hrs. Contrariwise, in an individual with primary amenorrhea at the

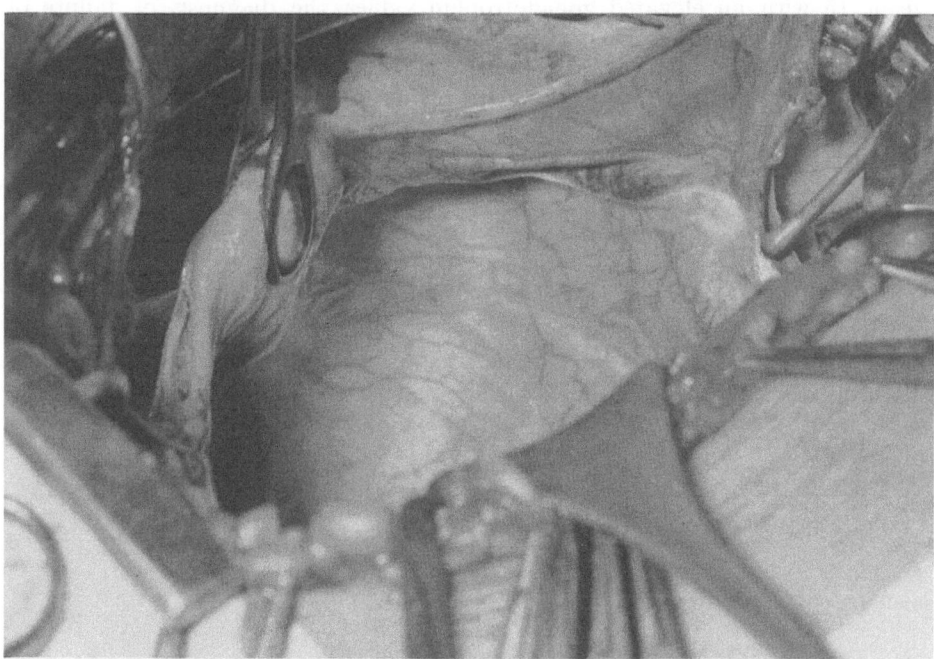

Fig. 15. Laparotomy findings on the same patient shown in Fig. 14. The streak on the right side is clearly shown

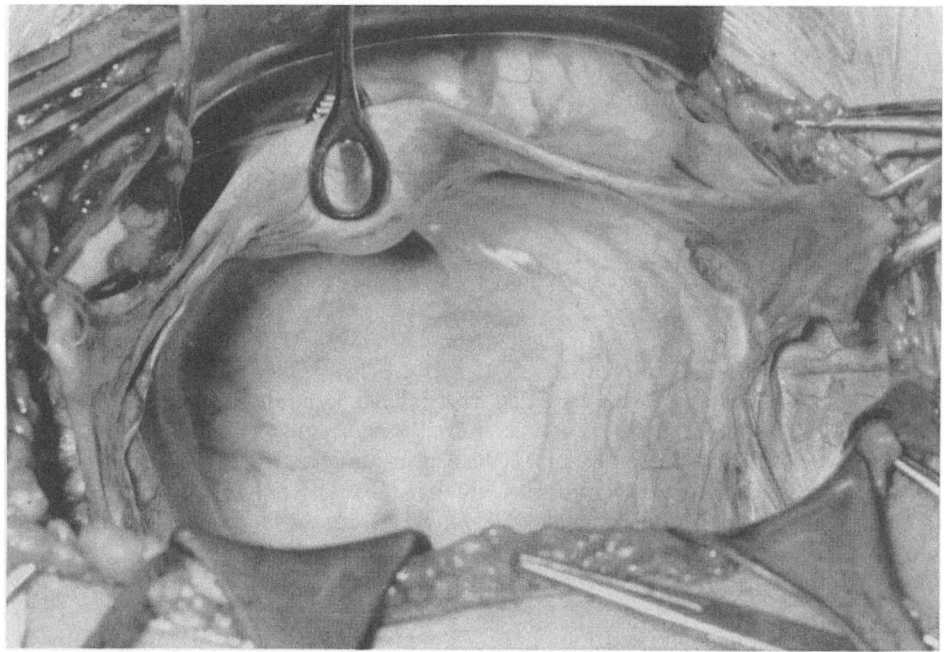

Fig. 16. Laparotomy findings on the same patient shown in Fig. 14. The left streak is clearly shown in this figure

age of 15 with an elevated gonadotrophin values the diagnosis of failure of ovarian development is almost a certainty.

Various endocrine determinations such as urinary 17-ketosteroids, 17-hydroxy-corticosteroids, and PBI are entirely within normal limits. Urinary excretion of estrogens is low and the maturation index and vaginal smear is shifted well to the left.

IV. True Hermaphroditism

I. General Considerations

By classic definition the state of true hermaphroditism may be considered to exist when both ovarian and testicular tissue has been demonstrated in the same patient (Figs. 17, 18, 19). Among all anomalies of sexual differentiation, true

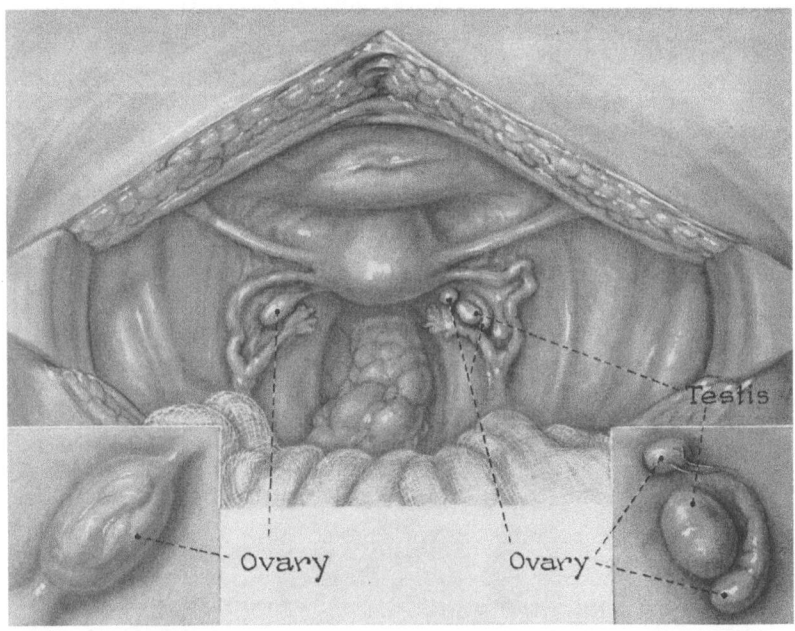

Fig. 17. Sketch of view at laparotomy of a patient with true hermaphroditism. (From: H. W. JONES, Jr., M. A. FERGUSON-SMITH and R. H. HELLER, 1965)

hermaphroditism has occupied a niche apart because of the still incomprehensible paradox of development of both testicular and ovarian tissue in the very organs which have traditionally been regarded as the Ultima Thule of sex identification. True hermaphroditism has recently acquired a new and critical fascination for an altogether unanticipated reason. It has been observed in the human that the Y chromosome carries genetic material which is normally responsible for testicular development and that this is active even when multiple X chromosomes are present. Thus, in the Klinefelter's syndrome a testis develops with up to 4 or more X's and only one Y. Conversely, with one general exception, a testis has not been observed to develop in the absence of the Y chromosome. The exceptions are those of true hermaphroditism and XX males, where in the majority of obser-ved cases, testicular tissue has developed in association with an XX sex chro-mosome complement.

2. Criteria for Diagnosis

As indicated above, it has been considered important to maintain the rigid criterion of a pathologic demonstration of both ovarian and testicular tissue before an individual can be accepted as a true hermaphrodite. A valid exception

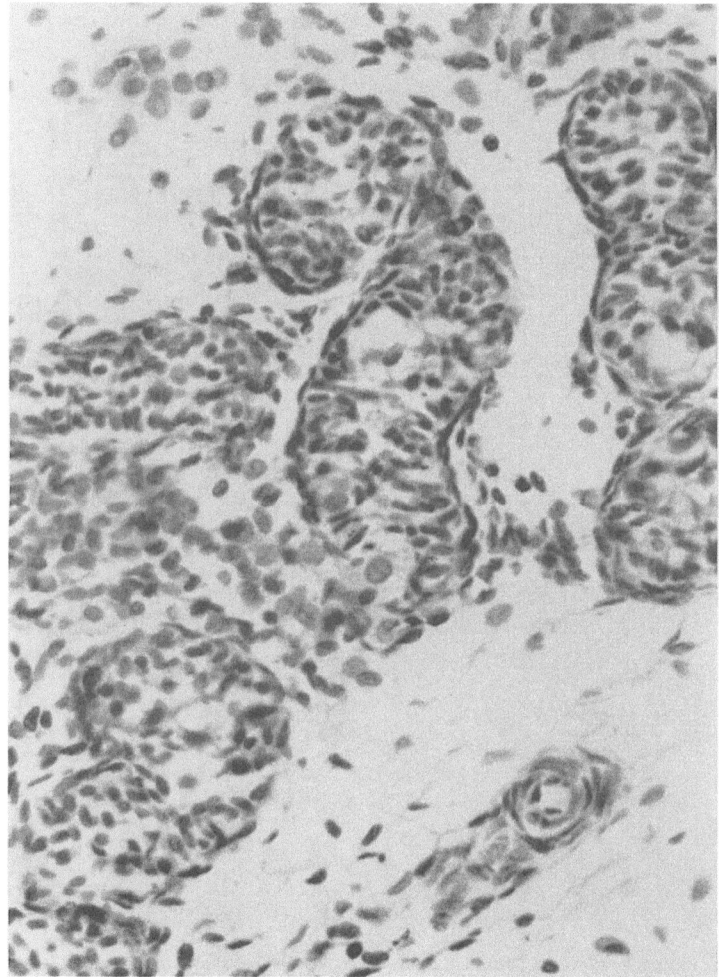

Fig. 18. Photomicrograph of testicular portion of right ovotestis of patient shown in Fig. 17 (H & E × 400). (From: H. W. JONES, Jr., M. A. FERGUSON-SMITH and R. H. HELLER, 1965)

to this rule may be made if there has been positive microscopic identification of ovarian tissue plus viable sperm in the ejaculate. It is important to emphasize that the ovarian identification must include specific recognition of ovogenesis. Ovarian stroma is not sufficient to identify an ovary. Although viable sperm may be considered as positive identification of testicular tissue, uterine bleeding cannot be considered as positive identification of ovarian tissue. This is because estrogenic stimulation may be from extra-ovarian sources and indeed can arise from testicular tissue itself.

3. Classification of True Hermaphroditism

The Klebs subdivision of true hermaphroditism recognized 16 possible groups. However, these groupings were made on theoretical considerations; many have not been recognized clinically and further experience has shown that groups have appeared which were not provided for in the classification. For these reasons it has seemed best not to attempt to use the Klebs classification but to describe

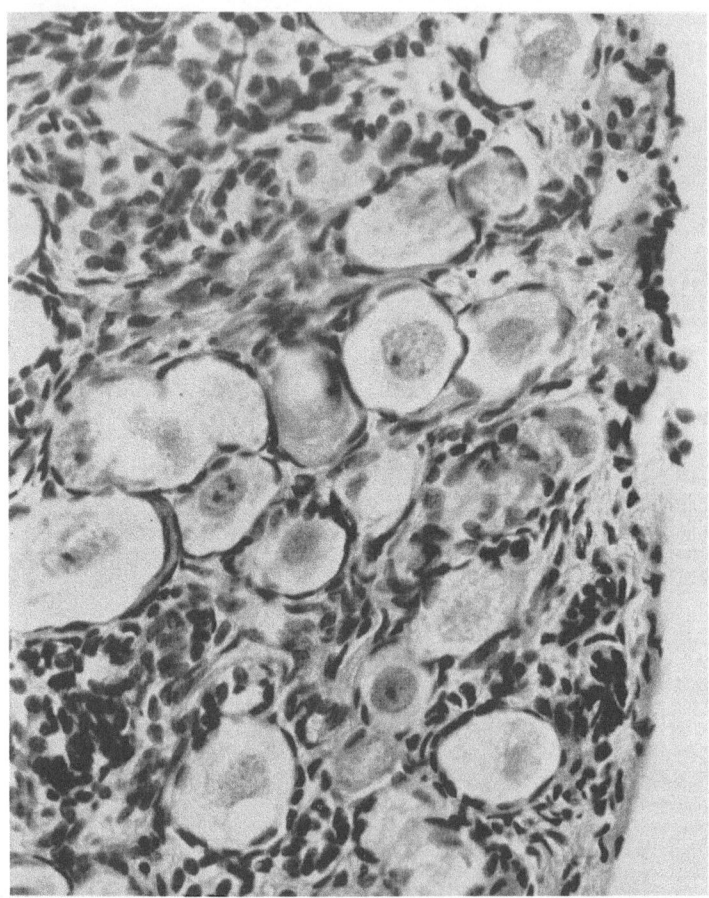

Fig. 19. Photomicrograph of ovarian portion of right ovotestis of patient shown in Fig. 17 (H & E × 400). (From: H. W. Jones, Jr., M. A. Ferguson-Smith and R. H. Heller, 1965)

6 groups, examples of which appear in the literature. This is an extension and an enlargement of Young's classification which itself was an attempted simplification of Klebs' groupings. It is as follows (Table 2):

Group I: alternating variety. In this group there is an ovary on one side and a testis on the other. In many instances the duct system on a given side corresponds in its development to the sex of the gonad on that side.

Group IIa: bilateral variety. In this group there is an ovotestis on each side.

Group IIb: bilateral variety. There is a separate ovary and testis on each side in this group.

Group IIIa: unilateral variety. In this group there is an ovary on one side and an ovotestis on the other.

Group IIIb: unilateral variety. An ovary is on one side and a separate testis and ovary on the other.

Group IIIc: unilateral variety. In this group an ovary is on one side and two separate ovotestes on the other.

Group IV: unilateral variety. A testis on one side and an ovotestis on the other form this group.

Group V: unilateral variety. This group is characterized by an ovary and testis on the same side but no gonad on the other.

Group VI: an incompletely studied group with an ovotestis on one side, but with the opposite gonad not examined.

Table 2

Group	Gonad One Side	Gonad Opposite Side
Alternating or Lateral Variety		
I	Ovary	Testis
Bilateral Variety		
IIa	Ovotestis	Ovotestis
IIb	Ovary Testis	Ovary Testis
Unilateral		
IIIa	Ovary	Ovotestis
IIIb	Ovary	Ovary Testis
IIIc	Ovary	Two ovotestes
IV	Testis	Ovotestis
V	Ovary Testis	No gonad
VI	Ovotestis	Not examined

4. Clinical Features

There are no exclusively characteristic features which clinically distinguish true hermaphroditism from other forms of intersexuality. This means that the diagnosis must be entertained in almost all forms of intersexuality with the exception of those with a continuing virilizing influence, as for example in congenital adrenal hyperplasia. On the other hand, a review of certain clinical features according to the groupings of true hermaphroditism brings out certain interesting characteristics about this group of patients. The following remarks are based on a study of 58 true hermaphrodites of Groups I to V, eliminating only Group VI, the incompletely studied cases (JONES and SCOTT, 1958).

The majority of true hermaphrodites are reared as boys and men. This, of course, means that the greater number of such hermaphrodites have rather masculine appearing external genitalia (Fig. 20). Of the 55 determinate cases, 39 were reared as men and if one eliminates Group II, that is the group with the most ovarian tissue, i.e. an ovary on one side and an ovotestis on the other, the preponderance of individuals reared as men is even more striking. In Group II, where the only virilizing influence was a unilateral ovotestis, half of the patients were reared as men and half as women. The accompanying table gives more information on this point (Table 3).

Almost all true hermaphrodites develop breasts. Of the 45 patients reviewed who were 14 years of age or older, 36 developed breasts. This could indeed, be a very helpful differential diagnostic point in distinguishing between male hermaphroditism and true hermaphroditism for no male hermaphrodites develop breasts except those with the testicular feminization syndrome. When the breasts develop, they usually do so at the expected time of puberty. Thus, if no breast development is present by the age of 14 years,one can be reasonably sure they will not develop thereafter (Table 4).

Many true hermaphrodites menstruate. The presence of absence of menstruation is determined in some measure by the development of the uterus, and there is a substantial group of true hermaphrodites who have rudimentary or no development of müllerian ducts. However, of those who have uteri which are developed well enough to menstruate, the great majority do menstruate as may be seen by the accompanying table (Table 4). This is especially interesting in Group IV where all patients with developed müllerian ducts menstruated in spite

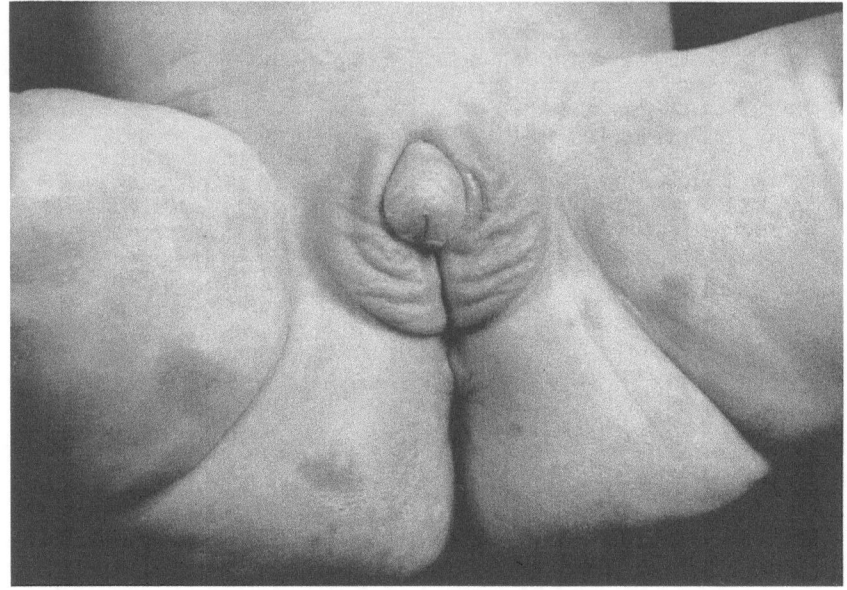

Fig. 20. Photograph of external genitalia of patient shown in Fig. 17. (From: H. W. Jones, Jr., M. A. Ferguson-Smith and R. H. Heller, 1965)

of the fact that their only ovarian tissue was in a unilateral ovotestis with a contralateral testis.

The development of the uterus from the müllerian ducts is of special interest not only because of its clinical importance in making it possible for such a patient to menstruate, but also because there is suggestive evidence from the true hermaphrodites as to the factors responsible for the normal development of the uterus. The detailed data are given in the accompanying table (Table 5). The striking fact from this table is that there are but 2 normal uteri out of 11 in Group II, the group consisting of individuals with bilateral ovotestes, whereas the proportion of normal uteri among the other groups is considerably higher. Furthermore, in this same Group II, 5 of the 11 uteri were of the bicornuate variety and the development of the müllerian ducts was completely inhibited in 4 of the 11 cases. The number of cases in the various groups is very small but these differences are quite striking. The implication from these data may be that the complete or partial inhibition of the müllerian ducts, as expressed by the absence or bicornuate development of the uterus, is apparently greatest when the inhibiting, virilizing local organizer effects are of bilateral origin.

Table 3. *Sex of Rearing*

Group		Determinate Cases	As Men	As Women
I	Ovary:testis	22	18	4
II	Ovotestis:ovotestis	10	7	3
III	Ovary:ovotestis	16	8	8
IV	Ovotestis:testis	5	4	1
V	Ovary Testis:O	2	2	0

Table 4. *Menstruation and Breast Development*

Group		No. of Cases over 14 yr. of Age	No. with Breast Development	No. with Menstruation	No. with Rudimentary or no Uteri
I	Ovary:testis	16	11	7	5
II	Ovotestis:ovotestis	7	5	2	4
III	Ovary:ovotestis	15	14	9	4
IV	Ovotestis:testis	5	4	4	1
V	Ovary Testis:O	1	1	0	0

Table 5. *Development of Uterus*

Group		Determinate Cases	Normal	Rudimentary	Bicornuate	Unicorn
I	Ovary:testis	22	10	7	2	3
II	Ovotestis:ovotestis	11	2	4	5	0
III	Ovary:ovotestis	16	10	6	0	0
IV	Ovotestis:testis	5	4	1	0	0
V	Ovary Testis:O	2	1	0	0	1

Table 6. *Relation of Karyotype to Type of True Hermaphrodite*

Karyotype	No. of cases	1	2	3	4	5	6
XX	24	5	6	6	4	2	1
XY	1		1				
XO/XY	1						1
XX/XXY	1	1					
XX/XXY/XXYYY	1				1		
XX/XY	1		1				

5. Sex Chromosome Complements

Through 1963, 29 patients with true hermaphroditism who also had karyotype analysis were reported in the literature. The pertinent data are presented in Table 6. A few case reports could not be used because of insufficient or conflicting data (JONES, FERGUSON-SMITH, HELLER, 1965).

The experience gained in this series clearly shows that the majority of patients with true hermaphroditism exhibit sex chromatin and karyotypes indistinguishable from those of the normal female (Figs. 21, 22). On the other hand, a few patients who cannot be distinguished from the others with true hermaphroditism have been reported to exhibit a variety of other karyotypes, mostly of a mosaic character.

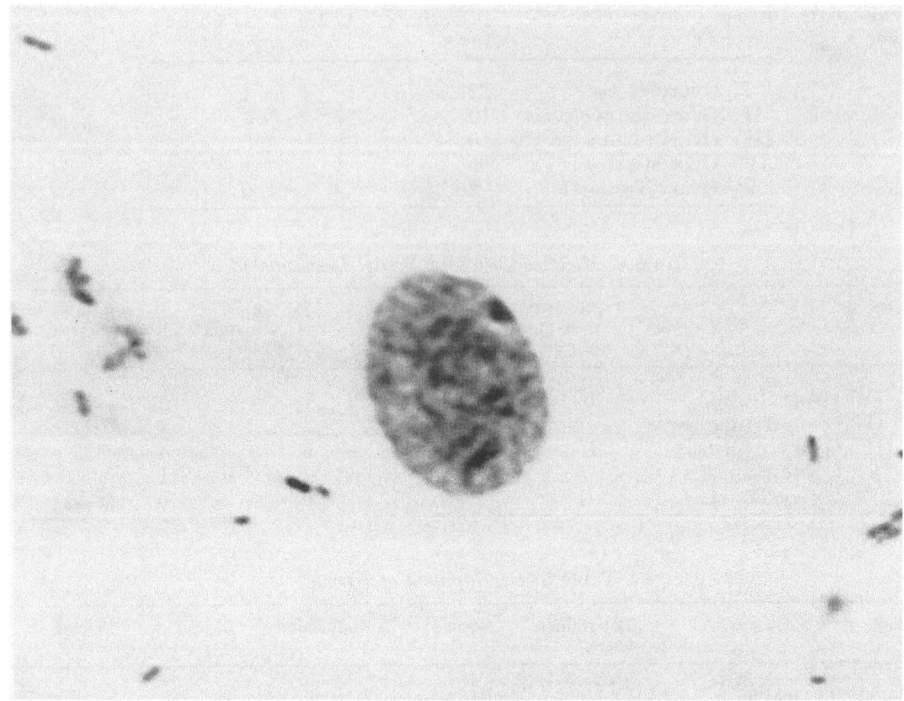

Fig. 21. Photomicrograph of cell from the buccal mucous membrane of the patient shown in Fig. 17. The sex chromatin body at 2 o'clock is entirely normal and indistinguishable from that of a normal female (acetic-orcein × 2500)

Examination of the location of the contradictory gonadal development in the 29 reported cases indicates that there is no recognizable relationship between that location and the karyotype (Table 6). This is brought out most clearly when those patients who have not been found to have mosaicism are considered. Thus, of the 24 XX patients, contradictory testicular development occurs in all possible combinations and in about the same proportion as expected from the known distribution of gonadal development in previous series (Table 7). Of special interest is the fact that the patient with XY sex chromosome complement was not of Group IV—i.e., with a testis on one side and an ovotestis on the other. The patients with mosaicism are too few for definitive analysis of this point, but each mosaic case has a different distribution of testicular tissue. It is regrettable that the pathologic description of the gonads in the mosaic cases seems to be less complete than might be desired; e.g., it would be valuable to know the relative amount of ovarian and testicular tissue, presence of germ cells, etc.

It seems of interest that the ovarian tissue in these cases is apparently normal from birth to adulthood. However, the testicular tissue in general appears to be relatively normal only in infancy. After puberty, germ cells begin to disappear and the adult testis invariably shows hyalinization and sclerosis of the tubules, with Leydig cell hyperplasia. This cannot simply be due to the fact that the gonads are intra-abdominal, as some are scrotal and, in other disorders (e.g., ectopic testes), germ cells survive the raised temperature (MACK et al., 1961). It may be that the germ cell degeneration is somehow associated with the proximity of the ovary or even with its estrogen production, for males who receive estrogen (as for prostatic cancer) show germ cell degeneration.

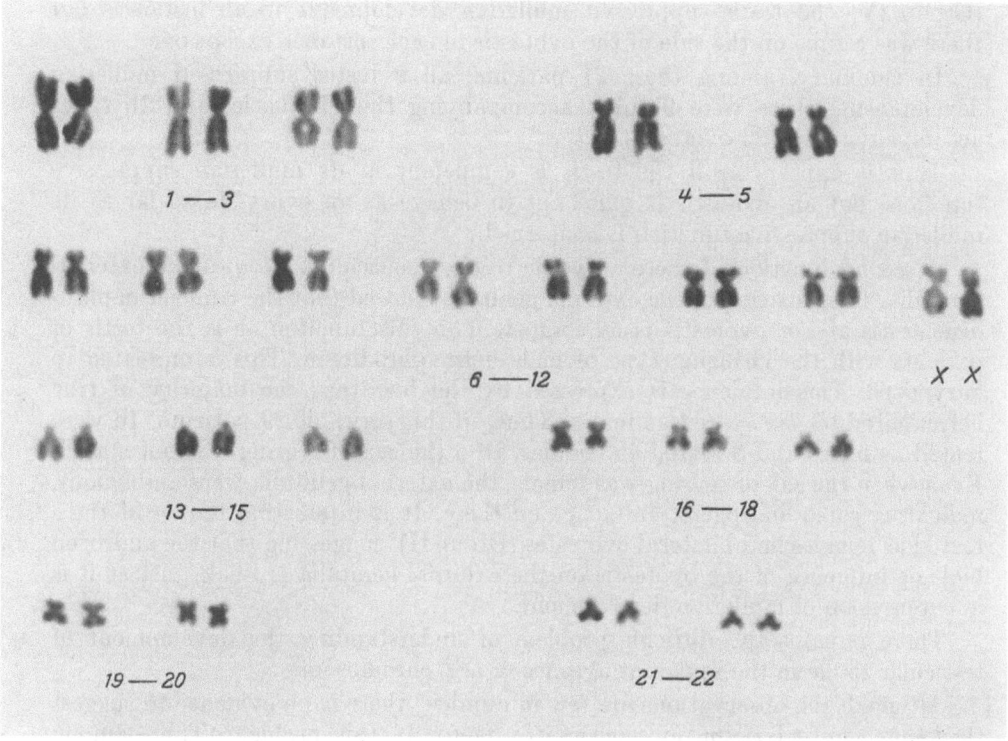

Fig. 22. A normal female karyotype of the same patient shown in Fig. 17. This karyotype is indistinguishable from that of a normal female (acetic-orcein × 4000). (From: H. W. JONES, Jr., M. A. FERGUSON-SMITH and R. H. HELLER, 1965)

Table 7. *Distribution of True Hermaphrodites According to Type*

Sources	No. of cases	Group of true hermaphrodite					
		1	2	3	4	5	6
JONES and SCOTT	66	23	12	16	5	2	8
Present series	24	5	6	6	4	2	1
Total	90	28	18	22	9	4	9

A study of testicular competence for müllerian duct suppression in the 24 reported true hermaphrodites with XX sex chromosome complement revealed that there was determinate (Groups V and VI not considered) information in 21 patients as follows.

In the 5 determinate lateral hermaphrodites (Group I), the müllerian duct was suppressed on the testicular side in all instances. A tube was present on the ovarian side in 4 instances and absent in one.

Curiously enough, in the 6 hermaphrodites with bilateral ovotestis (Group II) the tubes were mentioned as present bilaterally in 5, and the uterus, when mentioned was present although rudimentary and bicornuate in all except one.

In the 6 determinate patients with an ovary on one side and ovotestis on the other (Group III), the tube on the side of the ovary was present in all instances except one but the tube was present in 3 of the 6 instances on the side of the ovotestis.

In the 4 patients with a testis on one side and an ovotestis on the other (Group IV) the testis suppressed müllerian development in all instances but there was a tube on the side of the ovotestis in each instance except one.

In summary, among these 21 patients, all 9 testes suppressed müllerian development; there were 9 tubes accompanying the 11 ovaries and 16 tubes accompanying the 22 ovotestes.

Thus it appears as if the testis is competent in its müllerian suppressive functions but an ovotestis is quite apt to behave as an ovary, in so far as its müllerian suppressive function is concerned.

A second function of the embryonic testis is masculinization of the external genitalia. Examination of the external genitalia showed that the true hermaphroditic testis and/or ovotestis is as competent in this function as is the testis of patients with the virilizing type of male hermaphroditism. This is unrelated to karyotype. This influence is expressed by the fact that the majority of true hermaphrodites are reared as males. Thus, in this series of 29 patients, 16 were reared as males and 8 reared as females. In 5 the sex of rearing was not stated. Even when the sex of rearing was female, the external genitalia were ambiguous, indicating some androgenic influence on them. It is interesting that 6 of the 8 reared as females had bilateral ovotestes (Group II), suggesting that the androgen biologic influence of the ovotestis on the external genitalia is weak, just as it is in suppression of müllerian development.

There remains the difficult problem of understanding the development of testicular tissue in the apparent absence of a Y chromosome.

Although the observations are few in number, there is no evidence to suggest that unrecognized tissue mosaicism is a factor in this problem. Thus, among patients who have been found to have a normal female karyotype in the peripheral blood, 3 also had cultures with similar findings from the testicular portion of the gonads. Furthermore, in all 6 of our patients where a careful study was made, sex chromatin bodies were noted in the testicular portions of the gonads.

The question arises as to whether there has been developmental anatomic reversal of gonadal tissue by other than genetic factors. Such reversal has been observed under experimental conditions in a variety of vertebrates—fish (YAMAMOTO, 1963), amphibians (HUMPHREY, 1948), and birds (WILLIER et al., 1937).

There have been numerous unsuccessful attempts to achieve gonadal reversal in the mammal. However, only BURNS (1956) in the opossum and TURNER and ASAKAWA (1964) in the mouse, have achieved experimental reversal in these species. In the case of BURNS, the reversal was from testis to ovary by estradiol and in the case of TURNER and ASAKAWA, the reversal was from an explanted ovary to ovotestis by diffusion of an unknown substance from a testis.

To these experimental mammalian reversals should be added the classic reversal in the bovine freemartin. It was long ago shown by LILLIE (1922) that crossed placental circulation of male and female embryos resulted in the production of a sterile female. In some marked examples of this condition the medullary portion of the affected gonads was developed to such a degree that they could be called testes. Freemartins have been decribed in several mammalian species, a recent example being the lamb (ALEXANDER and WILLIAMS, 1964).

These experiments are perhaps pertinent to the situation with true hermaphroditism in the human in demonstrating that reversal may take place by the action of factors other than those involving genes. In this connection, it is of interest that in at least one of the 6 newly reported patients of JONES et al. (1965), the mother received a progestogenic steroid during early pregnancy.

That genetic influences may play a role in some patients with true hermaphroditism is clear from the families reported by CLAYTON et al. (1958) and from the case of WAXMAN et al. (1962)who found an interesting XX/XY mosaic, apparently the result of superfecundation.

In general, however, the clinical picture of true hermaphroditism does not fit in which the clinical situation of other examples of gross chromosomal anomalies. For example, there are very few patients with associated somatic anomalies, and mental retardation is almost unknown.

Thus, true hermaphroditism may have varied etiology, and it may be said that in the majority of patients an etiologic factor has escaped detection.

6. Principles of Treatment

The principles of treatment of patients who proved to be true hermaphrodites differ in no way from those for the treatment of hermaphroditism in general. These are expressed in the section on this subject, but can be summarized here by stating that the medical and surgical effort should be directed at removing contradictory organs and in reconstructing the external genitalia in keeping with the sex of rearing. The special problem in this group is to establish with certainty the character of the gonad. This is particularly difficult where an ovotestis concerned for its recognition by gross characteristics is notoriously inaccurate. It should be noted that in some instances the gonadal tissue of one sex was completely embedded within a gonadal structure primarily of the opposite sex.

V. Klinefelter's Syndrome

This syndrome was first described in 1942 by KLINEFELTER, REIFENSTEIN and ALBRIGHT. The clinical picture applies only to men and, as originally described, was characterized by small testes, azoospermia, relatively normal external genitalia, relatively normal somatic development, gynecomastia and high urinary excretion of gonadotrophins. Although the paper by KLINEFELTER et al. grouped such cases together for the first time as a definite clinical entity, individual case reports had appeared much earlier. In 1812, for example, BEDOR according to LEREBOULLET (1877) described two brothers, 21 and 24 years old, with bilateral gynecomastia and small testes. In retrospect these patients seem to be examples of Klinefelters's syndrome. One of the most complete but somewhat later reports was in an article by Dr. HENRY MEIGE, published in Paris in 1895. Interestingly enough, in the light of recent developments, MEIGE's patient was classified with several others who were examples of classic hermaphroditism (Fig. 23).

1. Clinical Characteristics

By definition the syndrome under discussion applies only to patients reared as males. The disease is not recognizable prior to puberty except by routine screening of newborn infants as will be described subsequently. Most patients from a clinical point of view come under observation between 16 and 40 years of age.

Early somatic development during infancy and childhood may be quite normal. Growth and muscular development also may be within the normal limits. Indeed, the majority of patients have a normal general appearance and no complaints referable to the disease which is often discovered in the course of a general physical examination or in a work-up for fertility.

26*

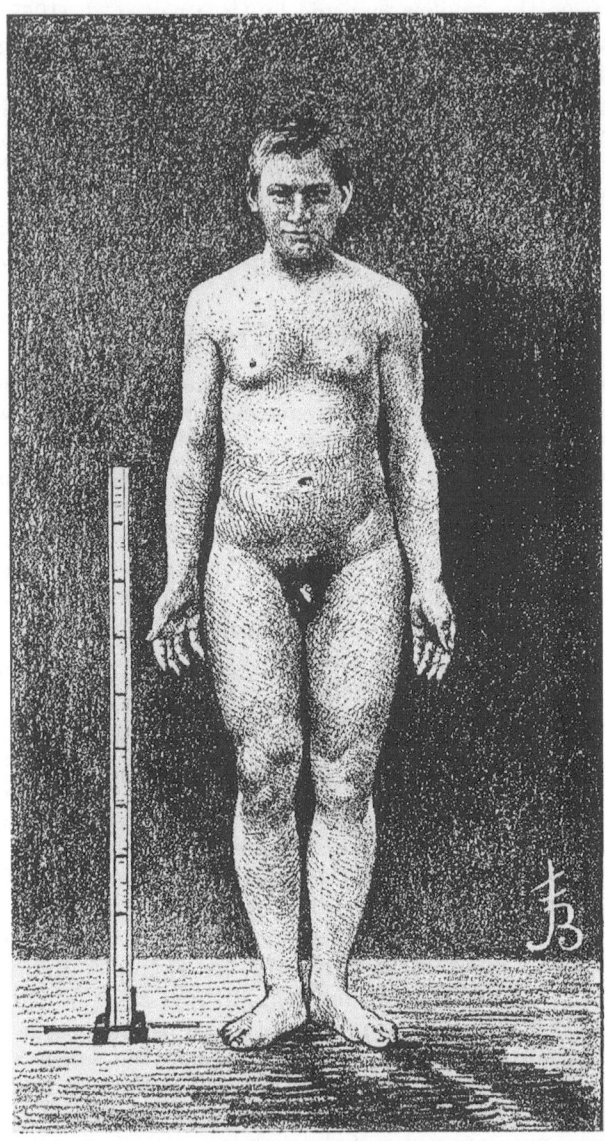

Fig. 23. Sketch of patient who probably was an example of Klinefelter's syndrome. Described in Paris by Henry Meige in 1895. (From: H. Meige, 1895)

In the original publication by Klinefelter et al., gynecomastia was considered to be a necessary part of the syndrome. However, Nelson and Heller (1945) and subsequently many other authors described patients without gynecomastia, but with other characteristics of the disease. When present, gynecomastia is bilateral and develops at or soon after the onset of puberty. The breasts usually show an increase in size over a period of several years and then remain stationary. In a few patients, breast enlargement has not been noted until 4 to 6 years after the onset of puberty. The gynecomastia is often the presenting complaint and represents one of the principal problems in the treatment of the

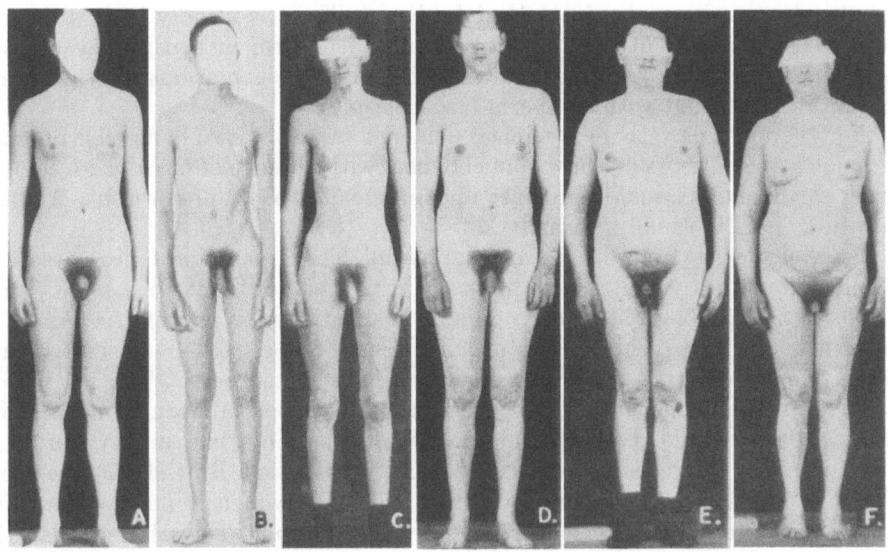

Fig. 24. Photograph of six patients with the syndrome. Taken from the original publication by KLINEFELTER et al. (From: H. F. KLINEFELTER, Jr., E. C. REIFENSTEIN, Jr. and F. ALBRIGHT, 1942)

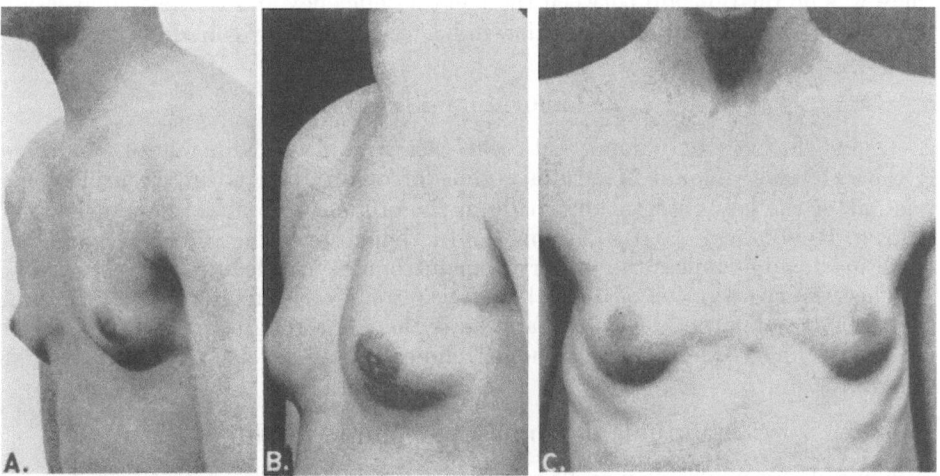

Fig. 25. Breast development of 3 patients with Klinefelter's syndrome. (From: H. F. KLINEFELTER, Jr., E. C. REIFENSTEIN, Jr. and F. ALBRIGHT, 1942)

disease. The breasts are not tender and no secretion is expressable from the nipples (Figs. 24, 25).

There is no history of delayed descent of the testes. The testes are always in the scrotum. Neither is there any history of testicular trauma or disease. Although a history of mumps orchitis can occasionally be enlisted, there has been no correlation of this disease with the syndrome.

The external genitalia are perfectly formed and in most patients are quite well developed. Growth and devolopment of the genitalia, however, in some instances may be impaired so that as adults the genitalia may be somewhat smaller than normal. In any case, erections and intercourse are usually satisfactory.

The testes, however, in contrast to the rest of the external genitalia, are often very small measuring about $1^1/_2$ by 1 by $^1/_2$ cm. There is usually little difference from patient to patient. The testes are normal or firm in consistency and are sensitive to pressure. As a rule, these patients have ejaculations with a normal amount of semen but with no sperm (azoospermia).

For the most part there is clinical evidence of adequate Leydig cell function although some patients do show eunuchoidism with absence of beard, high pitched voice, small phallus, small prostate and eunuchoid body proportions. Axillary and pubic hair is usually normal.

Psychological symptoms are often present. The actual incidence of serious psychological or psychiatric symptoms is extremely difficult to state as most studies of this problem have been done in institutions for psychiatric care. The seriousness of the psychiatric disturbance seems to be related to some degree to the number of extra X chromosomes as will be noted below. Thus, it is estimated that about one quarter of XXY patients are of subnormal intelligence, whereas BARR et al. (1962) found that 5 of 7 patients had an I.Q. of 35 or less. According to the study of PASQUALINI, VIDAL and BUR (1957) the mental deficiency does not measurably differ from that which is not accompanied by the other manifestations of Klinefelter's syndrome. They found that the picture was characterized by complete lack of initiative, interest and vivacity with slow awkward movements and activity strictly limited to immediate interests including patients with normal or near normal I.Q. They were of the opinion that there was no relationship between the mental deficiency and the testicular lesion but that the frequency of association indicated a common genetic origin.

2. Laboratory Findings

One of the extremely important clinical features of the Klinefelter's syndrome is the excessive amount of pituitary gonadotrophin excreted in the urine. This was one of the key observations made in the original communication of KLINE-FELTER, REIFENSTEIN and ALBRIGHT, and a diagnosis of the disorder cannot be made in the adult unless the urinary gonadotrophins are elevated.

The urinary excretion of neutral 17-ketosteroids varies from relatively normal to definitely subnormal levels. By and large there is a rough correlation between the degree of hypoleydigism as judged clinically and a lowered 17-ketosteroid excretion.

3. Pathological and Cytogenetic Findings

Klinefelter's syndrome may be regarded as a form of primary testicular failure due to a variety of causes which act during various periods of development. FERGUSON-SMITH et al. (1960) found that it was possible to classify those who fulfilled the key diagnostic criteria of Klinefelter's syndrome into three groups according to their testicular histopathology and chromosome count as follows: 1. *Chromatin negative individuals with postpubertal atrophy.* The microscopic appearance of the testes in this group suggests extensive secondary atrophy to what was at one stage a perfectly normal gonad. However, no relevant history of orchitis or injury can be obtained. The seminiferous tubules are shrunken, hyalinized and without epithelium. They contain large amounts of elastic fibers, indicating that atrophy had occurred after puberty. The Leydig cells are present in large numbers. These microscopic changes may follow such things as mumps orchitis with identical endocrinological consequences. Chromosomes studied in such patients have been rather sparse, but two such patients cited by FERGUSON-

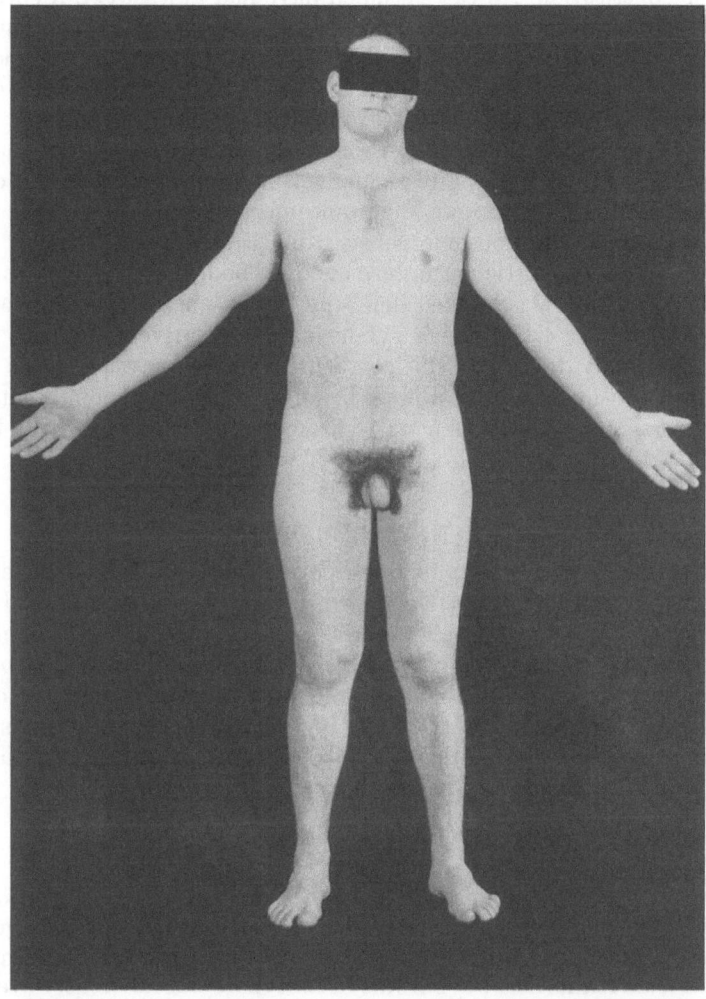

Fig. 26. Photograph of 25 year old normally well developed male. Note the external genitalia and the hands which evidence hard physical labor as a steel worker. The scars of the subcutaneous mastectomy are just visible. (From: H. W. JONES, Jr. and W. W. SCOTT)

SMITH showed normal male chromosomes. That such a type of patient may be the result of a genetic cause is illustrated by a family studied by REIFENSTEIN in 1947 where 9 out of 10 male members of 2 generations were affected. 2. *Chromatin negative individuals with absence of germ cells.* The testicular lesion in these patients if confined to absence of germinal cells and the tubules are lined almost entirely by Sertoli cells. There is peritubular fibrosis and rarely any tubular hyalinization. Leydig cells are diffusely increased. The etiology of this variety is obscure although it is possible that the lack of germ cells indicates some defect in embryogenesis. The chromosomes in this group of patients is entirely normal. The mean maternal age in a series of patients studied by FERGUSON-SMITH was 33 years which is as high as in a group of chromatin positive cases where this increased age has been considered an important factor predisposing to chromatin non-disjunction. 3. *Chromatin positive individuals with prepubertal testicular atrophy.* Males with chromatin positive nuclear sex are invariably found to have Klinefelter's syndrome. The nuclear sex anomaly reflects a basic genetic abnormality in sex chromosome constitution. All cases studied have had at least two X chromosomes and one Y chromosome. The commonest abnormality in the sex chromosome constitution is XXY but XXXY, XXYY XXXXY and XXXYY and mosaics of XX/XXY, XY/XXY, XY/XXXY, XXXY/XXXXY have all been described. It is interesting to note that in all examples except the XX/XXY mosaic a Y chromosome is present in all cells. From these patterns, it is obvious that the Y chromosome has a very strong testicular forming impulse which can operate in spite of the presence of up to four X chromosomes.

The abnormal sex chromosome constitution causes differentiation of an abnormal testis leading to testicular failure in adulthood. Before puberty and indeed at birth such testes show a marked deficiency or absence of germinal cells.

Chromatin positive Klinefelter's syndrome is a relatively common condition and many more individuals with this condition who are unaware of this disability and who never consult a physician must be circulating in the population at large. By nursery screening, the frequency of chromatin positive males has been estimated at 2.65 per 1000 live male births by MacLean (1961).

4. Treatment

The two principal complaints of these patients, infertility and gynecomastia, have no known specific therapy. No known pituitary preparation is effective in regenerating the hyalinized tubular epithelium and stimulating gametogenesis. Furthermore, no hormonal therapy is effective in treating the breast hypertrophy. When the breasts are a formidable psychological problem, plastic removal is an eminently satisfactory procedure (Fig. 26). In patients who have clinical symptoms of hypoleydigism substitutional therapy with testosterone in one form or another is an important physiologic and psychologic aid.

VI. Double X Males

A few adult men (SHAH et al., 1961; DE LA CHAPELLE et al., 1964; THERKELSEN, A. J., 1964) have been noted with slightly hypoplastic penis and very small testes but no other indication of abnormal sexual development. They were sterile and different from Klinefelter's syndrome in that there was no breast development. They are clinically very similar to the Del Castillo syndrome (1947). However, the double X males have a positive sex chromatin and a normal female karyotype. These are perhaps an extreme example of the sex reversal which partially occurs in true hermaphroditism.

The pediatric interest derives from their possible discovery by a routine buccal smear in a newborn.

VII. Multiple X Syndromes

The finding of more than one sex chromatin body in a cell indicates the presence of more than 2 X chromosomes in that particular cell (Fig. 3). In many patients such a finding is associated with mosaicism and the clinical picture is controlled by this fact, e.g. if one of the strains of the mosaicism is XO, gonadal agenesis is apt to occur.

However, there are some patients who appear not to have mosaicism but do have an abnormally increased number of X chromosomes in all cells. The most common form of such an abnormality results in patients with the triplo-X syndrome but patients with tetra X (CARR et al., 1961) and penta-X (KESAREE and WOOLLEY, 1963) have been reported.

The additional X chromosome does not appear to have a consistent effect on sexual differentiation. The external genitalia are quite female. A number of such patients have now been examined by laparotomy. There is clearly no consistent abnormality of the ovary. In a few cases, the number of follicles appears to be reduced and in at least one case the ovaries were very small and the ovarian stroma poorly differentiated. In about $1/5$ of the post-pubertal patients with the triplo-X syndrome, various degrees of amenorrhea were present or some irregularity in menstruation, but for the most part, the patients have a normal menstrual history and are of proved fertility.

The body proportion of these patients is quite normal.

Almost all patients with multiple X syndromes have had various degrees of mental retardation. Perhaps this is partly the result or the fact that such patients have been discovered by surveys in mental institutions, but the important clinical point is that infants with mental retardation should certainly have a buccal smear. A few patients have been reported with mongoloid features.

There has been some tendency for the maternal age to be somewhat increased as is in patients with mongolism.

The offspring of triplo-X patients have been uniformly normal. This is somewhat surprising, and it might be suspected, on theoretical grounds, that such individuals during meiosis should produce equal number of ova containing one and two X chromosomes. Fertilization of the abnormal XX ova should, therefore, give rise to XXX and XXY individuals. However, it seems to be that the fertile triplo-X cases exhibit selection in favor of normal ova or zygotes.

The diagnosis of this syndrome is made by discovering a high percent of cells in the buccal smear with a double sex chromatin and by finding 47 chromosomes with a karyotype showing an extra X chromosome in all cells cultured from the peripheral blood. It should be noted that in examination of the buccal smear some cells seem to have a single sex chromatin so that on the basis of the chromatin examination, one might suspect that there was an XX/XXX mosaicism. However, in the patients under discussion, only a single type of cell can be discovered by the chromosomes cultured from the peripheral blood. It is thought that the absence of the second chromatin in some of the somatic cells is due to the time the cell was examined in interphase and to the special orientation which prevented the 2 sex chromatin bodies being visible adjacent to the nuclear membrane. In this syndrome the number of cells containing either 1 or 2 sex chromatin bodies is very high, running from 60 to 80 percent or more as compared with an upper limit of about 40 in the normal human female (JOHNSTON, FERGUSON-SMITH, HANDMAKER, JONES and JONES, 1961).

VIII. Female Hermaphroditism Due to Congenital Adrenal Hyperplasia

1. Clinical Considerations

Female hermaphroditism due to congenital adrenal hyperplasia is now recognized as a clearly delineated clinical syndrome. It has, however, been only in the last decade since the discovery by WILKINS and co-workers (1950) that cortisone successfully arrested the process of virilization that the syndrome has been clearly understood.

The first clear record of the abnormality is found in the writing of CRECCHIO, who in 1865 described in detail the history and autopsy findings of such a patient, although CRECCHIO himself did not recognize the association of the adrenal enlargment with the general clinical picture. The exposition is a model of pathologic description and deserves a place with other classic descriptions of disease. This history of CRECCHIO's patient may be abstracted as follows: In June 1820 a woman in Naples gave birth to an infant who was identified by the midwife as female and who was given the name Josephine. At the age of 4 years a surgeon declared that the patient was a male with testicles in the abdomen. The child was then put into male attire and given the name of Joseph. This excited the raillery of the neighbors. At 12 years of age the patient become a valet. At 18 years Joseph had the voice of a man, a rapidly increasing beard and began to have adventures with women. This reassured his father, who had noticed the absence of nocturnal emissions. The following year his employers found it necessary to change his service because of his relations with the maid. In the meantime Joseph had contracted venereal disease twice. At the age of 25 years he fell in love with another maid, young and gentle, "qui lui rendait passion pour passion." Presents were exchanged, marriage was planned and Joseph was requested to produce his birth certificate. When he discovered that he had been pronounced at birth to be a female, he procrastinated. In the meantime his fiancee had become intimate with a new lover, "but Joseph conducted himself with dignity." Having been foiled in his matrimonial intentions, he became a drunkard and braggart of his numerous conquests. He smoked continually, turned against religion, wished to destroy the images of the saints and constantly frequented cabarets where he excelled in obscene stories. He lived to the age of 43 years.

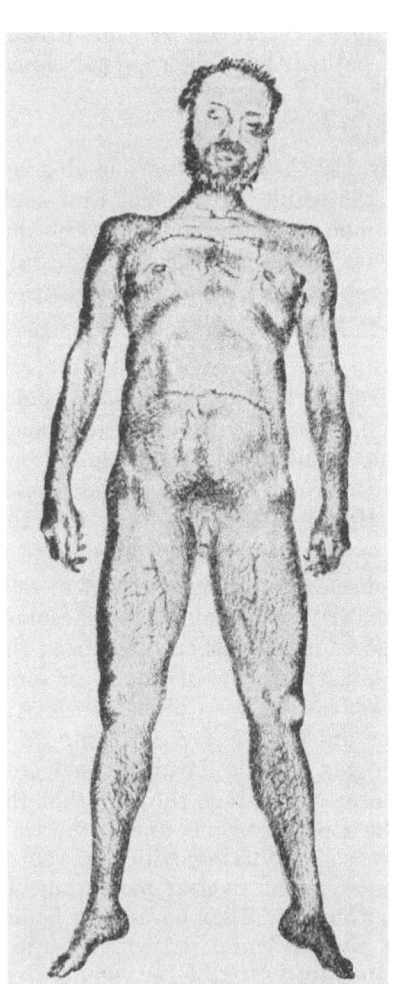

Fig. 27. Joseph Martzo, a sketch made postmortem. (From: H. W. JONES, Jr. and W. W. SCOTT, 1958)

The autopsy revealed the following data. Height 156 cm. Circumference of neck 38 cm; the chest 65 cm. and the circumference around the iliac crest 69 cm. The arms, hands and feet were small (Fig. 27). The penis was 6 cm long. It was measured 10 cm. on erection. The urinary meatus was at the base of the penis. There was no scrotum, nor were testicles palpable (Fig. 28). On opening the abdomen a normal uterus and tubes were discovered. The ovaries were found in their normal position and were elongated, smooth and showed no trace of corpora lutea and no irregularities. The microscope did not show any abnormality in these structures. The vagina opened into the urethra and measured 6.5 cm. in length and 4 cm. in circumference. The communication with the urethra was in the form of a valvule situated next to the prostatic portion. On each side was an opening which simulated the ejaculatory ducts into which a probe passed a short distance (Fig. 29). The adrenals were almost as large as the kidneys which were normal in size.

Since this original description by CRECCHIO, innumerable cases have been described, and the abnormality may now be recognized as the most common type of sexual aberration. Since the recognition of the fact that the exhibition

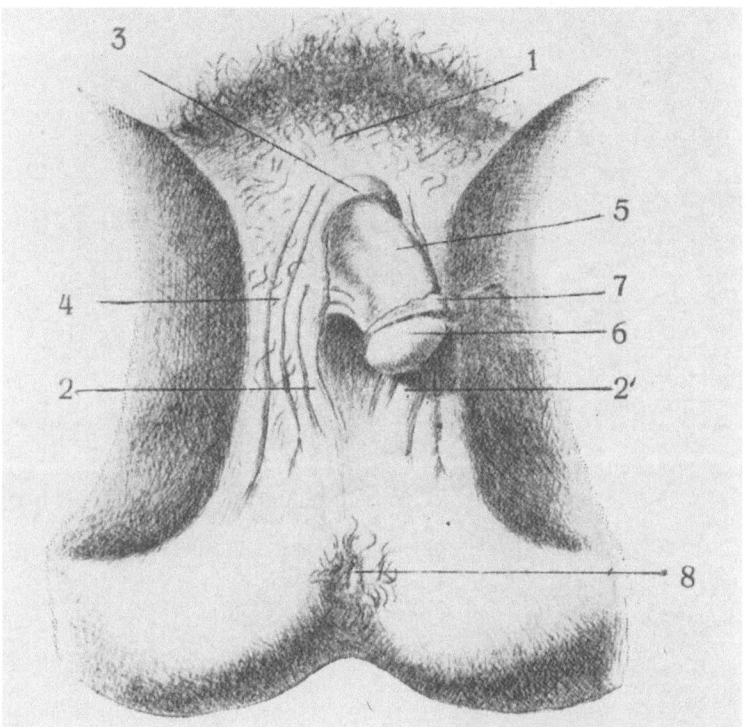

Fig. 28. Joseph Martzo, a sketch of external genitalia. (From: H. W. JONES, Jr. and W. W. SCOTT, 1958)

of cortisone or its derivatives can successfully arrest the process of virilization in these genetic females, it has become exceedingly important that the syndrome be recognized in infancy for if this is done, proper treatment by hormones and surgery will yield a girl and woman who can enjoy a relatively normal life.

If the diagnosis is not made in infancy, an unfortunate series of events occurs. Because the adrenals secrete an abnormally large amount of virilizing steroid, even during embryonic life, such infants are born with abnormal genitalia. In the fully developed case there is fusion of the scrotolabial folds and in rare instances the formation of a penile urethra. The clitoris is greatly enlarged so that it may be mistaken for a penis (Figs. 30, 31). There are, of course, no gonads palpable within the fused scrotolabial folds and their absence has sometimes given rise to the mistaken impression of male cryptorchidism. There is usually a single urinary meatus at the base of the phallus and the vagina enters the persistent urogenital sinus in a rather consistent fashion as will be detailed later (Fig. 32).

During infancy, provided there are no serious electrolyte disturbances, these children grow at a rate greater than normal so that for a time they greatly exceed the average of both height and weight. However, epiphyseal closure occurs early and full growth may be obtained by 10 years or thereabouts with the result that as adults these unfortunate people are much shorter than average (Fig. 33).

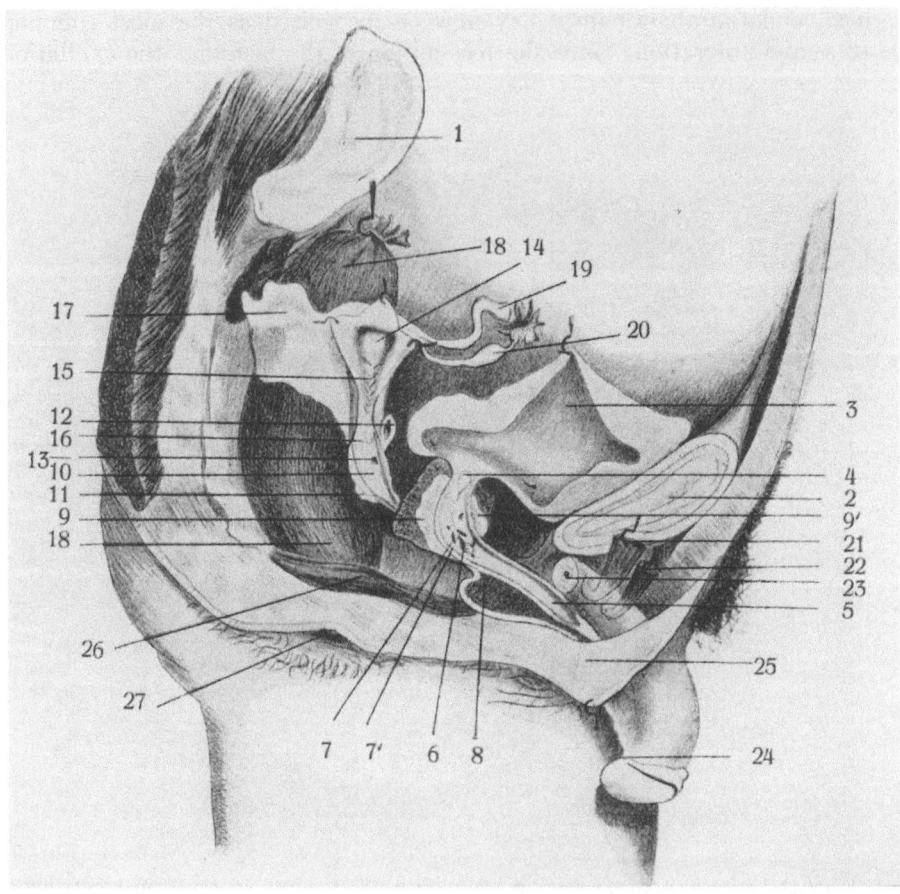

Fig. 29. Joseph Martzo, sagittal view of internal genitalia as discovered at autopsy. The uterus (#15) has been opened. The left tube (#19) and ovary (#20) are normal. The right tube and ovary are not shown in the sketch but were found to be similar to the left. The vaginal opening (#6) is shown in the prostatic urethra just in front of the prostate, but subsequent studies in many other cases have shown the orifice to be invariably more distal than this. However, it is noteworthy that it is clearly shown that the opening is not at the position of the prostatic utricle thus demonstrating even in this sketch of a century ago that the prostatic utricle may not be homologous with the vagina. (From: H. W. JONES, Jr. and W. W. SCOTT, 1958)

At an early age the process of virilization begins. Pubic hair may appear as early as 2 years but usually occurs at a somewhat later date. This is followed by the development of axillary hair and finally by the appearance of body hair and a beard which is often troublesome enough to require daily shaving. Acne may develop early. Puberty never appears. There is no breast development. Menstruation does not occur. During the entire process the urinary 17-ketosteroid excretion is always increased in relation to the patient's age.

Although our principal concern in this chapter is with females afflicted with this abnormality, it should be mentioned that adrenal hyperplasia of the adrenogenital

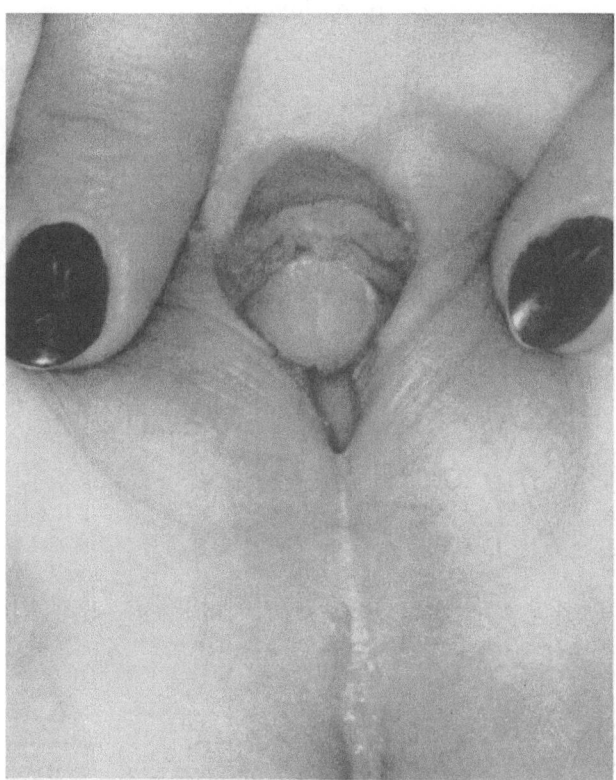

Fig. 30. Photograph of external genitalia of a patient with congenital adrenal hyperplasia. The enlarged clitoris and the fusion of the scrotolabial folds are easily seen. This gives rise to a single external orifice. (From: H. W. JONES, Jr. and W. W. SCOTT, 1958)

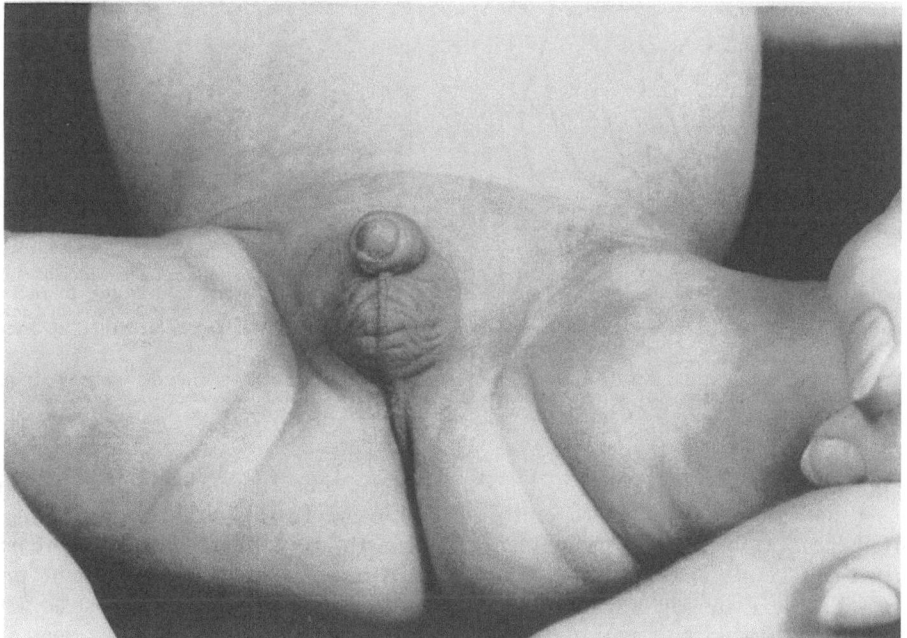

Fig. 31. Photograph of external genitalia of a female child with complete scrotolabial fusion giving genitalia which are indistinguishable from those of a normal male except for the absence of testes in the scrotum

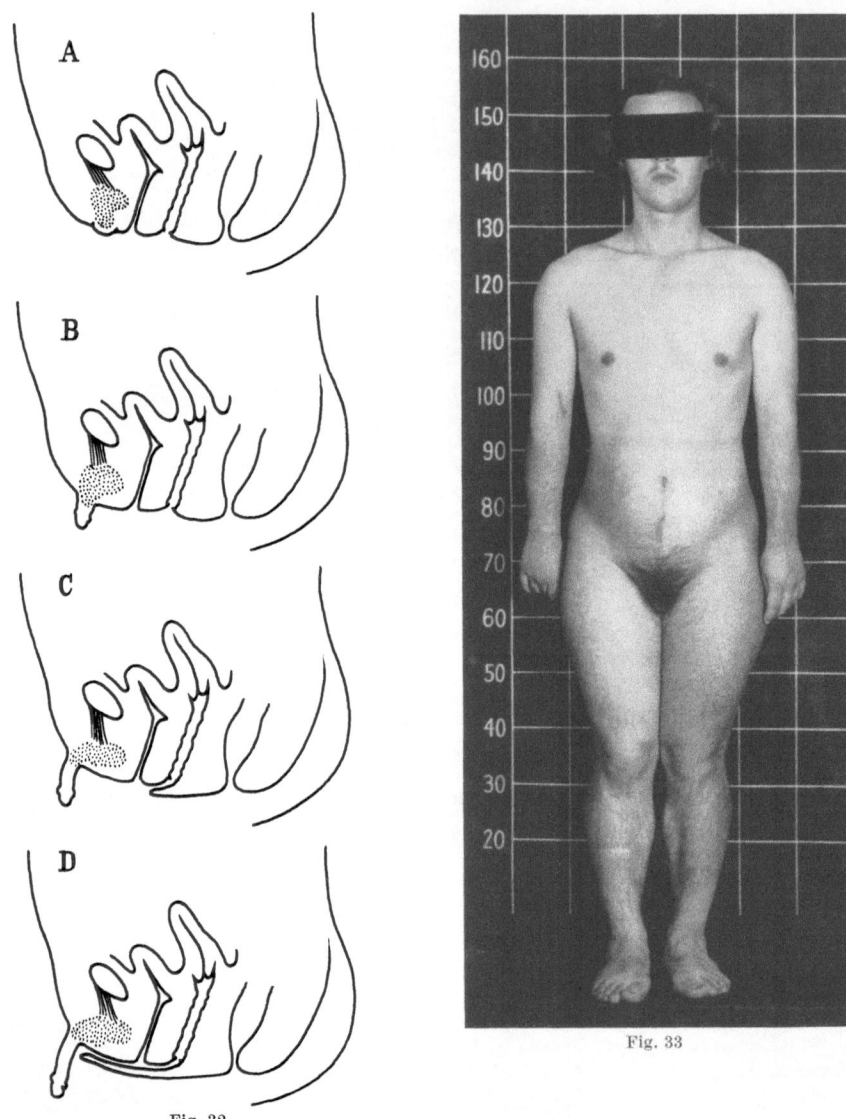

Fig. 32

Fig. 33

Fig. 32. Sketch of various stages of clitoral enlargement and scrotolabial fusion as seen in patients with congenital adrenal hyperplasia. a. normal female; b. slight enlargement of the clitoris; c. enlargement of the clitoris with some posterior fusion of the scrotolabial folds; d. enlarged clitoris with more fusion of the scrotolabial folds. In other cases there may be complete fusion giving a penile urethra as shown in Fig. 31. (From: H. W. JONES, Jr. and G. E. S. JONES, 1954)

Fig. 33. Photograph of a patient with untreated congenital adrenal hyperplasia showing the typical body proportions. Due to early epiphyseal closure, the trunk is of a normal adult height but the arms and legs are relatively too short. (From: H. W. JONES, Jr. and W. W. SCOTT, 1958)

type may also occur in males. Under these circumstances the clinical picture has been designated as macrogenitosomia praecox. In this situation the penis is often large at birth, but patients may reach $1^1/_2$ to 3 years of age before the growth of genitalia is sufficient to attract attention. Sexual development progresses rapidly and the sex organs attain adult size at an early age. Just as with the female, sexual hair and acne develop early and the voice becomes deep. In spite of the secondary sexual development it is important to recognize that the testes, which are usually present in the scrotum, remain small and immature and

spermatogenesis rarely occurs so that although the genitalia are of adult dimensions anatomically, impregnation is usually impossible. The somatic development
in the male corresponds to that of the female so that as children they exceed the
average in height and strength but, if untreated, as adults are stocky, muscular
individuals well below the average height.

Both male and female with this disorder, but especially the male, may have
the complicating problem of electrolyte imbalance. In infancy it is manifested by
vomiting, progressive loss of weight, dehydration and unless recognized promptly,
often by death. The condition is sometimes misdiagnosed as congenital pyloric
stenosis. The characteristic finding is an exceedingly low serum sodium and
carbon dioxide level and a high potassium.

A second complicating problem is hypertension. A few patients have this in
addition to the virilization.

In a study of 113 patients WILKINS (1961) found that 66 had simple virilization,
40 were complicated with salt loss and 7 were complicated by hypertension.

2. Adrenal Pathology

The pathologic changes in the adrenal gland were completely reviewed by
JONES and JONES (1954) based on a study of 15 specimens from female hermaphrodites. Of these, 7 were autopsy specimens and 8 were surgical specimens available
from the era when surgical excision of the adrenals was tried in an attempt to
alter the progress of the disease. Slides of paraffin blocks were examined in all
15 cases and in 7 instances, 6 of autopsy and 1 of surgery, formalin fixed material
was available for lipid study. The paraffin sections were stained not only with
hematoxylin and eosin but with Masson's trichrome method, iron hematoxylin
and Wilder's reticulum stain. The formalin fixed material was studied by staining
with Sudan III, Sudan Black B and by the Ashbel-Seligman method for ketosteroids, as well as examined for birefringence with polarized light.

DEANE and GREEP (1946) and others have maintained that ketosteroids react
positively to such lipid procedures as mentioned although no one test is specific
for steroids. A positive reaction to the entire group, however, has been considered
as presumptive evidence for the presence of ketosteroids. Other investigators,
on the contrary, have held that such methods were not specific enough for exact
ketosteroid localization. The findings about to be described must, therefore, be
interpreted in the light of the controversy. Fortunately the findings are of interest
and importance aside from their significance with respect to ketosteroids.

The autopsy material consisted of 7 patients, aged 7 and 13 days, 3 and 6
weeks, 3, 21 and 29 years. The surgical material consisted of 8 patients, aged
5 months, 6 years (3 cases), 7, 13, 16, 19 and 20 years. Each patient was a typical
female hermaphrodite with enlarged clitoris and a single external opening of a
urogenital sinus into which the vagina entered, except for 2 patients, aged 19 and
20, who had a somewhat less separate vaginal and urethral openings. Two had
been reared and named as males. None had menstruated. All were hirsute and
of typical stature. The 4 youngest autopsied children died of the disease, presumably as a result of disturbance of electrolyte metabolism and 1 died after an inter-
current acute illness of 18 hours, while under cortisone therapy. The other deaths
were note caused by the disorder. Adrenalectomy was performed in the surgical
cases in an attempt to alleviate the condition. Such a procedure is, of course,
no longer practiced.

The adrenals in all cases were greatly enlarged over the normal size for the age
(Fig. 34). The largest adrenal glands weighed 80 to 90 grams, respectively,

compared to a normal adult weight of about 5 grams. Grossly, the outer layers of the normal cortex are yellow and the inner, or reticular layer, is thin and light brown. In these cases the pathologist at the time of autopsy uniformly described the fresh adrenals as being composed of an outer pale or white zone in addition to a darker brown or reddish inner zone. In 3 of 6 instances it was commented that the cortex apparently was depleted of fat.

Microscopically, all cases showed some degree of hyperplasia of the reticular zone of the cortex. In general the degree of hyperplasia seemed to increase with age. It was most easily seen in adults where upward of 90 percent of the cortex was occupied by reticularis compared to a normal adult composition of from one-fourth to one-third of the cortical width. An estimation of the degree of reticular

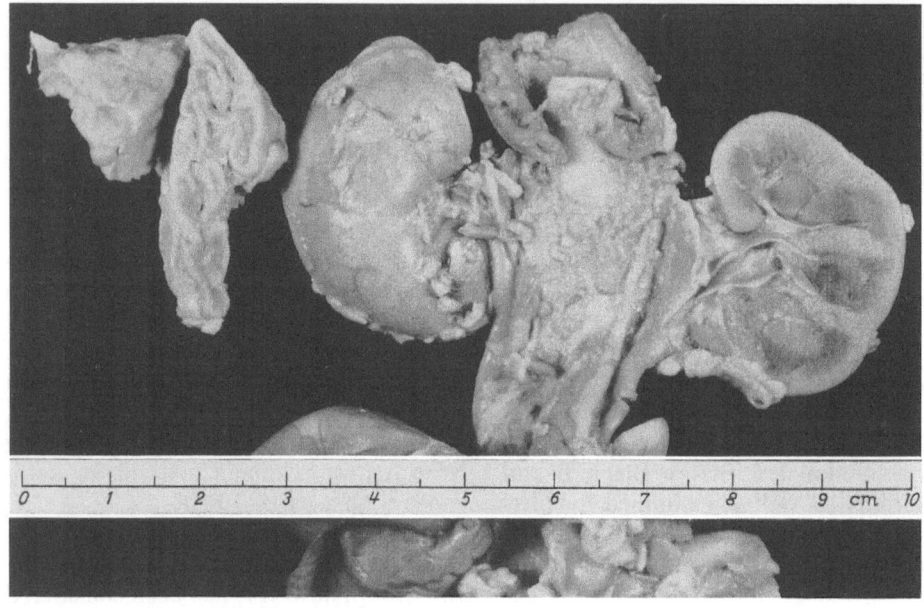

Fig. 34. Photograph of autopsy specimen showing huge adrenals. Patient was 6 weeks of age and typical female hermaphrodite. (From: H. W. JONES, Jr. and W. W. SCOTT, 1958)

hyperplasia in infancy and childhood depends upon comparison of the gland in question with the expected reticular width for the age. According to BLACKMAN (1946), and in agreement with other studies, the fetal reticular zone normally disappears by the end of the first month of postnatal life. The adult reticular zone, lying next to the fascicular zone and composed of cells smaller and with darker nuclei than those of the fetal reticular zone (hematoxylin and eosin) first appears as a thin layer at about 1 week of age. This layer gradually enlarges and becomes pigmented until it occupies its adult proportion of one-fourth to one-third of the cortical width at puberty. In infant hermaphrodites there was great hyperplasia of the adult reticular zone, and in all instances this zone occupied well over half the cortex, even in the patient who died at the age of 7 days. The hyperplasia of adult reticularis was, therefore, very great when compared with the expected reticular development at the corresponding age.

The fetal reticularis did not participate in the hyperplastic process. Fetal reticularis was present in the 5 patients under 5 months of age, but was less noticeable with increasing age until at the age of 5 months it was scarcely present. In the glands of very young patients, 17, 13 and 21 days old, it was not easy to be sure of the demarcation between fetal and adult reticularis but careful study, especially with the lipid preparations, resulted in a satisfactory diagnostic separation.

In addition to reticular hyperplasia the glands under discussion exhibited other abnormalities. In the 4 youngest infants who died and in the gland removed surgically from a patient aged 5 months, the glomerular layer, which is concerned with electrolyte metabolism, was practically absent although in 2 instances a few cells apparently belonging to the glomerulosa could be seen within the fibrous capsule. In other glands the glomerulosa seemed normally formed or actually hyperplastic. In the adult 29 years of age, the glomerular layer was about 30 cells deep compared to a normal of about 5. In 8 instances there was considered to be some degree of glomerular hyperplasia. It may be significant that the 4 infants with deficient glomerulosa died in Addisonian crisis and the child 5 months of age who had an adrenalectomy exhibited serious electrolyte imbalance.

The fascicular zone also exhibited marked variation. There was recognizable fasciculata in all instances. In all adults it was no more than 10 cells thick compared to a normal 30 to 50 cells and in some areas was absent. This variation in fasciculata in different areas of the same gland was characteristic for the majority of specimens. The fasciculata normally contains spongy, vacuolated cytoplasm when studied in the routine hematoxylin and the eosin section. In the cases under discussion the fasciculata was mainly composed of cells which exhibited homogeneous eosinophilic nonvacuolated cytoplasm. The fasciculata seemed to become less prominent with increasing age. It is interesting that in two instances in glands from patients who presented a slightly less severe form of the abnormality, the fasciculata approached normal and contained cells with more prominent vacuolization.

The studies with the formalin fixed material for lipid seemed most significant. Normally the reticularis contains very little lipid, the fasciculata contains abundant lipid and the glomerulosa less. In the hyperplastic glands there was substantial lipid in the reticularis of 5 of the 7 specimens. The fetal reticularis where present also reacted positively as is normal. The lipid was acetone soluble. The absence of lipid from the fasciculata together with its anatomic deformity may be of considerable importance in view of the probability that this layer is concerned with the production of the glucocorticoids. It is entirely likely that this abnormality of the fasciculata is the fundamental pathologic change in this syndrome, and that all other changes pathologically and endocrinologically and, therefore, clinically are the result of this fascicular abnormality.

It is important to note, however, that the significance of the absence of fascicular and glomerular lipid is somewhat obscured by the fact that in autopsy material adrenal lipid depletion is not uncommon in a variety of illnesses. We are led to believe, however, that the finding as recorded is probably significant, for a study of 25 other adrenals by the above method showed lipid in the reticularis in very few instances and only when there was considerable fascicular lipid. This finding is in confirmation of the more extensive and exhaustive study of AYRES, FIRMINGER and HAMILTON (1951) on adrenal lipid depletion in various disease states. Furthermore, in the one surgical specimen available for lipid staining, the finding was similar and very clear. The examination of adrenals surgically removed for various other conditions showed only minor fascicular lipid depletion. The findings, therefore, in the one gland removed surgically for hermaphroditism and available for study seems to make the above-mentioned autopsy observations more valid. Nevertheless, it is obvious conditions make the conclusion in regard to adrenal lipid distribution somewhat less than binding. The observation tallies so well,

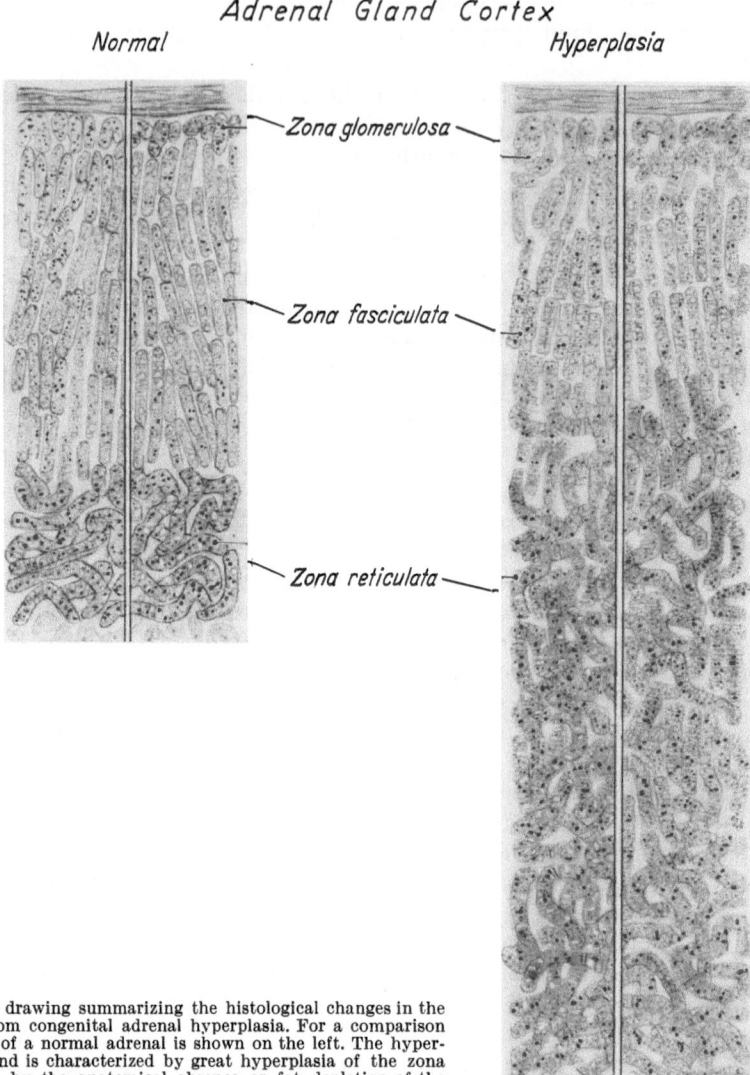

Adrenal Gland Cortex

Normal   Hyperplasia

Zona glomerulosa

Zona fasciculata

Zona reticulata

Fig. 35. A drawing summarizing the histological changes in the adrenal from congenital adrenal hyperplasia. For a comparison a drawing of a normal adrenal is shown on the left. The hyperplastic gland is characterized by great hyperplasia of the zona reticularis, by the anatomical absence or fat depletion of the zona fasciculata and by the difficulty of locating a zona glomerulosa. (From H. W. JONES, Jr. and G. E. S. JONES, 1954)

however, with the known endocrine aspects of the disease that it would note be unreasonable to suppose that it indeed accurately reflects the status in vivo.

An interesting exception to the lipid distribution just described occurred in the gland of the child who died within 18 hours of poisoning or of a fulminating infection after being under cortisone therapy for about 3 years. In this patient there was still recognizable reticular hyperplasia but to a much less degree than in untreated cases of corresponding severity. However, lipid was present in all cortical layers as demonstrated by Sudan Black B and Sudan III in contrast to the findings in untreated cases. However, there was much less lipid in the reticularis than in the untreated cases although there was still more than is normally seen especially by the Ashbel-Seligman technique.

One may summarize the adrenal changes by noting that the great adrenal enlargement described grossly seems to be due principally to a reticular hyperplasia and that this hyperplasia apparently becomes more marked with increasing age. In some instances the glomerulosa seems to participate in the hyperplasia, although in four glands of patients with fatal electrolyte depletion, the glomerulosa was absent. The fasciculata was greatly diminished in amount or entirely absent. Lipid studies showed absence of fascicular and glomerular lipid contrary to the normal situation while there was an abnormally strong lipid reaction in the reticularis in 5 of the 7 cases studied (Fig. 35).

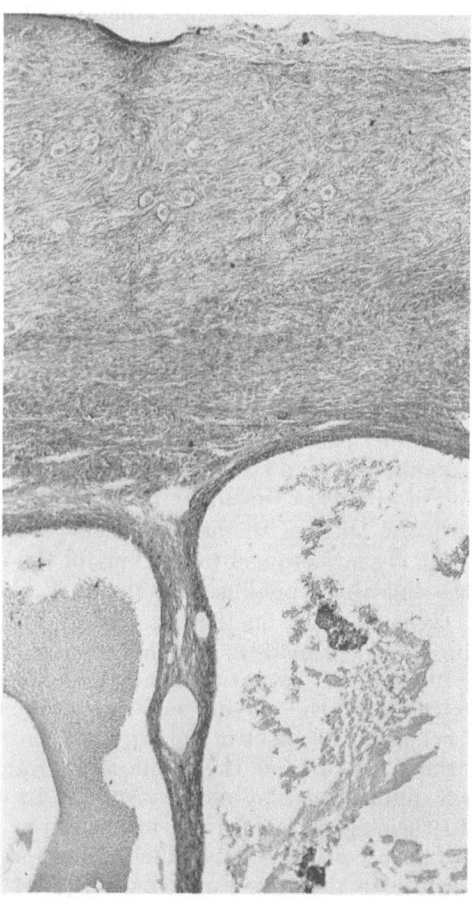

Fig. 36. Ovarian biopsy of a 19 year old untreated female hermaphrodite. The cortex contains numerous primordial follicles. Other follicles deep in the ovary have developed to the antrum stage. In no section of the ovary was there any evidence of ovulation (H & E ×35). (From: H. W. JONES, Jr. and G. E. S. JONES, 1954)

3. Ovary

In view of the absence of menstruation, the ovaries of these patients are of considerable interest. Ovarian tissue or sections were studied in 17 patients and reported by JONES and JONES in 1954. In infants the ovaries showed no recognizable change from normal. There were abundant primordial follicles, some of which had developed greater than 4 mm. in diameter. A few antrum follicles were seen. Atretic follicles were also noted. The ovarian stroma was normally sparse. The ovary in older untreated hermaphrodites became increasingly abnormal. In the teenage individuals there were also primordial, developing and antrum follicles but no sign of recent or previous ovulation (Fig. 36). The ovaries of 2 hermaphrodites aged 29 and 32 years were greatly abnormal, age considered. There were no primordial follicles and no developing atretic follicles. The ovarian cortex consisted entirely of stroma. In one of these ovaries there were a few structures suggesting very old corpora albicantia in spite of the absence of a history of any menstrual bleeding (Fig. 37). There was no sign of luteinization about the developing or atretic follicles in any case of marked hermaphroditism. In two patients showing a less severe degree of the abnormality, there was some luteinization. Primordial follicles could also be found. It is interesting that it was in these latter two cases that the adrenal sections contained some normal fasciculata.

The ovarian changes may be summarized by stating that in infants, children and teenagers there seemed to be normal follicular development to the antrum stage but no evidence of ovulation. As hermaphrodites aged there was less and less follicular activity and a disappearance of primordial follicles. It should be noted, however, that this disappearance must not be as complete as microscopically seems to be the case as cortisone therapy, even in adults, usually results in ovulatory menstruation after treatment of 4 to 6 months.

4. Developmental Anomalies of the Genital Tubercle and Urogenital Sinus Derivatives

The study of a relatively large number of cases of female hermaphroditism all due to congenital adrenal hyperplasia has indicated that the wolffian ducts atrophy in the usual manner. Furthermore, the müllerian derivatives are uniformly present but in an undeveloped state. There are, however, serious anomalies of the urogenital sinus derivatives including that part which unites with the genital tubercle to form the clitoris. In the past these anomalies have been described by urologists who are accustomed to thinking of the urethra as extending from the bladder to the end of the phallus. The vagina is, therefore, described as entering the urethra. Gynecologists, on the other hand, are used to considering the urethra as extending from the bladder for some 4 cm. to the vaginal vestibule. The circular muscule fibers along its entire length have an important sphincter action. Therefore, so think of the vagina entering this structure implies an important problem of urinary continence in the surgical correction of the anomaly. Past descriptions in the literature on the point of entrance of the vagina in relation to the urethral sphincter are by no means clear. In view of the fact that the lower part of both the vagina and female urethra arise from the urogenital sinus, there are embryologic considerations which underline the concern on this point. Normally, according to the review of GREENE (1944) the cranial portion of the urogenital sinus in the female gives rise to a portion of the bladder, the urethra, the paraurethral (Skene's) glands and Bartholin's gland. In the male, part of the bladder, the prostatic urethra, the prostate and Cowper's glands are derived from the cranial urogenital sinus. The caudal portion of the sinus in the female gives rise to the vaginal vestibule, the inner surface of the labia minora and the minor vestibular glands. In the male, it becomes the membranous urethra, the cavernous urethra and the para-urethral glands. The genital tubercle yields the clitoris in the female and the homologous structures of the male.

The description about to be made of the anomalies which arise from these structures is based upon a study of over 200 hermaphrodites with congenital adrenal hyperplasia and many cases of a less severe from of the anomaly in patients with primary amenorrhea who virilized at puberty. It should also be noted that women who at first glance appear normal may exhibit very mild forms of congenital adrenal hyperplasia and have minor anomalies of the external genitalia, such as an enlarged clitoris. The description is made in some detail because of the important surgical anatomical considerations involved in the reconstruction of the abnormal genitalia.

The phallus is composed of two lateral corpora cavernosa, but the corpus cavernosum urethrae is normally absent. The external urinary meatus is most often located at the base of the phallus (Fig. 30). However, an occasional case may be seen where the urethra does extend to the end of the clitoris (Fig. 31). The glans penis and the prepuce are present and indistinguishable from these

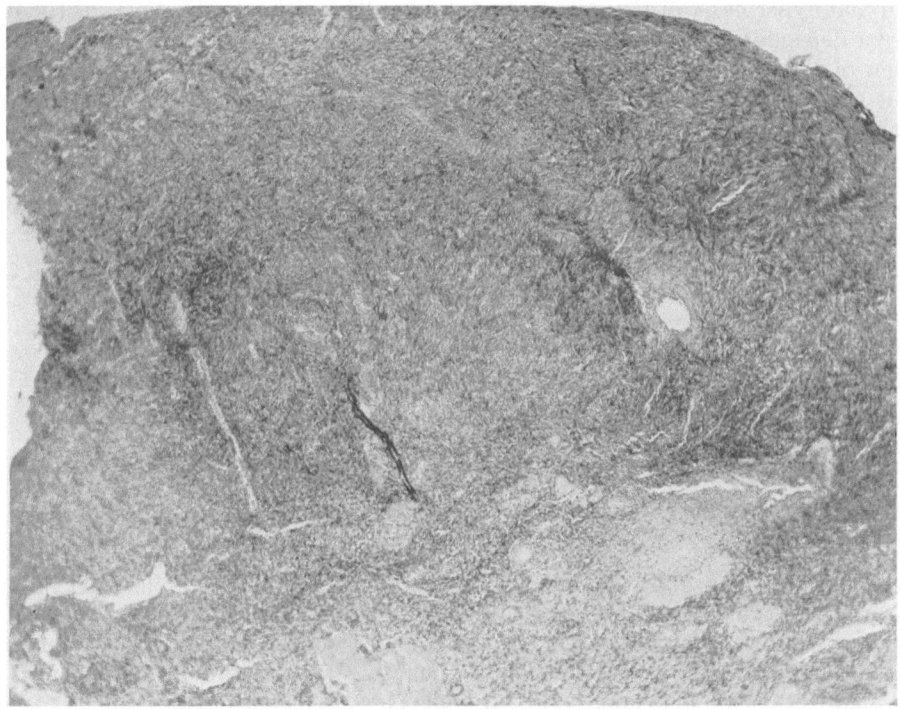

Fig. 37. Ovarian section of an untreated patient, age 29 years. The patient was a typical female hermaphrodite and examination of several sections of the ovary failed to discover any follicles (H & E × 40). (From: H. W. JONES, Jr. and G. E. S. JONES, 1954)

structures as developed in the male. The scrotolabial folds are characteristically fused in the midline giving a scroto-like appearance with a median perineal raphe although they seldom enlarge to normal scrotal size. No gonads are palpable within the scrotolabial folds. In circumstances where the anomaly is not so severe, as in patients with postnatal virilization, the fusion of the scrotolabial folds is not so complete and by gentle retraction it is often possible to locate not only the normally located external urinary meatus, but the orifice of the vagina.

An occasional case is encountered where no communication can be found between the urogenital sinus and the vagina. Careful endoscopic examination has shown that in no case has the vagina communicated with that portion of the urogenital sinus which gives rise to the female urethra in the case of a female or to the prostatic urethra in the case of the male. The vaginal communication was always in relation to the caudal urogenital sinus derivatives so that the sphincter mechanism fortunately is not involved and the anomalous communication is with that portion of the sinus yielding the vaginal vestibule in the female and the membranous urethra in the male. From the point of view of the gyneco-logist, it is much clearer to speak of the vagina and (female) urethra is entering a persistent urogenital sinus rather than to speak of the vagina as entering the (membranous male) urethra. It is to be noted that this conclusion casts some doubt on the embryologic significance of the prostatic utricle which is commonly stated to represent the homologue of the vagina in the normal male.

5. Hormonal Changes

Important and specific changes in the hormone milieu occur in congenital adrenal hyperplasia of the adrenogenital type. The ultimate diagnosis of the dis-order depends upon the demonstration of these abnormalities. From the practical view of diagnosis, the elevated urinary 17-ketosteroid value is the most important, but changes have been demonstrated in other important components as detailed below.

Table 8. *The 17-Ketosteroids Excretion of Untreated Patients with Adrenal Hyperplasia According to Age*

Age	No. of Patients	17-Ketosteroids Mg. in patients	Normal Value
0—6 mo.	19	1—10	0.5
7—12 mo.	3	3—10	0.5
1—5 yr.	28	4—30	1.0
6—9 yr.	9	11—40	2.0
10—15 yr.	6	16—50	10.0
15 yr.	12	21—80	15.0

From: BLIZZARD and WILKINS, A.M.A. Archives of Internal Medicine 100:731, 1957.

Urinary 17-Ketosteroids

The value of urinary neutral 17-ketosteroids is consistently elevated in this disorder. In evaluating a given result, consideration must be given to the age of the patient and collection of the specimen, as many of these patients are infants.

Normal values for adult females for urinary 17-ketosteroids vary from 5 to 12 mg/24 hrs. with an average of about 8 mg. Normal female children under 2 years of age normally excrete less than 1 mg. per day. From 2 to 8 years of age, values up to 2 mg. per day may be considered normal. From about 8 years to puberty there is a more or less linear rise to the adult level.

Values from 25 to 100 mg/24 hrs. for total neutral 17-ketosteroids are usually found in adults with virilizing adrenal hyperplasia. In infants up to 2 years of age any confirmed value over 1 mg. must be viewed as abnormal (Table 8).

The urinary 17-ketosteroids are apparently excretion products of both ovarian and adrenal androgens although in the normal female the principal fraction is adrenal in origin. BONGIOVANNI and CLAYTON (1954) made the interesting obser-vation that the 11-oxygenated neutral 17-ketosteroid fraction is relatively elevated with reference to the total 17-ketosteroid excretion in patients with adreno-cortical hyperplasia. As the 11-oxygenated group is of adrenal origin, this fits in with the anatomic changes in the adrenal.

EBERLEIN and BONGIOVANNI (1955) identified the elevated 17-ketosteroids as 11-ketoetiocholanolone and 11-hydroxyandrosterone by infrared analysis. It was reiterated that in normal individuals the 11-oxygenated compounds were found to constitute about 10 percent of the total neutral 17-ketosteroids whereas in patients with the adrenogenital syndrome 18 to 25 percent of the total were 11-oxygenated compounds.

Urinary Estrogens

The progressive virilization of female hermaphrodites due to adrenal hyperplasia would suggest that the estrogenic secretion of these patients was low. This opinion would seem to be further supported by the atrophic condition of both the ovarian follicular apparatus and the estrogen target organs. However, the determination of urinary estrogens both fluorometrically and biologically indicates that they are actually elevated. This must mean that the biologic effect of these estrogens is overridden by excessive androgen and that the estrogen is of adrenal origin. MIGEON and GARDNER (1952) determined the urinary estrogens in 11 female hermaphrodites by the fluorometric method of JAILER and found values up to 212 micrograms in 24 hrs. compared to the normal range of 25 to 50 mg/24 hrs. These elevated values were returned to normal after treatment with cortisone.

In addition to the elevation of estrogens measured fluorometrically, bioassay showed in one case a value equivalent to 4 μg. of estradiol. (Upper limit for normal for adults.) After treatment with cortisone, the bioassay value fell.

In further studies, MIGEON (1953) by counter current distribution and fluorometric measurement made the very important observation that there was a rather constant ratio between fractions resembling estrone, estradiol and estriol in patients with adrenal hyperplasia as compared with normal individuals. He furthermore showed that this ratio was unchanged when the total estrogen value was depressed by the administration of cortisone.

Urinary Pregnanetriol

BUTLER and MARRIAN (1937) first described and BONGIOVANNI (1953) greatly elaborated on the excretion of abnormally large amounts of 3-17-20 pregnanetriol in the urine of female hermaphrodites due to adrenal hyperplasia. This metabolite is not found in amounts over 1 to 2 mg. in normal adults. In severe adrenal cortical hyperplasia it may be found in amounts up to 50 mg/24 hrs. but in border line situations the value may approach normal. It may not be viewed as a "specific" metabolite for this condition as abnormally large amounts have been described in association with adrenal tumors and virilizing syndromes of the ovary.

One may speculate that the large quantity of pregnanetriol is the result of the metabolism of 17-hydroxyprogesterone to pregnanetriol. It has been postulated that the conversion of 17-hydroxyprogesterone to cortisone is one of the last steps in its biosynthesis. It may be assumed that the adrenal enzymatic defect in these patients results in the accumulation of large amounts of 17-hydroxyprogesterone which finds its way into the urine as pregnanetriol.

Blood 17-Hydroxycorticoids

ELY et al. (1953) using the method of NELSON and SAMUELS determined the blood 17-hydroxycorticoids in children with various diseases and found that in congenital adrenal hyperplasia consistently low values were obtained. In 5 patients levels of from 0 to 3 μg. percent were noted compared with normal values of from 4 to 10 μg. percent.

BONGIOVANNI et al. (1954) reported on the blood levels of 17-hydroxycorticoids in 9 patients with the adrenogenital syndrome. All but 2 of these revealed abnormally low levels, i.e. less than 1 per 100 ml. of plasma. These patients were

424 H. W. JONES:

stimulated by intravenous adrenocorticotropin (ACTH) and all but 2 failed to show the normal expected rise in the blood level of corticoids. Studies subsequently carried out with Bongiovanni's method for "conjugated" as well as for the "free" plasma 17-hydroxycorticosteroids indicated that these findings applied to both fractions.

Pituitary Fractions

With important changes in the steroid environment it would be expected that the pituitary fractions would show significant changes. Actually information on this point is sketchy although conclusive changes have been demonstrated.

SYDNOR et al. (1953) using the oxycellulose technique for the analysis of ACTH in blood, obtained a positive test in the blood of untreated children with the adrenogenital syndrome. The blood of normal children and cortisone treated pseudohermaphrodites did not give a positive reaction to this test.

The data on urinary gonadotrophins are scanty. MIGEON (1953) has presented information indicating that gonadotrophins are absent prior to therapy but appear after cortisone treatment.

Examination of ovaries of teenage female hermaphrodites (JONES and JONES, 1954) showed abundant follicular development to the antrum stage. However, there was no evidence of ovulation. This suggests that ICSH may be absent or low. We have been unable to demonstrate ICSH in the urine in 4 adult patients with congenital adrenal hyperplasia and are inclined to believe that this hormone is suppressed by the adrenal androgen.

6. Pathogenesis of Virilizing Adrenal Hyperplasia

With the pathologic and physiologic changes which have been demonstrated and discussed in the preceding sections, it is possible to formulate a hypothesis for the pathogenesis of adrenal hyperplasia. This has been attempted and discussed by a number of investigators. It was postulated by BARTTER et al. in 1951 that the primary disorder in the virilizing type of adrenal hyperplasia was an impairment of the production of Compound F (hydrocortisone). This fits in with the anatomic changes of the adrenal. HECHTER et al. (1951) on the basis of perfusion of beef adrenals suggested a step by step scheme for the conversion of cholesterol to Compound F (Fig. 38). This has been subject to some revision and debate in detail with regard to intermediate steps but the over-all concept is still valid. It has furthermore been pointed out that pregnanediol is an excretion product of progesterone and by analogy pregnanetriol might be expected to be an excretion product of 17-hydroxyprogesterone. In individuals with diseased adrenals the administrion of 17-hydroxyprogesterone will result in a great increase in the amount of excreted pregnanetriol. The fact that blood 17-hydroxycorticoids are low naturally suggests that there is a problem in the conversion of 17-hydroxyprogesterone into Compound F. Because of the difficulty in the synthesis of Compound F, there is, according to this theory, a compensatory increase in the secretion of pituitary ACTH. The increased ACTH stimulates to hyperplasia the only part of the adrenals capable of responding namely, the reticularis which is apparently the site of origin of the increased androgen and estrogen. It is furthermore postulated that the excessive amounts of androgen and estrogen which are produced by the adrenals inhibit the pituitary gonadotropin with the result that the ovary remains in a resting state (Figs. 39, 40). Although the evidence is not yet clear as to the exact pituitary hormones which are suppressed, there is suggestive evidence at the moment that FSH is probably inhibited as well as ICSH fraction. This is supported by the finding of WILKINS (1955) who showed

The Intersex States 425

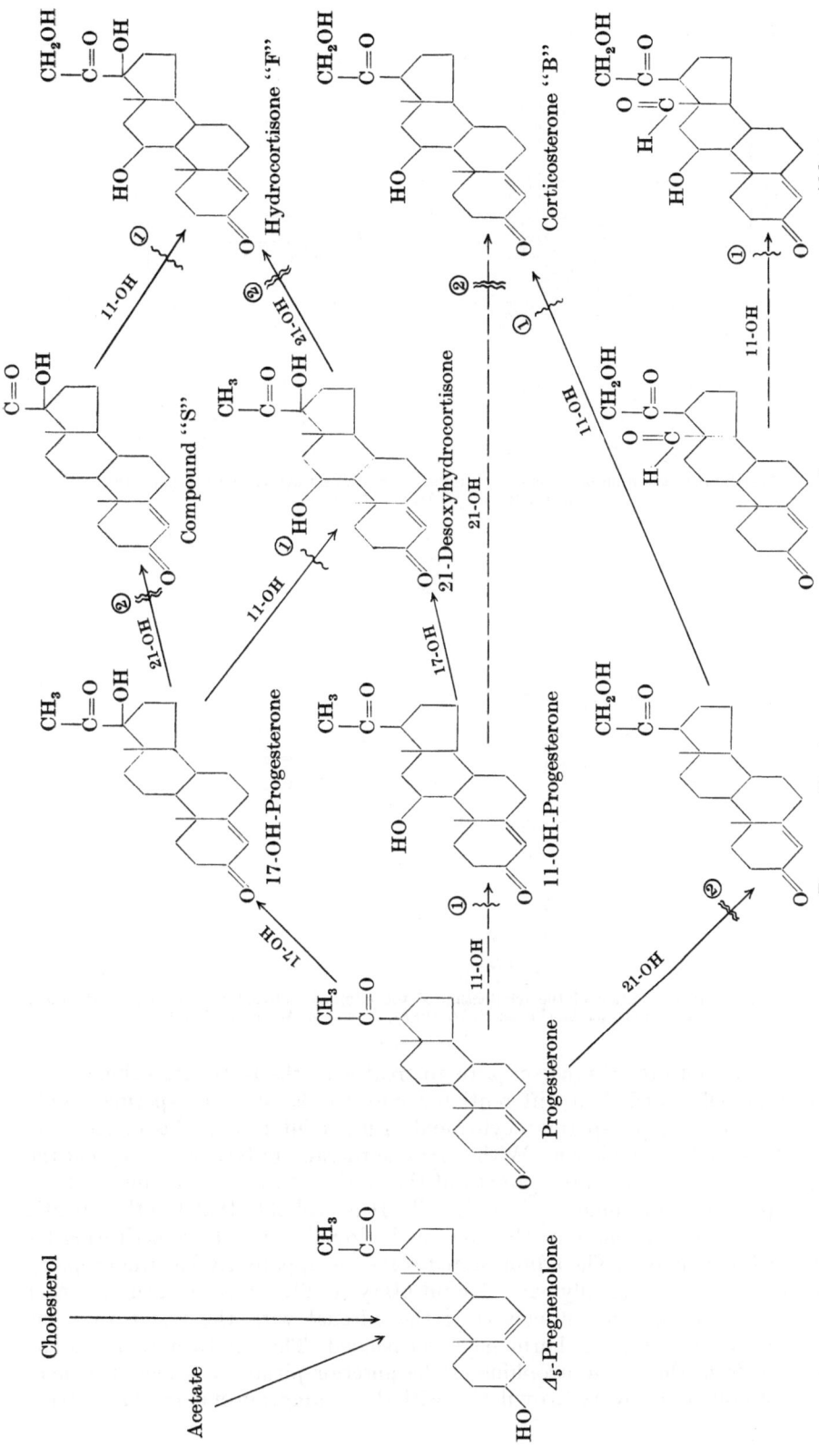

Fig. 38. This diagram shows a concept of the metabolic pathways of synthesis of 21-carbon adrenal steroids from acetate. (1) represents the block of hydroxylation of the 11-carbon atom which is present in the hypertensive type of virilizing adrenal hyperplasia; (2) represents the defective hydroxylation of the 21-carbon atom which is thought to be present in a majority of patients with this syndrome not of the hypertensive type. (From: R. M. Blizzard, G. Liddle, C. Migeon and L. Wilkins, 1959)

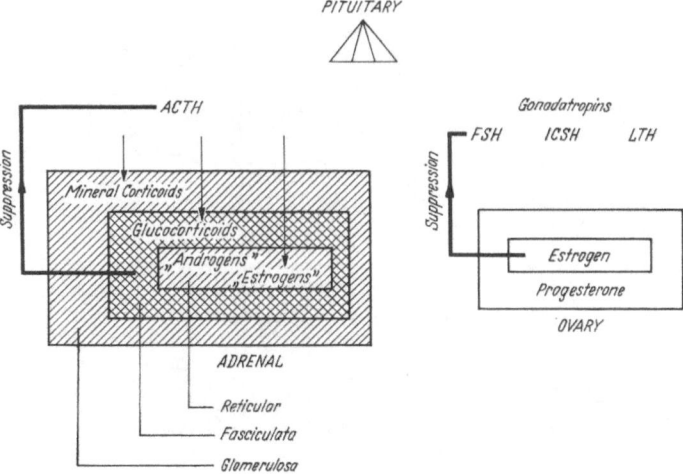

Fig. 39. A schematic representation of the normal pituitary, adrenal and ovarian relationships. (From: H. W. JONES, Jr. and W. W. SCOTT, 1958)

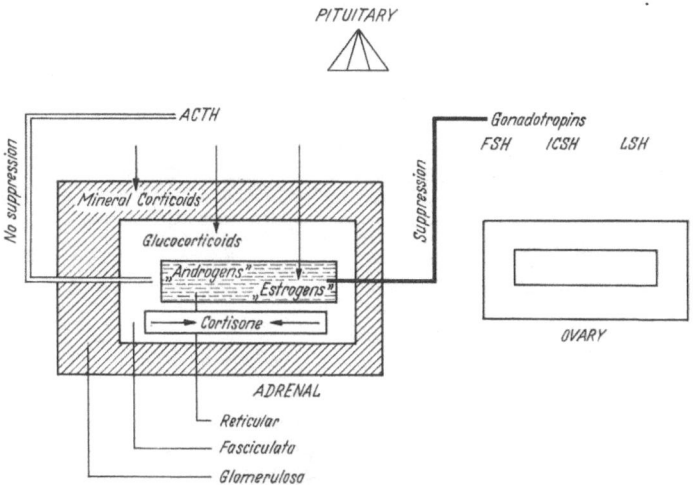

Fig. 40. A schematic representation of the relationship of the pituitary, adrenal and ovarian relationships in adrenal hyperplasia. (From: H. W. JONES, Jr. and W. W. SCOTT, 1958)

that in macrogenitosomia praecox prior to treatment the testicular tubules contained germ cells which had differentiated only to the stage of spermatogonia or rarely to the stage of spermatocytes and of great interest in this connection, no interstitial cells were present. With cortisone therapy in boys at or beyond the age of puberty there was rapid growth of the testis with active spermatogenesis and the appearance of numerous Leydig cells. It is probably that the therapeutic administration of cortisone puts the abnormal adrenal at rest by substituting for the absent Compound F. The administered cortisone apparently has the property of suppressing the abnormally excessive pituitary ACTH which in turn no longer excessively stimulates the reticularis of the adrenal with the result that the production of the virilizing hormone is suspended. This in turn removes the suppression from the gonadotrophins of the anterior pituitary which then may normally stimulate the ovary to ovulation with the resumption of normal menstrual cycles.

A report by BONGIOVANNI and EBERLEIN (1956) confirmed that most patients with adrenal hyperplasia have evidence of failure of the synthesis of Compound F (hydrocortisone) as indicated by low levels of plasma Porter-Silber chromogens, high urinary 17-ketosteroids and pregnanetriol. These findings may be interpreted to indicate a specific inability of the adrenal to hydroxylate the steroid nucleus at C-21. However, these investigators also reported the findings in an 8 year old female with marked virilization and hypertension. She was reared as a male. The urinary 17-ketosteroids were 20 to 47 mg/24 hrs. The pregnanetriol excretion of 6 to 7 mg/24 hrs., although elevated, was not as high as in other subjects. Surprisingly, the Porter-Silber chromogens were 24.7 to 26.2 mg. percent compared with a normal of about 7 to 9 mg. percent and compared with low Porter-Silber chromogens in most other cases of adrenal hyperplasia. Paper chromatography demonstrated that the substance responsible for the Porter-Silber chromogen reaction was not Compound F but Compound S which lacked an oxygen function at C-11. In addition the major metabolite in the urine of this subject was tetra-hydro S (pregnane-$3\alpha,17\alpha$-21α-triol-20 one). In contrast to the more usual defect described above this suggested an inability to hydroxylate at C-11 giving a metabolite which probably accounted for the hypertension. The entire picture including the hypertension was reversed by small doses of Compound F (hydrocortisone).

In discussing the pathogenesis of adrenal hyperplasia BONGIOVANNI and EBER-LEIN pointed out that crude extractions of urine of patients with adrenal hyperplasia contain normal or excess amounts of steroids which in the aggregate possess all the necessary molecular configuration of Compound F itself. Furthermore, Compound E (cortisone) can be metabolized by these patients in a relatively normal manner as evidenced by the appearance of the expected amounts of tetra-hydro E in the urine of treated patients. There is obviously a deficiency in the enzymes concerned with hydroxylation either at C-11 or C-21 but for the reasons just discussed the abnormal adrenal physiology is much more complicated than a simple deficiency of a single 11β-hydroxylase or a 21-hydroxylase.

Further study of other hypertension patients have confirmed the defect in C_{11}-hydroxylation leading to the formation of 11-desoxyhydrocortisone (Compound S) and desoxycorticosterone (DOC). The desoxycorticosterone is believed to be the cause of the hypertension.

Interestingly enough although hypertension patients are virilized the 17-ketosteroid excretion is at times not as elevated as might be expected when compared with patients with simple virilization.

There is no specific metabolic pattern in the salt losing form of the disorder. BLIZZARD et al. (1959) suggested that such children did not secrete normal amounts of aldosterone. It is postulated that there is a more complete defect of the 21-hydroxylation in the salt losers.

A few uniformly fatal cases of congenital adrenal hyperplasia have been found to have a 3β-hydroxysteroid dehydrogenase deficiency (BONGIOVANNI, A. M., 1962).

7. Diagnosis

The diagnosis of hermaphroditism due to congenital adrenal hyperplasia is to be suspected in any infant born with ambiguous or abnormal external genitalia. It is exceedingly important that the diagnosis be made at a very early age to prevent the undesirable disturbances of metabolism which may occur.

All patients with ambiguous external genitalia should have the benefit of the examination of the chromosomal characteristics by means of the buccal smear which is normally positive. In all instances of female hermaphroditism due to congenital hyperplasia the chromosomal arrangement has been entirely that of a normal female.

The critical determination is that of the urinary 17-ketosteroids. If these are elevated, the diagnosis must be either congenital adrenal hyperplasia or tumor. In the newborn, the latter is scarcely to be considered, for in Wilkins' survey of this question up to 1948, he found no tumors responsible for this syndrome at birth and so far as I am aware this is still true. However, in older children and in adults in the presence of elevated 17-ketosteroids the presence of tumor must be considered. One of the most satisfactory methods of making this differential diagnosis is to attempt to suppress the excess androgens by the administration of appropriate doses of dexamethasone. In an adult or an older child where the problem is likely to arise, a suitable dose of dexamethasone, preferably for 7 consecutive days is a satisfactory test procedure. In the presence of congenital adrenal hyperplasia there will be a marked suppression of the urinary 17-ketosteroids on the third to seventh days of the test to a value below 1 mg/24 hrs. while in the presence of tumor there will be either little effect or the 17-ketosteroids will rise.

A determination of the serum sodium, potassium and carbon dioxide combining power is also of importance to ascertain if serious electrolyte disturbances are present.

It should be mentioned at this point that it is no longer necessary to subject a hermaphrodite suspected of having congenital adrenal hyperplasia to a laparotomy for purposes of diagnosis of sex or to distinguish between tumor or hyperplasia. A female chromosomal sex arrangement plus the presence of elevated urinary 17-ketosteroids which are suppressed by dexamethasone are sufficient to establish the diagnosis.

8. Treatment

The treatment of female hermaphroditism due to congenital adrenal hyperplasia is partly endocrinologic and partly surgical. The endocrinological aspects are treated here and the surgical aspects subsequently in the section on therapy of intersexuality. In 1950 it was discovered that the administration of cortisone caused rapid diminution in the excretion of urinary 17-ketosteroids in congenital adrenal hyperplasia. This was followed by a demonstration that the virilizing manifestations of the disorder were suppressed and that the feminine sex characteristics developed. Although cortisone was the original steroid administered, it has been shown that various cortisone derivatives are equally effective, but are not necessarily any better. It is most satisfactory to begin treatment with relatively large doses of cortisone for 7 to 10 days to obtain rapid suppression of adrenal activity. In young infants the initial dose will approximate 25 mg. per day given intramuscularly and in older patients 100 mg. administered by the same route. After the output of ketosteroids has decreased to a low level the dose is reduced to the minimum amount required to maintain adequate suppression. This requires repeated measurements of the urinary 17-ketosteroids for the dosage determination must be on an individual basis. It seems to be somewhat easier to maintain uniform suppression with intramuscular rather than oral cortisone but because of the convenience, oral cortisone is frequently used. In the event that it is, it must be given daily. It is important to give it divided into a number of doses during the day whereas, if the intramuscular route is used, it may be administered every third or fourth day until suppression is obtained. The average amounts of cortisone for patients under 2 years of age for maintenance purposes varies from 8 to 12 mg. intramuscularly or 15 to 20 mg. orally. From the age of 2 to 6, the intramuscular dosage would be 16 to 25 mg. with an oral average of 25 to 50 mg. In patients over 6 years of age, it usually required an average of 25 mg. (i.m.) per day for maintenance and from 50 to 75 mg. per day by mouth.

It should be emphasized that it is of considerable importance that the ketosteroid output be suppressed to at least normal levels in order to obtain an adequate clinical response. This means that the ketosteroid output must be suppressed to 10 mg. or preferably even lower per day in adolescent patients and proportionately lower in infants. The observed effects under cortisone therapy depend upon the stage of the somatic development and virilization which the patient had attained before treatment was begun. If the diagnosis can be made at birth or before the age of 2 years and treatment started at this time, normal progress of growth and development takes place, puberty comes in the expected time, virilization is prevented and in general these patients enjoy a relatively normal existence. If treatment is begun after the age of 2, but before about 7 or 8, some sexual hair usually is developed already and bone age is usually considerably advanced with respect to chronologic age. The exhibition of cortisone usually stops further appearance of sex hair which, however, will seldom disappear; epiphyseal closure is prevented so that further growth is still possible. It is interesting that in some children in this age group, especially where the bone development is in the neighborhood of 11 years, there may be a gradual slow development of the nipples and breast, in effect, premature onset of puberty. In patients over the age of 7 or 8, bone development is usually far advanced and fusion either has occurred or is about to occur. Hair growth is already excessive, not only on the genital areas but also on the face, over the body and on the extremities. Treatment with cortisone checks the process of virilization and in most instances there is a gradual diminution in the noticeability of the hair, but with bone development well along there is usually rapid feminization including the onset of menstrual bleeding within a few weeks from the beginning of treatment. In patients who are above 20 years of age, the change in the hair is usually minimal although over a period of years, there is some thinning and softening of body hair as well as correction of any scalp baldness. There may be some delay in the onset of ovulatory bleeding. In one instance there was an interval of about 6 months before it appeared.

In the treatment of new born infants with congenital adrenal hyperplasia who have a defect of electrolyte regulation, it is usually necessary to administer additional sodium chloride in amounts of from 4 to 6 grams daily either orally or parenterally in addition to cortisone. Furthermore, desoxycorticosterone (DOCA) is usually required at the beginning. The dosage administered is entirely dependent upon the serum electrolytes which must be followed serially. The DOCA is conveniently supplied as pellets of 125 mg. each, but initial intramuscular therapy of 1 to 2 mg. per day is required. After regulation and on discharge of the patient 1 or 2 or rarely up to 5 pellets may be inserted, but after the initial insertion, additional DOCA pellets are seldom necessary, cortisone apparently being able to substitute for the loss of the salt regulating hormone.

The use of a 9α-fluorohydrocortisone in doses of 0.5—1.0 mg/day by mouth diminishes the use of DOCA as it has an important sodium retaining function. However, the tendency for hypertension to develop in patients so treated limits the therapeutic usefulness of the halogen substituted derivatives. In combination with cortisone 0.05—0.2 mg. 9α-fluorohydrocortisone can be used instead of DOCA to control sodium loss.

During acute illness and other stresses as well as during the operative period, additional cortisone is indicated. A doubling of the maintenance dose is usually adequate for this purpose.

In addition to the hormonal treatment of this disorder, surgical correction of the external genitalia is usually necessary. The surgical details of this will be described subsequently and will not be recorded here.

IX. Female Hermaphroditism without Progressive Masculinization

Female hermaphroditism due to congenital adrenal hyperplasia is a well known condition which has been described in the previous section. Fetal masculinization of the external genitalia with identical anatomical findings as in patients with congenital virilizing adrenal hyperplasia occurs also in females who do not have an adrenal abnormality. Unlike patients with the adrenogenital syndrome, these patients do not have elevated levels of urinary 17-ketosteroids, and as they grow older, they do not show precocious development, nor are they subject to metabolic difficulties as are sometimes associated with adrenal hyperplasia. At puberty, normal feminization with menstruation and ovulation occurs. The diagnosis of female hermaphroditism not due to adrenal abnormality depends upon the demonstration of a positive buccal smear and the finding of normal levels of 17-ketosteroids in the urine. With rather complete fusion of the scrotolabial folds, it is necessary to determine the exact relationship of the urogenital sinus to the urethra and vagina and to demonstrate the presence of a uterus by a rectal examination or observation of the cervix by endoscopy. When there is a high degree of masculinization, the differential diagnosis between this condition and true hermaphroditism may present an exceedingly troublesome situation which might require an exploratory laparotomy under some circumstances.

1. Classification

Patients with this problem may be seen under a variety of conditions as follows:
  I. Exogenous androgen
     a. maternal ingestion of androgen
     b. maternal androgenic tumor
 II. Idiopathic
III. Special or non-specific

2. Maternal Ingestion of Androgen

In 1958 it was first noted that masculinization of the female external genitalia occurred in patients whose mothers had been treated because of habitual or threatened abortion with certain steroid compounds (WILKINS, JONES, HOLMAN and STEMPFEL, 1958) (Figs. 41, 42, 43). Since the original report in 1958, several hundred instances of abnormality due to this cause have been reported. One of the largest series reported was that of WILKINS, who in 1960 was able to collect 91 patients who had had masculinization from mothers who had received various kinds of steroids (WILKINS, L., 1960). Although the information gathered in the 1960 paper is probably not contemporary with respect to the frequency of various drugs causing this abnormality, the difficulties reported were as follows. Ethinyl testosterone was taken by the mothers of 35 patients, the dose varying from 20 to 200 mg/day. This is sold under the trade name of Progestoral, Pranon, Lutocylol and others. Norlutin was involved in 35 instances, daily doses were from 10 to 40 mg. Enovid was noted once with a daily dose of 10 mg.; testosterone and other

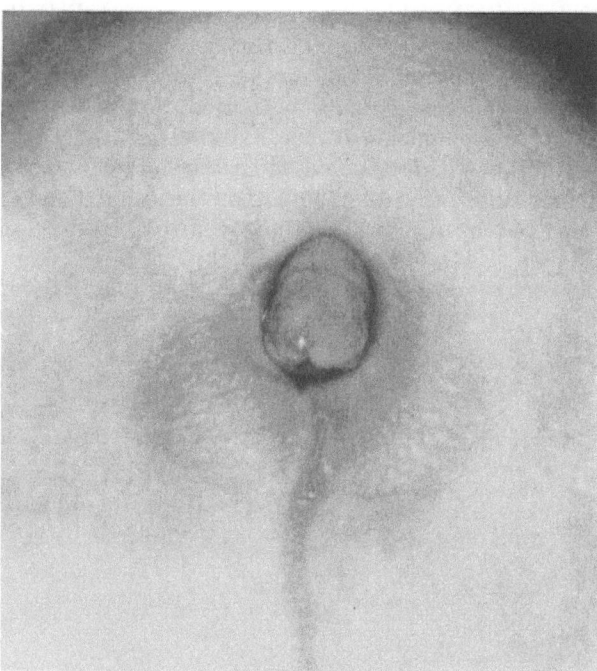

Fig. 41. Photograph of the external genitalia of an infant whose mother received ethinyl testosterone during the first few weeks of pregnancy. There is partial fusion of the scrotolabial folds and enlargement of the clitoris

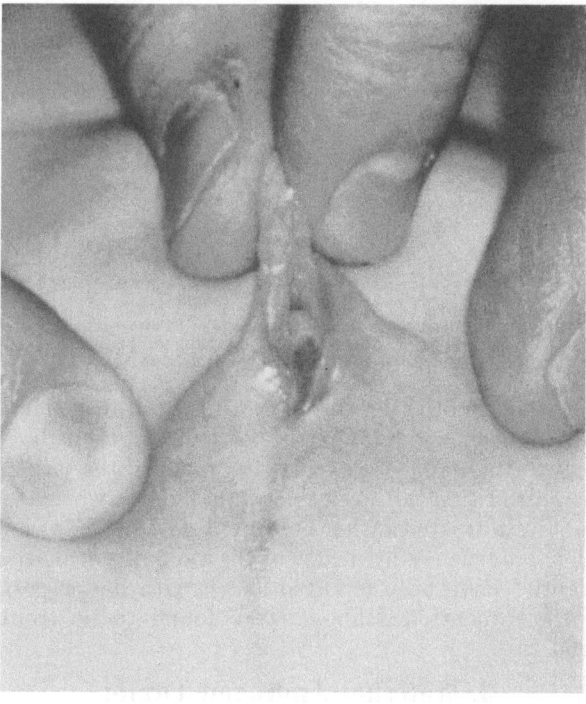

Fig. 42. Photograph of the same child shown in Fig. 41. As is characteristic in some instances, there is a good bit of fleshy enlargement when the masculinization of the external genitalia is due to maternal ingestion of androgenic steroids. (From: H. W. JONES, Jr. and W. W. SCOTT, 1958)

androgens were indicated 15 times in doses of 3 to 10 mg. daily. Two patients were observed who had received nothing but progesterone and diethylstilbesterol was noted in 4 instances. One of the puzzling things about this condition is its prevalence with respect to the number of mothers who take this drug. There have been very few efforts to express this in statistical terms. Ishizuka et al. (1962) reported 20 cases of masculinizing of the external genitalia among 888 female newborns whose mothers had received progestogens for at least 2 weeks with an incidence of masculinization of 2.25%. On the other hand, a much higher incidence was reported by Jacobson (1961) who observed 15 cases out of 81

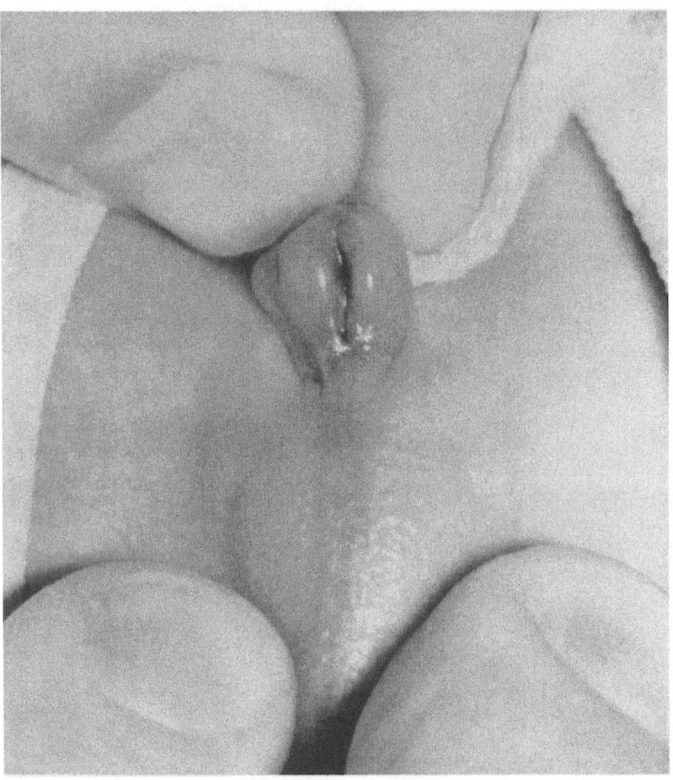

Fig. 43. Photograph of external genitalia of an infant whose mother received large doses of ethinyl testosterone early in pregnancy. There is complete fusion of the scrotolabial folds, the phallus is not tremendously enlarged but there is a complete penile urethra

females delivered whose mothers received Norlutin for an incidence of 18%. The reason for the difference in the 2 figures is not entirely clear, but probably arises from the fact that in the Japanese experience many of the patients were treated only for short periods of time with relatively small doses whereas in the Jacobson series the patients were treated rather continuously after it was determined they had a progestational deficiency by means of the vaginal smear. It should be noted that no abnormalities have been reported as yet with the progestational steroids marketed since those mentioned above were found to be troublesome in this regard about 1960.

3. Maternal Androgen Tumor

A very few patients have had masculinization of the external genitalia from virilizing ovarian tumors in the mother. Brentnall (1945) reported an often quoted case whose mother had an arrhenoblastoma which was removed post partum. A unique feature of this situation is the subsequent birth of a perfectly normal female child. JAVERT and FINN (1951) reported 2 female abortuses which were hermaphroditic and whose mothers had arrhenoblastomas. In 1964 when visiting in Lima, Peru, Dr. CARLOS MUNOZ showed me a child with an enlarged clitoris and fusion of the scrotolabial folds whose mother had had a tumor removed at the time of cesarean section. There also was an opportunity to view a microscopic section of the tumor which appeared to be that of an adrenal tumor of the ovary.

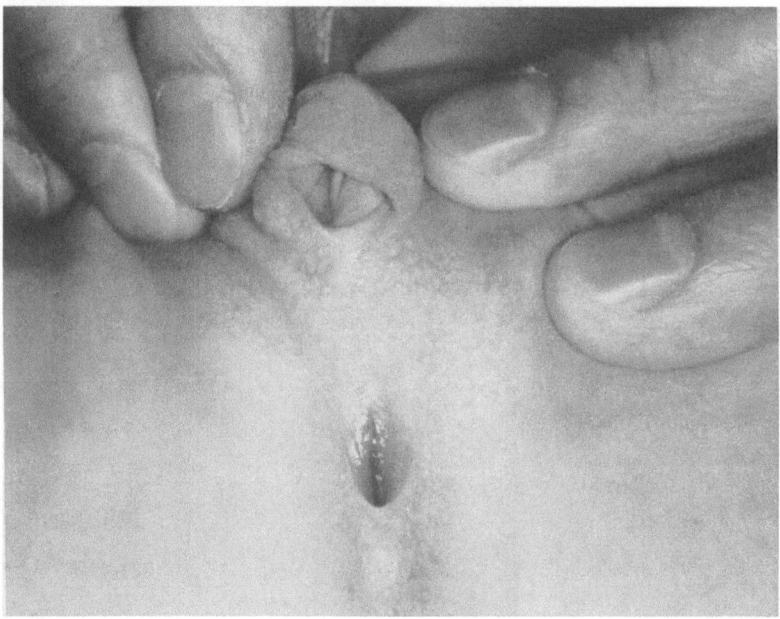

Fig. 44. Photograph of external genitalia of a 13 year old girl who had multiple deformities. The opening in the perineum formed by fusion of the scrotolabial folds leads to the vagina and the female urethral orifice. The apparent opening in the phallus which in the photograph appears to be an orifice is a blind pouch that leads down the phallus only for a distance of about a centimeter. Exploratory laparotomy revealed normal uterus, tubes and ovaries

4. Idiopathic

Perhaps a dozen patients who are otherwise normal females except for masculinized genitalia have been observed without an apparent source for the androgenic influence. Under this circumstance the differential diagnosis from true hermaphroditism is an exceedingly difficult point, and in the event of marked masculinization of the external genitalia, some consideration for an exploratory laparotomy must be entertained. Interestingly enough, in some instances the mothers of these patients have shown signs of temporary and often severe masculinization during pregnancy. In one such patient under our care very severe acne, deepening of the voice and hirsutism took place. All of these symptoms gradually cleared post partum. Temporary virilization with pregnancy sometimes to a very marked degree has been noted by other authors. FRIEDMAN et al. (1955) reported a patient in whom hirsutism, acne, deepening of the voice and enlargement of the nose occurred during pregnancy. Signs of virilization subsided a few months after the termination of the pregnancy. With the second pregnancy, some 15 months later, there was a recurrence of the signs and four days before labor the excretion of urinary 17-ketosteroids was 273 mg/24 hrs. Five weeks following delivery, the level had fallen to 13.6 mg. There was an enlargement of the ovary which, at laparotomy post partum, proved to be a dermoid cyst with a peculiar collection of lipoid cells but with no obvious androgenic tumor. Unfortunately in dealing with children born of mothers who might show some temporary virilization it is only in retrospect that the desirability of a study during pregnancy is apparent.

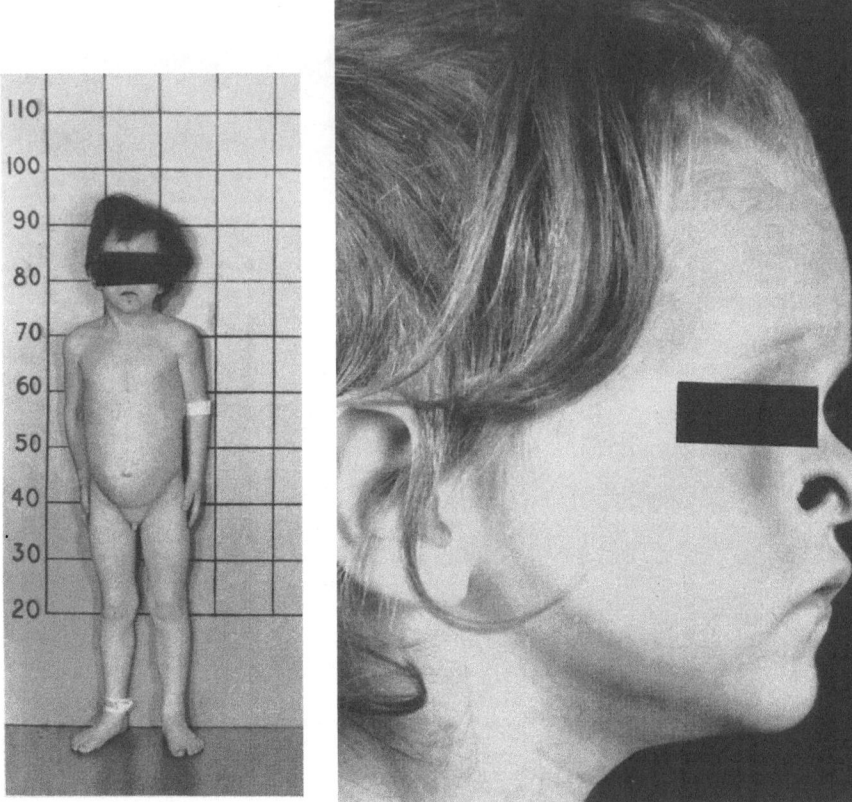

Fig. 45   Fig. 46

Fig. 45. Photograph of the same patient whose genitalia are shown in Fig. 44. At the age of 13 years, she is much shorter than normal

Fig. 46. Photograph of a nasal deformity of the same patient shown in Fig. 45. This photograph was taken when the patient was 7 years of age

5. Special or Non-Specific Female Hermaphroditism

In addition to the patients covered by the categories mentioned above, there remain a few individuals who have a positive sex chromatin, who have ovaries, who are hermaphroditic by virtue of masculinization of the external genitalia, but who also have a variety of other congenital anomalies. From a pathogenetic point of view, these patients are not entirely explained by an aberration of sexual development, but are apparently affected by a much more wide spread disorder during embryogenesis. JONES and SCOTT (1958) were able to collect 5 such patients and referred to them as "special" female hermaphrodites. CARPENTIER and POTTER (1959) studied in detail the generative tract of 48 stillborns all of whom had bilateral renal agenesis. They found among the females, as judged by positive sex chromatin, no example of an entirely normal genital tract. In addition to the hermaphroditic genitalia there were other anomalies (Figs. 44, 45, 46), such as anorectal malformation and developmental anomalies of the extremities. From a clinical view, these extreme cases are seldom encountered. However, some genetic female patients with diverse malformations with masculinization of the external genitalia or other anomalies of the generative tract would fall into this category. The etiology of this group is entirely unknown.

X. Male Hermaphroditism
1. General Considerations

Individuals with abnormal ectopic testes may have external genitalia so ambiguous at birth that the true sex is not recognizable. At puberty these individuals tend to masculinize or to feminize depending upon factors which will be discussed below. Thus, the adult habitus of these individuals may develop into a typical android form without breasts or to a normal female form with excellent breast development. In some instances the external genitalia may be indistinguishable from those of a normal female. In other patients the clirotis may be quite enlarged and in still other instances there is fusion of the labia in the midline resulting in what seems to be a hypospadic male. A vagina of varying depths may be present. A cervix, uterus and tubes may be developed to varying degrees although müllerian structures are most often absent. Mesonephric structures may be grossly visible or identifiable by microscopic examination of the appropriate tissue. Body hair may be of typical feminine distribution or masculine and of sufficient quantity to require plucking or shaving if the individual is reared as a female. In a special group the axillary and pubic hair is congenitally absent. In spite of a well developed uterus in some instances all cases so far reported have been amenorrheic in spite of the interesting theoretical possibility of uterine bleeding from endometrium stimulated by estrogen of testicular origin. There is no evidence of adrenal malfunction. In the feminized group and less frequently in the nonfeminized group there is a strong familial history. Male hermaphrodites reared as girls and women may be quite successfully married and have been well adjusted with their sex partner. Others, however, especially when there has been equivocation as to the best sex of rearing as infants, have as adults been less than attractive women and have been the unfortunate victims of indecisive therapy. Psychiatric studies have indicated that if the aim of therapy be an emotionally well-adjusted person, best results are obtained by directing endocrine, surgical and psychiatric measures toward improving the basic characteristics of the individual. Fortunately this point coincides with the surgical and endocrine possibilities in those reared as females, for current surgical techniques can produce with relatively greater ease more satisfactory feminine than masculine external genitalia. Furthermore the testes of male hermaphrodites are non-functional as far as spermatogenesis is concerned. The practicalities of the application of these problems and principles were brought out in an analysis by WILKINS of 47 male hermaphrodites with respect to the type of external genitalia and the effect of this on the sex of rearing (WILKINS, 1959). Among 47 patients 12 had feminine external genitalia, 17 ambiguous genitalia and 18 genitalia which simulated males. On this basis, only the 18 patients whose genitalia simulated the male were considered suitable candidates in infancy for male rearing. Therefore, 29, the sum of the ambiguous and feminine group were considered best reared as females. Thus, roughly only about $1/3$ of all male hermaphrodites are suitable for male rearing.

28*

2. Classification

All individuals under discussion in this section have unilateral or bilateral testicular tissue. They also have external genitalia which are female or ambiguous as to sex, they may develop feminine characteristics or have well developed müllerian ducts.

The morphology of the external genitalia, the endocrine potentiality of the patient as expressed by breast development, and the degree of development of the müllerian ducts may be used as convenient criteria of classification. With these facets in mind the following clinico-developmental classification of male hermaphrodites may be made:

A. Male hermaphrodites with ambiguous external genitalia who masculinize at puberty
  1. with rudimentary or no uterine (müllerian) structures with bilateral testes.
  2. with well formed uterine (müllerian) structures,
    a) with bilateral testes,
    b) with a single testis (asymmetrical gonadal differentiation).

B. Male hermaphrodites with female external genitalia who feminize at puberty.
  1. with rudimentary or no uterine (müllerian) structures.

3. Masculinizing Male Hermaphrodites

The sex of rearing preferred for this group of patients depends upon the judgment of all concerned at the time of birth. As has been noted above, two thirds of all such individuals are probably best reared as girls. It is to be hoped that with increased awareness of this problem more adequate consideration may be given to the proper selection of the sex of rearing. It should be emphasized that this does not imply that the sex of rearing will necessarily correspond to the chromatin or gonadal sex, but the recognition of the sex identification as a problem will allow the formulation of a suitable therapeutic plan in the neonatal period.

By definition, patients in this group tend to masculinize at puberty. This is manifested in a negative way in the failure of breast development or other development which is a feminine secondary sex charasteristic. In a positive way masculinization is indicated by hirsutism, lowering of the voice and increased size of phallus. In order to prevent these undesirable phenomena in patients reared as girls, it is always necessary to predict before puberty the expected type of secondary sexual characteristics — masculine or feminine — which will develop.

As a rule of thumb, it has been found that male hermaphrodites with entirely feminine external genitalia will feminize at puberty. However, if the genitalia are at all ambiguous such as exhibiting an enlarged clitoris, the probabilities are that the patient will virilize (Figs. 47, 48, 49).

Inguinal hernias are not uncommon and in several reported instances the mistake in sex identification was discovered during herniotomy.

The group with well developed müllerian structures (Group A, 2a) has no special clinical or endocrinological findings which separate it from the group of male hermaphrodites without müllerian structures (Group A, 1). However, there are some chromosomal differences as will be noted below. The presence of the müllerian structures serves as a convenient point of classification and has great implications with regard to therapy as a uterus provides the possibility of uterine bleeding with the administration of estrogen and from a surgical point of view, the presence of the uterus and the vagina makes the reconstruction of the external genitalia technically much easier.

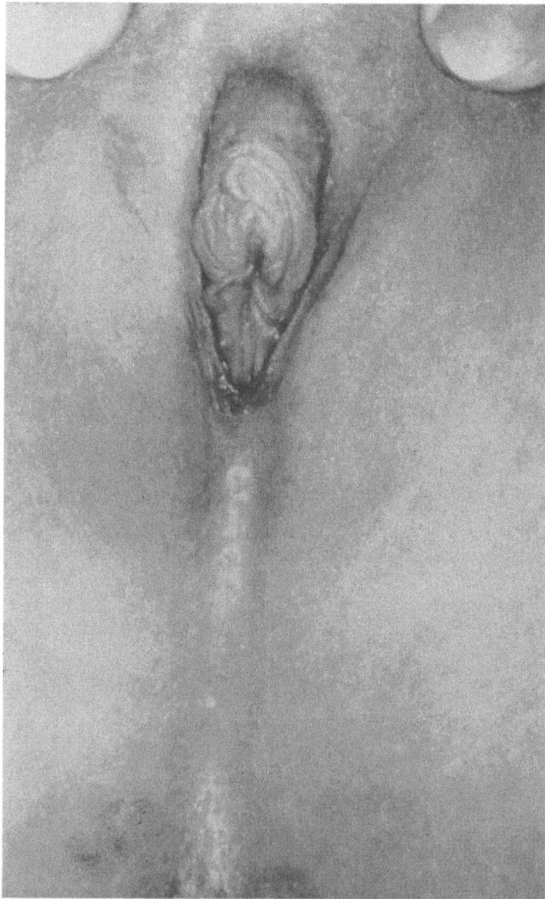

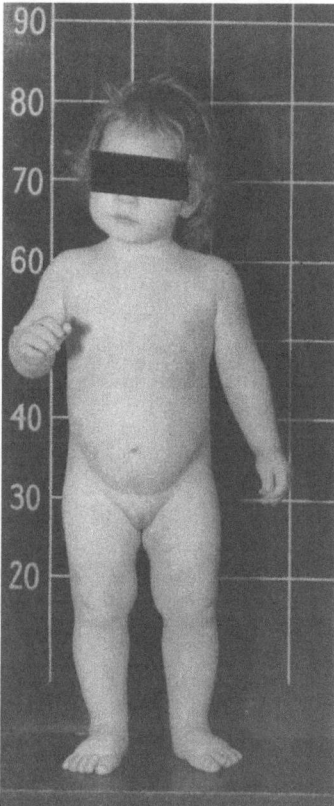

Fig. 47   Fig. 48

Fig. 47. Photograph of external genitalia of male hermaphrodite 3 years of age. The clitoris is enlarged; there is fusion of the scrotolabial folds; a single external orifice leads to a vagina and urethral orifice but this particular patient had no uterus. The urethra entered the urogenital sinus at the usual place for a female urethra

Fig. 48. Photograph of a 2 year old child with male hermaphroditism who was reared as a girl from birth

4. Asymmetrical Gonadal Differentiation

Asymmetrical gonadal differentiation (Group A, 2b) is an anomaly in which the patient has a unilateral testis and an indifferent contralateral gonadal streak. SOHVAL (1963) deserves credit for pointing out the homogeneity of this group and tentatively suggested the designation "mixed gonadal dysgenesis." We prefer to retain the designation "asymmetrical" to emphasize the asymmetrical disparity of development which characterizes the condition. Furthermore we have preferred to classify this group of patients under the heading of male hermaphrodites as opposed to a subheading under gonadal dysgenesis or agenesis in order to preserve the Klebs concept whereby the classification depends upon the histological character of the gonad. Regardless of the merit or demerits of histologic examination as a basis of classification, it is generally accepted to consider such patients under any other category would introduce a troublesome exception to this concept. Furthermore these patients fit well into this category from a view

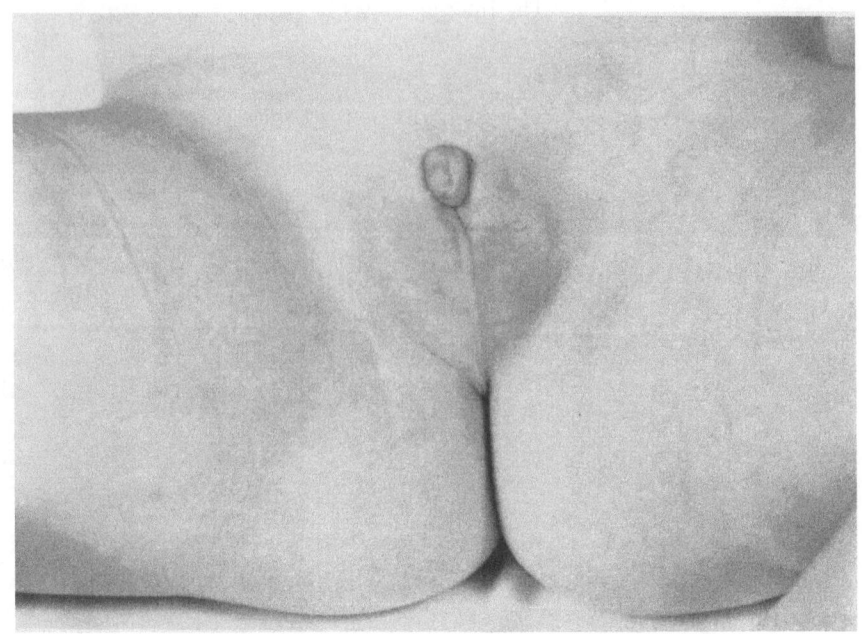

Fig. 49. Photograph of external genitalia of same child shown in Fig. 48

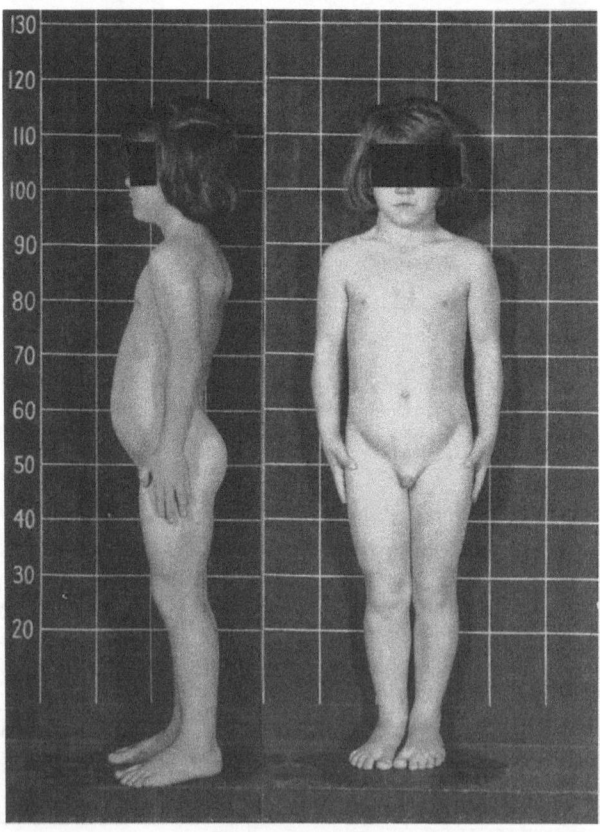

Fig. 50. Photograph, anterior and lateral, at the age of 8 years, of patient with asymmetrical gonadal differens-tiation. Note the short stature, the shield-like chest, the broad neck and the large phallus. (From: P.A. ZOURLAS-and H. W. JONES, Jr., 1965)

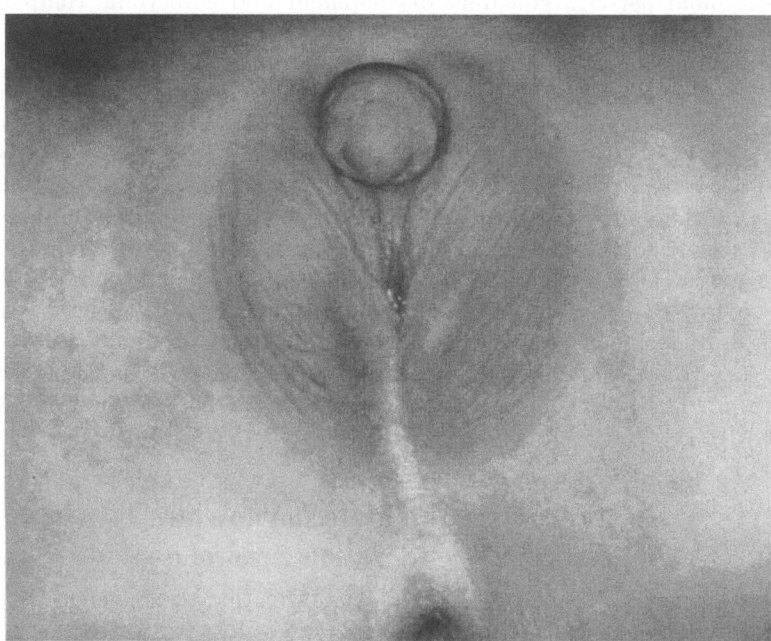

Fig. 51. External genitalia of same patient seen in Fig. 50. Phallus is enlarged and there is fusion of the scroto-labial folds with a single external meatus which leads both to the vagina and to the urethral meatus

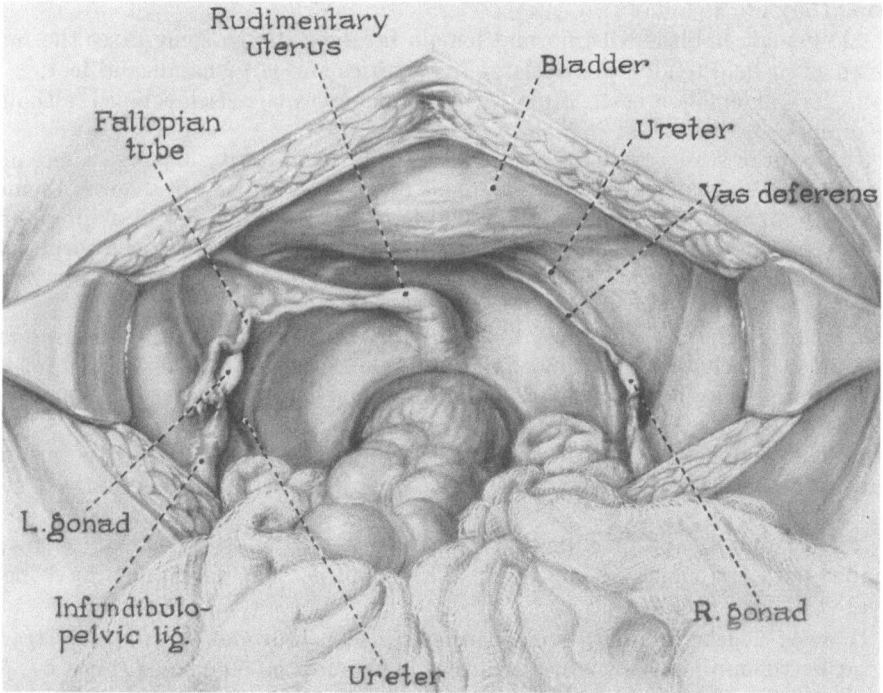

Fig. 52. Laparotomy findings in same patient shown in Fig. 50. Note the unicornuate uterus, with left tube and round ligament and no growth of the müllerian ducts on the right side. The right gonad was histologically an immature testis and the left gonad was composed only of fibrous tissue. (From: H. W. JONES, Jr. and W. W. SCOTT, 1958)

of chromosomal pattern, gonaduct development and embryonic competence in the testis.

To recapitulate and to clarify this syndrome under discussion, it may be said that it is characterized by patients who have unilateral testicular development with the opposite gonad represented by a gonadal streak or by no discoverable gonad at all (Figs. 50, 51, 52). The somatic features in this group of patients are essentially the same as with other male hermaphrodites except possibly for some increase in the frequency of short stature. Zourlas and Jones (1965b) reviewing 23 such patients found information on height available in 14 instances. In 11 of these 14 patients the height did not exceed 4'11" (150 cm.). A few additional patients exhibited some of the other somatic anomalies of Turner's syndrome although in no instance was there any mention made of the webbed neck.

It is easy to see that the short stature and the unilateral streak suggests that such cases might be considered as transitional between male hermaphrodites and patients with gonadal agenesis.

5. Feminizing Male Hermaphrodites
(Testicular Feminization Syndrome)

The most arresting syndrome in intersexuality is that of the genetic male with testes who resembles a normal appearing woman with excellent breast development. A large number of such cases have now been described. The syndrome was crystallized by Morris (1952) who was able to collect 76 cases and report 2 new ones. Since that time, three or four times as many cases have been reported. The description by Morris of the characteristics of the syndrome cannot be improved upon. They are as follows:

"1. Female habitus with normal female fat deposits. In some cases the build has an eunuchoid tendency with large extremities and large hands and feet.

2. Normal female breasts, often with a tendency to be overdeveloped, although the nipples are sometimes juvenile.

3. Absent or scanty axillary and pubic hair in the majority of cases. There may be a slight amount of vulvar hair. The hair on the head is that of a normal female without temporal recession, but the facial hair is more often absent as in a child.

4. Feminine external genitalia. The labia may be underdeveloped, especially the labia minora. The clitoris is normal or small. The vagina ends blindly, but is usually adequate for marital relations (Figs. 53, 54).

5. There is absence of internal genitalia except for rudimentary uterine and other anlage, including sometimes fallopian tubes or spermatic ducts, and for the gonads, which may be intra-abdominal or may lie along the course of the inguinal canal (Fig. 55).

6. The gonads consist largely of seminiferous tubules, usually without spermatogenesis, but in most cases with a marked increase in interstitial cells (Fig. 56).

7. Hormone assays in a limited number of cases suggest that these testes produce both estrogen and androgens. The pituitary gonadotrophins have been elevated in some instances."

There is a clear familial predisposition to this syndrome. It is always transmitted by the mother, never by the father. There is a 50-50 chance (Figs. 57, 58) that the males of such a family will be affected with this disorder.

Endocrine studies have been interesting if not clarifying. The 17-ketosteroid excretion above the age of puberty has been within the normal female range for

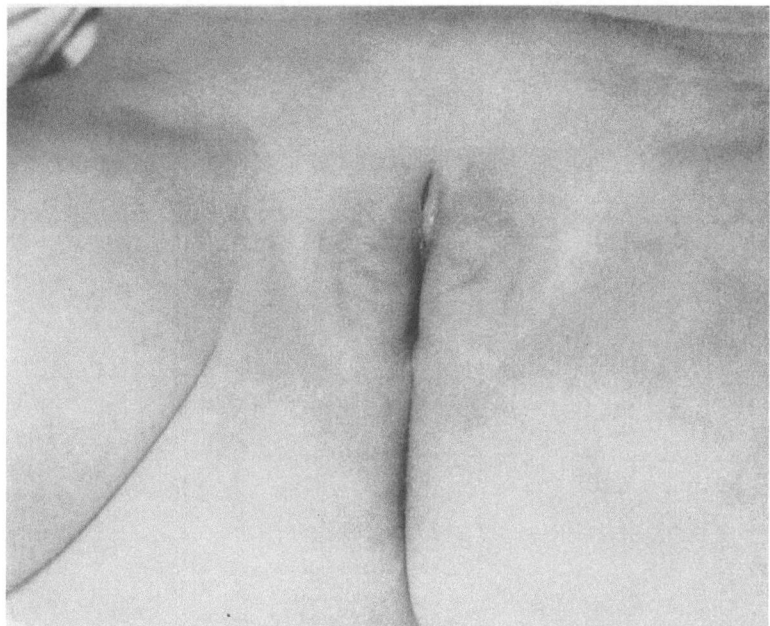

Fig. 53. Photograph of external genitalia of an 18 month old child with testicular feminization

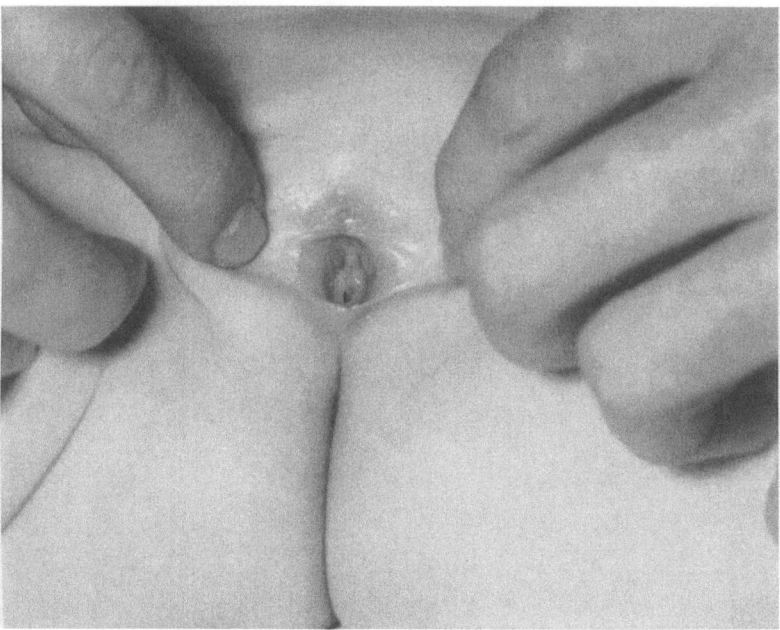

Fig. 54. Photograph of the same child shown in Fig. 53. The labia are retracted to show the normal vaginal
opening and the normal urethral opening. The vagina extended for a depth of 5 centimeters

the most part, but in some instances have been elevated. The fractionation of the
17-ketosteroids has indicated no peculiar distribution of the specific metabolites.

Biologic estrogen values and chemical estrogen values have been well within
the normal female range.

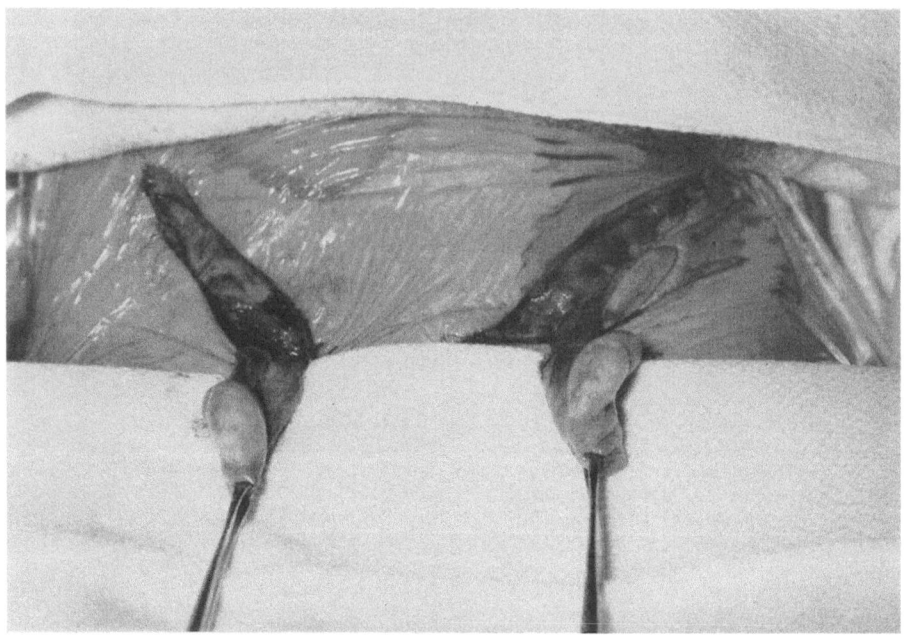

Fig. 55. Photograph of the testes in the same patient shown in Fig. 53. There were bilateral inguinal hernias which prompted the operation at this age

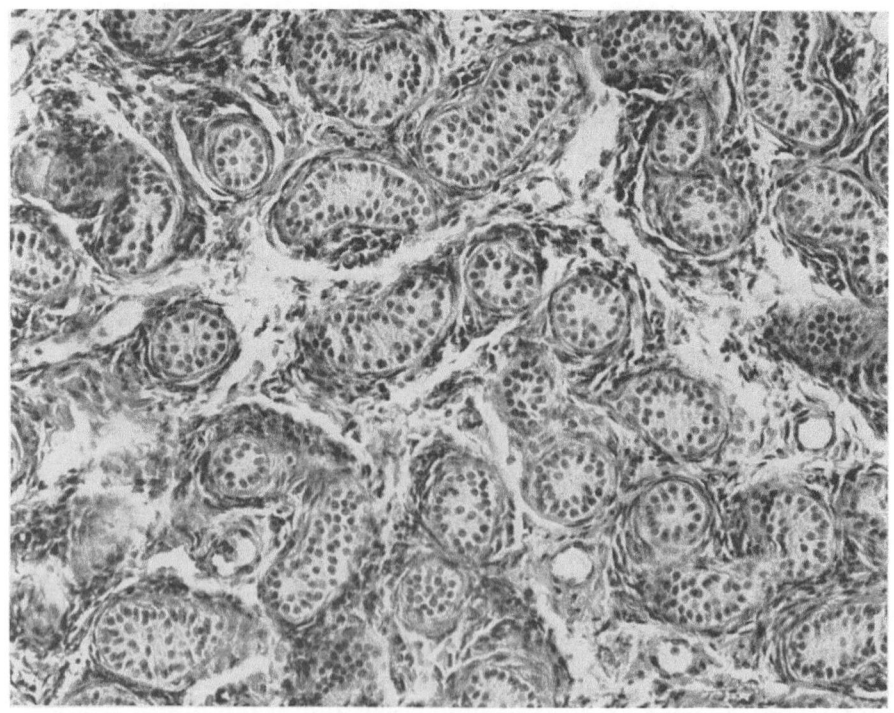

Fig. 56. Photomicrograph of the testes from the patient shown in Fig. 55. Seminiferous tubules are composed of a single type of cell. There is no evidence of cells of the spermatic series. Interstitial cells are much more prominent when the testes are removed above the age of puberty

The total gonadotrophins have been normal or in some cases elevated, a finding which is difficult to understand.

Following removal of the gonads there is a fall in the 17-ketosteroid excretion and the estrogen excretion and a rise in the total gonadotrophin value. This indicates that the testis is the site of at least some of the steroids which are keeping the gonadotrophins suppressed.

The histopathology of the gonads is of special interest and at one time it seemed likely that there was possibly a specific picture in such patients. However, at the present time it can be said that there is no special characteristic of the testes which would identify them as being associated with the testicular feminization

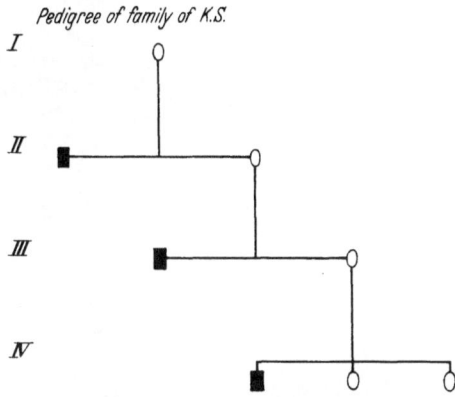

Fig. 57. Pedigree of the same patient in Fig. 54. This particular patient is the affected member in generation IV. Patients in generations II and III had the testes removed at about the age of 20 years

syndrome. For the most part the interstitial cells are well developed and very prominent above the age of puberty. The seminiferous tubules are quite healthy appearing with many cells of one kind (Sertoli). Germ cells are certainly not prominent and in many areas are completely absent. There is usually some hyalinization of the basement membrane of the tubules in adulthood. The infantile testis, of course, shows no development of the interstitial cells.

From the psychosexual point of view such patients are entirely feminine as they are reared as women, and there is usually no concern about sex identification at birth. The diagnosis is seldom made in infancy unless there is a familial history and a buccal smear of the female children is examined.

The pathogenesis of the syndrome is clearly understood, but MORRIS and MAHESH (1963) presented a considerable amount of data to reinforce the suggestion previously made by WILKINS that the syndrome may be the result of the failure of end organs to respond to normal androgenic stimuli. The principal evidence for this is that 1) no deficiency can be detected in the urinary excretion of androgenic metabolites of the testes and 2) there is failure of patients to masculinize when given exogenous androgens. Thus the syndrome can be very nicely explained in almost all its aspects on the assumption that there is some enzyme or series of enzymes missing which have to do with the biologic action of androgens with the result that androgens are unable to exert their action in embryonic or extrauterine life.

6. Sex Chromosomes in Male Hermaphroditism

The sex chromatin of all male hermaphrodites is negative. The relationship of the sex chromosomes to the various subdivisions of male hermaphroditism has been the subject of a series of papers by JONES and ZOURLAS (1965, 1965a, 1965b).

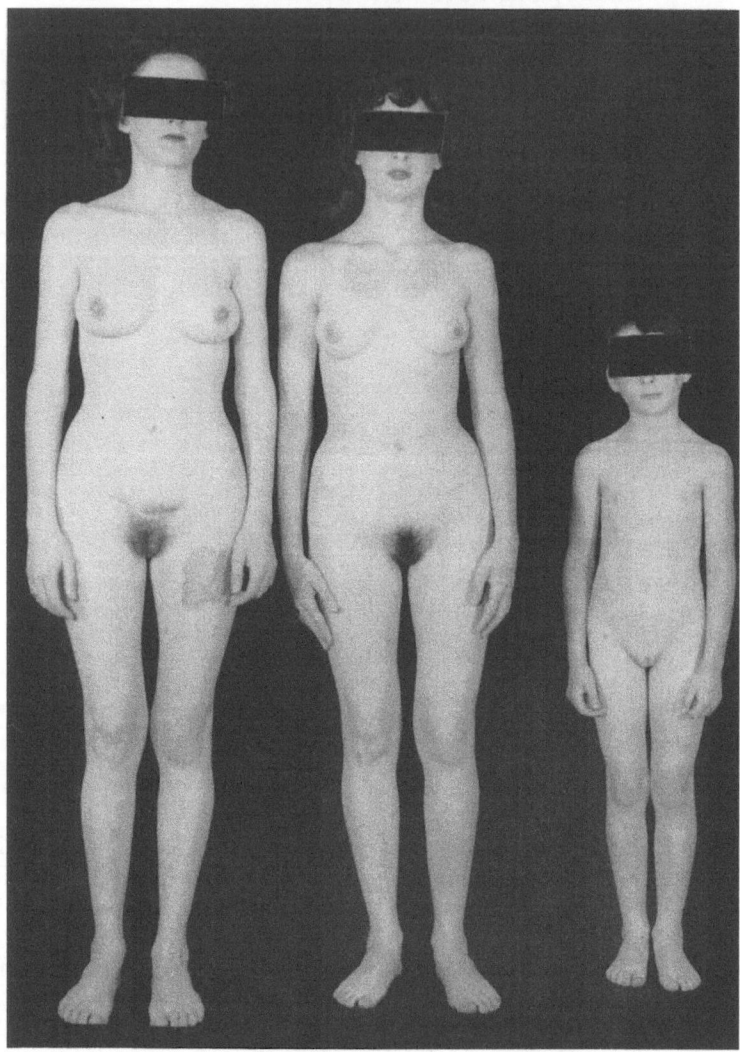

Fig. 58. Three sisters from the same family all affected with the testicular feminization syndrome. The two sisters to the left are over the age of puberty. Breast development is usually quite normal but the pubic hair is some-times very scanty

If the sex chromosome constitution be correlated with the clinico-developmental status of the patient, a provocative relationship becomes evident (Fig. 59). These data seem to indicate that a testis which has developed with all of its cells containing a Y chromosome is capable of suppressing müllerian development. However, the testis, which develops with some cells which contain a Y chromosome and some which do not, seems incapable of müllerian suppression. From a practical point of view this means that there is a close correlation between the presence or absence of müllerian structures and the presence or absence of chromosomal mosaicism for the Y chromosome. In those individuals who have müllerian development there is a very high probability of mosaicism, XO/XY being the most common, and in practically all cases with asymmetrical testicular development such mosaicism has been reported. On the other hand, in the absence of müllerian structures a normal male XY karyotype is usually found.

ZOURLAS and JONES (1965b) summarized the chromosome findings in such patients and found that all reported cases had an XY sex chromosome complement.

Type	Uterine development									
	Normal					None				
	No. Pts.	XY	XY/XO	Other		No. Pts.	XY	XY/XO	Other	
				No. Pts.	Pattern				No. Pts.	Pattern
Masculinizing bilateral testes	7	2	3	2	2	7	6	0	1	1
Masculinizing asymmetrical testicular development	23	5	12	6	4	0	0	0	0	0
Feminizing	0	0	0	0	0	25	25	0	0	0
Total	30	7	15	8	6	32	31	0	1	1

Fig. 59. Male hermaphrodites — sex chromosomes and uterine development

XI. Differential Diagnosis of Infants with Ambiguous Genitalia

If one has a careful history of maternal medication, has a sex chromatin determination or better still a full sex chromosome study, knows whether or not a uterus is present by rectal examination, has a 17-ketosteroid determination and

Diagnosis	Chromatin	History	Uterus	Anomalies	17-KS	Sex chromosomes
Adrenal hyperplasia	+	+	+	—	↑	XX
Maternal androgen	+	+	+	—	norm.	XX
Idiopathic masculinization	+	—	+	—	norm.	XX
True hermaphroditism	+ or —	—	+ or —	—	norm.	XX or other
Male hermaphroditism (masculinizing bilateral testes)	—	? —	+ or —	—	norm.	XY or other
Male hermaphroditism (masculinizing-asymmetrical gon. diff.)	—	—	+	—	norm.	XY/XO or other
Male hermaphroditism (feminizing)	—	+	—	—	norm.	XY
Streak gonad	+ or —	—	+	+ or —	norm.	XO or other

Fig. 60. Differential diagnosis of ambiguous external genitalia

some information about other congenital anomalies, it is quite possible to make a very accurate differential diagnosis in most patients with ambiguous genitalia (Fig. 60). An examination of the scheme in the figure discloses that an absolute diagnosis may be made with the various diagnostic aids recorded except for distinguishing idiopathic masculinization, the "special" female hermaphrodites and chromatin positive true hermaphroditism and for sometimes making the unimportant distinction between the 2 forms of masculinizing hermaphroditism. For these differentiations laparotomy is necessary, not only for diagnosis but also for therapy.

XII. The Treatment of Hermaphroditism

1. General Considerations

It has been shown that the sex of rearing is a primary consideration in the formation of the gender role of the individual. It has been demonstrated that it is much more important than such obvious signs of morphology of the external genitalia, the hormone dominence, or gonadal structure. Furthermore, it has also been shown that there may be serious psychiatric consequences from changing the sex of rearing after the age of infancy. It is seldom proper to advise a change of sex after infancy to conform to the gonadal structure, external genitalia and the like, but rather the physician should exert his efforts to complete the adjustment of the individual in the sex he finds himself. Happily at the present time most aberrations in sexual development are discovered in the newborn or in infancy or at the latest in early childhood, so that the problem of reassignment of sex is at a minimum and under ideal conditions should never occur.

Regardless of the time of treatment, and the earlier the better, for the surgeon this means reconstruction of the external genitalia to correspond to the sex of rearing. It also means the removal of any contradictory sex structures which may be functioning in a contradictory manner or which might in the future function in a contradictory manner. Specifically in our opinion, testes should always be removed from male hermaphrodites reared as women regardless of hormone production. In the case of the feminizing testes syndrome this thought is prompted by the instance of seminoma in these deformed retained testes.

In the case of virilized female hermaphrodites due to adrenal hyperplasia, the suppression of adrenal androgen production by the use of cortisone from an early age will result in completely female development including the onset of puberty at the expected time as well as normal menstruation and reproduction (See above). It is no longer necessary to surgically explore the abdomen and the internal genitalia in this well delineated syndrome and the surgical effort is confined entirely to reconstruction of the external genitalia along female lines.

It cannot be overemphasized that the use of cortisone must be carefully and properly supervised and the details of the endocrine therapy of this condition is covered in the section on adrenal hyperplasia.

Patients with gonadal agenesis or Turner's syndrome who are invariably reared as girls and women require the use of exogenous estrogen at the expected time of puberty. Other hermaphrodites reared as women who will not feminize also require the use of estrogen to promote the development of the female habitus and breasts. In patients with a well developed müllerian system, cyclic uterine bleeding may be produced although reproduction is, of course, out of the question. Estrogen is conveniently started at about the age of 18 and may be given at the rate of Stilbesterol 1 mg/day or its equivalent. In some patients after a period of time, this dosage has to be increased if additional breast development is desired. In patients without ovaries who have uteri and in male hermaphrodites in the same condition, cyclic uterine bleeding can often be induced by the cyclic use of estrogen for 20 days of each month. In other instance this seems inadequate to produce a sharp menstrual period and under this circumstance 20 days of estrogen may be followed by a few days of an oral progestogen or a single injection of progesterone which usually produces a very satisfactory crisp menstrual period.

Hermaphrodites reared as boys or men may require exogenous androgen as well as plastic reconstruction of the ambiguous external genitalia along male lines.

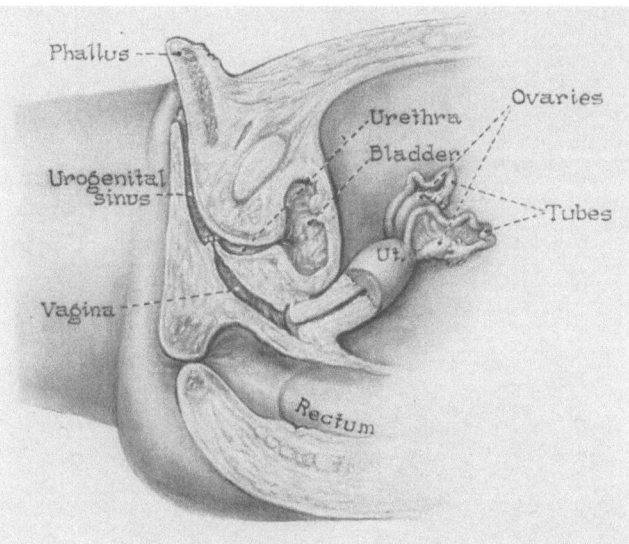

Fig. 61. Sagittal view showing relations of the urogenital sinus to the vagina and female urethra of an 18 month old patient with female hermaphroditism. (From: L. WILKINS, H. W. JONES, Jr., G. H. HOLMAN and R. S. STEMPFEL, Jr., 1958)

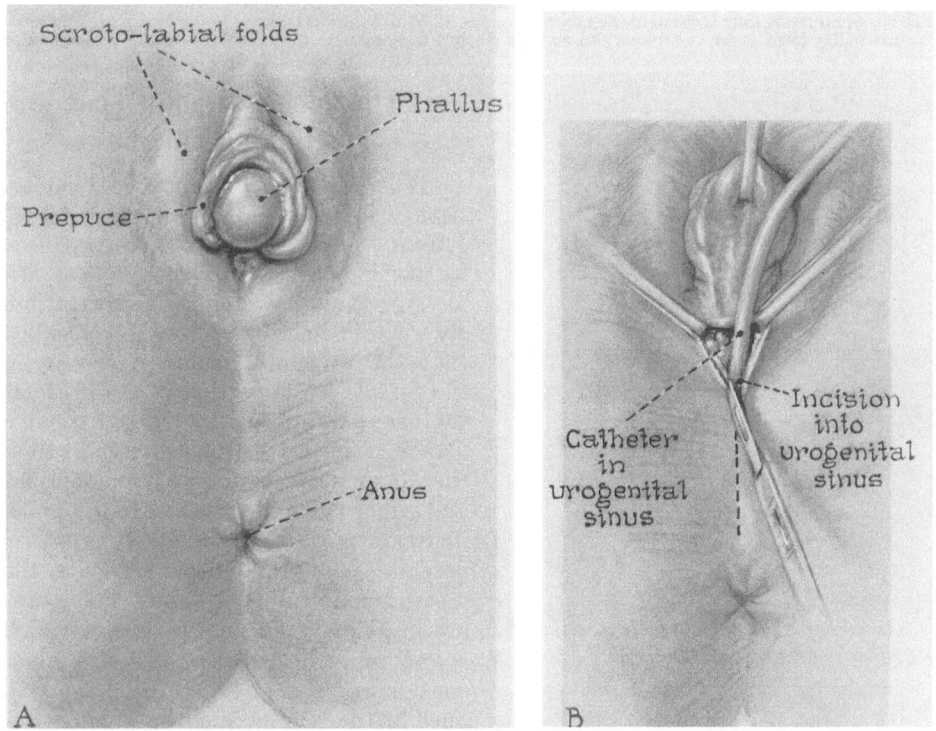

Fig. 62. A. Sketch of the external genitalia of an 18 month old patient with female hermaphroditism due to congenital adrenal hyperplasia. The operation is the same regardless of the etiology of the deformity. B. Beginning of the operation. Incision into the urogenital sinus. If the external meatus is large enough and the urogenital sinus will accommodate it, it is possible to introduce a catheter into the bladder through the urethra and besides this to introduce a sound into the vagina. When the structures are large enough, this maneuver greatly facilitates the operative procedure by assuring their identification. (From: H. W. JONES, Jr. and W. W. SCOTT, 1958)

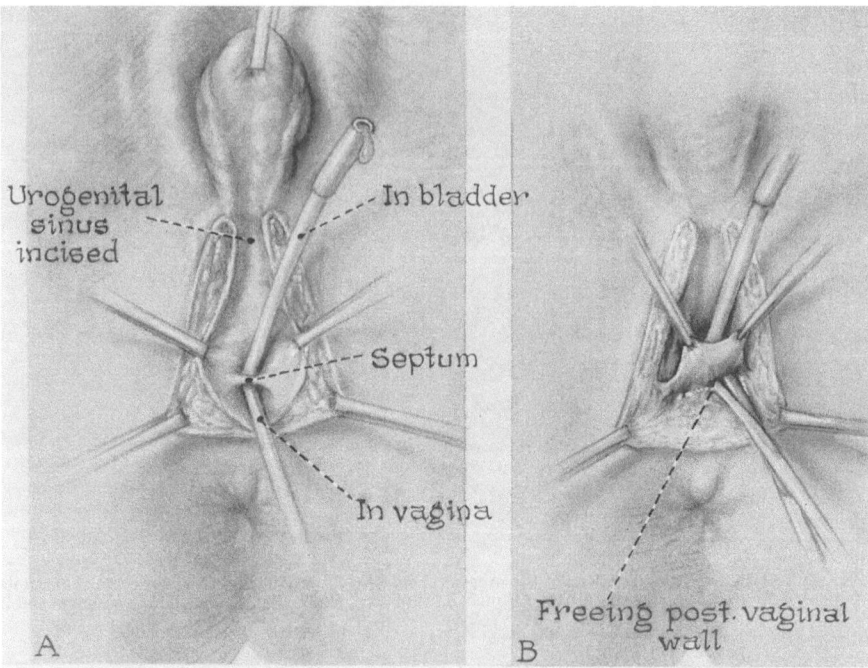

Fig. 63. A. Situation after incision of the urogenital sinus. B. With the glass catheter in the bladder the posterior vaginal wall is freed as far as necessary to make it possible to bring it to the skin edge without undue tension

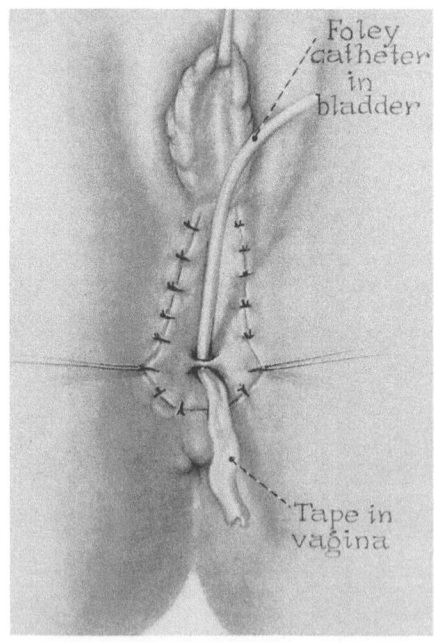

Fig. 64. The operative situation after the edges of the vagina are sutured to the skin and after the edges of the mucous membrane of the urogenital sinus are also sutured to the skin along the line of incision. (From: H. W. JONES, Jr. and W. W. SCOTT, 1958)

2. The Construction of Female External Genitalia

It is important to know and understand that the anomaly of the external genitalia in all hermaphrodites is caused by varying degrees of severity of essentially the same developmental aberration. By way of background it may be recalled that the urogenital sinus gives rise, in the female, to a portion of the bladder, urethra, paraurethral glands, Bartholin's glands, vaginal vestibule, lower vagina, the inner surface of the labia majora and minor vestibular glands. In the male, it yields part of the bladder, prostatic urethra, prostate, Cowper's glands, the membranous urethra and the paraurethral glands. The genital tubercle which is also concerned in the anomalies under consideration yield the clitoris, the prepuce in the female and the homologous structures in the male.

As these structures are concerned not only with the external appearance and with providing an entrance to the genital

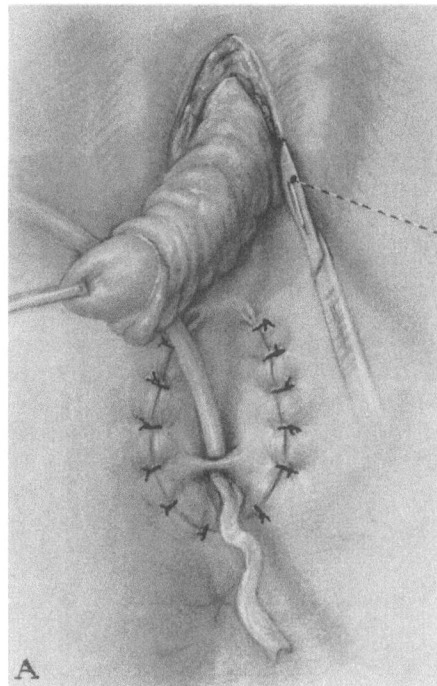

Incising around
phallus

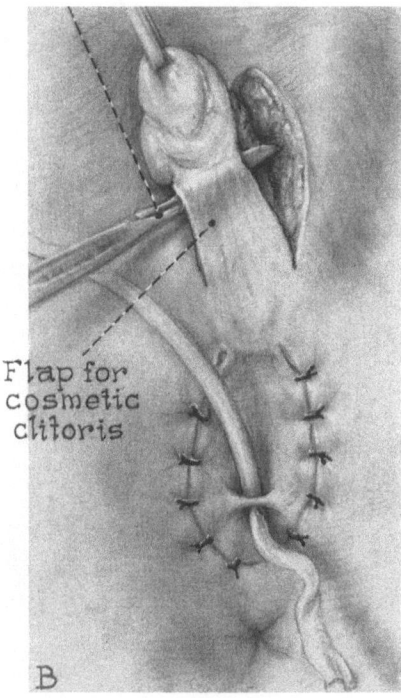

Flap for
cosmetic
clitoris

Fig. 65. A. The beginning of the elliptical incision around the base of the phallus. B. Drawing to show the shape of the incision to preserve the flap of mucous membrane along the ventral surface of the phallus. The mucous membrane is used to fashion a small cosmetic clitoris. (From: H. W. JONES, Jr. and W. W. SCOTT, 1958)

and urinary tract, but also with urinary control, it is of utmost importance for the gynecologic surgeon to know if the anomaly is uniform in detail regardless of the etiology of the hermaphroditic state and if the generative and urological structures can be surgically separated without damage to either.

The anatomical problem was studied in detail in a relatively large number of patients with anomalies due to congenital adrenal hyperplasia (JONES and JONES, 1954). It was concluded that in no case did the vagina communicate with that portion of the urogenital sinus that gives rise in the female to the entire urethra or

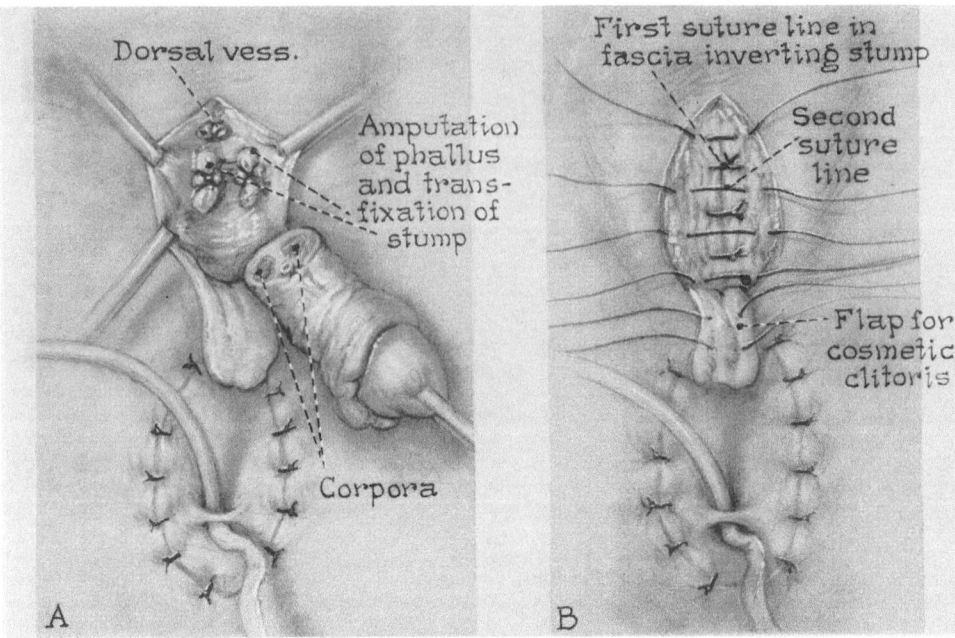

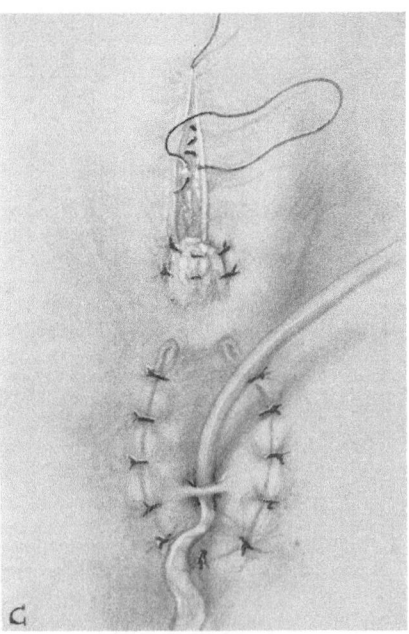

Fig. 66. A. The amputation of the phallus with transfixion of the stump. B. The stump is buried by drawing the fascia together in the midline. C. Final stages of the operation showing a tiny cosmetic clitoris. The size of this clitoris will vary from case to case and gives a remarkably normal appearance to the external genitalia especially after some pubic hair appears. (From: H. W. JONES, Jr. and W. W. SCOTT, 1958)

in the male to the prostatic urethra. The communication was always with the caudal urogenital sinus derivatives which in the female yield the vaginal vestibule and in the male the membranous urethra. The urinary sphincter mechanism was, therefore, fortunately not involved. Additinal studies of many hermaphroditic patients with various etiological backgrounds with masculine genitalia have shown that the anomaly was essentially the same regardless of genetic or other background (JONES and WILKINS, 1958 and 1960). The surgical procedures and techniques about to be described are, therefore, applicable to all hermaphrodites whose genitalia are to be surgically reconstructed to that of a female.

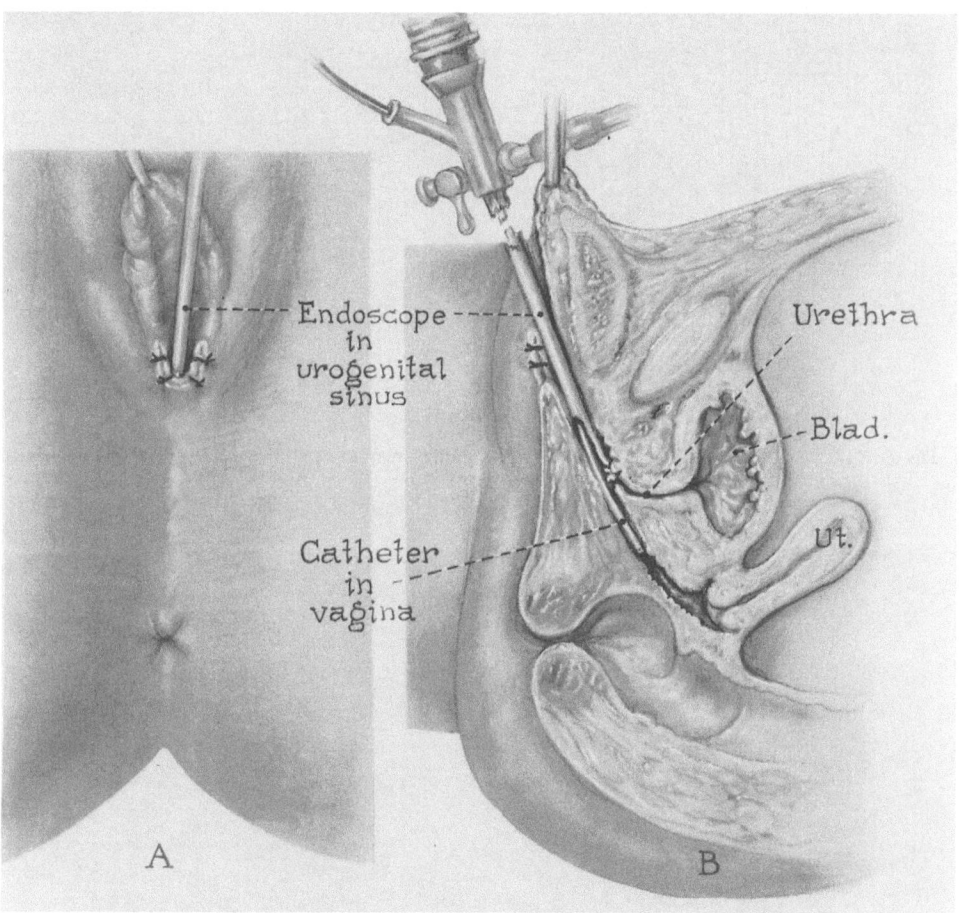

Fig. 67. The use of a McCarthy pan-endoscope with a small ureteral catheter to identify the vaginal communication with the urogenital sinus. (From: H. W. JONES, Jr. and L. WILKINS, 1961)

Prior to the surgical reconstruction, a careful study of the anomaly by endo-scopic examination or by roentgen visualization using a special catheter designed for this purpose will elucidate the exact anatomical arrangement. A rectal exam will usually reveal the presence or absence of a uterus beyond a reasonable doubt provided that the child can be examined under ideal conditions. The rectal examination as well as the endoscopic examination are so important that deep sedation of the child with barbiturates or with a general anesthesia must be carried out if necessary.

It is very desirable that surgical reconstruction of the genitalia be carried out at the earliest practical date. This is entirely on psychological grounds, not only

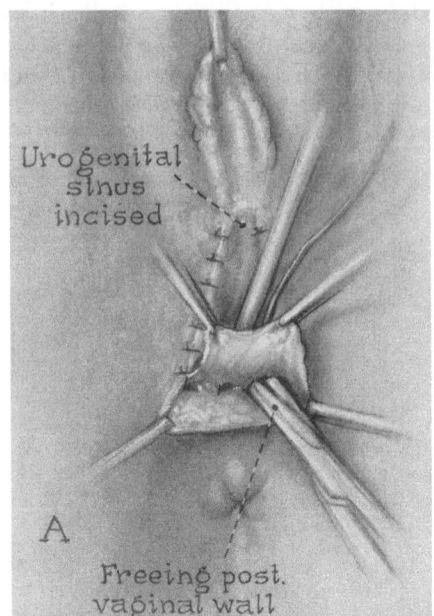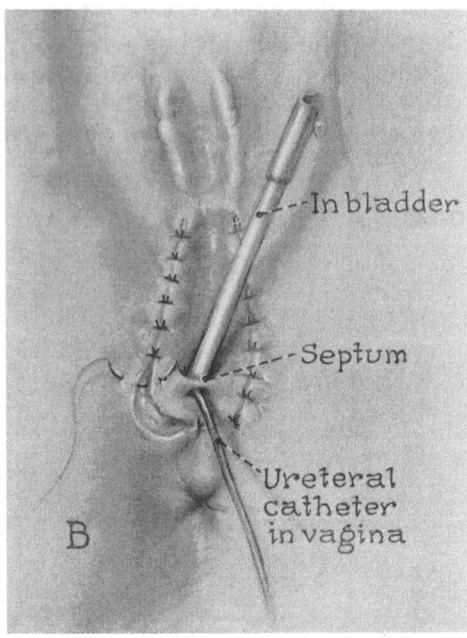

Fig. 68. With the ureteral catheter in place, the vaginal orifice may be identified and the operation completed in the usual manner. (From: W. H. JONES, Jr. and L. WILKINS, 1961)

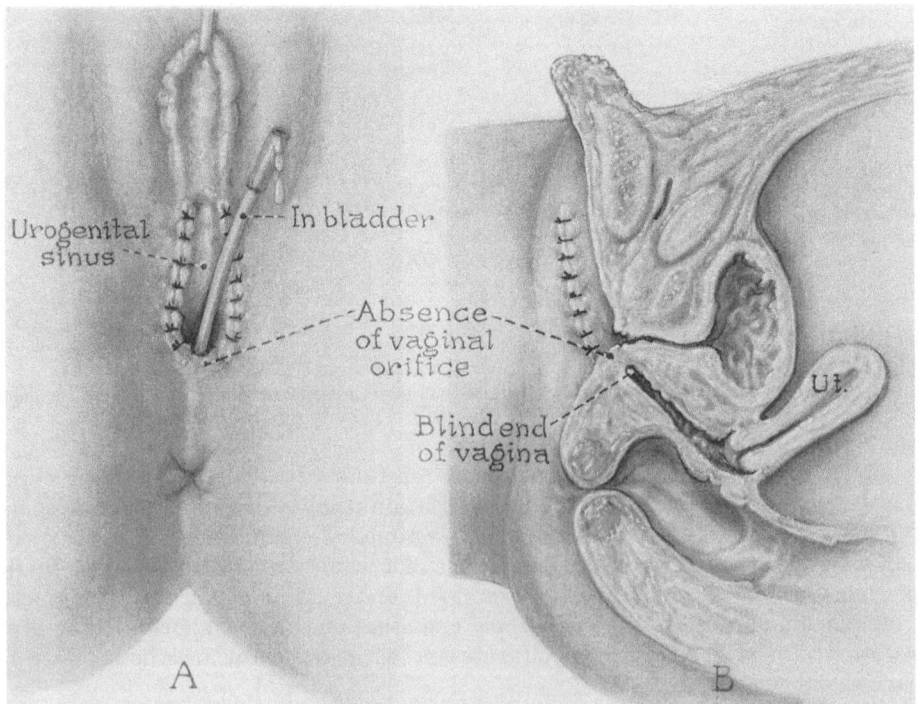

Fig. 69. Anterior-posterior and lateral drawing of an infant in whom there was no communication between the vagina and the urogenital sinus. (From: H. W. JONES, Jr. and L. WILKINS, 1961)

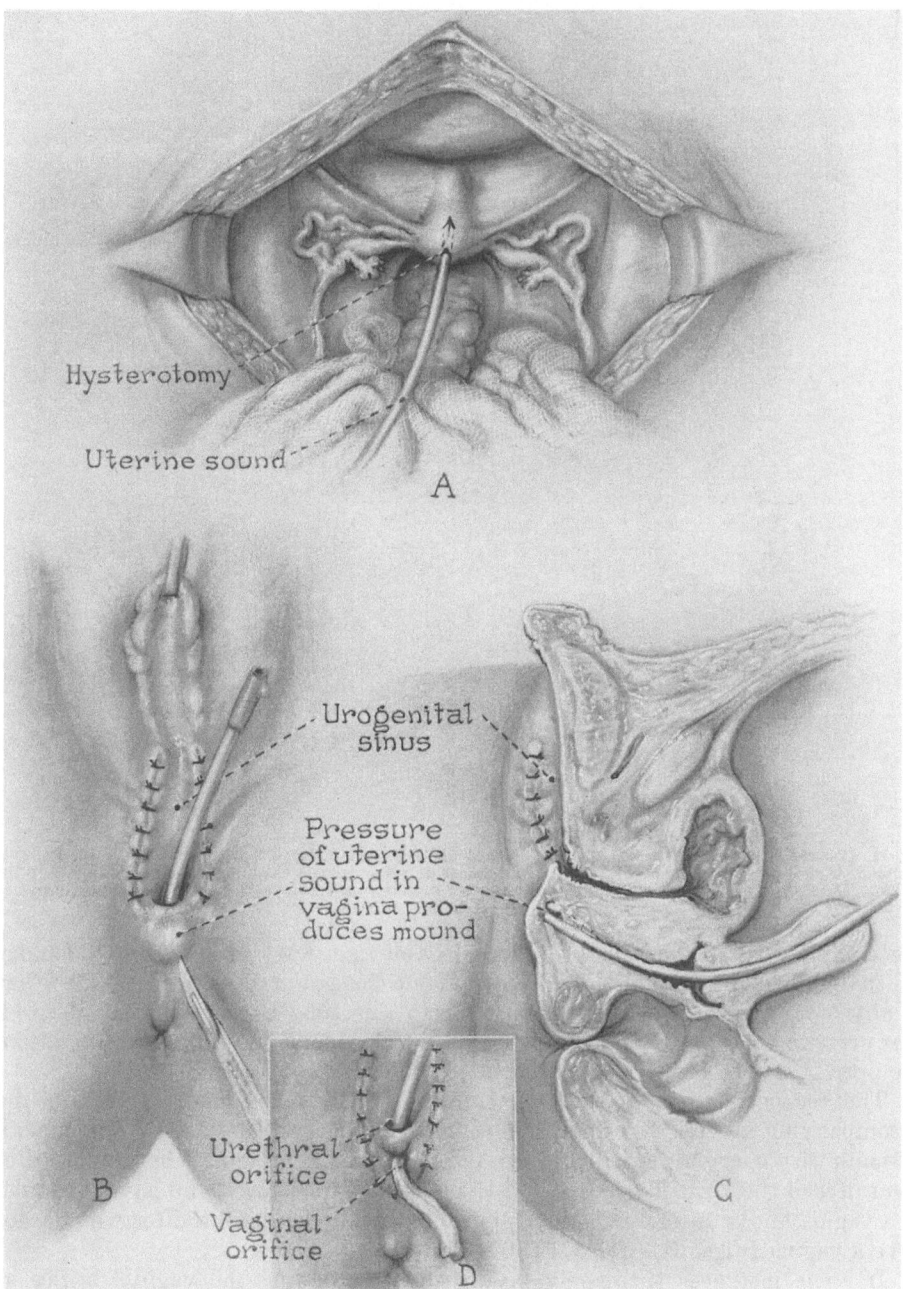

Fig. 70. Drawings to illustrate the use of the uterine sound to identify the vagina where there is lack of communication between the vagina and urogenital sinus. (From: H. W. JONES, Jr., and L. WILKINS, 1961)

for the patient but for the parents. The ease with which this can be done depends
entirely on the ease with which the communication between the vagina and the
urogenital sinus can be identified. In some instances this can easily be accomplished in the newborn in which case no hesitancy should be entertained in doing

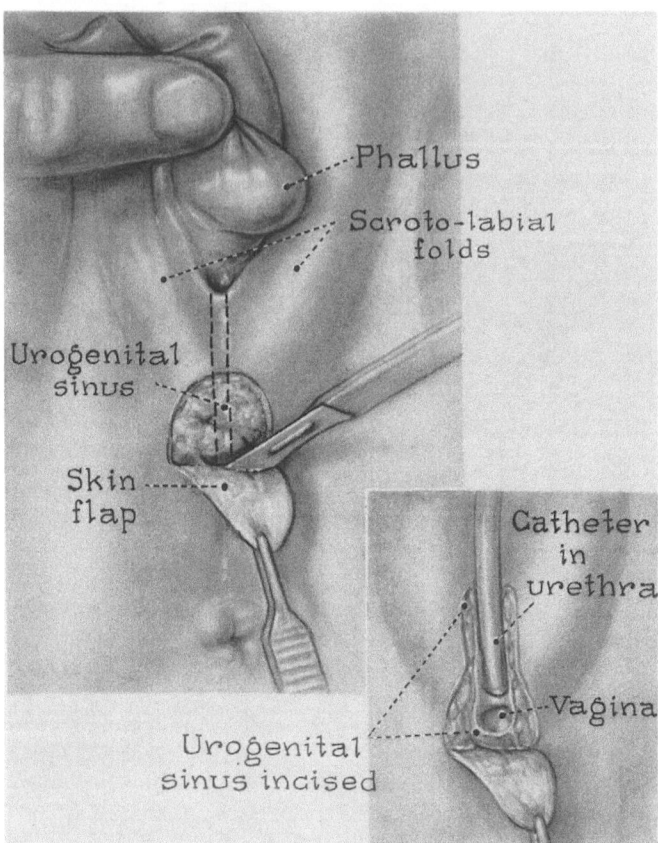

Fig. 71. Illustration of a modification of the standard operative procedure for reconstruction of the external genitalia. As shown in the illustration, the skin is developed as a flap whereas the urogenital sinus is incised in the midline. The skin flap is, therefore, useful in meeting the posterior aspect of the vagina if this should prove to be short

the complete reconstruction procedure in the first few weeks of life. Memory is usually developed in most children at about the age of 18—24 months, so that in any case an operative procedure prior to this time is greatly to be desired. However, if the vaginal opening cannot be identified by expert examination, the operation may have to be postponed.

The reconstruction procedure according to the technique outlined in the accompanying figures is applicable to most patients. This would include all patients with congenital virilizing adrenal hyperplasia, all female hermaphrodites regardless of the virilizing source and to male hermaphrodites who have a vagina, or a vagina and a uterus. It is not applicable to male hermaphrodites who do not have a vagina (Figs. 61, 62, 63, 64, 65, 66).

In some instances there may be difficulty in finding the vaginal orifice as mentioned above. Under this circumstance with an endoscope a small ureteral catheter can sometimes engage the vaginal opening even though it cannot be seen. In this circumstance the ureteral catheter is left in place so that it can be used as a guide to find the vagina (Figs. 67, 68).

In other instances the communication with the vagina and urogenital sinus may in fact not be present. This would correspond to an imperforate hymen in an otherwise normal genetic female. In this circumstance in a few instances it has been

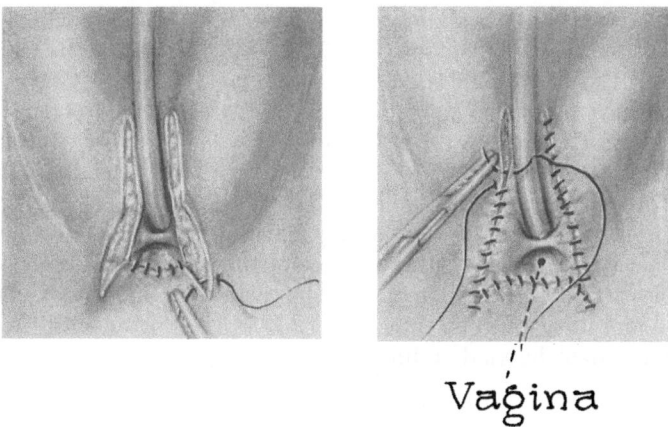

Fig. 72. Illustration of the method of placement of sutures when the skin flap procedure is used.

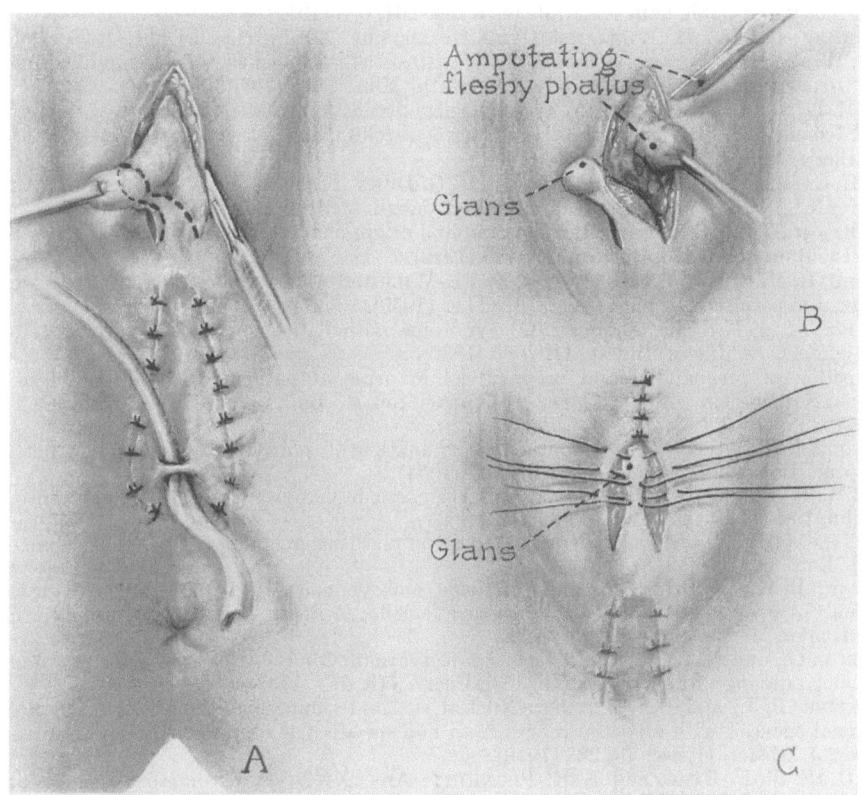

Fig. 73. Operative procedure carried out in a child with the enlargement of the phallus which was mostly of the fleshy variety.(F rom: H. W. JONES, Jr. and L. WILKINS, 1961)

necessary to identify the vagina by passing a uterine sound from above through the uterus. This is not applicable to newborn children and infants because the cavity of the uterus is not large enough to accept the uterine sound. This procedure should be reserved for children of an older age where it is necessary to establish vaginal communication because of the impending onset of menstruation (Figs. 69,

70). Recently we have on occasion modified our original incision somewhat as is shown in the accompanying diagrams to give a useful flap which somewhat enlarges the vaginal opening (Figs. 71, 72). This was originally suggested by FORTUNOFF and LATTIMER (1964).

In other situations where the phallus is not greatly enlarged or where the enlargement is mostly fleshy and not confined to the corpora, it is possible to remove the fleshy portion of the enlargement without actually removing the glans or the corpora cavernosa (Fig. 73).

Where there is no vagina, a vagina may be constructed as one would do with a genetic female who had congenital absence of the vagina. This procedure, however, cannot be carried out until the patient has attained full growth as the vaginal cavity created and lined by split-thickness graft will not grow with the individual and it, therefore, must be made adult size.

References

ALBRIGHT, F., P. H. SMITH, and R. FRASER: A syndrome characterized by primary ovarian-insufficiency and decreased stature; report of 11 cases with degression on hormonal control of axillary and pubic hair. Amer. J. med. Sci. 204, 625 (1942).

ALEXANDER, G., and D. WILLIAMS: Ovine freemartins. Nature (Lond.) 201, 1296 (1964).

AYRES, W. W., H. I. FIRMINGER, and P. K. HAMILTON: Birefringent and sudanophilic lipoids in adrenal cortex in disease and sudden death. Milit. Surg. 109, 503 (1951).

BARR, M. L., and L. F. BERTRAM: A morphological distinction between neurones of the males and females and the behavior of the nuclear satellite during accelerated nucleoprotein synthesis. Nature (Lond.) 163, 676 (1949).

— D. H. CARR, J. POZSONYI, R. A. WILSON, H. G. DUNN, T. S. JACOBSON, and J. R. MILLER: The XXXXY sex chromosome abnormality. Canad. Med. Assoc. J. 87, 891 (1962).

BLACKMAN jr., S. S.: Concerning the function and origin of the reticular zone of the adrenal cortex. Bull. Johns Hopk. Hosp. 78, 180 (1946).

BLIZZARD, R. M., G. LIDDLE, C. MIGEON, and L. WILKINS: Aldosterone excretion in virilizing adrenal hyperplasia. J. clin. Invest. 38, 1142 (1959).

BONGIOVANNI, A. M.: The adrenogenital syndrome with deficiency of 3 β-hydroxy-steroid dehydrogenase. J. clin. Invest. 41, 2086 (1962).

— Detection of pregnandiol and pregnantriol in urine of patients with adrenal hyperplasia. Suppression with cortisone; preliminary report. Bull. Johns Hopk. Hosp. 92, 244 (1953).

— A simplified method for the routine determination of pregnanediol and pregnanetriol in urine. Bull. Johns Hopk. Hosp. 94, 180 (1954).

—, and W. R. EBERLEIN: A defect in steroid biogenesis in man associated with hypertension. J. clin. Invest. 35, 693 (1956).

BRENTNALL, C. P.: A case of arrhenoblastoma complicating pregnancy. J. Obstet. Gynaec. Brit. Emp. 52, 235 (1945).

BURNS jr., R. K.: Transformation du testicule embryonnaire de l'Opossum en ovotestes ou en "ovaire" sous l'action de l'hormone femelle, le depropionate d'oestradiol. Arch. Anat. micr. Morph. exp. 45, 173 (1956).

BUTLER, C. G., and G. F. MARRIAN: Isolation of pregnane-3,17,20-triol from urine of women showing adrenogenital syndrome. J. biol. Chem. 119, 565 (1937).

CARPENTIER, P. J., and E. L. POTTER: Nuclear sex and genital malformations in 48 cases of renal agenesis with a special reference to non-specific female pseudohermaphroditism. Amer. J. Obstet. Gynec. 78, 235 (1959).

CARR, D. H., M. L. BARR, and E. R. PLUNKETT: An XXXX sex chromosome complex in two mentally defective females. Canad. med. Ass. J. 84, 131 (1961).

CHAPELLE, A. DE LA, H. HORTLING, M. NIEMI, and J. WENNSTRÖM: XX sex chromosomes in a human male. Acta med. scand. 175, Suppl. 412, 25 (1964).

CLAYTON, G. W., J. D. SMITH, and H. S. ROSENBERG: Familial true hermaphroditism in pre- and post-pubertal genetic females, hormonal and morphologic studies. J. clin. Endocr. 18, 1349 (1959).

CRECCHIO, L.: Ann. Hyg. 25, 178 (1865).

CREEVY, C. D.: Pseudohermaphroditism; report of 5 cases. Int. Sci. Digest 16, 195 (1933).

DEANE, H. W., and R. O. GREEP: A morphological and histochemical study of the rat's adrenal cortex after hypophysectomy with comments on the liver. Amer. J. Anat. 79, 117 (1946).

DEL CASTILLO, E. G., A. TRABUCCO, and F. A. DE LA BALZE: Syndrome produced by absence of the germinal epithelium without impairment of the Sertoli or Leydig cells. J. clin. Endocr. 7, 497 (1947).

EBERLEIN, W. R., and A. M. BONGIOVANNI: Partial characterization of urinary adreno-cortical steroids in adrenal hyperplasia. J. clin. Invest. 34, 1337 (1955).

FERGUSON-SMITH, M. A., B. LENNOX, J. S. S. STEWART, and W. S. MACK: Mem. Soc. Endocr. 7, 173 (1960).

FORTUNOFF, S., J. K. LATTIMER, and M. EDSON: Vaginoplasty technique for female pseudo-hermaphrodites. Surg. Gynec. Obstet. 118, 545 (1964).

FRIEDMAN, I. S., A. MACKLES, and I. DAICHMAN: Development of virilization during pregnancy. J. clin. Endocr. 15, 1281 (1955).

GARDNER, G. H., R. R. GREENE, and B. M. PECKHAM: Normal and cystic structures of broad ligament. Amer. J. Obstet. Gynec. 56, 1209 (1948).

GREENE, R. R.: Embryology of sexual structure and hermaphroditism. J. clin. Endocr. 4, 335 (1944).

HADDAD, H. M., and L. WILKINS: Congenital anomalies associated with gonadal aplasia; review of 55 cases. Pediatrics 23, 885 (1959).

HECHTER, O., A. ZAFFARONI, R. P. JACOBSEN, H. LEVY, R. W. JEANLOZ, V. SCHENKER, and G. PINCUS: The nature and the biogenesis of the adrenal secretory product. Recent Progr. Hormone Res. 6, 215 (1951).

HUMPHREY, R. R.: Reversal of sex in females of genotype WW in the Axolotl (Siredon or Ambystoma Mexicanum) and its bearing upon the role of the X chromosomes in the development of the testis. J. Zool. 109, 171 (1948).

ISHIZUKA, N., P. KAWASHIMA, T. NAKANISHI, T. SUGAWA, and Y. NISHIKAWA: Statistical observations on genital anomalies of newborns following the administration of progestins to their mothers. J. Jap. obstet. Gynaec. Soc. 9, 271 (1962).

JACOBSON, B. D.: Abortion: Its prediction and management. Fertil. and Steril. 12, 474 (1961).

JAVERT, C. T., and W. F. FINN: Arrhenoblastoma incidence of malignancy and relationship to pregnancy, to sterility and to treatment. Cancer (Philad.) 4, 60 (1951).

JOHNSTON, A. W., M. A. FERGUSON-SMITH, S. D. HANDMAKER, H. W. JONES, and G. E. S. JONES: The triple-X syndrome, clinical, pathological and chromosomal studies in 3 mentally retarded cases. Brit. med. J. 1961 II, 1046.

JONES Jr., H. W.: Clinical significance of anomalies of the sex chromosomes. Amer. J. Obstet. Gynec. 93, 335 (1965).

—, and G. E. S. JONES: The gynecological aspects of adrenal hyperplasia and allied disorders. Amer. J. Obstet. Gynec. 68, 1330 (1954).

—, and W. W. SCOTT: Hermaphroditism, genital anomalies and related endocrine disorders. Baltimore: Williams & Wilkins 1958.

—, and L. WILKINS: Gynecological surgery in 94 patients with intersexuality: Implications concerning the endocrine therapy of sexual differentiation. Amer. J. Obstet. Gynec. 82, 1142 (1961).

—, and P. A. ZOURLAS: Clinical, histologic and cytogenetic findings in male hermaphroditism. I. Male hermaphrodites with ambiguous external genitalia. Obstet. and Gynec. 25, 597 (1965).

—, M. A. FERGUSON-SMITH, and R. H. HELLER: The pathology and cytogenetics of gonadal agenesis. Amer. J. Obstet. Gynec. 87, 578 (1963).

— — — Pathological and cytogenetic findings in true hermaphroditism: A report of 6 cases and a review of twenty-three cases from the literature. Obstet. and Gynec. 25, 435 (1965).

KESAREE, N., and P. V. WOOLLEY: Phenotypic female with 49 chromosomes, presumably XXXXX: Case report. J. Pediat. 63, 1099 (1963).

KLEBS, E.: Handbuch der pathologischen Anatomie. Berlin: A. Hirschwald 1876.

KLINEFELTER jr., H. F., E. C. REIFENSTEIN jr., and F. ALBRIGHT: Syndrome characterized by gynecomastia, aspermatogenesis without A-Leydigism and increased excretion of follicle stimulating hormone. J. clin. Endocr. 2, 615 (1942).

LEREBOULLET, L.: Contribution à l'étude des atrophies testiculaires et des hypertrophies mammaires observées à la suite de certaines orchites (féminisme). Gaz. hebd. méd. 14, 533 (1877).

LILLIE, F. R.: The etiology of the freemartin. Vet. Rec. 2, 167 (1922).

LYON, M. F.: Sex chromatin and gene action in the mammalian X chromosome. Amer. J. hum. Genet. 14, 135 (1962).

MACK, W. S., L. S. SCOTT, A. M. FERGUSON-SMITH, and B. LENNOX: Ectopic testis and true undescendend testis: A histological comparison. J. Path. Bact. 82, 439 (1961).

MACLEAN, N., D. G. HARNDEN, and W. M. COURT BROWN: Abnormalities of sex chromosome constitution in newborn babies. Lancet 1961 II, 406.

MEIGE, H.: Infantilism, la feminism et l'hermaphrodite. Paris: G. Masson 1895.

Migeon, C. J.: Fractionation by countercurrent; distribution of urinary estrogens in normal individuals and in patients with hyperadrenocorticism. J. clin. Endocr. **13**, 674 (1953).

—, and L. I. Gardner: Urinary estrogens in hyperadrenocorticism; influence of cortisone, Compound F, Compound B and ACTH. J. clin. Endocr. **12**, 1513 (1952).

Money, J., J. G. Hampson, and J. L. Hampson: Hermaphroditism; recommendations concerning assignement of sex, change of sex and psychologic management. Bull. Johns Hopk. Hosp. **97**, 284 (1955a).

— — — An examination of some basic sexual concepts, the evidence of human hermaphroditism. Bull. Johns Hopk. Hosp. **97**, 301 (1955b).

Morris, J. M.: Syndrome of testicular feminization in male pseudohermaphrodites. Amer. J. Obstet. Gynec. **65**, 1192 (1953).

—, and V. Mahesh: Further observations on the syndrome "testicular feminization." Amer. J. Obstet. Gynec. **87**, 731 (1963).

Nelson, W. O., and C. G. Heller: Hyalinization of seminiferous tubules associated with normal or failing Leydig-cell function; discussion of relationship to eunuchoidism, elevated gonadotrophins, depressed 17-ketosteroids and estrogens. J. clin. Endocr. **5**, 1 (1945).

Pasqualini, R. Q., G. Vidal, and G. E. Bur: Psychopathology of Klinefelter's syndrome; review of thirty-one cases. Lancet **1957 II**, 164.

Reifenstein, Jr., E. C.: Hereditary familial hypogonadism. Proc. Am. Fed. Clin. Res. **3**, 86 (1947).

Richart, R. M., and K. Benirschke: Diagnosis of gonadal dysgenesis in newborn infants. Obstet. and Gynec. **15**, 621 (1960).

Sauramo, H.: Ovarian hilus cells in varying phases of life. Ann. Chir. Gynaec. Fenn. **51**, 108 (1962).

Shah, P. N., S. N. Naik, D. K. Mahajan, M. J. Dave, and J. C. Paymaster: A new variant of human intersex with discussion on the development aspects. Brit. med. J. **1961 II**, 474.

Sohval, A. R.: "Mixed gonadal dysgenesis" a variety of hermaphroditism. Amer. J. hum. Genet. **15**, 155 (1963).

Sydnor, K. L., V. C. Kelley, R. B. Raile, R. S. Ely, and G. Sayers: Blood adrenocorticotropin in children with congenital adrenal hyperplasia. Proc. Soc. exp. Biol. (N.Y.) **82**, 695 (1953).

Therkelsen, A. J.: Sterile male with the chromosome constitution 46 XX. Cytogenetics **3**, 207 (1964).

Tjio, J. H., and A. Levan: The chromosome number of man. Hereditas (Lund) **42**, 1 (1956).

Turner, C. D., and H. Asakawa: Experimental reversal of germ cells in ovaries of fetal mice. Science **143**, 1344 (1964).

Turner, H. H.: A syndrome of infantilism, congenital webbed neck and cubitus valgus. Endocrinology **23**, 566 (1938).

Varney, R. F., A. T. Kenyon, and F. C. Koch: An association of short stature retarded sexual development and high urinary gonadotrophin titers in women; ovarian dwarfism. J. clin. Endocr. **2**, 137 (1942).

Waxman, S. H., V. C. Kelley, S. M. Gartler, and B. Burt: Chromosome complement in a true hermaphrodite. Lancet **1962 II**, 161.

Wilkins, L.: Diagnosis and treatment of endocrine disorders in childhood and adolescence, ed. 2. Springfield, (Ill.): Ch. C. Thomas 1957.

— Masculinization of female fetus due to use of orally given progestins. J. Amer. med. Ass. **172**, 1028 (1960).

— Diagnosis and treatment of congenital virilizing adrenal hyperplasia. Postgrad. Med. **29**, 31 (1961).

—, and W. Fleischmann: Ovarian agenesis; pathology, associated clinical symptoms and bearing on theories of sex differentiation. J. clin. Endocr. **4**, 357 (1944).

—, and H. W. Jones jr.: Masculinization of the female fetus. Obstet. and Gynec. **11**, 355 (1958).

— — G. H. Holman, and R. S. Stempfel jr.: Masculinization of the female fetus associated with administration of oral and intramuscular progestins during gestation; non-adrenal female pseudohermaphroditism. J. clin. Endocr **18**, 559 (1958).

— R. A. Lewis, R. Klein, and E. Rosemberg: The suppression of androgen secretion by cortisone in a case of congenital adrenal hyperplasia. Bull. Johns Hopk. Hosp. 86, 249 (1950).

Willier, B. H., T. F. Gallagher, and F. C. Koch: The modification of sex development in the chick embryo by male and female sex hormone. Physiol. Zool. **10**, 101 (1937).

Wilson, K. M.: Correlation of external genitalia and sex glands in the human embryo. Contrib. Embryol. 18, 23 (1926).

Winiwarter, H. von: Recherches sur l'ovogenèse et l'organogenèse de l'ovaire des mammifères (Lapin et Homme). Arch. Biol. (Paris) **17**, 33 (1910).

Yamamoto, T.: Introduction of reversal in sex differentiation of XY zygotes in the Medaka Oryzias latipes. Genetics **48**, 293 (1963).

Author Index

Langworthy, O. R., and L. C.
Kolb 8, *49*
Lapides, J. 202, *220*, [1],
224, *241*
— C. R. Friend and E. P.
Ajemian [13], 237, *241*
Lascombes, G., see Neimann,
N. 58, *88*
Lathem, J. E., and K. H.
Smith *92*
Lattimer, J. K. 255, *283, 305*
— and M. Hubbart 261, *285*
— see Fortunoff, S. 456, *457*
— see Gleason, D. M. [10],
226, *241*
— see Landau, S. J. 202, *220*
— see Maloney, P. K., Jr.
199, *221*
— see Uson, A. C. 131, *164*,
194, 195, *222*
Lau, Fr. T., and R. B. Henline
96
Laughlin, V. C. *93*, 150, *163,
305*
Lauret, G., and A. Vigneron
49
Leadbetter, G. W., Jr. 217,
220
— J. H. Duxbury and J. R.
Dreyfuss 1, *49*
— and W. F. Leadbetter 45,
[6], 46, 47, *49*, 225, *241*,
246, 250, *282*
Leadbetter, W. F., see Leadbetter, G. W., Jr. 45, 46,
47, *49*, [6], 225, *241*, 246,
250, *282*
— see Parkhurst, E. C. 200,
221
— see Politano, V. A. *50*
Learmonth, J. R. 6, *49*
— and K. H. Watkins 185,
220
Lebedeff 82
Lebedev, A. P. *87*
Lefèvre, J., J. Sauvegrain,
Cl. Maitre, M. Savary and
M. Ethier *49*
Lehodey, F., see Neimann, N.
58, *88*
Leighton, P. W., see Fairley,
K. F. 64, *89*
Lemeh, C. N. *305*
Lenaghan, D. 103, 104, 107,
109, 119, *163*
Lenarduzzi, G. 83, *96*
Lennox, B., see Ferguson-
Smith, M. A. 406, *457*
— see Mack, W. S. 400, *457*
Leonetti, P., see Sorrentino, F.
211, *222*
Lepoutre, C. 171, *220*
Lereboullet, L. 403, *457*
Le Roy, A. 80, *95*
Leruitte 56, *88*

Lesbre, F. X. 348
— and F. Vigot *374*
Leslie, J. T., see Cook, C. E.
202, *219*
Levack, J. E. 111, 127, *163*
Levan, A., see Tjio, J. H.
375, *458*
Le Veen, H. H., see Ippolito,
J. J. *95*
Levi, D. 318, *343*
Levin, N. W., B. Rosenberg,
S. Zwi and F. P. Reid 84,
96
Levy, H., see Hechter, O.
424, *457*
Lewis, A. P. R., see Glaser,
K. H. 194, *219*
Lewis, L. *305*
Lewis, R. A., see Wilkins, L.
458
Lich, R., Jr., L. W. Howerton
and L. A. Davis *49*
Lichtenheld, F. R., see Mc
Cauley, R. T. 174, *221*
Liddle, G., see Blizzard, R. M.
425, 427, *456*
Light, I., see Dourmashkin,
R. L. *86*
Lillie, F. R. 402, *457*
Linberg 82
Lindau, see Byrnes 57
Lindsay, W. K., see Ross,
J. F. 309, *344*
Lindvall, Nila *96*
— see Ekstrom, T. 84, *95*
— see Lagergren, C. *96*
Linke, C. A., and J. H. Kiefer
305
Litzow, T. J., see Havens,
F. Z. *343*
Livaditis, A., K. Maurseth
and P. A. Skog 85, *96*,
124, *163*
Llanos, M. A. 65, *90*
Lodge, W. O., see Stewart,
M. J. *93*
Lodi, A. 354, *374*
Loechel, W. E., see Uhlen-
huth, E. de 5, *50*
Loquvam, G. S., see Hutch,
J. A. 17, *49*
Lorraine, H., see Dodson,
A. I. *284*
Loughran, A. M. 325, *343*
Lowen, W., and A. D. Smythe
83, *96*
Lowsley, O. S. 258, 259, *284,
305*
— and M. S. Curtis 59, *90*
— and R. Gutierrez 257, *283*
— and T. H. Johnston 200,
220
— and T. J. Kirwin 260,
262, *285*
Lubarch 62, *90*

Lucas, C. 72, *93*
Lund, A. J. 85, *96*, 102, 104,
163
Lurz, L. 188, *220*
Luschka, H. 173, *220*
Lynch, K. M. 60
— and R. R. Bradham 61,
90
Lyon, M. F. 375, *457*
Lyon, R. P., and D. R. Smith
244, 246, *281*
— and E. A. Tanagho 245,
281

Mac Alpine, J. B. 104, *163*
Mac Farlan, S. M., see Ringer,
M. G. 124, *164*
Mack, W. S., L. S. Scott,
A. M. Ferguson-Smith
and B. Lennox 400, *457*
— see Ferguson-Smith, M. A.
406, *457*
Mackellar, A., and F. D.
Stephens 187, 188, 189,
190, *221*
Mackenzie, L. L. 193, *221*
Mackles, A., see Friedman,
I. S. 433, *457*
Mac Lean, N., D. G. Harnden
and W. M. Court Brown
408, *457*
Macleod, M., see Edward, N.
92
Mac Millan, see Hoffmann 82
Maddock, W. G., see Anson,
B. J. *304*
Madisson, H. *93*
Magee, J. H., see Milam, J. H.
65, *90*
Magid, M. A., see Hejtmancik,
J. H. 193, *220*
Mahajan, D. K., see Shah,
P. N. 408, *458*
Mahesh, V., see Morris, J. M.
443, *458*
Maitre, Cl., see Lefèvre, J. *49*
Majnarich, G., see Malament
93
Makins, G. H. 216, *221*
Malament, Maxwell and
G. Majnarich *93*
Maletta, T. 281
Maloney, P. K., Jr., D. M.
Gleason and J. K. Latti-
mer 199, *221*
Mall, Fr. P., see Keibel, Fr. *96*
Malm, A., see Wulff, H. G. *95*
Marcel, J. E. *49*, 150, *163*
Marion, G. [4], 225, *241*, 319,
343
— and J. Perard *343*
Markee, J. E. 86
Marquardt, H. D. 85, *96*
Marrian, G. F., see Butler,
C. G. 423, *456*

Subject Index

Universitätsdruckerei H. Stürtz AG Würzburg